Cilia

A subject collection from *Cold Spring Harbor Perspectives in Biology*

Cilia

A subject collection from *Cold Spring Harbor Perspectives in Biology*

EDITED BY

Wallace Marshall

University of California, San Francisco

Renata Basto

Institut Curie

COLD SPRING HARBOR LABORATORY PRESS
Cold Spring Harbor, New York • www.cshlpress.org

Cilia

A subject collection from *Cold Spring Harbor Perspectives in Biology*
Articles online at www.cshperspectives.org

Executive Editor	Richard Sever
Managing Editor	Maria Smit
Senior Project Manager	Barbara Acosta
Permissions Administrator	Carol Brown
Production Editor	Diane Schubach
Production Manager/Cover Designer	Denise Weiss
Publisher	John Inglis

Front cover artwork: Scanning electron micrograph of cilia from olfactory sensory neurons on the apical surface of the mouse olfactory epithelium. (Figure kindly provided by Jeffrey R. Martens, Thomas H. Maren Professor and Chairman of the University of Florida, College of Medicine Department of Pharmacology & Therapeutics.)

Library of Congress Cataloging-in-Publication Data

Names: Marshall, Wallace F., editor. | Basto, Renata, editor.
Title: Cilia : a subject collection from Cold Spring Harbor perspectives in biology/edited by Wallace Marshall, University of California, San Francisco, and Renata Basto, Institut Curie, Paris.
Description: Cold Spring Harbor, New York : Cold Spring Harbor Laboratory Press, 2016. | Series: Cold Spring Harbor perspectives in biology | Includes bibliographical references and index.
Identifiers: LCCN 2016023671 | ISBN 9781621821410 (hardcover : alk. paper)
Subjects: LCSH: Cilia and ciliary motion.
Classification: LCC QP310.C5 C554 2016 | DDC 571.6/7--dc23
LC record available at https://lccn.loc.gov/2016023671

10   9   8   7   6   5   4   3   2   1

All World Wide Web addresses are accurate to the best of our knowledge at the time of printing.

For a complete catalog of all Cold Spring Harbor Laboratory Press publications, visit our website at www.cshlpress.org.

Contents

Contents

Preface

THE BASIC OUTLINES OF CELL BIOLOGY HAVE BEEN CLEAR SINCE AT LEAST the invention of the electron microscope, and the basic catalog of organelles has long been a standard part of high school biology texts. One would think, therefore, that ongoing research in cell biology would have little to tell us about the "big picture" of organelle functions, and would instead focus on increasingly fine molecular details of mechanism. Cilia, however, represent an example of an organelle whose existence has been known for well over a hundred years, but whose ubiquitous roles in physiology, development, and disease have only become apparent in the past two decades. This is particularly true of nonmotile primary cilia, which while found in almost all cells of the body, were long thought to be vestigial structures with no important functions. The discovery that these organelles play key roles in many different signaling pathways, and that defects in primary cilia can cause a wide spectrum of human diseases, came as a complete surprise. One might assume that the motile functions of cilia and flagella, being visually obvious, would have been much clearer, but even here there have been tremendous surprises during the past 20 years, such as the finding that ciliary motility plays a key role in left–right symmetry breaking. The complex range of ciliary functions is reflected in a complex internal architecture and proteome, with the list of ciliary proteins continuing to grow even as this volume goes to print. In a very few cases, the functions of these proteins are known, but in most cases, we are left with a list of parts whose individual functions are still not clear. Putting these parts together into a mechanistic understanding of ciliary function is going to require a concerted effort drawing on a wide range of approaches, including genomics, proteomics, structural biology, biophysics, and computation. It is our hope that, by collecting many of the key aspects of ciliary biology into a single volume with chapters provided by leaders in the field, we can help launch future integrative efforts to understand cilia at a mechanistic level.

We have tried to represent the major areas of ciliary biology, including diverse levels of questions from the molecular to the organismal, and representing many different model systems and viewpoints. This would have been much easier to do 10 years ago, but with the explosion of research on cilia, the field has grown to the point that we can no longer promise comprehensive coverage, although we have done the best that we can. The editors apologize in advance to readers who will read this volume looking for answers to specific questions and not find them. At least we hope the chapters in this volume, taken together, will provide an entry point for students, postdocs, and faculty who become interested in cilia and would like to get a snapshot of the current state of knowledge in the field. It is the nature of such a collection that it represents a picture of what is known, without necessarily making explicit what is not yet known. We encourage the reader to approach the volume as a whole with a view to identifying gaps in knowledge. Such gaps would be fruitful areas for future study.

The editors would like to thank the publishing team at Cold Spring Harbor Laboratory Press, especially Barbara Acosta, for keeping the whole project on track and for being extremely patient as deadlines approached and then passed. We also thank the members of our labs for our reduced accessibility while working on the book, and our colleagues in the cilia and flagella field for the infinite number of discussions we have had over the years, all of which fed into our selection of topics for the volume. We are especially grateful to the contributing authors for investing their priceless time on their contributions. We applaud them for recognizing the value of promoting the development of our field as a whole. Finally, we want to thank our families for putting up with us as we added this task onto our already-overloaded schedules.

WALLACE MARSHALL
RENATA BASTO

Axoneme Structure from Motile Cilia

Takashi Ishikawa[1,2]

[1]Laboratory of Biomolecular Research, Paul Scherrer Institute, 5232 Villigen PSI, Switzerland

[2]Department of Biology, ETH Zurich, 5232 Villigen PSI, Switzerland

Correspondence: takashi.ishikawa@psi.ch

The axoneme is the main extracellular part of cilia and flagella in eukaryotes. It consists of a microtubule cytoskeleton, which normally comprises nine doublets. In motile cilia, dynein ATPase motor proteins generate sliding motions between adjacent microtubules, which are integrated into a well-orchestrated beating or rotational motion. In primary cilia, there are a number of sensory proteins functioning on membranes surrounding the axoneme. In both cases, as the study of proteomics has elucidated, hundreds of proteins exist in this compartmentalized biomolecular system. In this article, we review the recent progress of structural studies of the axoneme and its components using electron microscopy and X-ray crystallography, mainly focusing on motile cilia. Structural biology presents snapshots (but not live imaging) of dynamic structural change and gives insights into the force generation mechanism of dynein, ciliary bending mechanism, ciliogenesis, and evolution of the axoneme.

Flagella and cilia are appendage-like organelles in eukaryotic cells. Cilia and flagella (these two terms are used interchangeably) are categorized into two classes by function and structure. One is motile cilia, which have dynein, a family of ATPase motor proteins, and generate motion, whereas primary cilia are nonmotile and play roles in sensory function and transportation. The axoneme is the main part of flagella and cilia and is located outside of the cell body (Fig. 1A–C). The part inside the cell that anchors cilia is called a basal body (Fig. 1D). These two regions are connected by a transition zone (TZ) (Fig. 1E). These three regions are connected continuously with microtubule cytoskeleton—normally as triplet microtubules in the basal body (Fig. 1D) and doublets in the axoneme (Figs. 1B,C, and 2) as well as in the TZ

(Fig. 1E). The axoneme consists of microtubule doublets (MTDs) and many other proteins encapsulated in the plasma membrane. The axoneme from most of the species has a 5- to 10-μm length and an ∼300-nm diameter.

The axoneme of motile cilia is a closed system—when it is isolated from the basal body and the cell body, it can still generate bending motion on addition of ATP. In this sense, eukaryotic flagella/cilia are different from bacterial flagella, which are driven by rotational motion of the base complex (Minamino et al. 2008). Structure and function of axonemes of motile cilia have been studied by genetics, physiology, and biochemistry for decades. Recently, progress in structural biology, genetic engineering, and proteomics has shed light on various new aspects of this organelle.

Cite this article as *Cold Spring Harb Perspect Biol* doi: 10.1101/cshperspect.a028076

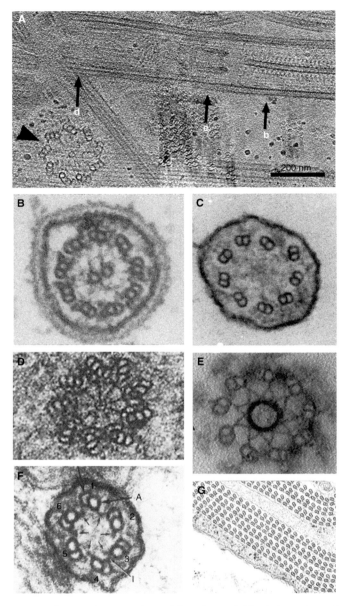

Figure 1. Axoneme structure from various species and locations of cilia. (*A*) Horizontal section from tomographic reconstruction of ice-embedded cilia including the basal body (centriole) and the transition zone (TZ) (M Hirono and T Ishikawa, unpubl.). A probasal body (a daughter centriole), which is positioned perpendicular to the basal body, is indicated by an arrowhead. (*B–G*) Electron micrographs of cross sections from plastic embedded cilia. (*B*) Typical 9+2 structure of the axoneme from *Chlamydomonas*. (Figure courtesy of Dr. Dennis Diener, Yale University.) (*C*) 9+0 structure from mouse nodal cilia. (Figure courtesy of Dr. Svetlana Markova, Dr. Dennis Diener, and Prof. Martina Brueckner, Yale University.) (*D*) Section of a centriole, showing microtubule triplets and the cartwheel (Berns et al. 1977). (*E*) The TZ from *Chlamydomonas* flagella. (Image modified from Awata et al. 2014.) (*F*) Unusual 6+0 structure (Schrevel and Besse 1975). (*G*) Abnormal axoneme with more than 100 microtubule doublets (MTDs) from the gall-midge fly *Asphondylia ruebsaameeni* Kertesz (Mencarelli et al. 2000). (Figure courtesy of Prof. Romano Dallai and Prof. Pietro Lupetti, University of Siena.) Corresponding position of cross sections in *B*, *D*, and *E* are indicated in *A* by b, d, and e, respectively.

Cite this article as *Cold Spring Harb Perspect Biol* doi: 10.1101/cshperspect.a028076

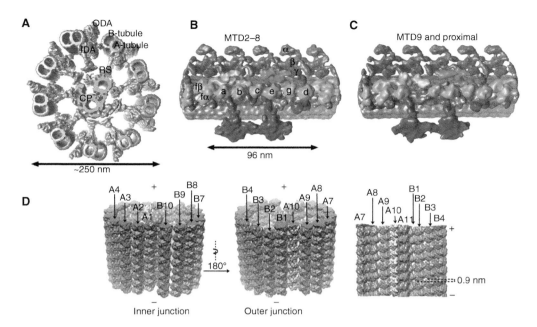

Figure 2. Overall structure of the 96-nm periodic unit from the axoneme. (*A*) Structure of the "9+2" axoneme based on cryoelectron tomography (cryo-ET). Outer and inner dynein arms (ODA and IDA, respectively), radial spokes (RS), A- and B-tubules from a microtubule doublet (MTD) as well as a central pair (CP) are indicated. (Figure modified from Bui and Ishikawa 2013.) (*B,C*) Enlarged views of one 96-nm periodic unit of regular MTDs. MTD2−8 in the distal region and MTD9 are shown (Bui et al. 2009, 2012). Dynein isoforms are indicated in *B*. (Image based on Bui et al. 2012.) MTD2−8 in the proximal region also lack dynein b. Red, ODA; cyan, IDA; purple, adjacent MTD; green, dynein regulatory complex (DRC); yellow, IC/LC (intermediate chain/light chain); blue, RS. Density maps are available as EMD2113-2130. (*D*) Molecular arrangement of tubulins in an MTD revealed by single particle cryoelectron microscopy (cryo-EM) (Maheshwari et al. 2015). Green, α-tubulin; blue, β-tubulin. Numbering of the protofilaments in A- and B-tubules is identical to that in Figure 3D.

Because cilia function in various tissues at respiratory tracts, lungs, kidneys, oviducts, sperm, brains, and embryos, defects of cilia cause diverse symptoms such as lung and kidney diseases, male and female infertility, mental disorder, and developmental abnormality. These diseases are called ciliopathy (Fliegauf et al. 2007). Ciliopathy takes place in some cases because of lack of the entire cilia in specific tissues and in other cases because of malfunction of ciliary motion. Therefore, research on both cilia formation (ciliogenesis) and cilia motility can have medical relevance.

More than 600 proteins were detected from the *Chlamydomonas* axoneme by mass spectroscopy−based proteomics (Pazour et al. 2005). *Chlamydomonas* is widely studied as a model

organism for cilia research, because its protein composition and physiological function and structure have much in common with other species, including humans. Major components of the axoneme in both *Chlamydomonas* and *Tetrahymena* are motor proteins, signal transduction proteins, and membrane proteins (Pazour et al. 2005; Smith et al. 2005), whereas components in primary cilia vary, depending on their functions (Liu et al. 2007; Mayer et al. 2009). It has not been established yet which among the more than 600 proteins of motile cilia (Pazour et al. 2005) are indispensable for bending motion. Microtubules and dynein motor proteins play central roles for force generation, driven by ATP-induced conformational change of dyneins (Figs. 3B and 4A). Dyneins are associated with a

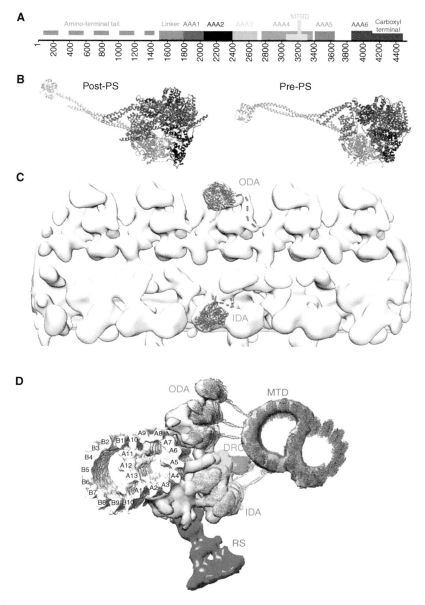

Figure 3. Dynein structure and arrangement in the axoneme. (*A*) Sequence motif of dynein. (*B*) Atomic structure of dynein in the post–power stroke (post-PS) conformation (PDB ID: 3VKH) (Kon et al. 2012) and in the pre–power stroke (pre-PS) conformation (4RH7) (Schmidt et al. 2014). In *A* and *B*, the color code is as follows: red, amino terminus and linker; blue, AAA1; black, AAA2; cyan, AAA3; green, AAA4; yellow, coiled-coil stalk (which is a part of AAA4); purple, AAA5; dark blue, AAA6 and carboxyl terminus. (*C*) Atomic models (3VKH) fitted to cryoelectron tomography (cryo-ET) structure (EMDB ID: 2117) (Bui et al. 2012) to show the front and back of dyneins in outer dynein arms (ODAs) and inner dynein arms (IDAs), respectively. The linker (red) is above the AAA ring (blue) in ODAs, whereas it is below the ring in IDAs. Yellow, stalks. (*D*) Atomic models in which density maps from cryo-ET (Bui et al. 2012) and single-particle analysis (ciliary microtubule doublet [MTD]) (Maheshwari et al. 2015) are fitted. Red, ODA; cyan, IDA; purple, adjacent MTD; green, dynein regulatory complex (DRC); yellow, intermediate and light chains (IC/LC); blue, radial spoke (RS).

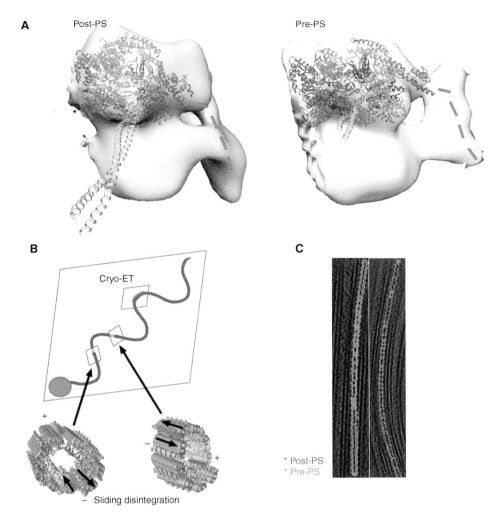

Figure 4. Dynein structural change induced by nucleotides. (*A*) Atomic model fitting to cryoelectron tomography (cryo-ET) (Ueno et al. 2014). The overall orientation of the linker at pre–power stroke (pre-PS) and post–power stroke (post-PS) condition (red) is the same as crystallography (Fig. 3B). The color code is the same as in Figure 3A,B. (*B*) Switching models based on sliding disintegration (Hayashi and Shingyoji 2008) and cryo-ET (Lin et al. 2014) of sea urchin sperm. Sliding disintegration occurs by sliding between two adjacent microtubule doublets (MTDs) (red in the surface-rendered models of two axonemes), whereas dyneins at the opposite side of the axoneme are relaxed (blue). In intact cilia, straight areas between principle and reverse bending (indicated by arrows) should be where sliding takes place. If sliding moves toward the opposite directions at these two sites, bending can happen. The sliding plane is perpendicular to the bending plane—switching occurs between the upper and lower sides of the bending plane. However, switching was found at the area of curvature by cryo-ET (Lin et al. 2014). Image classification of outer arm dynein structure showed localization of two dynein conformations at external and internal sides of curved cilia, suggesting switch within the bending plane. (*C*) Distribution of structure of outer dyneins along MTDs in the axoneme in the presence of the nonhydrolyzable ATP analog, ADP.Vi, show clustering, in which 10 to 20 dyneins in the post-PS conformation make a row (blue), followed by a row of dyneins in the pre-PS conformation (red). Under the same ADP.Vi concentration, almost all the dyneins turn to the pre-PS conformation without coexistence of multiple forms. This suggests cooperative conformational change of neighboring dyneins in the axoneme (Movassagh et al. 2010).

variety of intermediate chains (ICs), light chains (LCs), and other regulators (see review by King 2016).

Although one species of cytoplasmic dynein recognizes a variety of cargos with the help of adaptors, in cilia there are a number of axonemal dynein isoforms, similar to the case of kinesin. Axonemal dyneins form two complexes, inner and outer dynein arms (IDAs and ODAs, respectively), on MTDs. ODAs form either a heterodimer or heterotrimer. Although the functional difference between an ODA with dimer and ODA with trimer is not known, species with dimers are mostly unikont with calaxin as an associated calcium sensor and species with trimers are mostly bikont with LC4 as a sensor (Inaba 2015), suggesting their diversification during evolution. In *Chlamydomonas*, all the axonemal dyneins regularly functioning in cilia are identified and named as α, β, γ (ODAs) or a, b, c, d, e, fα, fβ, and g (IDAs) (Fig. 2B) (Yagi et al. 2009). They have different characteristics revealed by in vitro motility assay, such as velocity, ATPase, and duty ratio (Kagami and Kamiya 1992; Kamiya 2002). As described in detail later, all the species of dynein have fixed loci in the 96-nm periodic unit (Bui et al. 2008, 2012). It is not known whether they need to be in the proper positions to orchestrate bending motion in a designed manner or whether they only need to be on MTDs at high density to generate intense force. Judging from the beating of a deletion mutant, which lacks an entire ODA, with smaller amplitude but normal waveform (Kagami and Kamiya 1992), an ODA is an accelerator and not essential for waveform determination. Another important protein complex is the radial spoke (RS), which protrudes from an MTD toward the central pair (CP) singlet microtubules. In motile cilia, mutants lacking the RS (Ebersold et al. 1962; Huang et al. 1981; Yang et al. 2004) or CP (Starling and Randall 1971) are paralyzed under physiological beating conditions, and many IDA mutants have abnormal beating (summarized in Kamiya 2002), suggesting a signal transduction pathway from the CP and RS to the IDAs for bending motion. However, RS/CP mutants can be activated for motion in the presence of high ADP concentration,

indicating an alternative pathway, independent of RS/CP (Kamiya and Yagi 2014). Two pathways, one by CP/RS/IDA and the other by ODA, could either make the system robust by redundancy or enable finer regulation.

"9+2"

Most species—from the green algae *Chlamydomonas* to humans—share a similar ultrastructure of axoneme in motile cilia. Nine MTDs surround two singlet microtubules (the CP). This common architecture is called the "9+2" axonemal structure (Figs. 1B and 2A). An MTD consists of one complete cylindrical microtubule (A-tubule) and one incomplete tubule (B-tubule) attached on the A-tubule (Figs. 2A and 3D) and extends from a triplet microtubule in the basal body (centriole). Adjacent MTDs are linked by dyneins and dynein regulatory complex (DRC) (nexin), whereas RSs are between the MTD and the CP (Figs. 2A and 3D). Dyneins, RSs, and the DRC form a regular 96-nm periodicity along MTDs. This periodic length is determined by two coiled-coil proteins, FAP59 and FAP172, proved by extension of the periodicity induced by elongation of these proteins (Fig. 5A) (Oda et al. 2014a). Nine MTDs show pseudo ninefold symmetry, but not exact symmetry, as mentioned above (see the section Uneven Distribution of Axonemal Dynein Isoforms). Differently from cytoplasmic dynein, there are a number of axonemal dynein isoforms. These dyneins are composed of ODAs and IDAs (Fig. 2B). Within one 96-nm periodic unit, there are four ODAs with 24-nm spacing (Fig. 2B). One IDA consists of eight axonemal dynein molecules, whereas one ODA is either heterodimer or heterotrimer. Each proximal end of the axoneme (1 μm or less) has no dynein or RS.

MOTILE CILIA WITH EXCEPTIONAL STRUCTURE

There are also exceptional motile cilia. Nodal cilia and *Anguilla* (eel) sperm (Woolley 1997) are both motile, but lack the RS and CP ("9+0" with dynein) (Fig. 1C). Their conical or screw-

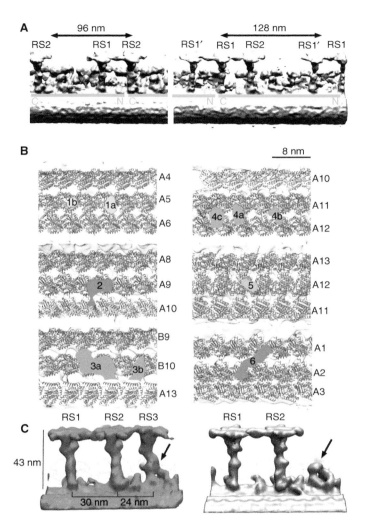

Figure 5. Rulers, microtubule inner proteins (MIPs), and radial spokes (RSs). (*A*) Location of FAP59 and FAP172, two coiled-coil proteins extending along the microtubule doublet (MTD) as proved by genetically tagged labels, which determine the periodic length, 96 nm. When amino-terminal subdomains of these proteins are duplicated by genetic engineering, the MTD has a 128-nm periodicity, allowing the third RS. Surface-rendered from EMD6108 and EMD6115 (Oda et al. 2014a). (*B*) Unidentified proteins binding to the inside of MTDs. Most of these MIPs bind two or more protofilaments. (From Maheshwari et al. 2015; reproduced, with permission, from the authors.) (*C*) Comparison of structure of RSs from *Tetrahymena* (*left*) and *Chlamydomonas* (*right*). The common structural motif at the base of RS3 is shown by arrows. (Modified from Pigino et al. 2011.)

like motion (Woolley 1998), instead of planar beating, likely originates from this "9+0" structure. Monogenean *Pseudodactylogyrus sp.* has "9+1" with nine MTDs, an RS, and, instead of a CP, one central singlet microtubule (Mollaret and Justine 1997). Although they are rare, there are even axonemes with different numbers of MTDs. Motile flagella from gametes of the gregarine *Lecdina tuzetae* and *Deplauxis hatti* have 6+0 and 3+0 structure, respectively (Fig. 1F) (Schrevel and Besse 1975; Prensier et al. 1980). Flagella from some insect sperms have large numbers of MTDs, such as Cecidomyiinae *Monarthoropalpus flavus*, in which more than 100 MTDs form laminae, instead of cylinders (Fig. 1G). They have only ODAs, which are structurally similar to those ODAs in "9+2" axonemes (Lupetti et al. 2005). Details of such insect sperm axonemes are reviewed in Dallai (2014). Except for these unusual flagella, all axonemes have nine MTDs. Nonmotile primary cilia have the "9+0" structure, without a CP, an RS, and dyneins. 3D reconstruction of primary cilia was done by cryoelectron tomography (cryo-ET) (Gilliam et al. 2012; Doroquez et al. 2014), but periodicity and other structural detail are still to be investigated. In this review, we will focus on the 3D structure of the "9+2" axoneme from motile cilia and flagella.

It is unclear why most species share the axonemal structure with nine doublets. Because nine MTDs stem from nine triplets in the basal body, ninefold symmetry is likely determined by centriolar structure, rather than ciliary structure. The mechanism to maintain the ninefold symmetry of centriole is still to be investigated, but it is surely linked to the SAS-6 protein, which builds a ninefold cartwheel in the centriole and also can form a 9-mer in vitro (Kitagawa et al. 2011; van Breugel et al. 2011). In *Chlamydomonas bld12* mutants, which are composed of eight to 12 MTDs and have abnormal motility, the CP is often missing in axomenes with eight MTDs, whereas two pairs of CPs are found in axonemes with more than nine MTDs, indicating that the CP is formed properly in an optimum space (Nakazawa et al. 2014). Although the diameter of the axoneme increases proportionally to the number of MTDs, the distance between MTDs is regular regardless of the number of MTDs (Nakazawa et al. 2014). This suggests that the proteins that contribute to the maintenance of axoneme structure, such as dyneins, RSs, and DRC proteins, are tuned to be accommodated in the "9+2" axoneme during evolution.

METHODS OF STUDY

Similar to studies of other cellular ultrastructures, the axoneme has been investigated by transmission electron microscopy (TEM), resin embedding and staining, freeze substitution of high-pressure frozen specimen, or freeze-fracture deep etch. These classical specimen preparations provide micrographs with high contrast, with which single-dynein molecules can be observed directly and many fine details of the axoneme can be revealed, such as arrangement of dyneins (Avolio et al. 1984; Goodenough and Heuser 1985a), RSs (Warner and Satir 1974; Goodenough and Heuser 1985b), and the DRC (Gardner et al. 1994). By these methods, periodicity of the axoneme, the T shape of the RS (Goodenough and Heuser 1985b), and dynein structure with a head and a tail domain (Goodenough and Heuser 1984) were visualized. By combining these classical techniques with image analysis, arrangement of dynein heavy chains in ODAs (Lupetti et al. 2005) and IDAs (Mastronarde et al. 1992) was modeled, which is consistent with our knowledge today. However, recent progress of cryo-ET brought a new scope to cilia studies. It enables 3D reconstruction, imaging conformation of individual molecules, and intermolecular network. By comparing deletion mutant structures, labeling by genetic tagging to add extra density to target molecules, their position can be located. Details of this method for the axoneme are described and findings revealed by this method are reviewed elsewhere (Bui and Ishikawa 2013; Ishikawa 2015).

One major reason for the enormous success of cryo-ET of the axoneme is its periodicity. To overcome the disadvantage of cryo-ET (i.e., poor signal-to-noise ratio [S/N]) and visualize molecules, many volumes, including the identical molecule with the same conformation (called subtomograms), must be computationally extracted, aligned three-dimensionally, and averaged. Structure along MTDs can be extracted and aligned, using 96-nm periodicity and ninefold symmetry, and is thus suitable for subtomogram averaging from cryotomography. Although another part of the cilia with less symmetry could also be a target to extract and average, until now successful 3D structural analysis is limited to exceptionally long stacks (therefore with periodicity) of centriolar SAS-6 cartwheels in *Trichonympha* (Guichard et al. 2012).

Cite this article as *Cold Spring Harb Perspect Biol* doi: 10.1101/cshperspect.a028076

A few component molecules, such as dynein and tubulin, are solved at atomic resolution by X-ray and electron crystallography, respectively (Nogales et al. 1998; Kon et al. 2012; Schmidt et al. 2012). Dyneins and MTDs were analyzed in vitro at intermediate resolution by single-particle cryoelectron microscopy (cryo-EM) analysis (Roberts et al. 2012, 2009; Maheshwari et al. 2015). Although the spatial resolution of in vivo cryo-ET even after subtomogram averaging is moderate (∼25 Å), the axoneme could be studied in greater molecular detail by fitting high-resolution in vitro structure to the in vivo structure by tomography (Roberts et al. 2012; Maheshwari et al. 2015).

OVERALL STRUCTURE OF THE "9+2" AXONEME

The triplet of the basal body consists of one complete microtubule (A-tubule) and two incomplete microtubules (B- and C-tubules) (Fig. 1D). The axonemal MTDs are extensions of A- and B-tubules of the basal body. Loss of the C-tubule happens at 300–500 nm from the proximal end of the basal body. However, the orientations of A- and B-tubules are slightly different between the doublet in the axoneme (Fig. 1B,C) and the triplet in the basal body (Fig. 1D). The factor to cause this twist has not been identified yet. The TZ, in which a characteristic star-shaped object (Diener et al. 2015) connects inside of MTDs, is located on an MTD (Fig. 1E) at the area close to the interface to the triplet. The point on the MTD where dyneins, RSs, and the CP start is located distal to the TZ. The structure of the area in between has not been analyzed. Axonemal dyneins, RSs, and DRCs are arranged with 96-nm periodicity (Mastronarde et al. 1992). Within this repeating unit, one set of IDAs, which contains eight inner arm dynein heavy chains, and four ODAs with 24-nm spacing (in which each ODA consists of two or three heavy chains), as well as ICs and LCs associated with the heavy chains, are found.

MTD STRUCTURE

The A-tubule of an MTD consists of 13 protofilaments, which are parallel to the axis of the tubule without twist, similar to the reconstituted 13-protofilament microtubule (Chrétien and Wade 1991). The arrangement of α- and β-tubulins in the A-tubule is also the same as the reconstituted microtubule—tubulin dimers form a left-handed stagger with one helical discontinuity between two adjacent protofilaments, called a B-lattice. Helical discontinuity occurs at the external side of the axoneme (Fig. 2D) (Maheshwari et al. 2015). The B-tubule is slightly larger in diameter than the A-tubule, corresponding to a 15-protofilament microtubule. However, although a 15-protofilament microtubule reconstituted from tubulin has helical twist (Chrétien and Wade 1991; Ray et al. 1993), protofilaments of the B-tubule are positioned straight—parallel to the axis of the MTD. This suggests a different mechanism of microtubule formation or stabilization of the MTD from reconstituted microtubules. Because this special structure is maintained after the removal of all the major proteins, such as dynein, RSs, and the DRC, proteins located inside an MTD may play an essential role for the tubulin arrangement in MTDs.

A number of proteins are found inside MTDs (Sui and Downing 2006). Many of them are common among nine MTDs from various species (Pigino et al. 2012), although some of them are specific. MTDs 1, 5, and 9 have a large complex named a beak, which occupies the inside of the B-tubule. Although its protein composition is not known, interestingly the beaks in MTD5 and MTD6 are missing in *mbo-1*, a backward swimming mutant (Segal et al. 1984). More inside proteins were found by cryo-EM tomography (Nicastro et al. 2011) and single-particle analysis (Fig. 4B) (Maheshwari et al. 2015). Some proteins inside MTDs have been identified and functionally characterized. Rib72, an EF-hand protein, was located in the region between the A- and B-tubules, called the ribbon region (Ikeda et al. 2003). The ribbon binds tektin, which forms a filament, as well as other EF-hand proteins, Rib74 and Rib85.5 (Linck et al. 2014). Other structures are not biochemically characterized yet. They connect adjacent protofilaments of A- and B-tubules (Fig. 5B), which could enable them to bundle protofilaments during the bending motion and keep their lattice arrangements different from those of reconstituted microtubules (Maheshwari et al. 2015).

AXONEMAL DYNEIN STRUCTURE

Dynein is a gigantic ATP-driven motor protein family with ∼4500 amino acids (Fig. 3A). The catalytic domain of dynein consists of ∼3000 amino acids, forming six AAA domains (Neuwald et al. 1999). A tail region with ∼1500 amino acids is at the amino terminus and responsible for cargo binding in cytoplasmic dynein and permanent (independent of ATP hydrolysis) binding to the A-tubule. Sequences of amino-terminal tail domains of dynein isoforms are diverse. Amino-terminal tails of axonemal dynein could be essential to determine the loci of dyneins, conformation of dynein arms, and connection between dyneins, RSs, and the DRC. There are two textbooks on dynein, edited by King (2011) and by Amos and Hirose (2012).

Among the six AAA domains of cytoplasmic dynein, one AAA domain (AAA1) is capable of ATP hydrolysis and three (AAA2, AAA3, and AAA4) bind nucleotides, whereas AAA5 and AAA6 do not have p-loops and therefore do not bind nucleotides. AAA3 hydrolyzes ATP (but less than AAA1) and is thought to be regulatory (Takahashi et al. 2004). AAA4 has a long coiled-coil protrusion called a stalk. The microtubule-binding domain (MTBD) at the tip of the stalk is responsible for ATP-dependent interaction to the microtubule (in the case of axonemal dynein, the B-tubule of the adjacent MTD). Atomic structures of the catalytic domain of cytoplasmic dynein in the absence of nucleotides (Schmidt et al. 2012), in the presence of ADP (Kon et al. 2012), and in the presence of ADP.Vi, which mimics the ADP.Pi state (Schmidt et al. 2014), are available. The first two are in the post-PS structure, whereas the ADP.Vi structure is in the pre-PS structure (Shimizu and Johnson 1983; Johnson 1985). Between the amino-terminal tail and AAA1 is an α-helix-rich linker domain (Carter et al. 2011; Kon et al. 2011). In the post-PS structure, the linker domain stems from AAA1 and extends across

the ring to dock on AAA4 in the ADP state (Kon et al. 2012) and on AAA5 in the apo state (Schmidt et al. 2012). In the pre-PS state, the linker kinks and bridges AAA1 and AAA2 (Schmidt et al. 2014).

Although no atomic structure of axonemal dynein has been solved, based on high homology with the sequence of catalytic domains of cytoplasmic and axonemal dyneins (47.7% homology, 26.4% identity) (Paschal et al. 1992), we assume that axonemal dynein has a structure similar to that of cytoplasmic dynein. Indeed, the cryo-EM structure of the inner arm dynein c at ∼20 Å resolution and its nucleotide-induced change are consistent with the high-resolution structure of cytoplasmic dynein (Roberts et al. 2009, 2012, 2013). On that basis, we can model axonemal dynein heads by fitting atomic models of cytoplasmic dynein to the density map of the axoneme from tomography (Lin et al. 2014; Ueno et al. 2014).

Orientation of dyneins in both ODAs and IDAs is similar along the longitudinal direction. In the 3D structure reconstructed from cryo-ET, the amino-terminal tail extends from the AAA ring toward the direction of the distal end of the axoneme, which corresponds to the plus end of the microtubule (Fig. 2B) (Bui et al. 2008). The stalks from two or three outer arm dyneins also show nearly the same orientation—downward when seen from the adjacent MTD with the proximal end left, but tilted ∼35° toward the proximal direction (yellow in Fig. 3C) (Bui et al. 2009, 2008), both with and without nucleotides. By fitting the atomic model of cytoplasmic dynein to the ODAs in the tomogram, the linker domain turns out to be above the AAA ring—the linker is located at the opposite side of the ring from the A-tubule, and the amino-terminal tail must bend and wrap one AAA domain to extend toward the A-tubule (red dotted line in Fig. 3C). Conformational change of axonemal dyneins induced by nucleotides also shows high coincidence with crystallography of cytoplasmic dynein. The linker extends from AAA1 toward AAA2 in the presence of ADP.Vi (Ueno et al. 2014). The orientation of the inner dynein heads is different from that of the outer dynein heads. Although inner dyneins have the amino termini extending toward the tip of cilia, similar to outer dyneins, the orientation of the stalk should be upward in Figure 3C to interact with the adjacent MTD (yellow in Fig. 3C). This means that the orientation of the inner dynein head is "flipped" with the linker down toward the A-tubule. Thus, the amino terminus should extend toward the A-tubule without wrapping the AAA domain (red dotted line in Fig. 3C).

DYNEIN STRUCTURAL CHANGE IN THE AXONEME AND BENDING MECHANISM

How are the conformational changes of the axonemal dyneins integrated into ciliary bending motion? This ultimate question will be decomposed into a few questions. How is the conformational change of each dynein in the axoneme? Is dynein activity in the axoneme correlated to the geometrical bending pattern? Do the dimeric axonemal dyneins (OADs and IADs) behave similarly to dimeric cytoplasmic dynein?

The switching model (switch model hypothesis)—in which one side of curved cilia has dyneins in one state of the force generation cycle and dyneins in the other side of the curved cilia are in another state to cause or maintain asymmetry and the two sides switch when the curvature flips—was proposed based on negative stain of cilia (Sale and Satir 1977; reviewed in Satir et al. 2014). The switching model was formulated by simulation (Sugino and Naitoh 1982) and is consistent with sliding disintegration studies, in which MTDs from enzymatically treated (i.e., the linkage between MTDs is disconnected) axonemes slide to split the axoneme into two parts (each consists of three to six MTDs) at the splitting plane perpendicular to the bending plane (arrows in Fig. 4B) (Hayashi and Shingyoji 2008). The direction of sliding disintegration indicates that dyneins on one MTD slide past each other (in Fig. 4B, MTDs at the red area in the surface model of the axonemes), whereas dyneins on the other side of the axoneme do not (in Fig. 4B, MTDs at the blue area). These MTDs are located on the same plane as the CP, suggesting the role of RS/CP for switching. On the contrary, cryo-ET showed

outer arm dyneins along MTDs at the external side of a curved part of native sea urchin sperm flagella are in one conformation, whereas dyneins at the internal side of curvature are in the other conformation (Fig. 4B, indicated by "cryo-ET") (Lin et al. 2014). Switching of sliding should take place in the straight area of bending cilia (indicated by arrows in Fig. 4B) in the model based on sliding disintegration, whereas in cryo-EM analysis conformational heterogeneity appeared at the area with curvature (Fig. 4C), suggesting these two experiments may observe different phenomena.

In the switching model, dynein activity must be suppressed locally. Indeed, ATPase activity of dynein in the axoneme is approximately five times lower than expected from microtubule-activated dynein ATPase (Maheshwari and Ishikawa 2012). Localization of suppression is supposed to occur either by a well-organized program (including 3D positioning of RS/CP) or by a spontaneous self-organization of dyneins; it could be the combination of both. Cryo-ET showed cooperative behavior of outer arm dyneins on MTDs—dynein conformations have a tendency to cluster, with 10 to 20 dyneins showing the same form in a row (Fig. 4C) (Movassagh et al. 2010). Dynein may sense mechanical strain from the conformational change of adjacent dyneins. This hypothesis is also consistent with force-triggered ciliary oscillation by a microneedle (Ishikawa and Shingyoji 2007).

Although the switching hypothesis presents a straightforward model to correlate molecular function of dynein and physiological bending, it must be examined from mechanical and kinetical aspects. Can outer and inner dyneins make as frequent ATPase as ciliary beating? ATPase turnover of microtubule-activated axonemal dyneins is two to five ATP hydrolyses per one dynein head (e.g., Shimizu et al. 1992, 2014). This means one dynein molecule needs ∼300 msec for turn over on average. Because frequency of ciliary beating is ∼50 Hz in wild-type and ∼25 Hz in ODA depleted *oda* mutants (Kamiya and Okamoto 1985), the ATPase cycle of axonemal dynein is too slow to make the transition from one curvature to the flipped curvature—it is unlikely that dyneins can turn from the pre-PS form to the post-PS form and come back to the pre-PS form in 20 msec, unless unknown factors drastically reduce the energy barrier to accelerate the ATPase cycle (however, it was shown that ATPase activity in the axoneme is not higher than in vitro) (Maheshwari and Ishikawa 2012). This suggests that, even if switching of MTD sliding is the mechanism of bending, it is not as simple as was thought. The hypothesis attributing the bending mechanism to asymmetrical distribution of conformations of dyneins is still attractive. However, dynein states in the real functional cilia will be more stochastic than the simplified switching of an entire row of dyneins.

UNEVEN DISTRIBUTION OF AXONEMAL DYNEIN ISOFORMS

It has been known that one of the nine MTDs lacks an outer arm in *Chlamydomonas*. The MTD without the ODA is numbered as MTD1 and is apposed (i.e., MTD1s of two cilia in one *Chlamydomonas* cell face each other) (Hoops and Witman 1983). Detailed analysis of *Chlamydomonas* flagella by cryo-ET proved that some inner dynein species are missing in parts of the axoneme (Bui et al. 2009). MTD1 and MTD9 lack dynein b. In MTD1, dynein c and e are also missing or fold abnormally (Bui et al. 2009). This indicates weaker sliding force at the inner side of the axoneme, suggesting that the mechanism generates an asymmetrical waveform. Dynein b is also missing in all of the MTDs in the proximal region (Bui et al. 2012), showing not only circumvention but also longitudinal asymmetry (Fig. 2B,C).

As immunolabeling and electron tomography showed recently, some dyneins are localized either in a proximal or a distal part of the axoneme. In *Chlamydomonas*, three dynein heavy chains, DHC3, DHC4, and DHC11, are localized at the proximal region ($< \sim 2$ μm from the basal body) (Yagi et al. 2009). They are called minor dyneins and replace major dyneins in the 96-nm unit—for example, DHC11 replaces dynein d (Bui et al. 2012).

In human cilia, replacement occurs in ODAs. DNAH5 is a homolog of *Chlamydomonas* dynein β and one cause of programmed cell death (PCD). It exists along the entire length of the cilia, whereas DNAH9 replaces DNAH5 partially but only at the distal region (Fliegauf et al. 2005); another outer arm dynein DNAH11 is localized at the proximal region (Dougherty et al. 2016). The mechanism and the meaning of the asymmetric distribution of dynein isoforms must be investigated further.

DYNEIN IC/LC

Although all the species of IADs and OADs are associated by ICs and LCs, structurally the most prominent and best analyzed are those of inner dynein f. ICs and LCs associated with the dynein f dimer form a large complex (40 nm in length along the MTD, 10 nm in width; yellow in Fig. 2B) and bridge the ODAs (by IC140 and IC2), dynein f, and other IADs (by IC138) (Heuser et al. 2012; Oda and Kikkawa 2013), and the connection extends further to the DRC (by the modifier of inner arms [MIA] complex) (Yamamoto et al. 2013) and RSs (Fig. 2B).

RSs AND THE CP

The repeating unit contains two or three RSs and one DRC. Each RS has a T shape and is of ~ 43 nm in height with a stalk and a head facing the CP. The first (RS1) and the second (RS2) are structurally similar, although not identical, suggesting similar components (Fig. 5C). RS1 and RS2 have branches between the stalk and the head (called a neck) (Pigino et al. 2011). At least 23 component proteins are identified from *Chlamydomonas* flagella (Yang et al. 2006). Comparison of the structures of deletion mutants enabled us to locate RSP1, 4, 6, 9, and 10 in the head, RSP2, 16, and 23 in the neck, and the rest in the stalk (Pigino et al. 2011). The head and the neck domains show pseudo twofold symmetry. It was an unexpected observation that symmetrical RS heads interact with the CP, which has clear polarity. Pseudo twofold symmetry can be explained as the sign of two preassembled subcomplexes proposed as two 12S complexes to be assembled into one 20S complex based on the L-shaped 12S complex from cytoplasm, whereas the entire RS is 20S (Diener et al. 2011). However, the assignment of RS head proteins by genetic tagging does not support the idea of twofold symmetrical arrangement of RS head proteins—RSP4 is located opposite from RSP6 (Oda et al. 2014c). When there are three RSs, the third RS (RS3) has a different structure than RS1 and RS2. *Chlamydomonas*, which has two RSs, has a short protrusion, which is at the same locus as RS3 and similar in shape to RS3. This suggests an evolutional origin of this RS3-like protrusion—either growth stopped or, once completed, RS3 was degenerated. Furthermore, mutants, which cause loss of RS1 or RS2, do not lose RS3. This indicates distinct protein components of RS1/2 and RS3 (Pigino et al. 2011; Pigino and Ishikawa 2012). The bases of RSs are connected to the tails of dyneins (Fig. 5C).

The CP consists of two singlet microtubules, composed of two sets of 13 protofilaments and various binding proteins with 32-nm periodicity. These decorating proteins form protrusions from the microtubules and linkers between two tubules (reviewed in Mitchell 2016). The CP seems straight in sperm flagella. This leads us to hypothesize that the planar bending motion of the "9+2" axoneme is defined by the CP. However, in *Chlamydomonas* and *Tetrahymena*, the CP twists shallowly (Mitchell and Nakatsugawa 2004; Pigino et al. 2012), although *Chlamydomonas* and *Tetrahymena* have planar waveforms. This hypothesis is still supported by the fact that the "9+0" axoneme from nodal cilia and eel sperm flagella makes a screw-like conical motion, instead of planar beating. The "6+0" axoneme from Gregarine (Schrevel and Besse 1975) makes a beating, but not planar, motion. Maybe the CP affects the waveform but with other influential factors. This mechanism must be further explored.

Interaction between RS heads and the CP will be central for ciliary regulation. Geometrically, there is a gap between them and it is difficult to build a model of direct protein interaction between RSs and the CP, taking twisting CP and RSs on the straight MTD into account. In addition, there is no predictable signal transduction subdomain in five RS head proteins.

Lack of chemical interaction and ellipsoidal sectional view of the CP leads us to build a hypothesis that mechanical pressure from the CP to the RS head induces regulation on the RS. A paralyzed mutant lacking one protrusion on the CP toward the RS head was rescued by recovering the volume with a genetically tagged (but completely different) protein from the CP protrusion on RS head proteins, strongly supporting this "mechanical interaction" hypothesis (Oda et al. 2014c).

LINKERS BETWEEN MTDs

There are linkers connecting adjacent MTDs in *Chlamydomonas* flagella. The DRC connects all nine adjacent pairs of MTDs. The DRC consists of at least 11 proteins (Bower et al. 2013) building bilobed structure (Heuser et al. 2009). Among them, DRC1, DRC2, and DRC4 are likely bundled coiled-coil proteins with the amino termini toward the adjacent B-tubule and the carboxyl termini anchored on the A-tubule (Oda et al. 2014b). Two more linkers between limited pairs of MTDs were found (Bui et al. 2009). IDL2 extends from near IC/LC of dynein f of MTDs 4, 5, and 9 to MTDs 5, 6, and 1. IDL3 exists only between MTD1 and MTD2 and is located close to the DRC.

Functions of these linkers are not known yet. Currently, it is unclear whether these linkers detach from MTDs during ciliary bending, which causes sliding of adjacent MTDs, or whether they keep binding and stretch. In the case when they keep bundling the adjacent MTDs, they might determine waveforms. IDL2 and IDL3 could limit sliding of MTDs 4–6 and 9–2. It could be the mechanism to limit ciliary motion planar (Bui et al. 2009).

CONCLUDING REMARKS

In this article, we reviewed what we have learned about ciliary axonemes through recent progress in structural biology, X-ray crystallography, cryo-EM single-particle analysis, cryo-ET, and 3D-image analysis. The in vitro and in vivo conformation of dyneins and the arrangement of tubulins in MTDs and the RS have recently been revealed. By combining structural biology with genetic engineering techniques, a number of features of dyneins and RSs were visualized. Structural analysis provides us insight not only into detailed 3D structure, molecular arrangement, and conformation of individual molecules but also into dynamics, cooperativity, and evolution.

ACKNOWLEDGMENTS

Our work is supported by grants from the Swiss National Science Foundation (NF3100A0-107540; NF31003A-125131/1; NF31003A-144035/1).

REFERENCES

*Reference is also in this collection.

Avolio J, Lebduska S, Sati P. 1984. Dynein arm substructure and the orientation of arm-microtubule attachments. *J Mol Biol* **173:** 389–401.

Awata J, Takada S, Standley C, Lechtreck KF, Bellvé KD, Pazour GJ, Fogarty KE, Witman GB. 2014. NPHP4 controls ciliary trafficking of membrane proteins and large soluble proteins at the transition zone. *J Cell Sci* **127:** 4714–4727.

Berns MW, Rattner JB, Brenner S, Meredith S. 1977. The role of the centriolar region in animal cell mitosis. A laser microbeam study. *J Cell Biol* **72:** 351–367.

Bower R, Tritschler D, Vanderwaal K, Perrone CA, Mueller J, Fox L, Sale WS, Porter ME. 2013. The N-DRC forms a conserved biochemical complex that maintains outer doublet alignment and limits microtubule sliding in motile axonemes. *Mol Biol Cell* **24:** 1134–1152.

Bui KH, Ishikawa T. 2013. 3D structural analysis of flagella/cilia by cryo-electron tomography. *Methods Enzymol* **524:** 305–323.

Bui KH, Sakakibara H, Movassagh T, Oiwa K, Ishikawa T. 2008. Molecular architecture of inner dynein arms in situ in *Chlamydomonas reinhardtii* flagella. *J Cell Biol* **183:** 923–932.

Bui KH, Sakakibara H, Movassagh T, Oiwa K, Ishikawa T. 2009. Asymmetry of inner dynein arms and inter-doublet links in *Chlamydomonas* flagella. *J Cell Biol* **186:** 437–446.

Bui KH, Yagi T, Yamamoto R, Kamiya R, Ishikawa T. 2012. Polarity and asymmetry in the arrangement of dynein and related structures in the *Chlamydomonas* axoneme. *J Cell Biol* **198:** 913–925.

Carter AP, Cho C, Jin L, Vale RD. 2011. Crystal structure of the dynein motor domain. *Science* **331:** 1159–1165.

Chrétien D, Wade RH. 1991. New data on the microtubule surface lattice. *Biol Cell* **71:** 161–174.

Dallai R. 2014. Overview on spermatogenesis and sperm structure of Hexapoda. *Arthropod Struct Dev* **43:** 257–290.

Diener DR, Yang P, Geimer S, Cole DG, Sale WS, Rosenbaum JL. 2011. Sequential assembly of flagellar radial spokes. *Cytoskeleton (Hoboken)* **68:** 389–400.

Diener DR, Lupetti P, Rosenbaum JL. 2015. Proteomic analysis of isolated ciliary transition zones reveals the presence of ESCRT proteins. *Curr Biol* **25:** 379–384.

Doroquez DB, Berciu C, Anderson JR, Sengupta P, Nicastro D. 2014. A high-resolution morphological and ultrastructural map of anterior sensory cilia and glia in *Caenorhabditis elegans*. *eLife* **3:** e01948.

Dougherty GW, Loges NT, Klinkenbusch JA, Olbrich H, Pennekamp P, Menchen T, Raidt J, Wallmeier J, Werner C, Westermann C, et al. 2016. DNAH11 localization in the proximal region of respiratory cilia defines distinct outer dynein arm complexes. *Am J Respir Cell Mol Biol* doi: 10.1165/rcmb.2015-0353OC.

Ebersold WT, Levine RP, Levine EE, Olmsted MA. 1962. Linkage maps in *Chlamydomonas reinhardi*. *Genetics* **47:** 531–543.

Fliegauf M, Olbrich H, Horvath J, Wildhaber JH, Zariwala MA, Kennedy J, Knowles MR, Omran H. 2005. Mislocalization of DNAH5 and DNAH9 in respiratory cells from patients with primary ciliary dyskinesia. *Am J Respir Crit Care Med* **171:** 1343–1349.

Fliegauf M, Benzing T, Omran H. 2007. When cilia go bad: Cilia defects and ciliopathies. *Nat Rev Mol Cell Biol* **8:** 880–893.

Gardner LC, O'Toole E, Perrone CA, Giddings T, Porter ME. 1994. Components of a "dynein regulatory complex" are located at the junction between the radial spokes and the dynein arms in *Chlamydomonas* flagella. *J Cell Biol* **127:** 1311–1325.

Gilliam JC, Chang JT, Sandoval IM, Zhang Y, Li T, Pittler SJ, Chiu W, Wensel TG. 2012. Three-dimensional architecture of the rod sensory cilium and its disruption in retinal neurodegeneration. *Cell* **151:** 1029–1041.

Goodenough U, Heuser J. 1984. Structural comparison of purified dynein proteins with in situ dynein arms. *J Mol Biol* **180:** 1083–1118.

Goodenough UW, Heuser JE. 1985a. Outer and inner dynein arms of cilia and flagella. *Cell* **41:** 341–342.

Goodenough UW, Heuser JE. 1985b. Substructure of inner dynein arms, radial spokes, and the central pair/projection complex of cilia and flagella. *J Cell Biol* **100:** 2008–2018.

Guichard P, Desfosses A, Maheshwari A, Hachet V, Dietrich C, Brune A, Ishikawa T, Sachse C, Gönczy P. 2012. Cartwheel architecture of *Trichonympha* basal body. *Science* **337:** 553.

Hayashi S, Shingyoji C. 2008. Mechanism of flagellar oscillation-bending-induced switching of dynein activity in elastase-treated axonemes of sea urchin sperm. *J Cell Sci* **121:** 2833–2843.

Heuser T, Raytchev M, Krell J, Porter ME, Nicastro D. 2009. The dynein regulatory complex is the nexin link and a major regulatory node in cilia and flagella. *J Cell Biol* **187:** 921–933.

Heuser T, Barber CF, Lin J, Krell J, Rebesco M, Porter ME, Nicastro D. 2012. Cryoelectron tomography reveals doublet-specific structures and unique interactions in the I1 dynein. *Proc Natl Acad Sci* **109:** E2067–E2076.

Hoops HJ, Witman GB. 1983. Outer doublet heterogeneity reveals structural polarity related to beat direction in *Chlamydomonas* flagella. *J Cell Biol* **97:** 902–908.

Huang B, Piperno G, Ramanis Z, Luck DJ. 1981. Radial spokes of *Chlamydomonas* flagella: Genetic analysis of assembly and function. *J Cell Biol* **88:** 80–88.

Ikeda K, Brown JA, Yagi T, Norrander JM, Hirono M, Eccleston E, Kamiya R, Linck RW. 2003. Rib72, a conserved protein associated with the ribbon compartment of flagellar A-microtubules and potentially involved in the linkage between outer doublet microtubules. *J Biol Chem* **278:** 7725–7734.

Inaba K. 2015. Calcium sensors of ciliary outer arm dynein: Functions and phylogenetic considerations for eukaryotic evolution. *Cilia* **4:** 6.

Ishikawa T. 2015. Cryo-electron tomography of motile cilia and flagella. *Cilia* **4:** 3.

Ishikawa R, Shingyoji C. 2007. Induction of beating by imposed bending or mechanical pulse in demembranated, motionless sea urchin sperm flagella at very low ATP concentrations. *Cell Struct Funct* **32:** 17–27.

Johnson KA. 1985. Pathway of the microtubule-dynein ATPase and the structure of dynein: A comparison with actomyosin. *Annu Rev Biophys Biophys Chem* **14:** 161–188.

Kagami O, Kamiya R. 1992. Translocation and rotation of microtubules caused by multiple species of *Chlamydomonas* inner-arm dynein. *J Cell Sci* **103:** 653–664.

Kamiya R. 2002. Functional diversity of axonemal dyneins as studied in *Chlamydomonas* mutants. *Int Rev Cytol* **219:** 115–155.

Kamiya R, Okamoto M. 1985. A mutant of *Chlamydomonas reinhardtii* that lacks the flagellar outer dynein arm but can swim. *J Cell Sci* **74:** 181–191.

Kamiya R, Yagi T. 2014. Functional diversity of axonemal dyneins as assessed by in vitro and in vivo motility assays of *Chlamydomonas* mutants. *Zoolog Sci* **31:** 633–644.

* King SM. 2016. Axonemal dynein arms. *Cold Spring Harb Perspect Biol* doi: 10.1101/cshperspect.a028100.

Kitagawa D, Vakonakis I, Olieric N, Hilbert M, Keller D, Olieric V, Bortfeld M, Erat MC, Flückiger I, Gönczy P, et al. 2011. Structural basis of the 9-fold symmetry of centrioles. *Cell* **144:** 364–375.

Kon T, Sutoh K, Kurisu G. 2011. X-ray structure of a functional full-length dynein motor domain. *Nat Struct Mol Biol* **18:** 638–642.

Kon T, Oyama T, Shimo-Kon R, Imamula K, Shima T, Sutoh K, Kurisu G. 2012. The 2.8 Å crystal structure of the dynein motor domain. *Nature* **484:** 345–350.

Lin J, Okada K, Raytchev M, Smith MC, Nicastro D. 2014. Structural mechanism of the dynein power stroke. *Nat Cell Biol* **16:** 479–485.

Linck R, Fu X, Lin J, Ouch C, Schefter A, Steffen W, Warren P, Nicastro D. 2014. Insights into the structure and function of ciliary and flagellar doublet microtubules: Tektins, Ca^{2+}-binding proteins and stable protofilaments. *J Biol Chem* **89:** 17427–17444.

Liu Q, Tan G, Levenkova N, Li T, Pugh EN, Rux JJ, Speicher DW, Pierce EA. 2007. The proteome of the mouse photoreceptor sensory cilium complex. *Mol Cell Proteomics* **6:** 1299–1317.

Lupetti P, Lanzavecchia S, Mercati D, Cantele F, Dallai R, Mencarelli C. 2005. Three-dimensional reconstruction of axonemal outer dynein arms in situ by electron tomography. *Cell Motil Cytoskeleton* **62:** 69–83.

Maheshwari A, Ishikawa T. 2012. Heterogeneity of dynein structure implies coordinated suppression of dynein motor activity in the axoneme. *J Struct Biol* **179:** 235–241.

Maheshwari A, Obbineni JM, Bui KH, Shibata K, Toyoshima YY, Ishikawa T. 2015. α- and β-tubulin lattice of the axonemal microtubule doublet and binding proteins revealed by single particle cryo-electron microscopy and tomography. *Structure* **23:** 1584–1595.

Mastronarde DN, O'Toole ET, McDonald KL, McIntosh JR, Porter ME. 1992. Arrangement of inner dynein arms in wild-type and mutant flagella of *Chlamydomonas*. *J Cell Biol* **118:** 1145–1162.

Mayer U, Küller A, Daiber PC, Neudorf I, Warnken U, Schnölzer M, Frings S, Möhrlen F. 2009. The proteome of rat olfactory sensory cilia. *Proteomics* **9:** 322–334.

Mencarelli C, Lupetti P, Rosetto M, Dallai R. 2000. Morphogenesis of the giant sperm axoneme in *Asphondylia ruebsaameni* Kertesz (Diptera, Cecidomyiidae). *Tissue Cell* **32:** 188–197.

Minamino T, Imada K, Namba K. 2008. Molecular motors of the bacterial flagella. *Curr Opin Struct Biol* **18:** 693–701.

* Mitchell DR. 2016. Evolution of cilia. *Cold Spring Harb Perspect Biol* doi: 10.1101/cshperspect.a028290.

Mitchell DR, Nakatsugawa M. 2004. Bend propagation drives central pair rotation in *Chlamydomonas reinhardtii* flagella. *J Cell Biol* **166:** 709–715.

Mollaret I, Justine JL. 1997. Immunocytochemical study of tubulin in the 9 + "1" sperm axoneme of a monogenean (Platyhelminthes), *Pseudodactylogyrus* sp. *Tissue Cell* **29:** 699–706.

Movassagh T, Bui KH, Sakakibara H, Oiwa K, Ishikawa T. 2010. Nucleotide-induced global conformational changes of flagellar dynein arms revealed by in situ analysis. *Nat Struct Mol Biol* **17:** 761–767.

Nakazawa Y, Ariyoshi T, Noga A, Kamiya R, Hirono M. 2014. Space-dependent formation of central pair microtubules and their interactions with radial spokes. *PLoS ONE* **9:** e110513.

Neuwald AF, Aravind L, Spouge JL, Koonin EV. 1999. Assembly, operation, and disassembly of protein complexes AAA: A class of chaperone-like ATPases associated with the assembly, operation, and disassembly of protein complexes. *Genome Res* **9:** 27–43.

Nicastro D, Fu X, Heuser T, Tso A, Porter ME, Linck RW. 2011. Cryo-electron tomography reveals conserved features of doublet microtubules in flagella. *Proc Natl Acad Sci* **108:** E845–853.

Nogales E, Wolf SG, Downing KH. 1998. Structure of the αβ tubulin dimer by electron crystallography. *Nature* **391:** 199–203.

Oda T, Kikkawa M. 2013. Novel structural labeling method using cryo-electron tomography and biotin-streptavidin system. *J Struct Biol* **183:** 305–311.

Oda T, Yanagisawa H, Kamiya R, Kikkawa M. 2014a. Cilia and flagella. A molecular ruler determines the repeat length in eukaryotic cilia and flagella. *Science* **346**: 857–860.

Oda T, Yanagisawa H, Kikkawa M. 2014b. Detailed structural and biochemical characterization of the nexin–dynein regulatory complex. *Mol Biol Cell* **26**: 294–304.

Oda T, Yanagisawa H, Yagi T, Kikkawa M. 2014c. Mechano-signaling between central apparatus and radial spokes controls axonemal dynein activity. *J Cell Biol* **204**: 807–819.

Paschal BM, Mikami A, Pfister KK, Vallee RB. 1992. Homology of the 74-kD cytoplasmic dynein subunit with a flagellar dynein polypeptide suggests an intracellular targeting function. *J Cell Biol* **118**: 1133–1143.

Pazour GJ, Agrin N, Leszyk J, Witman GB. 2005. Proteomic analysis of a eukaryotic cilium. *J Cell Biol* **170**: 103–113.

Pigino G, Ishikawa T. 2012. Axonemal radial spokes: 3D structure, function and assembly. *Bioarchitecture* **2**: 50–58.

Pigino G, Bui KH, Maheshwari A, Lupetti P, Diener D, Ishikawa T. 2011. Cryoelectron tomography of radial spokes in cilia and flagella. *J Cell Biol* **195**: 673–687.

Pigino G, Maheshwari A, Bui KH, Shingyoji C, Kamimura S, Ishikawa T. 2012. Comparative structural analysis of eukaryotic flagella and cilia from *Chlamydomonas*, *Tetrahymena*, and sea urchins. *J Struct Biol* **178**: 199–206.

Prensier G, Vivier E, Goldstein S, Schrével J. 1980. Motile flagellum with a "3 + 0" ultrastructure. *Science* **207**: 1493–1494.

Ray S, Meyhöfer E, Milligan RA, Howard J. 1993. Kinesin follows the microtubule's protofilament axis. *J Cell Biol* **121**: 1083–1093.

Roberts AJ, Numata N, Walker ML, Kato YS, Malkova B, Kon T, Ohkura R, Arisaka F, Knight PJ, Sutoh K, et al. 2009. AAA+ ring and linker swing mechanism in the dynein motor. *Cell* **136**: 485–95.

Roberts AJ, Malkova B, Walker ML, Sakakibara H, Numata N, Kon T, Ohkura R, Edwards TA, Knight PJ, Sutoh K, et al. 2012. ATP-driven remodeling of the linker domain in the dynein motor. *Struct Lond Engl* **20**: 1670–1680.

Roberts AJ, Kon T, Knight PJ, Sutoh K, Burgess SA. 2013. Functions and mechanics of dynein motor proteins. *Nat Rev Mol Cell Biol* **14**: 713–726.

Sale WS, Satir P. 1977. Direction of active sliding of microtubules in *Tetrahymena* cilia. *Proc Natl Acad Sci* **74**: 2045–2049.

Satir P, Heuser T, Sale WS. 2014. A structural basis for how motile cilia beat. *Bioscience* **64**: 1073–1083.

Schmidt H, Gleave ES, Carter AP. 2012. Insights into dynein motor domain function from a 3.3-Å crystal structure. *Nat Struct Mol Biol* **19**: 492–497.

Schmidt H, Zalyte R, Urnavicius L, Carter AP. 2014. Structure of human cytoplasmic dynein-2 primed for its power-er stroke. *Nature* **518**: 435–438.

Schrevel J, Besse C. 1975. A functional flagella with a 6 + 0 pattern. *J Cell Biol* **66**: 492–507.

Segal RA, Huang B, Ramanis Z, Luck DJ. 1984. Mutant strains of *Chlamydomonas reinhardtii* that move backwards only. *J Cell Biol* **98**: 2026–2034.

Shimizu T, Johnson KA. 1983. Presteady state kinetic analysis of vanadate-induced inhibition of the dynein ATPase. *J Biol Chem* **258**: 13833–13840.

Shimizu T, Hosoya N, Hisanaga S, Marchese-Ragona SP, Pratt MM. 1992. Activation of ATPase activity of 14S dynein from *Tetrahymena* cilia by microtubules. *Eur J Biochem* **206**: 911–917.

Shimizu Y, Sakakibara H, Kojima H, Oiwa K. 2014. Slow axonemal dynein e facilitates the motility of faster dynein c. *Biophys J* **106**: 2157–2165.

Smith JC, Northey JGB, Garg J, Pearlman RE, Siu KWM. 2005. Robust method for proteome analysis by MS/MS using an entire translated genome: Demonstration on the ciliome of *Tetrahymena thermophila*. *J Proteome Res* **4**: 909–919.

Starling D, Randall J. 1971. Flagella of temporary dikaryons of *Chlamydomonas reinhardii*. *Genet Res* **18**: 107.

Sugino K, Naitoh Y. 1982. Simulated cross-bridge patterns corresponding to ciliary beating in paramecium. *Nature* **295**: 609–611.

Sui H, Downing KH. 2006. Molecular architecture of axonemal microtubule doublets revealed by cryo-electron tomography. *Nature* **442**: 475–478.

Takahashi Y, Edamatsu M, Toyoshima YY. 2004. Multiple ATP-hydrolyzing sites that potentially function in cytoplasmic dynein. *Proc Natl Acad Sci* **101**: 12865–12869.

Ueno H, Bui KH, Ishikawa T, Imai Y, Yamaguchi T, Ishikawa T. 2014. Structure of dimeric axonemal dynein in cilia suggests an alternative mechanism of force generation. *Cytoskeleton (Hoboken)* **71**: 412–422.

van Breugel M, Hirono M, Andreeva A, Yanagisawa H, Yamaguchi S, Nakazawa Y, Morgner N, Petrovich M, Ebong IO, Robinson CV, et al. 2011. Structures of SAS-6 suggest its organization in centrioles. *Science* **331**: 1196–1199.

Warner FD, Satir P. 1974. The structural basis of ciliary bend formation. Radial spoke positional changes accompanying microtubule sliding. *J Cell Biol* **63**: 35–63.

Woolley DM. 1997. Studies on the eel sperm flagellum. I: The structure of the inner dynein arm complex. *J Cell Sci* **110**: 85–94.

Woolley DM. 1998. Studies on the eel sperm flagellum. 2: The kinematics of normal motility. *Cell Motil Cytoskeleton* **39**: 233–245.

Yagi T, Uematsu K, Liu Z, Kamiya R. 2009. Identification of dyneins that localize exclusively to the proximal portion of *Chlamydomonas* flagella. *J Cell Sci* **122**: 1306–1314.

Yamamoto R, Song K, Yanagisawa HA, Fox L, Yagi T, Wirschell M, Hirono M, Kamiya R, Nicastro D, Sale WS. 2013. The MIA complex is a conserved and novel dynein regulator essential for normal ciliary motility. *J Cell Biol* **201**: 263–278.

Yang P, Yang C, Sale WS. 2004. Flagellar radial spoke protein 2 is a calmodulin binding protein required for motility in *Chlamydomonas reinhardtii*. *Eukaryot Cell* **3**: 72–81.

Yang P, Diener DR, Yang C, Kohno T, Pazour GJ, Dienes JM, Agrin NS, King SM, Sale WS, Kamiya R, et al. 2006. Radial spoke proteins of *Chlamydomonas* flagella. *J Cell Sci* **119**: 1165–1174.

Posttranslational Modifications of Tubulin and Cilia

Dorota Wloga,[1] Ewa Joachimiak,[1] Panagiota Louka,[2] and Jacek Gaertig[2]

[1]Laboratory of Cytoskeleton and Cilia Biology, Department of Cell Biology, Nencki Institute of Experimental Biology, Polish Academy of Sciences, 02-093 Warsaw, Poland

[2]Department of Cellular Biology, University of Georgia, Athens, Georgia 30602

Correspondence: dwloga@nencki.gov.pl; jgaertig@uga.edu

Tubulin undergoes several highly conserved posttranslational modifications (PTMs) including acetylation, detyrosination, glutamylation, and glycylation. These PTMs accumulate on a subset of microtubules that are long-lived, including those in the basal bodies and axonemes. Tubulin PTMs are distributed nonuniformly. In the outer doublet microtubules of the axoneme, the B-tubules are highly enriched in the detyrosinated, polyglutamylated, and polyglycylated tubulin, whereas the A-tubules contain mostly unmodified tubulin. The nonuniform patterns of tubulin PTMs may functionalize microtubules in a position-dependent manner. Recent studies indicate that tubulin PTMs contribute to the assembly, disassembly, maintenance, and motility of cilia. In particular, tubulin glutamylation has emerged as a key PTM that affects ciliary motility through regulation of axonemal dynein arms and controls the stability and length of the axoneme.

TYPES OF CONSERVED TUBULIN PTMs

Tubulin undergoes several conserved posttranslational modifications (PTMs) (Janke 2014; Song and Brady 2015; Yu et al. 2015). The most studied tubulin PTMs and their responsible enzymes are summarized in Figure 1. These PTMs are enriched on the long-lived microtubules, including those of cilia and centrioles. Some tubulin PTMs appear to have coevolved with cilia (Janke et al. 2005; Shida et al. 2010). Although antibodies that recognize PTMs have been widely used to detect cilia (Magiera and Janke 2013), the functions of tubulin PTMs have started to emerge only recently. Within the cilium, there are striking differences in the levels of tubulin PTM among the specific microtubules and along the length or even around the circumference of the same microtubule (Fig. 2). According to the "tubulin code" model (Verhey and Gaertig 2007), tubulin PTMs form patterns of marks on the microtubules that locally influence various activities, such as the motility of motor proteins or severing factors. Although we focus on cilia, we will discuss key findings on the nonciliary microtubules as well, as it is likely that tubulin PTMs provide related if not identical functions inside and outside of cilia.

ACETYLATION OF K40 OF α-TUBULIN

This highly conserved PTM was discovered in cilia of *Chlamydomonas reinhardtii* (L'Hernault

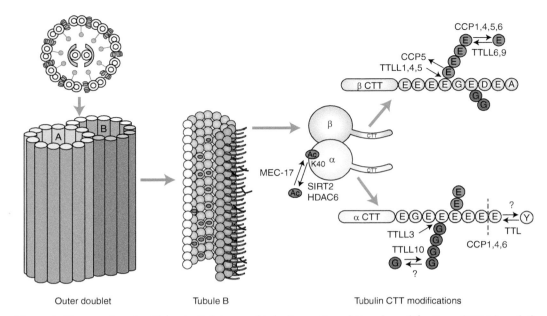

| Outer doublet | Tubule B | Tubulin CTT modifications |

Figure 1. Conserved and widely studied types of tubulin posttranslational modifications (PTMs) and the responsible enzymes that act in cilia. CTT, Carboxy-terminal tail; TTL(L), tubulin tyrosine ligase(-like); CCP, cytosolic carboxypeptidase.

and Rosenbaum 1983; LeDizet and Piperno 1987). Acetyl-K40 is the only known PTM that is located inside the microtubule lumen (Nogales et al. 1999; Soppina et al. 2012). Although mass spectrometry detected additional acetyl-Ks on tubulin (Choudhary et al. 2009; Chu et al. 2011; Liu et al. 2015), acetyl-K40 is the most abundant, if not exclusive in cilia (Akella et al. 2010). Acetyl-K40 marks long-lived microtubules, including those in the axonemes and basal bodies (Piperno and Fuller 1985). The enzyme that generates acetyl-K40, MEC-17/ATAT1 (Akella et al. 2010; Shida et al. 2010), has a catalytic domain homologous to that of the histone acetyltransferases (Friedmann et al. 2012; Kormendi et al. 2012; Li et al. 2012; Taschner et al. 2012; Davenport et al. 2014; Yuzawa et al. 2015), but its enzymatic rate is relatively slow because of "suboptimal" positions of the catalytic residues. Consequently, only long-lived microtubules carry acetyl-K40 (Szyk et al. 2014). Most of the axonemal α-tubulin has acetyl-K40 (Gaertig et al. 1995), and the PTM is present on both the central and outer microtubules (Piperno and Fuller 1985; Satish Tammana et al. 2013).

Tetrahymena mutants lacking MEC-17/ ATAT1 or expressing a K40R nonacetylatable α-tubulin have a normal gross phenotype but display altered sensitivities to tubulin-binding compounds (Gaertig et al. 1995; Akella et al. 2010). In *Chlamydomonas*, expression of a K40R α-tubulin did not detectably affect the gross phenotype (Kozminski et al. 1993). Surprisingly, in the apicomplexan *Toxoplasma gondii*, K40R α-tubulin-expressing parasites are viable only if a second mutation that confers oryzalin resistance (and likely changes the dynamics of microtubules) is cointroduced into α-tubulin. Furthermore, a depletion of MEC-17/ATAT1 in *Toxoplasma* inhibits mitosis and changes cell shape, indicating that the lack of acetylation affects both the nuclear and subpellicular microtubules (Varberg et al. 2016). In *Caenorhabditis elegans*, the acetyl-K40 α-tubulin is expressed exclusively in the mechanosensory (ciliated and nonciliated) neurons (Fukushige et al. 1999; Shida et al. 2010). Without MEC-17/ATAT1, the animals become touch-insensitive (Chalfie and Au 1989; Zhang et al. 2002; Akella et al. 2010; Shida et al. 2010), and

Cite this article as *Cold Spring Harb Perspect Biol* doi: 10.1101/cshperspect.a028159

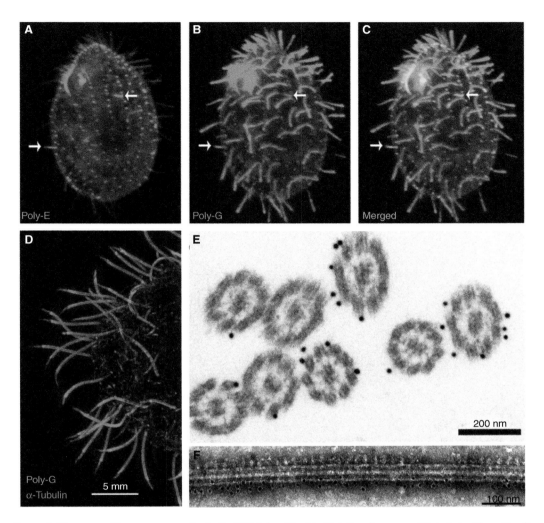

Figure 2. The nonuniform patterns of tubulin PTMs in cilia. (*A-C*) A *Tetrahymena* cell labeled with antibodies that recognize either polyglutamate (poly-E, green) (Shang et al. 2002) or polyglycine side chains (AXO49) (Bré et al. 1998). Note that the shorter assembling cilia (arrows) have a higher signal of polyglutamylation and a lower signal of polyglycylation as compared to the longer mature cilia on the same cell. These data suggest that during their assembly, cilia undergo remodeling of tubulin PTMs. (Figure adapted from Sharma et al. 2007; originally published in *Journal of Cell Biology.*) (*D*) A superresolution structured illumination microscopy (SR-SIM) image of the posterior end of a *Tetrahymena* cell, with cilia labeled with poly-G (green) (Duan and Gorovsky 2002) and anti-α-tubulin primary sequence antibodies 12G10 (red) (Jerka-Dziadosz et al. 1995). Note an absence of a signal of poly-G near the tips of cilia (distal segment) where the B-tubules are absent. (*E*) A postembedding immunogold image of cross sections of *Tetrahymena* cilia labeled with antipolyglycine AXO49 antibodies. Note the absence of a signal near the central microtubules. (From Wloga et al. 2009; adapted, with permission.) (*F*) An isolated doublet microtubule of *Chlamydomonas* labeled by poly-E antibodies with 10 nm colloidal gold and negatively stained with uranyl acetate. The A-tubule side can be identified by the presence of dynein arms (*top*), whereas the B-tubule has a smoother surface (*bottom*). Note that the gold particles are enriched along the B-tubule. (From Kubo et al. 2010; reprinted, with permission, from Elsevier © 2010.)

the neurites of the nonciliated touch receptor neurons contain fewer microtubules that have reduced diameters and lattice defects and are prone to buckling as the animal body bends (Cueva et al. 2012; Topalidou et al. 2012). A molecular dynamics simulation suggests that acetylation of K40 promotes the cohesion between protofilaments of microtubules (Cueva et al. 2012). Thus, it appears that acetyl-K40 has a microtubule lattice-stabilizing activity. Importantly, MEC-17/ATAT1 has additional functions beyond its acetyltransferase activity. The loss of touch sensation in the *C. elegans* mec-17 mutant animals (but not the defects in microtubule organization) can be rescued by expression of a catalytically inactive MEC17/ATAT1 (Topalidou et al. 2012). Also, the mammalian MEC-17/ATAT1 has a microtubule-destabilizing activity that is independent of its acetyltransferase activity (Kalebic et al. 2013a,b). Although siRNA depletion of MEC-17/ATAT1 delays the assembly of primary cilia (Shida et al. 2010), the mice lacking MEC-17/ATAT1 have not been reported to have cilia that are structurally defective or of improper length (Kalebic et al. 2013b; Kim et al. 2013).

ROLE OF ACETYL-K40 AND ASSOCIATED ENZYMES IN THE DISASSEMBLY OF CILIA

HDAC6 (Hubbert et al. 2003; Zhang et al. 2003) and SIRT2 (North et al. 2003) are two acetyl-K40 deacetylases that promote disassembly of primary cilia in mammalian cells. Knockdowns of either HDAC6 or SIRT2 increase the frequency of primary cilia or make them longer, whereas overexpressions lead to fewer or shorter cilia (Pugacheva et al. 2007; Yang et al. 2014; Zhou et al. 2014; Bangs et al. 2015; Ran et al. 2015). The cilia-shortening effect of overexpressed HDAC6 is counteracted by overexpression of a K40Q (acetyl-K mimic) α-tubulin indicating that HDAC6 acts through deacetylation of acetyl-K40. Thus, by deacetylating acetyl-K40, HDAC6 and SIRT2 could have a microtubule lattice-destabilizing effect (Cueva et al. 2012). Alternatively, deacetylation could enable another PTM on K40. In *Chlamydomonas*, α-tubulin (but not β-tubulin) undergoes lysine poly-

ubiquitination when the cilium disassembles (Huang et al. 2009). However, the effects of HDAC6 and SIRT2 are unlikely to be mediated solely by K40 ubiquitination, because poly-ubiquitinated α-tubulin has acetyl-K40 (Huang et al. 2009). HDAC6 also deacetylates cortactin, and this activity also shortens the cilium, by promoting actin polymerization around the base of the primary cilium (Ran et al. 2015); it is well documented that actin dynamics affects the length of cilia (discussed in Liang et al. 2016). The HDAC6 knockout mouse lacks ciliary phenotypes (Zhang et al. 2008), but the loss of HDAC6 could be covered by SIRT2.

ROLE OF ACETYL-K40 IN MOTOR-DRIVEN MOTILITY

Acetyl-K40 may influence motor proteins, also in cilia. Mice lacking MEC-17/ATAT1 are healthy, but the males have defective sperm cells whose flagella beat with a reduced bend amplitude (Kalebic et al. 2013b; Kim et al. 2013). Enzymatic acetylation of K40 increases, whereas deacetylation by SIRT2 decreases the rate of motility of axonemal outer arm dynein on microtubules in vitro, respectively (Alper et al. 2014). The effect of acetyl-K40 on the motility of axonemal dynein is surprising because K40 is located inside the microtubule lumen (Nogales et al. 1999; Soppina et al. 2012). However, changes in the microtubule lumen can affect the functionality of the external surface (Raff et al. 2008). Cryoelectron tomography failed to detect a difference between the tubulin dimers inside microtubules that are assembled of either acetylated or nonacetylated tubulin, but perhaps the resolution achieved (8 Å) was insufficient to detect a change that affects dynein (Howes et al. 2014).

In vitro, kinesin-1 binds more weakly and moves more slowly on mutant axonemes containing K40R α-tubulin, as compared to wild-type axonemes carrying acetyl-K40 (Reed et al. 2006). However, more recent studies that used enzymatically in vitro acetylated or deacetylated microtubules failed to detect an effect of acetyl-K40 on kinesin-1 motility in vitro (Walter et al. 2012; Kaul et al. 2014). In the study that reported a positive result for kinesin-1 (Reed et al.

2006), the nonacetylated microtubules were produced genetically and used in the form of isolated axonemes. The K40R mutation could have altered the composition of other axonemal proteins or the levels of other tubulin PTMs. One potential "reader" of the acetyl-K40 mark is a protein with a bromodomain, found in the flagellum of *Trypanosoma* (Alonso et al. 2014). Also, acetyl-K40 could affect the components of the lumen of outer doublets, microtubule inner proteins (MIPs) (Nicastro et al. 2011; Yanagisawa et al. 2014). Although the effects of acetyl-K40 on other tubulin PTMs have not been studied, a reverse relationship exists: The levels of acetyl-K40 increase in the axonemes of *Tetrahymena* mutants that are deficient in tubulin glycylation (Wloga et al. 2009). Such a PTM cross talk could compensate for an absence of acetyl-K40 and could explain the mild consequences of the loss of this PTM in some models.

DETYROSINATION OF α-TUBULIN

The carboxy-terminal Y on α-tubulin is removed by an unidentified carboxypeptidase resulting in detyrosination that occurs preferentially on microtubules (Kumar and Flavin 1981; Arce and Barra 1983). The tubulin tyrosine ligase (TTL) restores the carboxy-terminal Y on the unpolymerized tubulin (Schroder et al. 1985; Ersfeld et al. 1993; Szyk et al. 2011; Prota et al. 2013). Mice lacking TTL die shortly after birth, with defects in the organization of neurons (Erck et al. 2005). In cilia, detyrosinated tubulin is enriched on the B-tubules of outer doublets (Johnson 1998). The anterograde intraflagellar transport (IFT) trains that are driven by kinesin-2 move on the B-tubule, whereas the retrograde trains driven by IFT dynein move on the A-tubule (Kozminski et al. 1995; Pigino et al. 2009; Stepanek and Pigino 2016). In vitro, detyrosination increases the velocity and processivity of kinesin-2, suggesting that this PTM could stimulate the anterograde IFT (Sirajuddin et al. 2014). Detyrosination increases the landing rate of kinesin-1 on microtubules (Kaul et al. 2014). A combination of effects on the landing rate and processivity could explain the bias of kinesin-2-driven anterograde IFT trains for the B-tubule,

which in turn could mediate the collision-free trafficking between the base and the tip of the cilium (Stepanek and Pigino 2016). However, the B-tubule is also modified by polymodifications (see below). How the retrograde IFT trains could be biased for the A-tubule is less clear. The A-tubule has low levels of PTMs and its α-tubulin is mostly tyrosinated (Johnson 1998). The yeast cytoplasmic dynein, a motor similar to IFT dynein, is not sensitive to the status of the carboxy-terminal Y (Sirajuddin et al. 2014).

GLUTAMYLATION

Glutamylation is an addition of a side-chain peptide composed of Es to the γ-carboxyl group of an E in the protein primary sequence (Eddé et al. 1990). Glutamylation occurs on multiple E sites within the carboxy-terminal tail domains (CTTs) of α- and β-tubulin (Redeker et al. 1991, 1992). The CTTs are flexible domains that are likely to form a negatively charged "polymer brush" on the microtubule (Roll-Mecak 2015). A related PTM, tubulin glycylation, occurs within the same cluster of Es (Redeker et al. 1994) and the two polymodifications are in a competition (Rogowski et al. 2009; Wloga et al. 2009). In *Tetrahymena*, a substitution of multiple Es (to charge-conserving but nonmodifiable Ds) on the CCT of β-tubulin, causes a failure in the axoneme assembly, including its shortness, loss of the central microtubules, and loss of the B-tubules (Xia et al. 2000; Thazhath et al. 2002). A substitution of all Es on the CTT of α-tubulin is well tolerated (Wloga et al. 2008). The exact placement of the modifiable Es on tubulin is not important; for example, the severe axoneme assembly defect caused by E to D substitutions on β-tubulin is rescued by replacement of the CTT of α-tubulin with the corresponding sequence of the β-tubulin CTT (Xia et al. 2000; Duan and Gorovsky 2002).

Glutamylation is generated by a subset of enzymes related to TTL, TTL-like proteins (or TTLLs) (Janke et al. 2005). The glutamylating (glutamylase) TTLLs use a region adjacent to the enzymatic core as a microtubule-binding domain and preferably modify microtubules over unpolymerized tubulin (Garnham et al.

2015). There are several glutamylase TTLL subtypes, which differ in the levels of chain-initiating and elongating activities, on either α- or β-tubulin, but the exact enzymatic profiles vary greatly among the TTLL orthologs or even paralogs (van Dijk et al. 2007; Wloga et al. 2008; Suryavanshi et al. 2010). The catalytic core of the mammalian TTLL7 is flexible enough to allow the microtubule-docked enzyme to perform both chain initiation and elongation reactions on β-tubulin (Mukai et al. 2009; Garnham et al. 2015), but other TTLLs are active mostly in either initiation or elongation (van Dijk et al. 2007; Wloga et al. 2008; Suryavanshi et al. 2010). Some TTLLs have an autonomous activity, whereas others need associated activators (Janke et al. 2005). TTLLs are enriched in cilia and basal bodies (van Dijk et al. 2007; Wloga et al. 2008; Suryavanshi et al. 2010). In mammalian and zebrafish cells, a centrosomal protein, CEP41, is required for targeting of TTLL6 to cilia (Lee et al. 2012). In *Chlamydomonas* cilia, TTLL9 is transported by IFT, in a complex with an adapter, FAP234 (Kubo et al. 2014).

Trimming of the polyglutamate side chains on tubulin (deglutamylation) is accomplished by the zinc carboxypeptidases: CCP1, -4, -5, and -6 (Kimura et al. 2010; Rogowski et al. 2010; Tort et al. 2014). CCPs also remove the terminal E from the end of detyrosinated α-tubulin, creating the "Δ2 α-tubulin" (Rogowski et al. 2010) also in cilia (Paturie-Lafanechère et al. 1994). Polyglutamylation also occurs on nontubulin proteins (van Dijk et al. 2008), and the same TTLL can glutamylate both a tubulin and a nontubulin substrate (van Dijk et al. 2007; Xia et al. 2016). CCPs also deglutamylate nontubulin proteins (Rogowski et al. 2010; Ye et al. 2014). Thus, caution is needed in the interpretation of the effects of manipulations of TTLLs or CCPs. In *Tetrahymena*, glutamylated tubulin is present on the outer doublets but absent from the central pair (Suryavanshi et al. 2010). However, the central microtubules are glutamylated in the sperm of *Drosophila melanogaster* and *Lytechinus pictus* (Hoyle et al. 2008). In motile cilia, within the outer doublet, glutamylation is enriched on the B-tubule (Fig. 2F) (Lechtreck and Geimer 2000; Kubo et al. 2010;

Suryavanshi et al. 2010). It is not clear to what extent this enrichment is a result of targeting of TTLLs or CCPs or a consequence of the saturation of most of the A-tubule surface with protein complexes such as dynein arms and radial spokes (Ishikawa 2015).

GLUTAMYLATION AND CILIARY MOTILITY

Loss of glutamylating TTLL subtypes (TTLL1, TTLL5, TTLL6, TTLL9) either disturbs the waveform or reduces the beat frequency or causes stalling of motile cilia (Ikegami et al. 2010; Kubo et al. 2010; Suryavanshi et al. 2010; Pathak et al. 2011; Bosch Grau et al. 2013; Konno et al. 2016); in most of these studies, the motility defects are not associated with gross structural defects in the axoneme (with some exceptions, see below). In the mouse, loss of function of TTLL1 phenocopies primary ciliary dyskinesia associated with reduced motility of respiratory cilia (Ikegami et al. 2010; Vogel et al. 2010). Among the axonemal dynein arms that drive ciliary motility, the inner dynein arms (IDAs) are mainly involved with the ciliary waveform, whereas the outer dynein arms (ODAs) control the beat frequency (Kamiya 2002). The reduction in the beat frequency in the absence of TTLL6, a β-tubulin elongase (Suryavanshi et al. 2010; Pathak et al. 2011; Bosch Grau et al. 2013), indicates that glutamylation regulates ODAs. Of key importance is that in *Chlamydomonas*, a mutant lacking ODAs has residual ciliary motility (supported by the IDAs alone), but a double mutant lacking TTLL9 glutamylase and ODAs is completely paralyzed. This indicates that tubulin glutamylation is essential for the waveform control by IDAs (Kubo et al. 2010). A depletion of CCP5 in zebrafish leads to hyperglutamylation of ciliary microtubules and a reduction in the beat amplitude of pronephric cilia (Pathak et al. 2014). Surprisingly, in vitro, in the axonemes lacking ODAs that were treated with ATP, the velocity of sliding of microtubules increases in the absence of either TTLL9 or TTLL6 (Kubo et al. 2010; Suryavanshi et al. 2010). This leads to a counterintuitive model that the IDA activity on microtubules is inhibited by tubulin glutamylation. There are several subtypes of IDAs, each occupying a different position within the 96-nm repeat on the A-tubule (Bui et al. 2009; Barber et al. 2012). In *Chlamydomonas*, a loss of TTLL9 specifically affects the function of dynein e, an IDA subspecies (Kubo et al. 2012). This effect could involve positive charge patches in the microtubule-biding domain of dynein e that could interact with the negatively charged polyglutamate side chains (Kubo et al. 2012). Glutamylation of tubulin on the B-tubule could increase the time spent by dynein e in a microtubule-bound state. This could increase the cohesion between the adjacent microtubule doublets and affect the engagement of other dyneins with the B-tubule (Kubo et al. 2012). Alternatively, the dynein e/B-tubule links could increase the microtubule curvature by generating a drag force on the microtubules that are simultaneously pushed by other "faster" dyneins (Kotani et al. 2007; Suryavanshi et al. 2010). In addition to the effects on dynein e, TTLL9 also affects the axoneme integrity by influencing the DRC-nexin/B-tubule links between the adjacent doublets (Kubo et al. 2012; Alford et al. 2016).

GLUTAMYLATION IN CILIARY ASSEMBLY AND MICROTUBULE STABILITY

Glutamylating TTLLs are important for the assembly or stability of ciliary microtubules, most notably in the mammalian sperm. The males of mice lacking TTLL1 or its partner subunit Pgs1 are infertile and produce sperm with truncated axonemes (Campbell et al. 2002; Regnard et al. 2003; Vogel et al. 2010). In the TTLL5-null mouse, the sperm axonemes frequently lack a single doublet, number 4 (Lee et al. 2013). Similarly, a deletion of TTLL9 leads to a loss of the distal portion of doublet 7 in the sperm axoneme (Konno et al. 2016).

The sperm axoneme defects could result from either a lack of proper assembly or instability. In the *Tetrahymena* cells overproducing TTLL6, the cytoplasmic microtubules undergo hyperglutamylation and excessive stabilization, associated with increased K40-acetylation and resistance to depolymerization (Wloga et al. 2010). Tubulin glutamylation could increase

the microtubule stability by promoting the binding of stabilizing factors. Centriole and spindle-associated protein (CSAP) colocalizes with the polyglutamylated microtubules, including those in cilia, centrioles, and the mitotic spindle (Backer et al. 2012). Depletion of CSAP destabilizes microtubules of the mitotic spindle (Ohta et al. 2015). In zebrafish, a depletion of CSAP causes defects in the left–right asymmetry in association with impaired beating of cilia inside Kupffer's vesicle (Backer et al. 2012). Thus, CSAP could be a tubulin code reader that recognizes and stabilizes polyglutamylated microtubules, including those present in motile cilia.

The axoneme structure defects could result from a suboptimal IFT (Hou et al. 2007; Dave et al. 2009). The anterograde IFT trains driven by kinesin-2 move on the B-tubule of the outer doublets (Kozminski et al. 1995; Pigino et al. 2009; Stepanek and Pigino 2016), which is enriched in tubulin glutamylation (Lechtreck and Geimer 2000; Kubo et al. 2010; Suryavanshi et al. 2010). Polyglutamylation of β-tubulin increases the processivity and speed of kinesin-2 on microtubules in vitro (Sirajuddin et al. 2014). The velocity of Osm3 kinesin-2 increases on the hyperglutamylated axonemes in cilia of *C. elegans* lacking CCP1 (O'Hagan et al. 2011). In zebrafish, defects in ciliogenesis caused by depletions of either fleer/DYF1/IFT70 or IFT88 IFT train subunits are partially rescued by a depletion of CCP5 (Pathak et al. 2014), arguing that increased tubulin glutamylation can improve a suboptimal IFT. However, the axoneme assembly defects caused by deficiencies in TTLLs could also be caused by a lack of glutamylation of nontubulin proteins. In addition to the sperm defects (Lee et al. 2013), a loss of TTLL5 also causes retinal dystrophy (Sergouniotis et al. 2014), but this phenotype is mainly, if not entirely, caused by a lack of glutamylation of RPGR, a protein that functions in the connecting cilium of the photoreceptor (Sun et al. 2016).

ROLE OF GLUTAMYLATION IN THE DISASSEMBLY AND NEGATIVE LENGTH REGULATION OF CILIA

Excessive tubulin glutamylation may destabilize and shorten cilia. In *C. elegans*, despite a stimulatory effect on the anterograde IFT velocity described above, a loss of CCP1 causes an age-dependent degeneration of cilia, including a loss of the B-tubules (O'Hagan et al. 2011). A loss of CCP1 in the (pcd) mouse causes degeneration of the Purkinje neurons, bulb mitral cells, and retina photoreceptors and male sterility (Greer and Shepherd 1982; Fernandez-Gonzalez et al. 2002; Rogowski et al. 2010). Because all of the cell types affected in the pcd mouse are ciliated, these phenotypes can at least in part be caused by a degeneration of cilia. Consistently, the mammalian CCP1 is a positive regulator of the length of the primary cilium (Kim et al. 2010). In Tetrahymena, overexpression of TTLL6 glutamylase causes rapid shortening and fragmentation of axonemes and a loss of B-tubules (Wloga et al. 2010), which phenocopies overexpression of katanin (Sharma et al. 2007). The depolymerizing effect of TTLL6 is axoneme-specific; in the same cells, nonciliary microtubules undergo stabilization (Wloga et al. 2010). This argues that hyperglutamylation does not have a microtubule disassembly-promoting effect per se but rather that it activates axoneme-associated microtubule depolymerizers. Specifically, hyperglutamylation could activate katanin to sever the axoneme (O'Hagan and Barr 2012). There is substantial evidence that tubulin glutamylation stimulates katanin and the structurally similar spastin (Lu et al. 2004; Sharma et al. 2007; Roll-Mecak and Vale 2008; Lacroix et al. 2010; Valenstein and Roll-Mecak 2016). Katanin subunits are present inside cilia where they associate with the outer doublets (Dymek et al. 2004; Sharma et al. 2007).

Recent observations in Chlamydomonas provide further evidence that links tubulin glutamylation to axoneme stability and depolymerization and ultimately to the length of the cilium. It has been puzzling that Chlamydomonas mutants that lack repeated components on the A-tubule (IDAs, ODAs, DRC-nexin) often have short cilia (Huang et al. 1979; LeDizet and Piperno 1995; Yamamoto et al. 2010). A loss of function of either TTLL9 or the associated FAP234 reduces tubulin glutamylation in cilia and, surprisingly, partially rescues the length defect of the IDA, ODA, and DRC-nexin mu-

tants (Kubo et al. 2015). Thus, tubulin glutamylation has a strong shortening effect on an axoneme that has incomplete A-tubule-associated protein complexes. Importantly, a loss of either FAP234 or TTLL9 alone mildly elongates cilia and decreases the rate of ciliary disassembly (Kubo et al. 2015). In the absence of TTLL9, other TTLLs could provide the IFT-promoting function of tubulin glutamylation (discussed above). It therefore appears that the repeated protein complexes bound to the A-tubule stabilize the axoneme and are in competition with tubulin glutamylation that destabilizes the axoneme. It is tempting to speculate that the shortening of the axoneme is mediated by tubulin glutamylation that activates ciliary katanin.

Surprisingly, a loss of katanin causes shortening of cilia and a failure to assemble the central pair (Dymek et al. 2004; Sharma et al. 2007; Dymek and Smith 2012), a counterintuitive phenotype for a microtubule-severing factor. However, in vitro, the severing activity of another microtubule depolymerizer, spastin, is dependent on the number of glutamates per tubulin, with a small number stimulating spastin and a large number inhibiting spastin (Valenstein and Roll-Mecak 2016). The levels of tubulin polyglutamylation detected by antibodies are higher in the assembling cilia than in the steady-state cilia, which also have a higher level of polyglycylation (Fig. 1A–C) (Adoutte et al. 1991; Sharma et al. 2007; Bosch Grau et al. 2013). It is possible that the polyglutamate side chains undergo remodeling as the cilium matures, including their shortening (by deglutamylases) and replacement by the polyglycine chains (see below). Such a PTM remodeling could modulate the possible multiple activities of katanin at different stages of ciliogenesis.

TUBULIN GLYCYLATION

Although all ciliated species have tubulin glutamylation (Janke et al. 2005), most, but not all, have tubulin glycylation (Adoutte et al. 1985; Bré et al. 1996). Glycylation is generated by a subset of TTLLs: TTLL3 is an initiase (Rogowski et al. 2009; Wloga et al. 2009), whereas TTLL10 is an elongase (Ikegami et al. 2008; Ikegami and

Cite this article as Cold Spring Harb Perspect Biol doi: 10.1101/cshperspect.a028159

Setou 2009; Rogowski et al. 2009). One exception is *D. melanogaster* in which polyglycylated tubulin is generated by TTLL3 alone (Rogowski et al. 2009). Curiously, in the ciliated cells of primates (including humans), the side chain is limited to a single G (monoglycylation), because of a mutation that inactivates the catalytic center of TTLL10 (Kann et al. 1998; Million et al. 1999; Rogowski et al. 2009). *Tetrahymena* cells with a knockout of all six paralogs of TTLL3 grow poorly and have slightly shorter, but ultrastructurally normal cilia that, unlike wild-type cilia, fail to elongate in the presence of paclitaxel, indicating a subtle defect in the tubulin turnover in the axoneme (Wloga et al. 2009). In zebrafish, a morpholino knockdown of TTLL3 glycylase causes reversals in the direction of ciliary beating in the pronephron. A depletion of TTLL6 glutamylase reduces the beat frequency (Pathak et al. 2011). Surprisingly, a depletion of both TTLL3 and TTLL6 leads to high penetrance ciliary phenotypes, including the body axis curvature and pronephric cysts, associated with reduced ciliary beat amplitude and even paralysis of pronephric cilia (Pathak et al. 2011). The axonemes depleted in both TTLL3 and TTLL6 often lack the B-tubules and have misplaced doublets and ectopic central singlets (Pathak et al. 2011). The severe assembly defects in the zebrafish depleted in TTLL3 and TTLL6 are reminiscent of the profound failure in the assembly of the axoneme in *Tetrahymena* mutants with substitutions in the modifiable Es on the CTT of β-tubulin (Xia et al. 2000; Thazhath et al. 2002). Thus, tubulin glycylation and glutamylation could both compete and cooperate in providing essential functions for axoneme assembly and motility. Polyglycine side chains are predicted to have an extended conformation, whereas the polyglutamate side chains may form swollen coils because of increased charge repulsion (Roll-Mecak 2015), supporting a view that the two polymodifications have different biological effects. In the mouse and *Drosophila*, tubulin glycylation could be important on its own. In *Drosophila*, a depletion of TTLL3 with RNAi causes male sterility associated with a failure in the axoneme assembly (Rogowski et al. 2009). In the murine ependy-

mal cells, a depletion of TTLL3 destabilizes already assembled motile cilia (Bosch Grau et al. 2013) and reduces the frequency of primary cilia in the colon epithelium (Rocha et al. 2014). Because a loss of glycylation leads to an increase in glutamylation (Rogowski et al. 2009; Wloga et al. 2009), it will be of interest to address whether, in the mouse, the defects in cilia caused by depletions of TTLL3 are mediated by tubulin hyperglutamylation.

CONCLUSION

It is increasingly convincing that tubulin PTMs play multiple important roles in cilia. The tubulin code model, originally inspired by the studies on histone tail modifications, seems useful in exploring the functions of microtubule PTMs. Many microtubule interactors, such as the motors, +TIPs, and even the tubulin code generators themselves (MEC-17/ATAT1 [Szyk et al. 2014] and TTLL7 [Garnham et al. 2015]) either walk unidirectionally or undergo a one-dimensional diffusion along the microtubule surface or inside its lumen. Thus, microtubule interactors can sense and respond to the PTM marks in a processive manner and therefore can be seen as genuine readers of the tubulin code. Inside the axoneme, the nonuniform patterns of PTM marks (Fig. 2) may be critical for the confinement of the anterograde and retrograde IFT streams to separate tubules of the outer doublets, and for the generation of the bend curvature in motile cilia. Although the evidence behind the importance of the tubulin code is growing steadily, a major challenge is in linking the phenotypes produced by manipulations of tubulin PTM enzymes to tubulin modifications. This task is difficult because of the complexity of the PTM pathways, which include nontubulin modification targets and nonenzymatic activities of the tubulin PTM enzymes. Despite these challenges, exploring the cilium has proved productive in generating ideas about the mechanisms and the biological effects of the tubulin code. The same studies have greatly added to the knowledge about how cilia assemble, maintain their length, and move.

ACKNOWLEDGMENTS

The work in the laboratory of D.W. is currently supported by funds from the National Science Center (Harmonia 6, 2014/14/M/NZ3/00511) and the European Molecular Biology Organization (EMBO) (IG No. 2331). E.J. is supported by a Polish Ministry of Science and Higher Education Grant (N N303 817840). Work in the laboratory of J.G. is supported by the University of Georgia. P.L. is supported by a predoctoral fellowship from the American Heart Association.

REFERENCES

Adoutte A, Claisse M, Maunoury R, Beisson J. 1985. Tubulin evolution: Ciliate-specific epitopes are conserved in the ciliary tubulin in metazoa. *J Mol Evol* **22:** 220–229.

Adoutte A, Delgado P, Fleury A, Levilliers N, Lainé MC, Marty MC, Boisvieux-Ulrich E, Sandoz D. 1991. Microtubule diversity in ciliated cells: Evidence for its generation by post-translational modification in the axonemes of *Paramecium* and quail oviduct cells. *Biol Cell* **71:** 227–245.

Akella JS, Wloga D, Kim J, Starostina NG, Lyons-Abbott S, Morrissette NS, Dougan ST, Kipreos ET, Gaertig J. 2010. MEC-17 is an α-tubulin acetyltransferase. *Nature* **467:** 218–222.

Alford LM, Stoddard D, Li JH, Hunter EL, Tritschler D, Bower R, Nicastro D, Porter ME, Sale WS. 2016. The nexin link and B-tubule glutamylation maintain the alignment of outer doublets in the ciliary axoneme. *Cytoskeleton (Hoboken)* **73:** 331–340.

Alonso VL, Villanova GV, Ritagliati C, Machado Motta MC, Cribb P, Serra EC. 2014. *Trypanosoma cruzi* bromodomain factor 3 binds acetylated α-tubulin and concentrates in the flagellum during metacyclogenesis. *Eukaryot Cell* **13:** 822–831.

Alper JD, Decker F, Agana B, Howard J. 2014. The motility of axonemal dynein is regulated by the tubulin code. *Biophys J* **107:** 2872–2880.

Arce CA, Barra HS. 1983. Association of tubulinyl-tyrosine carboxypeptidase with microtubules. *FEBS Lett* **157:** 75–78.

Backer CB, Gutzman JH, Pearson CG, Cheeseman IM. 2012. CSAP localizes to polyglutamylated microtubules and promotes proper cilia function and zebrafish development. *Mol Biol Cell* **23:** 2122–2130.

Bangs FK, Schrode N, Hadjantonakis AK, Anderson KV. 2015. Lineage specificity of primary cilia in the mouse embryo. *Nat Cell Biol* **17:** 113–122.

Barber CF, Heuser T, Carbajal-Gonzalez BI, Botchkarev VV Jr, Nicastro D. 2012. Three-dimensional structure of the radial spokes reveals heterogeneity and interactions with dyneins in *Chlamydomonas* flagella. *Mol Biol Cell* **23:** 111–120.

Bosch Grau M, Gonzalez Curto G, Rocha C, Magiera MM, Marques Sousa P, Giordano T, Spassky N, Janke C. 2013. Tubulin glycylases and glutamylases have distinct functions in stabilization and motility of ependymal cilia. *J Cell Biol* **202:** 441–451.

Bré MH, Redeker V, Quibell M, Darmanaden-Delome J, Bressac C, Cosson J, Huitore P, Schmitte JM, Rossier J, Johnson T, et al. 1996. Axonemal tubulin polyglycylation probed with two monoclonal antibodies: Widespread evolutionary distribution, appearance during spermatozoan maturation and possible function in motility. *J Cell Sci* **109:** 727–738.

Bré MH, Redeker V, Vinh J, Rossier J, Levilliers N. 1998. Tubulin polyglycylation: Differential posttranslational modification of dynamic cytoplasmic and stable axonemal microtubules in *Paramecium*. *Mol Biol Cell* **9:** 2655–2665.

Bui KH, Sakakibara H, Movassagh T, Oiwa K, Ishikawa T. 2009. Asymmetry of inner dynein arms and inter-doublet links in *Chlamydomonas* flagella. *J Cell Biol* **186:** 437–446.

Campbell PK, Waymire KG, Heier RL, Sharer C, Day DE, Reimann H, Jaje JM, Friedrich GA, Burmeister M, Bartness TJ, et al. 2002. Mutation of a novel gene results in abnormal development of spermatid flagella, loss of intermale aggression and reduced body fat in mice. *Genetics* **162:** 307–320.

Chalfie M, Au M. 1989. Genetic control of differentiation of the *Caenorhabditis elegans* touch receptor neurons. *Science* **243:** 1027–1033.

Choudhary C, Kumar C, Gnad F, Nielsen ML, Rehman M, Walther TC, Olsen JV, Mann M. 2009. Lysine acetylation targets protein complexes and co-regulates major cellular functions. *Science* **325:** 834–840.

Chu CW, Hou F, Zhang J, Phu L, Loktev AV, Kirkpatrick DS, Jackson PK, Zhao Y, Zou H. 2011. A novel acetylation of β-tubulin by San modulates microtubule polymerization via down-regulating tubulin incorporation. *Mol Biol Cell* **22:** 448–456.

Cueva JG, Hsin J, Huang KC, Goodman MB. 2012. Posttranslational acetylation of α-tubulin constrains protofilament number in native microtubules. *Curr Biol* **22:** 1066–1074.

Dave D, Wloga D, Sharma N, Gaertig J. 2009. DYF-1 Is required for assembly of the axoneme in *Tetrahymena thermophila*. *Eukaryot Cell* **8:** 1397–1406.

Davenport AM, Collins LN, Chiu H, Minor PJ, Sternberg PW, Hoelz A. 2014. Structural and functional characterization of the α-tubulin acetyltransferase MEC-17. *J Mol Biol* **426:** 2605–2616.

Duan J, Gorovsky MA. 2002. Both carboxy terminal tails of α and β tubulin are essential, but either one will suffice. *Curr Biol* **12:** 313–316.

Dymek EE, Smith EF. 2012. PF19 encodes the p60 catalytic subunit of katanin and is required for assembly of the flagellar central apparatus in *Chlamydomonas*. *J Cell Sci* **125:** 3357–3366.

Dymek EE, Lefebvre PA, Smith EF. 2004. PF15p is the *Chlamydomonas* homologue of the katanin p80 subunit and is required for assembly of flagellar central microtubules. *Eukaryot Cell* **3:** 870–879.

Eddé B, Rossier J, Le Caer JP, Desbruyères E, Gros F, Denoulet P. 1990. Posttranslational glutamylation of α-tubulin. *Science* **247:** 83–85.

Erck C, Peris L, Andrieux A, Meissirel C, Gruber AD, Vernet M, Schweitzer A, Saoudi Y, Pointu H, Bosc C, et al. 2005. A vital role of tubulin-tyrosine-ligase for neuronal organization. *Proc Natl Acad Sci* **102:** 7853–7858.

Ersfeld K, Wehland J, Plessmann U, Dodemont H, Gerke V, Weber K. 1993. Characterization of the tubulin-tyrosine ligase. *J Cell Biol* **120:** 725–732.

Fernandez-Gonzalez A, La Spada AR, Treadaway J, Higdon JC, Harris BS, Sidman RL, Morgan JI, Zuo J. 2002. *Purkinje cell degeneration (pcd)* phenotypes caused by mutations in the axotomy-induced gene, *Nna1. Science* **295:** 1904–1906.

Friedmann DR, Aguilar A, Fan J, Nachury MV, Marmorstein R. 2012. Structure of the α-tubulin acetyltransferase, αTAT1, and implications for tubulin-specific acetylation. *Proc Natl Acad Sci* **109:** 19655–19660.

Fukushige T, Siddiqui ZK, Chou M, Culotti JG, Gogonea CB, Siddiqui SS, Hamelin M. 1999. MEC-12, an α-tubulin required for touch sensitivity in *C. elegans. J Cell Sci* **112:** 395–403.

Gaertig J, Cruz MA, Bowen J, Gu L, Pennock DG, Gorovsky MA. 1995. Acetylation of lysine 40 in α-tubulin is not essential in *Tetrahymena thermophila. J Cell Biol* **129:** 1301–1310.

Garnham CP, Vemu A, Wilson-Kubalek EM, Yu I, Szyk A, Lander GC, Milligan RA, Roll-Mecak A. 2015. Multivalent microtubule recognition by tubulin tyrosine ligase-like family glutamylases. *Cell* **161:** 1112–1123.

Greer CA, Shepherd GM. 1982. Mitral cell degeneration and sensory function in the neurological mutant mouse Purkinje cell degeneration (PCD). *Brain Res* **235:** 156–161.

Hou Y, Qin H, Follit JA, Pazour GJ, Rosenbaum JL, Witman GB. 2007. Functional analysis of an individual IFT protein: IFT46 is required for transport of outer dynein arms into flagella. *J Cell Biol* **176:** 653–665.

Howes SC, Alushin GM, Shida T, Nachury MV, Nogales E. 2014. Effects of tubulin acetylation and tubulin acetyltransferase binding on microtubule structure. *Mol Biol Cell* **25:** 257–266.

Hoyle HD, Turner FR, Raff EC. 2008. Axoneme-dependent tubulin modifications in singlet microtubules of the *Drosophila* sperm tail. *Cell Motil Cytoskeleton* **65:** 295–313.

Huang B, Piperno G, Luck DJ. 1979. Paralyzed flagella mutants of *Chlamydomonas reinhardtii.* Defective for axonemal doublet microtubule arms. *J Biol Chem* **254:** 3091–3099.

Huang K, Diener DR, Rosenbaum JL. 2009. The ubiquitin conjugation system is involved in the disassembly of cilia and flagella. *J Cell Biol* **186:** 601–613.

Hubbert C, Guardiola A, Shao R, Kawaguchi Y, Ito A, Yoshida M, Wang XF, Yao T-P. 2003. Identification of HDAC6 as a microtubule-associated deacetylase. *Nature* **417:** 455–458.

Ikegami K, Setou M. 2009. TTLL10 can perform tubulin glycylation when co-expressed with TTLL8. *FEBS Lett* **583:** 1957–1963.

Ikegami K, Horigome D, Mukai M, Livnat I, Macgregor GR, Setou M. 2008. TTLL10 is a protein polyglycylase that can modify nucleosome assembly protein 1. *FEBS Lett* **582:** 1129–1134.

Ikegami K, Sato S, Nakamura K, Ostrowski LE, Setou M. 2010. Tubulin polyglutamylation is essential for airway ciliary function through the regulation of beating asymmetry. *Proc Natl Acad Sci* **107:** 10490–10495.

Ishikawa T. 2015. Cryo-electron tomography of motile cilia and flagella. *Cilia* **4:** 3.

Janke C. 2014. The tubulin code: Molecular components, readout mechanisms, and functions. *J Cell Biol* **206:** 461–472.

Janke C, Rogowski K, Wloga D, Regnard C, Kajava AV, Strub JM, Temurak N, van Dijk J, Boucher D, van Dorsselaer A, et al. 2005. Tubulin polyglutamylase enzymes are members of the TTL domain protein family. *Science* **308:** 1758–1762.

Jerka-Dziadosz M, Jenkins LM, Nelsen EM, Williams NE, Jaeckel-Williams R, Frankel J. 1995. Cellular polarity in ciliates: Persistence of global polarity in a disorganized mutant of *Tetrahymena thermophila* that disrupts cytoskeletal organization. *Dev Biol* **169:** 644–661.

Johnson KA. 1998. The axonemal microtubules of the *Chlamydomonas* flagellum differ in tubulin isoform content. *J Cell Sci* **111:** 313–320.

Kalebic N, Martinez C, Perlas E, Hublitz P, Bilbao-Cortes D, Fiedorczuk K, Andolfo A, Heppenstall PA. 2013a. Tubulin acetyltransferase αTAT1 destabilizes microtubules independently of its acetylation activity. *Mol Cell Biol* **33:** 1114–1123.

Kalebic N, Sorrentino S, Perlas E, Bolasco G, Martinez C, Heppenstall PA. 2013b. αTAT1 is the major α-tubulin acetyltransferase in mice. *Nat Commun* **4:** 1962.

Kamiya R. 2002. Functional diversity of axonemal dyneins as studied in *Chlamydomonas* mutants. *Int Rev Cytol* **219:** 115–155.

Kann ML, Prigent Y, Levilliers N, Bré MH, Fouquet JP. 1998. Expression of glycylated tubulin during the differentiation of spermatozoa in mammals. *Cell Motil Cytoskeleton* **41:** 341–352.

Kaul N, Soppina V, Verhey KJ. 2014. Effects of α-tubulin K40 acetylation and detyrosination on kinesin-1 motility in a purified system. *Biophys J* **106:** 2636–2643.

Kim J, Lee JE, Heynen-Genel S, Suyama E, Ono K, Lee K, Ideker T, Aza-Blanc P, Gleeson JG. 2010. Functional genomic screen for modulators of ciliogenesis and cilium length. *Nature* **464:** 1048–1051.

Kim GW, Li L, Gorbani M, You L, Yang XJ. 2013. Mice lacking α-tubulin acetyltransferase 1 are viable but display α-tubulin acetylation deficiency and dentate gyrus distortion. *J Biol Chem* **288:** 20334–20350.

Kimura Y, Kurabe N, Ikegami K, Tsutsumi K, Konishi Y, Kaplan OI, Kunitomo H, Iino Y, Blacque OE, Setou M. 2010. Identification of tubulin deglutamylase among *Caenorhabditis elegans* and mammalian cytosolic carboxypeptidases (CCPs). *J Biol Chem* **285:** 22936–22941.

Konno A, Ikegami K, Konishi Y, Yang HJ, Abe M, Yamazaki M, Sakimura K, Yao I, Shiba K, Inaba K, et al. 2016. Doublet 7 shortening, doublet 5-preferential poly-Glu reduction, and beating stall of sperm flagella in *Ttll9⁻/⁻* mice. *J Cell Sci* **129:** 2757–2766.

Kormendi V, Szyk A, Piszczek G, Roll-Mecak A. 2012. Crystal structures of tubulin acetyltransferase reveal a conserved catalytic core and the plasticity of the essential N terminus. *J Biol Chem* **287:** 41569–41575.

Kotani N, Sakakibara H, Burgess SA, Kojima H, Oiwa K. 2007. Mechanical properties of inner-arm dynein-f (dynein I1) studied with in vitro motility assays. *Biophys J* **93:** 886–894.

Kozminski KG, Diener DR, Rosenbaum JL. 1993. High level expression of nonacetylatable α-tubulin in *Chlamydomonas reinhardtii*. *Cell Motil Cytoskel* **25:** 158–170.

Kozminski KG, Beech PL, Rosenbaum JL. 1995. The *Chlamydomonas* kinesin-like protein FLA10 is involved in motility associated with the flagellar membrane. *J Cell Biol* **131:** 1517–1527.

Kubo T, Yanagisawa HA, Yagi T, Hirono M, Kamiya R. 2010. Tubulin polyglutamylation regulates axonemal motility by modulating activities of inner-arm dyneins. *Curr Biol* **20:** 441–445.

Kubo T, Yagi T, Kamiya R. 2012. Tubulin polyglutamylation regulates flagellar motility by controlling a specific inner-arm dynein that interacts with the dynein regulatory complex. *Cytoskeleton (Hoboken)* **69:** 1059–1068.

Kubo T, Yanagisawa HA, Liu Z, Shibuya R, Hirono M, Kamiya R. 2014. A conserved flagella-associated protein in *Chlamydomonas*, FAP234, is essential for axonemal localization of tubulin polyglutamylase TTLL9. *Mol Biol Cell* **25:** 107–117.

Kubo T, Hirono M, Aikawa T, Kamiya R, Witman GB. 2015. Reduced tubulin polyglutamylation suppresses flagellar shortness in *Chlamydomonas*. *Mol Biol Cell* **26:** 2810–2822.

Kumar N, Flavin M. 1981. Preferential action of a brain detyrosinolating carboxypeptidase on polymerized tubulin. *J Biol Chem* **256:** 7678–7680.

Lacroix B, van Dijk J, Gold ND, Guizetti J, Aldrian-Herrada G, Rogowski K, Gerlich DW, Janke C. 2010. Tubulin polyglutamylation stimulates spastin-mediated microtubule severing. *J Cell Biol* **189:** 945–954.

Lechtreck KF, Geimer S. 2000. Distribution of polyglutamylated tubulin in the flagellar apparatus of green flagellates. *Cell Motil Cytoskeleton* **47:** 219–235.

LeDizet M, Piperno G. 1987. Identification of an acetylation site of *Chlamydomonas* α-tubulin. *Proc Natl Acad Sci* **84:** 5720–5724.

LeDizet M, Piperno G. 1995. The light chain p28 associates with a subset of inner dynein arm heavy chains in *Chlamydomonas* axonemes. *Mol Biol Cell* **6:** 697–711.

Lee JE, Silhavy JL, Zaki MS, Schroth J, Bielas SL, Marsh SE, Olvera J, Brancati F, Iannicelli M, Ikegami K, et al. 2012. CEP41 is mutated in Joubert syndrome and is required for tubulin glutamylation at the cilium. *Nat Genet* **44:** 193–199.

Lee GS, He Y, Dougherty EJ, Jimenez-Movilla M, Avella M, Grullon S, Sharlin DS, Guo C, Blackford JA Jr, Awasthi S, et al. 2013. Disruption of *Ttll5/Stamp* gene (tubulin tyrosine ligase-like protein 5/SRC-1 and TIF2-associated modulatory protein gene) in male mice causes sperm malformation and infertility. *J Biol Chem* **288:** 15167–15180.

L'Hernault SW, Rosenbaum JL. 1983. *Chlamydomonas* α-tubulin is posttranslationally modified in the flagella during flagellar assembly. *J Cell Biol* **97:** 258–263.

Li W, Zhong C, Li L, Sun B, Wang W, Xu S, Zhang T, Wang C, Bao L, Ding J. 2012. Molecular basis of the acetyltransferase activity of MEC-17 towards α-tubulin. *Cell Res* **22:** 1707–1711.

Liang Y, Meng D, Zhu B, Pan J. 2016. Mechanism of ciliary disassembly. *Cell Mol Life Sci* **73:** 1787–1802.

Liu N, Xiong Y, Li S, Ren Y, He Q, Gao S, Zhou J, Shui W. 2015. New HDAC6-mediated deacetylation sites of tubulin in the mouse brain identified by quantitative mass spectrometry. *Sci Rep* **5:** 16869.

Lu C, Srayko M, Mains PE. 2004. The *Caenorhabditis elegans* microtubule-severing complex MEI-1/MEI-2 katanin interacts differently with two superficially redundant β-tubulin isotypes. *Mol Biol Cell* **15:** 142–150.

Magiera MM, Janke C. 2013. Investigating tubulin post-translational modifications with specific antibodies. *Methods Cell Biol* **115:** 247–267.

Million K, Larcher J-C, Laokili J, Bourguigon D, Marano F, Tournier F. 1999. Polyglutamylation and polyglycylation of α- and β-tubulins during in vitro ciliated cell differentiation of human respiratory epithelial cells.

Mukai M, Ikegami K, Sugiura Y, Takeshita K, Nakagawa A, Setou M. 2009. Recombinant mammalian tubulin polyglutamylase TTLL7 performs both initiation and elongation of polyglutamylation on β-tubulin through a random sequential pathway. *Biochemistry* **48:** 1084–1093.

Nicastro D, Fu X, Heuser T, Tso A, Porter ME, Linck RW. 2011. Cryo-electron tomography reveals conserved features of doublet microtubules in flagella. *Proc Natl Acad Sci* **108:** E845–853.

Nogales E, Whittaker M, Milligan RA, Downing KH. 1999. High-resolution model of the microtubule. *Cell* **96:** 79–88.

North BJ, Marshall BL, Borra MT, Denu JM, Verdin E. 2003. The human Sir2 ortholog, SIRT2, is an NAD$^+$-dependent tubulin deacetylase. *Mol Cell* **11:** 437–444.

O'Hagan R, Barr MM. 2012. Regulation of tubulin glutamylation plays cell-specific roles in the function and stability of sensory cilia. *Worm* **1:** 155–159.

O'Hagan R, Piasecki BP, Silva M, Phirke P, Nguyen KC, Hall DH, Swoboda P, Barr MM. 2011. The tubulin deglutamylase CCPP-1 regulates the function and stability of sensory cilia in *C. elegans*. *Curr Biol* **21:** 1685–1694.

Ohta S, Hamada M, Sato N, Toramoto I. 2015. Polyglutamylated tubulin binding protein C1orf96/CSAP is involved in microtubule stabilization in mitotic spindles. *PLoS ONE* **10:** e0142798.

Pathak N, Austin CA, Drummond IA. 2011. Tubulin tyrosine ligase-like genes *ttll3* and *ttll6* maintain zebrafish cilia structure and motility. *J Biol Chem* **286:** 11685–11695.

Pathak N, Austin-Tse CA, Liu Y, Vasilyev A, Drummond IA. 2014. Cytoplasmic carboxypeptidase 5 regulates tubulin glutamylation and zebrafish cilia formation and function. *Mol Biol Cell* **25:** 1836–1844.

Paturie-Lafanechère L, Manier M, Trigault N, Pirollet F, Mazarguil H, Job D. 1994. Accumulation of δ-2-tubulin, a major tubulin variant that cannot be tyrosinated, in neuronal tissues and in stable microtubule assemblies. *J Cell Sci* **107:** 1529–1543.

Pigino G, Cantele F, Vannuccini E, Lanzavecchia S, Paccagnini E, Lupetti P. 2009. Electron-tomographic analysis of intraflagellar transport particle trains in situ. *J Cell Biol* **187:** 135–148.

Piperno G, Fuller MT. 1985. Monoclonal antibodies specific for an acetylated form of α-tubulin recognize the antigen in cilia and flagella from a variety of organisms. *J Cell Biol* **101:** 2085–2094.

Prota AE, Magiera MM, Kuijpers M, Bargsten K, Frey D, Wieser M, Jaussi R, Hoogenraad CC, Kammerer RA, Janke C, et al. 2013. Structural basis of tubulin tyrosination by tubulin tyrosine ligase. *J Cell Biol* **200:** 259–270.

Pugacheva EN, Jablonski SA, Hartman TR, Henske EP, Golemis EA. 2007. HEF1-dependent Aurora A activation induces disassembly of the primary cilium. *Cell* **129:** 1351–1363.

Raff EC, Hoyle HD, Popodi EM, Turner FR. 2008. Axoneme β-tubulin sequence determines attachment of outer dynein arms. *Curr Biol* **18:** 911–914.

Ran J, Yang Y, Li D, Liu M, Zhou J. 2015. Deacetylation of α-tubulin and cortactin is required for HDAC6 to trigger ciliary disassembly. *Sci Rep* **5:** 12917.

Redeker V, Le Caer JP, Rossier J, Promé JC. 1991. Structure of the polyglutamyl side chain posttranslationally added to α-tubulin. *J Biol Chem* **266:** 23461–23466.

Redeker V, Melki R, Promé D, Le Caer JP, Rossier J. 1992. Structure of tubulin C-terminal domain obtained by subtilisin treatment. The major α- and β-tubulin isotypes from pig brain are glutamylated. *FEBS Lett* **313:** 185–192.

Redeker V, Levilliers N, Schmitter JM, Le Caer JP, Rossier J, Adoutte A, Bré MH. 1994. Polyglycylation of tubulin: A post-translational modification in axonemal microtubules. *Science* **266:** 1688–1691.

Reed NA, Cai D, Blasius L, Jih GT, Meyhofer E, Gaertig J, Verhey KJ. 2006. Microtubule acetylation promotes kinesin-1 binding and transport. *Curr Biol* **16:** 2166–2172.

Regnard C, Fesquet D, Janke C, Boucher D, Desbruyères E, Koulakoff A, Insina C, Travo P, Edde B. 2003. Characterization of PGs1, a subunit of a protein complex co-purifying with tubulin polyglutamylase. *J Cell Sci* **116:** 4181–4190.

Rocha C, Papon L, Cacheux W, Marques Sousa P, Lascano V, Tort O, Giordano T, Vacher S, Lemmers B, Mariani P, et al. 2014. Tubulin glycylases are required for primary cilia, control of cell proliferation and tumor development in colon. *EMBO J* **33:** 2247–2260.

Rogowski K, Juge F, van Dijk J, Wloga D, Strub JM, Levilliers N, Thomas D, Bre MH, Van Dorsselaer A, Gaertig J, et al. 2009. Evolutionary divergence of enzymatic mechanisms for posttranslational polyglycylation. *Cell* **137:** 1076–1087.

Rogowski K, van Dijk J, Magiera MM, Bosc C, Deloulme JC, Bosson A, Peris L, Gold ND, Lacroix B, Grau MB, et al. 2010. A family of protein-deglutamylating enzymes associated with neurodegeneration. *Cell* **143:** 564–578.

Roll-Mecak A. 2015. Intrinsically disordered tubulin tails: Complex tuners of microtubule functions? *Semin Cell Dev Biol* **37:** 11–19.

Roll-Mecak A, Vale RD. 2008. Structural basis of microtubule severing by the hereditary spastic paraplegia protein spastin. *Nature* **451:** 363–367.

Satish Tammana TV, Tammana D, Diener DR, Rosenbaum J. 2013. Centrosomal protein CEP104 (*Chlamydomonas* FAP256) moves to the ciliary tip during ciliary assembly. *J Cell Sci* **126:** 5018–5029.

Schroder HC, Wehland J, Weber K. 1985. Purification of brain tubulin-tyrosine ligase by biochemical and immunological methods. *J Cell Biol* **100:** 276–281.

Sergouniotis PI, Chakarova C, Murphy C, Becker M, Lenassi E, Arno G, Lek M, MacArthur DG, Consortium UCE, Bhattacharya SS, et al. 2014. Biallelic variants in TTLL5, encoding a tubulin glutamylase, cause retinal dystrophy. *Am J Hum Genet* **94:** 760–769.

Shang Y, Li B, Gorovsky MA. 2002. *Tetrahymena thermophila* contains a conventional γ tubulin that is differentially required for the maintenance of different microtubule organizing centers. *J Cell Biol* **158:** 1195–1206.

Sharma N, Bryant J, Wloga D, Donaldson R, Davis RC, Jerka-Dziadosz M, Gaertig J. 2007. Katanin regulates dynamics of microtubules and biogenesis of motile cilia. *J Cell Biol* **178:** 1065–1079.

Shida T, Cueva JG, Xu Z, Goodman MB, Nachury MV. 2010. The major α-tubulin K40 acetyltransferase αTAT1 promotes rapid ciliogenesis and efficient mechanosensation. *Proc Natl Acad Sci* **107:** 21517–21522.

Sirajuddin M, Rice LM, Vale RD. 2014. Regulation of microtubule motors by tubulin isotypes and post-translational modifications. *Nat Cell Biol* **16:** 335–344.

Song Y, Brady ST. 2015. Post-translational modifications of tubulin: Pathways to functional diversity of microtubules. *Trends Cell Biol* **25:** 125–136.

Soppina V, Herbstman JF, Skiniotis G, Verhey KJ. 2012. Luminal localization of α-tubulin K40 acetylation by cryo-EM analysis of fab-labeled microtubules. *PLoS ONE* **7:** e48204.

Stepanek L, Pigino G. 2016. Microtubule doublets are double-track railways for intraflagellar transport trains. *Science* **352:** 721–724.

Sun X, Park JH, Gumerson J, Wu Z, Swaroop A, Qian H, Roll-Mecak A, Li T. 2016. Loss of RPGR glutamylation underlies the pathogenic mechanism of retinal dystrophy caused by TTLL5 mutations. *Proc Natl Acad Sci* **113:** E2925–E2934.

Suryavanshi S, Edde B, Fox LA, Guerrero S, Hard R, Hennessey T, Kabi A, Malison D, Pennock D, Sale WS, et al. 2010. Tubulin glutamylation regulates ciliary motility by altering inner dynein arm activity. *Curr Biol* **20:** 435–440.

Szyk A, Deaconescu G, Piszczek G, Roll-Mecak A. 2011. Tubulin tyrosine ligase structure reveals adaptation of an ancient fold to bind and modify tubulin. *Nat Struct Mol Biol* **18:** 1250–1258.

Szyk A, Deaconescu AM, Spector J, Goodman B, Valenstein ML, Ziolkowska NE, Kormendi V, Grigorieff N, Roll-Mecak A. 2014. Molecular basis for age-dependent microtubule acetylation by tubulin acetyltransferase. *Cell* **157:** 1405–1415.

Taschner M, Vetter M, Lorentzen E. 2012. Atomic resolution structure of human α-tubulin acetyltransferase bound to acetyl-CoA. *Proc Natl Acad Sci* **109:** 19649–19654.

Thazhath R, Liu C, Gaertig J. 2002. Polyglycylation domain of β-tubulin maintains axonemal architecture and affects cytokinesis in *Tetrahymena*. *Nat Cell Biol* **4:** 256–259.

Topalidou I, Keller C, Kalebic N, Nguyen KC, Somhegyi H, Politi KA, Heppenstall P, Hall DH, Chalfie M. 2012. Enzymatic and non-enzymatic activities of the tubulin acetyltransferase MEC-17 are required for microtubule organization and mechanosensation in *C. elegans. Curr Biol* **22:** 1057–1065.

Tort O, Tanco S, Rocha C, Bieche I, Seixas C, Bosc C, Andrieux A, Moutin MJ, Aviles FX, Lorenzo J, et al. 2014. The cytosolic carboxypeptidases CCP2 and CCP3 catalyze posttranslational removal of acidic amino acids. *Mol Biol Cell* **25:** 3017–3027.

Valenstein ML, Roll-Mecak A. 2016. Graded control of microtubule severing by tubulin glutamylation. *Cell* **164:** 911–921.

van Dijk J, Rogowski K, Miro J, Lacroix B, Eddé B, Janke C. 2007. A targeted multienzyme mechanism for selective microtubule polyglutamylation. *Mol Cell* **26:** 437–448.

van Dijk J, Miro J, Strub JM, Lacroix B, van Dorsselaer A, Edde B, Janke C. 2008. Polyglutamylation is a post-translational modification with a broad range of substrates. *J Biol Chem* **283:** 3915–3922.

Varberg JM, Padgett LR, Arrizabalaga G, Sullivan WJ. 2016. TgATAT-mediated α-tubulin acetylation is required for division of the protozoan parasite *Toxoplasma gondii. mSphere* **1:** e00088–e00015.

Verhey KJ, Gaertig J. 2007. The tubulin code. *Cell Cycle* **6:** 2152–2160.

Vogel P, Hansen G, Fontenot G, Read R. 2010. Tubulin tyrosine ligase-like 1 deficiency results in chronic rhinosinusitis and abnormal development of spermatid flagella in mice. *Vet Pathol* **47:** 703–712.

Walter WJ, Beranek V, Fischermeier E, Diez S. 2012. Tubulin acetylation alone does not affect kinesin-1 velocity and run length in vitro. *PLoS ONE* **7:** e42218.

Wloga D, Rogowski K, Sharma N, Van Dijk J, Janke C, Edde B, Bre MH, Levilliers N, Redeker V, Duan J, et al. 2008. Glutamylation on α-tubulin is not essential but affects the assembly and functions of a subset of microtubules in *Tetrahymena thermophila. Eukaryot Cell* **7:** 1362–1372.

Wloga D, Webster DM, Rogowski K, Bre MH, Levilliers N, Jerka-Dziadosz M, Janke C, Dougan ST, Gaertig J. 2009. TTLL3 is a tubulin glycine ligase that regulates the assembly of cilia. *Dev Cell* **16:** 867–876.

Wloga D, Dave D, Meagley J, Rogowski K, Jerka-Dziadosz M, Gaertig J. 2010. Hyperglutamylation of tubulin can either stabilize or destabilize microtubules in the same cell. *Eukaryot Cell* **9:** 184–193.

Xia L, Hai B, Gao Y, Burnette D, Thazhath R, Duan J, Bré MH, Levilliers N, Gorovsky MA, Gaertig J. 2000. Polyglycylation of tubulin is essential and affects cell motility and division in *Tetrahymena thermophila. J Cell Biol* **149:** 1097–1106.

Xia P, Ye B, Wang S, Zhu X, Du Y, Xiong Z, Tian Y, Fan Z. 2016. Glutamylation of the DNA sensor cGAS regulates its binding and synthase activity in antiviral immunity. *Nat Immunol* **17:** 369–378.

Yamamoto R, Hirono M, Kamiya R. 2010. Discrete PIH proteins function in the cytoplasmic preassembly of different subsets of axonemal dyneins. *J Cell Biol* **190:** 65–71.

Yanagisawa HA, Mathis G, Oda T, Hirono M, Richey EA, Ishikawa H, Marshall WF, Kikkawa M, Qin H. 2014. FAP20 is an inner junction protein of doublet microtubules essential for both the planar asymmetrical waveform and stability of flagella in *Chlamydomonas. Mol Biol Cell* **25:** 1472–1483.

Yang Y, Ran J, Liu M, Li D, Li Y, Shi X, Meng D, Pan J, Ou G, Aneja R, et al. 2014. CYLD mediates ciliogenesis in multiple organs by deubiquitinating Cep70 and inactivating HDAC6. *Cell Res* **24:** 1342–1353.

Ye B, Li C, Yang Z, Wang Y, Hao J, Wang L, Li Y, Du Y, Hao L, Liu B, et al. 2014. Cytosolic carboxypeptidase CCP6 is required for megakaryopoiesis by modulating Mad2 polyglutamylation. *J Exp Med* **211:** 2439–2454.

Yu I, Garnham CP, Roll-Mecak A. 2015. Writing and reading the tubulin code. *J Biol Chem* **290:** 17163–17172.

Yuzawa S, Kamakura S, Hayase J, Sumimoto H. 2015. Structural basis of cofactor-mediated stabilization and substrate recognition of the α-tubulin acetyltransferase αTAT1. *Biochem J* **467:** 103–113.

Zhang Y, Ma C, Delohery T, Nasipak B, Foat BC, Bounoutas A, Bussemaker HJ, Kim SK, Chalfie M. 2002. Identification of genes expressed in *C. elegans* touch receptor neurons. *Nature* **418:** 331–335.

Zhang Y, Li N, Caron C, Matthias G, Hess D, Khochbin S, Matthias P. 2003. HDAC-6 interacts with and deacetylates tubulin and microtubules in vivo. *EMBO J* **22:** 1168–1179.

Zhang Y, Kwon S, Yamaguchi T, Cubizolles F, Rousseaux S, Kneissel M, Cao C, Li N, Cheng HL, Chua K, et al. 2008. Mice lacking histone deacetylase 6 have hyperacetylated tubulin but are viable and develop normally. *Mol Cell Biol* **28:** 1688–1701.

Zhou X, Fan LX, Li K, Ramchandran R, Calvet JP, Li X. 2014. SIRT2 regulates ciliogenesis and contributes to abnormal centrosome amplification caused by loss of polycystin-1. *Hum Mol Genet* **23:** 1644–1655.

Axonemal Dynein Arms

Stephen M. King

Department of Molecular Biology and Biophysics, University of Connecticut Health Center, Farmington, Connecticut 06030-3305

Correspondence: king@uchc.edu

Axonemal dyneins form the inner and outer rows of arms associated with the doublet microtubules of motile cilia. These enzymes convert the chemical energy released from adenosine triphosphate (ATP) hydrolysis into mechanical work by causing the doublets to slide with respect to each other. Dyneins form two major groups based on the number of heavy-chain motors within each complex. In addition, these enzymes contain other components that are required for assembly of the complete particles and/or for the regulation of motor function in response to phosphorylations status, ligands such as Ca^{2+}, changes in cellular redox state and which also apparently monitor and respond to the mechanical state or curvature in which any given motor finds itself. It is this latter property, which is thought to result in waves of motor function propagating along the axoneme length. Here, I briefly describe our current understanding of axonemal dynein structure, assembly, and organization.

Motile cilia[1] are organelles involved in the transport of fluids (such as mucus flow in the lungs) and in the directed movement of individual cells (e.g., sperm). Reactivation of demembranated organelles has clearly shown that all the components necessary to generate and propagate waveforms are built into the microtubular axoneme, which can thus be considered a solid-state (or hard-wired) motility system requiring only an external energy source to function. Indeed, in vitro, these structures even retain the ability to alter their waveform or beat frequency in response to alterations in added signaling factors. Although the path by which signals are propagated through the axonemal superstructure is complex, ultimately the motile behavior is generated by two rows of dynein arms that are permanently attached to the A-tubule of one microtubule doublet and transiently interact in an adenosine triphosphate (ATP)-dependent manner with the B-tubule of an adjacent doublet. This generates a sliding force that is converted to bending by additional structures (including the nexin–dynein regulatory complex) that link the doublets circumferentially. In this article, I briefly describe the basic composition of these dynein complexes and how they are preassembled in cytoplasm and trafficked into the ciliary shaft. I also discuss how they are incorporated into the axoneme, their patterning within the superstructure, the mechanisms by which they are regulated to generate specific ciliary beat patterns, and the general consequences of their dysfunction in mammals.

---

[1] In some organisms, such as *Chlamydomonas*, this organelle is commonly referred to as a flagellum. However, here I use the term cilium throughout for clarity.

As dyneins are highly complex macro-molecular systems containing, in some cases, >20 different protein components, and as they have been studied over many years in numerous different model systems, their nomenclature has become extraordinarily complex and highly confusing. Here, I use the nomenclature scheme used in *Chlamydomonas*; to translate this to orthologous components in other organisms, the reader is referred to Hom et al. (2011), which presents a comprehensive taxonomic guide to axonemal dynein nomenclature.

COMPOSITION OF DYNEIN ARMS

The dynein arms that power ciliary motility may be classified into two distinct groups based on the number of heavy-chain motor units that they contain. These motors are ∼4500 residues and consist of an amino-terminal region required for assembly, followed by a motor unit consisting of a hexameric ring of nonidentical AAA^+ domains (Fig. 1A,B). The microtubule-binding region is located at the tip of a coiled coil that emanates from AAA4 and also interacts with a second coiled-coil segment (termed the buttress) that derives from AAA5. Finally, there is a carboxy-terminal domain that associates across the plane of the AAA ring and may be involved in regulating motor activity.

The first dynein class consists of single motors that are associated with a molecule of actin and either a specific light chain (termed p28 in *Chlamydomonas*; dyneins a, c, and d) or the Ca^{2+}-binding protein centrin (dyneins b, e, and g); dynein d also has two additional components p38 and p44. These motors form a subset of the inner dynein arms and are arrayed in a complex manner both along the axonemal length and potentially even on distinct outer doublets. In addition to this major group of monomeric dyneins, there are also several "minor" dyneins that are present in greatly reduced quantity (Yagi et al. 2009). In *Chlamydomonas*, one of these (termed DHC3) is estimated to be almost 100 kDa larger than other dyneins because of a series of insertions within the AAA^+ ring domain. Although little is known about the composition or role of these minor isoforms,

their discrete localization in the region of the axoneme near the ciliary base supports the idea that they may be involved in bend initiation. *Chlamydomonas* mutants defective for actin are actually viable and the lack of this protein leads to the failure of most of these monomeric heavy-chain dyneins to assemble (Kato-Minoura et al. 1997). However, it has been observed that two particular motors are still incorporated into the axonemal superstructure as they are able to use an actin-related protein that becomes up-regulated in the mutant strain (Kato-Minoura et al. 1998). The p28 light chain is also absolutely required for the assembly of those dyneins with which it associates (LeDizet and Piperno 1995).

The second dynein class includes both the outer arm (Fig. 1C) and an additional inner arm termed I1 or f (see Table 1). These motors are built in a manner related to that of canonical cytoplasmic dynein and the dynein that powers retrograde intraflagellar transport (IFT). In all of these complexes, two heavy-chain motors (or three in the case of outer arms in organisms such as *Chlamydomonas* and *Tetrahymena*) associate together via their amino-terminal regions. Cytoplasmic and IFT dyneins are formed from homodimers of heavy chains. However, for axonemal dyneins these multimotor enzymes contain nonidentical heavy chains encoded by different genes that show distinct motor and ATPase properties; indeed, the individual motor units within the same complex are even regulated by different signaling inputs. The heavy-chain amino-terminal segments also bind an additional subcomplex consisting of two WD-repeat intermediate chains and light chains belonging to three conserved classes (namely, DYNLL1/2 or LC8, DYNLT1 or Tctex1, and DYNLRB1/2, roadblock, or LC7) (King and Patel-King 1995b; Harrison et al. 1998; Bowman et al. 1999). This subcomplex is absolutely required for motor integrity and assembly and, for example, mutants defective for the outer arm WD-repeat intermediate chains are completely unable to assemble these structures into the axoneme (Mitchell and Kang 1991; Wilkerson et al. 1995). The outer arms from metazoans also contain an additional in-

Cite this article as *Cold Spring Harb Perspect Biol* doi: 10.1101/cshperspect.a028100

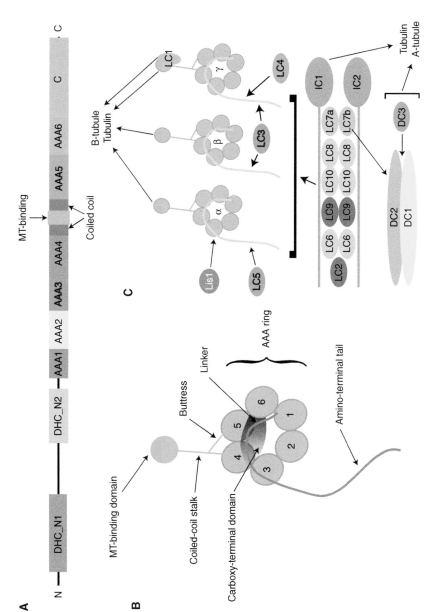

Figure 1. Organization of axonemal dynein motors. (*A*) Linear map of a generic ∼4500-residue dynein heavy chain. The amino-terminal region contains two conserved segments found in nearly all these motors: DHC_N1 is involved in heavy-chain assembly, whereas DHC_N2 represents the linker that traverses the plane of the AAA ring and is key to the power stroke mechanism. This is followed by the hexameric AAA ring and carboxy-terminal domain. (From King 2012a; modified, with permission, from the author.) (*B*) Model of a single dynein heavy chain showing how the various domains are arranged. (From King 2010; modified, with permission, from the author.) (*C*) Diagram showing the interactions of the core components that comprise the *Chlamydomonas* outer dynein arm. The heavy chains interact with a series of proteins that provide regulatory inputs (LCs1, 3, 4, and 5, and Lis1). They also bind to the intermediate-chain/light-chain complex, which in turn associates with the trimeric docking complex. The general properties of all these components are indicated in Table 1. MT, Microtubule. (From King 2012b; modified, with permission, from the author.)

termediate chain that consists of an amino-terminal thioredoxin module followed by two or three nucleoside diphosphate kinase catalytic cores (Ogawa et al. 1996; Padma et al. 2001). This observation suggested that nucleotides other than ATP might be important for axonemal dynein function. However, to date, there is no experimental evidence to support this notion. Furthermore, these multi-heavy-chain motors also contain additional members of the LC8 and Tctex1 light-chain classes not found in cytoplasmic dynein, and also other light-chain components that allow the motors to sense and be regulated by external factors including redox poise, Ca^{2+} levels, and the mechanical state or local curvature experienced by the motors themselves. For example, the *Chlamydomonas* outer arm contains two bona fide thioredoxins, a calmodulin homolog and a leucine-rich repeat protein thought to be involved in sensing curvature that is actually the most highly conserved outer arm dynein-specific component.

CYTOPLASMIC PREASSEMBLY AND TRAFFICKING OF DYNEIN ARM COMPLEXES

Dynein heavy chains are massive proteins consisting of ~4500 residues. Consequently, their synthesis in the cytoplasm takes a considerable time; >13 min per molecule for a eukaryotic ribosome operating at a rate of 5.6 residues incorporated per sec (Ingolia et al. 2011) (BNID107952). During this time, there is clearly massive potential for protein aggregation because of exposure of hydrophobic core residues, and this is likely one reason why cells have evolved a complex series of cytoplasmic factors that are now known to be required for stable synthesis and maintenance of these proteins in cytoplasm before their transport into the ciliary compartment. These factors include LRRC50 (ODA7 in *Chlamydomonas*) (Duquesnoy et al. 2009), multiple PIH domain proteins (Omran et al. 2008; Yamamoto et al. 2010), and their interacting partners such as Dyx1c1 (Tarkar et al. 2013) that associate with Hsp70 and Hsp90 and are presumably playing a chaperone role in helping fold dynein components. Cur-

rently, there is little information on the precise functional interplay among all of these molecules. Various dynein components are then specifically assembled to form discrete subcomplexes (such as that formed by the WD-repeat intermediate chains and a series of light chains) that are then associated with the heavy chains and regulatory light chains to yield complete dynein arm particles (Fowkes and Mitchell 1998). For the outer arm, recent studies show that the ODA5/ODA10 proteins are needed both in the cytoplasm to generate fully assembly-competent dynein particles and also in the axoneme in which, surprisingly, ODA10 assembles onto the proximal region of doublet #1, which, in *Chlamydomonas*, is the only doublet that lacks outer dynein arms (Dean and Mitchell 2013). What role it plays there is uncertain at present. Inner arm dyneins appear to concentrate at the base of growing cilium where they can presumably be readily and rapidly injected through the ciliary gate at the transition zone into the ciliary compartment proper. Surprisingly, at least in *Chlamydomonas*, outer arm dyneins do not concentrate in this region but rather are found spread throughout the cytoplasmic volume, although why this should be remains unclear.

During growth of the cilium, the cell must transport axonemal dyneins from the cytoplasm into the ciliary shaft where they can then be assembled. There appear to be specific adaptors that are necessary to attach the different arms to the IFT machinery so they can be transported (Ahmed et al. 2008). However, once in the ciliary matrix, they must presumably be kept in an inactive state until fully assembled, otherwise they would likely interact with microtubule doublets and traffic toward the microtubule minus end at the ciliary base. One intriguing assembly factor (known as FBB18 in *Chlamydomonas*) appears to be present in the matrix (Austin-Tse et al. 2013) and could potentially be involved in keeping dyneins inactive until they have incorporated into the ciliary superstructure. As the various axonemal dyneins are present in different amounts and also differentially arranged along the axonemal length, the cell likely also has a sorting system that delivers these motors in the correct proportions and at the appropriate time during axoneme assembly. For example, motors only present in the distal region are presumably transported into the growing structure only during the latter phases of assembly. However, how this sorting might be achieved is completely unknown.

ASSEMBLING DYNEIN ARMS IN THE AXONEMAL SUPERSTRUCTURE

The basic axonemal building block is a 96-nm repeat that contains four outer arms, one inner arm I1/f, and six monomeric inner arms in addition to two complete radial spokes and a third truncated spoke (Fig. 2). Also included in the repeat is the N-DRC regulatory complex (Heuser et al. 2009) and the MIA complex (Yamamoto et al. 2013), which links the N-DRC to inner arm I1/f. Although the outer arms all appear to be identical in composition, one of the four in each repeat is physically connected through a linker structure to inner arm I1/f; it is this linker that is thought to provide a pathway for regulatory signals from the central pair/radial spokes to reach the outer arms. A key question in axoneme assembly is how this complex arrangement is patterned such that all the dynein arms are correctly located on the outer doublet microtubules. What is clear is that the mechanism(s) involved are robust as ectopic localization of dynein arms at inappropriate sites is never observed. In contrast, when purified dyneins are rebound to brain microtubules in vitro, they are capable of associating all around the microtubule circumference (Haimo and Fenton 1988). For the inner arm system, recent studies have revealed that a pair of extended coiled-coil proteins FAP59 and FAP172 (CCDC39 and CCDC40, in mammals) play a key role (Oda et al. 2014). These proteins asso-

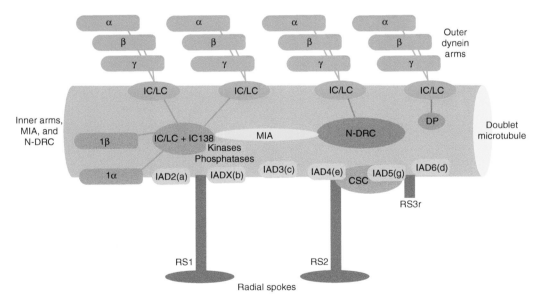

Figure 2. Arrangement of dyneins in the 96-nm axonemal repeat. Each axonemal repeat unit consists of four outer dynein arms arranged with a spacing of 24 nm, a single inner arm I1/f, and six different monomeric heavy-chain inner arms. Linkages between the outer arms and other components vary on different microtubule doublets and also on the position along the axoneme. The N-DRC regulatory complex is linked to inner arm I1/f through the MIA complex, which provides a physical pathway for regulatory signals. Location of the radial spokes, the calmodulin and spoke-associated complex (CSC), and a distal projection that can interact with the last outer arm in the repeat are also shown. (From King 2013; reprinted, with permission, from the author under a Creative Commons License.)

ciate together and span the repeat distance on the doublet microtubule surface. They provide defined docking sites for individual inner arms at specific locations along the repeat. For outer arms to interact with the doublet microtubules, an additional trimeric docking complex consisting of two coiled-coil proteins (DC1 and DC2) and a calmodulin homolog (DC3) is needed (Takada and Kamiya 1994). However, although this complex is necessary for outer arm association with the doublets, it does not appear sufficient to define the site of axonemal incorporation. Recently, an additional protein (CCDC103) has also been found to be necessary for outer arm insertion into the superstructure (Panizzi et al. 2012). This protein has rather remarkable biophysical properties including a very high T_m and the ability to form dimers and higher order oligomers that are resistant to (or reform following) heating in the presence of SDS and reducing agents. In vitro, CCDC103 self-assembles into polymeric structures that can bind in linear arrays along the microtubule surface with a 12-nm periodicity (King and Patel-King 2015). Thus, one possibility is that CCDC103 polymers assemble during axonemal growth and provide a high-affinity track along the outer doublets with which the outer arms/ docking complex can then associate.

DYNEIN MOTOR FUNCTION AND REGULATION

Both rows of dynein arms are capable of generating motive force (Paschal et al. 1987; Vale and Toyoshima 1988). When analyzed in vitro, individual dynein HC motors show discrete rates of microtubule translocation, different ATPase activities; and some (mainly the monomeric inner arms) also generate torque such that the gliding microtubules rotate as they move (Kagami and Kamiya 1992). Indeed, the outer arms increase cilia-generated propulsive force output by fourfold, and, in *Chlamydomonas*, beat frequency is reduced by approximately half in mutants unable to assemble outer arms (Kamiya and Okamoto 1985; Mitchell and Rosenbaum 1985). Furthermore, cilia retain the ability to essentially double their force output when placed under viscous load; this requires the outer arms (Yagi et al. 2005; Patel-King and King 2009). At least one mechanism involved in this response, is directed trafficking of the Lis1 regulatory factor into the ciliary matrix when load is encountered (Rompolas et al. 2012). For cytoplasmic dynein, Lis1 acts as a clutch to disengage the ATPase and microtubule-binding activities allowing this motor to drag large cargoes, such as nuclei, without releasing the microtubule (McKenney et al. 2010). In the cilium, Lis1 has been found to associate with the α heavy chain within the outer arm and to increase the intrinsic beat frequency of reactivated cell models; that is, it enhances force output per unit time (Rompolas et al. 2012). In contrast, the inner arm system provides the exquisite control necessary to actually generate and propagate specific waveforms (Kamiya et al. 1991). Signals derived from the central pair/radial spokes and from the N-DRC are propagated directly through physical connections to the various inner arm dynein motors (Yamamoto et al. 2013). It is thought that this leads to dynein activation on individual doublets at appropriate intervals. The inner arm system is also controlled by the direct phosphorylation of the IC138 intermediate chain associated with inner arm I1/f. When phosphorylated by casein kinase 1 or protein kinase A, this dynein shows slow microtubule gliding. However, dephosphorylation by PP1 and/or PP2A leads to faster sliding (Smith and Sale 1992; Alford et al. 2012). Ultimately, this regulatory pathway is controlled by signals emanating from the central pair microtubule complex.

To propagate a bend, dyneins along a given doublet must be sequentially activated so that a wave of motor activity transits the structure usually from base to tip. Current models based on direct physical activation of mutant cilia lacking particular axonemal structures propose that there are actually two distinct mechanosensory systems (Hayashibe et al. 1997). One involving the central pair and radial spokes that impacts the inner arms, and second distinct system that activates the outer arms. This second system is thought to involve the highly conserved leucine-rich repeat protein LC1 (Benash-

Table 1. Polypeptide composition of *Chlamydomonas* axonemal dyneins containing multiple heavy chains

Dynein	Protein	Mass (kDa)	Description
Outer arm	α HC	504	ATPase, microtubule motor; unusual amino-terminal domain consisting of kelch repeats and immunoglobulin-like folds; not essential for outer arm assembly; interacts with Lis1 to increase propulsive force output
	β HC	520	ATPase, microtubule motor; required for outer arm assembly; provides most of the power output
	γ HC	513	ATPase, microtubule motor; required for outer arm assembly; appears to be the key regulatory node integrating multiple signaling inputs
	IC1	76	WD-repeat protein, binds microtubules in an ATP-independent manner; required for outer arm assembly
	IC2	63	WD-repeat protein; required for outer arm assembly
	Lis1	37	Entry into cilium is regulated in response to perceived load; binds the α HC to increase propulsive force output
	LC1	22	Leucine-rich repeat protein; one copy binds to the microtubule-binding domain of the γ HC; also binds microtubules directly; proposed to act as a sensor of axonemal curvature
	LC2	16	Tctex2; binds the ICs and LC6
	LC3	17	Thioredoxin associated with amino-terminal region of β HC
	LC4	18	Calmodulin homolog; specifically binds Ca^{2+} and controls structural rigidity of γ HC amino-terminal region
	LC5	14	Thioredoxin associated with amino-terminal region of α HC
	LC6	14	*Chlamydomonas*-specific variant of LC8
	LC7a	12	Related to mammalian roadblock/DYNLRB; associates with ICs
	LC7b	11	Related to mammalian roadblock/DYNLRB; associates with ICs and interacts with docking complex and LC3
	LC8	10	Extraordinarily highly conserved; acts as a dimerization hub and is found in a large number of macromolecular complexes; interacts with ICs
	LC9	13	Dimeric Tctex1 homolog; associates with ICs
	LC10	12	LC8 variant; apparently outer arm specific; in mammals, the ortholog (DNAL4) is very highly expressed suggesting additional roles
Inner arm I1/f	1α HC	523	ATPase, microtubule motor; required for inner arm assembly
	1β HC	510	ATPase, microtubule motor; required for inner arm assembly
	IC140	110	WD-repeat protein; required for inner arm assembly
	IC138	111	WD-repeat protein; phosphorylated by casein kinase 1 and protein kinase A; regulates motor activity
	IC97	90	Ortholog of mammalian LAS1 that is implicated in lung tumor formation; interacts with tubulin and IC140/IC138
	FAP120	32	Contains ankyrin repeats
	Tctex1	12	Dimer, associates with ICs; LC9 homolog
	Tctex2b	14	Related to LC2
	LC7a	12	Related to mammalian roadblock/DYNLRB; associates with ICs; also present in outer arm dynein
	LC7b	11	Related to mammalian roadblock/DYNLRB; associates with ICs; also present in outer arm dynein
	LC8	10	Extraordinarily highly conserved; acts as a dimerization hub and is found in a large number of macromolecular complexes; interacts with ICs; also present in outer arm dynein

Cite this article as *Cold Spring Harb Perspect Biol* doi: 10.1101/cshperspect.a028100

ski et al. 1999; Wu et al. 2000) that has recently been shown to bind the microtubule-binding domain of the γ heavy chain (Ichikawa et al. 2015). When mutant forms of this protein are expressed in a wild-type background, they yield dominant negative effects that have dramatic consequences on swimming velocity and on the ability of the cilia to maintain hydrodynamic coupling (Patel-King and King 2009; Rompolas et al. 2010).

Many organisms are able to change their ciliary beat pattern in response to specific signals. In *Chlamydomonas*, switching between forward motion using an asymmetric or ciliary beat and backward movement driven by a symmetric sinusoidal waveform is mediated by altering intraciliary Ca^{2+} (Bessen et al. 1980). This switch from forward to backward motion occurs in two steps: first the cilia stop beating as Ca^{2+} increases to $\sim 10^{-5}$ M and then they restart with the different waveform as levels increase further. It is this latter step that requires the outer dynein arms as mutant cilia lacking these structures either just stop beating or show a highly aberrant waveform transition at high Ca^{2+}. This suggests that there may be two Ca^{2+} sensors necessary to mediate the complete response. The most likely candidate for the outer arm Ca^{2+} sensor that leads to waveform conversion is LC4, which is an EF-hand protein shown to specifically interact with Ca^{2+} (King and Patel-King 1995a). This protein binds to two distinct segments of the γ heavy-chain amino-terminal region and, in the absence of Ca^{2+}, acts as a rigid strut maintaining a bent orientation in this heavy-chain domain (Sakato et al. 2007). However, when Ca^{2+} increases, LC4 appears to detach from one site on the heavy-chain amino-terminal region and instead interacts directly with IC1. This allows the γ heavy-chain amino-terminal domain to adopt conformations not attainable in the absence of the ligand. How this structural change might then manifest itself as an alteration in waveform is still not clear. The other known *Chlamydomonas* outer arm Ca^{2+}-binding protein is the docking complex DC3 component (Casey et al. 2003a). However, this protein can bind both Ca^{2+} and Mg^{2+} and its absence does not

abrogate waveform conversion (Casey et al. 2003b). There is also the possibility that soluble calmodulin present in the ciliary matrix transiently interacts with the two IQ motifs located on the γ heavy-chain amino-terminal region as Ca^{2+} increases.

Outer arm dyneins also contain thioredoxins either as individual light chains (as in *Chlamydomonas*) or as part of a modular intermediate chain (metazoans) that also has functional nucleoside diphosphate kinase catalytic cores (Ogawa et al. 1996; Patel-King et al. 1996; Padma et al. 2001). These thioredoxin units are redox-active, and it was subsequently found that beat frequency can be controlled by alterations in cellular redox state (Wakabayashi and King 2006). This response appears to be mediated through the formation of different mixed disulfides in vivo. Furthermore, in vitro experiments revealed that changes in the redox state of these dynein-associated thioredoxin moieties also affected motility. Indeed, the beat frequency of reactivated cell models can be controlled simply by manipulating the GSH:GSSG ratio in glutathione redox buffers. In *Chlamydomonas*, both the α and β heavy chains have associated thioredoxins (LC5 and LC3, respectively), and it was found that lack of the α heavy chain and LC5 resulted in cell models that could not respond to redox changes presumably caused by disruption of an intradynein redox cascade (Wakabayashi and King 2006).

CONSEQUENCES OF DYNEIN ARM DYSFUNCTION

In *Chlamydomonas*, lack of outer dynein arms results in a drop in ciliary beat frequency to approximately half that of wild-type, whereas lack of inner arms alters the normal waveform by decreasing the angular size of the principal bend. However, apart from slow motility, there appear to be no other consequences for these unicellular organisms, at least in the laboratory. In contrast, in vertebrates (including humans), ciliary motility and the consequent fluid flow powered by axonemal dyneins, play key roles in development, reproduction, and organismal homeostasis (Fliegauf et al. 2007; Becker-Heck et al. 2012). Reduction of ciliary beat frequency caused by mutations within axonemal dyneins or their cytoplasmic assembly factors leads to primary ciliary dyskinesia (PCD), which includes multiple phenotypes of varying severity depending on the precise mutational target (Becker-Heck et al. 2012). In general, patients with PCD show highly reduced fertility because of poor sperm motility in males and reduction in beat frequency of the cilia lining the fallopian tubes in females that are involved in oocyte movement toward the uterus. Dysfunctional cilia also result in sometimes-severe bronchial problems caused by the inability of the lungs to clear secreted mucus, which acts to protect the tissue from inhaled pollutants and pathogens. Cilia also line the brain ventricles and in many vertebrates ependymal ciliary dysfunction leads to hydrocephalus. One other major consequence of ciliary immotility is that ∼50% of patients with PCD will show "situs inversus," in which the position of the internal organs is reversed. Several forms are known including partial reversals within either the abdominal or thoracic cavities. This is caused during embryogenesis as ciliary beating at the embryonic node is required to set up left–right asymmetry. In the absence of ciliary motility, the embryo breaks symmetry randomly resulting in a 50% chance of incorrect "situs." In general, patients with "situs inversus totalis" are asymptomatic. However, in some cases, the embryo fails to setup a left–right axis and this leads to heterotaxy and severe heart deformations, which can require surgical intervention for survival. Also of note is that defects in the dynein that power retrograde IFT also have severe phenotypic consequences. In humans, dysfunction of this motor complex results in developmental problems including severe (often fatal) skeletal abnormalities such as those observed for Jeune's asphyxiating thoracic dystrophy and short-rib polydactyly (Schmidts et al. 2013).

CONCLUDING REMARKS

Axonemal dyneins represent a highly divergent and complex class of motors whose ensemble activity results in exquisitely controlled ciliary beating. Although much has been learned about these motors in terms of their composition, location, and in vitro properties, there remain many outstanding questions concerning how they are assembled in cytoplasm, specifically trafficked into the cilium, and then located at precise sites within the axonemal superstructure. Furthermore, even though high-resolution structures are now available for dynein motor domains there is still much to be understood about the precise arrangement and function of the many other components that comprise these motors.

ACKNOWLEDGMENTS

S.M.K.'s laboratory is supported by Grant GM051293 from the National Institutes of Health (NIH).

REFERENCES

Ahmed N, Gao C, Lucker B, Cole D, Mitchell D. 2008. ODA16 aids axonemal outer row dynein assembly through an interaction with the intraflagellar transport machinery. *J Cell Biol* **183:** 313–322.

Alford L, Wirschell M, Yamamoto R, Sale WS. 2012. Control of axonemal inner dynein arms. In *Dyneins: Structure, biology and disease* (ed. King SM), pp. 313–335. Elsevier, Waltham, MA.

Austin-Tse C, Halbritter J, Zariwala MA, Gilberti ReM, Gee HY, Hellman N, Pathak N, Liu Y, Panizzi JR, Patel-King RS, et al. 2013. Zebrafish ciliopathy screen plus human mutational analysis identifies *C21orf59* and *CCDC65* defects as causing primary ciliary fyskinesia. *Am J Hum Genet* **93:** 672–686.

Becker-Heck A, Loges NT, Omran H. 2012. Dynein dysfunction as a cause of primary ciliary dyskinesia and other ciliopathies. In *Dyneins: Structure, biology and disease* (ed. King SM), pp. 603–627. Elsevier, Waltham, MA.

Benashski SE, Patel-King RS, King SM. 1999. Light chain 1 from the *Chlamydomonas* outer dynein arm is a leucine-rich repeat protein associated with the motor domain of the γ heavy chain. *Biochemistry* **38:** 7253–7264.

Bessen M, Fay RB, Witman GB. 1980. Calcium control of waveform in isolated flagellar axonemes of *Chlamydomonas*. *J Cell Biol* **86:** 446–455.

Bowman AB, Patel-King RS, Benashski SE, McCaffery JM, Goldstein LS, King SM. 1999. *Drosophila* roadblock and *Chlamydomonas* LC7: A conserved family of dynein-associated proteins involved in axonal transport, flagellar motility, and mitosis. *J Cell Biol* **146:** 165–180.

Casey D, Inaba K, Pazour G, Takada S, Wakabayashi K, Wilkerson C, Kamiya R, Witman G. 2003a. DC3, the 21-kD subunit of the outer dynein arm-docking complex (ODA-DC), is a novel EF-hand protein important for assembly of both the outer arm and the ODA-DC. *Mol Biol Cell* **14**: 3650–3663.

Casey D, Yagi T, Kamiya R, Witman G. 2003b. DC3, the smallest subunit of the *Chlamydomonas* flagellar outer dynein arm-docking complex, is a redox-sensitive calcium-binding protein. *J Biol Chem* **278**: 42652–42659.

Dean AB, Mitchell DR. 2013. *Chlamydomonas* ODA10 is a conserved axonemal protein that plays a unique role in outer dynein arm assembly. *Mol Biol Cell* **24**: 3689–3696.

Duquesnoy P, Escudier E, Vincensini L, Freshour J, Bridoux AM, Coste A, Deschildre A, de Blic J, Legendre M, Montantin G, et al. 2009. Loss-of-function mutations in the human ortholog of *Chlamydomonas reinhardtii* ODA7 disrupt dynein arm assembly and cause primary ciliary dyskinesia. *Am J Hum Genet* **85**: 890–896.

Fliegauf M, Benzing T, Omran H. 2007. When cilia go bad: Cilia defects and ciliopathies. *Nat Rev Mol Cell Biol* **8**: 880–893.

Fowkes ME, Mitchell DR. 1998. The role of preassembled cytoplasmic complexes in assembly of flagellar dynein subunits. *Mol Biol Cell* **9**: 2337–2347.

Haimo LT, Fenton RD. 1988. Interaction of *Chlamydomonas* dynein with tubulin. *Cell Motil Cytoskeleton* **9**: 129–139.

Harrison A, Olds-Clarke P, King SM. 1998. Identification of the *t* complex-encoded cytoplasmic dynein light chain Tctex1 in inner arm I1 supports the involvement of flagellar dyneins in meiotic drive. *J Cell Biol* **140**: 1137–1147.

Hayashibe K, Shingyoji C, Kamiya R. 1997. Induction of temporary beating in paralyzed flagella of *Chlamydomonas* mutants by application of external force. *Cell Motil Cytoskeleton* **37**: 232–239.

Heuser T, Raytchev M, Krell J, Porter ME, Nicastro D. 2009. The dynein regulatory complex is the nexin link and a major regulatory node in cilia and flagella. *J Cell Biol* **187**: 921–933.

Hom E, Witman GB, Harris EH, Dutcher SK, Kamiya R, Mitchell DR, Pazour GJ, Porter ME, Sale WS, Wirschell M, et al. 2011. A unified taxonomy for ciliary dyneins. *Cytoskeleton* **68**: 555–565.

Ichikawa M, Saito K, Yanagisawa HA, Yagi T, Kamiya R, Yamaguchi S, Yajima J, Kushida Y, Nakano K, Numata O, et al. 2015. Axonemal dynein light chain-1 locates at the microtubule-binding domain of the γ heavy chain. *Mol Biol Cell* **26** 4236–4247.

Ingolia N, Lareau L, Weissman J. 2011. Ribosome profiling of mouse embryonic stem cells reveals the complexity and dynamics of mammalian proteomes. *Cell* **147**: 789–802.

Kagami O, Kamiya R. 1992. Translocation and rotation of microtubules caused by multiple species of *Chlamydomonas* inner-arm dynein. *J Cell Sci* **103**: 653–664.

Kamiya R, Okamoto M. 1985. A mutant of *Chlamydomonas reinhardtii* that lacks the flagellar outer dynein arm but can swim. *J Cell Sci* **74**: 181–191.

Kamiya R, Kurimoto E, Muto E. 1991. Two types of *Chlamydomonas* flagellar mutants missing different components of inner-arm dynein. *J Cell Biol* **112**: 441–447.

Kato-Minoura T, Hirono M, Kamiya R. 1997. *Chlamydomonas* inner-arm dynein mutant, *ida5*, has a mutation in an actin-encoding gene. *J Cell Biol* **137**: 649–656.

Kato-Minoura T, Uryu S, Hirono M, Kamiya R. 1998. Highly divergent actin expressed in a *Chlamydomonas* mutant lacking the conventional actin gene. *Biochem Biophys Res Commun* **251**: 71–76.

King SM. 2010. Axonemal dyneins winch the cilium. *Nat Struct Mol Biol* **17**: 673–674.

King SM. 2012a. Composition and assembly of axonemal dyneins. In *Dyneins: Structure, biology and disease* (ed. King SM), pp. 209–243. Elsevier, Waltham, MA.

King SM. 2012b. Integrated control of axonemal dynein AAA$^+$ motors. *J Struct Biol* **179**: 222–228.

King SM. 2013. A solid-state control system for dynein-based ciliary/flagellar motility. *J Cell Biol* **201**: 173–175.

King SM, Patel-King RS. 1995a. Identification of a Ca^{2+}-binding light chain within *Chlamydomonas* outer arm dynein. *J Cell Sci* **108**: 3757–3764.

King SM, Patel-King RS. 1995b. The M_r = 8,000 and 11,000 outer arm dynein light chains from *Chlamydomonas* flagella have cytoplasmic homologues. *J Biol Chem* **270**: 11445–11452.

King SM, Patel-King RS. 2015. The oliogmeric outer arm dynein assembly factor CCDC103 is tightly integrated within the ciliary axoneme and exhibits periodic binding to microtubules. *J Biol Chem* **290**: 7388–7401.

LeDizet M, Piperno G. 1995. *ida4-1*, *ida4-2*, and *ida4-3* are intron splicing mutations affecting the locus encoding p28, a light chain of *Chlamydomonas* axonemal inner dynein arms. *Mol Biol Cell* **6**: 713–723.

McKenney RJ, Vershinin M, Kunwar A, Vallee RB, Gross SP. 2010. LIS1 and NudE induce a persistent dynein force-producing state. *Cell* **141**: 304–314.

Mitchell DR, Kang Y. 1991. Identification of *oda6* as a *Chlamydomonas* dynein mutant by rescue with the wild-type gene. *J Cell Biol* **113**: 835–842.

Mitchell DR, Rosenbaum JL. 1985. A motile *Chlamydomonas* flagellar mutant that lacks outer dynein arms. *J Cell Biol* **100**: 1228–1234.

Oda T, Yanagisawa H, Kamiya R, Kikkawa M. 2014. A molecular ruler determines the repeat length in eukaryotic cilia and flagella. *Science* **346**: 857–860.

Ogawa K, Takai H, Ogiwara A, Yokota E, Shimizu T, Inaba K, Mohri H. 1996. Is outer arm dynein intermediate chain 1 multifunctional? *Mol Biol Cell* **7**: 1895–1907.

Omran H, Kobayashi D, Olbrich H, Tsukahara T, Loges NT, Hagiwara H, Zhang Q, Leblond G, O'Toole E, Hara C, et al. 2008. Ktu/PF13 is required for cytoplasmic pre-assembly of axonemal dyneins. *Nature* **456**: 611–616.

Padma P, Hozumi A, Ogawa K, Inaba K. 2001. Molecular cloning and characterization of a thioredoxin/nucleoside diphosphate kinase related dynein intermediate chain from the ascidian, *Ciona intestinalis*. *Gene* **275**: 177–183.

Panizzi J, Becker-Heck A, Castleman V, Al-Mutairi D, Liu Y, Loges NT, Pathak N, Austin-Tse C, Sheridan E, Schmidts M, et al. 2012. *CCDC103* mutations cause primary ciliary dyskinesia by disrupting assembly of ciliary dynein arms. *Nat Genet* **44**: 714–719.

Paschal BM, King SM, Moss AG, Collins CA, Vallee RB, Witman GB. 1987. Isolated flagellar outer arm dynein translocates brain microtubules in vitro. *Nature* **330**: 672–674.

Patel-King RS, King SM. 2009. An outer arm dynein light chain acts in a conformational switch for flagellar motility. *J Cell Biol* **186:** 283–295.

Patel-King RS, Benashki SE, Harrison A, King SM. 1996. Two functional thioredoxins containing redox-sensitive vicinal dithiols from the *Chlamydomonas* outer dynein arm. *J Biol Chem* **271:** 6283–6291.

Rompolas P, Patel-King RS, King SM. 2010. An outer arm dynein conformational switch is required for metachronal synchrony of motile cilia in planaria. *Mol Biol Cell* **21:** 3669–3679.

Rompolas P, Patel-King RS, King SM. 2012. Association of Lis1 with outer arm dynein is modulated in response to alterations in flagellar motility. *Mol Biol Cell* **23:** 3554–3656.

Sakato M, Sakakibara H, King SM. 2007. *Chlamydomonas* outer arm dynein alters conformation in response to Ca^{2+}. *Mol Biol Cell* **18:** 3620–3634.

Schmidts M, Arts HH, Bongers EMHF, Yap A, Oud MM, Antony D, Duijkers L, Emes RD, Stalker J, Yntema JBL, et al. 2013. Exome sequencing identifies *DYNC2H1* mutations as a common cause of asphyxiating thoracic dystrophy (Jeune syndrome) without major polydactyly, renal or retinal involvement. *J Med Genet* **50:** 309–323.

Smith EF, Sale WS. 1992. Regulation of dynein-driven microtubule sliding by the radial spokes in flagella. *Science* **257:** 1557–1559.

Takada S, Kamiya R. 1994. Functional reconstitution of *Chlamydomonas* outer dynein arms from α–β and γ subunits: Requirement of a third factor. *J Cell Biol* **126:** 737–745.

Tarkar A, Loges NT, Slagle CE, Francis R, Dougherty GW, Tamayo JV, Shook B, Cantino M, Schwartz D, Jahnke C,

et al. 2013. DYX1C1 is required for axonemal dynein assembly and ciliary motility. *Nat Genet* **45:** 995–1003.

Vale RD, Toyoshima YY. 1988. Rotation and translocation of microtubules in vitro induced by dyneins from *Tetrahymena* cilia. *Cell* **52:** 459–469.

Wakabayashi K, King SM. 2006. Modulation of *Chlamydomonas reinhardtii* flagellar motility by redox poise. *J Cell Biol* **173:** 743–754.

Wilkerson CG, King SM, Koutoulis A, Pazour GJ, Witman GB. 1995. The 78,000 M_r intermediate chain of *Chlamydomonas* outer arm dynein is a WD-repeat protein required for arm assembly. *J Cell Biol* **129:** 169–178.

Wu H, Maciejewski MW, Marintchev A, Benashski SE, Mullen GP, King SM. 2000. Solution structure of a dynein motor domain associated light chain. *Nat Struct Biol* **7:** 575–579.

Yagi T, Minoura I, Fujiwara A, Saito R, Yasunaga T, Hirono M, Kamiya R. 2005. An axonemal dynein particularly important for flagellar movement at high viscosity: Implications from a new *Chlamydomonas* mutant deficient in the dynein heavy chain gene DHC9. *J Biol Chem* **280:** 41412–41420.

Yagi T, Uematsu K, Liu Z, Kamiya R. 2009. Identification of dyneins that localize exclusively to the proximal portion of *Chlamydomonas* flagella. *J Cell Sci* **122:** 1306–1314.

Yamamoto R, Hirono M, Kamiya R. 2010. Discrete PIH proteins function in the cytoplasmic preassembly of different subsets of axonemal dyneins. *J Cell Biol* **190:** 65–71.

Yamamoto R, Song K, Yanagisawa H, Fox L, Yagi T, Wirschell M, Hirono M, Kamiya R, Nicastro S, Sale WS. 2013. The MIA complex is a conserved and novel dynein regulator essential for normal ciliary motility. *J Cell Biol* **201:** 263–278.

The Central Apparatus of Cilia and Eukaryotic Flagella

Thomas D. Loreng and Elizabeth F. Smith

Department of Biological Sciences, Dartmouth College, Hanover, New Hampshire 03755

Correspondence: elizabeth.f.smith@dartmouth.edu

The motile cilium is a complex organelle that is typically comprised of a 9+2 microtubule skeleton; nine doublet microtubules surrounding a pair of central singlet microtubules. Like the doublet microtubules, the central microtubules form a scaffold for the assembly of protein complexes forming an intricate network of interconnected projections. The central microtubules and associated structures are collectively referred to as the central apparatus (CA). Studies using a variety of experimental approaches and model organisms have led to the discovery of a number of highly conserved protein complexes, unprecedented high-resolution views of projection structure, and new insights into regulation of dynein-driven microtubule sliding. Here, we review recent progress in defining mechanisms for the assembly and function of the CA and include possible implications for the importance of the CA in human health.

Motile cilia and eukaryotic flagella are characterized by the canonical "9+2" arrangement of microtubules in which nine doublet microtubules surround a central pair of singlet microtubules. These central microtubules and their associated protein projections are collectively referred to as the central apparatus (CA) (Fig. 1). Although there are a few exceptions to the 9+2 arrangement of microtubules, the CA is remarkably well conserved throughout eukaryotes and is thought to have been present in cilia of the last common eukaryotic ancestor (Mitchell 2004). Since researchers first discovered the nearly crystalline arrangement of proteins that form the axoneme, they have tried to answer two fundamental questions: How do these structures assemble? What is their role in motility? Answering these questions has

required a sophisticated array of structural, biochemical, genetic, and functional approaches. Here, we review new insights into the structure and composition of the CA, possible mechanisms for assembly, and the role of the CA in regulating ciliary and flagellar motility.

CENTRAL APPARATUS STRUCTURE

Unlike the nine doublet microtubules, the microtubules of the CA are not continuous with the basal body. The central microtubules extend from a region near the transition zone, elongate beyond the doublet microtubules, and end in a capping structure at the tip of the cilium (Ringo 1967; Dentler and Rosenbaum 1977; Dentler 1984). This cap has not been biochemically characterized, but structurally it contains two

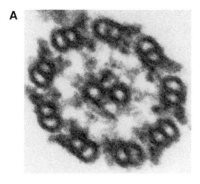

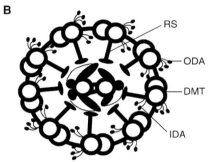

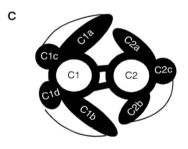

Figure 1. Central apparatus (CA) structure. (*A*) Transmission electron microscopy (TEM) image of an axoneme transverse section from *Chlamydomonas reinhardtii*. The central microtubules and their associated projections are surrounded by the outer doublet microtubules. (*B*) Diagram of axoneme transverse section seen in *A*. (*C*) Diagram of CA with projections labeled. DMT, Doublet microtubule; IDA, inner dynein arm; ODA, outer dynein arm; RS, radial spoke.

major components: a "bead" attached to the membrane, and two plates attached to the central microtubules (Dentler 1984). Each microtubule of the CA is structurally and biochemically distinct (reviewed below) and are referred to as C1 and C2. The portions of the central microtubules that extend beyond the outer dou-

blets are devoid of protein projections (Ringo 1967). In addition to protein projections associated with the surface of the microtubules, the Nicastro laboratory has shown that the internal lumen of the C2 microtubule of *Chlamydomonas* contains two small densities (Carbajal-Gonzalez et al. 2013), which they refer to as microtubule inner proteins, or MIPs, named MIP-C2a and MIP-C2b. MIP-C2a repeats every 16 nm and is 50 kDa in size, whereas MIP-C2b has an 8-nm periodicity and is only 35 kDa in size. These proteins are not seen in the flagella of sea urchin *Strongylocentrotus purpuratus* sperm, possibly indicating a specialization unique to *Chlamydomonas* cilia (Carbajal-Gonzalez et al. 2013).

Advances in electron microscopy (EM) in the 1960s and 1970s opened the door to our current understanding of CA structure. Using both *Chlamydomonas* and *Tetrahymena*, early work identified structural differences in the CA projections (Chasey 1969; Hopkins 1970). The microtubule with longer projections was termed C1, and the other C2. The longest projections on C1 are designated C1a and C1b and have a 16-nm repeat period along the length of C1. The prominent projections on C2 (C2a and C2b) have the same 16-nm repeat period. Later work using the newly developed rapid-freeze deep-etch method of EM revealed additional complexities including a sheath-like structure surrounding the CA (Goodenough and Heuser 1985). Advances in digital imaging and image-averaging techniques led to the discovery of smaller projections on both central microtubules: C2c repeating every 16 nm, and C1c and C1d that repeat every 32 nm (Mitchell and Sale 1999; Mitchell 2003b). The central microtubules are held together by a bridging structure that repeats at 16-nm intervals and is likely involved in CA stability (see section on Composition of the Central Apparatus).

Recent application of cryo-electron tomography (cryo-ET) has provided unprecedented views of CA structure and revealed four entirely new protein projections termed C1e, C1f, C2d, and C2e (Fig. 2) (Carbajal-Gonzalez et al. 2013). These new views indicate that the projections actually form a continuous network sur-

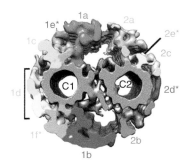

Figure 2. Detailed structural elements of the central apparatus (CA). Isosurface renderings derived from cryo-electron tomography (cryo-ET) of the *Chlamydomonas* CA viewed in transverse section. The projections form an interconnected network surrounding the two singlet microtubules with connections between C1a and C2a and C1b and C2b. The bridge structure connecting the two microtubules is in gray. MIP-C2a is highlighted within the lumen of the C2 microtubule. The projections labeled C1e, C1f, C2d, and C2e likely correspond to the material previously described as a sheath. (From Carbajal-Gonzalez et al. 2013; reprinted, with permission, from Wiley.)

rounding the two central microtubules and may explain the sheath structure observed in previous studies (Carbajal-Gonzalez et al. 2013).

Finally, the entire CA structure is known to twist in a left-handed helix (Omoto and Kung 1980; Kamiya 1982; Goodenough and Heuser 1985; Mitchell and Nakatsugawa 2004). This helix contains one 360° twist over the length of the *Chlamydomonas* cilium, whereas in *Paramecium* the number of twists per cilium length varies (Omoto and Kung 1980; Kamiya 1982). CA twist in *Chlamydomonas* occurs with or without RS heads, indicating that twist is not due to CA contact with the RSs (Mitchell and Nakatsugawa 2004). In addition, CAs extruded from the axoneme retain a helical structure (Kamiya 1982). The function of this twist is unknown, but it may have a role in regulating motility.

COMPOSITION OF THE CENTRAL APPARATUS

Our current understanding of the composition of the projections is largely based on biochemical approaches and study of *Chlamydomonas*

mutants (for example, Witman et al. 1978; Adams et al. 1981). In one class of mutants, the entire CA fails to assemble (*pf15*, *pf18*, *pf19*, and *pf20*); these mutants have immotile flagella. The identity of the *PF18* gene is unknown; however, the other three loci have been studied extensively. PF20 is a WD repeat-containing protein localized to the bridge structure connecting the two central microtubules and is hypothesized to stabilize the central tubules (Smith and Lefebvre 1997). *PF15* and *PF19*, encode the p80 and p60 subunits, respectively, of the microtubule-severing protein katanin (Dymek et al. 2004; Dymek and Smith 2012) (see section on Assembly of the Central Apparatus for further discussion of PF15 and PF19).

The second class of *Chlamydomonas* mutants lack subsets of CA projections and includes *pf6*, *cpc1*, and *pf16*. The C1a projection is lacking from *pf6* flagella (Dutcher et al. 1984). *PF6* encodes a large (240-kDa) protein that contains numerous proline-rich domains (Rupp et al. 2001). Additional biochemical studies led to the discovery of a PF6-containing complex that includes calmodulin (CaM) and four other proteins (Table 1). The PF6 protein likely plays a scaffolding role in the assembly of this complex to form C1a (Wargo et al. 2005).

The C1b projection is lacking in *cpc1* flagella (Mitchell and Sale 1999). These mutant flagella often lack portions of the C2b projection as well, possibly indicating structural connections between these two projections (Mitchell and Sale 1999; Zhang and Mitchell 2004). CPC1 is a 265-kDa protein composed of an EF-hand domain and an adenylate kinase domain. Along with CPC1, the C1b projection contains four other identified proteins including the chaperone protein HSP70 and the glycolytic enzyme enolase, which were identified through cosedimentation in sucrose gradients (Table 1) (Mitchell et al. 2005).

Mutations in *PF16* cause the disassembly of the C1 microtubule when flagella are demembranated (Dutcher et al. 1984). PF16 is a 57-kDa protein composed of eight continuous armadillo repeat motifs (Smith and Lefebvre 1996, 2000). Armadillo repeats are involved in protein–protein interactions (Peifer et al. 1994),

Table 1. Proteins associated with the central apparatus

Structure/ localization	Protein	Molecular mass (kDa)	Accession number	Description	References
C1a projection	PF6[a]	240	AAK38270	Contains alanine-proline rich domains; ASH domain	Rupp et al. 2001; Wargo et al. 2005
	C1a-86	86	AAZ31187	Contains a PKA RII-like LRR domain	Wargo et al. 2005
	C1a-34	34	AAZ31186	Contains a coiled-coil region	Wargo et al. 2005
	C1a-32	32	AAZ31185	Similar to C1a-34	Wargo et al. 2005
	C1a-18	18	AAZ31184	Contains MORN domains	Wargo et al. 2005
	Calmodulin	18	AAA33083	Calcium-binding protein	Wargo et al. 2005
C1b projection	CPC1	265	AAT40992	Contains EF-hand domain; adenylate kinase domains	Zhang and Mitchell 2004; Mitchell et al. 2005
	C1b-350 (FAP42)	350	EDP00757	Contains five guanylate kinase domains; one adenylate kinase domain	Mitchell et al. 2005
	C1b-135 (FAP69)	135	EDP06190	Armadillo repeat-containing protein	Mitchell et al. 2005
	HSP70	78	P25840	Chaperone/heat shock protein	Zhang and Mitchell 2004; Mitchell et al. 2005
	Enolase	56	P13683	Glycolytic enzyme	Mitchell et al. 2005
C1d projection	Pcdp1 (FAP221)[a]	100	ADD85929	Calmodulin-binding protein	DiPetrillo and Smith 2010
	FAP54	318	AFG30957	Contains no known functional domains	DiPetrillo and Smith 2010
	FAP46	289	AFG30956	Contains no known functional domains	DiPetrillo and Smith 2010
	FAP74	204	ADD85930	Contains no known functional domains	DiPetrillo and Smith 2010
	C1d-87	87	EDP09774	WD repeat-containing protein	Brown et al. 2012
C1 microtubule	PF16[a]	57	AAC49169	Armadillo repeat-containing protein	Dutcher et al. 1984; Smith and Lefebvre 1996
	PP1c	35	AAD38850	Phosphatase	Yang et al. 2000
	C1 kinesin	110		Kinesin-like protein	Fox et al. 1994
C1–C2 bridge	PF20[a]	63	AAB41727	WD repeat-containing protein	Smith and Lefebvre 1997
C2b projection	Hydin[a]	540	EDP09735	Contains four ASH domains	Lechtreck and Witman 2007
C2c projection	KLP1	96	P46870	Kinesin-like protein	Bernstein et al. 1994; Yokoyama et al. 2004

Continued

Cite this article as *Cold Spring Harb Perspect Biol* doi: 10.1101/cshperspect.a028118

Table 1. *Continued*

Structure/ localization	Protein	Molecular mass (kDa)	Accession number	Description	References
Other	Katanin p60	60	AAF12877	Catalytic subunit of katanin; AAA ATPase	Dymek and Smith 2012
	Katanin p80	80	EDP00085	Regulatory subunit of katanin; WD repeat-containing protein	Dymek et al. 2004

[a]These proteins have mammalian homologs that are discussed in the section on Central Apparatus in Mammalian Health and Human Disease.

suggesting that PF16 forms complexes that stabilize the C1 microtubule. The *pf16* mutant has been invaluable for defining the composition of C1 and C2. Proteins that remain associated with the axoneme in a *pf16* mutant are good candidates for C2-associated proteins.

Biochemical approaches have revealed the identities of additional CA-associated proteins. Experiments designed to identify axonemal CaM interacting proteins led to the discovery of a second CaM-containing complex of five proteins associated with C1 (Table 1) (DiPetrillo and Smith 2010). When expression of complex components is reduced, the C1d projection fails to assemble and the cells show uncoordinated flagellar movement (DiPetrillo and Smith 2010). A strain with a mutation in a single complex component has the same phenotype (Brown et al. 2012). Furthermore, the C1d protein FAP221 is homologous to the mouse primary ciliary dyskinesia protein 1 (see section on Central Apparatus in Mammalian Health and Human Disease) (Lee et al. 2008; DiPetrillo and Smith 2010).

Two other proteins are associated with the C1 microtubule. Protein phosphatase 1 was identified in cilia from *Paramecium* and *Chlamydomonas* (Friderich et al. 1992; Yang et al. 2000). In addition, a 110-kDa kinesin has been localized to C1 through the use of mutants and polyclonal antibodies (Fox et al. 1994; Johnson et al. 1994). Localization of these proteins to specific projections has not been determined.

Two proteins have been assigned to C2 projections: KLP1 and Hydin. The study that led to the discovery of the 110-kDa C1 kinesin, also led to the identification of KLP1 (Fox et al. 1994;

Johnson et al. 1994). KLP1 is tightly bound to the axoneme in a *pf16* mutant, thereby localizing it to the C2 microtubule (Bernstein et al. 1994). KLP1 is a member of the kinesin-9 family of proteins and when its expression is knocked down, the C2c projection and a portion of the C2b projection fail to assemble (Yokoyama et al. 2004).

Although many studies in *Chlamydomonas* have preceded the identification of mammalian homologs (see section on Central Apparatus in Mammalian Health and Human Disease), a study in mammals led to the discovery of Hydin. Mice that have the *hy3* mutation in the Hydin gene show hydrocephalus (Davy and Robinson 2003). Furthermore, similar mutations in trypanosomes lead to motility defects (Broadhead et al. 2006). Knockdown of Hydin in *Chlamydomonas* causes the failure of the C2b projection to assemble and, in some instances, the destabilization of the C1b and C2c projections as well (Lechtreck and Witman 2007). As noted above, the destabilization of the C1b projection due to mutations in CPC1 leads to the reciprocal destabilization of the C2b projection (Mitchell and Sale 1999; Yokoyama et al. 2004; Lechtreck and Witman 2007). Based on its size, Hydin may act as a scaffold for other CA-associated proteins.

ASSEMBLY OF THE CENTRAL APPARATUS

Mutants that fail to assemble the entire CA have contributed to our understanding of CA assembly. Although PF20 has been implicated in stabilizing the central microtubules, katanin may play a direct role in the formation of the central microtubules. Katanin is a hexameric protein composed of an AAA ATPase catalytic subunit (p60, encoded by the *PF19* gene) and a regulatory subunit with WD repeat domains (p80, encoded by the *PF15* gene) (Hartman et al. 1998; Dymek et al. 2004; Dymek and Smith 2012). Mutations in either subunit yield a central pairless phenotype. Katanin severs microtubules by the binding of its p60 hexamer to tubulin, whereas the p80 subunit potentially targets and enhances this activity (Hartman et al. 1998; McNally et al. 2000). The *pf19* mutation inhibits the microtubule-severing activity of the p60 subunit (Dymek and Smith 2012). Therefore, severing activity is necessary for the assembly of the central tubules. It seems paradoxical that the assembly of microtubules requires microtubule-severing activity; however, there is precedent for katanin's involvement in the formation of noncentrosomal microtubule arrays in plants and neurons (Karabay et al. 2004; Yu et al. 2005; Gardiner and Marc 2011). Recall that the central microtubules are not nucleated from the basal body and are also noncentrosomal microtubules.

Several other proteins are linked to central microtubule assembly. Studies in *Drosophila* identified the basal body protein Bld10p/ CEP135 as important for assembly. Bld10p is localized to the central cartwheel structure of the basal body (Carvalho-Santos et al. 2012). In *Drosophila*, a single microtubule exists before the formation of the central microtubules and is stabilized by Bld10p. Presumably, the stabilization of this microtubule is necessary for the eventual formation of two central microtubules (Carvalho-Santos et al. 2012). It is unknown whether the requirement for Bld10p in CA assembly is unique to *Drosophila*.

RNA interference has revealed that γ-tubulin and its associated ring complex are necessary for CA assembly in trypanosomes (McKean et al. 2003; Zhou and Li 2015). In addition, in *Tetrahymena*, central pair formation depends on the glycylation domain of β-tubulin and if this domain is mutated, cilia are short and fail to form central microtubules (Thazhath et al. 2004). Three possible roles for glycylation have

been proposed. First, glycylation may be necessary for intraflagellar transport (IFT)-mediated transport of CA proteins (Thazhath et al. 2004). Short doublet microtubules, but not central pair microtubules, assemble in *Tetrahymena* IFT mutants and in sea urchin eggs in which IFT was inhibited by antibodies (Brown et al. 1999, 2003). Perhaps glycylation is necessary for tubulin's association with anterograde IFT particles. Second, glycylation may be necessary for central microtubule nucleation at the transition zone (Thazhath et al. 2004). Finally, β-tubulin glycylation may stabilize the growing distal ends of the central microtubules or mediates associations with cap-specific proteins (Thazhath et al. 2004).

Why has evolution favored a pair of singlet central microtubules? The answer may be as simple as space dependence. *Chlamydomonas pf14* mutants lack RSs; therefore, the center of the axoneme contains more empty space than in wild-type. Mutant *pf14* axonemes have been observed to contain two pairs of central microtubules with identical and correct polarities and complete protein projections (Lechtreck et al. 2013). Another example includes mutations in *BLD12*. Bld12p is the *Chlamydomonas* homolog of Sas6 in *Caenorhabditis elegans* and forms the cartwheel structure of the basal body that establishes ninefold symmetry (Nakazawa et al. 2007). Mutations in *BLD12* cause misshaped basal bodies that can manifest as an expansion of the axonemal diameter; in these cases, a corresponding increase in the number of CAs is observed (Nakazawa et al. 2014). These results provide evidence that the number of CAs that assemble may be determined by the allowable space within the axoneme. A key question remains: Why is the CA typically a pair of microtubules as opposed to a single or three or more microtubules?

Assembly and targeting of the CA projections to the central microtubules is also an active area of research. Several ciliary structures, including dynein arm and RS components, form preassembly complexes in the cytoplasm that are then transported into the cilium (Fowkes and Mitchell 1998; Diener et al. 2011; Viswanadha et al. 2014). Do CA proteins assemble

Cite this article as *Cold Spring Harb Perspect Biol* doi: 10.1101/cshperspect.a028118

in preassembly complexes before targeting to the axoneme? Proteins are transported into the cilium in one of two ways: IFT or diffusion. Are CA projection proteins transported through IFT or diffusion?

These questions have recently been studied in *Chlamydomonas* flagellar regeneration and dikaryon rescue experiments performed in the Lechtreck laboratory (Lechtreck et al. 2013; Wren et al. 2013). Observations of flagellar regeneration have revealed time-dependent assembly in which the projections assemble onto the central microtubules before dynein arm assembly, despite the observation that the central microtubules form after the outer doublets. Interestingly, the direction of assembly depends on whether the cell is in the process of de novo flagellar assembly (regeneration) or if the CA is assembling in an assembled central pairless axoneme (dikaryon rescue). In regenerating flagella, central microtubule growth begins proximal to the transition zone and continues distally. However, in dikaryon rescue experiments using *pf15* or *pf19* cells, subdistal microtubule assembly occurred with projection proteins assembling tip to base. These observations indicate that the CA is capable of assembling independent of other structures in the axoneme and that microtubule nucleation can occur from either distal or proximal regions of the flagellum. This laboratory has also shown that PF16 uses IFT to enter the cilium suggesting that other CA proteins may as well (Wren et al. 2013).

CENTRAL APPARATUS FUNCTION

The high degree of CA conservation in motile cilia implies that there is selective pressure for motile cilia to retain the 9+2 microtubule structure. However, is the CA actually necessary for motility? There are examples of motile cilia that lack a CA. For instance, the mature male gametes of the alga *Lithodesmium undulatum* contain a 9+0 microtubule arrangement and are fully motile (Manton 1966; Carvalho-Santos et al. 2011). The protozoan *Lecudina tuzetae* and the arthropod *Acerentomon microrhinus* have more extreme versions of alternate yet mo-

tile axonemal structure with 6+0 and 14+0, respectively (Schrevel and Besse 1975; Dallai et al. 2010; Carvalho-Santos et al. 2011). In addition, the nodal cilia present during mammalian development have a 9+0 structure and show a more rotational motility rather than the more planar symmetric and asymmetric waveforms typical of other flagella and cilia (Nonaka et al. 1998).

Examples of motile cilia that lack a CA are rare. Furthermore, in organisms with motile cilia that assemble a CA, the CA is required for motility. As noted above, central pairless mutants of *Chlamydomonas* show complete flagellar paralysis. In addition, mutations in which CA projections are absent also show aberrant flagellar motility. The *pf6* mutant flagella lacking C1a twitch ineffectively (Rupp et al. 2001) and mutants missing only the carboxy terminus of PF6 are paralyzed as well (Goduti and Smith 2012). The flagella of *cpc1* mutants beat at 40 Hz instead of 60 Hz and have axonemes that switch waveforms when exposed to high concentrations of calcium (Mitchell and Sale 1999). Mutations affecting C1d cause slow swimming. In addition, the flagella of these cells are uncoordinated and beat at a frequency of ∼30% that of wild-type (DiPetrillo and Smith 2010, 2011). Hydin mutants missing C2b are locked in a position in which one flagellum is oriented along the cell body, whereas the other extends away from it in a "hands up/hands down" conformation and mice missing this projection show cilia stalling (Lechtreck and Witman 2007; Lechtreck et al. 2008). Finally, *pf16* flagella are lacking only three proteins yet are completely paralyzed (Dutcher et al. 1984). These findings are not limited to *Chlamydomonas*. Defects in the mammalian CA, like in Hydin mutants, have also been shown to affect motility (see section on Central Apparatus in Mammalian Health and Human Disease).

Extragenic suppressor mutations that restore motility to paralyzed CA mutants (and RS defective mutants) without restoring the missing CA structures have provided important genetic evidence of a mechanism for the CA's role in regulating motility (Huang et al. 1982). The suppressor mutations are found in components of the inner and outer dynein arms along with the N-DRC and suggest a regulatory pathway by which the CA regulates dynein activity through the RSs and regulatory complexes located on the doublet microtubules (Huang et al. 1982; Piperno et al. 1992, 1994; Porter et al. 1992, 1994). However, it is important to note that the CA and RS mutants carrying suppressor mutations produce symmetric waveforms of reduced amplitude and beat frequency (Brokaw et al. 1982). This implicates the CA and RSs in the production of proper waveforms necessary for forward swimming.

Despite flagellar paralysis, dynein arms of CA mutants retain their force-generating properties. Using a sliding disintegration assay in which microtubule sliding is uncoupled from flagellar bending, dynein activity can be quantified as microtubule sliding velocity (Summers and Gibbons 1971; Okagaki and Kamiya 1986). Using this assay, *pf18* and *pf15* mutant axonemes have been shown to have reduced sliding velocities compared with wild-type (Smith 2002b). Evidently, dynein activity is reduced in these mutants. Reduced sliding velocities are also observed for *pf16* mutant axonemes showing the importance of the C1 microtubule in regulating dynein. Combining the sliding assay with structural studies has shown that the position of the CA correlates with the position of active doublet sliding (Yoshimura and Shingyoji 1999; Wargo and Smith 2003).

The sliding disintegration assay has also revealed transduction pathways that include second messengers, and kinases and phosphatases anchored to the axoneme. Several studies have shown a link between the CA and regulation of inner dynein arms through axonemal kinases and phosphatases (Smith and Sale 1992; Howard et al. 1994; Habermacher and Sale 1996; Yang and Sale 2000; Kikushima 2009). These studies have been reviewed in Wirschell et al. (2011). A relationship between calcium regulation of motility and the CA has also been shown. For instance, increased intraflagellar calcium concentrations can increase sliding velocity in central pairless mutants, but not *pf16* axonemes (Smith 2002a). Studies of sea urchin sperm also show the importance of calcium signaling

through the CA in the localized sliding of specific subsets of doublet microtubules (Nakano et al. 2003). Work in *Chlamydomonas* indicates that these changes are mediated by axoneme-associated CaM (Smith 2002a) and that CaM is associated with both the C1a and C1d projections (Wargo et al. 2005; DiPetrillo and Smith 2009, 2010). These studies provide evidence of a role for the CA in calcium regulation of motility.

Although these studies provide a strong correlation between the CA and regulating microtubule sliding, they do not provide a mechanism for how this regulation is converted to complex ciliary waveforms. Given the location of the CA in relation to the dynein arms, CA signals would likely be mediated by the RSs. This idea was first supported by structural studies of *Elliptio* gill cilia in which the RS heads make transient contact with the projections of the CA, causing a tilt in the RSs relative to the longitudinal axis of the cilium (Warner and Satir 1974). RS–CA interactions may also occur in protozoan organisms whose CA rotates during bend propagation (Omoto and Kung 1980; Omoto et al. 1999; Mitchell 2003a). Even though this rotation has been shown to be a passive response to flagellar bending, rotation and twist of the CA may provide important positional cues that are transmitted via the spokes (Mitchell and Nakatsugawa 2004).

These studies raise the question of whether the CA provides biochemical or mechanical cues or both to regulate microtubule sliding. What is the relationship between specific protein projections and this regulation? Of the CA projections, C1d is preferentially oriented toward areas of active microtubule sliding (Wargo and Smith 2003; DiPetrillo and Smith 2011). Furthermore, axonemes with expanded diameters show preferential binding between the RSs and specific C1 projections; one site near C1a and one near C1b (Nakazawa et al. 2014). Removing C1a and C1b reveals weaker, yet still seemingly specific, interactions between RS heads and the CA (Nakazawa et al. 2014). These studies indicate that there is specificity in RS–CA interactions; however, a recent study suggests otherwise. By engineering epitope tags on

RS heads, Kikkawa's group rescued motility in *pf6* mutants lacking C1a (Oda et al. 2014). The tag extended the RS across the gap creating an artificial interaction that reconstituted motility. These results suggest CA regulation may not require specific CA–RS interactions but may rely on nonspecific mechanosignaling. Further research is required to reconcile these seemingly conflicting studies.

Additional mechanisms for CA regulation of motility have been proposed from mathematical modeling approaches applied to beating cilia and flagella. Mathematical modeling of sperm flagellar bending led to a "geometric clutch" model (Lindemann 1994; Lindemann and Kanous 1995). In this model, the formation of dynein cross-bridges is limited by the space between doublet microtubules. Therefore, only a subset of dynein can be active at any given time. When cross-bridges form, microtubule sliding is induced causing axonemal distortions that result in dynein cross bridges on one side of the axoneme to release and dynein cross-bridges on the opposite side of the axoneme to form (Lindemann and Mitchell 2007). Bending of the flagellum occurs due to switching of cross-bridge formation and release. In this model, the CA and RSs transmit a transverse force during the switch point (Lindemann 2003). However, the force-bearing capacity of the CA is unknown and must be determined to understand how the t-force is distributed in the axoneme (Lindemann 2007). For a recent summary of this model, see Lindemann and Lesich (2015).

One surprising contribution the CA may make in regulating motility is maintaining stable intraciliary ATP levels. The CPC1 protein contains an adenylate kinase domain. CPC1-defective mutants have 64% the beat frequency of wild-type flagella, potentially due to a lack of ATP production from recycled ADP. When mutant axonemes are saturated with ATP in flagellar reactivation assays, beat frequency is returned to 90% that of wild-type (Zhang and Mitchell 2004). Reactivation of beating despite the lack of CPC1 in these flagella indicates that there are potentially other adenylate kinases to be discovered in the axoneme.

THE CENTRAL APPARATUS IN MAMMALIAN HEALTH AND HUMAN DISEASE

One of the most astounding findings in cell biology in the past few decades is the identification of diseases that are caused by ciliary defects known collectively as ciliopathies (Braun and Hildebrandt 2016). These defects can be in either motile or nonmotile cilia. Here, we focus on proteins in motile cilia that when defective cause disease in mammals.

In mammalian motile cilia, three proteins with *Chlamydomonas* homologs are believed to form a complex: Spag16 (PF20), Spag6 (PF16), and Spag17 (PF6) (Zhang et al. 2005). Each member of this putative "interactome" is essential for proper ciliary motility and mammalian health. The mammalian homolog of PF20, Spag16, occurs in two isoforms: Spag16L, localized to the cilium and Spag16S, localized to the nucleus (Zhang et al. 2006). Spag16 is essential for male fertility by regulating sperm flagellar motility (Zhang et al. 2006). Further studies identified Spag16 as essential for human fertility as well (Zhang et al. 2007b). The mammalian homolog of PF16, Spag6 (Neilson et al. 1999; Zhang et al. 2002), is essential for mammalian health; mice with mutations in Spag6 show hydrocephalus, respiratory distress, infertility, and die 8 weeks after birth (Sapiro et al. 2002). Furthermore, Spag6-deficient mice also have reduced ciliary beat frequency in tracheal epithelial cells and reduced numbers of cilia in both the trachea and brain ependymal cells (Teves et al. 2014). The PF6 homolog, Spag17, is also essential for motility and linked to severe health problems in mice. Spag17-deficient mice show skeletal abnormalities and have enlarged brain ventricles, respiratory distress, and die in an extremely short 12 hours after birth (Teves et al. 2013, 2015). These phenotypes are exacerbated in mice that have both Spag6 and Spag17 mutations (Zhang et al. 2007a). To date, there is no evidence that these three proteins form a complex in *Chlamydomonas*.

Additional mammalian homologs of *Chlamydomonas* proteins include the protein FAP221 associated with C1d, which is mammalian Pcdp1, a protein associated with primary ciliary dyskinesia in mice (Lee et al. 2008). More recently, mammalian homologs of other members of C1d have been identified. Mutations in one of these proteins, CFAP54, lead to symptoms associated with primary ciliary dyskinesia such as sperm motility defects and an accumulation of mucus in the lungs (McKenzie et al. 2015). These findings further implicate this small projection as being essential for ciliary motility. Finally, the C2b protein Hydin was first characterized in mice showing hydrocephalus (Davy and Robinson 2003). This hydrocephalus is not due to defects in ciliary length or density in the brain, which are normal in Hydin mutants. Instead, the brain cilia lack C2b causing a motility defect termed "cilia stalling"; this leads to a lack of fluid flow in the brain and hydrocephalus (Lechtreck et al. 2008). (For more information about CA proteins in mammals, see Teves et al. 2016.)

These findings can be extended beyond mouse models to humans. Mutations in projection proteins like Spag16L can cause male infertility, and recessive mutations in Hydin have been linked to primary ciliary dyskinesia in human patients (Zhang et al. 2007b; Olbrich et al. 2012). In addition, there are several accounts of patients with primary ciliary dyskinesia that lack the entire CA (Bautista-Harris et al. 2000; Stannard et al. 2004). Often, the cilia of these patients will show the transposition of one outer doublet microtubule to the center of the axoneme perhaps to function as a CA (Smallman and Gregory 1986; Chilvers et al. 2003; Burgoyne et al. 2014). This particular observation is especially interesting for showing the importance of the CA for the regulation of proper ciliary motility in humans. The locations of the mutations in these patients have not been published. Therefore, it is not known whether they correspond to previously identified CA proteins.

QUESTIONS FOR THE FUTURE

By applying a combination of biochemical, structural, and genetic approaches to understanding how motile cilia assemble and function, investigators continue to peel away the layers of complexity associated with the CA. As we make headway identifying the components of essential complexes, our challenge is to map these complexes onto the intricate CA structures revealed by cryo-ET. In addition, we are only beginning to define the molecular and mechanical mechanisms that couple the CA to regulation of ciliary motility. We need to combine our knowledge of the composition and structure of the CA projections with functional assays that allow for computational and mathematical modeling approaches to elucidate the mechanisms of the CA in regulating motility, including the role of second messengers and other signaling molecules. Finally, despite the intriguing discovery that katanin may play a role in nucleating the central microtubules, there are many questions that remain to be answered about the nucleation of the central microtubules and the assembly of associated structures.

REFERENCES

*Reference is also in this collection.

Adams GM, Huang B, Piperno G, Luck DJ. 1981. Central-pair microtubular complex of *Chlamydomonas* flagella: Polypeptide composition as revealed by analysis of mutants. *J Cell Biol* **91:** 69–76.

Bautista-Harris G, Julia-Serda G, Rodriguez de Castro F, Santana-Benitez I, Cabrera-Navarro P. 2000. Absence of central microtubules and transposition in the ciliary apparatus of three siblings. *Respiration* **67:** 449–452.

Bernstein M, Beech PL, Katz SG, Rosenbaum JL. 1994. A new kinesin-like protein (Klp1) localized to a single microtubule of the *Chlamydomonas* flagellum. *J Cell Biol* **125:** 1313–1326.

* Braun DA, Hildebrandt F. 2016. Ciliopathies. *Cold Spring Harb Perspect Biol* doi: 10.1101/cshperspect.a028191.

Broadhead R, Dawe HR, Farr H, Griffiths S, Hart SR, Portman N, Shaw MK, Ginger ML, Gaskell SJ, McKean PG, et al. 2006. Flagellar motility is required for the viability of the bloodstream trypanosome. *Nature* **440:** 224–227.

Brokaw CJ, Luck DJ, Huang B. 1982. Analysis of the movement of *Chlamydomonas* flagella: The function of the radial-spoke system is revealed by comparison of wild-type and mutant flagella. *J Cell Biol* **92:** 722–732.

Brown JM, Marsala C, Kosoy R, Gaertig J. 1999. Kinesin-II is preferentially targeted to assembling cilia and is required for ciliogenesis and normal cytokinesis in *Tetrahymena*. *Mol Biol Cell* **10:** 3081–3096.

Brown JM, Fine NA, Pandiyan G, Thazhath R, Gaertig J. 2003. Hypoxia regulates assembly of cilia in suppressors of *Tetrahymena* lacking an intraflagellar transport subunit gene. *Mol Biol Cell* **14:** 3192–3207.

Cite this article as *Cold Spring Harb Perspect Biol* doi: 10.1101/cshperspect.a028118

Brown JM, Dipetrillo CG, Smith EF, Witman GB. 2012. A FAP46 mutant provides new insights into the function and assembly of the C1d complex of the ciliary central apparatus. *J Cell Sci* **125:** 3904–3913.

Burgoyne T, Lewis A, Dewar A, Luther P, Hogg C, Shoemark A, Dixon M. 2014. Characterizing the ultrastructure of primary ciliary dyskinesia transposition defect using electron tomography. *Cytoskeleton* **71:** 294–301.

Carbajal-Gonzalez BI, Heuser T, Fu X, Lin J, Smith BW, Mitchell DR, Nicastro D. 2013. Conserved structural motifs in the central pair complex of eukaryotic flagella. *Cytoskeleton* **70:** 101–120.

Carvalho-Santos Z, Azimzadeh J, Pereira-Leal JB, Bettencourt-Dias M. 2011. Evolution: Tracing the origins of centrioles, cilia, and flagella. *J Cell Biol* **194:** 165–175.

Carvalho-Santos Z, Machado P, Alvarez-Martins I, Gouveia SM, Jana SC, Duarte P, Amado T, Branco P, Freitas MC, Silva ST, et al. 2012. BLD10/CEP135 is a microtubule-associated protein that controls the formation of the flagellum central microtubule pair. *Dev Cell* **23:** 412–424.

Chasey D. 1969. Observations on the central pair of microtubules from the cilia of *Tetrahymena pyriformis*. *J Cell Sci* **5:** 453–458.

Chilvers MA, Rutman A, O'Callaghan C. 2003. Ciliary beat pattern is associated with specific ultrastructural defects in primary ciliary dyskinesia. *J Allergy Clin Immunol* **112:** 518–524.

Dallai R, Mercati D, Bu Y, Yin YW, Callaini G, Riparbelli MG. 2010. The spermatogenesis and sperm structure of *Acerentomon microrhinus* (Protura, Hexapoda) with considerations on the phylogenetic position of the taxon. *Zoomorphology* **129:** 61–80.

Davy BE, Robinson ML. 2003. Congenital hydrocephalus in *hy3* mice is caused by a frameshift mutation in *Hydin*, a large novel gene. *Hum Mol Genet* **12:** 1163–1170.

Dentler WL. 1984. Attachment of the cap to the central microtubules of *Tetrahymena* cilia. *J Cell Sci* **66:** 167–173.

Dentler WL, Rosenbaum JL. 1977. Flagellar elongation and shortening in *Chlamydomonas*. III: Structures attached to the tips of flagellar microtubules and their relationship to the directionality of flagellar microtubule assembly. *J Cell Biol* **74:** 747–759.

Diener DR, Yang P, Geimer S, Cole DG, Sale WS, Rosenbaum JL. 2011. Sequential assembly of flagellar radial spokes. *Cytoskeleton* **68:** 389–400.

DiPetrillo C, Smith E. 2009. Calcium regulation of ciliary motility analysis of axonemal calcium-binding proteins. *Method Cell Biol* **92:** 163–180.

DiPetrillo CG, Smith EF. 2010. Pcdp1 is a central apparatus protein that binds Ca^{2+}-calmodulin and regulates ciliary motility. *J Cell Biol* **189:** 601–612.

DiPetrillo CG, Smith EF. 2011. The Pcdp1 complex coordinates the activity of dynein isoforms to produce wild-type ciliary motility. *Mol Biol Cell* **22:** 4527–4538.

Dutcher SK, Huang B, Luck DJ. 1984. Genetic dissection of the central pair microtubules of the flagella of *Chlamydomonas reinhardtii*. *J Cell Biol* **98:** 229–236.

Dymek EE, Smith EF. 2012. PF19 encodes the p60 catalytic subunit of katanin and is required for assembly of the flagellar central apparatus in *Chlamydomonas*. *J Cell Sci* **125:** 3357–3366.

Dymek EE, Lefebvre PA, Smith EF. 2004. PF15p is the *Chlamydomonas* homologue of the Katanin p80 subunit and is required for assembly of flagellar central microtubules. *Eukaryotic Cell* **3:** 870–879.

Fowkes ME, Mitchell DR. 1998. The role of preassembled cytoplasmic complexes in assembly of flagellar dynein subunits. *Mol Biol Cell* **9:** 2337–2347.

Fox LA, Sawin KE, Sale WS. 1994. Kinesin-related proteins in eukaryotic flagella. *J Cell Sci* **107:** 1545–1550.

Friderich G, Klumpp S, Russell CB, Hinrichsen RD, Kellner R, Schultz JE. 1992. Purification, characterization and structure of protein phosphatase 1 from the cilia of *Paramecium tetraurelia*. *Eur J Biochem* **209:** 43–49.

Gardiner J, Marc J. 2011. *Arabidopsis thaliana*, a plant model organism for the neuronal microtubule cytoskeleton? *J Exp Botany* **62:** 89–97.

Goduti DJ, Smith EF. 2012. Analyses of functional domains within the PF6 protein of the central apparatus reveal a role for PF6 sub-complex members in regulating flagellar beat frequency. *Cytoskeleton* **69:** 179–194.

Goodenough UW, Heuser JE. 1985. Substructure of inner dynein arms, radial spokes, and the central pair/projection complex of cilia and flagella. *J Cell Biol* **100:** 2008–2018.

Habermacher G, Sale WS. 1996. Regulation of flagellar dynein by an axonemal type-1 phosphatase in *Chlamydomonas*. *J Cell Sci* **109:** 1899–1907.

Hartman JJ, Mahr J, McNally K, Okawa K, Iwamatsu A, Thomas S, Cheesman S, Heuser J, Vale RD, McNally FJ. 1998. Katanin, a microtubule-severing protein, is a novel AAA ATPase that targets to the centrosome using a WD40-containing subunit. *Cell* **93:** 277–287.

Hopkins JM. 1970. Subsidiary components of the flagella of *Chlamydomonas reinhardtii*. *J Cell Sci* **7:** 823–839.

Howard DR, Habermacher G, Glass DB, Smith EF, Sale WS. 1994. Regulation of *Chlamydomonas* flagellar dynein by an axonemal protein kinase. *J Cell Biol* **127:** 1683–1692.

Huang B, Ramanis Z, Luck DJ. 1982. Suppressor mutations in *Chlamydomonas* reveal a regulatory mechanism for flagellar function. *Cell* **28:** 115–124.

Johnson KA, Haas MA, Rosenbaum JL. 1994. Localization of a kinesin-related protein to the central pair apparatus of the *Chlamydomonas reinhardtii* flagellum. *J Cell Sci* **107:** 1551–1556.

Kamiya R. 1982. Extrusion and rotation of the central-pair microtubules in detergent-treated *Chlamydomonas* flagella. *Prog Clin Biol Res* **80:** 169–173.

Karabay A, Yu W, Solowska JM, Baird DH, Baas PW. 2004. Axonal growth is sensitive to the levels of katanin, a protein that severs microtubules. *J Neuroscience* **24:** 5778–5788.

Kikushima K. 2009. Central pair apparatus enhances outer-arm dynein activities through regulation of inner-arm dyneins. *Cell Motil Cytoskeleton* **66:** 272–280.

Lechtreck KF, Witman GB. 2007. *Chlamydomonas reinhardtii* hydin is a central pair protein required for flagellar motility. *J Cell Biol* **176:** 473–482.

Lechtreck KF, Delmotte P, Robinson ML, Sanderson MJ, Witman GB. 2008. Mutations in *Hydin* impair ciliary motility in mice. *J Cell Biol* **180:** 633–643.

Lechtreck KF, Gould TJ, Witman GB. 2013. Flagellar central pair assembly in *Chlamydomonas reinhardtii*. *Cilia* **2:** 15.

Lee L, Campagna DR, Pinkus JL, Mulhern H, Wyatt TA, Sisson JH, Pavlik JA, Pinkus GS, Fleming MD. 2008. Primary ciliary dyskinesia in mice lacking the novel ciliary protein Pcdp1. *Mol Cell Biol* **28:** 949–957.

Lindemann CB. 1994. A model of flagellar and ciliary functioning which uses the forces transverse to the axoneme as the regulator of dynein activation. *Cell Motil Cytoskeleton* **29:** 141–154.

Lindemann CB. 2003. Structural–functional relationships of the dynein, spokes, and central-pair projections predicted from an analysis of the forces acting within a flagellum. *Biophys J* **84:** 4115–4126.

Lindemann CB. 2007. The geometric clutch as a working hypothesis for future research on cilia and flagella. *Ann NY Acad Sci* **1101:** 477–493.

Lindemann CB, Kanous KS. 1995. "Geometric clutch" hypothesis of axonemal function: Key issues and testable predictions. *Cell Motil Cytoskeleton* **31:** 1–8.

Lindemann CB, Lesich KA. 2015. The geometric clutch at 20: Stripping gears or gaining traction? *Reproduction* **150:** R45–R53.

Lindemann CB, Mitchell DR. 2007. Evidence for axonemal distortion during the flagellar beat of *Chlamydomonas*. *Cell Motil Cytoskeleton* **64:** 580–589.

Manton I. 1966. Observations on the fine structure of the male gamete of the marine centric diatom *Lithodesmium undulatum*. *J R Microscopical Soc* **85:** 119–134.

McKean PG, Baines A, Vaughan S, Gull K. 2003. γ-Tubulin functions in the nucleation of a discrete subset of microtubules in the eukaryotic flagellum. *Curr Biol* **13:** 598–602.

McKenzie CW, Craige B, Kroeger TV, Finn R, Wyatt TA, Sisson JH, Pavlik JA, Strittmatter L, Hendricks GM, Witman GB, et al. 2015. CFAP54 is required for proper ciliary motility and assembly of the central pair apparatus in mice. *Mol Biol Cell* **26:** 3140–3149.

McNally KP, Bazirgan OA, McNally FJ. 2000. Two domains of p80 katanin regulate microtubule severing and spindle pole targeting by p60 katanin. *J Cell Sci* **113:** 1623–1633.

Mitchell DR. 2003a. Orientation of the central pair complex during flagellar bend formation in *Chlamydomonas*. *Cell Motil Cytoskeleton* **56:** 120–129.

Mitchell DR. 2003b. Reconstruction of the projection periodicity and surface architecture of the flagellar central pair complex. *Cell Motil Cytoskeleton* **55:** 188–199.

Mitchell DR. 2004. Speculations on the evolution of 9+2 organelles and the role of central pair microtubules. *Biol Cell* **96:** 691–696.

Mitchell DR, Nakatsugawa M. 2004. Bend propagation drives central pair rotation in *Chlamydomonas reinhardtii* flagella. *J Cell Biol* **166:** 709–715.

Mitchell DR, Sale WS. 1999. Characterization of a *Chlamydomonas* insertional mutant that disrupts flagellar central pair microtubule-associated structures. *J Cell Biol* **144:** 293–304.

Mitchell BF, Pedersen LB, Feely M, Rosenbaum JL, Mitchell DR. 2005. ATP production in *Chlamydomonas reinhardtii* flagella by glycolytic enzymes. *Mol Biol Cell* **16:** 4509–4518.

Nakano I, Kobayashi T, Yoshimura M, Shingyoji C. 2003. Central-pair-linked regulation of microtubule sliding by calcium in flagellar axonemes. *J Cell Sci* **116:** 1627–1636.

Nakazawa Y, Hiraki M, Kamiya R, Hirono M. 2007. SAS-6 is a cartwheel protein that establishes the 9-fold symmetry of the centriole. *Curr Biol* **17:** 2169–2174.

Nakazawa Y, Ariyoshi T, Noga A, Kamiya R, Hirono M. 2014. Space-dependent formation of central pair microtubules and their interactions with radial spokes. *PLoS ONE* **9:** e110513.

Neilson LI, Schneider PA, Van Deerlin PG, Kiriakidou M, Driscoll DA, Pellegrini MC, Millinder S, Yamamoto KK, French CK, Strauss JF III. 1999. cDNA cloning and characterization of a human sperm antigen (SPAG6) with homology to the product of the *Chlamydomonas PF16* locus. *Genomics* **60:** 272–280.

Nonaka S, Tanaka Y, Okada Y, Takeda S, Harada A, Kanai Y, Kido M, Hirokawa N. 1998. Randomization of left–right asymmetry due to loss of nodal cilia generating leftward flow of extraembryonic fluid in mice lacking KIF3B motor protein. *Cell* **95:** 829–837.

Oda T, Yanagisawa H, Yagi T, Kikkawa M. 2014. Mechanosignaling between central apparatus and radial spokes controls axonemal dynein activity. *J Cell Biol* **204:** 807–819.

Okagaki T, Kamiya R. 1986. Microtubule sliding in mutant *Chlamydomonas* axonemes devoid of outer or inner dynein arms. *J Cell Biol* **103:** 1895–1902.

Olbrich H, Schmidts M, Werner C, Onoufriadis A, Loges NT, Raidt J, Banki NF, Shoemark A, Burgoyne T, Al Turki S, et al. 2012. Recessive *HYDIN* mutations cause primary ciliary dyskinesia without randomization of left–right body asymmetry. *Am J Hum Genet* **91:** 672–684.

Omoto CK, Kung C. 1980. Rotation and twist of the central-pair microtubules in the cilia of *Paramecium*. *J Cell Biol* **87:** 33–46.

Omoto CK, Gibbons IR, Kamiya R, Shingyoji C, Takahashi K, Witman GB. 1999. Rotation of the central pair microtubules in eukaryotic flagella. *Mol Biol Cell* **10:** 1–4.

Peifer M, Berg S, Reynolds AB. 1994. A repeating amino acid motif shared by proteins with diverse cellular roles. *Cell* **76:** 789–791.

Piperno G, Mead K, Shestak W. 1992. The inner dynein arms I2 interact with a "dynein regulatory complex" in *Chlamydomonas* flagella. *J Cell Biol* **118:** 1455–1463.

Piperno G, Mead K, LeDizet M, Moscatelli A. 1994. Mutations in the "dynein regulatory complex" alter the ATP-insensitive binding sites for inner arm dyneins in *Chlamydomonas* axonemes. *J Cell Biol* **125:** 1109–1117.

Porter ME, Power J, Dutcher SK. 1992. Extragenic suppressors of paralyzed flagellar mutations in *Chlamydomonas reinhardtii* identify loci that alter the inner dynein arms. *J Cell Biol* **118:** 1163–1176.

Porter ME, Knott JA, Gardner LC, Mitchell DR, Dutcher SK. 1994. Mutations in the SUP-PF-1 locus of *Chlamydomonas reinhardtii* identify a regulatory domain in the β-dynein heavy chain. *J Cell Biol* **126:** 1495–1507.

Ringo DL. 1967. Flagellar motion and fine structure of the flagellar apparatus in *Chlamydomonas*. *J Cell Biol* **33:** 543–571.

Rupp G, O'Toole E, Porter ME. 2001. The *Chlamydomonas* PF6 locus encodes a large alanine/proline-rich polypeptide that is required for assembly of a central pair projection and regulates flagellar motility. *Mol Biol Cell* **12**: 739–751.

Sapiro R, Kostetskii I, Olds-Clarke P, Gerton GL, Radice GL, Strauss IJ. 2002. Male infertility, impaired sperm motility, and hydrocephalus in mice deficient in sperm-associated antigen 6. *Mol Cell Biol* **22**: 6298–6305.

Schrevel J, Besse C. 1975. A functional flagella with a 6+0 pattern. *J Cell Biol* **66**: 492–507.

Smallman LA, Gregory J. 1986. Ultrastructural abnormalities of cilia in the human respiratory tract. *Hum Pathol* **17**: 848–855.

Smith EF. 2002a. Regulation of flagellar dynein by calcium and a role for an axonemal calmodulin and calmodulin-dependent kinase. *Mol Biol Cell* **13**: 3303–3313.

Smith EF. 2002b. Regulation of flagellar dynein by the axonemal central apparatus. *Cell Motil Cytoskeleton* **52**: 33–42.

Smith EF, Lefebvre PA. 1996. PF16 encodes a protein with armadillo repeats and localizes to a single microtubule of the central apparatus in *Chlamydomonas* flagella. *J Cell Biol* **132**: 359–370.

Smith EF, Lefebvre PA. 1997. *PF20* gene product contains WD repeats and localizes to the intermicrotubule bridges in *Chlamydomonas* flagella. *Mol Biol Cell* **8**: 455–467.

Smith EF, Lefebvre PA. 2000. Defining functional domains within PF16: A central apparatus component required for flagellar motility. *Cell Motil Cytoskeleton* **46**: 157–165.

Smith EF, Sale WS. 1992. Regulation of dynein-driven microtubule sliding by the radial spokes in flagella. *Science* **257**: 1557–1559.

Stannard W, Rutman A, Wallis C, O'Callaghan C. 2004. Central microtubular agenesis causing primary ciliary dyskinesia. *Am J Respir Crit Care Med* **169**: 634–637.

Summers KE, Gibbons IR. 1971. Adenosine triphosphate-induced sliding of tubules in trypsin-treated flagella of sea-urchin sperm. *Proc Natl Acad Sci* **68**: 3092–3096.

Teves ME, Zhang Z, Costanzo RM, Henderson SC, Corwin FD, Zweit J, Sundaresan G, Subler M, Salloum FN, Rubin BK, et al. 2013. Sperm-associated antigen-17 gene is essential for motile cilia function and neonatal survival. *Am J Respir Cell Mol Biol* **48**: 765–772.

Teves ME, Sears PR, Li W, Zhang Z, Tang W, van Reesema L, Costanzo RM, Davis CW, Knowles MR, Strauss JF III, et al. 2014. Sperm-associated antigen 6 (SPAG6) deficiency and defects in ciliogenesis and cilia function: Polarity, density, and beat. *PLoS ONE* **9**: e107271.

Teves ME, Sundaresan G, Cohen DJ, Hyzy SL, Kajan I, Maczis M, Zhang Z, Costanzo RM, Zweit J, Schwartz Z, et al. 2015. *Spag17* deficiency results in skeletal malformations and bone abnormalities. *PLoS ONE* **10**: e0125936.

Teves ME, Nagarkatti-Gude DR, Zhang Z, Strauss JF III. 2016. Mammalian axoneme central pair complex proteins: Broader roles revealed by gene knockout phenotypes. *Cytoskeleton* **73**: 3–22.

Thazhath R, Jerka-Dziadosz M, Duan J, Wloga D, Gorovsky MA, Frankel J, Gaertig J. 2004. Cell context-specific effects of the β-tubulin glycylation domain on assembly and size of microtubular organelles. *Mol Biol Cell* **15**: 4136–4147.

Viswanadha R, Hunter EL, Yamamoto R, Wirschell M, Alford LM, Dutcher SK, Sale WS. 2014. The ciliary inner dynein arm, I1 dynein, is assembled in the cytoplasm and transported by IFT before axonemal docking. *Cytoskeleton* **71**: 573–586.

Wargo MJ, Smith EF. 2003. Asymmetry of the central apparatus defines the location of active microtubule sliding in *Chlamydomonas* flagella. *Proc Natl Acad Sci* **100**: 137–142.

Wargo MJ, Dymek EE, Smith EF. 2005. Calmodulin and PF6 are components of a complex that localizes to the C1 microtubule of the flagellar central apparatus. *J Cell Sci* **118**: 4655–4665.

Warner FD, Satir P. 1974. The structural basis of ciliary bend formation. Radial spoke positional changes accompanying microtubule sliding. *J Cell Biol* **63**: 35–63.

Wirschell M, Yamamoto R, Alford L, Gokhale A, Gaillard A, Sale WS. 2011. Regulation of ciliary motility: Conserved protein kinases and phosphatases are targeted and anchored in the ciliary axoneme. *Arch Biochem Biophys* **510**: 93–100.

Witman GB, Plummer J, Sander G. 1978. *Chlamydomonas* flagellar mutants lacking radial spokes and central tubules. Structure, composition, and function of specific axonemal components. *J Cell Biol* **76**: 729–747.

Wren KN, Craft JM, Tritschler D, Schauer A, Patel DK, Smith EF, Porter ME, Kner P, Lechtreck KF. 2013. A differential cargo-loading model of ciliary length regulation by IFT. *Curr Biol* **23**: 2463–2471.

Yang P, Sale WS. 2000. Casein kinase I is anchored on axonemal doublet microtubules and regulates flagellar dynein phosphorylation and activity. *J Biol Chem* **275**: 18905–18912.

Yang P, Fox L, Colbran RJ, Sale WS. 2000. Protein phosphatases PP1 and PP2A are located in distinct positions in the *Chlamydomonas* flagellar axoneme. *J Cell Sci* **113**: 91–102.

Yokoyama R, O'Toole E, Ghosh S, Mitchell DR. 2004. Regulation of flagellar dynein activity by a central pair kinesin. *Proc Natl Acad Sci* **101**: 17398–17403.

Yoshimura M, Shingyoji C. 1999. Effects of the central pair apparatus on microtubule sliding velocity in sea urchin sperm flagella. *Cell Struct Funct* **24**: 43–54.

Yu W, Solowska JM, Qiang L, Karabay A, Baird D, Baas PW. 2005. Regulation of microtubule severing by katanin subunits during neuronal development. *J Neurosci* **25**: 5573–5583.

Zhang H, Mitchell DR. 2004. Cpc1, a *Chlamydomonas* central pair protein with an adenylate kinase domain. *J Cell Sci* **117**: 4179–4188.

Zhang Z, Sapiro R, Kapfhamer D, Bucan M, Bray J, Chennathukuzhi V, McNamara P, Curtis A, Zhang M, Blanchette-Mackie EJ, et al. 2002. A sperm-associated WD repeat protein orthologous to *Chlamydomonas* PF20 associates with Spag6, the mammalian orthologue of *Chlamydomonas* PF16. *Mol Cell Biol* **22**: 7993–8004.

Zhang Z, Jones BH, Tang W, Moss SB, Wei Z, Ho C, Pollack M, Horowitz E, Bennett J, Baker ME, et al. 2005. Dissecting the axoneme interactome: The mammalian orthologue of *Chlamydomonas* PF6 interacts with sperm-associated antigen 6, the mammalian orthologue of *Chlamydomonas* PF16. *Mol Cell Proteomics* **4:** 914–923.

Zhang Z, Kostetskii I, Tang W, Haig-Ladewig L, Sapiro R, Wei Z, Patel AM, Bennett J, Gerton GL, Moss SB, et al. 2006. Deficiency of SPAG16L causes male infertility associated with impaired sperm motility. *Biol Reprod* **74:** 751–759.

Zhang Z, Tang W, Zhou R, Shen X, Wei Z, Patel AM, Povlishock JT, Bennett J, Strauss JF 3rd. 2007a. Accelerated mortality from hydrocephalus and pneumonia in mice with a combined deficiency of SPAG6 and SPAG16L reveals a functional interrelationship between the two central apparatus proteins. *Cell Motil Cytoskeleton* **64:** 360–376.

Zhang Z, Zariwala MA, Mahadevan MM, Caballero-Campo P, Shen X, Escudier E, Duriez B, Bridoux AM, Leigh M, Gerton GL, et al. 2007b. A heterozygous mutation disrupting the *SPAG16* gene results in biochemical instability of central apparatus components of the human sperm axoneme. *Biol Reprod* **77:** 864–871.

Zhou Q, Li Z. 2015. γ-Tubulin complex in *Trypanosoma brucei*: Molecular composition, subunit interdependence and requirement for axonemal central pair protein assembly. *Mol Microbiol* **98:** 667–680.

Cite this article as *Cold Spring Harb Perspect Biol* doi: 10.1101/cshperspect.a028118

Radial Spokes—A Snapshot of the Motility Regulation, Assembly, and Evolution of Cilia and Flagella

Xiaoyan Zhu, Yi Liu, and Pinfen Yang

The Department of Biological Sciences, Marquette University, Milwaukee, Wisconsin 53201

Correspondence: pinfen.yang@marquette.edu

Propulsive forces generated by cilia and flagella are used in events that are critical for the thriving of diverse eukaryotic organisms in their environments. Despite distinctive strokes and regulations, the majority of them adopt the 9+2 axoneme that is believed to exist in the last eukaryotic common ancestor. Only a few outliers have opted for a simpler format that forsakes the signature radial spokes and the central pair apparatus, although both are unnecessary for force generation or rhythmicity. Extensive evidence has shown that they operate as an integral system for motility control. Recent studies have made remarkable progress on the radial spoke. This review will trace how the new structural, compositional, and evolutional insights pose significant implications on flagella biology and, conversely, ciliopathy.

The intricate radial spokes (RSs) and central pair apparatus (CP) in the 9+2 axoneme have attracted the interest of many. By all accounts, flagellar motility is contingent on their direct interactions, and thus they are often mentioned in the same context. Although motile cilia and flagella in some cell types or some organisms naturally lack both RSs and the CP, for the majority that possess the 9+2 axoneme, they prove to be critical for motility. Genetic defects afflicting either structure will result in a spectrum of dyskinesia, ranging from paralysis to jerking to a mixed population of motile and dysmotile cilia.

As mentioned in Loreng and Smith (2016), the CP is substantially more complex than the RS. The synchronized spinning of the CP with rhythmic beating in certain cilia (Omoto and Witman 1981) has especially fueled much imagination. In contrast, the RS is relatively simple. Nonetheless, the simplicity presented a distinctive opportunity for asking questions fundamental to flagella biology. Vigorous efforts in combination with brilliant ideas and new technologies in recent years have shed considerable insight on RSs from diverse organisms, including humans. Although many earlier RS studies used *Chlamydomonas*, emerging evidence showed species-specific differences in RSs and experimental advantages. This review will narrate the converging discoveries that unveil RSs' structures, molecular interactions, functions and divergence among different organisms, and the implications. The in-depth discussion of motility control mechanism, radial spoke proteins (RSPs) and RS mutants in *Chlamydo-*

Cite this article as *Cold Spring Harb Perspect Biol* doi: 10.1101/cshperspect.a028126

monas could be found in a previous review (Yang and Smith 2008).

MORPHOLOGY AND PERIODICITY OF RADIAL SPOKES

Physical attributes of axonemal components are integral to the mechanics of motile cilia and flagella. RSs were named after their radial pattern in the axoneme cross sections (Ishikawa 2016). Individual RSs rendered by conventional electron microscopy (EM) often were hard to discern or varied substantially. Recent cryo-electron tomography (ET) that bypassed fixations and metal decoration finally resolves the discrepancies. Interestingly, RSs from diverse organisms actually bear striking resemblances but indeed with distinctions. A side view of the *Trypanosome* axoneme (Koyfman et al. 2011) in which the CP is extruded (Fig. 1A) reveals the typical radial pattern of RSs, with the thinner spoke stalk docking to each outer doublet and the enlarged head projecting toward the CP. In a lateral view of *Chlamydomonas* axoneme (Fig. 1B), there are two RSs—RS1 and RS2—in each 96-nm repeat, contrary to three RSs for most organisms. The missing third RS—RS3—is replaced by a stand-in stub, RS3S (Pigino et al. 2011; Barber et al. 2012; Lin et al 2014). The distances among the three RSs adhere to a multiplication of 8 nm—the length of $\alpha\beta$-tubulins—namely 32, 24, and 40 nm.

As revealed by negative-stained splayed axoneme (Witman et al. 1978), RS1 and RS2 appear as a T shape, which, as the view rotates outward, turns into a Y shape with an enlarged head connected to a neck region by two arms (Fig. 1C). The top view shows spoke heads comprised of two asymmetric modules arranged in a twofold rotational symmetry (Fig. 1D), presumably each connected to the neck by one arm. At higher resolutions, the spoke head in humans (Fig. 1E) or sea urchins appeared smaller and more compact, lacking the tendril-like extensions in protist spoke heads. Notably, RS3 has a rather different morphology, with an asymmetric head (Fig. 1E) and a kinked stalk. Furthermore, RS3 seems unaffected in primary cilia dyskinesia (PCD) patients with defective

spoke head proteins (Lin et al. 2014). The biochemical bases underlying the distinctive RS3 morphology are unknown.

The molecular context at the base of each RS is also distinct, suggesting that the task of the three RSs are not identical. RS1 base is adjacent to the base of the two-headed inner dynein arm I1, whose base module contacts the MIA complex named after the deficiencies in *modifier of inner arm* (*mia*) mutants (King and Dutcher 1997; Yamamoto et al. 2013). *mia* mutants, albeit containing all I1 components, are slow swimmers deficient in phototaxis as I1 mutants (see the section Chemical Signaling for a detailed discussion). RS2 and RS3/RS3S are, respectively, adjacent to the front and back of the nexin–dynein regulatory complex (N-DRC) and perhaps the calmodulin- and spoke-associated complex (CSC) (Gardner et al. 1994; Dymek and Smith 2007; Heuser et al. 2009, 2012; Urbanska et al. 2015). This arrangement is consistent with distinct inner dynein anomalies in *mia* mutants (King and Dutcher 1997) and N-DRC mutants (Piperno et al. 1992). These precise structural relationships are predetermined by a complex coined as a molecular ruler that is, surprisingly, comprised of only two coiled-coil paralogous proteins (FAP59 and FAP172) (Oda et al. 2014a). Defects in their encoding genes in *Chlamydomonas* mutants (*pf7* and *pf8*) result in diminished inner dynein arms and N-DRC. Interestingly, the 32- to 64-nm RS periodicity is replaced by a 32-nm only periodicity. Lengthening these ruler proteins also alters the periodicity of RSs or their adjacent ensemble. It is interpreted that RS periodicity is founded on the molecular ruler that recruits N-DRC and inner dynein arms, which in turn inhibits the inherent ability of RSs in binding to outer doublets.

RADIAL SPOKE PROTEINS IN *CHLAMYDOMONAS*

The repertoire of RS mutants (Huang et al. 1981), the RSP 2D map (Piperno et al. 1981), nuclear transformation (Diener et al. 1990), and extraction of RS particles (Yang et al. 2001) set the stage for the comprehensive identifications of RSPs in *Chlamydomonas*, *Ciona* (Satouh et al.

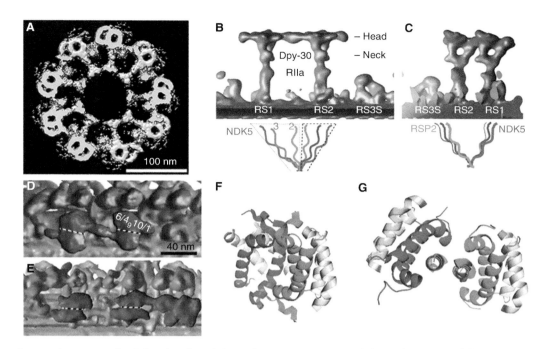

Figure 1. Structures of radial spokes. (*A*–*E*) Cryo-electron tomogram renditions of axonemes. (*A*) Cross section from *Trypanosome brucei* in which the central pair apparatus was extruded offered an unobstructed view of the entire radial spoke. The central pair apparatus is absent in this sample (electron microscopy [EM] Data Bank [EMDB] ID 5302; Koyfman et al. 2011). (*B,C*) Lateral view of the T-shaped twin radial spokes, RS1 and RS2, in one 96-nm repeat of *Chlamydomonas* flagella and a counterclockwise rotation view to reveal the bifurcation of the neck region (EMDB ID 1941; Pigino et al. 2011). The *bottom* panels depict predicted positions of the helices from RSP2, RSP3, and NDK5 (pink, green, and blue lines) that form the two arms at the neck and tether to two head modules. Spoke HSP40 at this position is not illustrated for clarity. (*D,E*) *Top* view of a 96-nm repeat in a *Chlamydomonas* flagellum (EMDB ID 5845; Oda et al. 2014b) and human respiratory cilium (EMDB ID 5950; Lin et al. 2014). Each spoke head has two modules arranged in rotary symmetry. Numbers in *D* indicate predicted locations of RSPs. Note humans have triplet RSs, with RS3 strikingly different from RS1 and RS2. The images were generated using the EM Navigator at the Protein Data Bank Japan (PDBj) (pdbj.org). (*F,G*) Crystallographic structures predicting the assembly process of the spokeneck region with Dpy-30 domains from RSP2 (pink ribbons) and NDK5 (blue ribbons) and amphipathic helices from a RSP3 dimer (green ribbons). (*F*) Crystallography of tetramers of the Dpy-30 domain that is comprised of two X helical bundles. The domain exists as a homodimer in solution. In crystals, staggering of the two dimers partially exposes the binding cleft for an amphipathic helix in their interacting partners. The polypeptides are from Dpy-30 protein. The illustration is generated from PDB ID 3G36 (Wang et al. 2009). This may account for the direct association of RSP2 and NDK5 human orthologs. (*G*) The face-to-face orientation of the two dimeric Dpy-30 domains binding to an amphipathic helix from Ash2 in the Set1-like histone methyltransferase complex. The illustration is generated from PDB ID 4RIQ (Tremblay et al. 2014). This may represent the conformational change as the RSP2/NDK5 complex associates with RSP3's amphipathic helix.

2005; Satouh and Inaba 2009), and other organisms. RSPs were first defined based on the missing 17 proteins in the spokeless axonemes of *Chlamydomonas* mutant, *pf14* (Huang et al. 1981). Extracted RS particles contain these 17 RSPs, tubulins, and at least six more polypep-

tides. Two are ubiquitous—calmodulin and LC8, which are present in multiple protein complexes—and therefore are not absent in *pf14*. Together, these 19 proteins reside in the Y-shaped RS complex. The rest are likely tethered to the RS base and coextracted when microtubules are disrupted by the KI buffer. All *Chlamydomonas* RSPs have been identified except RSP13 (Yang et al. 2006). RSP13 may be orthologous to CMUB116 in *Ciona* RSs, a conserved protein with both a calmodulin-binding domain and a ubiquitin domain. There is no evidence yet that the content in RS1 and RS2 differs. Only those that were studied further will be discussed below.

ORGANIZATION OF RSPs

Based on partial deficiencies in a few *Chlamydomonas* RS mutants, the 17 RSPs were assigned to separate regions—five (RSP 4, 6, 1, 10, and 9) in the head, three (RSP2, HSP40, and RSP23) at the neck, and the rest at the stalk (Huang et al. 1981; Patel-King et al. 2004; Yang et al. 2006) (Fig. 1B–E). The comprehensive data showed that the largely symmetric RS1 and RS2 contained many homodimers and paired paralogs with an identical domain. For example, in the spoke head, RSP4 and 6 are paralogs expressed from duplicated genes (Curry et al. 1992), whereas RSP1 and 10 are homologs with membrane occupation and recognition nexus (MORN) repeats. Analysis of recombinant proteins and tagging indicate that each spoke head contains two copies of each protein that is inherently dimeric (Kohno et al. 2011; Oda et al. 2014b). Furthermore, the RIIa domains in RSP11 and RSP7, and the Dpy-30 domains in RSP2 and RSP23 (NDK5) in the stalk are dimerization and docking (DD) domains of striking similar tertiary structures (Sivadas et al. 2012). Finally, RSP3 that docks the RS to outer doublets also exists as a homodimer (Diener et al. 1993; Wirschell et al. 2008).

RIIa-containing RSPs showed the physiological relevance of the binding of RSP3 to the RIIa-containing RII in the cAMP-dependent protein kinase (PKA) (Gaillard et al. 2001) in an in vitro assay credited for the discoveries of many PKA-anchoring proteins (AKAPs). As such, RSP3 is often referred to as an AKAP. Deletion mutagenesis (Sivadas et al. 2012) and structural tagging (Oda et al. 2014b) confirmed that an RII-binding amphipathic helix in RSP3 binds to the RIIa domain in RSP7 and RSP11,

which otherwise bears no other PKA signature features (Fig. 1B). This led to a prediction that some "AKAPs"—including RSP3 in *Chlamydomonas*—anchors proteins with a RIIa domain that tethers functional modules irrelevant to PKA, and the term of "AKAPs" should be reserved for those that anchor PKA holoenzyme in vivo (see detailed discussion in the section Evolution of Radial Spokes). In addition, RSP3 harbors another amphipathic helix that binds Dpy-30 domains in RSP2 and NDK5. Chemical cross-linking suggests further association of NDK5 with RSP1, and RSP2 with RSP4/RSP6 (Kohno et al. 2011).

Taken together, these observations support a model that one RSP3 dimer serves as a scaffold to anchor the rest of the RSPs. The conserved region of RSP3 likely defines the RS length that is strictly conserved (Gaillard et al. 2001). Its amino-terminal region docks RSs to the outer doublets, whereas its carboxyl terminus directs toward the spoke head (Sivadas et al. 2012; Oda et al. 2014b). This orientation positions the RIIa-containing RSP7 and RSP11 toward the outer doublets and the two Dpy-30-containing RSP2 and NDK5 toward the spoke head (Fig. 1B). As such, the carboxy-terminal helices in RSP2, RSP3, and NDK5 may form the arms that bifurcate at the neck and secure two spoke head modules directionally (Fig. 1B,C, bottom panel). For the spoke head, we designate that RSP6 is peripheral to RSP4, and RSP1 is peripheral to RSP10 (Fig. 1D), because both RSP6 and RSP1 are not required for spoke head assembly in *Chlamydomonas* (Wei et al. 2010; X Zhu, unpubl.); and RSP10 could be chemically cross-linked into a homodimer (Kohno et al. 2011). RSP9 is required for spoke head assembly but there is insufficient data to predict its location.

EVOLUTION OF RADIAL SPOKES

In spite of the similar Y-shaped morphology, RSs diverged evolutionarily—toward simplicity—perhaps accelerated by gene duplication, redundancy, and the versatility of DD domains. Consistent with partial redundancy of spoke head proteins, the Basic Local Alignment Search Tool (BLAST) search indicates that RSP1 and its predicted partner, NDK5, are absent in *Tetrahymena* and other ciliates, whereas *Ciona intestinalis* as well as sea urchins only have one gene for RSP4/6 (Table 1) (P Yang, unpubl.). This is substantiated by the presence of only one RSP4/6 and one MORN protein in purified *Ciona* RSs (Satouh and Inaba 2009). This may be also true for humans, although humans have genes for RSP4, 6, 1, and 10. The abundance of the transcripts (Kott et al. 2013) or express sequence tags (ESTs) for RSP1 (RSPH1) and RSPH10 are opposite in airway and testis. Similarly, RSPH4 transcripts are far more abundant than RSPH6 transcripts. This raises the possibility that the spoke head in human cilia may contain homodimers of RSPH4 and RSPH1, forsaking RSPH6 and RSPH10, whereas RSPH1 is absent in sperm flagella. As such, the notion that certain defective RS genes will not affect fertility and thus will be more prevalent (Onoufriadis et al. 2014) should be entertained.

Simplification also reflects in the sizes of RSP1, 2, 3, and NDK5 in the head/neck region (Table 1). These proteins from *Tetrahymena* and *Chlamydomonas* have a long carboxy-terminal extension that is absent in their metazoan coun-terparts. For instance, contrary to *Chlamydomonas* RSP2's 738 aa residues, human RSP2 counterparts—DYDC1 and DYDC2—are only ~170-aa long, encompassing only the region required for rhythmic beating, including the Dpy-30 domain for docking to RSP3, and the flanking helices perhaps for forming the bifurcated arms that interact with RSP4/6 (Fig. 1B,C) (Gopal et al. 2012). As the extensions are predicted to be near the spoke head, absence of the extension is consistent with metazoans' smaller, compact spoke head. Although these carboxy-terminal tails are dispensable, RSP2s and NDK5s have calmodulin-binding motifs. *Chlamydomonas* cells lacking RSP2's tail cannot steer properly under bright light (Gopal et al. 2012). This suggests that the calmodulin-binding extension acts on the spoke head region (Fig. 1B,C) to coordinate flagellar motility as this photosynthetic organism navigates illuminated environments. Such measures apparently are unnecessary for most organisms.

The RS epitomizes the versatility of DD domains and their amphipathic helix partners (Gopal et al. 2012; Sivadas et al. 2012). The DD domains bring to molecular complexes

Table 1. Divergence of radial spoke proteins (RSPs) in the head and neck region of four representative species, as shown by their sizes and selective absence

RSP	Species			
	C. r.	T. t.	C. i.	H. s.
RSP4	465	463 486	527	717
RSP6	459	493	N/D	612
RSP1	814	N/D	300	309
RSP10	216	227 221	N/D	870
RSP9	269	296 463	276	276
RSP2 (DYDC1,2)	738	628	169	177
RSP23 (NDK5)	586	N/D	257	212
RSP3	516	691 950 764	336	380 ($^{N'}$160)

RSPs in the head region are shaded in gray. C.r., *Chlamydomonas reinhardtii*; T. t., *Tetrahymena thermophile*; C. i., *Ciona intestinalis*; H. s., *Homo sapiens*. The polypeptides were identified based on a BLAST search against *Chlamydomonas* RSPs. Those from C. r. and C. i. were also confirmed by protein biochemistry of isolated RS particles (Yang et al. 2006; Satouh and Inaba 2009). Note that some RSPs in the protists are far larger, while human RSP3 has an additional 160-aa fragment at the amino terminus ($^{N'}$160) (Jivan et al. 2009). N/D, Not detectable in the genome or RS proteome.

their tethered effector moieties—for cNMP signaling in the case of RII in PKA, for calcium-sensing such as RSP7, for nucleoside metabolism such as NDK5, and for assembly such as DYDC2 and RSP11. The tethered moieties could further duplicate or combine, as in *Chlamydomonas*' RSP2 and NDK5. Mammals particularly have multiple genes that encode RIIa-containing proteins, most of which are enriched in testis. But they share little semblance to each other, RSP7, or RSP11 except the RIIa domain (e.g., Newell et al. 2008). The divergence of DD domain proteins may allow airway and testis to build slightly different RSs and perhaps CPs that also contains an RIIa-binding protein (Gaillard et al. 2001; Rao et al. 2016). Some of them bind to "AKAPs" in the fibrous sheath of sperm flagella (e.g., Fiedler et al. 2012). As such, bona fide RIIa partners of RSP3 and any AKAPs may differ in each cell type, and should be determined empirically. Although RSP3 does not bind PKA's RII in *Chlamydomonas*, it potentially could in animals' motile cilia or flagella as a bona fide AKAP. Different RIIa partners may explain opposite effects of PKA on *Chlamydomonas* flagella and other cilia and flagella (Hasegawa et al. 1987). It is worthwhile to point out that RSP3 indeed possesses a key feature of AKAPs, the scaffold of molecular complexes, to integrate proteins of diverse functions for flagellar motility control. Although *Chlamydomonas* RSP3 is not a typical PKA-anchoring AKAP, it still anchors calcium-sensing molecules perhaps for calcium-related motility control and structural proteins for transducing mechanical feedback.

RS ASSEMBLY

RSPs are assembled into RSs in two phases. Fractionations of cell body extracts and the flagellar membrane matrix suggest that the cell body only makes partially furbished RSs that sediment as 12S particles in the sucrose gradient (Qin et al. 2004). After delivery into flagella by intraflagellar transport (IFT), 12S precursors are converted into 20S mature RSs. The direct interaction of human orthologs of NDK5 and RSP2 at the neck (e.g., Rual et al. 2005) suggests

that they assemble first, perhaps mediated by the Dpy-30 domain that was crystallized as two partially staggered dimeric dimers (Fig. 1F). Interactions with amphipathic helices in the RSP3 dimer during the assembly of the 12S RS precursors might reposition the two sets of dimers into a face-to face orientation (Fig. 1G). This structure-based interpretation is concordant with the detection of RS subparticles in the RSP3 mutant (Diener et al. 2011).

The distinct sedimentation coefficients of 12S and 20S RS particles are more likely caused by disparateness in shapes than masses. While the Y-shaped mature RSs are about 32 nm by 42 nm, the 12S RS is a Γ-shape of 20 nm by 28 nm in negative-stained EM (Diener et al. 2011). The precursors lack the two strictly conserved small dimeric proteins—HSP40 and LC8—that are involved in the conformational change during RS assembly. Most organisms have a multitude of HSP40 genes. Typically, HSP40s cooperate with other chaperones to mediate protein folding, whereas some may act solitarily. Flagella that lack HSP40 jerk incessantly although the other RSPs appear normal (Yang et al. 2008). EM revealed irregular tilt or a lower bifurcated neck of RSs. These observations support a model that the V-shaped dimeric HSP40, transported differently from 12S precursors, bind to the polypeptides at the neck region at the flagellar tip where final assembly occurs (Johnson and Rosenbaum 1992). LC8 dimer, present in numerous protein complexes including multiheaded dyneins and RSs, typically brings two polypeptides in a complex together. But in RSs, a stack of LC8 dimers bring together the amino terminus of an RSP3 dimer in RS precursors to form a rod at the RS base, promoting RSP3's phosphorylation and docking to the outer doublets (Yang et al. 2009; Gupta et al. 2012).

RSP3 phosphorylation is tightly linked to RS docking. RSP3 is hypophosphorylated in RS precursors and in mutants with reduced RSs, such as *fla14* that lacks LC8 and *pf27* (Yang and Yang 2006; Gupta et al. 2012), suggesting that the RSs in *pf27*, as in *fla14*, are not assembled via a normal process that involves phosphorylation. The *PF27* gene has not been identified but likely encodes a protein outside of flagella (Huang et al. 1981). Interestingly, RSs in *pf27* preferentially distribute toward the base of flagella (Alford et al. 2013). It is proposed that *PF27* is an adaptor linking spoke precursors to IFT.

RS OUTER DOUBLET CONNECTIONS

Much progress has been made in elucidating the distinct structural linkages of the three RSs. Three proteins (CaM-IP2, IP3, and IP4) cosediment with the 20S RS fraction but were not abundant enough for identification. Instead, they were identified in the pull-down of the anticalmodulin antibody. Knockdown of CSC proteins impaired RS2, RS3S, N-DRC, and a single-headed dynein, e (Dymek et al. 2011; Heuser et al. 2012), suggesting that the CSC is among the base of RS2 and RS3, N-DRC, and the nearby dynein. Consistent with this, the antibody for a CSC protein inhibits microtubule sliding velocity as the reduced microtubule sliding velocity of *pf14* axonemes. And yet the seemingly normal CSC in *pf14* axonemes indicates that the CSC assembly is independent of the rest of RSs (Dymek and Smith 2007).

A pull-down of the spoke stalk recovered the CSC and another protein that associated with RS2, FAP206 (Gupta et al. 2012). FAP206 is reduced in *pf14* axonemes (Lin et al. 2011). Knockout of *Tetrahymena* FAP206 also impaired RS2 specifically, and indicated that FAP206 linked the proximal end of RS2 base to the single-headed dynein, c (Vasudevan et al. 2015). Therefore, FAP206 and the CSC appear to link the base of RS2 to different single-headed dyneins aligned in tandem. The biochemical evidence for direct coupling among RS1, the MIA complex, and I1 has not yet emerged.

THE SPECTRUM OF MOTILITY ANOMALIES OF RS MUTANTS

It is important to note that not all RS mutants are immotile. Although *Chlamydomonas* RS mutants are known for paralyzed flagella, they actually show a spectrum of motility anomalies, depending on the defective RSP, the nature of mutations, and, surprisingly, the culture conditions. They are paralyzed if RS-CP contacts are absent because they lack the entire RS, the spoke head, or spoke head and neck. Flagella lacking spoke HSP40 jerk constantly. In contrast, the paralysis of the spoke head mutant *pf26* (RSP6) is conditional. In the healthy log phase cultures at room temperature, the motility and the other spoke head proteins appear normal (Huang et al. 1981; Wei et al. 2010). However, in stationary phase cultures, *pf26* cells are largely paralyzed, lacking nearly all spoke head proteins in flagella. In an intermediate condition, a culture will contain swimmers, immotile cells, and cells with jerky (uncoordinated) flagella. Therefore, although RSP26 is dispensable in the laboratory, it is presumably needed in *Chlamydomonas*' natural environment in which the temperature and nutrient availability are expected to fluctuate. Similar motility phenotypes are shown by *pf25* that lacks RSP11 (Yang and Yang 2006) and partially rescued RS mutants (Gupta et al. 2012).

Similar scenarios were noted in other species. *Tetrahymena* FAP206 knockout could swim despite diminished RS2. Its velocity is reduced, attributed to altered waveform, likely because of the deficit of inner dynein arms (Vasudevan et al. 2015) and uncoordinated beat as expected of minor RS deficits. Unexpectedly, cilia of human RS PCD patients beat rhythmically with helical waveform—amid asynchronous beating ones in some cases—despite lacking spoke heads in RS1 and RS2 or most of RSs (Burgoyne et al. 2014; Jeanson et al. 2015). The helical movement is attributed to the presence of RS3, because *Chlamydomonas* flagella with same genetic defect and lacking RS3 are immotile. However, one should bear in mind that natural 9+0 cilia and flagella generate rhythmic beating with helical waveform without any RS (Nonaka et al. 2002). Therefore, phenotype polymorphism, cell conditions, and evolutionary divergence should be considered in phenotyping RS mutants.

ROLES OF RSS

It will be easier to appreciate RSs if one considers the 9+2 cilia and flagella as reversible biological nanomachines.

Generation of Planar Waveform

This is not absolute. The 9+0 motile cilia is rare, such as that found in the node or certain eel and horseshoe crab sperm (Gibbons et al. 1985; Ishijima et al. 1988). As such, they connote an inferior perception. However, these sperm swim rapidly with a 9+0 flagellum of a high beat frequency. Importantly, the typical waveform is helical, similar to the movement of the respiratory cilia of RS PCD patients (e.g., Castleman et al. 2009; Jeanson et al. 2015) and reactivated axonemes of paralyzed *Chlamydomonas* RS or CP mutants under altered conditions (e.g., Yagi and Kamiya 2000; and see Loreng and Smith 2016).

In contrast, 9+2 cilia usually display planar waveform of a large amplitude, but retain the ability to generate helical waveform. For *Chlamydomonas*, power strokes are largely planar, but recovery strokes are helical. The planar waveform with large amplitude appears to be crucial for respiratory cilia to generate sufficient propulsive forces, because the helical movement of cilia in RS PCD patients does not offer sufficient mucociliary clearance despite a normal beat frequency. An exception is the 9+2 axoneme of *Trypanosome* that generates 3D bihelical movements, perhaps because of the paraflagellar rod, a cytoskeletal system unique to kinoplastid parasites (Portman and Gull 2010; Koyfman et al. 2011). Interestingly, helical and planar waveforms of both 9+0 and 9+2 flagella could be switched by changes in physical properties of the aqueous environment, such as surface tension and viscosity (Ishijima et al. 1988; Woolley and Vernon 2001). Therefore, nine outer doublets are inherently switchable to various waveforms.

We envisage that the RS-CP system is a device for swift reordering of the activation sequence of motors in the nine outer doublets. The resulting high-frequency beating of planar waveform is hydrodynamically favorable in most circumstances. But such movement may not be appropriate for a select few. The low-frequency helical beat of 9+0 nodal cilia may be just right for establishing a gradient of morphogens that direct the development of left–right asymmetry (Nonaka et al. 2002; Shinohara et al. 2015). By the same token, 3D-bihelical movement is what *Trypanosome* adopt to navigate in the viscous bloodstream.

Coordination of Molecular Motors by Mechanical Feedback

Diverse evidence suggests that the RS-CP system overrides the inherent helical movement of the outer doublet bundle by serving as a venue for mechanical feedback, perhaps of forces or tensions. Although *Chlamydomonas* RS or CP mutants have paralyzed flagella, their axonemes are able to vibrate in reactivation conditions, suggesting that dynein motors are active but are not coordinated (Yagi et al. 1994). An additional mutation in outer dynein or N-DRC partially rescues the paralysis (e.g., Huang et al. 1982; Porter et al. 1994; Rupp et al. 1996). This leads to the prediction that the RS-CP system coordinates the activation of dynein motors via N-DRC—to generate planar waveform in most cases.

Generation and propagation of a planar waveform are a result of alternate sliding of opposing subsets of outer doublets at any particular moment. The selection appears determined by the CP with asymmetric projections (see Loreng and Smith 2016). The geometric clutch model (Lindemann and Lesich 2015) offers an explicit explanation of mechanical feedback. It posits that the CP and RSs, which contact intermittently during each beat cycle (Warner and Satir 1974), transmit transverse forces developed from flagellar bend, which in turn differentially alter the distances among doublets confined by the base, tip, and N-DRC links. Like the action of an automobile clutch, increasing and reducing interdoublet distances would, respectively, disengage and engage distinct subsets of dynein motors, leading to bend propagation and rhythmic beating. This theory is concordant with elliptic versus circular cross sections of motile and immotile axonemes, and oscillation of axoneme diameters during rhythmic beating (Warner 1978; Sakakibara et al. 2004; Lindemann and Mitchell 2007; Yang et al. 2008).

Mechanical feedback is also supported by perturbations that paralyze motile flagella or rescue the paralyzed. For example, simply removing HSP40 at a spoke neck resulted in jerky flagella with uncoordinated bend initiation, propagation, and switching, presumably because of weakening of RS rigidity necessary for force transmission (Yang et al. 2008). Conversely, forced bending could jolt paralyzed flagella to beat rhythmically for a brief period (Hayashibe et al. 1997). Finally, adding a large tag to the spoke head rescued paralyzed flagella of a CP mutant, probably complementing a missing CP projection to restore structural contact necessary for mechanical signaling (Oda et al. 2014b). Suppressor mutations may alter N-DRC or motors as to bypasses the defective mechanical feedback, or return toward the 9+0 machinery.

Chemical Signaling

The RS-CP system is integral to the modulation of cilia and flagella motility by calcium, calmodulin, cyclic nucleotides, and phosphorylation. Consistent with this, the calcium-dependent waveform changes appeared missing in 9+0 sperm flagella (Gibbons et al. 1985). Contrary to the expected universal principle of mechanical feedback, the mechanisms of chemical signaling may differ substantially for *Chlamydomonas*, which use two relatively short flagella to steer, and for sperm's long flagella, which are activated by Wnt signaling (Koch et al. 2015).

The biflagellate *Chlamydomonas* is known to show two calcium-dependent responses: a waveform switch when encountering barriers or stimulated by intense white light (photoshock), and phototaxis in an environment with differential light intensity (Yang and Smith 2008). The RS-CP system is proposed to be involved in the former because of the symmetric waveform of the suppressor of RS mutants (Brokaw et al. 1982). While this is likely, calcium signaling pathways are far more complicated given more than 80 proteins in algal flagella are capable of binding calcium or calmodulin (Pazour et al. 2005), including multiple ones residing in the

structure corridor from the CP to dynein motors (see Loreng and Smith 2016). Indeed, the mutant lacking the calmodulin-binding region in RSP2 is capable of normal swimming and both light-induced responses, but cannot steer under bright light (Gopal et al. 2012). It is likely that motility control is via concerted actions of multiple calcium sensors, rather than a handful of calcium switches. RSs may be involved in this via the RSPs with calcium- and calmodulin-binding motifs. If calmodulin binding of *Tetrahymena* RSP4 and RSP6 (Ueno et al. 2006) proves physiologically relevant, these two conserved molecules are appealing candidates for universal major calcium switch.

Independent approaches strongly suggest that RSs also regulate dynein motors via phosphorylation—of the inner dynein arm, I1. While RS mutant flagella are paralyzed, their axoneme can undergo interdoublet sliding albeit with a reduced velocity (Witman et al. 1978; Smith and Sale 1992). The inhibition stems from phosphorylation mediated by kinases such as PKA and CK1 (Howard et al. 1994; Yang and Sale 2000) when RSs are defective, and could be antagonized by calcium-independent phosphatases (Habermacher and Sale 1996). The key substrate of these enzymes is the intermediate chain IC138 at I1 base, near RS1 (Habermacher and Sale 1997). IC138 hyperphosphorylation as well as defects in I1 and the MIA complex correlate with reduced sliding velocity (Yamamoto et al. 2013) and, interestingly, ineffective phototaxis (King and Dutcher 1997; Okita et al. 2005). Therefore, I1 is a key effector that enables the calcium-dependent dominance of one flagellum for turning the trajectory during phototaxis. RS may operate to suppress IC138 phosphorylation that may otherwise hinder such responses.

Maintenance of 9+2 Axoneme Structural Stability

This role is not as well known as motility control, and is less obvious in *Chlamydomonas* than in some other organisms. First of all, RSs nudge the CP to its central location (e.g., Sivadas et al. 2012). The CP shifts from the central location when RSs are defective. In addition, RSs enhance the stability of the CP and the entire axoneme. Cilia EMs from RS PCD patients often show an array of structural anomalies such as two CPs and transposition—the replacement of the CP by an outer doublet (e.g., Castleman et al. 2009; Kott et al. 2013). Cilia tomography suggests that with defective RSs, the CP actually is assembled but tends to disassemble, leading to transposition (Burgoyne et al. 2014). *Tetrahymena* axonemes with diminished RS2 are also susceptible to deformation (Vasudevan et al. 2015). Eel 9+0 axoneme is exceptionally fragile when the membrane is removed (Gibbons et al. 1985). Conversely, the perceived structural stability of spokeless axonemes in *Chlamydomonas* RS mutant may be caused by multiple factors (LeDizet and Piperno 1995; Kubo et al. 2015). Nonetheless, the structural instability surfaces as short flagella of double mutants lacking both RSs and the CP, likely caused by an increasing disassembly rate.

While all major axonemal complexes may confer stability to the axoneme that is built on inherently unstable microtubules (Kubo et al. 2015), we take the liberty to speculate that the RS-CP system is particularly so by filling up the hollow center, and by the transient contact that dissipates forces and tensions, which the axoneme endures as it repetitively sweeps through fluid of relatively high viscosity. This may explain the nearly identical dimensions of RSs and thus the diameter of 9+2 axonemes across all sources. Interestingly, the ratio of the dimensions of the CP and RS versus the CP in cross sections is strikingly similar to the golden ratio (φ). Perhaps the 9+2 axonemal platform of the last common ancestor of eukaryotes is as good as it gets. Much of the subsequent changes are all but individual preferences.

FUTURE DIRECTIONS

Much about RSs is now known, but key questions remain. First and foremost is how the RS-CP system is involved in calcium/calmodulin-signaled motility changes and, second, is regarding phosphorylation. It is still unclear how RS defects enhance phosphorylation of dynein mediated by PKA and CK1 that also regulate critical events occurring at the basal body area in diverse organisms (Briscoe and Therond 2013). Although *Chlamydomonas*, like plants, do not have the typical PKA tetramer common in animals, PKA activity is indisputable (Hasegawa et al 1987; Howard et al. 1994; Yang and Sale 2000). Identification of PKA in this organism and elucidation of how anchored PKA and CK1 phosphorylate IC138 is central to elucidate RS-mediated regulation. Related, but not identical, to this are the questions about how RSPs become phosphorylated and whether the phosphorylation is a consequence or a cause of RS assembly. The third is RS assembly, likely via spoke-unique mechanisms because of drastic differences between spoke precursor and mature RSs, the multiple chaperone-like spoke subunits, and subunits of a low stoichiometry (Piperno et al. 1981). The fourth is characterization of RS3. Aside from satisfying curiosity, the findings may lead to new insight of RS evolution—how *Chlamydomonas* managed to lose RS3, and what are the expense and benefits to retain a distinct RS3. Last, but not least, is to elucidate RSs of mammals and phenotype RS PCD patients. With the advance of technologies and reagents (Lin et al. 2014; Frommer et al. 2015), the challenges inherent to these systems (Papon et al. 2010) will be overcome and the findings will complement the limitations of simple model systems.

ACKNOWLEDGMENTS

We thank Dr. James Courtright (Marquette University) for his advice and editing.

REFERENCES

*Reference is also in this collection.

Alford LM, Mattheyses AL, Hunter EL, Lin H, Dutcher SK, Sale WS. 2013. The *Chlamydomonas* mutant pf27 reveals novel features of ciliary radial spoke assembly. *Cytoskeleton (Hoboken)* **70**: 804–818.

Barber CF, Heuser T, Carbajal-Gonzalez BI, Botchkarev VV Jr, Nicastro D. 2012. Three-dimensional structure of the radial spokes reveals heterogeneity and interactions with dyneins in *Chlamydomonas* flagella. *Mol Biol Cell* **23**: 111–120.

Cite this article as *Cold Spring Harb Perspect Biol* doi: 10.1101/cshperspect.a028126

Briscoe J, Therond PP. 2013. The mechanisms of Hedgehog signalling and its roles in development and disease. *Nat Rev Mol Cell Biol* **14:** 416–429.

Brokaw CJ, Luck DJ, Huang B. 1982. Analysis of the movement of *Chlamydomonas* flagella: The function of the radial-spoke system is revealed by comparison of wild-type and mutant flagella. *J Cell Biol* **92:** 722–732.

Burgoyne T, Lewis A, Dewar A, Luther P, Hogg C, Shoemark A, Dixon M. 2014. Characterizing the ultrastructure of primary ciliary dyskinesia transposition defect using electron tomography. *Cytoskeleton (Hoboken)* **71:** 294–301.

Castleman VH, Romio L, Chodhari R, Hirst RA, de Castro SC, Parker KA, Ybot-Gonzalez P, Emes RD, Wilson SW, Wallis C, et al. 2009. Mutations in radial spoke head protein genes RSPH9 and RSPH4A cause primary ciliary dyskinesia with central-microtubular-pair abnormalities. *Am J Hum Genet* **84:** 197–209.

Curry AM, Williams BD, Rosenbaum JL. 1992. Sequence analysis reveals homology between two proteins of the flagellar radial spoke. *Mol Cell Biol* **12:** 3967–3977.

Diener DR, Curry AM, Johnson KA, Williams BD, Lefebvre PA, Kindle KL, Rosenbaum JL. 1990. Rescue of a paralyzed-flagella mutant of *Chlamydomonas* by transformation. *Proc Natl Acad Sci* **87:** 5739–5743.

Diener DR, Ang LH, Rosenbaum JL. 1993. Assembly of flagellar radial spoke proteins in *Chlamydomonas*: Identification of the axoneme binding domain of radial spoke protein 3. *J Cell Biol* **123:** 183–190.

Diener DR, Yang P, Geimer S, Cole DG, Sale WS, Rosenbaum JL. 2011. Sequential assembly of flagellar radial spokes. *Cytoskeleton (Hoboken)* **68:** 389–400.

Dymek EE, Smith EF. 2007. A conserved CaM- and radial spoke associated complex mediates regulation of flagellar dynein activity. *J Cell Biol* **179:** 515–526.

Dymek EE, Heuser T, Nicastro D, Smith EF. 2011. The CSC is required for complete radial spoke assembly and wild-type ciliary motility. *Mol Biol Cell* **22:** 2520–2531.

Fiedler SE, Sisson JH, Wyatt TA, Pavlik JA, Gambling TM, Carson JL, Carr DW. 2012. Loss of ASP but not ROPN1 reduces mammalian ciliary motility. *Cytoskeleton (Hoboken)* **69:** 22–32.

Frommer A, Hjeij R, Loges NT, Edelbusch C, Jahnke C, Raidt J, Werner C, Wallmeier J, Grosse-Onnebrink J, Olbrich H, et al. 2015. Immunofluorescence analysis and diagnosis of primary ciliary dyskinesia with radial spoke defects. *Am J Respir Cell Mol Biol* **53:** 563–573.

Gaillard AR, Diener DR, Rosenbaum JL, Sale WS. 2001. Flagellar radial spoke protein 3 is an A-kinase anchoring protein (AKAP). *J Cell Biol* **153:** 443–448.

Gardner LC, O'Toole E, Perrone CA, Giddings T, Porter ME. 1994. Components of a "dynein regulatory complex" are located at the junction between the radial spokes and the dynein arms in *Chlamydomonas* flagella. *J Cell Biol* **127:** 1311–1325.

Gibbons BH, Baccetti B, Gibbons IR. 1985. Live and reactivated motility in the 9+0 flagellum of Anguilla sperm. *Cell Motil* **5:** 333–350.

Gopal R, Foster KW, Yang P. 2012. The DPY-30 domain and its flanking sequence mediate the assembly and modulation of flagellar radial spoke complexes. *Mol Cell Biol* **32:** 4012–4024.

Gupta A, Diener DR, Sivadas P, Rosenbaum JL, Yang P. 2012. The versatile molecular complex component LC8 promotes several distinct steps of flagellar assembly. *J Cell Biol* **198:** 115–126.

Habermacher G, Sale WS. 1996. Regulation of flagellar dynein by an axonemal type-1 phosphatase in *Chlamydomonas*. *J Cell Sci* **109:** 1899–1907.

Habermacher G, Sale WS. 1997. Regulation of flagellar dynein by phosphorylation of a 138-kD inner arm dynein intermediate chain. *J Cell Biol* **136:** 167–176.

Hasegawa E, Hayashi H, Asakura S, Kamiya R. 1987. Stimulation of in vitro motility of *Chlamydomonas* axonemes by inhibition of cAMP-dependent phosphorylation. *Cell Motil Cytoskeleton* **8:** 302–311.

Hayashibe K, Shingyoji C, Kamiya R. 1997. Induction of temporary beating in paralyzed flagella of *Chlamydomonas* mutants by application of external force. *Cell Motil Cytoskeleton* **37:** 232–239.

Heuser T, Raytchev M, Krell J, Porter ME, Nicastro D. 2009. The dynein regulatory complex is the nexin link and a major regulatory node in cilia and flagella. *J Cell Biol* **187:** 921–933.

Heuser T, Dymek EE, Lin J, Smith EF, Nicastro D. 2012. The CSC connects three major axonemal complexes involved in dynein regulation. *Mol Biol Cell* **23:** 3143–3155.

Howard DR, Habermacher G, Glass DB, Smith EF, Sale WS. 1994. Regulation of *Chlamydomonas* flagellar dynein by an axonemal protein kinase. *J Cell Biol* **127:** 1683–1692.

Huang B, Piperno G, Ramanis Z, Luck DJ. 1981. Radial spokes of *Chlamydomonas* flagella: Genetic analysis of assembly and function. *J Cell Biol* **88:** 80–88.

Huang B, Ramanis Z, Luck DJ. 1982. Suppressor mutations in *Chlamydomonas* reveal a regulatory mechanism for flagellar function. *Cell* **28:** 115–124.

Ishijima S, Sekiguchi K, Hiramoto Y. 1988. Comparative study of the beat patterns of American and Asian horseshoe crab sperm: Evidence for a role of the central pair complex in forming planar waveforms in flagella. *Cytoskeleton* **9:** 264–270.

* Ishikawa T. 2016. Axoneme structure from motile cilia. *Cold Spring Harb Perspect Biol* doi: 10.1101/cshperspect. a028076.

Jeanson L, Copin B, Papon JF, Dastot-Le Moal F, Duquesnoy P, Montantin G, Cadranel J, Corvol H, Coste A, Desir JM. 2015. RSPH3 mutations cause primary ciliary dyskinesia with central-complex defects and a near absence of radial spokes. *Am J Hum Genet* **97:** 153–162.

Jivan A, Earnest S, Juang YC, Cobb MH. 2009. Radial spoke protein 3 is a mammalian protein kinase A-anchoring protein that binds ERK1/2. *J Biol Chem* **284:** 29437–29445.

Johnson KA, Rosenbaum JL. 1992. Polarity of flagellar assembly in *Chlamydomonas*. *J Cell Biol* **119:** 1605–1611.

King SJ, Dutcher SK. 1997. Phosphoregulation of an inner dynein arm complex in *Chlamydomonas reinhardtii* is altered in phototactic mutant strains. *J Cell Biol* **136:** 177–191.

Koch S, Acebron SP, Herbst J, Hatiboglu G, Niehrs C. 2015. Post-transcriptional Wnt signaling governs epididymal sperm maturation. *Cell* **163:** 1225–1236.

Kohno T, Wakabayashi K, Diener DR, Rosenbaum JL, Kamiya R. 2011. Subunit interactions within the *Chlamydomonas* flagellar spokehead. *Cytoskeleton (Hoboken)* **68:** 237–246.

Kott E, Legendre M, Copin B, Papon JF, Dastot-Le Moal F, Montantin G, Duquesnoy P, Piterboth W, Amram D, Bassinet L, et al. 2013. Loss-of-function mutations in RSPH1 cause primary ciliary dyskinesia with central-complex and radial-spoke defects. *Am J Hum Genet* **93:** 561–570.

Koyfman AY, Schmid MF, Gheiratmand L, Fu CJ, Khant HA, Huang D, He CY, Chiu W. 2011. Structure of *Trypanosoma brucei* flagellum accounts for its bihelical motion. *Proc Natl Acad Sci* **108:** 11105–11108.

Kubo T, Hirono M, Aikawa T, Kamiya R, Witman GB. 2015. Reduced tubulin polyglutamylation suppresses flagellar shortness in *Chlamydomonas*. *Mol Biol Cell* **26:** 2810–2822.

LeDizet M, Piperno G. 1995. The light chain p28 associates with a subset of inner dynein arm heavy chains in *Chlamydomonas* axonemes. *Mol Biol Cell* **6:** 697–711.

Lin J, Tritschler D, Song K, Barber CF, Cobb JS, Porter ME, Nicastro D. 2011. Building blocks of the nexin–dynein regulatory complex in *Chlamydomonas* flagella. *J Biol Chem* **286:** 29175–29191.

Lin J, Yin W, Smith MC, Song K, Leigh MW, Zariwala MA, Knowles MR, Ostrowski LE, Nicastro D. 2014. Cryo-electron tomography reveals ciliary defects underlying human RSPH1 primary ciliary dyskinesia. *Nat Commun* **5:** 5727.

Lindemann CB, Lesich KA. 2015. The geometric clutch at 20: Stripping gears or gaining traction? *Reproduction* **150:** R45–R53.

Lindemann CB, Mitchell DR. 2007. Evidence for axonemal distortion during the flagellar beat of *Chlamydomonas*. *Cell Motil Cytoskeleton* **64:** 580–589.

* Loreng TD, Smith EF. 2016. The central apparatus of cilia and eukaryotic flagella. *Cold Spring Harb Perspect Biol* doi: 10.1101/cshperspect.a028118.

Newell AE, Fiedler SE, Ruan JM, Pan J, Wang PJ, Deininger J, Corless CL, Carr DW. 2008. Protein kinase A RII-like (R2D2) proteins exhibit differential localization and AKAP interaction. *Cell Motil Cytoskeleton* **65:** 539–552.

Nonaka S, Shiratori H, Saijoh Y, Hamada H. 2002. Determination of left–right patterning of the mouse embryo by artificial nodal flow. *Nature* **418:** 96–99.

Oda T, Yanagisawa H, Kamiya R, Kikkawa M. 2014a. A molecular ruler determines the repeat length in eukaryotic cilia and flagella. *Science* **346:** 857–860.

Oda T, Yanagisawa H, Yagi T, Kikkawa M. 2014b. Mechanosignaling between central apparatus and radial spokes controls axonemal dynein activity. *J Cell Biol* **204:** 807–819.

Okita N, Isogai N, Hirono M, Kamiya R, Yoshimura K. 2005. Phototactic activity in *Chlamydomonas* "non-phototactic" mutants deficient in Ca^{2+}-dependent control of flagellar dominance or in inner-arm dynein. *J Cell Sci* **118:** 529–537.

Omoto CK, Witman GB. 1981. Functionally significant central-pair rotation in a primitive eukaryotic flagellum. *Nature* **290:** 708–710.

Onoufriadis A, Shoemark A, Schmidts M, Patel M, Jimenez G, Liu H, Thomas B, Dixon M, Hirst RA, Rutman A, et al. 2014. Targeted NGS gene panel identifies mutations in RSPH1 causing primary ciliary dyskinesia and a common mechanism for ciliary central pair agenesis due to radial spoke defects. *Hum Mol Genet* **23:** 3362–3374.

Papon JF, Coste A, Roudot-Thoraval F, Boucherat M, Roger G, Tamalet A, Vojtek AM, Amselem S, Escudier E. 2010. A 20-year experience of electron microscopy in the diagnosis of primary ciliary dyskinesia. *Eur Respir J* **35:** 1057–1063.

Patel-King RS, Gorbatyuk O, Takebe S, King SM. 2004. Flagellar radial spokes contain a Ca^{2+}-stimulated nucleoside diphosphate kinase. *Mol Biol Cell* **15:** 3891–3902.

Pazour GJ, Agrin N, Leszyk J, Witman GB. 2005. Proteomic analysis of a eukaryotic cilium. *J Cell Biol* **170:** 103–113.

Pigino G, Bui KH, Maheshwari A, Lupetti P, Diener D, Ishikawa T. 2011. Cryoelectron tomography of radial spokes in cilia and flagella. *J Cell Biol* **195:** 673–687.

Piperno G, Huang B, Ramanis Z, Luck DJ. 1981. Radial spokes of *Chlamydomonas* flagella: Polypeptide composition and phosphorylation of stalk components. *J Cell Biol* **88:** 73–79.

Piperno G, Mead K, Shestak W. 1992. The inner dynein arms I2 interact with a "dynein regulatory complex" in *Chlamydomonas*. *J Cell Biol* **118:** 1455–1463.

Porter ME, Knott JA, Gardner LC, Mitchell DR, Dutcher SK. 1994. Mutations in the SUP-PF-1 locus of *Chlamydomonas reinhardtii* identify a regulatory domain in the beta-dynein heavy chain. *J Cell Biol* **126:** 1495–1507.

Portman N, Gull K. 2010. The paraflagellar rod of kinetoplastid parasites: From structure to components and function. *Int J Parasitol* **40:** 135–148.

Qin H, Diener DR, Geimer S, Cole DG, Rosenbaum JL. 2004. Intraflagellar transport (IFT) cargo: IFT transports flagellar precursors to the tip and turnover products to the cell body. *J Cell Biol* **164:** 255–266.

Rao VG, Sarafdar RB, Chowdhury TS, Sivadas P, Yang P, Dongre PM, D'Souza JS. 2016. Myc-binding protein orthologue interacts with AKAP240 in the central pair apparatus of the *Chlamydomonas* flagella. *BMC Cell Biol* **17:** 24.

Rual JF, Venkatesan K, Hao T, Hirozane-Kishikawa T, Dricot A, Li N, Berriz GF, Gibbons FD, Dreze M, Ayivi-Guedehoussou N, et al. 2005. Towards a proteome-scale map of the human protein–protein interaction network. *Nature* **437:** 1173–1178.

Rupp G, O'Toole E, Gardner LC, Mitchell BF, Porter ME. 1996. The sup-pf-2 mutations of *Chlamydomonas* alter the activity of the outer dynein arms by modification of the gamma-dynein heavy chain. *J Cell Biol* **135:** 1853–1865.

Sakakibara HM, Kunioka Y, Yamada T, Kamimura S. 2004. Diameter oscillation of axonemes in sea-urchin sperm flagella. *Biophys J* **86:** 346–352.

Satouh Y, Inaba K. 2009. Proteomic characterization of sperm radial spokes identifies a novel spoke protein with an ubiquitin domain. *FEBS Lett* **583:** 2201–2207.

Satouh Y, Padma P, Toda T, Satoh N, Ide H, Inaba K. 2005. Molecular characterization of radial spoke subcomplex containing radial spoke protein 3 and heat shock protein 40 in sperm flagella of the ascidian *Ciona intestinalis*. *Mol Biol Cell* **16:** 626–636.

Shinohara K, Chen D, Nishida T, Misaki K, Yonemura S, Hamada H. 2015. Absence of radial spokes in mouse node cilia is required for rotational movement but confers ultrastructural instability as a trade-off. *Dev Cell* **35:** 236–246.

Sivadas P, Dienes JM, St Maurice M, Meek WD, Yang P. 2012. A flagellar A-kinase anchoring protein with two amphipathic helices forms a structural scaffold in the radial spoke complex. *J Cell Biol* **199:** 639–651.

Smith EF, Sale WS. 1992. Regulation of dynein-driven microtubule sliding by the radial spokes in flagella. *Science* **257:** 1557–1559.

Tremblay V, Zhang P, Chaturvedi CP, Thornton J, Brunzelle JS, Skiniotis G, Shilatifard A, Brand M, Couture JF. 2014. Molecular basis for DPY-30 association to COMPASS-like and NURF complexes. *Structure* **22:** 1821–1830.

Ueno H, Iwataki Y, Numata O. 2006. Homologues of radial spoke head proteins interact with Ca^{2+}/calmodulin in Tetrahymena cilia. *J Biochem* **140:** 525–533.

Urbanska P, Song K, Joachimiak E, Krzemien-Ojak L, Koprowski P, Hennessey T, Jerka-Dziadosz M, Fabczak H, Gaertig J, Nicastro D, et al. 2015. The CSC proteins FAP61 and FAP251 build the basal substructures of radial spoke 3 in cilia. *Mol Biol Cell* **26:** 1463–1475.

Vasudevan KK, Song K, Alford LM, Sale WS, Dymek EE, Smith EF, Hennessey T, Joachimiak E, Urbanska P, Wloga D, et al. 2015. FAP206 is a microtubule-docking adapter for ciliary radial spoke 2 and dynein c. *Mol Biol Cell* **26:** 696–710.

Wang X, Lou Z, Dong X, Yang W, Peng Y, Yin B, Gong Y, Yuan J, Zhou W, Bartlam M, et al. 2009. Crystal structure of the C-terminal domain of human DPY-30-like protein: A component of the histone methyltransferase complex. *J Mol Biol* **390:** 530–537.

Warner FD. 1978. Cation-induced attachment of ciliary dynein cross-bridges. *J Cell Biol* **77:** R19–R26.

Warner FD, Satir P. 1974. The structural basis of ciliary bend formation. Radial spoke positional changes accompanying microtubule sliding. *J Cell Biol* **63:** 35–63.

Wei M, Sivadas P, Owen HA, Mitchell DR, Yang P. 2010. *Chlamydomonas* mutants display reversible deficiencies in flagellar beating and axonemal assembly. *Cytoskeleton (Hoboken)* **67:** 71–80.

Wirschell M, Zhao F, Yang C, Yang P, Diener D, Gaillard A, Rosenbaum JL, Sale WS. 2008. Building a radial spoke: Flagellar radial spoke protein 3 (RSP3) is a dimer. *Cell Motil Cytoskeleton* **65:** 238–248.

Witman GB, Plummer J, Sander G. 1978. *Chlamydomonas* flagellar mutants lacking radial spokes and central tubules. *J Cell Biol* **76:** 729–747.

Woolley DM, Vernon GG. 2001. A study of helical and planar waves on sea urchin sperm flagella, with a theory of how they are generated. *J Exp Biol* **204:** 1333–1345.

Yagi T, Kamiya R. 2000. Vigorous beating of *Chlamydomonas* axonemes lacking central pair/radial spoke structures in the presence of salts and organic compounds. *Cell Motil Cytoskeleton* **46:** 190–199.

Yagi T, Kamimura S, Kamiya R. 1994. Nanometer scale vibration in mutant axonemes of *Chlamydomonas*. *Cell Motil Cytoskeleton* **29:** 177–185.

Yamamoto R, Song K, Yanagisawa HA, Fox L, Yagi T, Wirschell M, Hirono M, Kamiya R, Nicastro D, Sale WS. 2013. The MIA complex is a conserved and novel dynein regulator essential for normal ciliary motility. *J Cell Biol* **201:** 263–278.

Yang P, Sale WS. 2000. Casein kinase I is anchored on axonemal doublet microtubules and regulates flagellar dynein phosphorylation and activity. *J Biol Chem* **275:** 18905–18912.

Yang P, Smith EF. 2008. The flagellar radial spokes. In *The Chlamydomonas source book* (ed. Witman GB), pp. 207–231. Academic, New York.

Yang C, Yang P. 2006. The flagellar motility of *Chlamydomonas* pf25 mutant lacking an AKAP-binding protein is overtly sensitive to medium conditions. *Mol Biol Cell* **17:** 227–238.

Yang P, Diener DR, Rosenbaum JL, Sale WS. 2001. Localization of calmodulin and dynein light chain LC8 in flagellar radial spokes. *J Cell Biol* **153:** 1315–1326.

Yang P, Diener DR, Yang C, Kohno T, Pazour GJ, Dienes JM, Agrin NS, King SM, Sale WS, Kamiya R, et al. 2006. Radial spoke proteins of *Chlamydomonas* flagella. *J Cell Sci* **119:** 1165–1174.

Yang C, Owen HA, Yang P. 2008. Dimeric heat shock protein 40 binds radial spokes for generating coupled power strokes and recovery strokes of 9+2 flagella. *J Cell Biol* **180:** 403–415.

Yang P, Yang C, Wirschell M, Davis S. 2009. Novel LC8 mutations have disparate effects on the assembly and stability of flagellar complexes. *J Biol Chem* **284:** 31412–31421.

The Intraflagellar Transport Machinery

Michael Taschner and Esben Lorentzen

Department of Structural Cell Biology, Max-Planck-Institute of Biochemistry, D-82152 Martinsried, Germany

Correspondence: taschner@biochem.mpg.de; lorentze@biochem.mpg.de

Eukaryotic cilia and flagella are evolutionarily conserved organelles that protrude from the cell surface. The unique location and properties of cilia allow them to function in vital processes such as motility and signaling. Ciliary assembly and maintenance rely on intra-flagellar transport (IFT), the bidirectional movement of a multicomponent transport system between the ciliary base and tip. Since its initial discovery more than two decades ago, considerable effort has been invested in dissecting the molecular mechanisms of IFT in a variety of model organisms. Importantly, IFT was shown to be essential for mammalian development, and defects in this process cause a number of human pathologies known as ciliopathies. Here, we review current knowledge of IFT with a particular emphasis on the IFT machinery and specific mechanisms of ciliary cargo recognition and transport.

The main structural component of cilia and flagella (interchangeable terms) is an internal microtubule (MT)-based axoneme, which gives the organelle its characteristic elongated shape (Fig. 1). The axoneme is templated from the basal body (BB), a modified centriole anchored to the plasma membrane via transition fibers. In motile cilia, the axonemal MTs are densely decorated with additional multisubunit complexes, such as outer dynein arms (ODAs), inner dynein arms (IDAs), and radial spokes (RS). These motility complexes act together in a highly coordinated fashion to achieve ciliary beating (Lindemann and Lesich 2010). Early studies in the green-alga *Chlamydomonas reinhardtii* (*Cr*) and other protozoan species demonstrated that new axonemal subunits are added to the ciliary tip (Rosenbaum and Child 1967; Johnson and Rosenbaum 1992), and the assembly site is thus continuously moving farther away from the site of protein synthesis in the cytoplasm during flagellar growth. Together with the fact that diffusion-limiting protein complexes localize to the transition zone (Reiter et al. 2012), this necessitates an active transport mechanism (i.e., intraflagellar transport [IFT]), to deliver building blocks (e.g., tubulin, ODAs, and IDAs) from the flagellar base to the tip. IFT was first observed in paralyzed *C. reinhardtii* flagella using differential interference contrast (DIC) microscopy in the early 1990s (Kozminski et al. 1993), and subsequent work has improved the visualization of IFT using a variety of methods not only in the green alga (Engel et al. 2009a), but also in other model systems such as *Caenorhabditis elegans* (Hao et al. 2009), *Tetrahymena thermophila* (Jiang et al. 2015), *Trypanosoma brucei* (Santi-Rocca et al. 2015), and mammalian cells (Williams et al. 2014; Ishikawa and Marshall 2015). In *Chlamydomonas*, dense particles were seen to move continuously from the base to the tip (anterograde

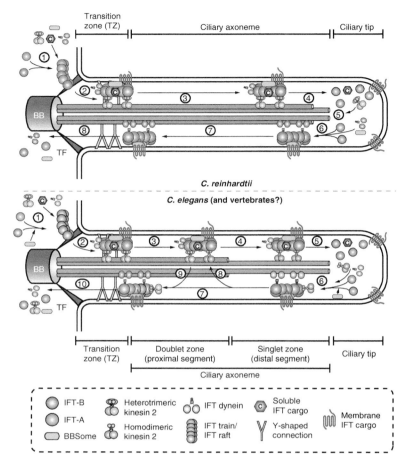

Figure 1. Schematic overview of the main steps during intraflagellar transport (IFT). Because of important differences in various organisms (mainly regarding IFT motors), the individual steps are shown for both *Chlamydomonas reinhardtii* (*top*; only one anterograde motor) and *Caenorhabditis elegans* (*bottom*; two anterograde motors). (*Top*) (1) In *C. reinhardtii*, IFT trains are assembled from IFT-A and IFT-B particles at the ciliary base around the transition fibers and bind to the anterograde motor, the retrograde motor (as a cargo), soluble and membrane cargos, as well as the BBSome. (2) The trains enter the cilium, and (3) move processively toward the ciliary tip (anterograde IFT). (4) At the tip, the IFT trains are remodeled, cargo is unloaded, and the anterograde motor is inactivated (by phosphorylation). (5) The inactive heterotrimeric kinesin 2 motor exits the flagellum independently of retrograde IFT. (6) Retrograde IFT trains assemble at the ciliary tip with active IFT dynein linking them to the ciliary axoneme. (7) Processive retrograde IFT returns the trains and associated proteins back to the ciliary base, and (8) the trains exit flagella and get disassembled. (*Bottom*) (1) In *C. elegans*, IFT-A and IFT-B particles also form trains at the ciliary base, and this step is assisted by the BBSome complex. (2) After binding to cargos (retrograde motor, soluble cargos, membrane cargos, BBSome), the heterotrimeric kinesin 2 motor transports the trains through the transition zone (characterized by Y-shaped connectors linking the microtubule (MT) doublets to the ciliary membrane). (3) Along the ciliary proximal segment ("handover zone") heterotrimeric kinesin 2 gradually dissociates from the trains and is replaced by homodimeric kinesin 2 (OSM-3). (4) Along the ciliary distal segment, the trains are exclusively moved by OSM-3. (5) At the tip the trains are remodeled, cargo is unloaded, and the anterograde motor is inactivated. (6) Retrograde trains assemble, which contain activated IFT dynein and inactivated OSM-3 as a retrograde cargo. (7) Retrograde IFT returns the trains back to the ciliary base. (8) Along the proximal segment, OSM-3 is gradually unloaded, and (9) inactive heterotrimeric kinesin 2 is picked up for transport back to the base. (10) Trains exit the cilium and are disassembled. TF, Transition fiber; BB, basal body.

Cite this article as *Cold Spring Harb Perspect Biol* doi: 10.1101/cshperspect.a028092

IFT) at a speed of ∼2.5 μm/sec and back from the tip to the base at a faster velocity of ∼4 μm/sec (retrograde IFT). A follow-up study correlated these particles with large electron-dense structures (IFT trains) forming tight contacts between the outer axonemal MT doublets and the overlying flagellar membrane (Kozminski et al. 1995). An elegant study in the green alga using a combined fluorescence and electron microscopy approach showed that anterograde and retrograde IFT trains run on the B- and A-tubules of the axonemal MT doublets, respectively, explaining why these complexes do not crash into each other inside cilia (Stepanek and Pigino 2016). Efforts to identify and characterize the biochemical components of the IFT machinery (most importantly, molecular motors and IFT complex subunits) made use of several already characterized mutants in various model organisms, such as flagellar assembly (*fla*) mutants in *C. reinhardtii* (Huang et al. 1977; Adams et al. 1982), or chemosensory (*che*), osmotic avoidance (*osm*), dye-filling (*dyf*), and dauer-formation (*daf*) mutants in *C. elegans* (Inglis et al. 2007).

Movement of the IFT particles in *Chlamydomonas* flagella was shown to depend on the FLA10 protein, thus identifying the anterograde IFT motor as heterotrimeric kinesin II (Cole et al. 1993; Walther et al. 1994). IFT trains ceased to move after shifting temperature-sensitive *fla10-1* mutant cells to the restrictive temperature; the IFT trains disappeared, and the flagella resorbed owing to a requirement of continuous IFT for *Chlamydomonas* flagellar maintenance (Kozminski et al. 1995). A more complex situation was observed in *C. elegans*, where heterotrimeric kinesin II mutants are still able to assemble neuronal sensory cilia because of the presence of homodimeric kinesin II (OSM-3) as a second anterograde IFT motor. These two motor complexes function redundantly to build the ciliary middle segment containing axonemal MT doublets, but OSM-3 functions alone to build the distal singlet zone (Snow et al. 2004). A recent study investigated the relationship between the two anterograde motors in more detail and showed that rather than transporting IFT trains together as previously suggested, the two different kinesin motors localize to different regions of the cilium (heterotrimeric kinesin II more proximal, OSM-3 more distal) (Prevo et al. 2015). Despite the presence of an OSM3 homolog in vertebrates (Kif17), heterotrimeric kinesin II is the main motor complex for ciliary assembly in mice (Nonaka et al. 1998; Marszalek et al. 1999) and zebrafish (Zhao et al. 2012), but Kif17 is required for the proper formation of specific subtypes of the organelle (Insinna et al. 2009; Zhao et al. 2012).

Cytoplasmic dynein 2 was identified as the retrograde IFT motor that carries the IFT-complex (and cargos/turnover products) back from the ciliary tip. Inactivation of subunits in this complex leads to short cilia with bulges accumulating IFT proteins in a variety of model systems (Pazour et al. 1998; Porter et al. 1999; Signor et al. 1999; Perrone et al. 2003; Hou et al. 2004; May et al. 2005). The process of IFT can be subdivided into several distinct phases (see Fig. 1). IFT particles and trains are first assembled at the ciliary base, where they also bind to ciliary cargos followed by processive anterograde transport to the ciliary tip. At the tip, the ciliary cargos are unloaded, and the IFT particles are remodeled and switched to a retrograde mode by inactivating the anterograde and activating the retrograde motor. Finally, IFT particles move back to the base, where they are recycled for subsequent rounds of transport. Although a general outline of the IFT cycle is in place, the molecular mechanisms of most steps of this process remain poorly understood.

THE DISCOVERY OF THE IFT COMPLEX

The key to biochemical purification and characterization of endogenous IFT complexes from *C. reinhardtii* was the temperature-sensitive kinesin II mutant (*fla10-1*), in which the activity of the anterograde heterotrimeric kinesin II motor could be abrogated by a shift to the nonpermissive temperature. Two independent studies used this strain to identify a large protein complex, which was only present in the flagella at the permissive temperature (Piperno and Mead 1997; Cole et al. 1998). Although Piperno and Mead (1997) identified 13 proteins as members of this complex, Cole et al. (1998) reported 15 proteins and showed that the complex dissociated at increased ionic strength into two biochemically distinct subcomplexes, IFT-A and IFT-B. The components of the IFT complex in *C. reinhardtii* were named according to their apparent molecular weight, as judged by their migration in sodium dodecyl sulfate polyacrylamide gel electrophoresis(SDS-PAGE). In the rest of this paper, we will use this *Chlamydomonas* nomenclature for the proteins even when referring to work carried out in other systems in which the factors have different names (Table 1). Immunofluorescence and immunogold electron microscopy showed that these proteins were highly enriched at the flagellar base but were also detectable as punctuate stainings along the ciliary axoneme (for examples, see Deane et al. 2001; Pedersen et al. 2005; Hou et al. 2007; Fan et al. 2010). This subcellular

distribution is also characteristic of IFT proteins in other model systems (for example, see Blacque and Leroux 2006; Tsao and Gorovsky 2008a; Follit et al. 2009; Huet et al. 2014).

Additional IFT proteins have since then been identified (Piperno et al. 1998; Wang et al. 2009; Fan et al. 2010; Ishikawa et al. 2014), and, currently, six proteins are believed to form the IFT-A subcomplex (IFT144, 140, 139, 122, 121, and 43), whereas 16 subunits are found in IFT-B (IFT172, 88, 81, 80, 74, 70, 57, 56, 54, 52, 46, 38, 27, 25, 22, and 20). Analysis of mutants deficient for those factors showed that IFT-A mutants typically yielded ciliary defects similar to cytoplasmic dynein 2 mutants (stumpy organelles filled with IFT material) (for example, see Piperno et al. 1998; Blacque et al. 2006; Tran et al. 2008; Iomini et al. 2009), whereas IFT-B mutants usually had severe ciliogenesis defects (examples in Pazour et al. 2000; Tsao and Gorovsky 2008a;

Table 1. Nomenclature of IFT proteins in several model organisms

| Complex | General | \multicolumn{5}{c}{Alternative name in other organisms (if different)} |
		Chlamydomonas reinhardtii	*Trypanosoma brucei*	*Caenorhabditis elegans*	*Danio rerio*	Mammals
IFT-B						
IFT-B1	IFT88	-	-	OSM-5	Polaris	Polaris/Tg737
	IFT81	-	-	-	-	-
	IFT74	-	-	-	-	-
	IFT70	FAP259	PIFTB2	DYF-1	Fleer	TTC30A/B
	IFT56	DYF-13	PIFTC3	DYF-13	-	TTC26
	IFT52	BLD1	-	OSM-5	-	NGD5
	IFT46	-	-	DYF-6	-	-
	IFT27	-	-	(Absent)	-	RabL4
	IFT25	FAP232	-	(Absent)	-	HSPB11
	IFT22	FAP9	-	IFTA-2	-	RabL5
IFT-B2	IFT172	-	-	OSM-1	-	SLB
	IFT80	-	-	CHE-2	-	WDR56
	IFT57	-	-	CHE-13	-	Hippi
	IFT54	FAP116	-	DYF-11	Elipsa	Traf3IP1/MIP-T3
	IFT38	FAP22	PIFTA1	DYF-3	Qilin	Cluap1
	IFT20	-	-	-	-	-
IFT-A						
Core	IFT144	-	-	DYF-2	-	WDR19
	IFT140	-	-	CHE-11	-	WDTC2
	IFT122	FAP80	-	DAF-10	-	WDR10
Noncore	IFT139	-	-	-	-	THM1/TTC21B
	IFT121	-	PIFTD4	IFTA-1	-	WDR35
	IFT43	-	-	-	-	C14ORF179

Cite this article as *Cold Spring Harb Perspect Biol* doi: 10.1101/cshperspect.a028092

Adhiambo et al. 2009), comparable to kinesin II mutants (with a few important exceptions, see below). These results showed that IFT-A and IFT-B were not only biochemically distinct complexes, but suggested that they also have separate functions in retrograde and anterograde transport of cargos, respectively. Although this simplistic view may be true to a large extent, several exceptions have since been reported, such as the importance of the IFT-A factors for ciliary protein import (Mukhopadhyay et al. 2010) or the requirement for IFT-B proteins in ciliary cargo export (Keady et al. 2012; Eguether et al. 2014; Huet et al. 2014). Importantly, IFT mutations in mammals were shown to cause severe pathologies (ciliopathies) (for review, see Fliegauf et al. 2007) and affect cellular signaling pathways (Huangfu et al. 2003; Wang et al. 2006). These discoveries highlighted the importance of IFT in particular (and cilia in general) in human development, tissue homeostasis, and disease, and led to an increased interest in understanding cilium formation and function.

Bioinformatic analysis of IFT proteins revealed well-known protein–protein interaction motifs (Taschner et al. 2012), such as tetratricopeptide repeats (TPRs), WD-40 repeats, and coiled coils, consistent with their assembly into a large macromolecular complex and their putative binding to hundreds of ciliary proteins found in ciliary and flagellar proteomes (Ostrowski et al. 2002; Pazour et al. 2005; Liu et al. 2007). Nevertheless, assays to identify direct interactions involving IFT proteins using bacterial coexpressions/pulldowns and yeast two-hybrid analyses only gave limited insights. Recently, the recombinant reconstitution and biochemical characterization of IFT subcomplexes (primarily from *C. reinhardtii*) has improved our understanding especially of the IFT-B complex, and has led to a first glimpse on high-resolution structures of IFT proteins.

THE ARCHITECTURE OF THE IFT-B COMPLEX

The IFT-B Core (IFT-B1)

Some of the first insights into the organization of the 16-subunit IFT-B complex was obtained when Lucker et al. (2005) showed that in *Chla-*

mydomonas, several subunits formed a salt-stable subcomplex (called IFT-B core), and others (peripheral subunits) dissociated at an NaCl concentration of 300 mM. The originally identified IFT-B core contained six proteins (IFT88, 81, 74, 52, 46, and 27), but four additional members in the green alga were identified in subsequent studies, namely, IFT70 (Fan et al. 2010), IFT25 (Lechtreck et al. 2009b; Wang et al. 2009), IFT22 (Wang et al. 2009), and IFT56 (Ishikawa et al. 2014). Homologs of these factors have been identified in many other model systems, with the exceptions being *C. elegans,* which lacks IFT27 and IFT25, and *Drosophila*, which lacks IFT81, 74, 27, 25, and 22 (Cole and Snell 2009). Although most IFT-B core factors are obligatory for ciliogenesis, some factors have more specialized roles in cilium formation and function (see below).

Interaction studies of IFT-B core proteins using yeast two-hybrid analysis or bacterial coexpressions/pulldowns suggested direct interactions between IFT27 and IFT25 (Follit et al. 2009; Wang et al. 2009; Bhogaraju et al. 2011), IFT70 and IFT52 (Zhao and Malicki 2011; Howard et al. 2013), IFT70 and IFT46 (Fan et al. 2010), IFT81 and IFT74 (Lucker et al. 2005; Kobayashi et al. 2007), IFT88, IFT52, and IFT46 (Lucker et al. 2010), and, finally, IFT56 and IFT46 (Swiderski et al. 2014). These initial insights into subunit interactions paved the way for the recombinant reconstitution of several IFT-B core subcomplexes using *Chlamydomonas* proteins (Taschner et al. 2011) as well as a nine-subunit IFT-B core lacking only IFT56 (Taschner et al. 2014). In addition, most of the direct protein–protein interactions were mapped to individual domains in IFT proteins (Taschner et al. 2011, 2014), leading to an initial interaction map of the nine-subunit IFT-B core with domain resolution (Fig. 2). Several of these interactions appear to occur through composite interfaction interfaces because they require the presence of more than two proteins, providing an explanation for why they were not observed in yeast two-hybrid assays. IFT81, 74, 52, and 46, for example, only bind when preassembled IFT81/74 and IFT52/46 complexes are used. Similarly, IFT22 interacts with a minimal

IFT81/74 complex, but not with the individual IFT81 or IFT74 fragments (Taschner et al. 2014). Salt-stability tests of the nonameric IFT-B core complex confirmed previously published results (Lucker et al. 2005), but showed that this complex was stable at NaCl concentrations >2 M in vitro, suggesting substantial hydrophobic interactions between subunits (Taschner et al. 2014). This idea was confirmed by several crystal structures of IFT-B core subcomplexes. The first IFT subcomplex structure obtained was that of Chlamydomonas IFT27/25 (Bhogaraju et al. 2011), in which the atypical small GTPase IFT27 binds to IFT25 via an unusually long carboxy-terminal helix through hydrophobic contacts. This hydrophobicity is the likely explanation for why IFT27 can only be made recombinantly in the presence of IFT25 (Bhogaraju et al. 2011) and for the instability of endogenous IFT27 in mouse cells lacking IFT25 (Keady et al. 2012). Interestingly, the ciliate T. thermophila (Tt) does not have an IFT25 homolog, but does have an IFT27 protein that lacks the hydrophobic residues. TtIFT27, unlike the algal and mouse proteins, can be produced recombinantly in a soluble form (M Taschner and E Lorentzen, unpubl.).

The high-resolution structure of the IFT70/52 complex is also in agreement with the high salt stability of the IFT-B core complex. The tetratricopeptide repeat (TPR)-containing superhelical protein IFT70 was shown to wrap tightly in two full turns around a short (40 residues) and largely unstructured, proline-rich peptide of IFT52 that can be considered part of the hydrophobic core required for proper folding of the IFT70 protein (Taschner et al. 2014). Finally, a complex between the carboxy-terminal domains of TtIFT52 and TtIFT46 also revealed the presence a large hydrophobic surface between the two proteins (Taschner et al. 2014), and that this carboxy-terminal region of IFT46 is necessary for proper folding of the IFT52 carboxy-terminal domain.

The Peripheral IFT-B Complex (IFT-B2)

Several peripheral subunits (IFT172, 80, 57, and 20) were reported to be loosely associated with the IFT-B core complex (Lucker et al. 2005), and

the two additional peripheral factors IFT54 (Kunitomo and Iino 2008; Omori et al. 2008; Berbari et al. 2011) and IFT38 (Ou et al. 2005b; Inglis et al. 2007; Pasek et al. 2012) were identified later. With respect to protein–protein interactions, IFT20 was shown to bind to the carboxy-terminal coiled-coil region of IFT54 in pulldowns (Omori et al. 2008; Follit et al. 2009) and to IFT57 in yeast-two-hybrid assays (Baker et al. 2003), suggesting the formation of a ternary complex between these three subunits. We could recently confirm the presence of a stable IFT54/20 complex of Chlamydomonas, but found that IFT57 binds strongly to IFT38 and only weakly to IFT20 (Taschner et al. 2016). The IFT57/38 and IFT54/20 complexes interact with each other to form a heterotetramer, although this interaction appears to be much weaker than the interactions holding together the IFT54/20 and IFT57/38 heterodimers (Taschner et al. 2016).

Surprisingly, we found that the six peripheral IFT-B subunits form a stable complex in the absence of the IFT-B1 complex and show that IFT57/38 is a central component that directly interacts with IFT172, IFT80, and IFT54/20 (Taschner et al. 2016). These results show that the IFT-B complex is composed of two distinct subcomplexes that we refer to as IFT-B1 (the IFT-B core) and IFT-B2 (for the complex containing the proteins formerly referred to as peripheral IFT-B proteins) (Fig. 2). This conclusion, based on the analysis of purified Chlamydomonas proteins, was recently confirmed using a "visible immunoprecipitation" (VIP) approach with human proteins (Katoh et al. 2016), as well as in human cells in a large-scale proteomic study (Boldt et al. 2016). With regard to salt stability, it should be noted that five of the six IFT-B2 proteins are stable at NaCl concentrations of 1 M (Taschner et al. 2016), with IFT172 being the only weakly attached subunit, confirming previously published data (Cole et al. 1998; Pedersen et al. 2005).

The presence of two stable subcomplexes within IFT-B raised the question about how they interact with each other to form the full IFT-B complex. Mixing of IFT-B1 and IFT-B2 led to the reconstitution of a nearly complete 15-subunit IFT-B complex (missing only

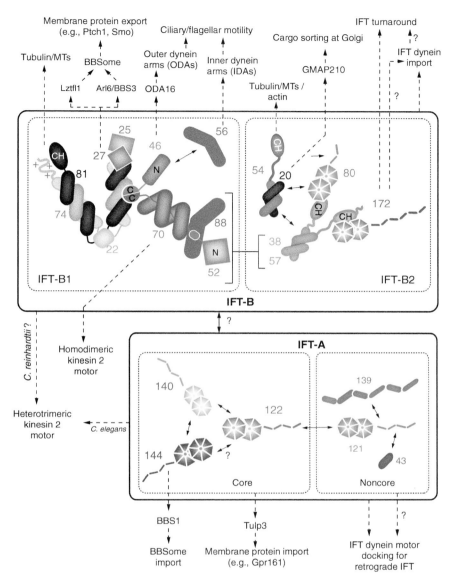

Figure 2. Interactions within intraflagellar transport (IFT) proteins and interaction between IFT proteins/complexes and ciliary motor/cargo proteins (see text for details). MT, Microtubule; *C. reinhardtii*, *Chlamydomonas reinhardtii*; *C. elegans*, *Caenorhabditis elegans*.

IFT56), and pulldown analyses between recombinant parts showed that the minimal requirement for this binding is a preformed IFT88/52N subcomplex (on the IFT-B1 side) that directly contacts the IFT57/38 subcomplex in IFT-B2 (Fig. 2) (Taschner et al. 2016). The necessity for the presence of four distinct proteins explains why such an interaction was previously missed in yeast-two-hybrid analysis and similar experiments. Although our understanding of IFT-B complex architecture has significantly improved in recent years, the relative position of proteins and ciliary cargo-binding sites within the IFT complex are still unknown. Further studies using not only X-ray crystallography but also single-particle cryo-EM will be necessary to gain further insights into the ciliary assembly machinery.

FUNCTIONS OF IFT-B PROTEINS IN IFT-B COMPLEX FORMATION AND STABILITY

The specific functions of individual IFT-B proteins in ciliogenesis are still largely enigmatic. The reason for this is that null mutants of most IFT-B factors typically result in IFT complex destabilization and thus a strong ciliogenesis defect, masking the specific contribution of the missing protein to IFT regulation and/or cargo transport. IFT52, for example, is indispensable for IFT-B complex formation because of its central location within the complex and interaction with at least five other IFT-B subunits (Fig. 2). Consistently, the *ift52/bld1* mutant in *Chlamydomonas* is unable to assemble flagella and shows strong destabilization of several IFT-B1 subunits (Richey and Qin 2012), with the exception of the IFT27/25 subcomplex, which forms independently of IFT-B1, and IFT46 (Brazelton et al. 2001). Short and malformed flagella are formed in the *ift46* mutant, in which not only the IFT-B1 protein IFT81, but also the IFT-B2 proteins IFT57 and IFT20 were destabilized, indicating an interdependence of the IFT-B1 and B2 subcomplexes for stability (Hou et al. 2007). Interestingly, the cellular levels of the weakly attached IFT-B2 protein IFT172 were not substantially affected, but a strong increase was observed for IFT-A proteins. Quantification of transcript levels showed an upregulation of mRNAs for all of the examined IFT-factors, which led to the model that *ift46* cells try to compensate for the loss of the IFT-B1 complex by upregulating expression of IFT genes. This upregulation leads to an increase in the abundance of IFT proteins that are not dependent on the presence of an intact IFT-B complex for stability (such as IFT-A components). However, IFT proteins whose stability depends on IFT46 are likely degraded in the absence of IFT46 and thus less abundant despite the up-regulated mRNA levels (Hou et al. 2007). Interestingly, the authors also identified an *ift46* suppressor mutant in which the carboxy-terminal IFT46 region was expressed, and found that in these cells the levels of IFT-B proteins were stabilized, which is in agreement with the fact that this domain is required for IFT-B1 complex stability (Fig. 2) (Lucker et al. 2010; Taschner et al. 2011, 2014). Intriguingly, the higher levels of IFT-A proteins did not return to wild-type levels in this strain for unknown reasons (Hou et al. 2007). A similar destabilization of IFT-B1 (IFT81 and IFT46) and IFT-B2 (IFT57 and IFT20) proteins as well as a slight increase in an IFT-A protein (IFT139) were recently observed in an IFT74 (*ift74-2*) *Chlamydomonas* mutant (Brown et al. 2015). IFT88 is not required for the stability of the IFT-B1 complex (Pazour et al. 2000; Richey and Qin 2012; Taschner et al. 2014), but as indicated above is part of the interface between IFT-B1 and IFT-B2 (Boldt et al. 2016; Taschner et al. 2016). Consistently, the IFT-B complex formed in the *ift88* mutant is smaller (Richey and Qin 2012), and IFT57 (IFT-B2 component) but not IFT81 (IFT-B1 component) is destabilized (Pazour et al. 2000). Again, the abundance of an IFT-A protein was increased (Pazour et al. 2000). Finally, knockdown of IFT70 in *Chlamydomonas* also led to a destabilization of IFT-B1 and an increase in IFT-A proteins (Fan et al. 2010).

Taken together, the observed destabilization of several factors within the IFT-B complex in the absence of a particular protein creates a serious problem for the dissection of the individual functions of a particular IFT component, and more subtle alterations compatible with IFT complex assembly (e.g., deletions of IFT protein domains dispensable for complex formation) will be necessary to obtain a detailed understanding of the contribution of individual factors to IFT.

FUNCTIONS OF IFT-B PROTEINS IN CARGO TRANSPORT, IFT REGULATION, AND SIGNALING

IFT46 and IFT56 Function in the Transport of Motility Factors

Despite the difficulties listed above, some examples of specific cargo transport functions of IFT proteins are available in the literature, and these discoveries were possible because the mutants were—at least to some extent—able to form cilia and flagella for further analyses. For instance, IFT46 has been linked to the specific transport of outer dynein arms (ODAs) into *Chlamydomonas* flagella. ODAs were previously shown to depend on IFT for axonemal assembly (Piperno et al. 1996) and to interact with the IFT complex (Qin et al. 2004). The short axonemes of *ift46* mutant flagella lacked ODAs; and, in a suppressor strain (expressing the carboxy-terminal IFT46 domain), overall flagellar assembly, but not axonemal ODA deficiency, was rescued (Hou et al. 2007). This implies a function of the IFT46 amino-terminal domain, which is not required for IFT-B1 complex stability (Taschner et al. 2014), in the transport of ODAs as cargos. In agreement with this notion, a direct interaction between IFT46 and the ODA assembly factor ODA16 was detected both by yeast-two-hybrid analysis as well as by pull-downs using recombinantly expressed proteins (Ahmed et al. 2008), but further examination of amino-terminal and carboxy-terminal IFT46 domains was not published. IFT46 knockdown in zebrafish was recently shown to lead to a similar shortening of cilia in various organs and to several phenotypes indicative of motility defects (e.g., left–right asymmetry defects), but analysis of the axonemal ultrastructure did not show an obvious ODA assembly defect (Lee et al. 2015).

In contrast to the ODA assembly defect in *ift46* mutants, the flagella in this mutant had no obvious deficiency in inner dynein arms (IDAs) (Hou et al. 2007). Transport of these motor complexes was shown to specifically require IFT56/TTC26 (Ishikawa et al. 2014). *ift56* mutant cells have normal levels of IFT and can assemble flagella of nearly wild-type length but display motility defects, and proteomic analysis of isolated *ift56* flagella showed a deficiency in certain types of IDAs (in addition to other motility-related components). A direct interaction between IDAs and IFT56 was not reported, but the authors speculate that IFT56 might bind IDAs via an IDA chaperone (TWI1) (Ishikawa et al. 2014). IDA deficiencies and ciliary motility defects were also observed in a mouse mutant (*hop* mouse) of IFT56 (Bryan 1983), and

knockdown of the zebrafish homolog similarly led to shorter cilia with abnormal motility (Ishikawa et al. 2014). Additional phenotypes in the *hop* mouse could not be explained by defects in motile cilia, and a recent study showed that nonmotile primary cilia in these mice have a specific defect in Hedgehog (Hh) signaling (Swiderski et al. 2014).

IFT27 and IFT25 Function in BBSome Transport and Regulate Hedgehog (Hh) Signaling

Another IFT-B protein with a specific function in Hh signaling is the atypical small GTPase IFT27 (Keady et al. 2012; Eguether et al. 2014). It was first reported that the IFT25 knockout mouse, unlike most other IFT-B mutant mice, neither had a detectable defect in ciliogenesis, nor did cilia on those cells have any obvious structural abnormalities (Keady et al. 2012). However, several phenotypes hinted toward defective Hh signaling, which requires coordinated ciliary import and export of signaling factors such as the membrane proteins Patched1 (Ptch1) and Smoothened (Smo), depending on the absence or presence of Hh ligand. Both Ptch1 and Smo accumulated in the cilia of *ift25*$^{-/-}$ MEFs, hinting at a role for IFT25 in the ciliary export of these proteins. However, IFT27 was destabilized in mutant cells, and so the defect could not be attributed specifically to the loss of IFT25 (Keady et al. 2012). A subsequent study showed that the IFT27 knockout mouse, despite having wild-type levels of IFT25, also had a similar export defect of Ptch1 and Smo (Eguether et al. 2014). Furthermore, the mutant cilia accumulated the multisubunit BBSome complex (Eguether et al. 2014; Mick et al. 2015), a well-known adapter for the transport of ciliary membrane proteins (Loktev et al. 2008; Jin et al. 2010). The authors explained their findings by a defective link between the IFT complex (IFT27) and the BBSome via the intermediate protein Lztfl1 (Seo et al. 2011), which leads to the accumulation of the BBSome and its membrane cargos (Ptch1 and Smo) in mutant cilia (Eguether et al. 2014). A similar requirement for IFT27 in

BBSome removal from cilia was reported in a separate study, but the authors attributed this to a function of the protein in the stabilization of the small GTPase Arl6/BBS3 required for assembly of the BBSome membrane coat inside cilia (Liew et al. 2014). GTP-bound Arl6/BBS3 recruits the BBSome components (via interaction with BBS1) to membranes to initiate the formation of a planar membrane coat (Jin et al. 2010; Mourão et al. 2014). After Arl6 GTP hydrolysis at the ciliary tip, the transient stabilization of the nucleotide-free form of Arl6 through binding to IFT27 was proposed to be essential for the formation of a BBSome coat and efficient retrograde movement and ciliary export (Liew et al. 2014). Mutations in IFT27 have been found in patients with Bardet–Biedl syndrome (BBS), strengthening the functional link between this GTPase and the BBSome (Aldahmesh et al. 2014).

The functions of the IFT27/25 complex in unicellular eukaryotes have not been investigated in similar detail, but the presence of a BBSome in *Chlamydomonas* and other single-celled organisms implies that a similar mechanism could be at work there for export of membrane cargos. IFT27 in *T. brucei* was reported to participate in retrograde IFT via import of the IFT-A complex and the retrograde dynein motor (Huet et al. 2014). This was not observed in mammalian cells and may represent a species-specific effect in *Trypanosoma*.

IFT172 Functions in IFT Turnaround at the Ciliary Tip

IFT172 is mutated in a temperature-sensitive mutant strain of *C. reinhardtii* (*fla11^{ts}*) (Pedersen et al. 2005). These cells have normal levels of anterograde and retrograde IFT at the permissive temperature (Iomini et al. 2001), but accumulate IFT proteins near the flagellar tip after a shift to restrictive conditions (Iomini et al. 2001; Pedersen et al. 2005, 2006) This suggests that, in the absence of functional IFT172, the transition from anterograde to retrograde IFT is compromised. The MT plus-end binding protein EB1 localizes to the flagellar tip in *Chlamydomonas* (Pedersen et al. 2003), interacts

with IFT172 in coimmunoprecipitation experiments (Pedersen et al. 2005), and the flagellar localization of EB1 is abolished in the *fla11^{ts}* strain at the restrictive temperature (Pedersen et al. 2003). Although these results would suggest a requirement for IFT172 in the anterograde transport of EB1, flagellar tip localization of EB1 does not depend on IFT (Pedersen et al. 2003) and has recently been shown to occur by a diffusion and capture mechanism (Harris et al. 2016). The underlying molecular mechanism by which IFT172 promotes IFT turnaround at the tip and the potential role for EB1 in this process remain elusive.

A function of IFT172 in the switch between anterograde and retrograde transport was also implied in a study using the ciliate *T. thermophila* where recombination can be used to create specific gene knockouts (Dave et al. 2009). Deletion of IFT172 abolished ciliogenesis, showing that this factor is essential for anterograde IFT (Tsao and Gorovsky 2008a). Wild-type cilia could be restored by rescue with full-length IFT172, but several IFT proteins accumulated at the ciliary tip when a part of the carboxyl terminus was missing (interestingly the same region that contains the *Chlamydomonas fla11^{ts}* mutation). The authors speculated that the carboxy-terminal region of the protein interacts with EB1 at the tip, and that such an interaction is required for efficient regulation of IFT turnaround (Tsao and Gorovsky 2008a).

IFT20 Is a Golgi-Associated Protein Involved in Membrane Protein Sorting to the Cilium

Unlike other IFT proteins, IFT20 not only localizes to the cilium but also to the Golgi complex in mammalian cells (Follit et al. 2006, 2008), where it is directly tethered to the Golgi protein GMAP210/Trip11 (Follit et al. 2008). Both IFT20 and GMAP210 are essential for embryonic development (Follit et al. 2008; Jonassen et al. 2008), and it was proposed that they act together in the sorting of membrane cargos destined for the cilium. Indeed, reduced levels of certain ciliary membrane proteins were observed after moderate knockdown of IFT20 (Follit et al. 2006) and in the cilia of embryonic kidney cells

Cite this article as *Cold Spring Harb Perspect Biol* doi: 10.1101/cshperspect.a028092

from GMAP210 knockout mice (Follit et al. 2008). Whether IFT20 also functions in Golgi to cilium transport in unicellular organisms such as *Chlamydomonas* remains to be shown.

IFT81/74 and IFT54 Bind αβ-Tubulin Dimers and MT

The most abundant IFT cargo is the αβ-tubulin heterodimer, the main component of the MTs of the ciliary axoneme. About 350,000 tubulin dimers are required to construct a full-length flagellum in *Chlamydomonas* (Bhogaraju et al. 2014), and this process takes ∼90 min to be complete (Rosenbaum et al. 1969). Tubulin was shown to move along the axoneme with IFT particles in *C. reinhardtii* (Craft et al. 2015) and *C. elegans* (Hao et al. 2011), and a first insight into how it is recognized as a cargo came from studies on the amino-terminal domains of IFT81 and IFT74. These regions are not required for stability of the IFT–B1 complex (Bhogaraju et al. 2013; Taschner et al. 2014), but instead form a tubulin-binding module. The amino-terminal IFT81 region was crystallized and displays the fold of a calponin-homology (CH) domain, known to recognize tubulin/MTs (Wei et al. 2007; Ciferri et al. 2008; Bhogaraju et al. 2013). Although the IFT81 CH-domain itself can bind to tubulin dimers, it does so with low affinity. In complex with the amino terminus of IFT74, however, this affinity is significantly increased (Bhogaraju et al. 2013). The current model is that the IFT81 CH-domain recognizes tubulin using conserved surface-exposed basic residues, and the highly positively charged amino terminus of IFT74 locks the cargo to the complex by electrostatic interactions with the negatively charged carboxy-terminal tubulin tails (E-hooks) (see Janke and Bulinski 2011). IFT81 proteins with mutations in the CH domain move normally by IFT in *T. brucei* flagella, but are unable to rescue the ciliogenesis defect cause by siRNA-mediated depletion of IFT81 in human cells (Bhogaraju et al. 2013). In *Chlamydomonas*, the expression of an IFT74 version lacking the amino-terminal positively charged region required for high-affinity tubulin binding did not abolish ciliogenesis, but slowed

down ciliogenesis (Brown et al. 2015). The same was shown to be true also for cells carrying mutations in the CH-domain of IFT81, but when IFT81N and IFT74N were mutated in combination the cells were able to assemble only very short flagella (Kubo et al. 2016).

Several lines of evidence imply that IFT81/74 may not be the only tubulin-binding site in the IFT complex. First, mutations in *C. elegans* IFT81 and IFT74 do not disrupt formation of sensory cilia (Kobayashi et al. 2007). Second, *Drosophila melanogaster* does not contain homologs of these proteins (Cole and Snell 2009), but nevertheless assembles sensory cilia by IFT (e.g., Sarpal et al. 2003). Last, the presence of only one dedicated tubulin-binding site in the IFT complex is in theory not sufficient to explain the fast initial kinetics of ciliogenesis in *Chlamydomonas* after deflagellation (Bhogaraju et al. 2014). Interestingly, three other IFT proteins within the IFT-B2 complex (IFT57, IFT54, and IFT38) contain amino-terminal CH domains just like IFT81 (Taschner et al. 2012; Schou et al. 2013), but two of those (IFT57 and IFT38) are engaged in intra-IFT interactions with IFT172 and IFT80 and do not bind tubulin (see Fig. 2) (Taschner et al. 2016). Recent results indicate that the CH domain of IFT54 is able to bind to tubulin with a similar affinity as IFT81/74, and the crystal structure revealed important residues for this interaction (Taschner et al. 2016). Two separate tubulin-binding sites within the IFT complex would be sufficient to explain the flagellar regeneration kinetics in *Chlamydomonas*, and future experiments should provide further insights into the transport of this important IFT cargo.

Although it is tempting to speculate that the IFT81/74 amino-terminal module and the IFT54 CH domain function in the transport of soluble tubulin dimers as an IFT cargo, alternative/additional functions of these domains cannot be excluded at this moment. Interaction with polymerized MTs has been reported both for IFT81N/74N (Bhogaraju et al. 2013) and IFT54 (Ling and Goeddel 2000; Taschner et al. 2016); and although it seems counterproductive for a complex that moves processively along MT tracks to also bind to them, a transient MT interaction could be involved in the localization of IFT complexes to the basal body region or may represent an intermediate stage in IFT turnaround at the ciliary tip. Alternatively, the MT binding function of these proteins could be linked to a nonciliary role in modulating the overall stability of cellular microtubules, as recently described for IFT54/TRAF3IP1 in mammalian cells (Berbari et al. 2011; Bizet et al. 2015).

ARCHITECTURE OF THE IFT-A COMPLEX

Although the IFT-A complex contains significantly fewer proteins than the IFT-B counterpart, much less information is available on IFT-A architecture and the specific functions of IFT-A subunits. Bioinformatic analysis predicted that four of the six IFT-A proteins (IFT144, 140, 122, and 121) share domain organization with membrane coat proteins (with amino-terminal WD40 β-propellers and an α-solenoid carboxy-terminal tail) (Taschner et al. 2012), suggesting that they evolved from a protocoatomer protein present in an early ancestral cell (Jékely and Arendt 2006; van Dam et al. 2013). A study using human cell lines showed that three of the IFT-A components (IFT144, IFT140, and IFT122) form a subcomplex independent of the remaining three subunits, and this subcomplex was termed IFT-A core (Mukhopadhyay et al. 2010). Experimental evidence regarding how these proteins interact is still missing, but their structural similarity to members of the coat protein complex I (COPI) makes it tempting to speculate that they form a trimeric complex via their amino-terminal β propellers (Faini et al. 2013). A subsequent study in *Chlamydomonas* using yeast two-hybrid and coexpression/pulldown assays confirmed the existence of this "core" complex and reported additional direct interactions within the core (IFT144/140, IFT140/122), between the remaining noncore subunits (IFT121/43, IFT139/121), as well as between the two subcomplexes (IFT122/121) (Behal et al. 2011). Our own efforts to express IFT-A proteins and to reconstitute this complex confirmed that IFT139, IFT121, and IFT43 form a stable complex and demonstrated

that IFT121 interacts with both IFT43 and IFT139 (M Taschner and E Lorentzen, unpubl.).

Functions of IFT-A Proteins in IFT-A Complex Stability and Cargo Transport

As mentioned earlier, mutations in IFT-A components typically result in stumpy flagella that accumulate IFT-B proteins similarly to what is observed in mutants of retrograde IFT motor subunits. This has been observed in model systems such as *C. reinhardtii* (for example, see Piperno et al. 1998; Iomini et al. 2009), *C. elegans* (for example, see Blacque et al. 2006; Efimenko et al. 2006), *T. thermophila* (Tsao and Gorovsky 2008b), *D. melanogaster* (Lee et al. 2008), and mouse (for example, see Tran et al. 2008; Cortellino et al. 2009; Liem et al. 2012). As described for the IFT-B complex, the similar mutant phenotypes of IFT-A mutants may be a result of IFT-A (sub)complex disruption. Mutation of IFT121 in *Chlamydomonas* destabilizes the other "noncore" IFT-A protein IFT139, leaving the core unaffected (Behal et al. 2011), similar to what was observed in mammalian cells (Mukhopadhyay et al. 2010). In an *ift122* mutant (with a mutation in an IFT-A core component), the other two core members were unaffected, but the noncore proteins IFT139 and IFT121 were depleted (Behal et al. 2011). It thus seems that the noncore proteins are more dependent on the core proteins rather than the other way around. Some analyses also suggested that although the IFT-A core can move to the ciliary tip independently of the other subunits the return of this core to the base of the cilium depends on IFT139, 121, and 43 (Mukhopadhyay et al. 2010; Bredrup et al. 2011).

Most stumpy cilia observed in IFT-A mutant mice are associated with ectopic, ligand-independent activation of the Hh pathway (for example, see Tran et al. 2008; Qin et al. 2011; Liem et al. 2012), in strong contrast to IFT-B mutants, which typically display defects in pathway activation (for examples, see Huangfu et al. 2003; Rix et al. 2011; Keady et al. 2012; Eguether et al. 2014; Swiderski et al. 2014). Interestingly, a mutant in the retrograde motor displayed similar short and bulgy cilia, but was deficient in Hh signaling (Ocbina et al. 2011), showing that the overactivation of the pathway cannot be explained solely by a defect in retrograde IFT. Indeed, the IFT-A core complex was shown to be important for the ciliary import of the protein TULP3 (Mukhopadhyay et al. 2010), a member of the Tubby protein family (Mukhopadhyay and Jackson 2011), which in turn mediates ciliary import of G-protein-coupled receptors (GPCRs), among which is the negative Hh regulator Gpr161 (Mukhopadhyay et al. 2013). Detailed information about which of the three IFT-A core proteins mediates the recognition of TULP3 is currently unavailable.

An example of a confirmed direct interaction between a specific IFT-A protein and a cargo came from studies involving the BBSome. IFT144 was shown to directly interact with BBS1 in mammalian cells, and hypomorphic mutations in *C. elegans* in IFT144/DYF-2 or in BBS1 that disrupt this interaction lead to exclusion of the BBSome from cilia and a defective return of IFT-B proteins from the ciliary tip (Wei et al. 2012). This was another example of the requirement of the IFT-A complex for ciliary cargo (i.e., the BBSome) import and provided important clues on the function of the BBSome in regulation of IFT particle assembly at the ciliary base and tip (see below).

ASSOCIATION BETWEEN IFT-A AND IFT-B COMPLEXES AND THE FORMATION OF IFT TRAINS

No specific direct interactions between IFT-A and IFT-B proteins have so far been demonstrated, and biochemical evidence suggests that such an interaction, if it occurs, is weak compared to intra-IFT-B and intra-IFT-A interactions. The initial characterization of the IFT complex showed that, although IFT-A and IFT-B cosedimented in sucrose density gradients at low ionic strength, a slight increase in NaCl concentration led to their dissociation (Cole et al. 1998). Coimmunoprecipitation attempts between subunits of the two complexes led to contradictory results. Although some studies were unable to detect coprecipiation of IFT-A and IFT-B subunits (for example, see Cole et al.

1998; Baker et al. 2003; Follit et al. 2009; Lechtreck et al. 2009b; Fan et al. 2010), other studies did detect an interaction (Qin et al. 2004; Pedersen et al. 2005), although coprecipitation seemed to be dependent on the antibodies and pulldown conditions used. High-throughput pulldowns and mass-spectrometry analysis did not detect direct interactions between IFT-A and IFT-B (Boldt et al. 2016), but a recent study using *Chlamydomonas* provided evidence that the amino-terminal part of the IFT-B1 component IFT74 could be linking IFT-B and IFT-A. In a mutant strain missing this IFT74 region, the IFT-A complex was unable to enter the flagellum, and IFT-B components accumulated at a tip as a result (Brown et al. 2015).

As mentioned earlier, anterograde IFT in *C. elegans* uses two separate kinesin motors, namely, the slow heterotrimeric and faster homodimeric kinesin II (OSM-3) (Snow et al. 2004; Prevo et al. 2015). Another study showed that in BBS mutant worms (*bbs-7* and *bbs-8*) the IFT-A complex moves at the slow speed of heterotrimeric kinesin II, whereas IFT-B is faster and seems to be transported by OSM-3. The authors concluded that instead of being tightly bound to each other IFT-A and B are physically held together by the BBSome (Ou et al. 2005a). Such a model would, however, not work for *Chlamydomonas*, in which the BBSome was shown to be substoichiometric to the IFT proteins (Lechtreck et al. 2009a). More information about the exact role of the BBSome in IFT complex formation was obtained later, again in work carried out in the nematode model. Hypomorphic mutations in BBS-1 and IFT144/DYF2 allow the BBSome to assemble IFT-A and -B complexes at the ciliary base and anterograde IFT-trains are properly formed, but the BBSome is no longer transported to the tip as an IFT cargo. According to the author's model, after disassembly of trains at the tip, the BBSome is not available to attach IFT-B particles to retrograde trains, leading to their accumulation at this location (Wei et al. 2012). In the *bbs-7* and *bbs-8* mutants used in the previous study (Ou et al. 2005a), BBSome function is also disrupted at the ciliary base, leading to dissociation of IFT-A and IFT-B already in the anterograde direction. Taken together, the data suggest that the BBSome mediates IFT-A and -B assembly at the base and tip of *C. elegans* sensory cilia.

In the early seminal studies of IFT in *Chlamydomonas*, IFT material was observed to oligomerize into long (several 100-nm) electron-dense IFT trains (also known as IFT rafts) (Kozminski et al. 1993, 1995). The most detailed view of IFT trains was published in an electron-tomography study in plastic-embedded *Chlamydomonas* flagella (Pigino et al. 2009). A structural building block (most likely corresponding to the IFT complex and associated cargos) was seen to form a dimer that repeated every 40 nm along the longitudinal axis of the flagellum, implying that the IFT complexes not only make longitudinal but also lateral contacts (Pigino et al. 2009). While it was initially speculated that these long assemblies represented IFT trains moving in the anterograde direction (Pigino et al. 2009), it now became clear that they, in fact, are stalled trains, and that both anterograde and retrograde trains are significantly shorter and display a different periodicity (Stepanek and Pigino 2016; Vannuccini et al. 2016). The molecular basis of IFT train formation is still not understood, and it is not clear whether the IFT complex itself is able to form trains or whether association with motors, cargos, or other accessory factors is required. Regardless of the exact mechanism of IFT train formation, it appears to be a highly regulated process as the length of IFT trains is adjusted according to flagellar length in *Chlamydomonas* (Engel et al. 2009b). It is currently not known whether IFT train formation is evolutionarily conserved in all ciliated organisms and is a requirement for ciliogenesis.

ASSOCIATION BETWEEN IFT COMPLEXES AND IFT MOTORS

Several regulated interactions between the IFT complexes and the motors are necessary to ensure proper cycling of IFT trains between the ciliary base and tip. Kinesin II motor(s) power(s) anterograde movement of complexes carrying inactive dynein 2 as a cargo. After reaching the tip, kinesin inactivation and dynein activa-

tion are a prerequisite for retrograde IFT. Interaction studies between IFT complex proteins and motors have so far not been very conclusive and, in some cases, contradictory (Baker et al. 2003; Qin et al. 2004; Follit et al. 2009; Liang et al. 2014). A direct interaction between heterotrimeric kinesin II and IFT20 was proposed for the mouse proteins based on a yeast two-hybrid assay (Baker et al. 2003), but this could not be confirmed in an independent study (Follit et al. 2009), and might have been a false positive interaction caused by coexpression of two coiled-coil proteins in the absence of their native binding partners. However, the IFT-B mutant phenotypes, often observed to have a general assembly defect, make it very likely that IFT-B attaches to the kinesin motor for anterograde transport. This notion was confirmed by studies in *Chlamydomonas* (Liang et al. 2014) and mouse (Baker et al. 2003), which found specific coimmunoprecipiations between IFT-B and heterotrimeric kinesin II. In the green alga, a specific phosphorylation event on FLA8 (kinesin II motor subunit, a homolog of Kif3b) was shown to disrupt the interaction between the motor and IFT-B and thus to be important for IFT regulation (Liang et al. 2014). At the ciliary tip, this phosphorylation leads to dissociation and inactivation of the anterograde motor, and inhibition of phosphorylation results in IFT complex accumulation at the tip (Liang et al. 2014). After dissociation, the kinesin motor is able to exit the *Chlamydomonas* flagellum independently of retrograde transport (Pedersen et al. 2006; Engel et al. 2012), indicating that it returns to the ciliary base via diffusion. This is in contrast to the retrograde dynein motor that needs heterotrimeric kinesin II for transport to the tip, and so it is a bona fide anterograde IFT cargo attached to IFT-B potentially via IFT172 (Pedersen et al. 2006; Williamson et al. 2011). Retrograde IFT seems to depend on docking of the dynein complex to IFT-A (Williamson et al. 2011), consistent with the similar flagellar phenotypes of IFT-A and dynein mutants (Pazour et al. 1998; Piperno et al. 1998; Porter et al. 1999; Perrone et al. 2003; Iomini et al. 2009).

Apart from the fact that anterograde IFT relies on two distinct kinesin motors in

C. elegans (Snow et al. 2004), another difference in this model system is the apparent attachment of heterotrimeric kinesin II to IFT-A instead of IFT-B (Ou et al. 2005a). Recently, a study showed that the heterotrimeric kinesin motor has an important function in importing IFT trains past the transition zone and into the ciliary compartment proper, after which the trains are "handed over" along the proximal segment to the faster homodimeric OSM-3 motor that transports them along the distal segment to the ciliary tip (Prevo et al. 2015). In contrast to the situation in *Chlamydomonas*, both kinesin motors in the nematode are retrograde IFT cargos (Prevo et al. 2015). Regarding direct attachment points between motors and cargos, it was found that Dyf-1 in *C. elegans* (the worm homolog of the IFT-B1 protein IFT70) is necessary for binding and/or activation of the OSM-3 motor, because in *dyf-1* mutants the IFT trains are moved only by heterotrimeric kinesin II, and the OSM-3 dependent distal singlet segment fails to form (Ou et al. 2005a). A direct interaction between the homologous proteins in mammals was suggested (Howard et al. 2013).

CONCLUDING REMARKS

Since the discovery of intraflagellar transport more than two decades ago, our understanding of this process has improved considerably. The identities of the main components of the IFT machinery are known, the architectures of these complexes are emerging, and our knowledge of cargo binding and IFT regulation is increasing. Nevertheless, much remains to be deciphered before we fully comprehend this transport mechanism. How do the 22 IFT proteins come together to form a multisubunit IFT complex in three dimensions? How do IFT complexes oligomerize into long IFT trains? What are the specific roles of the individual IFT subunits in cargo transport? How is cargo loading onto IFT trains regulated? What is the exact mechanism of motor attachment/activation, and how are these steps regulated to ensure unperturbed bidirectional IFT movement? A multidisciplinary approach to the study of IFT should provide answers to these questions.

ACKNOWLEDGMENTS

We apologize to our colleagues whose work could not be mentioned in this review because of space limitations. We thank the members of the Lorentzen laboratory for critical reading of the manuscript.

REFERENCES

Adams GM, Huang B, Luck DJ. 1982. Temperature-sensitive, assembly-defective flagella mutants of *Chlamydomonas reinhardtii*. *Genetics* **100**: 579–586.

Adhiambo C, Blisnick T, Toutirais G, Delannoy E, Bastin P. 2009. A novel function for the atypical small G protein Rab-like 5 in the assembly of the trypanosome flagellum. *J Cell Sci* **122**: 834–841.

Ahmed NT, Gao C, Lucker BF, Cole DG, Mitchell DR. 2008. ODA16 aids axonemal outer row dynein assembly through an interaction with the intraflagellar transport machinery. *J Cell Biol* **183**: 313–322.

Aldahmesh MA, Li Y, Alhashem A, Anazi S, Alkuraya H, Hashem M, Awaji AA, Sogaty S, Alkharashi A, Alzahrani S, et al. 2014. *IFT27*, encoding a small GTPase component of IFT particles, is mutated in a consanguineous family with Bardet–Biedl syndrome. *Hum Mol Genet* **23**: 3307–3315.

Baker SA, Freeman K, Luby-Phelps K, Pazour GJ, Besharse JC. 2003. IFT20 links kinesin II with a mammalian intraflagellar transport complex that is conserved in motile flagella and sensory cilia. *J Biol Chem* **278**: 34211–34218.

Behal RH, Miller MS, Qin H, Lucker B, Jones A, Cole DG. 2011. Subunit interactions and organization of the *Chlamydomonas reinhardtii* intraflagellar transport complex A. *J Biol Chem* **287**: 11689–11703.

Berbari NF, Kin NW, Sharma N, Michaud EJ, Kesterson RA, Yoder BK. 2011. Mutations in Traf3ip1 reveal defects in ciliogenesis, embryonic development, and altered cell size regulation. *Dev Biol* **360**: 66–76.

Bhogaraju S, Taschner M, Morawetz M, Basquin C, Lorentzen E. 2011. Crystal structure of the intraflagellar transport complex 25/27. *EMBO J* **30**: 1907–1918.

Bhogaraju S, Cajanek L, Fort C, Blisnick T, Weber K, Taschner M, Mizuno N, Lamla S, Bastin P, Nigg EA, et al. 2013. Molecular basis of tubulin transport within the cilium by IFT74 and IFT81. *Science* **341**: 1009–1012.

Bhogaraju S, Weber K, Engel BD, Lechtreck KF, Lorentzen E. 2014. Getting tubulin to the tip of the cilium: One IFT train, many different tubulin cargo-binding sites? *Bioessays* **36**: 463–467.

Bizet AA, Becker-Heck A, Ryan R, Weber K, Filhol E, Krug P, Halbritter J, Delous M, Lasbennes MC, Linghu B, et al. 2015. Mutations in TRAF3IP1/IFT54 reveal a new role for IFT proteins in microtubule stabilization. *Nat Commun* **6**: 8666.

Blacque OE, Leroux MR. 2006. Bardet-Biedl syndrome: An emerging pathomechanism of intracellular transport. *Cell Mol Life Sci* **63**: 2145–2161.

Blacque OE, Li C, Inglis PN, Esmail MA, Ou G, Mah AK, Baillie DL, Scholey JM, Leroux MR. 2006. The WD repeat-containing protein IFTA-1 is required for retrograde intraflagellar transport. *Mol Biol Cell* **17:** 5053–5062.

Boldt K, van Reeuwijk J, Lu Q, Koutroumpas K, Nguyen TM, Texier Y, van Beersum SE, Horn N, Willer JR, Mans DA, et al. 2016. An organelle-specific protein landscape identifies novel diseases and molecular mechanisms. *Nat Commun* **7:** 11491.

Brazelton WJ, Amundsen CD, Silflow CD, Lefebvre PA. 2001. The *bld1* mutation identifies the *Chlamydomonas osm-6* homolog as a gene required for flagellar assembly. *Curr Biol* **11:** 1591–1594.

Bredrup C, Saunier S, Oud MM, Fiskerstrand T, Hoischen A, Brackman D, Leh SM, Midtbø M, Filhol E, Bole-Feysot C, et al. 2011. Ciliopathies with skeletal anomalies and renal insufficiency due to mutations in the IFT-A gene *WDR19*. *Am J Hum Genet* **89:** 634–643.

Brown JM, Cochran DA, Craige B, Kubo T, Witman GB. 2015. Assembly of IFT trains at the ciliary base depends on IFT74. *Curr Biol* **25:** 1583–1593.

Bryan JH. 1983. Abnormal cilia in a male-sterile mutant mouse. *Virchows Arch A Pathol Anat Histopathol* **400:** 77–86.

Ciferri C, Pasqualato S, Screpanti E, Varetti G, Santaguida S, Reis Dos G, Maiolica A, Polka J, De Luca JG, De Wulf P, et al. 2008. Implications for kinetochore–microtubule attachment from the structure of an engineered Ndc80 complex. *Cell* **133:** 427–439.

Cole DG, Snell WJ. 2009. SnapShot: Intraflagellar transport. *Cell* **137:** 784–784.e1.

Cole DG, Chinn SW, Wedaman KP, Hall K, Vuong T, Scholey JM. 1993. Novel heterotrimeric kinesin-related protein purified from sea urchin eggs. *Nature* **366:** 268–270.

Cole DG, Diener DR, Himelblau AL, Beech PL, Fuster JC, Rosenbaum JL. 1998. *Chlamydomonas* kinesin-II-dependent intraflagellar transport (IFT): IFT particles contain proteins required for ciliary assembly in *Caenorhabditis elegans* sensory neurons. *J Cell Biol* **141:** 993–1008.

Cortellino S, Wang C, Wang B, Bassi MR, Caretti E, Champeval D, Calmont A, Jarnik M, Burch J, Zaret KS, et al. 2009. Defective ciliogenesis, embryonic lethality and severe impairment of the Sonic Hedgehog pathway caused by inactivation of the mouse complex A intraflagellar transport gene *Ift122/Wdr10*, partially overlapping with the DNA repair gene *Med1/Mbd4*. *Dev Biol* **325:** 225–237.

Craft JM, Harris JA, Hyman S, Kner P, Lechtreck KF. 2015. Tubulin transport by IFT is upregulated during ciliary growth by a cilium-autonomous mechanism. *J Cell Biol* **208:** 223–237.

Dave D, Wloga D, Gaertig J. 2009. Manipulating ciliary protein-encoding genes in *Tetrahymena thermophila*. *Methods Cell Biol* **93:** 1–20.

Deane JA, Cole DG, Seeley ES, Diener DR, Rosenbaum JL. 2001. Localization of intraflagellar transport protein IFT52 identifies basal body transitional fibers as the docking site for IFT particles. *Curr Biol* **11:** 1586–1590.

Efimenko E, Blacque OE, Ou G, Haycraft CJ, Yoder BK, Scholey JM, Leroux MR, Swoboda P. 2006. *Caenorhabditis elegans* DYF-2, an orthologue of human WDR19, is a component of the intraflagellar transport machinery in sensory cilia. *Mol Biol Cell* **17:** 4801–4811.

Eguether T, San Agustin JT, Keady BT, Jonassen JA, Liang Y, Francis R, Tobita K, Johnson CA, Abdelhamed ZA, Lo CW, et al. 2014. IFT27 links the BBSome to IFT for maintenance of the ciliary signaling compartment. *Dev Cell* **31:** 279–290.

Engel BD, Lechtreck KF, Sakai T, Ikebe M, Witman GB, Marshall WF. 2009a. Total internal reflection fluorescence (TIRF) microscopy of *Chlamydomonas* flagella. *Methods Cell Biol* **93:** 157–177.

Engel BD, Ludington WB, Marshall WF. 2009b. Intraflagellar transport particle size scales inversely with flagellar length: Revisiting the balance-point length control model. *J Cell Biol* **187:** 81–89.

Engel BD, Ishikawa H, Wemmer KA, Geimer S, Wakabayashi KI, Hirono M, Craige B, Pazour GJ, Witman GB, Kamiya R, et al. 2012. The role of retrograde intraflagellar transport in flagellar assembly, maintenance, and function. *J Cell Biol* **199:** 151–167.

Faini M, Beck R, Wieland FT, Briggs JAG. 2013. Vesicle coats: Structure, function, and general principles of assembly. *Trends Cell Biol* **23:** 279–288.

Fan ZC, Behal RH, Geimer S, Wang Z, Williamson SM, Zhang H, Cole DG, Qin H. 2010. *Chlamydomonas* IFT70/CrDYF-1 is a core component of IFT particle complex B and is required for flagellar assembly. *Mol Biol Cell* **21:** 2696–2706.

Fliegauf M, Benzing T, Omran H. 2007. When cilia go bad: Cilia defects and ciliopathies. *Nat Rev Mol Cell Biol* **8:** 880–893.

Follit JA, Tuft RA, Fogarty KE, Pazour GJ. 2006. The intraflagellar transport protein IFT20 is associated with the Golgi complex and is required for cilia assembly. *Mol Biol Cell* **17:** 3781–3792.

Follit JA, San Agustin JT, Xu F, Jonassen JA, Samtani R, Lo CW, Pazour GJ. 2008. The Golgin GMAP210/TRIP11 anchors IFT20 to the Golgi complex. *PLoS Genet* **4:** e1000315.

Follit JA, Xu F, Keady BT, Pazour GJ. 2009. Characterization of mouse IFT complex B. *Cell Motil Cytoskeleton* **66:** 457–468.

Hao L, Acar S, Evans J, Ou G, Scholey JM. 2009. Analysis of intraflagellar transport in *C. elegans* sensory cilia. *Methods Cell Biol* **93:** 235–266.

Hao L, Thein M, Brust-Mascher I, Civelekoglu-Scholey G, Lu Y, Acar S, Prevo B, Shaham S, Scholey JM. 2011. Intraflagellar transport delivers tubulin isotypes to sensory cilium middle and distal segments. *Nat Cell Biol* **13:** 790–798.

Harris JA, Liu Y, Yang P, Kner P, Lechtreck KF. 2016. Single-particle imaging reveals intraflagellar transport–independent transport and accumulation of EB1 in *Chlamydomonas* flagella. *Mol Biol Cell* **27:** 295–307.

Hou Y, Pazour GJ, Witman GB. 2004. A dynein light intermediate chain, D1bLIC, is required for retrograde intraflagellar transport. *Mol Biol Cell* **15:** 4382–4394.

Hou Y, Qin H, Follit JA, Pazour GJ, Rosenbaum JL, Witman GB. 2007. Functional analysis of an individual IFT protein: IFT46 is required for transport of outer dynein arms into flagella. *J Cell Biol* **176:** 653–665.

Howard PW, Jue SF, Maurer RA. 2013. Interaction of mouse TTC30/DYF-1 with multiple intraflagellar transport complex B proteins and KIF17. *Exp Cell Res* **319:** 2275–2281.

Huang B, Rifkin MR, Luck DJ. 1977. Temperature-sensitive mutations affecting flagellar assembly and function in *Chlamydomonas reinhardtii*. *J Cell Biol* **72:** 67–85.

Huangfu D, Liu A, Rakeman AS, Murcia NS, Niswander L, Anderson KV. 2003. Hedgehog signalling in the mouse requires intraflagellar transport proteins. *Nature* **426:** 83–87.

Huet D, Blisnick T, Perrot S, Bastin P. 2014. The GTPase IFT27 is involved in both anterograde and retrograde intraflagellar transport. *eLife* **3:** e02419.

Inglis PN, Ou G, Leroux MR, Scholey JM. 2007. The sensory cilia of *Caenorhabditis elegans*. *WormBook* doi: 10.1895/wormbook.1.126.2.

Insinna C, Humby M, Sedmak T, Wolfrum U, Besharse JC. 2009. Different roles for KIF17 and kinesin II in photoreceptor development and maintenance. *Dev Dyn* **238:** 2211–2222.

Iomini C, Babaev-Khaimov V, Sassaroli M, Piperno G. 2001. Protein particles in *Chlamydomonas* flagella undergo a transport cycle consisting of four phases. *J Cell Biol* **153:** 13–24.

Iomini C, Li L, Esparza JM, Dutcher SK. 2009. Retrograde intraflagellar transport mutants identify complex A proteins with multiple genetic interactions in *Chlamydomonas reinhardtii*. *Genetics* **183:** 885–896.

Ishikawa H, Marshall WF. 2015. Efficient live fluorescence imaging of intraflagellar transport in mammalian primary cilia. *Methods Cell Biol* **127:** 189–201.

Ishikawa H, Ide T, Yagi T, Jiang X, Hirono M, Sasaki H, Yanagisawa H, Wemmer KA, Stainier DY, Qin H, et al. 2014. TTC26/DYF13 is an intraflagellar transport protein required for transport of motility-related proteins into flagella. *eLife* **3:** e01566.

Janke C, Bulinski JC. 2011. Post-translational regulation of the microtubule cytoskeleton: Mechanisms and functions. *Nat Rev Mol Cell Biol* **12:** 773–786.

Jékely G, Arendt D. 2006. Evolution of intraflagellar transport from coated vesicles and autogenous origin of the eukaryotic cilium. *Bioessays* **28:** 191–198.

Jiang YY, Lechtreck K, Gaertig J. 2015. Total internal reflection fluorescence microscopy of intraflagellar transport in *Tetrahymena thermophila*. *Methods Cell Biol* **127:** 445–456.

Jin H, White SR, Shida T, Schulz S, Aguiar M, Gygi SP, Bazan JF, Nachury MV. 2010. The conserved Bardet–Biedl syndrome proteins assemble a coat that traffics membrane proteins to cilia. *Cell* **141:** 1208–1219.

Johnson KA, Rosenbaum JL. 1992. Polarity of flagellar assembly in *Chlamydomonas*. *J Cell Biol* **119:** 1605–1611.

Jonassen JA, San Agustin J, Follit JA, Pazour GJ. 2008. Deletion of IFT20 in the mouse kidney causes misorientation of the mitotic spindle and cystic kidney disease. *J Cell Biol* **183:** 377–384.

Katoh Y, Terada M, Nishijima Y, Takei R, Nozaki S, Hamada H, Nakayama K. 2016. Overall architecture of the intraflagellar transport (IFT)-B complex containing Cluap1/IFT38 as an essential component of the IFT-B peripheral subcomplex. *J Biol Chem* **291:** 10962–10975.

Keady BT, Samtani R, Tobita K, Tsuchya M, San Agustin JT, Follit JA, Jonassen JA, Subramanian A, Lo CW, Pazour GJ. 2012. IFT25 links the signal-dependent movement of Hedgehog components to intraflagellar transport. *Dev Cell* **22:** 940–951.

Kobayashi T, Gengyo-Ando K, Ishihara T, Katsura I, Mitani S. 2007. IFT-81 and IFT-74 are required for intraflagellar transport in *C. elegans*. *Genes Cells* **12:** 593–602.

Kozminski KG, Johnson KA, Forscher P, Rosenbaum JL. 1993. A motility in the eukaryotic flagellum unrelated to flagellar beating. *Proc Natl Acad Sci* **90:** 5519–5523.

Kozminski KG, Beech PL, Rosenbaum JL. 1995. The *Chlamydomonas* kinesin-like protein FLA10 is involved in motility associated with the flagellar membrane. *J Cell Biol* **131:** 1517–1527.

Kubo T, Brown JM, Bellve K, Craige B, Craft JM, Fogarty K, Lechtreck KF, Witman GB. 2016. The IFT81 and IFT74 N-termini together form the major module for intraflagellar transport of tubulin. *J Cell Sci* doi: 10.1242/jcs.187120.

Kunitomo H, Iino Y. 2008. *Caenorhabditis elegans* DYF-11, an orthologue of mammalian Traf3ip1/MIP-T3, is required for sensory cilia formation. *Genes Cells* **13:** 13–25.

Lechtreck KF, Johnson EC, Sakai T, Cochran D, Ballif BA, Rush J, Pazour GJ, Ikebe M, Witman GB. 2009a. The *Chlamydomonas reinhardtii* BBSome is an IFT cargo required for export of specific signaling proteins from flagella. *J Cell Biol* **187:** 1117–1132.

Lechtreck KF, Luro S, Awata J, Witman GB. 2009b. HA-tagging of putative flagellar proteins in *Chlamydomonas reinhardtii* identifies a novel protein of intraflagellar transport complex B. *Cell Motil Cytoskeleton* **66:** 469–482.

Lee E, Sivan-Loukianova E, Eberl DF, Kernan MJ. 2008. An IFT-A protein is required to delimit functionally distinct zones in mechanosensory cilia. *Curr Biol* **18:** 1899–1906.

Lee MS, Hwang KS, Oh HW, Ji-Ae K, Kim HT, Cho HS, Lee JJ, Yeong Ko J, Choi JH, Jeong YM, et al. 2015. IFT46 plays an essential role in cilia development. *Dev Biol* **400:** 248–257.

Liang Y, Pang Y, Wu Q, Hu Z, Han X, Xu Y, Deng H, Pan J. 2014. FLA8/KIF3B phosphorylation regulates kinesin-II interaction with IFT-B to control IFT entry and turnaround. *Dev Cell* **30:** 585–597.

Liem KF, Ashe A, He M, Satir P, Moran J, Beier D, Wicking C, Anderson KV. 2012. The IFT-A complex regulates Shh signaling through cilia structure and membrane protein trafficking. *J Cell Biol* **197:** 789–800.

Liew GM, Ye F, Nager AR, Murphy JP, Lee JS, Aguiar M, Breslow DK, Gygi SP, Nachury MV. 2014. The intraflagellar transport protein IFT27 promotes BBSome exit from cilia through the GTPase ARL6/BBS3. *Dev Cell* **31:** 265–278.

Lindemann CB, Lesich KA. 2010. Flagellar and ciliary beating: The proven and the possible. *J Cell Sci* **123:** 519–528.

Ling L, Goeddel DV. 2000. MIP-T3, a novel protein linking tumor necrosis factor receptor-associated factor 3 to the microtubule network. *J Biol Chem* **275:** 23852–23860.

Liu Q, Tan G, Levenkova N, Li T, Pugh EN, Rux JJ, Speicher DW, Pierce EA. 2007. The proteome of the mouse photoreceptor sensory cilium complex. *Mol Cell Proteomics* **6:** 1299–1317.

Loktev AV, Zhang Q, Beck JS, Searby CC, Scheetz TE, Bazan JF, Slusarski DC, Sheffield VC, Jackson PK, Nachury MV. 2008. A BBSome subunit links ciliogenesis, microtubule stability, and acetylation. *Dev Cell* **15:** 854–865.

Lucker BF, Behal RH, Qin H, Siron LC, Taggart WD, Rosenbaum JL, Cole DG. 2005. Characterization of the intraflagellar transport complex B core: Direct interaction of the IFT81 and IFT74/72 subunits. *J Biol Chem* **280:** 27688–27696.

Lucker BF, Miller MS, Dziedzic SA, Blackmarr PT, Cole DG. 2010. Direct interactions of intraflagellar transport complex B proteins IFT88, IFT52, and IFT46. *J Biol Chem* **285:** 21508–21518.

Marszalek JR, Ruiz-Lozano P, Roberts E, Chien KR, Goldstein LS. 1999. Situs inversus and embryonic ciliary morphogenesis defects in mouse mutants lacking the KIF3A subunit of kinesin-II. *Proc Natl Acad Sci* **96:** 5043–5048.

May SR, Ashique AM, Karlen M, Wang B, Shen Y, Zarbalis K, Reiter J, Ericson J, Peterson AS. 2005. Loss of the retrograde motor for IFT disrupts localization of Smo to cilia and prevents the expression of both activator and repressor functions of Gli. *Dev Biol* **287:** 378–389.

Mick DU, Rodrigues RB, Leib RD, Adams CM, Chien AS, Gygi SP, Nachury MV. 2015. Proteomics of primary cilia by proximity labeling. *Dev Cell* **35:** 497–512.

Mourão A, Nager AR, Nachury MV, Lorentzen E. 2014. Structural basis for membrane targeting of the BBSome by ARL6. *Nat Struct Mol Biol* **21:** 1035–1041.

Mukhopadhyay S, Jackson PK. 2011. The tubby family proteins. *Genome Biol* **12:** 225.

Mukhopadhyay S, Wen X, Chih B, Nelson CD, Lane WS, Scales SJ, Jackson PK. 2010. TULP3 bridges the IFT-A complex and membrane phosphoinositides to promote trafficking of G protein-coupled receptors into primary cilia. *Genes Dev* **24:** 2180–2193.

Mukhopadhyay S, Wen X, Ratti N, Loktev A, Rangell L, Scales SJ, Jackson PK. 2013. The ciliary G-protein-coupled receptor Gpr161 negatively regulates the Sonic Hedgehog pathway via cAMP signaling. *Cell* **152:** 210–223.

Nonaka S, Tanaka Y, Okada Y, Takeda S, Harada A, Kanai Y, Kido M, Hirokawa N. 1998. Randomization of left-right asymmetry due to loss of nodal cilia generating leftward flow of extraembryonic fluid in mice lacking KIF3B motor protein. *Cell* **95:** 829–837.

Ocbina PJR, Eggenschwiler JT, Moskowitz I, Anderson KV. 2011. Complex interactions between genes controlling trafficking in primary cilia. *Nat Genet* **43:** 547–553.

Omori Y, Zhao C, Saras A, Mukhopadhyay S, Kim W, Furukawa T, Sengupta P, Veraksa A, Malicki J. 2008. *Elipsa* is an early determinant of ciliogenesis that links the IFT particle to membrane-associated small GTPase Rab8. *Nat Cell Biol* **10:** 437–444.

Ostrowski LE, Blackburn K, Radde KM, Moyer MB, Schlatzer DM, Moseley A, Boucher RC. 2002. A proteomic analysis of human cilia: Identification of novel components. *Mol Cell Proteomics* **1:** 451–465.

Ou G, Blacque OE, Snow JJ, Leroux MR, Scholey JM. 2005a. Functional coordination of intraflagellar transport motors. *Nature* **436:** 583–587.

Ou G, Qin H, Rosenbaum JL, Scholey JM. 2005b. The PKD protein qilin undergoes intraflagellar transport. *Curr Biol* **15:** R410–R411.

Pasek RC, Berbari NF, Lewis WR, Kesterson RA, Yoder BK. 2012. Mammalian Clusterin associated protein 1 is an evolutionarily conserved protein required for ciliogenesis. *Cilia* **1:** 20.

Pazour GJ, Wilkerson CG, Witman GB. 1998. A dynein light chain is essential for the retrograde particle movement of intraflagellar transport (IFT). *J Cell Biol* **141:** 979–992.

Pazour GJ, Dickert BL, Vucica Y, Seeley ES, Rosenbaum JL, Witman GB, Cole DG. 2000. *Chlamydomonas IFT*88 and its mouse homologue, polycystic kidney disease gene *tg737*, are required for assembly of cilia and flagella. *J Cell Biol* **151:** 709–718.

Pazour GJ, Agrin N, Leszyk J, Witman GB. 2005. Proteomic analysis of a eukaryotic cilium. *J Cell Biol* **170:** 103–113.

Pedersen LB, Geimer S, Sloboda RD, Rosenbaum JL. 2003. The microtubule plus end-tracking protein EB1 is localized to the flagellar tip and basal bodies in *Chlamydomonas reinhardtii*. *Curr Biol* **13:** 1969–1974.

Pedersen LB, Miller MS, Geimer S, Leitch JM, Rosenbaum JL, Cole DG. 2005. *Chlamydomonas* IFT172 is encoded by *FLA11*, interacts with CrEB1, and regulates IFT at the flagellar tip. *Curr Biol* **15:** 262–266.

Pedersen LB, Geimer S, Rosenbaum JL. 2006. Dissecting the molecular mechanisms of intraflagellar transport in *Chlamydomonas*. *Curr Biol* **16:** 450–459.

Perrone CA, Tritschler D, Taulman P, Bower R, Yoder BK, Porter ME. 2003. A novel dynein light intermediate chain colocalizes with the retrograde motor for intraflagellar transport at sites of axoneme assembly in *Chlamydomonas* and mammalian cells. *Mol Biol Cell* **14:** 2041–2056.

Pigino G, Geimer S, Lanzavecchia S, Paccagnini E, Cantele F, Diener DR, Rosenbaum JL, Lupetti P. 2009. Electron-tomographic analysis of intraflagellar transport particle trains in situ. *J Cell Biol* **187:** 135–148.

Piperno G, Mead K. 1997. Transport of a novel complex in the cytoplasmic matrix of *Chlamydomonas* flagella. *Proc Natl Acad Sci* **94:** 4457–4462.

Piperno G, Mead K, Henderson S. 1996. Inner dynein arms but not outer dynein arms require the activity of kinesin homologue protein KHP1(FLA10) to reach the distal part of flagella in *Chlamydomonas*. *J Cell Biol* **133:** 371–379.

Piperno G, Siuda E, Henderson S, Segil M, Vaananen H, Sassaroli M. 1998. Distinct mutants of retrograde intraflagellar transport (IFT) share similar morphological and molecular defects. *J Cell Biol* **143:** 1591–1601.

Porter ME, Bower R, Knott JA, Byrd P, Dentler W. 1999. Cytoplasmic dynein heavy chain 1b is required for flagellar assembly in *Chlamydomonas*. *Mol Biol Cell* **10:** 693–712.

Prevo B, Mangeol P, Oswald F, Scholey JM, Peterman EJG. 2015. Functional differentiation of cooperating kinesin-2 motors orchestrates cargo import and transport in *C. elegans* cilia. *Nat Cell Biol* **17:** 1536–1545.

Qin H, Diener DR, Geimer S, Cole DG, Rosenbaum JL. 2004. Intraflagellar transport (IFT) cargo: IFT transports flagellar precursors to the tip and turnover products to the cell body. *J Cell Biol* **164:** 255–266.

Qin J, Lin Y, Norman RX, Ko HW, Eggenschwiler JT. 2011. Intraflagellar transport protein 122 antagonizes Sonic Hedgehog signaling and controls ciliary localization of pathway components. *Proc Natl Acad Sci* **108:** 1456–1461.

Reiter JF, Blacque OE, Leroux MR. 2012. The base of the cilium: Roles for transition fibres and the transition zone in ciliary formation, maintenance and compartmentalization. *EMBO Rep* **13:** 608–618.

Richey EA, Qin H. 2012. Dissecting the sequential assembly and localization of intraflagellar transport particle complex B in *Chlamydomonas. PLoS ONE* **7:** e43118.

Rix S, Calmont A, Scambler PJ, Beales PL. 2011. An *Ift80* mouse model of short rib polydactyly syndromes shows defects in hedgehog signalling without loss or malformation of cilia. *Hum Mol Genet* **20:** 1306–1314.

Rosenbaum JL, Child FM. 1967. Flagellar regeneration in protozoan flagellates. *J Cell Biol* **34:** 345–364.

Rosenbaum JL, Moulder JE, Ringo DL. 1969. Flagellar elongation and shortening in *Chlamydomonas.* The use of cycloheximide and colchicine to study the synthesis and assembly of flagellar proteins. *J Cell Biol* **41:** 600–619.

Santi-Rocca J, Chenouard N, Fort C, Lagache T, Olivo-Marin JC, Bastin P. 2015. Imaging intraflagellar transport in trypanosomes. *Methods Cell Biol* **127:** 487–508.

Sarpal R, Todi SV, Sivan-Loukianova E, Shirolikar S, Subramanian N, Raff EC, Erickson JW, Ray K, Eberl DF. 2003. *Drosophila* KAP interacts with the kinesin II motor subunit KLP64D to assemble chordotonal sensory cilia, but not sperm tails. *Curr Biol* **13:** 1687–1696.

Schou KB, Andersen JS, Pedersen LB. 2013. A divergent calponin homology (NN-CH) domain defines a novel family: Implications for evolution of ciliary IFT complex B proteins. *Bioinformatics* **30:** 899–902.

Seo S, Zhang Q, Bugge K, Breslow DK, Searby CC, Nachury MV, Sheffield VC. 2011. A novel protein LZTFL1 regulates ciliary trafficking of the BBSome and Smoothened. *PLoS Genet* **7:** e1002358.

Signor D, Wedaman KP, Orozco JT, Dwyer ND, Bargmann CI, Rose LS, Scholey JM. 1999. Role of a class DHC1b dynein in retrograde transport of IFT motors and IFT raft particles along cilia, but not dendrites, in chemosensory neurons of living *Caenorhabditis elegans. J Cell Biol* **147:** 519–530.

Snow JJ, Ou G, Gunnarson AL, Walker MRS, Zhou HM, Brust-Mascher I, Scholey JM. 2004. Two anterograde intraflagellar transport motors cooperate to build sensory cilia on *C. elegans* neurons. *Nat Cell Biol* **6:** 1109–1113.

Stepanek L, Pigino G. 2016. Microtubule doublets are double-track railways for intraflagellar transport trains. *Science* **352:** 721–724

Swiderski RE, Nakano Y, Mullins RF, Seo S, Bánfi B. 2014. A mutation in the mouse *ttc26* gene leads to impaired hedgehog signaling. *PLoS Genet* **10:** e1004689.

Taschner M, Bhogaraju S, Vetter M, Morawetz M, Lorentzen E. 2011. Biochemical mapping of interactions within the intraflagellar transport (IFT) B core complex: IFT52 binds directly to four other IFT-B subunits. *J Biol Chem* **286:** 26344–26352.

Taschner M, Bhogaraju S, Lorentzen E. 2012. Architecture and function of IFT complex proteins in ciliogenesis. *Differentiation* **83:** S12–S22.

Taschner M, Kotsis F, Braeuer P, Kuehn EW, Lorentzen E. 2014. Crystal structures of IFT70/52 and IFT52/46 provide insight into intraflagellar transport B core complex assembly. *J Cell Biol* **207:** 269–282.

Taschner M, Weber K, Mourão A, Vetter M, Awasthi M, Stiegler M, Bhogaraju S, Lorentzen E. 2016. Intraflagellar transport proteins 172, 80, 57, 54, 38, and 20 form a stable tubulin-binding IFT-B2 complex. *EMBO J* **35:** 773–790.

Tran PV, Haycraft CJ, Besschetnova TY, Turbe-Doan A, Stottmann RW, Herron BJ, Chesebro AL, Qiu H, Scherz PJ, Shah JV, et al. 2008. THM1 negatively modulates mouse sonic hedgehog signal transduction and affects retrograde intraflagellar transport in cilia. *Nat Genet* **40:** 403–410.

Tsao CC, Gorovsky MA. 2008a. Different effects of *Tetrahymena* IFT172 domains on anterograde and retrograde intraflagellar transport. *Mol Biol Cell* **19:** 1450–1461.

Tsao CC, Gorovsky MA. 2008b. *Tetrahymena* IFT122A is not essential for cilia assembly but plays a role in returning IFT proteins from the ciliary tip to the cell body. *J Cell Sci* **121:** 428–436.

van Dam TJP, Townsend MJ, Turk M, Schlessinger A, Sali A, Field MC, Huynen MA. 2013. Evolution of modular intraflagerllar transport from a coatomer-like progenitor. *Proc Natl Acad Sci* **110:** 6943–6948.

Vannuccini E, Paccagnini E, Cantele F, Gentile M, Dini D, Fino F, Diener D, Mencarelli C, Lupetti P. 2016. Two classes of short intraflagellar transport trains with different 3D structures are present in *Chlamydomonas* flagella. *J Cell Sci* **129:** 2064–2074.

Walther Z, Vashishtha M, Hall JL. 1994. The *Chlamydomonas FLA10* gene encodes a novel kinesin-homologous protein. *J Cell Biol* **126:** 175–188.

Wang Q, Pan J, Snell WJ. 2006. Intraflagellar transport particles participate directly in cilium-generated signaling in *Chlamydomonas. Cell* **125:** 549–562.

Wang Z, Fan ZC, Williamson SM, Qin H. 2009. Intraflagellar transport (IFT) protein IFT25 is a phosphoprotein component of IFT complex B and physically interacts with IFT27 in *Chlamydomonas. PLoS ONE* **4:** e5384.

Wei RR, Al-Bassam J, Harrison SC. 2007. The Ndc80/HEC1 complex is a contact point for kinetochore-microtubule attachment. *Nat Struct Mol Biol* **14:** 54–59.

Wei Q, Zhang Y, Li Y, Zhang Q, Ling K, Hu J. 2012. The BBSome controls IFT assembly and turnaround in cilia. *Nat Cell Biol* **14:** 950–957.

Williams CL, McIntyre JC, Norris SR, Jenkins PM, Zhang L, Pei Q, Verhey K, Martens JR. 2014. Direct evidence for BBSome-associated intraflagellar transport reveals distinct properties of native mammalian cilia. *Nat Commun* **5:** 5813.

Williamson SM, Silva DA, Richey E, Qin H. 2011. Probing the role of IFT particle complex A and B in flagellar entry and exit of IFT-dynein in *Chlamydomonas. Protoplasma* **249:** 851–856.

Zhao C, Malicki J. 2011. Nephrocystins and MKS proteins interact with IFT particle and facilitate transport of selected ciliary cargos. *EMBO J* **30:** 2532–2544.

Zhao C, Omori Y, Brodowska K, Kovach P, Malicki J. 2012. Kinesin-2 family in vertebrate ciliogenesis. *Proc Natl Acad Sci* **109:** 2388–2393.

Open Sesame: How Transition Fibers and the Transition Zone Control Ciliary Composition

Francesc R. Garcia-Gonzalo[1] and Jeremy F. Reiter[2]

[1]Departamento de Bioquímica, Facultad de Medicina, and Instituto de Investigaciones Biomédicas Alberto Sols UAM-CSIC, Universidad Autónoma de Madrid, 28029 Madrid, Spain

[2]Department of Biochemistry and Biophysics, and Cardiovascular Research Institute, University of California, San Francisco, San Francisco, California 94158

Correspondence: francesc.garcia@uam.es; jeremy.reiter@ucsf.edu

Cilia are plasma membrane protrusions that act as cellular propellers or antennae. To perform these functions, cilia must maintain a composition distinct from those of the contiguous cytosol and plasma membrane. The specialized composition of the cilium depends on the ciliary gate, the region at the ciliary base separating the cilium from the rest of the cell. The ciliary gate's main structural features are electron dense struts connecting microtubules to the adjacent membrane. These structures include the transition fibers, which connect the distal basal body to the base of the ciliary membrane, and the Y-links, which connect the proximal axoneme and ciliary membrane within the transition zone. Both transition fibers and Y-links form early during ciliogenesis and play key roles in ciliary assembly and trafficking. Accordingly, many human ciliopathies are caused by mutations that perturb ciliary gate function.

Cilia are born from the docking of a mother centriole to a membrane. The interaction of centriole and membrane depends on the centriolar distal appendages, whose tips anchor to the membrane. In the context of ciliogenesis, these membrane-anchored distal appendages are often referred to as transition fibers (TFs) (Fig. 1). After this early ciliogenic step, the nine concentric microtubule triplets in the centriole, now referred to as the basal body, act as templates onto which the nine axonemal microtubule doublets are constructed. Because axoneme growth and disassembly occur only at its distal tip, the basal-most axoneme is the first

to form. This proximal region of the cilium, known as the transition zone (TZ), begins at the distal end of the basal body and is defined by the presence of Y-links, bifid fibers connecting each TZ microtubule doublet to the adjacent membrane (Fig. 1). Perhaps because of their tethering to microtubules, ciliary gate membranes are particularly resistant to detergent extraction and harbor specialized proteinaceous structures like the ciliary necklace (Fig. 1) (Gilula and Satir 1972; Horst et al. 1987).

The ciliary gate sits at the juncture between the basal body and axoneme, between the plas-

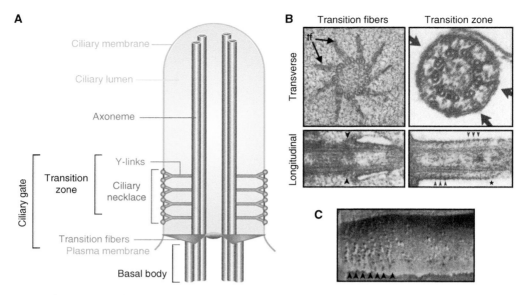

Figure 1. Ultrastructure of the ciliary gate. (*A*) Schematic showing the different parts of a cilium, with emphasis on the ciliary gate comprised by the transition fibers and the transition zone. (From Reiter et al. 2012; adapted, with permission, from the authors.) (*B*) Electron microscopic images of the transition fibers and transition zone. *Left* panels show transition fibers in transverse. (*Top*, from O'Toole et al. 2003; adapted, with permission, from the American Society for Cell Biology © 2003; and longitudinal [*bottom*] views from Tateishi et al. 2013; reprinted under a Creative Commons License [Attribution–Noncommercial–Share Alike 3.0 Unported license, as described at creativecommons.org/licenses/by-nc-sa/3.0].) Arrows point to transition fibers (tf). *Right* panels show transverse and longitudinal views of transition zones from rat photoreceptor connecting cilia (From data in Horst et al. 1987; adapted, with permission, from the authors.) Arrows point to Y-links. Asterisk marks the distal end of the transition zone (TZ). (*C*) Freeze-fracture etching of the ciliary necklace of a rat tracheal cilium. (Image adapted from data in Gilula and Satir 1972.) Arrowheads point to beads in the ciliary necklace.

ma membrane and ciliary membrane, and between the cytosol and ciliary lumen (Fig. 1), and thus is in a position to regulate proteins entering and exiting the cilium. A decade ago, little was known about this region other than its ultrastructure. The last 10 years, however, have changed that: proteomes and interactomes of the ciliary gate and several of its components have been generated, superresolution microscopy has helped localize several gate components, loss of function genetic studies have elucidated functional modules, and many ciliary gate genes have been linked to specific human ciliopathies. We are now at the point where an integrated understanding of how the ciliary gate controls ciliary composition can be anticipated. Below, we review the architecture, composition, function, and disease involvement of each ciliary

gate region. We conclude by identifying outstanding questions in the field.

TRANSITION FIBERS

Transition Fiber Architecture

Basal bodies start their lives as mother centrioles, distinguished from daughter centrioles by the distal appendages through which they dock to the vesicles that are eventually remodeled into the ciliary membrane (Sorokin 1962; Lu et al. 2015). When associated with the ciliary membrane or its precursors, centriolar distal appendages are referred to as transition fibers (TFs) or alar sheets. Although the former term has gained general acceptance, the latter is evocative of their shape: wing-like trapezoidal sheets

(Anderson 1972). There are nine TFs, each of which emerges from the distal portion of the C-tubule of the basal body and ends at an electron dense knob at the most proximal ciliary membrane (Fig. 1) (Gibbons and Grimstone 1960; Ringo 1967). Cross sections suggest that there is ~60 nm between adjacent TFs, enough so that a ribosome, but not a vesicle, might pass between them (Nachury et al. 2010). Because the basal body lumen is obstructed by centrin-2-containing electron dense structures (Fisch and Dupuis-Williams 2011), these spaces between TFs are likely to be the main passageways for macromolecules entering and exiting the cilium.

The TF is comprised of at least five proteins (Graser et al. 2007; Sillibourne et al. 2011; Joo et al. 2013; Tanos et al. 2013). Hierarchical localization analysis shows that Cep83 (also called Ccdc41) is required for the localization of Cep89 (also called Ccdc123 or Cep123) and Sclt1, and the latter is required for localizing Fbf1 and Cep164 (Tanos et al. 2013). Many of these core TF components are required for basal body docking to a membrane and ciliogenesis (Schmidt et al. 2012; Tanos et al. 2013; Joo et al. 2013; Tateishi et al. 2013; Burke et al. 2014; Ye et al. 2014). Another protein, Chibby (Cby1), is recruited to TFs by Cep164 and extends toward the proximal TZ, but is dispensable for at least some forms of ciliogenesis (Voronina et al. 2009; Burke et al. 2014; Lee et al. 2014). In turn, TF localization of all the above proteins requires components of the distal centriole, including Ofd1 and C2cd3, two proteins that, despite their shared role in building the distal appendages, exert opposite effects on centriole length (Singla et al. 2010; Thauvin-Robinet et al. 2014; Ye et al. 2014), and Odf2 (Cenexin), a Cby1 interactor required for the formation of both distal and subdistal centriolar appendages (Ishikawa et al. 2005; Steere et al. 2012; Tateishi et al. 2013). Superresolution microscopy confirms the localization of the above proteins to the TFs, and suggests that Cep89 and Cep164 localize along the length of TFs (Fig. 2A) (Sillibourne et al. 2011; Tanos et al. 2013; Lee et al. 2014; Yang et al. 2015). Despite these advances, how different TF components are spatially ar-

ranged to give rise to these structures remains to be determined.

Transition Fiber Functions

An early step in ciliogenesis is the recruitment of small vesicles to the distal appendages of a mother centriole, the distal appendage vesicles (DAVs). The DAVs fuse to form a larger ciliary vesicle that caps the distal centriole in a process that depends on EHD1 and EHD3 (Lu et al. 2015). Expansion of this ciliary vesicle gives rise to the ciliary membrane, which is assembled in parallel with the axoneme. Axoneme outgrowth is associated with removal from the distal mother centriole of CP110, a protein with both positive and negative roles in early ciliogenesis (Spektor et al. 2007; Tsang et al. 2008; Kobayashi et al. 2011; Yadav et al. 2016). Axoneme formation also involves the construction of the TZ and recruitment of intraflagellar transport (IFT) particles, the microtubule motor-driven complexes that mediate ciliary transport and assembly (Lu et al. 2015).

Disrupting distal appendage formation or function (e.g., interfering with Ofd1, C2cd3, Odf2, Cep164, or Cep83) blocks ciliogenesis and, in all cases where it has been examined, prevents DAV formation (Singla et al. 2010; Schmidt et al. 2012; Joo et al. 2013; Tanos et al. 2013; Tateishi et al. 2013; Burke et al. 2014; Thauvin-Robinet et al. 2014; Ye et al. 2014). Moreover, TFs are essential to remove CP110 from the distal centriole and recruit TZ and IFT proteins to the ciliary base (Deane et al. 2001; Schmidt et al. 2012; Joo et al. 2013; Tanos et al. 2013; Čajánek and Nigg 2014; Ye et al. 2014).

Cep164 is a key TF component that promotes ciliogenesis through several mechanisms. First, facilitated by Cby1, Cep164 interacts with Rabin8 to activate Rab8 and promote growth of the ciliary membrane (Knödler et al. 2010; Westlake et al. 2011; Feng et al. 2012; Schmidt et al. 2012; Burke et al. 2014; Lu et al. 2015). Second, Cep164 triggers axoneme extension by recruiting Tau tubulin kinase 2 (Ttbk2) (Goetz et al. 2012; Čajánek and Nigg 2014). For Cep164 to recruit Ttbk2, the phosphoinositide PI(4)P, which binds to Cep164 and inhibits its inter-action with Ttbk2, must first be depleted near the centriole by the PI(4)P 5-kinase PIPKIγ (Xu et al. 2016). Successful Ttbk2 recruitment by Cep164 leads to the removal of CP110 from the distal centriole and the recruitment of IFT components to TFs (Goetz et al. 2012; Čajánek and Nigg 2014). In the absence of Cep164, IFT-B component IFT88 is not recruited to the basal body, preventing ciliogenesis (Schmidt et al. 2012; Čajánek and Nigg 2014). Besides Cep164, other TF proteins also participate in recruiting IFT particles. For example, Fbf1 and Cep83 both bind IFT-B proteins (IFT54 and IFT20, respectively), help recruit them to the TFs, and participate in their crossing into the cilium (Wei et al. 2013; Ye et al. 2014). TF proteins are also needed to recruit TZ proteins to the forming cilium (Schmidt et al. 2012; Tanos et al. 2013; Lu et al. 2015).

Complementing their roles in ciliogenesis, TFs also function in regulating protein traffick-ing to mature cilia. For example, their role as docking stations for IFT particles en route to cilia holds true both during ciliogenesis and af-ter cilia are constructed (Deane et al. 2001; Wei et al. 2013). Likewise, Inpp5e, a lipid-modified phosphoinositide phosphatase involved in both ciliogenesis and ciliary trafficking, is recruited to the ciliary base by interacting with Cep164 (Humbert et al. 2012; Chávez et al. 2015; Gar-cia-Gonzalo et al. 2015; Xu et al. 2016).

Thus, TFs recruit a variety of proteins bound for the cilium, but whether TFs actively promote the movement of these proteins through the ciliary gate is unclear. The entry of IFT particles into cilia is not constant: larger amounts of IFT proteins enter the cilium after longer periods of accumulation at the ciliary base, a behavior similar to that of avalanches (Ludington et al. 2013). Whether this ava-lanche-like behavior depends on accumulation of IFT particles on the TFs remains unclear. TFs (or the spaces between them) might also play a role in the size-dependent entry of soluble pro-teins into cilia (Kee et al. 2012; Breslow et al. 2013; Lin et al. 2013). The TFs and TZ are good candidates for imparting sieve-like properties to the ciliary gate, but whether either or both do so

is unclear. While the trafficking of soluble pro-teins across the ciliary gate will be described more fully below, TFs affect this process by per-forming critical roles in both ciliogenesis and protein recruitment to the ciliary base.

TRANSITION ZONE

Transition Zone Architecture

Distal to the TFs, the axoneme becomes closely apposed to the ciliary membrane. This apposi-tion is at least partly mediated by Y-links connecting each microtubule doublet of the ax-oneme to the ciliary membrane (Fig. 1) (Wil-liams et al. 2011; Jensen et al. 2015). In trans-verse cross sections, Y-links appear as Y-shaped fibers with one end attached to the interface between the A and B tubules, and the other two ends docked to the ciliary membrane (Fig. 1). In three dimensions, Y-links were pro-posed to resemble individual champagne glass-es, and more recent electron tomographic re-constructions of *Caenorhabditis elegans* Y-links indicate that they form continuous sheets spanning the entire TZ length (Gilula and Satir 1972; Schouteden et al. 2015; Lambacher et al. 2016). Freeze-fracture electron microscopy of the transition zone membrane reveals circum-ferential bead-like intramembranous particles collectively resembling a necklace (Fig. 1) (Gi-lula and Satir 1972). In longitudinal sections, the lipid bilayer contacting the Y-links contains similar beads, suggesting that the ciliary neck-lace is physically associated with the outermost portion of the Y-links (Horst et al. 1987). The necklace need not consist of a single ring of beads: strand number appears to vary with cilia type, ranging from one in some fibroblasts to around 40 in retinal rod cells, whose connecting cilium constitutes an extended TZ (Horst et al. 1987; Fisch and Dupuis-Williams 2011). Neck-lace strand number correlates with the proxi-modistal extent of the Y-link region, further suggesting that there is an intimate connection between the two (Fisch and Dupuis-Williams 2011). However, the precise relationship be-tween individual Y-links and necklace beads re-mains unclear, as is the question of whether necklace strands are separate stacked rings or turns of a single spiral spanning the TZ (Reiter et al. 2012; Lambacher et al. 2016).

Many TZ proteins have been identified, most of which have been linked to ciliopathies. Several interactome studies have identified two biochemically distinct TZ protein complexes (Chih et al. 2011; Dowdle et al. 2011; Garcia-Gonzalo et al. 2011; Sang et al. 2011; Roberson et al. 2015). One of these, the NPHP complex, is mostly involved in the ciliopathy nephronophthisis (NPHP) and includes Nphp1, Nphp4, and Rpgrip1l (Mollet et al. 2005; Arts et al. 2007; Sang et al. 2011). A second complex, involved mostly in Meckel (MKS) and Joubert (JBTS) syndromes and referred to as the MKS complex, includes the three Tectonic proteins (Tctn1, 2, 3), the three B9 domain proteins (Mks1, B9d1, B9d2), the coiled-coil proteins Cc2d2a and Cep290, Ahi1 and the transmembrane proteins Tmem67, Tmem216, Tmem17, Tmem231, Tmem107, and possibly others such as Tmem237 and Tmem218 (Chih et al. 2011; Garcia-Gonzalo et al. 2011; Huang et al. 2011; Sang et al. 2011; Christopher et al. 2012; Barker et al. 2014; Roberson et al. 2015; Lambacher et al. 2016; Li et al. 2016; Shylo et al. 2016).

The MKS and NPHP complexes interact with each other through a network of looser connections (Fig. 3). An important hub in this network is Cep290, which is part of the MKS complex but also binds Nphp5, a basal body and TZ protein that associates with two NPHP complex components (Schäfer et al. 2008; Sang et al. 2011; Barbelanne et al. 2013; Barbelanne et al. 2015; Gupta et al. 2015). Inversin is another such hub, linking the MKS and NPHP complexes to the inversin/Nphp3/Nek8/Anks6/Anks3 complex (Sang et al. 2011; Hoff et al. 2013; Leettola et al. 2014; Czarnecki et al. 2015; Yakulov et al. 2015). Nevertheless, because this latter complex mostly localizes to the inversin compartment, immediately distal from the TZ, it is likely that some of these interactions occur at different times or involve different pools of the same protein (Shiba et al. 2009, 2010; Sang et al. 2011; Czarnecki et al. 2015).

Additional candidate TZ components identified in a proteomic analysis of isolated *Chlamydomonas* TZs include several endosomal sorting complex required for transport (ESCRT) proteins (Diener et al. 2015). Because ESCRT proteins mediate vesicle budding from the cell surface in other contexts, they may perform a similar role in cilia (Olmos and Carlton 2016). Consistently, cilia shed extracellular vesicles that, at least in *Chlamydomonas*, play functionally important roles (Hogan et al. 2009; Bakeberg et al. 2011; Wood et al. 2013; Maguire et al. 2015). ESCRT proteins also mediate the shedding of midbodies during cytokinesis, a process that topologically resembles autotomy, the shedding of *Chlamydomonas* flagella through microtubule severing at the TZ (Quarmby 2004; Olmos and Carlton 2016).

Other proteins and lipids may cooperate with the established TZ complexes. For example, septins 2 and 7, and possibly others, are part of a ring at or near the TZ that regulates protein access to cilia (Hu et al. 2010; Kim et al. 2010; Fliegauf et al. 2014). In addition to a distinct protein composition, the TZ domain may also have a distinct lipid composition: the phosphoinositide lipid $PI(4,5)P_2$ is restricted to the proximal ciliary membrane by the ciliary enzyme Inpp5e (Chávez et al. 2015; Garcia-Gonzalo et al. 2015; Jensen et al. 2015).

Biochemical and genetic approaches have been useful for identifying TZ complexes and their functions, and now superresolution microscopy is beginning to elucidate how they are organized to explain how the TZ controls ciliary composition (Fig. 2B–F) (Lee et al. 2014; Yang et al. 2015; Lambacher et al. 2016). For example, one form of superresolution microscopy, stimulated emission depletion (STED) microscopy, has shown that MKS (Tmem67, Ahi1) and NPHP (Rpgrip1l) complex proteins comprise discontinuous rings transverse to the ciliary axis with a pattern and diameter consistent with ciliary necklace localization (Fig. 2C) (Lee et al. 2014; Lambacher et al. 2016). Along the ciliary axis, MKS (Tctn2, Tmem67, Mks1) and NPHP (Rpgrip1l) complex proteins localize at the same level, 100–200 nm above the basal body in RPE-1 cells, matching the location of Y-links and ciliary necklace defined by electron microscopy (Fig.

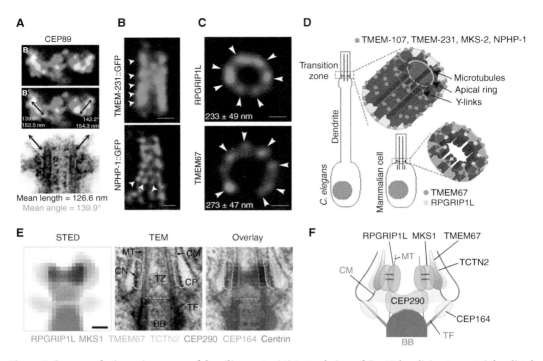

Figure 2. Superresolution microscopy of the ciliary gate. (*A*) Lateral view of Cep89 localizing to centriolar distal appendages, as observed using photoactivated localization microscopy (PALM). A comparable electron microscopic image is shown at the bottom. Black arrows indicate distal appendages. Red lines indicate the angle at which distal appendages emerge from the centriole. (From Sillibourne et al. 2011; adapted, with permission, from John Wiley and Sons © 2011.) (*B*) Longitudinal views of *Caenorhabditis elegans* sensory cilia transition zones with the indicated proteins localized using stimulated emission depletion (STED) microscopy. Arrowheads indicate axial and radial periodicity of TMEM-231::GFP and NPHP-1::GFP, respectively. Scale bars, 200 nm. (Panel *B* is from Lambacher et al. 2016; adapted, with permission, from Nature Publishing Group © 2015.) (*C*) Transverse STED images of cilia from human renal proximal tubule epithelial cells (RPTECs) stained for the indicated proteins. Arrowheads indicate radial periodicity. Ring diameters indicated in each image. Scale bars, 100 nm. (Panel *C* is from Lambacher et al. 2016; adapted, with permission, from Nature Publishing Group © 2015.) (*D*) Schematic of transition zone (TZ) protein localization in nematode and mammalian cilia. (Panel *D* is from Lambacher et al. 2016; adapted, with permission, from Nature Publishing Group © 2015.) (*E*) Summary of STED data from human retinal pigment epithelium (RPE) cell cilia. *Left* panel shows overlayed STED signals, colored as indicated. The *center* panel shows an electron microscopic image for reference in which ciliary gate structures are indicated. The *right* panel shows merge. Scale bar, 200 nm. (*F*) Schematic of the data shown in E. (Images E and F are from Yang et al. 2015; reprinted courtesy of Nature Publishing Group under Creative Commons CC-BY Licensing.) MT, Microtubule, CN, ciliary necklace; TF, transition fiber; CP, ciliary pocket; CM, ciliary membrane; BB, basal body.

2E−F) (Yang et al. 2015). As expected, given their transmembrane domains, Tctn2 and Tmem67 rings have the same diameter as that of the ciliary membrane, whereas Mks1, which can bind lipids (our unpublished data), has a slightly smaller diameter. The Rpgrip1l ring has an even smaller diameter, similar to that of the axoneme (Fig. 2C,E) (Yang et al. 2015; Lam-

bacher et al. 2016). Thus, these initial observations suggest that the NPHP complex is proximal to the transition zone microtubule sides, consistent with the tubulin-binding ability of both Nphp1 and 4, whereas the MKS complex is mostly associated with the ciliary membrane, as previously predicted from the domain structures of these proteins (Fig. 3) (Otto et al. 2003;

Mollet et al. 2005; Garcia-Gonzalo and Reiter 2012). Interestingly, whereas Tmem67 and Rpgrip1l form a single ring in the cilia of a human kidney cell line, longitudinal views of several *C. elegans* TZ proteins (Tmem107, Tmem231, Tmem216, and Nphp1) display a periodic pattern that is highly reminiscent of ciliary necklace strands or their underlying Y-links (Fig. 2B–D) (Lambacher et al. 2016). Consistent with these proteins being stably anchored to microtubules or other ciliary structures, fluorescence recovery after photobleaching (FRAP) of *C. elegans* MKS complex proteins (Tmem107, Tmem216, Cc2d2a) indicates that these proteins do not enter the TZ once it has been assembled (Lambacher et al. 2016).

STED microscopy of mammalian Cep290 positions it more proximally and centrally than Rpgrip1l, suggesting that it may localize to the TZ axoneme (Fig. 2E–F) (Yang et al. 2015). This may also be the case for the *C. elegans* ortholog of Cep290, which localizes near the TZ axoneme and is required to form the central cylinder, a structure on the lumenal side of the microtubule doublets (Schouteden et al. 2015). In contrast, *Chlamydomonas* Cep290 localizes to and is involved in the formation of Y-links (Craige et al. 2010). Whether these divergent observations reflect species-specific or cilia-type-specific differences, or whether Cep290, a large protein, can span the TZ axoneme and Y-links, remains to be clarified.

Roles of the Transition Zone in Ciliogenesis

Loss of function studies of TZ components have revealed two major roles for the TZ: ciliogenesis and the control of ciliary composition. Mouse *Nphp1* and *Nphp4* mutants, encoding core components of the NPHP complex, show only mild ciliogenic defects affecting two types of highly specialized cilia, photoreceptor cilia and sperm flagella (Jiang et al. 2008, 2009; Won et al. 2011). In contrast, many mutations in genes encoding MKS complex components cause severe ciliogenesis defects in mice, typically leading to embryonic lethality (Weatherbee et al. 2009; Chih et al. 2011; Dowdle et al. 2011; Garcia-Gonzalo et al. 2011; Sang et al. 2011). Despite the severity

of the associated embryological defects, many of these MKS mutations do not abrogate ciliogenesis, although this varies by cell type. For instance, *Tctn1* mutant embryos lack nodal cilia, have few and dysmorphic neural tube cilia, but display normal rates of ciliogenesis in the limb bud and perineural mesenchyme (Garcia-Gonzalo et al. 2011). In *C. elegans*, in which only select sensory neurons possess cilia, ciliogenesis is not affected by disruption of either NPHP or MKS proteins but is strongly perturbed when both an NPHP and an MKS complex component are disrupted (Williams et al. 2008, 2010, 2011; Huang et al. 2011; Jensen et al. 2015; Yee et al. 2015). Electron microscopy in these MKS NPHP double mutants reveals a ciliary gate wherein neither transition fibers nor Y-links are present, leading to membrane detachment from the ciliary axoneme (Huang et al. 2011; Williams et al. 2011; Jensen et al. 2015; Yee et al. 2015). This functional redundancy between the MKS and NPHP complexes is recapitulated in mice, where MKS NPHP double-mutant limbs display more severe disruption of ciliogenesis and patterning defects than those of single MKS mutants (Yee et al. 2015).

Interestingly, the above analysis does not apply to Rpgrip1l, which, despite being part of the biochemical NPHP complex, genetically behaves more like a member of the MKS complex (Sang et al. 2011; Williams et al. 2011). Indeed, *Rpgrip1l* mouse mutants die embryonically with a phenotype highly reminiscent of those seen for other MKS genes such as *Tctn1*, *Tctn2*, *B9d1*, or *Cc2d2a*, and for double mouse mutants affecting both MKS and NPHP module components (Vierkotten et al. 2007; Dowdle et al. 2011; Garcia-Gonzalo et al. 2011; Sang et al. 2011; Yee et al. 2015). Consistently, human *NPHP1* and *NPHP4* mutations have only been reported to cause nephronophthisis, whereas *RPGRIP1L* mutations can also cause more severe ciliopathies such as Meckel and Joubert syndromes (Arts et al. 2007; Delous et al. 2007; Wolf et al. 2007). In *C. elegans*, *Mks-5* is the single ortholog of both *RPGRIP1L* and *RPGRIP1* (an *RPGRIP1L* paralog, mutations in which cause Leber congenital amaurosis, a severe retinal ciliopathy) (Williams et al. 2011;

Li 2014; Jensen et al. 2015). Remarkably, although *C. elegans Mks-5* mutant TZs completely lack Y-links and fail to localize all known nematode MKS and NPHP module components to the TZ, these mutants still form largely normal ciliary axonemes (Jensen et al. 2015). Ciliary axonemes, however, are disrupted in *Mks-5 Nphp-4* double mutants, indicating that MKS-5 behaves genetically as an MKS module component (Williams et al. 2011). Unlike MKS complex proteins, *C. elegans* NPHP-1 and NPHP-4 localize not only to TZ but also to TFs, and it may be their function at the TFs that supports ciliogenesis in *Mks-5* mutants (Jensen et al. 2015). In single *Nphp-4* mutants, which also lack NPHP-1 in their cilia, TFs, Y-links and ciliary axonemes are present, although both the Y-links and the ciliary axonemes display defects apparent using electron microscopy (Winkelbauer et al. 2005; Jauregui et al. 2008; Jensen et al. 2015; Yee et al. 2015). Together, these data suggest that, in nematodes (but not in many mammalian cell types), ciliogenesis can be supported by TFs alone and that, hence, the TZ functions mainly in the control of cilia composition. How MKS-5/Rpgrip1l affects the MKS complex without being part of it remains unclear. Among the known Rpgrip1l interactors that could mediate effects on the MKS complex are CSPP and Nek4, both of which affect ciliogenesis and localize at or near the TZ (Patzke et al. 2010; Coene et al. 2011).

As mentioned above, vertebrate TZ proteins are required for ciliogenesis in a tissue-specific manner. The TZ forms early in ciliogenesis, soon after membrane docking of the basal body and before axoneme extension (Williams et al. 2011; Lu et al. 2015). Although it is unclear how TZ defects impair subsequent axoneme elongation, one possibility is that a damaged TZ interferes with IFT particle trafficking. Intriguingly, *C. elegans Mks-5* mutants display higher IFT speeds than controls in the proximal axoneme but lower speeds in the more distal axoneme, even though overall IFT particle flux remains unaltered (Jensen et al. 2015). The MKS-5-dependent slowing of IFT progress in the proximal axoneme could reflect an interaction between TZ and IFT proteins that restrains IFT speed; the TZ-dependent increase in IFT speed in the distal cilium may reflect a TZ-dependent change in the composition or regulation of the IFT particles (Boldt et al. 2011; Zhao and Malicki 2011). Perhaps TZ-dependent regulation of IFT composition or cargo is critical for some forms of ciliogenesis.

Functional ciliogenic interactions have also been found between TZ proteins (of both the MKS and NPHP complexes) and components of the BBSome, an IFT-associated complex (Nachury et al. 2007; Wei et al. 2012; Williams et al. 2014; Yee et al. 2015). For instance, limb bud cells from mice lacking either Tctn1 or Bbs1 are highly ciliated, whereas cells lacking both proteins grow virtually no cilia (Yee et al. 2015). Because the BBSome associates and may travel with IFT particles, one possible explanation for these results is that TZ proteins and the BBSome have overlapping functions in facilitating IFT-dependent ciliogenesis (Lechtreck et al. 2009; Wei et al. 2012; Williams et al. 2014). For instance, both the TZ and the BBSome could promote loading of ciliogenic cargo, such as tubulins, onto IFT particles (Craft et al. 2015). Alternatively, the BBSome could be required for IFT particles to enter the cilium across a defective TZ region. How ciliary trafficking complexes, such as IFT particles and the BBSome, interact with ciliary-gate components, including those of the transition zone, to facilitate ciliogenesis remains to be elucidated.

Roles of the Transition Zone in Soluble Protein Trafficking

Soluble proteins not associated with the membrane can enter the cilium by diffusion or active transport. Because the ciliary gate behaves as a size-exclusion filter, only small proteins can enter the cilium by diffusion (Kee et al. 2012; Breslow et al. 2013; Lin et al. 2013). Using a chemically inducible diffusion trap at cilia (C-IDTc), proteins with Stokes radii of up to 8 nm were seen to enter cilia with influx rates decreasing exponentially with increasing Stokes radii (Lin et al. 2013). These data, which were similar in

Cite this article as *Cold Spring Harb Perspect Biol* doi: 10.1101/cshperspect.a028134

both fibroblasts (NIH-3T3) and kidney tubule cells (IMCD3), are consistent with the presence of a molecular sieve spanning 1.4 μm along the axis of the ciliary base and with a mean mesh radius of 8 nm (Lin et al. 2013). Another study using both C-IDTc and in vitro reconstitution of soluble trafficking across the ciliary gate of IMCD3 cells detected no ciliary entry for proteins with Stokes radii above 4.5 nm (Breslow et al. 2013). However, ciliary entry in this study was monitored for 6–10 min, which, according to the study by Lin et al. (2013), corresponds to the time a 6- to 7-nm protein needs to reach half-maximal accumulation within cilia (Breslow et al. 2013). Hence, the data in these two studies are fairly consistent with one another when time is accounted for. The same can be

said of a third study in which 10 kDa (1.9 nm) but not 40 kDa (4.8 nm) dextran was found inside cilia 5 min after microinjection into the cell body of RPE cells (Kee et al. 2012). Last, a fourth study showing that trimeric GFP (Baum et al. 2014) can readily accumulate in the outer segments of transgenic frog photoreceptors can be explained by the fact that trimeric GFP (5.5 nm), when measured, had been moving across the connecting cilium for a long time (Najafi et al. 2012). Therefore, the ciliary gate appears to behave as a molecular sieve-like barrier, but the exact location and molecular makeup of this barrier are unknown. The data from the Lin et al. (2013) model (1.4-μm-long barrier with 8 nm mean mesh radius) suggests that the entire ciliary gate stretching from the TFs

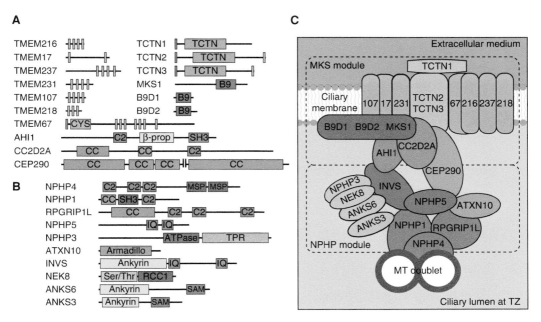

Figure 3. The transition zone interactome. (*A*) Domain structure of human MKS module proteins. Orange rectangles, transmembrane helices; purple rectangles, signal peptides; TCTN. Tectonic domains (also known as DUF1619); CYS, cysteine-rich domain; CC, coiled-coil; SH3, Src homology-3 motif; β-prop, β-propeller domain. With the exception of CEP290, the schematic lengths scale with amino acid number. (Panel *A* from Garcia-Gonzalo and Reiter 2012; adapted, with permission, from the authors.) (*B*) Domain structure of human NPHP module proteins, including those of the inversin subcomplex. MSP, major sperm protein domain; CC, coiled coil; SH3, Src-homology-3 motif; IQ, IQ calmodulin-binding motif; TPR, tetratricopeptide repeats; Armadillo, Armadillo-like domain; Ankyrin, ankyrin repeats; Ser/Thr, serine/threonine kinase domain; RCC1, RCC1-like domain; SAM, sterile α motif. (Panel *B* from Garcia-Gonzalo and Reiter 2012; adapted, with permission, from the authors.) (*C*) Model of the MKS and NPHP protein interaction network. TMEM proteins are only indicated by their number. The TCTN and B9 subcomplexes are shown as compact structures with the individual protein names indicated inside.

through the TZ acts as a sieve, and that the pore size is smaller than the distances between adjacent TFs or Y-links, possibly because of a non-electron dense meshwork spanning those gaps.

Unlike diffusion, which only small soluble proteins can use to efficiently enter cilia, microtubule motor-dependent transport into and out of cilia is unaffected by cargo size, and depends rather on the ability of the cargo to associate with motor complexes (reviewed in Garcia-Gonzalo and Reiter 2012). For ciliary entry, ciliary proteins can associate with heterotrimeric kinesin-2 (comprised of Kif3a, Kif3b, and Kap) or with monomeric Kif17, whereas ciliary exit is mediated by cytoplasmic dynein 2 (Verhey et al. 2011). Heterotrimeric kinesin-2, and many of its associated IFT proteins, are required for ciliogenesis, complicating investigation of how these complexes use force and protein–protein interactions to negotiate passage across the TZ. Apart from tubulins, some of the soluble cargos carried by these kinesin-IFT complexes in *Chlamydomonas* include the outer dynein arms and the partially preassembled radial spoke complexes, two large macromolecular assemblies required for flagellar motility (Qin et al. 2004; Hou et al. 2007; Bhogaraju et al. 2013; Craft et al. 2015). In contrast to trimeric kinesin-2, Kif17 is not required for ciliogenesis and its ciliary entry requires its interactions with both importin-β2 and Rab23 (Dishinger et al. 2010; Leaf and Von Zastrow 2015; Lim and Tang 2015). The requirement for importin-β2, together with the possible presence of some nucleoporins at the ciliary base and a Ran-GTP gradient across it, has revealed parallels between transport into cilia and transport into nuclei (Dishinger et al. 2010; Kee et al. 2012; Breslow et al. 2013; Takao et al. 2014). The extent of these parallels remains an area of investigation, and has been reviewed elsewhere (Garcia-Gonzalo and Reiter 2012; Reiter et al. 2012).

Roles of the Transition Zone in Membrane Trafficking

Some plasma membrane proteins cannot reach the ciliary base because they are anchored to the cytoskeleton, as is the case of cortical ac-tin-bound podocalyxin (Francis et al. 2011). Many of the plasma membrane proteins that do reach the ciliary base are excluded from the cilium by the TZ, as seen by their presence within cilia only on TZ disruption (Chih et al. 2011; Williams et al. 2011; Jensen et al. 2015; Roberson et al. 2015; Yee et al. 2015). How the TZ membrane prevents the ciliary accumulation of such membrane-associated proteins is unclear but may relate to a low fluidity of the TZ membrane associated with the abundance of membrane-associated anchors attached to the underlying microtubule cytoskeleton (as reviewed in Nachury et al. 2010). As discussed above, the TZ membrane contains the ciliary necklace, which may contain, among others, the many transmembrane proteins of the MKS complex (e.g., Tmem67, Tmem231, Tmem216, Tmem17, Tmem107, Tctn2, and Tctn3) (Fig. 3). Because septin2 is required for the TZ localization of some of these MKS complex proteins, it is possible that the ciliary septin ring adjoins the ciliary necklace, where it could stabilize MKS complex protein localization (Chih et al. 2011). Consistent with this possibility, both septin2 and B9d1 (an MKS complex protein requiring septin2 for its localization) are required for the ciliary gate to act as a membrane diffusion barrier, slowing down the rate at which transmembrane proteins cross the ciliary gate (Hu et al. 2010; Chih et al. 2011). More extensive reviews on the functions of septins in cilia can be found elsewhere (Hu and Nelson 2011; Garcia-Gonzalo and Reiter 2012; Reiter et al. 2012).

Despite the membrane diffusion barrier at the ciliary gate, many membrane-associated proteins still enter and exit cilia. As with cytosolic proteins, association with trafficking complexes, including microtubule motors and cargo adaptor complexes such as IFT-B, Tulp3-IFT-A, and the BBSome, can facilitate the localization of these proteins to the cilium (Nachury et al. 2007; Jin et al. 2010; Mukhopadhyay et al. 2010; Garcia-Gonzalo and Reiter 2012; Mukhopadhyay et al. 2013; Williams et al. 2014). Membrane-associated cargo proteins can interact with these trafficking complexes via ciliary localization sequences (CLSs). For instance, multiple ciliary G-protein-coupled receptors (GPCRs) rely on CLSs in their third intracellular loops that directly bind to the BBSome complex, which is itself associated with IFT particles, perhaps explaining why the Tulp3-IFTA complex is also required for the cilia localization of GPCRs (Berbari et al. 2008a,b; Jin et al. 2010; Mukhopadhyay et al. 2010, 2013; Loktev and Jackson 2013; Williams et al. 2014). Ciliary localization of a lipid-anchored ciliary protein, RP2, requires binding to importin-β2, which associates with the Kif17 motor (Dishinger et al. 2010; Hurd et al. 2011). Importin-β1 localizes to cilia and interacts with the ciliary form of Crumbs3, Crb3-CLPI, suggesting that it may also work in ciliary targeting (Fan et al. 2007). Kif17 seems to use ankyrin G as an adaptor to mediate the ciliary localization of cyclic nucleotide-gated channels in photoreceptors (Jenkins et al. 2006; Kizhatil et al. 2009).

In addition to using trafficking complexes to enter cilia across the TZ, membrane-associated proteins use trafficking complexes to exit cilia across the TZ. An important example of this involves IFT27, which dissociates from the IFT-B complex inside cilia and acts via Lztfl1 and Arl6 to recruit the BBSome and promote the removal of Hh-mediators Patched 1 and Smoothened (Seo et al. 2011; Eguether et al. 2014; Liew et al. 2014). Another Hh-mediator, Gpr161, accumulates in cilia lacking the BBSome component Bbs1, suggesting that Gpr161 also requires the BBSome to exit cilia (Yee et al. 2015). The BBSome, acting as an IFT cargo, also mediates the ciliary exit of other proteins, such as *Chlamydomonas* phospholipase D, indicating that the BBSome may have evolutionarily ancient roles in promoting the departure of proteins from cilia (Lechtreck et al. 2009, 2013).

Unlike integral membrane proteins, peripheral membrane proteins can be transiently solubilized by interaction with lipid-binding proteins that themselves cross the ciliary gate, thereby circumventing the TZ membrane when entering the cilium (Milenkovic et al. 2009). Two such transporter proteins, Unc119b and Pde6d, mediate ciliary targeting of myristoylated and prenylated cargo, respectively (Wright

et al. 2011; Humbert et al. 2012; Wätzlich et al. 2013; Thomas et al. 2014; Zhang et al. 2014; Lee and Seo 2015; Zhang et al. 2015). Unc119b recognizes the myristoylated amino-terminal residues of both Nphp3 and Cystin, thereby protecting them from interaction with the lipid bilayer (Wright et al. 2011). Only when bound to cargo can Unc119b enter the cilium across the TZ, where it encounters Arl3, a small GTP-binding protein that binds to and induces the dissociation of the Unc119-cargo complex, thereby allowing the myristoylated cargo to associate with the ciliary membrane (Wright et al. 2011; Zhang et al. 2011; Nakata et al. 2012). In turn, Arl3 activity is regulated by an Arl3-GAP (RP2) and an Arl3-GEF (Arl13b), which regulate its association with GDP or GTP and, therefore, its activity (Wright et al. 2011; Gotthardt et al. 2015). Similarly, Pde6d binds prenylated cargo such as Inpp5e, Grk1, and Pde6, whose intraciliary release depends on RPGR bringing together Arl3·GTP and the loaded cargo receptor (Baehr 2014; Gotthardt et al. 2015; Lee and Seo 2015). Intriguingly, double *Unc119 Pde6d* mouse knockouts show a partial rescue of the *Pde6d* mutant phenotype, suggesting that these trafficking systems antagonize each other at some level (Zhang et al. 2014). Despite these fascinating advances, how the lipid-binding proteins Unc119b and Pde6d themselves cross the ciliary gate remains to be elucidated. Other ciliary proteins, such as Arl13b or Calflagin/FCaBP, require palmitoylation or dual acylation (myristoylation and palmitoylation) of their amino-termini to enter cilia, but how these lipid modifications mediate their ciliary entry is not yet clear (Emmer et al. 2009, 2010; Cevik et al. 2010; Maric et al. 2011).

In summary, much is known about how membrane proteins use their CLSs to associate with specialized ciliary trafficking complexes that cross the ciliary gate, but little is known about how these complexes actually surmount the ciliary necklace, Y-links, septin ring, and other hurdles they may encounter along the way. Besides TZ-negotiating CLSs, ciliary proteins may also contain CLSs that function in the earlier step of bringing them to the ciliary base from other cellular locales. These CLSs have been reviewed elsewhere (Garcia-Gonzalo and Reiter 2012). In at least one case, photoreceptor cilia CLSs can even be dispensable, as the connecting cilium is the default destination of Golgi-derived membrane proteins (Baker et al. 2008; Gospe et al. 2010). Whether such bulk transport is a feature of other types of cilia remains to be seen.

Other Roles of the Transition Zone

Beyond ciliogenesis and trafficking, the TZ may play additional roles in cell morphogenesis and signaling. For example, in *C. elegans* sensory neurons, TZ proteins are required for anchoring the TZ at the tips of dendrites to the extracellular matrix, a process that occurs before axonemes sprout from those TZs (Schouteden et al. 2015). The presence of ESCRT proteins, involved in membrane budding elsewhere in the cell, may reflect a TZ role in ciliary vesicle shedding or in autotomy (ciliary shedding) (Quarmby et al. 2004; Diener et al. 2015). TZ proteins may also be involved in Wnt signaling. Tmem67 and the orphan receptor tyrosine kinase Ror2 form a TZ complex that binds Wnt5a, a noncanonical Wnt ligand (Abdelhamed et al. 2015). This mechanism may account for the overactivity of canonical Wnt signaling in *Tmem67* mutant cells and tissues (Adams et al. 2012; Abdelhamed et al. 2013, 2015; Leightner et al. 2013). Similar effects are observed in the absence of Tmem67-interactor and fellow MKS complex member Tmem216, and in the absence of Tmem237, whose *C. elegans* ortholog also behaves as a component of the MKS module (Valente et al. 2010; Huang et al. 2011). Forced expression of either Tmem67, Tmem216, or Tmem237 can suppress overactivation of canonical Wnt signaling in *Tmem67*-null cells, suggesting that these proteins have overlapping functions in restricting Wnt signaling (Huang et al. 2011). Another MKS complex protein, Ahi1, binds β-catenin and helps it translocate into the nucleus, and this activation is impaired by the presence of cilia, which sequester Ahi1 away from β-catenin (Lancaster et al. 2009, 2011). Whether Tmem67/216/237 exert their effects through Ahi1, or by affecting the β-catenin repressor Jade-1 via Nphp4, remains unclear (Borgal et al. 2012). These TZ mechanisms may provide an explanation as to why mutations that block ciliation increase responsiveness to canonical Wnt signals (Gerdes et al. 2007; Corbit et al. 2008).

CILIARY GATE DISEASES

The genes encoding the components of the ciliary gate are mutated in a diverse range of human ciliopathies (Table 1). Despite having been identified only recently, many TF and distal basal body components are associated with ciliopathies. Mutations in three of these genes (*SCLT1*, *OFD1*, and *C2CD3*) cause orofaciodigital syndrome, characterized by craniofacial dysmorphology and polydactyly (Ferrante et al. 2001, 2006; Adly et al. 2014; Thauvin-Robinet et al. 2014). Ofd1 and C2cd3 interact and colocalize in the distal centriole, where they have antagonistic influences on centriole length and are required for TF formation and ciliogenesis (Singla et al. 2010; Thauvin-Robinet et al. 2014; Ye et al. 2014). Similarly, Sclt1 is a TF protein required for ciliary assembly (Tanos et al. 2013). Mutations in two other TF genes, *CEP164* and *CEP83*, can cause nephronophthisis (NPHP), a cystic kidney disease, sometimes accompanied by other signs such as intellectual disability (Chaki et al. 2012; Failler et al. 2014). Both proteins are required for ciliogenesis and Cep164 can also participate in the DNA damage response, a function it can share with other NPHP proteins (Graser et al. 2007; Chaki et al. 2012; Tanos et al. 2013; Daly et al. 2016). As noted above, *C. elegans* orthologs of the NPHP proteins Nphp1 and Nphp4 localize in both TFs and TZ (Hildebrandt et al. 1997; Mollet et al. 2002; Otto et al. 2002; Jensen et al. 2015). However, whether the mammalian orthologs are also present or function at TFs is unclear. An individual with a homozygous deletion covering a portion of *CEP89* displayed impaired cognitive and neuronal function, but this may be caused by a mitochondrial and not a TF role of Cep89 (van Bon et al. 2013). No

Cite this article as *Cold Spring Harb Perspect Biol* doi: 10.1101/cshperspect.a028134

Table 1. Ciliary-gate diseases

Human gene	Loc	MKS	JBTS	NPHP	SLSN	OFD	BBS	LCA
MKS1	TZ	MKS1	-				BBS13	
TMEM216	TZ	MKS2	JBTS2					
TMEM67	TZ	MKS3	JBTS6	NPHP11				
CEP290	TZ	MKS4	JBTS5	NPHP6	SLSN6		BBS14	LCA10
RPGRIP1L	TZ	MKS5	JBTS7	NPHP8				
CC2D2A	TZ	MKS6	JBTS9					
NPHP3	IC	MKS7	-	NPHP3				
TCTN2	TZ	MKS8	JBTS24					
B9D1	TZ	MKS9	-					
B9D2	TZ	MKS10	-					
TMEM231	TZ	MKS11	JBTS20			OFD3		
AHI1	TZ	-	JBTS3					
NPHP1	TZ	-	JBTS4	NPHPI	SLSN1			
OFD1	DC	-	JBTS10			OFD1		
TCTN1	TZ	-	JBTS13					
TMEM237	TZ	-	JBTS14					
TMEM138	TZ		JBTS16					
TCTN3	TZ	-	JBTS18			OFD4		
INVS	IC	-	-	NPHP2				
NPHP4	TZ	-		NPHP4	SLSN4			
NPHP5	TZ	-	-	NPHP5	SLSN5			
NEK8	IC	-	-	NPHP9				
CEP164	TF	-	-	NPHP15				
ANKS6	IC	-	-	NPHP16				
CEP83	TF	-	-	NPHP1 8				
C2CD3	DC	-	-	NPHP18		OFDM		

Genes in the left column encode proteins mentioned in the text. As shown in the second column, these proteins localize (Loc) to the transition zone (TZ), transition fibers (TF), distal centriole (DC), or inversin compartment (IC). See the text for details and references. Columns three through nine indicate which of the genes cause, when mutated, the following ciliopathies: Meckel–Gruber syndrome (MKS), Joubert syndrome (JBTS), nephronophthisis (NPHP), Senior–Løken syndrome (SLSN), orofaciodigital syndrome (OFD), Bardet–Biedl syndrome (BBS), and Leber congenital amaurosis (LCA). The numbers correspond to the locus number for each disease. The information in this table was compiled from www.omim.org (last accessed January 30, 2016).

ciliopathic mutations have yet been identified for *FBF1*, *ODF2*, and *CBY1*.

Human pathogenic mutations have been identified for most of the genes encoding TZ proteins. The MKS complex derives its name from Meckel–Gruber syndrome (MKS), an extremely severe ciliopathy, leading to perinatal lethality and characterized by occipital encephalocele, polycystic kidneys, and polydactyly (Hildebrandt et al. 2011). MKS can arise from mutations in *MKS1*, *TMEM216*, *TMEM67*, *CEP290*, *RPGRIP1L*, *CC2D2A*, *NPHP3*, *TCTN2*, *B9D1*, *B9D2*, *TMEM231*, *KIF14*, *TCTN3*, and *TMEM107* (Thomas et al. 2012; Shaheen et al. 2013; Filges et al. 2014; Valente et al. 2014;

Roberson et al. 2015; Shaheen et al. 2015). Of these, all except *RPGRIP1L*, *NPHP3*, and *KIF14* encode identified components of the MKS complex (Chih et al. 2011; Dowdle et al. 2011; Garcia-Gonzalo et al. 2011; Sang et al. 2011; Roberson et al. 2015; Lambacher et al. 2016). Some of these genes (i.e., *TMEM216*, *TMEM67*, *CEP290*, *RPGRIP1L*, *CC2D2A*, *TCTN2*, *TCTN3*, and *TMEM231*) are also mutated in a milder ciliopathy, Joubert syndrome (JBTS), defined by a specific cerebellar malformation, the molar tooth sign, but usually also including polydactyly (Valente et al. 2014; Huppke et al. 2015). The differences in phenotype and severity of these two recessive

Mendelian syndromes suggest that MKS results from homozygous null mutations in these genes, whereas JBTS results from the involvement of at least one hypomorphic allele (Delous et al. 2007; Mougou-Zerelli et al. 2009; Iannicelli et al. 2010). Hence, MKS may result from more severe dysfunction of the ciliary gate and ciliogenesis than JBTS (Garcia-Gonzalo et al. 2011). Alternatively, coinherited genetic modifiers may account for the phenotypic differences (Roberson et al. 2015). TZ genes are also mutated in NPHP, as is the case for *NPHP1*, *NPHP4*, *RPGRIP1L*, *TMEM67*, and *CEP290* (Hurd and Hildebrandt 2011). Despite the extensive overlap between MKS-, JBTS-, and NPHP-associated genes, the extent to which these syndromes arise from disruption of different functions of the TZ ciliary-gating mechanism is unclear. TZ proteins are mutated in yet other ciliopathies, like OFD, further extending the phenotypic consequences of perturbed ciliary gate function (Thomas et al. 2012; Roberson et al. 2015). As the discrete cellular functions of TZ proteins become clearer, a better understanding will emerge of how different perturbations of their functions act alone or in combination to give rise to diverse ciliopathies.

CONCLUDING REMARKS

The last decade has seen great advances in our molecular understanding of the ciliary gate, and the pace of discovery only seems to be increasing. An already visible trend in the field includes the widespread application of superresolution microscopy, structural biology, and reconstitution to obtain high-resolution maps of ciliary gate structures such as TFs, Y-links, and ciliary necklaces (Fig. 2). As this information starts to be integrated, a more detailed picture of ciliary gate dynamics and function will emerge. Another trend is an increasing understanding of how the molecular ciliary composition and structure differs between tissues and species. The lipid composition and function of the transition zone is only starting to be uncovered: recent studies have revealed the presence of a $PI(4,5)P_2$ gradient at the ciliary base that affects trafficking of ciliary membrane proteins and is

dependent on TZ proteins (Chávez et al. 2015; Garcia-Gonzalo et al. 2015; Jensen et al. 2015). The next years should see further developments regarding phosphoinositides and other lipids at the ciliary gate, such as the ones that may bestow it with lipid raft-like properties (Emmer et al. 2010; Follit et al. 2010); and, of course, in this exciting and fast-moving field, one should always expect the unexpected: new ciliary gate functions, new exceptions for previously established rules, and new ways in which the ciliary gate controls the diverse functions of motile and signaling cilia.

REFERENCES

Abdelhamed ZA, Wheway G, Szymanska K, Natarajan S, Toomes C, Inglehearn C, Johnson CA. 2013. Variable expressivity of ciliopathy neurological phenotypes that encompass Meckel–Gruber syndrome and Joubert syndrome is caused by complex de-regulated ciliogenesis, Shh and Wnt signalling defects. *Hum Mol Genet* **22:** 1358–1372.

Abdelhamed ZA, Natarajan S, Wheway G, Inglehearn CF, Toomes C, Johnson CA, Jagger DJ. 2015. The Meckel–Gruber syndrome protein TMEM67 controls basal body positioning and epithelial branching morphogenesis in mice via the non-canonical Wnt pathway. *Dis Model Mech* **8:** 527–541.

Adams M, Simms RJ, Abdelhamed Z, Dawe HR, Szymanska K, Logan CV, Wheway G, Pitt E, Gull K, Knowles MA, et al. 2012. A meckelin-filamin A interaction mediates ciliogenesis. *Hum Mol Genet* **21:** 1272–1286.

Adly N, Alhashem A, Ammari A, Alkuraya FS. 2014. Ciliary genes TBC1D32/C6orf170 and SCLT1 are mutated in patients with OFD type IX. *Hum Mutat* **35:** 36–40.

Anderson RG. 1972. The three-dimensional structure of the basal body from the rhesus monkey oviduct. *J Cell Biol* **54:** 246–265.

Arts HH, Doherty D, van Beersum SE, Parisi MA, Letteboer SJ, Gorden NT, Peters TA, Märker T, Voesenek K, Kartono A, et al. 2007. Mutations in the gene encoding the basal body protein RPGRIP1L, a nephrocystin-4 interactor, cause Joubert syndrome. *Nat Genet* **39:** 882–888.

Baehr W. 2014. Membrane protein transport in photoreceptors: The function of PDEδ: The Proctor lecture. *Invest Ophthalmol Vis Sci* **55:** 8653–8666.

Bakeberg JL, Tammachote R, Woollard JR, Hogan MC, Tuan HF, Li M, van Deursen JM, Wu Y, Huang BQ, Torres VE, et al. 2011. Epitope-tagged Pkhd1 tracks the processing, secretion, and localization of fibrocystin. *J Am Soc Nephrol* **22:** 2266–2277.

Baker SA, Haeri M, Yoo P, Gospe SM III, Skiba NP, Knox BE, Arshavsky VY. 2008. The outer segment serves as a default destination for the trafficking of membrane proteins in photoreceptors. *J Cell Biol* **183:** 485–498.

Barbelanne M, Song J, Ahmadzai M, Tsang WY. 2013. Pathogenic NPHP5 mutations impair protein interaction

Cite this article as *Cold Spring Harb Perspect Biol* doi: 10.1101/cshperspect.a028134

with Cep290, a prerequisite for ciliogenesis. *Hum Mol Genet* **22:** 2482–2494.

Barbelanne M, Hossain D, Chan DP, Peränen J, Tsang WY. 2015. Nephrocystin proteins NPHP5 and Cep290 regulate BBSome integrity, ciliary trafficking and cargo delivery. *Hum Mol Genet* **24:** 2185–2200.

Barker AR, Renzaglia KS, Fry K, Dawe HR. 2014. Bioinformatic analysis of ciliary transition zone proteins reveals insights into the evolution of ciliopathy networks. *BMC Genomics* **15:** 531.

Baum M, Erdel F, Wachsmuth M, Rippe K. 2014. Retrieving the intracellular topology from multi-scale protein mobility mapping in living cells. *Nat Commun* **5:** 4494.

Berbari NF, Lewis JS, Bishop GA, Askwith CC, Mykytyn K. 2008a. Bardet–Biedl syndrome proteins are required for the localization of G protein–coupled receptors to primary cilia. *Proc Natl Acad Sci* **105:** 4242–4246.

Berbari NF, Johnson AD, Lewis JS, Askwith CC, Mykytyn K. 2008b. Identification of ciliary localization sequences within the third intracellular loop of G protein–coupled receptors. *Mol Biol Cell* **19:** 1540–1547.

Bhogaraju S, Cajanek L, Fort C, Blisnick T, Weber K, Taschner M, Mizuno N, Lamla S, Bastin P, Nigg EA, et al. 2013. Molecular basis of tubulin transport within the cilium by IFT74 and IFT81. *Science* **341:** 1009–1012.

Boldt K, Mans DA, Won J, van Reeuwijk J, Vogt A, Kinkl N, Letteboer SJ, Hicks WL, Hurd RE, Naggert JK, et al. 2011. Disruption of intraflagellar protein transport in photoreceptor cilia causes Leber congenital amaurosis in humans and mice. *J Clin Invest* **121:** 2169–2180.

Borgal L, Habbig S, Hatzold J, Liebau MC, Dafinger C, Sacarea I, Hammerschmidt M, Benzing T, Schermer B. 2012. The ciliary protein nephrocystin-4 translocates the canonical Wnt regulator Jade-1 to the nucleus to negatively regulate β-catenin signaling. *J Biol Chem* **287:** 25370–25380.

Breslow DK, Koslover EF, Seydel F, Spakowitz AJ, Nachury MV. 2013. An in vitro assay for entry into cilia reveals unique properties of the soluble diffusion barrier. *J Cell Biol* **203:** 129–147.

Burke MC, Li FQ, Cyge B, Arashiro T, Brechbuhl HM, Chen X, Siller SS, Weiss MA, O'Connell CB, Love D, et al. 2014. Chibby promotes ciliary vesicle formation and basal body docking during airway cell differentiation. *J Cell Biol* **207:** 123–137.

Čajánek L, Nigg EA. 2014. Cep164 triggers ciliogenesis by recruiting Tau tubulin kinase 2 to the mother centriole. *Proc Natl Acad Sci* **111:** E2841–E2850.

Cevik S, Hori Y, Kaplan OI, Kida K, Toivenon T, Foley-Fisher C, Cottell D, Katada T, Kontani K, Blacque OE. 2010. Joubert syndrome Arl13b functions at ciliary membranes and stabilizes protein transport in *Caenorhabditis elegans. J Cell Biol* **188:** 953–969.

Chaki M, Airik R, Ghosh AK, Giles RH, Chen R, Slaats GG, Wang H, Hurd TW, Zhou W, Cluckey A, et al. 2012. Exome capture reveals ZNF423 and CEP164 mutations, linking renal ciliopathies to DNA damage response signaling. *Cell* **150:** 533–548.

Chávez M, Ena S, Van Sande J, de Kerchove d'Exaerde A, Schurmans S, Schiffmann SN. 2015. Modulation of ciliary phosphoinositide content regulates trafficking and sonic Hedgehog signaling output. *Dev Cell* **34:** 338–350.

Chih B, Liu P, Chinn Y, Chalouni C, Komuves LG, Hass PE, Sandoval W, Peterson AS. 2011. A ciliopathy complex at the transition zone protects the cilia as a privileged membrane domain. *Nat Cell Biol* **14:** 61–72.

Christopher KJ, Wang B, Kong Y, Weatherbee SD. 2012. Forward genetics uncovers Transmembrane protein 107 as a novel factor required for ciliogenesis and Sonic hedgehog signaling. *Dev Biol* **368:** 382–392.

Coene KL, Mans DA, Boldt K, Gloeckner CJ, van Reeuwijk J, Bolat E, Roosing S, Letteboer SJ, Peters TA, Cremers FP, et al. 2011. The ciliopathy-associated protein homologs RPGRIP1 and RPGRIP1L are linked to cilium integrity through interaction with Nek4 serine/threonine kinase. *Hum Mol Genet* **20:** 3592–3605.

Corbit KC, Shyer AE, Dowdle WE, Gaulden J, Singla V, Chen MH, Chuang PT, Reiter JF. 2008. Kif3a constrains β-catenin-dependent Wnt signalling through dual ciliary and non-ciliary mechanisms. *Nat Cell Biol* **10:** 70–76.

Craft JM, Harris JA, Hyman S, Kner P, Lechtreck KF. 2015. Tubulin transport by IFT is upregulated during ciliary growth by a cilium-autonomous mechanism. *J Cell Biol* **208:** 223–237.

Craige B, Tsao CC, Diener DR, Hou Y, Lechtreck KF, Rosenbaum JL, Witman GB. 2010. CEP290 tethers flagellar transition zone microtubules to the membrane and regulates flagellar protein content. *J Cell Biol* **190:** 927–940.

Czarnecki PG, Gabriel GC, Manning DK, Sergeev M, Lemke K, Klena NT, Liu X, Chen Y, Li Y, San Agustin JT, et al. 2015. ANKS6 is the critical activator of NEK8 kinase in embryonic situs determination and organ patterning. *Nat Commun* **6:** 6023.

Daly OM, Gaboriau D, Karakaya K, King S, Dantas TJ, Lalor P, Dockery P, Krämer A, Morrison CG. 2016. Gene-targeted CEP164-deficient cells show a ciliation defect with intact DNA repair capacity. *J Cell Sci* doi: 10.1242/jcs.186221.

Deane JA, Cole DG, Seeley ES, Diener DR, Rosenbaum JL. 2001. Localization of intraflagellar transport protein IFT52 identifies basal body transitional fibers as the docking site for IFT particles. *Curr Biol* **11:** 1586–1590.

Delous M, Baala L, Salomon R, Laclef C, Vierkotten J, Tory K, Golzio C, Lacoste T, Besse L, Ozilou C, et al. 2007. The ciliary gene RPGRIP1L is mutated in cerebello-oculorenal syndrome (Joubert syndrome type B) and Meckel syndrome. *Nat Genet* **39:** 875–881.

Diener DR, Lupetti P, Rosenbaum JL. 2015. Proteomic analysis of isolated ciliary transition zones reveals the presence of ESCRT proteins. *Curr Biol* **25:** 379–384.

Dishinger JF, Kee HL, Jenkins PM, Fan S, Hurd TW, Hammond JW, Truong YN, Margolis B, Martens JR, Verhey KJ. 2010. Ciliary entry of the kinesin-2 motor KIF17 is regulated by importin-β2 and RanGTP. *Nat Cell Biol* **12:** 703–710.

Dowdle WE, Robinson JF, Kneist A, Sirerol-Piquer MS, Frints SGM, Corbit KC, Zaghloul NA, van Lijnschoten G, Mulders L, Verver DE, et al. 2011. Disruption of a ciliary B9 protein complex causes Meckel syndrome. *Am J Hum Genet* **89:** 94–110.

Eaton S. 2008. Multiple roles for lipids in the Hedgehog signalling pathway. *Nat Rev Mol Cell Biol* **9:** 437–445.

Eguether T, San Agustin JT, Keady BT, Jonassen JA, Liang Y, Francis R, Tobita K, Johnson CA, Abdelhamed ZA, Lo

CW, et al. 2014. IFT27 links the BBSome to IFT for maintenance of the ciliary signaling compartment. *Dev Cell* **31:** 279–290.

Emmer BT, Souther C, Toriello KM, Olson CL, Epting CL, Engman DM. 2009. Identification of a palmitoyl acyltransferase required for protein sorting to the flagellar membrane. *J Cell Sci* **122:** 867–874.

Emmer BT, Maric D, Engman DM. 2010. Molecular mechanisms of protein and lipid targeting to ciliary membranes. *J Cell Sci* **123:** 529–536.

Failler M, Gee HY, Krug P, Joo K, Halbritter J, Belkacem L, Filhol E, Porath JD, Braun DA, Schueler M, et al. 2014. Mutations of CEP83 cause infantile nephronophthisis and intellectual disability. *Am J Hum Genet* **94:** 905–914.

Fan S, Fogg V, Wang Q, Chen XW, Liu CJ, Margolis B. 2007. A novel Crumbs3 isoform regulates cell division and ciliogenesis via importin β interactions. *J Cell Biol* **178:** 387–398.

Feng S, Knödler A, Ren J, Zhang J, Zhang X, Hong Y, Huang S, Peränen J, Guo W. 2012. A Rab8 guanine nucleotide exchange factor-effector interaction network regulates primary ciliogenesis. *J Biol Chem* **287:** 15602–15609.

Ferrante MI, Giorgio G, Feather SA, Bulfone A, Wright V, Ghiani M, Selicorni A, Gammaro L, Scolari F, Woolf AS, et al. 2001. Identification of the gene for oral-facial-digital type I syndrome. *Am J Hum Genet* **68:** 569–576.

Ferrante MI, Zullo A, Barra A, Bimonte S, Messaddeq N, Studer M, Dollé P, Franco B. 2006. Oral-facial-digital type I protein is required for primary cilia formation and left–right axis specification. *Nat Genet* **38:** 112–117.

Filges I, Nosova E, Bruder E, Tercanli S, Townsend K, Gibson WT, Röthlisberger B, Heinimann K, Hall JG, Gregory-Evans CY, et al. 2014. Exome sequencing identifies mutations in KIF14 as a novel cause of an autosomal recessive lethal fetal ciliopathy phenotype. *Clin Genet* **86:** 220–228.

Fisch C, Dupuis-Williams P. 2011. Ultrastructure of cilia and flagella—Back to the future! *Biol Cell* **103:** 249–270.

Fliegauf M, Horvath J, von Schnakenburg C, Olbrich H, Müller D, Thumfart J, Schermer B, Pazour GJ, Neumann HP, Zentgraf H, et al. 2006. Nephrocystin specifically localizes to the transition zone of renal and respiratory cilia and photoreceptor connecting cilia. *J Am Soc Nephrol* **17:** 2424–2433.

Fliegauf M, Kahle A, Häffner K, Zieger B. 2014. Distinct localization of septin proteins to ciliary sub-compartments in airway epithelial cells. *Biol Chem* **395:** 151–156.

Follit JA, Li L, Vucica Y, Pazour GJ. 2010. The cytoplasmic tail of fibrocystin contains a ciliary targeting sequence. *J Cell Biol* **188:** 21–28.

Francis SS, Skafianos J, Lo B, Mellman I. 2011. A hierarchy of signals regulates entry of membrane proteins into the ciliary membrane domain in epithelial cells. *J Cell Biol* **193:** 219–233.

Garcia-Gonzalo FR, Reiter JF. 2012. Scoring a backstage pass: Mechanisms of ciliogenesis and ciliary access. *J Cell Biol* **197:** 697–709.

Garcia-Gonzalo FR, Corbit KC, Sirerol-Piquer MS, Ramaswami G, Otto EA, Seol AD, Bennett CL, Robinson JF, Josifova DJ, Garcia-Verdugo JM, et al. 2011. A transition

zone complex regulates mammalian ciliogenesis and ciliary membrane composition. *Nat Genet* **43:** 776–784.

Garcia-Gonzalo FR, Phua SC, Roberson EC, Garcia G III, Abedin M, Schurmans S, Inoue T, Reiter JF. 2015. Phosphoinositides regulate ciliary protein trafficking to modulate Hedgehog signaling. *Dev Cell* **34:** 400–409.

Gerdes JM, Liu Y, Zaghloul NA, Leitch CC, Lawson SS, Kato M, Beachy PA, Beales PL, DeMartino GN, Fisher S, et al. 2007. Disruption of the basal body compromises proteasomal function and perturbs intracellular Wnt response. *Nat Genet* **39:** 1350–1360.

Gibbons IR, Grimstone AV. 1960. On flagellar structure in certain flagellates. *J Biophys Biochem Cytol* **7:** 697–716.

Gilula NB, Satir P. 1972. The ciliary necklace. A ciliary membrane specialization. *J Cell Biol* **53:** 494–509.

Goetz SC, Liem KF Jr, Anderson KV. 2012. The spinocerebellar ataxia-associated gene Tau tubulin kinase 2 controls the initiation of ciliogenesis. *Cell* **151:** 847–858.

Gospe SM III, Baker SA, Arshavsky VY. 2010. Facilitative glucose transporter Glut1 is actively excluded from rod outer segments. *J Cell Sci* **123:** 3639–3644.

Gotthardt K, Lokaj M, Koerner C, Falk N, Giessl A, Wittinghofer A. 2015. A G-protein activation cascade from Arl13B to Arl3 and implications for ciliary targeting of lipidated proteins. *eLife* **4:** e11859.

Graser S, Stierhof YD, Lavoie SB, Gassner OS, Lamla S, Le Clech M, Nigg EA. 2007. Cep164, a novel centriole appendage protein required for primary cilium formation. *J Cell Biol* **179:** 321–330.

Gupta GD, Coyaud É, Gonçalves J, Mojarad BA, Liu Y, Wu Q, Gheiratmand L, Comartin D, Tkach JM, Cheung SW, et al. 2015. A dynamic protein interaction landscape of the human centrosome–cilium interface. *Cell* **163:** 1484–1499.

Hildebrandt F, Otto E, Rensing C, Nothwang HG, Vollmer M, Adolphs J, Hanusch H, Brandis M. 1997. A novel gene encoding an SH3 domain protein is mutated in nephronophthisis type 1. *Nat Genet* **17:** 149–153.

Hildebrandt F, Benzing T, Katsanis N. 2011. Ciliopathies. *New Engl J Med* **364:** 1533–1543.

Hoff S, Halbritter J, Epting D, Frank V, Nguyen TM, van Reeuwijk J, Boehlke C, Schell C, Yasunaga T, Helmstädter M, et al. 2013. ANKS6 is a central component of a nephronophthisis module linking NEK8 to INVS and NPHP3. *Nat Genet* **45:** 951–956.

Hogan MC, Manganelli L, Woollard JR, Masyuk AI, Masyuk TV, Tammachote R, Huang BQ, Leontovich AA, Beito TG, Madden BJ, et al. 2009. Characterization of PKD protein-positive exosome-like vesicles. *J Am Soc Nephrol* **20:** 278–288.

Horst CJ, Forestner DM, Besharse JC. 1987. Cytoskeletal-membrane interactions: A stable interaction between cell surface glycoconjugates and doublet microtubules of the photoreceptor connecting cilium. *J Cell Biol* **105:** 2973–2987.

Hou Y, Qin H, Follit JA, Pazour GJ, Rosenbaum JL, Witman GB. 2007. Functional analysis of an individual IFT protein: IFT46 is required for transport of outer dynein arms into flagella. *J Cell Biol* **176:** 653–665.

Cite this article as *Cold Spring Harb Perspect Biol* doi: 10.1101/cshperspect.a028134

Hu Q, Nelson WJ. 2011. Ciliary diffusion barrier: The gatekeeper for the primary cilium compartment. *Cytoskeleton (Hoboken)* **68:** 313–324.

Hu Q, Milenkovic L, Jin H, Scott MP, Nachury MV, Spiliotis ET, Nelson WJ. 2010. A septin diffusion barrier at the base of the primary cilium maintains ciliary membrane protein distribution. *Science* **329:** 436–439.

Huang L, Szymanska K, Jensen VL, Janecke AR, Innes AM, Davis EE, Frosk P, Li C, Willer JR, Chodirker BN, et al. 2011. TMEM237 is mutated in individuals with a Joubert syndrome related disorder and expands the role of the TMEM family at the ciliary transition zone. *Am J Hum Genet* **89:** 713–730.

Humbert MC, Weihbrecht K, Searby CC, Li Y, Pope RM, Sheffield VC, Seo S. 2012. ARL13B, PDE6D, and CEP164 form a functional network for INPP5E ciliary targeting. *Proc Natl Acad Sci* **109:** 19691–19696.

Huppke P, Wegener E, Böhrer-Rabel H, Bolz HJ, Zoll B, Gärtner J, Bergmann C. 2015. Tectonic gene mutations in patients with Joubert syndrome. *Eur J Hum Genet* **23:** 616–620.

Hurd TW, Hildebrandt F. 2011. Mechanisms of nephronophthisis and related ciliopathies. *Nephron Exp Nephrol* **118:** e9–e14.

Hurd TW, Fan S, Margolis BL. 2011. Localization of retinitis pigmentosa 2 to cilia is regulated by Importin β2. *J Cell Sci* **124:** 718–726.

Iannicelli M, Brancati F, Mougou-Zerelli S, Mazzotta A, Thomas S, Elkhartoufi N, Travaglini L, Gomes C, Ardissino GL, Bertini E, et al. 2010. Novel TMEM67 mutations and genotype-phenotype correlates in meckelin-related ciliopathies. *Hum Mutat* **31:** E1319–E1331.

Ishikawa H, Kubo A, Tsukita S, Tsukita S. 2005. Odf2-deficient mother centrioles lack distal/subdistal appendages and the ability to generate primary cilia. *Nat Cell Biol* **7:** 517–524.

Jauregui AR, Nguyen KC, Hall DH, Barr MM. 2008. The *Caenorhabditis elegans* nephrocystins act as global modifiers of cilium structure. *J Cell Biol* **180:** 973–988.

Jenkins PM, Hurd TW, Zhang L, McEwen DP, Brown RL, Margolis B, Verhey KJ, Martens JR. 2006. Ciliary targeting of olfactory CNG channels requires the CNGB1b subunit and the kinesin-2 motor protein, KIF17. *Curr Biol* **16:** 1211–1216.

Jensen VL, Li C, Bowie RV, Clarke L, Mohan S, Blacque OE, Leroux MR. 2015. Formation of the transition zone by Mks5/Rpgrip1L establishes a ciliary zone of exclusion (CIZE) that compartmentalises ciliary signalling proteins and controls PIP2 ciliary abundance. *EMBO J* **34:** 2537–2556.

Jiang ST, Chiou YY, Wang E, Lin HK, Lee SP, Lu HY, Wang CK, Tang MJ, Li H. 2008. Targeted disruption of Nphp1 causes male infertility due to defects in the later steps of sperm morphogenesis in mice. *Hum Mol Genet* **17:** 3368–3379.

Jiang ST, Chiou YY, Wang E, Chien YL, Ho HH, Tsai FJ, Lin CY, Tsai SP, Li H. 2009. Essential role of nephrocystin in photoreceptor intraflagellar transport in mouse. *Hum Mol Genet* **18:** 1566–1577.

Jin H, White SR, Shida T, Schulz S, Aguiar M, Gygi SP, Bazan JF, Nachury MV. 2010. The conserved Bardet–Biedl syndrome proteins assemble a coat that traffics membrane proteins to cilia. *Cell* **141:** 1208–1219.

Joo K, Kim CG, Lee MS, Moon HY, Lee SH, Kim MJ, Kweon HS, Park WY, Kim CH, Gleeson JG, et al. 2013. CCDC41 is required for ciliary vesicle docking to the mother centriole. *Proc Natl Acad Sci* **110:** 5987–5992.

Kee HL, Dishinger JF, Lynne Blasius T, Liu CJ, Margolis B, Verhey KJ. 2012. A size-exclusion permeability barrier and nucleoporins characterize a ciliary pore complex that regulates transport into cilia. *Nat Cell Biol* **14:** 431–437.

Kim SK, Shindo A, Park TJ, Oh EC, Ghosh S, Gray RS, Lewis RA, Johnson CA, Attie-Bittach T, Katsanis N, et al. 2010. Planar cell polarity acts through septins to control collective cell movement and ciliogenesis. *Science* **329:** 1337–1340.

Kizhatil K, Baker SA, Arshavsky VY, Bennett V. 2009. Ankyrin-G promotes cyclic nucleotide-gated channel transport to rod photoreceptor sensory cilia. *Science* **323:** 1614–1617.

Knödler A, Feng S, Zhang J, Zhang X, Das A, Peränen J, Guo W. 2010. Coordination of Rab8 and Rab11 in primary ciliogenesis. *Proc Natl Acad Sci* **107:** 6346–6351.

Kobayashi T, Tsang WY, Li J, Lane W, Dynlacht BD. 2011. Centriolar kinesin Kif24 interacts with CP110 to remodel microtubules and regulate ciliogenesis. *Cell* **145:** 914–925.

Lambacher NJ, Bruel AL, van Dam TJ, Szymanska K, Slaats GG, Kuhns S, McManus GJ, Kennedy JE, Gaff K, Wu KM, et al. 2016. TMEM107 recruits ciliopathy proteins to subdomains of the ciliary transition zone and causes Joubert syndrome. *Nat Cell Biol* **18:** 122–131.

Lancaster MA, Louie CM, Silhavy JL, Sintasath L, Decambre M, Nigam SK, Willert K, Gleeson JG. 2009. Impaired Wnt-β-catenin signaling disrupts adult renal homeostasis and leads to cystic kidney ciliopathy. *Nat Med* **15:** 1046–1054.

Lancaster MA, Schroth J, Gleeson JG. 2011. Subcellular spatial regulation of canonical Wnt signalling at the primary cilium. *Nat Cell Biol* **13:** 700–707.

Leaf A, Von Zastrow M. 2015. Dopamine receptors reveal an essential role of IFT-B, KIF17, and Rab23 in delivering specific receptors to primary cilia. *eLife* **4:** e06996

Lechtreck KF, Johnson EC, Sakai T, Cochran D, Ballif BA, Rush J, Pazour GJ, Ikebe M, Witman GB. 2009. The *Chlamydomonas reinhardtii* BBSome is an IFT cargo required for export of specific signaling proteins from flagella. *J Cell Biol* **187:** 1117–1132.

Lechtreck KF, Brown JM, Sampaio JL, Craft JM, Shevchenko A, Evans JE, Witman GB. 2013. Cycling of the signaling protein phospholipase D through cilia requires the BBSome only for the export phase. *J Cell Biol* **201:** 249–261.

Lee JJ, Seo S. 2015. PDE6D binds to the C-terminus of RPGR in a prenylation-dependent manner. *EMBO Rep* **16:** 1581–1582.

Lee YL, Santé J, Comerci CJ, Cyge B, Menezes LF, Li FQ, Germino GG, Moerner WE, Takemaru K, Stearns T. 2014. Cby1 promotes Ahi1 recruitment to a ring-shaped domain at the centriole–cilium interface and facilitates proper cilium formation and function. *Mol Biol Cell* **25:** 2919–2933.

Leettola CN, Knight MJ, Cascio D, Hoffman S, Bowie JU. 2014. Characterization of the SAM domain of the PKD-related protein ANKS6 and its interaction with ANKS3. *BMC Struct Biol* **14:** 17.

Leightner AC, Hommerding CJ, Peng Y, Salisbury JL, Gainullin VG, Czarnecki PG, Sussman CR, Harris PC. 2013. The Meckel syndrome protein meckelin (TMEM67) is a key regulator of cilia function but is not required for tissue planar polarity. *Hum Mol Genet* **22:** 2024–2040.

Li T. 2014. Leber congenital amaurosis caused by mutations in RPGRIP1. *Cold Spring Harb Perspect Med.* **5:** a017384.

Li C, Jensen VL, Park K, Kennedy J, Garcia-Gonzalo FR, Romani M, De Mori R, Bruel AL, Gaillard D, Doray B, et al. 2016. MKS5 and CEP290 dependent assembly pathway of the ciliary transition zone. *PLoS Biol* **14:** e1002416.

Liew GM, Ye F, Nager AR, Murphy JP, Lee JS, Aguiar M, Breslow DK, Gygi SP, Nachury MV. 2014. The intraflagellar transport protein IFT27 promotes BBSome exit from cilia through the GTPase ARL6/BBS3. *Dev Cell* **31:** 265–278.

Lim YS, Tang BL. 2015. A role for Rab23 in the trafficking of Kif17 to the primary cilium. *J Cell Sci* **128:** 2996–3008.

Lin YC, Niewiadomski P, Lin B, Nakamura H, Phua SC, Jiao J, Levchenko A, Inoue T, Rohatgi R, Inoue T. 2013. Chemically-inducible diffusion trap at cilia (C-IDTc) reveals molecular sieve-like barrier. *Nat Chem Biol* **9:** 437–443.

Loktev AV, Jackson PK. 2013. Neuropeptide Y family receptors traffic via the Bardet-Biedl syndrome pathway to signal in neuronal primary cilia. *Cell Rep* **5:** 1316–1329.

Lu Q, Insinna C, Ott C, Stauffer J, Pintado PA, Rahajeng J, Baxa U, Walia V, Cuenca A, Hwang YS, et al. 2015. Early steps in primary cilium assembly require EHD1/EHD3-dependent ciliary vesicle formation. *Nat Cell Biol* **17:** 228–240.

Ludington WB, Wemmer KA, Lechtreck KF, Witman GB, Marshall WF. 2013. Avalanche-like behavior in ciliary import. *Proc Natl Acad Sci* **110:** 3925–3930.

Maguire JE, Silva M, Nguyen KC, Hellen E, Kern AD, Hall DH, Barr MM. 2015. Myristoylated CIL-7 regulates ciliary extracellular vesicle biogenesis. *Mol Biol Cell* **26:** 2823–2832.

Maric D, McGwire BS, Buchanan KT, Olson CL, Emmer BT, Epting CL, Engman DM. 2011. Molecular determinants of ciliary membrane localization of *Trypanosoma cruzi* flagellar calcium-binding protein. *J Biol Chem* **286:** 33109–33117.

Milenkovic L, Scott MP, Rohatgi R. 2009. Lateral transport of Smoothened from the plasma membrane to the membrane of the cilium. *J Cell Biol* **187:** 365–374.

Mollet G, Salomon R, Gribouval O, Silbermann F, Bacq D, Landthaler G, Milford D, Nayir A, Rizzoni G, Antignac C, et al. 2002. The gene mutated in juvenile nephronophthisis type 4 encodes a novel protein that interacts with nephrocystin. *Nat Genet* **32:** 300–305.

Mollet G, Silbermann F, Delous M, Salomon R, Antignac C, Saunier S. 2005. Characterization of the nephrocystin/nephrocystin-4 complex and subcellular localization of nephrocystin-4 to primary cilia and centrosomes. *Hum Mol Genet* **14:** 645–656.

Mougou-Zerelli S, Thomas S, Szenker E, Audollent S, Elkhartoufi N, Babarit C, Romano S, Salomon R, Amiel J, Esculpavit C, et al. 2009. CC2D2A mutations in Meckel and Joubert syndromes indicate a genotype–phenotype correlation. *Hum Mutat* **30:** 1574–1582.

Mukhopadhyay S, Wen X, Chih B, Nelson CD, Lane WS, Scales SJ, Jackson PK. 2010. TULP3 bridges the IFT-A complex and membrane phosphoinositides to promote trafficking of G protein-coupled receptors into primary cilia. *Genes Dev* **24:** 2180–2193.

Mukhopadhyay S, Wen X, Ratti N, Loktev A, Rangell L, Scales SJ, Jackson PK. 2013. The ciliary G-protein-coupled receptor Gpr161 negatively regulates the Sonic hedgehog pathway via cAMP signaling. *Cell* **152:** 210–223.

Myers BR, Sever N, Chong YC, Kim J, Belani JD, Rychnovsky S, Bazan JF, Beachy PA. 2013. Hedgehog pathway modulation by multiple lipid binding sites on the smoothened effector of signal response. *Dev Cell* **26:** 346–357.

Nachury MV, Loktev AV, Zhang Q, Westlake CJ, Peränen J, Merdes A, Slusarski DC, Scheller RH, Bazan JF, Sheffield VC, et al. 2007. A core complex of BBS proteins cooperates with the GTPase Rab8 to promote ciliary membrane biogenesis. *Cell* **129:** 1201–1213.

Nachury MV, Seeley ES, Jin H. 2010. Trafficking to the ciliary membrane: How to get across the periciliary diffusion barrier? *Annu Rev Cell Dev Biol* **26:** 59–87.

Najafi M, Maza NA, Calvert PD. 2012. Steric volume exclusion sets soluble protein concentrations in photoreceptor sensory cilia. *Proc Natl Acad Sci* **109:** 203–208.

Nakata K, Shiba D, Kobayashi D, Yokoyama T. 2012. Targeting of Nphp3 to the primary cilia is controlled by an N-terminal myristoylation site and coiled-coil domains. *Cytoskeleton (Hoboken)* **69:** 221–234.

Olmos Y, Carlton JG. 2016. The ESCRT machinery: New roles at new holes. *Curr Opin Cell Biol* **38:** 1–11.

O'Toole ET, Giddings TH, McIntosh JR, Dutcher SK. 2003. Three-dimensional organization of basal bodies from wild-type and δ-tubulin deletion strains of *Chlamydomonas reinhardtii*. *Mol Biol Cell* **14:** 2999–3012.

Otto E, Hoefele J, Ruf R, Mueller AM, Hiller KS, Wolf MT, Schuermann MJ, Becker A, Birkenhäger R, Sudbrak R, et al. 2002. A gene mutated in nephronophthisis and retinitis pigmentosa encodes a novel protein, nephroretinin, conserved in evolution. *Am J Hum Genet* **71:** 1161–1167.

Otto EA, Schermer B, Obara T, O'Toole JF, Hiller KS, Mueller AM, Ruf RG, Hoefele J, Beekmann F, Landau D, et al. 2003. Mutations in INVS encoding inversin cause nephronophthisis type 2, linking renal cystic disease to the function of primary cilia and left–right axis determination. *Nat Genet* **34:** 413–420.

Patzke S, Redick S, Warsame A, Murga-Zamalloa CA, Khanna H, Doxsey S, Stokke T. 2010. CSPP is a ciliary protein interacting with Nephrocystin 8 and required for cilia formation. *Mol Biol Cell* **21:** 2555–2567.

Qin H, Diener DR, Geimer S, Cole DG, Rosenbaum JL. 2004. Intraflagellar transport (IFT) cargo: IFT transports flagellar precursors to the tip and turnover products to the cell body. *J Cell Biol* **164:** 255–266.

Quarmby LM. 2004. Cellular deflagellation. *Int Rev Cytol* **233:** 47–91.

Reiter JF, Blacque OE, Leroux MR. 2012. The base of the cilium: Roles for transition fibres and the transition zone

Cite this article as *Cold Spring Harb Perspect Biol* doi: 10.1101/cshperspect.a028134

in ciliary formation, maintenance and compartmentalization. *EMBO Rep* **13**: 608–618.

Ringo DL. 1967. Flagellar motion and fine structure of the flagellar apparatus in *Chlamydomonas*. *J Cell Biol* **33**: 543–571.

Roberson EC, Dowdle WE, Ozanturk A, Garcia-Gonzalo FR, Li C, Halbritter J, Elkhartoufi N, Porath JD, Cope H, Ashley-Koch A, et al. 2015. TMEM231, mutated in orofaciodigital and Meckel syndromes, organizes the ciliary transition zone. *J Cell Biol* **209**: 129–142.

Rohatgi R, Snell WJ. 2010. The ciliary membrane. *Curr Opin Cell Biol* **22**: 541–546.

Sang L, Miller JJ, Corbit KC, Giles RH, Brauer MJ, Otto EA, Baye LM, Wen X, Scales SJ, Kwong M, et al. 2011. Mapping the NPHP-JBTS-MKS protein network reveals ciliopathy disease genes and pathways. *Cell* **145**: 513–528.

Schäfer T, Pütz M, Lienkamp S, Ganner A, Bergbreiter A, Ramachandran H, Gieloff V, Gerner M, Mattonet C, Czarnecki PG, et al. 2008. Genetic and physical interaction between the NPHP5 and NPHP6 gene products. *Hum Mol Genet* **17**: 3655–3662.

Schmidt KN, Kuhns S, Neuner A, Hub B, Zentgraf H, Pereira G. 2012. Cep164 mediates vesicular docking to the mother centriole during early steps of ciliogenesis. *J Cell Biol* **199**: 1083–1101.

Schouteden C, Serwas D, Palfy M, Dammermann A. 2015. The ciliary transition zone functions in cell adhesion but is dispensable for axoneme assembly in *C. elegans*. *J Cell Biol* **210**: 35–44.

Seo S, Zhang Q, Bugge K, Breslow DK, Searby CC, Nachury MV, Sheffield VC. 2011. A novel protein LZTFL1 regulates ciliary trafficking of the BBSome and Smoothened. *PLoS Genet* **7**: e1002358.

Shaheen R, Ansari S, Mardawi EA, Alshammari MJ, Alkuraya FS. 2013. Mutations in TMEM231 cause Meckel–Gruber syndrome. *J Med Genet* **50**: 160–162.

Shaheen R, Almoisheer A, Faqeih E, Babay Z, Monies D, Tassan N, Abouelhoda M, Kurdi W, Al Mardawi E, Khalil MM, et al. 2015. Identification of a novel MKS locus defined by TMEM107 mutation. *Hum Mol Genet* **24**: 5211–5218.

Shiba D, Yamaoka Y, Hagiwara H, Takamatsu T, Hamada H, Yokoyama T. 2009. Localization of Inv in a distinctive intraciliary compartment requires the C-terminal ninein-homolog-containing region. *J Cell Sci* **122**: 44–54.

Shiba D, Manning DK, Koga H, Beier DR, Yokoyama T. 2010. Inv acts as a molecular anchor for Nphp3 and Nek8 in the proximal segment of primary cilia. *Cytoskeleton (Hoboken)* **67**: 112–119.

Shylo NA, Christopher KJ, Iglesias A, Daluiski A, Weatherbee SD. 2016. TMEM107 is a critical regulator of ciliary protein composition and is mutated in orofaciodigital syndrome. *Hum Mutat* **37**: 155–159.

Sillibourne JE, Specht CG, Izeddin I, Hurbain I, Tran P, Triller A, Darzacq X, Dahan M, Bornens M. 2011. Assessing the localization of centrosomal proteins by PALM/STORM nanoscopy. *Cytoskeleton (Hoboken)* **68**: 619–627.

Singla V, Romaguera-Ros M, Garcia-Verdugo JM, Reiter JF. 2010. Ofd1, a human disease gene, regulates the length and distal structure of centrioles. *Dev Cell* **18**: 410–424.

Sorokin S. 1962. Centrioles and the formation of rudimentary cilia by fibroblasts and smooth muscle cells. *J Cell Biol* **15**: 363–377.

Spektor A, Tsang WY, Khoo D, Dynlacht BD. 2007. Cep97 and CP110 suppress a cilia assembly program. *Cell* **130**: 678–690.

Steere N, Chae V, Burke M, Li FQ, Takemaru K, Kuriyama R. 2012. A Wnt/β-catenin pathway antagonist Chibby binds Cenexin at the distal end of mother centrioles and functions in primary cilia formation. *PLoS ONE* **7**: e41077.

Stinchcombe JC, Randzavola LO, Angus KL, Mantell JM, Verkade P, Griffiths GM. 2015. Mother centriole distal appendages mediate centrosome docking at the immunological synapse and reveal mechanistic parallels with ciliogenesis. *Curr Biol* **25**: 3239–3244.

Takao D, Dishinger JF, Kee HL, Pinskey JM, Allen BL, Verhey KJ. 2014. An assay for clogging the ciliary pore complex distinguishes mechanisms of cytosolic and membrane protein entry. *Curr Biol* **24**: 2288–2294.

Tanos BE, Yang HJ, Soni R, Wang WJ, Macaluso FP, Asara JM, Tsou MF. 2013. Centriole distal appendages promote membrane docking, leading to cilia initiation. *Genes Dev* **27**: 163–168.

Tateishi K, Yamazaki Y, Nishida T, Watanabe S, Kunimoto K, Ishikawa H, Tsukita S. 2013. Two appendages homologous between basal bodies and centrioles are formed using distinct Odf2 domains. *J Cell Biol* **203**: 417–425.

Thauvin-Robinet C, Lee JS, Lopez E, Herranz-Pérez V, Shida T, Franco B, Jego L, Ye F, Pasquier L, Loget P, et al. 2014. The oral-facial-digital syndrome gene C2CD3 encodes a positive regulator of centriole elongation. *Nat Genet* **46**: 905–911.

Thomas S, Legendre M, Saunier S, Bessières B, Alby C, Bonnière M, Toutain A, Loeuillet L, Szymanska K, Jossic F, et al. 2012. TCTN3 mutations cause Mohr–Majewski syndrome. *Am J Hum Genet* **91**: 372–378.

Thomas S, Wright KJ, Le Corre S, Micalizzi A, Romani M, Abhyankar A, Saada J, Perrault I, Amiel J, Litzler J, et al. 2014. A homozygous PDE6D mutation in Joubert syndrome impairs targeting of farnesylated INPP5E protein to the primary cilium. *Hum Mutat* **35**: 137–146.

Tsang WY, Bossard C, Khanna H, Peränen J, Swaroop A, Malhotra V, Dynlacht BD. 2008. CP110 suppresses primary cilia formation through its interaction with CEP290, a protein deficient in human ciliary disease. *Dev Cell* **15**: 187–197.

Valente EM, Logan CV, Mougou-Zerelli S, Lee JH, Silhavy JL, Brancati F, Iannicelli M, Travaglini L, Romani S, Illi B, et al. 2010. Mutations in TMEM216 perturb ciliogenesis and cause Joubert, Meckel and related syndromes. *Nat Genet* **42**: 619–625.

Valente EM, Rosti RO, Gibbs E, Gleeson JG. 2014. Primary cilia in neurodevelopmental disorders. *Nat Rev Neurol* **10**: 27–36.

van Bon BW, Oortveld MA, Nijtmans LG, Fenckova M, Nijhof B, Besseling J, Vos M, Kramer JM, de Leeuw N, Castells-Nobau A, et al. 2013. CEP89 is required for mitochondrial metabolism and neuronal function in man and fly. *Hum Mol Genet* **22**: 3138–3151.

Verhey KJ, Dishinger J, Kee HL. 2011. Kinesin motors and primary cilia. *Biochem Soc Trans* **39:** 1120–1125.

Vierkotten J, Dildrop R, Peters T, Wang B, Rüther U. 2007. Ftm is a novel basal body protein of cilia involved in Shh signalling. *Development* **134:** 2569–2577.

Voronina VA, Takemaru K, Treuting P, Love D, Grubb BR, Hajjar AM, Adams A, Li FQ, Moon RT. 2009. Inactivation of Chibby affects function of motile airway cilia. *J Cell Biol* **185:** 225–233.

Wätzlich D, Vetter I, Gotthardt K, Miertzschke M, Chen YX, Wittinghofer A, Ismail S. 2013. The interplay between RPGR, PDEδ and Arl2/3 regulate the ciliary targeting of farnesylated cargo. *EMBO Rep* **14:** 465–472.

Weatherbee SD, Niswander LA, Anderson KV. 2009. A mouse model for Meckel syndrome reveals Mks1 is required for ciliogenesis and Hedgehog signaling. *Hum Mol Genet* **18:** 4565–4575.

Wei Q, Zhang Y, Li Y, Zhang Q, Ling K, Hu J. 2012. The BBSome controls IFT assembly and turnaround in cilia. *Nat Cell Biol* **14:** 950–957.

Wei Q, Xu Q, Zhang Y, Li Y, Zhang Q, Hu Z, Harris PC, Torres VE, Ling K, Hu J. 2013. Transition fibre protein FBF1 is required for the ciliary entry of assembled intraflagellar transport complexes. *Nat Commun* **4:** 2750.

Westlake CJ, Baye LM, Nachury MV, Wright KJ, Ervin KE, Phu L, Chalouni C, Beck JS, Kirkpatrick DS, Slusarski DC, et al. 2011. Primary cilia membrane assembly is initiated by Rab11 and transport protein particle II (TRAPPII) complex-dependent trafficking of Rabin8 to the centrosome. *Proc Natl Acad Sci* **108:** 2759–2764.

Williams CL, Winkelbauer ME, Schafer JC, Michaud EJ, Yoder BK. 2008. Functional redundancy of the B9 proteins and nephrocystins in Caenorhabditis elegans ciliogenesis. *Mol Biol Cell* **19:** 2154–2168.

Williams CL, Masyukova SV, Yoder BK. 2010. Normal ciliogenesis requires synergy between the cystic kidney disease genes MKS-3 and NPHP-4. *J Am Soc Nephrol* **21:** 782–793.

Williams CL, Li C, Kida K, Inglis PN, Mohan S, Semenec L, Bialas NJ, Stupay RM, Chen N, Blacque OE, et al. 2011. MKS and NPHP modules cooperate to establish basal body/transition zone membrane associations and ciliary gate function during ciliogenesis. *J Cell Biol* **192:** 1023–1041.

Williams CL, McIntyre JC, Norris SR, Jenkins PM, Zhang L, Pei Q, Verhey K, Martens JR. 2014. Direct evidence for BBSome-associated intraflagellar transport reveals distinct properties of native mammalian cilia. *Nat Commun* **5:** 5813.

Winkelbauer ME, Schafer JC, Haycraft CJ, Swoboda P, Yoder BK. 2005. The *C. elegans* homologs of nephrocystin-1 and nephrocystin-4 are cilia transition zone proteins involved in chemosensory perception. *J Cell Sci* **118:** 5575–5587.

Wolf MTF, Saunier S, O'Toole JF, Wanner N, Groshong T, Attanasio M, Salomon R, Stallmach T, Sayer JA, Waldherr R, et al. 2007. Mutational analysis of the RPGRIP1L gene in patients with Joubert syndrome and nephronophthisis. *Kidney Int* **72:** 1520–1526.

Won J, Marín de Evsikova C, Smith RS, Hicks WL, Edwards MM, Longo-Guess C, Li T, Naggert JK, Nishina PM. 2011. NPHP4 is necessary for normal photoreceptor ribbon synapse maintenance and outer segment formation, and for sperm development. *Hum Mol Genet* **20:** 482–496.

Wood CR, Huang K, Diener DR, Rosenbaum JL. 2013. The cilium secretes bioactive ectosomes. *Curr Biol* **23:** 906–911.

Wright KJ, Baye LM, Olivier-Mason A, Mukhopadhyay S, Sang L, Kwong M, Wang W, Pretorius PR, Sheffield VC, Sengupta P, et al. 2011. An ARL3-UNC119-RP2 GTPase cycle targets myristoylated NPHP3 to the primary cilium. *Genes Dev* **25:** 2347–2360.

Xu Q, Zhang Y, Wei Q, Huang Y, Hu J, Ling K. 2016. Phosphatidylinositol phosphate kinase PIPKIγ and phosphatase INPP5E coordinate initiation of ciliogenesis. *Nat Commun* **7:** 10777.

Yadav SP, Sharma NK, Liu C, Dong L, Li T, Swaroop A. 2016. Centrosomal protein CP110 controls maturation of mother centriole during cilia biogenesis. *Development* **143:** 1491–1501.

Yakulov TA, Yasunaga T, Ramachandran H, Engel C, Müller B, Hoff S, Dengjel J, Lienkamp SS, Walz G. 2015. Anks3 interacts with nephronophthisis proteins and is required for normal renal development. *Kidney Int* **87:** 1191–1200.

Yang TT, Su J, Wang WJ, Craige B, Witman GB, Tsou MF, Liao JC. 2015. Superresolution pattern recognition reveals the architectural map of the ciliary transition zone. *Sci Rep* **5:** 14096.

Ye X, Zeng H, Ning G, Reiter JF, Liu A. 2014. C2cd3 is critical for centriolar distal appendage assembly and ciliary vesicle docking in mammals. *Proc Natl Acad Sci* **111:** 2164–2169.

Yee LE, Garcia-Gonzalo FR, Bowie RV, Li C, Kennedy JK, Ashrafi K, Blacque OE, Leroux MR, Reiter JF. 2015. Conserved genetic interactions between ciliopathy complexes cooperatively support ciliogenesis and ciliary signaling. *PLoS Genet* **11:** e1005627.

Zhang H, Liu XH, Zhang K, Chen CK, Frederick JM, Prestwich GD, Baehr W. 2004. Photoreceptor cGMP phosphodiesteraseδ subunit (PDEδ) functions as a prenyl-binding protein. *J Biol Chem* **279:** 407–413.

Zhang H, Li S, Doan T, Rieke F, Detwiler PB, Frederick JM, Baehr W. 2007. Deletion of PrBP/δ impedes transport of GRK1 and PDE6 catalytic subunits to photoreceptor outer segments. *Proc Natl Acad Sci* **104:** 8857–8862.

Zhang H, Constantine R, Vorobiev S, Chen Y, Seetharaman J, Huang YJ, Xiao R, Montelione GT, Gerstner CD, Davis MW, et al. 2011. UNC119 is required for G protein trafficking in sensory neurons. *Nat Neurosci* **14:** 874–880.

Zhang H, Frederick JM, Baehr W. 2014. Unc119 gene deletion partially rescues the GRK1 transport defect of Pde6d$^{-/-}$ cones. *Adv Exp Med Bio.* **801:** 487–493.

Zhang H, Hanke-Gogokhia C, Jiang L, Li X, Wang P, Gerstner CD, Frederick JM, Yang Z, Baehr W. 2015. Mistrafficking of prenylated proteins causes retinitis pigmentosa 2. *FASEB J* **29:** 932–942.

Zhao C, Malicki J. 2011. Nephrocystins and MKS proteins interact with IFT particle and facilitate transport of selected ciliary cargos. *EMBO J* **30:** 2532–2544.

Transition Zone Migration: A Mechanism for Cytoplasmic Ciliogenesis and Postaxonemal Centriole Elongation

Tomer Avidor-Reiss, Andrew Ha, and Marcus L. Basiri

University of Toledo, Department of Biological Sciences, Toledo, Ohio 43606

Correspondence: tomer.avidorreiss@utoledo.edu

The cilium is an elongated and continuous structure that spans two major subcellular domains. The cytoplasmic domain contains a short centriole, which serves to nucleate the main projection of the cilium. This projection, known as the axoneme, remains separated from the cytoplasm by a specialized gatekeeping complex within a ciliary subdomain called the transition zone. In this way, the axoneme is compartmentalized. Intriguingly, however, this general principle of cilium biology is altered in the sperm cells of many animals, which instead contain a cytoplasmic axoneme domain. Here, we discuss the hypothesis that the formation of specialized sperm giant centrioles and cytoplasmic cilia is mediated by the migration of the transition zone from its typical location as part of a structure known as the annulus and examine the intrinsic properties of the transition zone that may facilitate its migratory behavior.

This review specifically addresses a phenomenon unique to certain ciliated cell types whereby a ciliary substructure known as the transition zone migrates distally from its original position between the centriole and the cilium during the process of axoneme assembly. We begin our discussion with a general introduction to cilium biology that is primarily intended for those who are less familiar with the transition zone and other fundamental ciliary substructures. Next, we build on this foundation by introducing the relatively recent observation that, in certain sperm cell types, the transition zone and a related structure known as the annulus migrates away from the centriole and its original position at the base of the ciliary compartment. We then consider the ways in which transition zone migration may contribute to centriole elongation, and we end our discussion by hypothesizing the mechanism by which transition zone migration enables cytoplasmic ciliogenesis.

CENTRIOLE, CILIUM, AND TRANSITION ZONE

Centriole, Centrosome, Basal Body, and Cilium Are Related Terms

The centriole lives a double life. In dividing cells, the centriole is located centrally, where it functions to organize and maintain dense layers of protein complexes referred to as pericentriolar material around its periphery to assemble the centrosome and participate in mitotic activities. Once mitosis ends, the centriole sheds the majority of its centrosomal complexes and travels to the cell membrane where it produces a cilium during interphase ($G_{0/1}$). If the cell cycle continues to mitosis, the cilium is disassembled and the centriole returns to the center of the cell where it again nucleates centrosomal complexes ($S-G_2$). Therefore, the centrosome and cilium are distinct functional and morphological manifestations of the same organelle, shifting back and forth throughout the cell cycle to accomplish separate cellular responsibilities. Of note, the centriole is historically referred to as the "basal body" while it is a component of the cilium. For simplicity, we will use the term centriole in the context of both the centrosome and cilium.

A Centriole and a Cilium Form a Continuous Microtubule-Based Structure Dynamic at One End

In broad terms, the cilium is comprised of two parts: a cytoplasmic segment characterized by the centriole, and a second longer segment that projects from the cell surface via a structure known as the axoneme. Although, these ciliary substructures are often regarded separately, the axoneme is not structurally discrete from the centriole but is instead formed as a continuous extension of the centriole's distal end that gains compositional and functional distinction by incorporating axoneme-specific proteins (Fig. 1A). Importantly, a common and continuous polymer of microtubules forms both structures. At its proximal segment, the centriole is comprised of a ninefold radial array of triplet microtubules, by convention referred to as the "A,"

"B," and "C" microtubules (Fig. 1A). This radial arrangement continues distally along the centriole and throughout the axoneme via extension of the "A" and "B" microtubules. Considering this relationship, the centriole can therefore be regarded as both the template and the cytoplasmic anchor for the axoneme.

Because microtubules have an inherent polarity, the centriole and axoneme are also polarized, with the microtubule "minus" end found at the proximal end of the centriole and the "plus" end found at the distal end of the axoneme. In general, microtubule minus ends are stabilized by complexes formed from γ-tubulin (γ-TuRC) (Moritz et al. 1995). Consistently, it appears that the proximal end of centriolar "A" microtubules are also stabilized by a γ-TuRC complex (Guichard et al. 2010). On the other hand, however, microtubule "plus" ends are dynamic, displaying the capacity for both dramatic extension and collapse (Mitchison and Kirschner 1984; Desai and Mitchison 1997). Similarly, axonemal microtubules are highly dynamic at their growing "plus" ends, enabling them to form cilia of distinct lengths (Johnson and Rosenbaum 1992).

Typical Centriole Elongation Precedes Axoneme Assembly

Although the capacity for length variability is an inherent property of microtubule-based struc-

---

Figure 1. Models of transition zone migration in centriole formation and ciliogenesis. (*A*) In a typical cell, the centriole forms in the cytoplasm. At its base, the centriole is comprised of triplet microtubules known as the A, B, and C microtubules, with only the A and B microtubules extending to its tip. The centriole then either associates with a vesicle that fuses with the plasma membrane (not shown) or migrates directly to the membrane for docking. There, the microtubules at the distal end of the centriole elongate to form an axoneme, which is surrounded by a specialized membrane known as the ciliary membrane. At the base of the cilium, this membrane forms a pocket known as the ciliary pocket, and the axoneme is embedded in a network of proteins that serves as a ciliary gate known as the transition zone. (*B*) During *Drosophila* spermatogenesis, ciliogenesis begins in the premeiotic diploid spermatocytes and is completed in the postmeiotic haploid spermatids. In the spermatocyte, the centriole docks to the plasma membrane and forms a short cilium. After the axoneme is initiated, both the centriole and the cilium elongate. Presumably, the centriole grows as a result of transition zone migration along the axoneme. In the spermatid, the cilium continues to grow and the transition zone continues to migrate along the axoneme, exposing axonemal microtubules to the cytoplasm. (*C*) During mammalian spermatogenesis, ciliogenesis starts and is completed in spermatids. In the spermatid, the centriole docks to the plasma membrane and forms a full-length cilium. After the cilium is formed, the transition zone and annulus migrate along the axoneme. Proteins involved in transition zone migration and annulus formation and migration are indicated in *B* and *C*.

Cite this article as *Cold Spring Harb Perspect Biol* doi: 10.1101/cshperspect.a028142

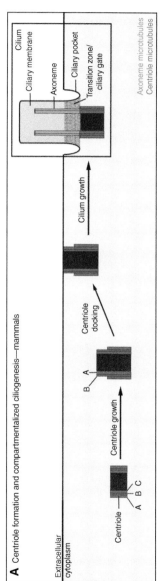

A Centriole formation and compartmentalized ciliogenesis—mammals

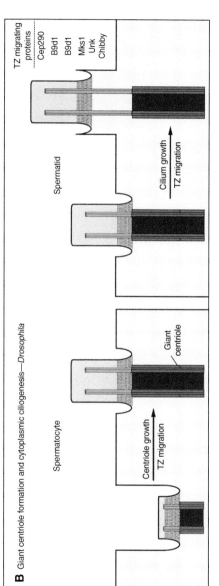

B Giant centriole formation and cytoplasmic ciliogenesis—*Drosophila*

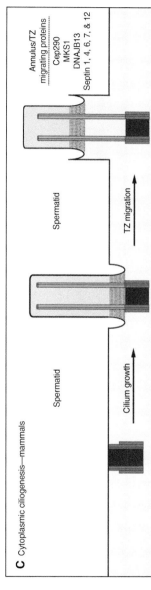

C Cytoplasmic ciliogenesis—mammals

Figure 1. (*Legend on facing page.*)

tures, centriole length is strictly regulated, with precise lengths consistently observed within similar settings. Still, this characteristic length can vary between contexts and is dependent on the organism, cell type, and stage of the cell cycle. Centriole elongation occurs via a regulated program that proceeds before axoneme assembly. For example, mature mammalian centrioles typically measure ~500 nm in length, whereas centrioles of *Drosophila* early embryos are ~200 nm long (Gonzalez et al. 1998). In mammals, a centriole reaches its final length in a stepwise process occurring across almost two cell cycles. Initially, an ~200-nm-long procentriole is made from a centriolar scaffold called the cartwheel during S phase. The centriole continues to elongate throughout G_2 and M to reach ~500 nm in length. Typically, the conclusion of centriole elongation is marked by the incorporation of distal appendage proteins. Once a cell with a mature centriole is committed for ciliogenesis, the centriole's distal "A" and "B" microtubules elongate to form the axoneme. In most cells, centriole length is thought to become locked following the onset of axoneme assembly, remaining stable not only throughout subsequent phases of ciliogenesis but also throughout the life of the mature functional cilium. Therefore, to summarize, centriole length is tightly controlled, displaying little variability within its specific biological context, and final centriole length is usually achieved before the initiation of axoneme assembly.

The Typical Cilium Is Compartmentalized by a Transition Zone

As described above, the axoneme can be regarded as an uninterrupted extension of the microtubule-based architecture of the centriole, upon which is layered a unique and defining complement of axoneme-specific proteins. Thus, various mechanisms must ensure that the centriole and axoneme faithfully establish and maintain their distinct compositions, and therefore functions. One such mechanism is an inherent consequence of the stepwise assembly process directing centriole and axoneme growth. As the centriole grows distally, centriolar proteins are initially deposited on the microtubule scaffold. This process is similarly mirrored during axoneme assembly, with the deposition of axonemal proteins occurring instead. This differential deposition occurring at the level of the centriole and axoneme is intimately tied to the activity of both a ciliary trafficking mechanism known as intraflagellar transport (IFT) (Cole et al. 1998; Rosenbaum and Witman 2002) and a molecular gatekeeping complex within a structure known as the transition zone, which resides at the interface of the cytoplasmic centriole and the axonemal compartment (Craige et al. 2010; Garcia-Gonzalo et al. 2011). Together, IFT and the transition zone ciliary gate cooperate to maintain the centriole and axoneme within distinct cellular compartments. The spatial separation of the centriole and axoneme encourages the differential deposition of specific proteins onto otherwise structurally similar substrates, and axoneme compartmentalization is generally thought to be an essential prerequisite for normal ciliogenesis.

In a typical cilium, the transition zone is found just distal to the centriole and represents the most proximal segment of the cilium. Within this region, the axoneme becomes anchored to the membrane of the cilium, which takes on a characteristic bend known as the ciliary pocket (Fig. 1A). The transition zone is composed of a complex arrangement of both cytoplasmic and membrane-bound proteins that are thought to both connect the axoneme to the base of the ciliary membrane and to establish an active sieve that gates ciliary entry and exit at both the level of the cytoplasm and membrane (Garcia-Gonzalo et al. 2011). As a result of its multiple activities, the transition zone is compositionally diverse. For example, mass spectrometry of the centrosome–cilium interface using proximity-dependent biotinylation and capture has recently shown that at least 22 distinct proteins may comprise the mammalian transition zone (Gupta et al. 2015). In addition to cilium-specific proteins, the transition zone also contains multiple proteins of the Septin family (Chih et al. 2012), members of the nuclear import machinery such as Ran-GTP, importins and nucleoporins (Kee et al. 2012), and ESCRT proteins (Diener et al. 2015). To add to this complexity, the transition zone functions as an intricate network, with extensive protein–protein interactions occurring both within the transition zone and with outside partners (Sang et al. 2011; Williams et al. 2011). Nevertheless, although the various roles of individual transition zone proteins remain poorly understood, proteomic and genetic studies have shown that transition zone proteins segregate into distinct functional modules defined by core interacting partners and related mutant phenotypes. Accordingly, these modules have been named by the clinical manifestations associated with mutations of their respective components—namely, nephronopthisis (NPHP module) and Meckel–Gruber syndrome (MKS module) (Sang et al. 2011; Williams et al. 2011). Still, it remains unclear which components of the transition zone are either directly or indirectly involved in axoneme compartmentalization.

Besides its interactions with the axoneme and ciliary membrane, the transition zone is also intimately associated with the distal end of the centriole. For example, just proximal to the transition zone, centriolar structures known as the distal appendages are involved in attaching the centriole to the cell membrane and docking various complexes for ciliary transport (Deane et al. 2001). Furthermore, transition zone proteins also directly interact with components of the main centriole body (Gupta et al. 2015). Thus, the transition zone is generally regarded as both spatially anchored to and functionally dependent on the distal end of the centriole (Wang et al. 2013). Accordingly, like centriole elongation, transition zone formation generally occurs before axoneme formation, thereby ensuring that axoneme assembly occurs entirely within a separate cellular compartment.

Sperm Cells Have Specialized Centrioles and Cilia

Until now, we have discussed generalized features of ciliogenesis that occur in most species and cell types. To briefly reiterate, we described that centriole elongation occurs before the initiation of axoneme assembly and that final centriole length is usually ≤ 500 μm. Next, we discussed that axoneme assembly occurs entirely within a separate cellular compartment protected by a gatekeeping mechanism within the transition zone. Intriguingly, however, sperm cells—which are motile by means of a modified cilium (often referred to as the flagellum)—often infringe on these general rules of cilium biology, showing two important distinctions: (1) the centriole is longer and (2) a portion of the axoneme is exposed to the cytoplasm (Fig. 2). These unique centrioles, which occur in some insect species and in non-passerine birds, can reach 2–5 μm in length (up to 10 times typical centriole length) and are referred to as giant centrioles. Also, unlike typical fully compartmentalized cilia, the proximal segment of the axoneme in such species is exposed to the cytoplasm (Fig. 1). Nevertheless, electron microscopy studies have shown that in these species, the distal end of the cilium still forms what appears to be a typical membrane-bound axoneme (Fig. 2B).

Together, these observations have led to the distinction between two broad types of cilia, which are classified as either "cytoplasmic" or "compartmentalized" (Avidor-Reiss et al. 2004; Avidor-Reiss and Leroux 2015). Although this unique ciliary phenomena occurring in spermatids has been observed for many decades, little is known about how or why spermatid ciliogenesis forms giant centrioles and cytoplasmic axonemes. However, one adaptation that appears to be central to these processes is the disengagement of the transition zone from the centriole and its successive distal migration throughout axoneme elongation (Basiri et al. 2014).

CENTRIOLE ELONGATION AND TRANSITION ZONE MIGRATION

Giant Centrioles Are Found In Sperm Cells of Some Insects

Giant centrioles were originally described in insects using electron microscopy, first by Friedländer and Wahrman (1966) in neuropteran meiosis and shortly after by Phillips (1967) in

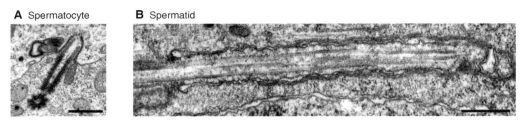

A Spermatocyte   **B** Spermatid

C Early spermatid

Ana1, Cep290, Acetyl-tubulin

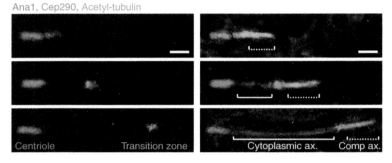

Centriole   Transition zone

Cytoplasmic ax.   Comp ax.

D Intermediate spermatid

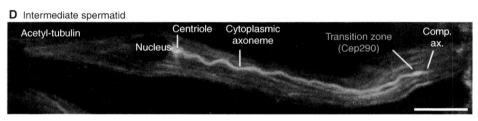

Acetyl-tubulin

Nucleus

Centriole   Cytoplasmic axoneme

Transition zone (Cep290)

Comp. ax.

Figure 2. Transition zone migration in *Drosophila melanogaster*. (*A*) Electron micrograph showing two giant centrioles, one of which forms a short cilium. Scale bar, 500 nm. (*B*) Electron micrograph showing a spermatid cytoplasmic and compartmentalized axoneme segment. The ciliary pocket is observed at the base of the compartmentalized axoneme. Scale bar, 500 nm. (*C*) Light microscopy demonstrating transition zone migration in early spermatids. The centriole is labeled by Ana1-tdTomato (red), transition zone is labeled by Cep290-GFP (green), and axoneme (ax) is labeled with antiacetylated-tubulin (cyan). Scale bar, 1 μm. (*D*) A full-length intermediate spermatid with centriole, cytoplasmic axoneme, transition zone, and compartmentalized axoneme. Scale bar, 5 μm. (*D*, From Basiri et al. 2014; reprinted, with permission, from Elsevier © 2014.)

sperm cells of *Sciara*. Later, atypically long centrioles were found in other insects including *Drosophila melanogaster* (Tates 1971), in which they were studied in the most detail. In *Drosophila*, during early spermatogenesis (from stem cells until early spermatocytes), mitotic centrioles display the typical short length representative of other tissue types (Tates 1971). Shortly afterward, these short centrioles dock to the plasma membrane and form a cilium. However, unlike most cell types in which centriole elongation terminates on attaching to

the membrane, the centrioles of *Drosophila* spermatocytes and early spermatids unexpectedly continue to elongate extensively after membrane attachment and axoneme nucleation. During this stage of elongation, the centriole grows fourfold, ultimately reaching a length of ~1.8 μm while attached to the plasma membrane (Tates 1971; Basiri et al. 2013). Simultaneously, the ciliary axoneme slowly begins to grow from the distal end of the centriole to provide a platform for transition zone assembly. This modest axoneme growth occurs throughout most of the spermatocyte differentiation program and does not extend beyond the region defined by transition zone markers at the level of light microscopy (Tates 1971; Riparbelli et al. 2013; Basiri et al. 2014).

Transition Zone Migration in *Drosophila* Spermatocytes May Enable Postaxonemal Centriole Elongation

Typically, the transition zone is regarded as a static structure whose position does not change after its assembly. However, a static model of the transition zone poses problems for postaxonemal centriole elongation. So, how can a centriole elongate after forming a cilium? Theoretically, there are three possibilities: by elongating at its microtubule "plus" end, its "minus" end, or at some point within. Based on our current understanding of the rules governing microtubule assembly, it is unlikely that centrioles elongate via the latter two options. Considering the synonymous microtubule-based scaffolding of both the centriole and the axoneme as well as the capacity of the transition zone to migrate, it is tempting to speculate that distal displacement of the transition zone would expose bare, immature axoneme to the cytoplasm where it would be transformed into a centriole. Such a transformation could be accomplished by incorporating centriolar proteins to the freshly exposed "A" and "B" microtubules of the axoneme and concurrent extension of centriolar "C" microtubules (Fig. 1C).

Giant Centrioles Are Also Present in Some Vertebrate Taxa

Similar to *Drosophila* and other insect species, the centrioles of some non-passerine birds are also considerably longer than in most species and cell types. In fact, in ostrich, the sperm distal centriole elongates to become giant in the spermatid stage, outgrowing *Drosophila* giant centrioles to ultimately reach 3 μm in length (Soley 1994). However, there are important distinctions between *Drosophila* and ostrich giant centrioles. For example, unlike *Drosophila* in

which spermatid giant centrioles are present in the background of cytoplasmic ciliogenesis, ostrich giant centrioles occur during an otherwise typical compartmentalized ciliogenesis. Furthermore, the distal centriole in ostrich grows to its final "giant" length before docking to the plasma membrane, and no centriole elongation occurs after the initiation of axoneme formation (Soley 1994). Thus, giant centrioles can also form outside of cytoplasmic ciliogenesis and transition zone migration, although in such cases final giant centriole length must be achieved before membrane docking.

CYTOPLASMIC CILIOGENESIS AND TRANSITION ZONE MIGRATION

The Axoneme Is Exposed to the Cytoplasm in Some Cells

Although a general theme of cilium biology is that the axoneme resides within a distinct compartment isolated from the cytoplasm via the transition zone, the exclusion of the axoneme from the cytoplasm is not an absolute requirement for ciliogenesis and examples of cytoplasmic axonemes are not uncommon (Avidor-Reiss and Leroux 2015). These "cytoplasmic cilia" can occur via three different mechanisms. Primary cytoplasmic ciliogenesis is found in mammalian sperm cells and is characterized by the exposure of the mature axoneme to the cytoplasm only after it is first fully formed within a compartmentalized cilium. In secondary cytoplasmic ciliogenesis, which is found in insect sperm cells, immature axoneme is exposed to the cytoplasm after its initial polymerization within a small compartment referred to as the ciliary cap. Tertiary cytoplasmic ciliogenesis is found in the microgametes of some protists such as plasmodium and is characterized by both the assembly and maturation of the axoneme in the cytoplasm. In all three types of cytoplasmic ciliogenesis, the axoneme eventually becomes associated with the cell membrane to form a tight and slender structure according to cell-type-specific processes.

In primary and secondary cytoplasmic ciliogenesis, the exposure of the axoneme to the cytoplasm requires the displacement of the attachment point between the axoneme and the ciliary membrane. This membrane attachment occurs at an electron-dense structure, known as the annulus in mammals (Fawcett et al. 1970) or the ring centriole in insects (Phillips 1970). For clarification, please note that the ring centriole is not a true microtubule-based centriole, but instead only resembles a centriole by its location at the base of a compartmentalized cilium. Similar to the transition zone of typical compartmentalized cilia, the annulus is essential for compartmentalizing the ciliary membrane by preventing exchange with the cell membrane (Caudron and Barral 2009). Recently, the annulus and ring centriole have been shown to contain typical transition zone components such as Septins and MKS module proteins (Ihara et al. 2005; Basiri et al. 2014).

Transition Zone Migration in *D. melanogaster* Spermatids Forms the Cytoplasmic Cilium

Earlier, we described the modest axoneme elongation that occurs in the context of giant centriole growth and transition zone assembly in *Drosophila* spermatocytes. However, in *Drosophila*, axoneme growth is not initiated in earnest until the conclusion of meiosis. Here, axoneme assembly involves a dramatic elongation program in which axoneme growth occurs in parallel to spermatid morphological differentiation, ultimately yielding an axoneme that is ~2 mm in length. This exaggerated spermatid length seems to be the manifestation of evolutionarily pressures favoring increased spermatid length (Joly et al. 2004). Interestingly, although this extreme axoneme growth is required for the formation of motile spermatids, spermatid elongation occurs independently of axoneme assembly and appears to be driven by a mitochondrial mechanism (Noguchi et al. 2011). Transmission electron microscopy of *Drosophila* spermatids performed in the 1970s were the first to show that in this context of dramatic axoneme growth, the point of association of the ciliary membrane with the axoneme becomes displaced distally, ultimately exposing the majority of the axoneme to the cytoplasm

(Tates 1971; Tokuyasu 1975). Although this region of axoneme membrane association contains morphological features reminiscent of a typical compartmentalized cilium, the axoneme it encloses is structurally and compositionally immature (Fig. 2B) and only represents the final 2 μm of what will eventually become an ~2-mm-long cytoplasmic axoneme.

In *Drosophila* spermatids, axoneme elongation is accompanied by transition zone disengagement from the distal end of the centriole. As the axoneme is assembled, the transition zone continuously migrates in close association with the axoneme's growing end. Throughout this process, the migrating transition zone remains consistently localized ~2 μm proximal to the very tip of the axoneme and remains coupled to the base of the ciliary pocket, which is pulled distally in concert with spermatid elongation. Thus, the migrating transition zone of *Drosophila* spermatids is perfectly situated to compartmentalize a cilium-like structure housing the growing axoneme tip. This unique centriole-detached structure, which is referred to as the ciliary cap, seems to contain the minimum elements required for axoneme compartmentalization. For example, immunostaining *Drosophila* spermatids with anti-acetyl-tubulin under a protocol that selectively permeabilizes the cell membrane without disrupting membrane of the ciliary cap allows for labeling of the entire cytoplasmic segment of the axoneme while excluding any labeling inside of the ciliary cap. Consistently, mutations of transition zone proteins such as Cep290 compromise the integrity of the ciliary cap compartment, allowing for staining of both the cytoplasmic and ciliary cap axoneme segments following selective permeabilization of the cell membrane without disrupting the ciliary cap membrane. Interestingly, mutation of *Drosophila* Cep290 also disrupts spermatid axoneme assembly, suggesting that compartmentalization of the site of axoneme assembly within the ciliary cap is essential for proper ciliogenesis in this system (Basiri et al. 2014).

In this context, two simultaneous and discrete processes characterize axoneme assembly in *Drosophila* spermatids. Within the ciliary cap compartment distal to the transition zone, axonemal microtubules are extended via patterned tubulin polymerization, the template for which was originally established in the centriole (Riparbelli et al. 2013). As the transition zone migrates, it leaves behind bare axonemal microtubules assembled in the appropriate architecture. Simultaneously, as this "naked" axoneme becomes exposed to the cytoplasm, axoneme-specific proteins become incorporated into its structure, representing the second phase of axoneme assembly (Tates 1971; Tokuyasu 1974). These proteins, which include dyneins and other axonemal proteins, are essential for sperm tail function in motility. Thus, axoneme assembly during spermatid cytoplasmic ciliogenesis occurs via compartmentalized polymerization and subsequent cytoplasmic maturation, with the migrating transition zone separating the two processes.

The Annulus Is a Septin-Based Ring Structure Found in Mammalian Spermatozoa

The mammalian sperm tail is structurally divided into four parts from proximal to distal: the connecting piece, midpiece, principal piece, and end piece. The connecting piece contains the centriole as well as other specialized structures such as a modified PCM assembly known as the striated columns. In some instances, the midpiece contains a cytoplasmic axoneme segment and mitochondria (Phillips 1974); we refer to this arrangement as an "axonemal midpiece" (Fig. 3A). However, in many fish and non-passerine bird species, the connecting piece and the midpiece are combined and contain the centriole, striated columns, and mitochondria without a cytoplasmic axoneme segment (Mattei 1988; Soley 1994), an arrangement that we refer to as a "centriolar midpiece" (Fig. 3A). Apart from this distinction, however, the junction of the midpiece and principal piece in most vertebrate species is morphologically similar, and is marked by a distinct ciliary pocket with a ring-like annulus (Fig. 3B,C). Thus, the base of the midpiece resembles the base of a typical compartmentalized cilium.

The annulus was originally characterized ultrastructurally as comprised of closely packed filaments attached to the membrane of the ciliary pocket (Fig. 3C) (Fawcett 1970). More recently, molecular studies have shown that the annulus contains several members of the Septin family of polymerizing GTP-binding proteins including Septin 1, 4, 6, 7, and 12 (Ihara et al. 2005; Steels et al. 2007). Although septin family proteins play diverse roles in many cell types, they have been shown to be particularly essential for spermatid development and motility. For example, loss-of-function mutations in Septin 12 and Septin 4 abolish the annulus and result in a marked kink at the midpiece–principal piece junction in spermatozoa, resulting in male infertility in both mice and humans (Kuo et al. 2012). Despite these morphological defects, however, these mutants displayed a fully formed midpiece (Ihara et al. 2005; Kissel et al. 2005; Kwitny et al. 2010), suggesting that the annulus is dispensable for structurally delineating the midpiece–principal piece boundary. Still, Septin 4 knockout mice show defects in the compartmentalization of membrane proteins in the midpiece, suggesting that the annulus functions as a membrane diffusion barrier to prevent the exchange of membrane components between the midpiece and principal piece (Kwitny et al. 2010). Thus, the annulus displays a similar structural and functional role to the transition zone in compartmentalizing the distal ciliary membrane.

Annulus Migration Happens during Spermiogenesis in Mammals and Some Birds during Flagellum Formation

In many vertebrates, the annulus is formed during the early stages of spermiogenesis after the centrioles dock to the plasma membrane, but before the axoneme initiates its assembly (Holstein and Roosen-Runge 1981). Initially, the annulus is located at the distal end of the centriole, later departing from this position and migrating in concert with axoneme growth. Annulus migration then stops at a precise location that is species-specific and remains in this position in the mature sperm. As a result of this migration,

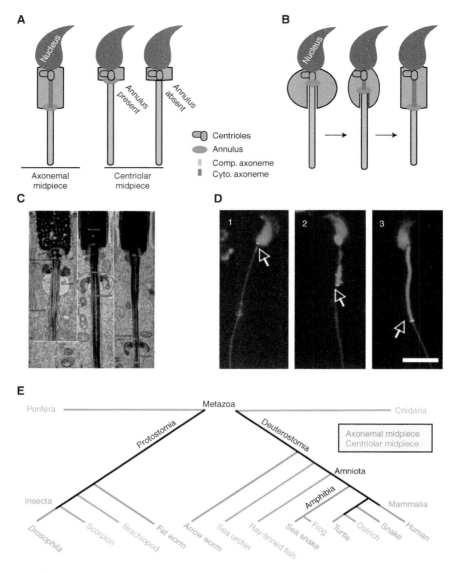

Figure 3. Annulus migration occurs in mammalian spermatids. (*A*) Sperm can be classified into three types based on the location and presence of the annulus: sperm with an annulus separating cytoplasmic and compartmentalized axoneme segments (axonemal midpiece, *left*), sperm with an annulus separating the centriole from a fully compartmentalized axoneme (centriolar midpiece with annulus), and sperm lacking an annulus but containing a centriolar midpiece and a fully compartmentalized axoneme within the principal piece (centriolar midpiece without annulus). (*B*) Illustration of annulus migration in mammalian spermatids. (*C*) Longitudinal sections of *Macaca mulatta* spermatids demonstrating annulus migration during spermiogenesis. The annulus is indicated by a red dotted circle. (*C*, Reprinted from data in Fawcett et al. 1970.) (*D*) Septin 4 localization during mouse spermiogenesis (white arrow). The annulus indicated by Septin 4 antibody (green) is found near the nucleus (blue) in early stage elongating spermatids (*1*). The annulus then begins to migrate toward the growing end of the axoneme to form the midpiece (red) (*2*). Finally, the annulus reaches the distal end of the midpiece to define the midpiece-principal piece junction in mature sperm (*3*). (*D*, From Guan et al. 2009; reprinted under the terms of the Creative Commons Attribution License.) Scale bar, 10 μm. (*E*) Distribution of spermatids containing either a centriolar midpiece or an axonemal midpiece across animal phylogeny. (Data based on Afzelius 1955; Silveira and Porter 1964; Sato et al. 1967; Reger and Cooper 1968; Mattei 1988; Hess et al. 1991; Mita and Nakamura 1992; Medina 1994; Dallai et al. 1995; Iomini and Justine 1997; Reunov and Klepal 2003; Scheltinga et al. 2003; Smita et al. 2004; Al-Dokhi et al. 2007, 2010; Vignoli et al. 2008; Lipke et al. 2009.)

Cite this article as *Cold Spring Harb Perspect Biol* doi: 10.1101/cshperspect.a028142

the midpiece contains both the centriole and a cytoplasmic segment of axoneme of defined length (Fig. 3C) (Nagano 1962; Fawcett et al. 1970; Phillips 1974; Gunawardana and Scott 1977). In mice, this process of annulus migration is nicely shown by the marker Septin 4 (Fig. 3D) (Guan et al. 2009). Aside from Septins, the annulus has also been shown to be marked by the co-chaperone DNAJB13 (Guan et al. 2009) in a manner independent of Septin 4 during early spermiogenesis, as well as the SLC26 family protein testis anion transporter 1 (Tat1) in mature sperm (Toure et al. 2007).

In some animals, such as the primitive non-passerine birds, ostrich and rhea, the annulus does not migrate but instead maintains a fixed position throughout spermiogenesis (Soley 1994). In such cases, the annulus is formed when the centriole pair docks to the plasma membrane, ultimately creating a midpiece that is only composed of a centriole without a cytoplasmic axoneme segment (centriolar midpiece) (Fig. 3A) (Soley 1994). Considering this, the annulus is likely to have a role that is independent of transition zone migration.

Surprisingly, there are also animals in which electron microscopy does not identify an annulus. In these cases, the midpiece is only composed of a centriole, similar to spermatozoa in which the annulus maintains a fixed position (Jamieson et al. 1995; Lovas et al. 2012). Although untested, it is possible that these spermatids do in fact contain an annulus-like septin-based structure that is not apparent via electron microscopy.

Considering their similarities, it is likely that the migrating transition zone (the ring centrioles of insects) and the annulus are distinct names for the same ciliary structure. Both structures migrate away from the centriole during axoneme growth (Guan et al. 2009; Basiri et al. 2014). Furthermore, both contain transition zone proteins and are essential for compartmentalizing the two homologous segments of the sperm, namely, the *Drosophila* ciliary cap and the vertebrate sperm principle piece (Kissel et al. 2005; Basiri et al. 2014). One important distinction, however, is the timing of migration; annulus migration occurs after cilium forma-

tion, whereas transition zone migration in *Drosophila* occurs during axoneme growth. It remains untested whether the *Drosophila* migrating transition zone also contains a septin-based component that may contribute to ciliary cap compartmentalization.

Transition Zone Migration May Be Motor-Driven or Exchange-Driven

The mechanism of transition zone migration has not yet been characterized. Because migration occurs along the axoneme, one can imagine that microtubule-based motors are involved. Alternatively, migration may occur via a treadmilling mechanism in which the transition zone is disassembled at the "minus" end and reassembled at the "plus" end.

Studies on the ciliary microtubule-binding protein Cep162 may provide insight into the mechanisms of migration. In RPE1 cells, siRNA depletion of Cep162 (QN1/KIAA1009), which normally localizes at the centriole just beyond the distal appendages, was shown to prevent the recruitment of various transition zone proteins, thereby blocking the initiation of ciliogenesis (Wang et al. 2013). Interestingly, expression of mutant forms of Cep162 that are unable to bind the distal centriole but with intact axoneme binding capacity was shown to result in the distal accumulation of transition zone proteins including Cep290, TMEM67, TCTN1, and RPGRIP1L at the ciliary tip. Cep162 interacts with Cep290 via its axoneme recognition domain, and although Cep290 can independently associate with the distal centriole, its microtubule binding capacity in mitotic centrosomal complexes is dependent on Cep162. Thus, the Cep290–Cep162 interaction appears to be required for its transition zone association with axoneme microtubules, and untethering the transition zone from the distal centriole may permit an inherent constitutive transition zone migratory capacity to exert itself.

Cep162 is an ATPase that contains structural similarity with the kinesin microtubule plus-end motor family (Leon et al. 2006). Because this domain was not disrupted in the centriole-binding mutant (Wang et al. 2013), it is tempting to speculate that untethering Cep162 and, therefore indirectly, the transition zone, may allow for active transition zone migration via a Cep162 kinesin-type activity. Interestingly, the *Drosophila* ortholog of Cep162 (CG42699, via BLAST reciprocal best-hit) (data not shown) is enriched in the *Drosophila* testis (Chintapalli et al. 2007), and multiple sequence alignment of mammalian Cep162 and CG42699 (using Clustal Omega) (data not shown) shows that *Drosophila* does not seem to contain the amino-terminal region Cep162 that is essential for centriole binding in mammals (Wang et al. 2013). Nevertheless, aberrant displacement of the transition zone to the distal axoneme tip has not been reported in mutations of any other ciliary protein in compartmentalized cilia.

On the other hand, one can also speculate that transition zone migration occurs not via a motorized active mechanism but instead by successive rounds of displacement and replacement on the growing axonemal microtubules in a "treadmilling" manner. Consistent with this, previous studies have shown that transition zone proteins are in a dynamic equilibrium with a cytoplasmic reserve, rapidly cycling in between transition zone and cytoplasmic fractions (Craige et al. 2010). In this way, diffusion of fixed transition zone complexes back into the cytoplasm could allow for the reintegration of new transition zone proteins at more distal locations on the growing axoneme.

Transition Zone Migration May Reflect an Inherent Elasticity in the Ciliary Gate

Investigation of animals from distinct phylogenic groups that possess flagellated sperm reveals that spermiogenesis always begins with a centriole and an immature compartmentalized cilium. However, sperm development ultimately forms either an axonemal midpiece or a centriolar midpiece. Of these, spermatozoa with a centriolar midpiece are considered to represent the ancestral (primitive) form in animal evolution (Baccetti 1982). Interestingly, however, a review of the literature reveals that phylogenetic clades often contain species of both types, with some bearing a centriolar midpiece and others

an axonemal midpiece (Fig. 3E). Stated differently, the appearance and divergence of the axonemal midpiece from the primitive centriolar type does not occur at a single clear bifurcation in metazoan evolutionary history, but instead occurs seemingly indiscriminately and in manifold throughout deuterostome and protostome phylogeny. This observation suggests that transition zone migration can be acquired quickly via minor evolutionary changes, and that the capacity for migration is an intrinsic feature of spermatid cilia.

In light of this capacity, we speculate that transition zone migration may reflect an inherent elasticity in the ciliary gate that may be necessary to maintain cilium compartmentalization under external forces, even in traditional fully compartmentalized cilia. In *Drosophila* spermatids, mitochondria grow dramatically in the cytoplasm and provide the mechanical force required for spermatid elongation (Noguchi et al. 2011). As the mitochondria elongate, they advance the cell membrane in the direction of axoneme growth. Because the transition zone is anchored to the advancing cell membrane, the ciliary gate may adapt to these forces by migrating relative to the axoneme. However, forces impacting the transition zone are not exclusive to insect spermatids. Because cilia project from cell surfaces and are often motile, they are inherently exposed to significant internal and external forces. For example, in mammalian primary cilia exposed to fluid flow, force is predicted to be highest at the ciliary base where the transition zone is localized (Rydholm et al. 2010; Young et al. 2012). Therefore, even in compartmentalized cilia, the ciliary gate is expected to be elastic and show a dynamic association with the surrounding ciliary architecture to adapt to stresses that would otherwise compromise the ciliary compartment.

ACKNOWLEDGMENTS

This work is supported by Grant 1121176 (Division of Molecular and Cellular Biosciences [MCB]) from the National Science Foundation and R01GM098394 from the National Institute of General Medical Sciences.

REFERENCES

Afzelius BA. 1955. The fine structure of the sea urchin spermatozoa as revealed by the electron microscope. *Z Zellforsch Mikrosk Anat* **42:** 134–148.

Al-Dokhi O, Al-Onazee Y, Mubarak M. 2007. Fine structure of the epididymal sperm of the snake *Eryx jayakari* (Squamata, Reptilia). *Int J Zool Res* **3:** 1–13.

Avidor-Reiss T, Leroux MR. 2015. Shared and distinct mechanisms of compartmentalized and cytoplasmic ciliogenesis. *Curr Biol* **25:** R1143–R1150.

Avidor-Reiss T, Maer AM, Koundakjian E, Polyanovsky A, Keil T, Subramaniam S, Zuker CS. 2004. Decoding cilia function: Defining specialized genes required for compartmentalized cilia biogenesis. *Cell* **117:** 527–539.

Baccetti B. 1982. The evolution of the sperm tail. *Symp Soc Exp Biol* **35:** 521–532.

Basiri ML, Blachon S, Chim YC, Avidor-Reiss T. 2013. Imaging centrosomes in fly testes. *J Vis Exp* e50938.

Basiri ML, Ha A, Chadha A, Clark NM, Polyanovsky A, Cook B, Avidor-Reiss T. 2014. A migrating ciliary gate compartmentalizes the site of axoneme assembly in *Drosophila* spermatids. *Curr Biol* **24:** 2622–2631.

Caudron F, Barral Y. 2009. Septins and the lateral compartmentalization of eukaryotic membranes. *Dev Cell* **16:** 493–506.

Chih B, Liu P, Chinn Y, Chalouni C, Komuves LG, Hass PE, Sandoval W, Peterson AS. 2012. A ciliopathy complex at the transition zone protects the cilia as a privileged membrane domain. *Nat Cell Biol* **14:** 61–72.

Chintapalli VR, Wang J, Dow JA. 2007. Using FlyAtlas to identify better *Drosophila melanogaster* models of human disease. *Nat Genet* **39:** 715–720.

Cole DG, Diener DR, Himelblau AL, Beech PL, Fuster JC, Rosenbaum JL. 1998. Chlamydomonas kinesin-II-dependent intraflagellar transport (IFT): IFT particles contain proteins required for ciliary assembly in *Caenorhabditis elegans* sensory neurons. *J Cell Biol* **141:** 993–1008.

Craige B, Tsao CC, Diener DR, Hou Y, Lechtreck KF, Rosenbaum JL, Witman GB. 2010. CEP290 tethers flagellar transition zone microtubules to the membrane and regulates flagellar protein content. *J Cell Biol* **190:** 927–940.

Dallai R, Afzelius BA, Witalinski W. 1995. The axoneme of the spider spermatozoon. *Boll Zool* **62:** 335–338.

Dallai R, Mercati D, Bu Y, Yin YW, Callaini G, Riparbelli MG. 2010. The spermatogenesis and sperm structure of *Acerentomon microrhinus* (Protura, Hexapoda) with considerations on the phylogenetic position of the taxon. *Zoomorphology* **129:** 61–80.

Deane JA, Cole DG, Seeley ES, Diener DR, Rosenbaum JL. 2001. Localization of intraflagellar transport protein IFT52 identifies basal body transitional fibers as the docking site for IFT particles. *Curr Biol* **11:** 1586–1590.

Desai A, Mitchison TJ. 1997. Microtubule polymerization dynamics. *Annu Rev Cell Dev Biol* **13:** 83–117.

Diener DR, Lupetti P, Rosenbaum JL. 2015. Proteomic analysis of isolated ciliary transition zones reveals the presence of ESCRT proteins. *Curr Biol* **25:** 379–384.

Fawcett DW. 1970. A comparative view of sperm ultrastructure. *Biol Reprod* **2:** 90–127.

Fawcett DW, Eddy EM, Phillips DM. 1970. Observations on the fine structure and relationships of the chromatoid body in mammalian spermatogenesis. *Biol Reprod* **2:** 129–153.

Friedlander M, Wahrman J. 1966. Giant centrioles in neuropteran meiosis. *J Cell Sci* **1:** 129–144.

Garcia-Gonzalo FR, Corbit KC, Sirerol-Piquer MS, Ramaswami G, Otto EA, Noriega TR, Seol AD, Robinson JF, Bennett CL, Josifova DJ, et al. 2011. A transition zone complex regulates mammalian ciliogenesis and ciliary membrane composition. *Nat Genet* **43:** 776–784.

Gonzalez C, Tavosanis G, Mollinari C. 1998. Centrosomes and microtubule organisation during *Drosophila* development. *J Cell Sci* **111:** 2697–2706.

Guan J, Kinoshita M, Yuan L. 2009. Spatiotemporal association of DNAJB13 with the annulus during mouse sperm flagellum development. *BMC Dev Biol* **9:** 23.

Guichard P, Chretien D, Marco S, Tassin AM. 2010. Procentriole assembly revealed by cryo-electron tomography. *EMBO J* **29:** 1565–1572.

Gunawardana VK, Scott MG. 1977. Ultrastructural studies on the differentiation of spermatids in the domestic fowl. *J Anat* **124:** 741–755.

Gupta GD, Coyaud E, Goncalves J, Mojarad BA, Liu Y, Wu Q, Gheiratmand L, Comartin D, Tkach JM, Cheung SW, et al. 2015. A dynamic protein interaction landscape of the human centrosome–cilium interface. *Cell* **163:** 1484–1499.

Hess RA, Thurston RJ, Gist DH. 1991. Ultrastructure of the turtle spermatozoon. *Anat Rec* **229:** 473–481.

Holstein AF, Roosen-Runge EC. 1981. *Atlas of human spermatogenesis*. Grosse, Berlin.

Ihara M, Kinoshita A, Yamada S, Tanaka H, Tanigaki A, Kitano A, Goto M, Okubo K, Nishiyama H, Ogawa O, et al. 2005. Cortical organization by the septin cytoskeleton is essential for structural and mechanical integrity of mammalian spermatozoa. *Dev Cell* **8:** 343–352.

Iomini C, Justine JL. 1997. Spermiogenesis and spermatozoon of *Echinostoma caproni* (Platyhelminthes, Digenea): Transmission and scanning electron microscopy, and tubulin immunocytochemistry. *Tissue Cell* **29:** 107–118.

Jamieson BG, Koehler L, Todd BJ. 1995. Spermatozoal ultrastructure in three species of parrots (aves, Psittaciformes) and its phylogenetic implications. *Anat Rec* **241:** 461–468.

Johnson KA, Rosenbaum JL. 1992. Polarity of flagellar assembly in *Chlamydomonas*. *J Cell Biol* **119:** 1605–1611.

Joly D, Korol A, Nevo E. 2004. Sperm size evolution in *Drosophila*: Inter- and intraspecific analysis. *Genetica* **120:** 233–244.

Kee HL, Dishinger JF, Blasius TL, Liu CJ, Margolis B, Verhey KJ. 2012. A size-exclusion permeability barrier and nucleoporins characterize a ciliary pore complex that regulates transport into cilia. *Nat Cell Biol* **14:** 431–437.

Kissel H, Georgescu MM, Larisch S, Manova K, Hunnicutt GR, Steller H. 2005. The *Sept4* septin locus is required for sperm terminal differentiation in mice. *Dev Cell* **8:** 353–364.

Kuo YC, Lin YH, Chen HI, Wang YY, Chiou YW, Lin HH, Pan HA, Wu CM, Su SM, Hsu CC, et al. 2012. *SEPT12* mutations cause male infertility with defective sperm annulus. *Hum Mutat* **33:** 710–719.

Kwitny S, Klaus AV, Hunnicutt GR. 2010. The annulus of the mouse sperm tail is required to establish a membrane diffusion barrier that is engaged during the late steps of spermiogenesis. *Biol Reprod* **82:** 669–678.

Leon A, Omri B, Gely A, Klein C, Crisanti P. 2006. QN1/KIAA1009: A new essential protein for chromosome segregation and mitotic spindle assembly. *Oncogene* **25:** 1887–1895.

Lipke C, Meinecke-Tillmann S, Meyer W, Meinecke B. 2009. Preparation and ultrastructure of spermatozoa from green poison frogs, *Dendrobates auratus*, following hormonal induced spermiation (Amphibia, Anura, Dendrobatidae). *Anim Reprod Sci* **113:** 177–186.

Lovas EM, Filippich LJ, Johnston SD. 2012. Spermiogenesis in the Australian cockatiel *Nymphicus hollandicus*. *J Morphol* **273:** 1291–1305.

Mattei X. 1988. The flagellar apparatus of spermatozoa in fish. Ultrastructure and evolution. *Biol Cell* **63:** 151–158.

Medina A. 1994. Spermiogenesis and sperm structure in the shrimp *Parapenaeus longirostris* (Crustacea: Dendrobranchiata): Comparative aspects among decapods. *Marine Biol* **119:** 449–460.

Mita M, Nakamura M. 1992. Ultrastructural study of an endogenous energy substrate in spermatozoa of the sea urchin *Hemicentrotus pulcherrimus*. *Biol Bull* **182:** 298–304.

Mitchison T, Kirschner M. 1984. Dynamic instability of microtubule growth. *Nature* **312:** 237–242.

Moritz M, Braunfeld MB, Sedat JW, Alberts B, Agard DA. 1995. Microtubule nucleation by γ-tubulin-containing rings in the centrosome. *Nature* **378:** 638–640.

Nagano T. 1962. Observations on the fine structure of the developing spermatid in the domestic chicken. *J Cell Biol* **14:** 193–205.

Noguchi T, Koizumi M, Hayashi S. 2011. Sustained elongation of sperm tail promoted by local remodeling of giant mitochondria in *Drosophila*. *Curr Biol* **21:** 805–814.

Phillips DM. 1967. Giant centriole formation in *Sciara*. *J Cell Biol* **33:** 73–92.

Phillips DM. 1970. Insect sperm: Their structure and morphogenesis. *J Cell Biol* **44:** 243–277.

Phillips DM. 1974. *Spermiogenesis*. Academic, New York.

Reger JF, Cooper DP. 1968. Studies on the fine structure of spermatids and spermatozoa from the millipede Polydesmus sp. *J Ultrastruct Res* **23:** 60–70.

Reunov A, Klepal W. 2003. Ultrastructural study of spermatogenesis in *Phoronopsis harmeri* (Lophophorata, Phoronida). *Helgol Mar Res* **58:** 1–10.

Riparbelli MG, Cabrera OA, Callaini G, Megraw TL. 2013. Unique properties of *Drosophila* spermatocyte primary cilia. *Biol Open* **2:** 1137–1147.

Rosenbaum JL, Witman GB. 2002. Intraflagellar transport. *Nat Rev Mol Cell Biol* **3:** 813–825.

Rydholm S, Zwartz G, Kowalewski JM, Kamali-Zare P, Frisk T, Brismar H. 2010. Mechanical properties of primary cilia regulate the response to fluid flow. *Am J Physiol Renal Physiol* **298:** F1096–1102.

Sang L, Miller JJ, Corbit KC, Giles RH, Brauer MJ, Otto EA, Baye LM, Wen X, Scales SJ, Kwong M, et al. 2011. Mapping the NPHP-JBTS-MKS protein network reveals ciliopathy disease genes and pathways. *Cell* **145:** 513–528.

Sato M, Oh M, Sakodas K. 1967. Electron microscopic study of spermatogenesis in the lung fluke (*Paragonimus miyazakii*). *Z Zellforsch Mikrosk Anat* **77:** 232–243.

Scheltinga DM, Wilkinson M, Jamieson BG, Oommen OV. 2003. Ultrastructure of the mature spermatozoa of caecilians (Amphibia: Gymnophiona). *J Morphol* **258:** 179–192.

Silveira M, Porter KR. 1964. The spermatozoids of flatworms and their microtubular systems. *Protoplasma* **59:** 240–265.

Smita M, George JM, Girija R, Akbarsha MA, Oommen OV. 2004. Spermiogenesis in caecilians *Ichthyophis tricolor* and *Uraeotyphlus cf. narayani* (Amphibia: Gymnophiona): Analysis by light and transmission electron microscopy. *J Morphol* **262:** 484–499.

Soley JT. 1994. Centriole development and formation of the flagellum during spermiogenesis in the ostrich (*Struthio camelus*). *J Anat* **185:** 301–313.

Steels JD, Estey MP, Froese CD, Reynaud D, Pace-Asciak C, Trimble WS. 2007. Sept12 is a component of the mammalian sperm tail annulus. *Cell Motil Cytoskeleton* **64:** 794–807.

Tates AD. 1971. Cytodifferentiation during spermatogenesis in *Drosophila melanogaster*: An electron microscope study. Rijksuniversiteit de Leiden, Leiden, Netherlands.

Tokuyasu KT. 1974. Dynamics of spermiogenesis in *Drosophila melanogaster*. IV: Nuclear transformation. *J Ultrastruct Res* **48:** 284–303.

Tokuyasu KT. 1975. Dynamics of spermiogenesis in *Drosophila melanogaster*. VI: Significance of "onion" nebenkern formation. *J Ultrastruct Res* **53:** 93–112.

Toure A, Lhuillier P, Gossen JA, Kuil CW, Lhote D, Jegou B, Escalier D, Gacon G. 2007. The testis anion transporter 1 (Slc26a8) is required for sperm terminal differentiation and male fertility in the mouse. *Hum Mol Genet* **16:** 1783–1793.

Vignoli V, Klann AE, Michalik P. 2008. Spermatozoa and sperm packages of the European troglophylous scorpion *Belisarius xambeui* Simon, 1879 (Troglotayosicidae, Scorpiones). *Tissue Cell* **40:** 411–416.

Wang WJ, Tay HG, Soni R, Perumal GS, Goll MG, Macaluso FP, Asara JM, Amack JD, Tsou MF. 2013. CEP162 is an axoneme-recognition protein promoting ciliary transition zone assembly at the cilia base. *Nat Cell Biol* **15:** 591–601.

Williams CL, Li C, Kida K, Inglis PN, Mohan S, Semenec L, Bialas NJ, Stupay RM, Chen N, Blacque OE, et al. 2011. MKS and NPHP modules cooperate to establish basal body/transition zone membrane associations and ciliary gate function during ciliogenesis. *J Cell Biol* **192:** 1023–1041.

Young YN, Downs M, Jacobs CR. 2012. Dynamics of the primary cilium in shear flow. *Biophys J* **103:** 629–639.

Cite this article as *Cold Spring Harb Perspect Biol* doi: 10.1101/cshperspect.a028142

Primary Cilia and Coordination of Receptor Tyrosine Kinase (RTK) and Transforming Growth Factor β (TGF-β) Signaling

Søren T. Christensen, Stine K. Morthorst, Johanne B. Mogensen, and Lotte B. Pedersen

Department of Biology, University of Copenhagen, DK-2100 Copenhagen OE, Denmark

Correspondence: stchristensen@bio.ku.dk; lbpedersen@bio.ku.dk

Since the beginning of the millennium, research in primary cilia has revolutionized our way of understanding how cells integrate and organize diverse signaling pathways during vertebrate development and in tissue homeostasis. Primary cilia are unique sensory organelles that detect changes in their extracellular environment and integrate and transmit signaling information to the cell to regulate various cellular, developmental, and physiological processes. Many different signaling pathways have now been shown to rely on primary cilia to function properly, and mutations that lead to ciliary dysfunction are at the root of a pleiotropic group of diseases and syndromic disorders called ciliopathies. In this review, we present an overview of primary cilia-mediated regulation of receptor tyrosine kinase (RTK) and transforming growth factor β (TGF-β) signaling. Further, we discuss how defects in the coordination of these pathways may be linked to ciliopathies.

Cellular signaling pathways form complex networks that converge signals from many different receptors to regulate a range of cellular and physiological processes during development and in tissue homeostasis (Lage et al. 2010; Kirouac et al. 2012). These networks operate in a spatiotemporal manner to balance the capacity of cells to perceive and transmit signals and process information. It is now evident that primary cilia take center stage in the ability of cells to register extracellular cues and coordinate the activity of multiple signaling pathways (Satir et al. 2010), which when defective may lead to ciliopathies (Hildebrandt et al. 2011; Waters and Beales 2011; Norris and Grimes 2012).

Cilia are membrane-bound, microtubule (MT)-based organelles that extend from a modified centriole (basal body) at the surface of many eukaryotic cells ranging from single-celled organisms, such as *Chlamydomonas* and *Trypanosoma*, to complex multicellular organisms, such as human (Satir and Christensen 2007). The MT core of the cilium, the axoneme, can vary in structure and composition among cell types and organisms, and cilia are, therefore, classified based on axoneme structure and presence or absence of proteins important for motility (i.e., inner and outer dynein arms, nexins and radial spokes). Motile cilia typically have axonemes comprised by nine outer doublets of MTs surrounding a central MT pair (9+2 cilia) and are usually 5 to 10 μm long and present in large numbers per cell, such as in the epithelium of the tracheal and nasal cavity, the oviduct, and the brain ventricles (Satir and Christensen 2007). Flagella, which are present in a single copy in sperm cells and one or more copies in many protists, may be longer than cilia but are otherwise similar in axonemal construction. Primary cilia, on the other hand, are thought to be nonmotile, usually display axonemes with 9+0 MT configuration, and are present in a single copy on the surface of most nondividing vertebrate cell types, depending on their lineage and differentiation stage (Blitzer et al. 2011; Bangs et al. 2015). The typical primary cilium ranges in length from about 1 to 20 μm with a diameter of roughly 200 nm, similar to that of motile cilia.

The basal body of primary cilia is derived from the centrosomal mother centriole (Fig. 1) (Paintrand et al. 1992; Kobayashi and Dynlacht 2011), whereas in multiciliated cells basal body formation requires the production of many new centrioles (Al et al. 2014). Importantly, primary cilia can be found in different shapes and structures, depending on their function, which primarily is associated with the ability to organize and integrate receptors and cellular signaling pathways during development and in tissue homeostasis (Satir and Christensen 2007). Modified primary cilia comprise those present at dendritic endings of sensory neurons, including the outer segment of photoreceptor cells and cilia on olfactory sensory neurons. As an example, the outer segment of rod photoreceptors corresponds to the ciliary shaft of a prototypic

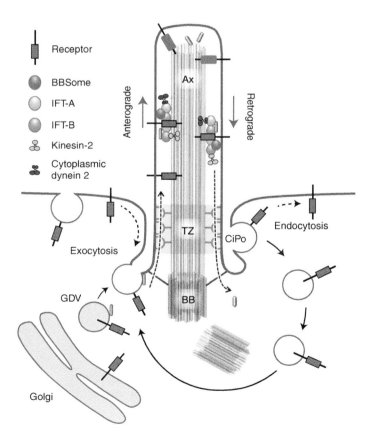

Figure 1. Overview on trafficking pathways involved in ciliary assembly, homeostasis, and signaling. IFT-A/B, Intraflagellar transport complexes A/B; Ax, axoneme; TZ, ciliary transition zone; BB, basal body; GDV, Golgi-derived vesicle; CiPo, ciliary pocket. Please see text for further details and references.

cilium and is composed of a cylinder-shaped stack of membrane disks that detect photons through the activation of the dim light receptor Rhodopsin (Roepman and Wolfrum 2007). Motile cilia in mammalian cells also have sensory capacities, such as those in the oviduct epithelium, in which transient receptor potential (TRP) ion channels as well as angiopoietin and progesterone receptors may localize to the cilia in a temporal manner during ovulation (Teilmann and Christensen 2005; Teilmann et al. 2005; Teilmann et al. 2006). Further, members of the bitter taste receptors in the class A/rhodopsin family of G-protein-coupled receptors (GPCRs) localize to motile cilia in the mammalian airways to support sinonasal innate immunity and prevent lung inflammation (Shah et al. 2009).

For most classes of cilia, axonemes are assembled and maintained by intraflagellar transport (IFT). IFT is characterized by kinesin-2 and cytoplasmic dynein 2-mediated movement of trains of IFT particles, with associated ciliary cargo, from the ciliary base toward the tip and back along the axonemal outer doublet MTs (Fig. 1) (Pedersen et al. 2008; Bhogaraju et al. 2013; Lechtreck 2015). Importantly, the IFT system also participates in ciliary transport and compartmentalization of selected membrane receptors and other signaling proteins. For example, during Sonic hedgehog (SHH) signaling retrograde IFT and the BBSome (a complex of eight Bardet–Biedl syndrome proteins) (Nachury et al. 2007; Loktev et al. 2008) mediate ciliary export of the SHH receptor Patched1 (PTCH1) and the class F GPCR Smoothened (SMO) (Keady et al. 2012; Eguether et al. 2014), as well as ciliary trafficking and processing of the GLI transcription factors (Haycraft et al. 2005), which are the main downstream effectors of the pathway. This dynamic IFT-mediated compartmentalization of signaling proteins is fundamental for ciliary function, exemplified by the fact that mutations that specifically inhibit the association of signaling receptors (e.g., PTCH1 and SMO) with the BBSome or retrograde IFT machinery lead to disease-associated signaling defects (Seo et al. 2011; Keady et al. 2012; Marion et al. 2012b;

Eguether et al. 2014; Schaefer et al. 2014). For a recent review on IFT and regulation of ciliary signaling, see Mourão et al. (2016). In addition to IFT-mediated compartmentalization of ciliary signaling components, the transition zone (TZ) located between the basal body and cilium proper also plays an essential role in establishment and maintenance of the cilium as a compartmentalized signaling organelle by regulating selective passage of proteins and lipids into and out of the cilium (Fig. 1) (Garcia-Gonzalo et al. 2011; Szymanska and Johnson 2012; Jensen et al. 2015; Takao and Verhey 2015). Ultrastructurally, the TZ is characterized by champagne-glass-like structures that connect the axonemal outer doublet MTs to the ciliary/TZ membrane and which may organize rows of intramembrane particles known as the ciliary necklace (Gilula and Satir 1972). Furthermore, the TZ membrane appears to be enriched in condensed lipids that may function as a fence to separate the ciliary membrane from the adjacent periciliary membrane domain (Vieira et al. 2006). Importantly, a large number of genes that are mutated in ciliopathies such as nephronophthisis (NPHP) and Meckel–Gruber syndrome (MKS) code for proteins that localize to the TZ (Reiter et al. 2012; Takao and Verhey 2015), highlighting the functional importance of this region.

The region between the TZ membrane and the plasma membrane, referred to as the periciliary membrane, is often infolded to produce a ciliary pocket (CiPo) that comprises an active site for exocytosis and clathrin-mediated endocytosis (CME) of ciliary components (Fig. 1) (Benmerah 2013). These processes are essential not only for ciliary membrane formation and maintenance, but also for regulation of trafficking and activation/inactivation of ciliary receptors and downstream signaling components (Dwyer et al. 2001; Field and Carrington 2009; Hu et al. 2007; Kaplan et al. 2012; Clement et al. 2013a; Bauss et al. 2014).

In summary, multiple components along the centrosome–cilium axis contribute to the establishment and maintenance of the cilium as a dynamic, compartmentalized signaling organelle, which in a spatiotemporally controlled manner can detect, integrate, and transmit a range of different signals to regulate fundamental processes at various levels from cell to organism. Consequently, mutations that impair the structure or function of either subdomain along the centrosomal–ciliary axis are associated with diseases and syndromic disorders (ciliopathies) that include cystic diseases of kidney, pancreas and liver, blindness, congenital heart disease, craniofacial and skeletal patterning defects, neurological disorders and obesity, which may be manifested in pleiotropic syndromes such BBS, NPHP, and MKS as well as Alström (AS) and Joubert (JS) syndromes (Hildebrandt et al. 2011; Koefoed et al. 2014; Valente et al. 2014). In the following, we first present a brief, general overview of signaling pathways that are known to be coordinated by primary cilia. Next, we provide a more in-depth description of cilia-mediated regulation of RTK and TGF-β signaling, including a discussion on how defects in the coordination of these pathways may be linked to ciliopathies.

OVERVIEW OF SIGNALING IN PRIMARY CILIA

The sensory capacity of primary cilia is reflected by the enriched localization of specified receptors in the ciliary membrane, where ligand binding activates downstream signaling events for transmission of information to the cell. Some of the best described examples of ciliary receptors include the 12 transmembrane SHH receptor PTCH1, which when bound to ligand exits the cilium concomitantly with ciliary entry of SMO, thereby initiating a series of downstream events to activate GLI-mediated target gene expression during development and in the adult (Mukhopadhyay and Rohatgi 2014; Pedersen et al. 2016), class A and B GPCRs that regulate various behavioral responses and tissue homeostasis (Schou et al. 2015; Hilgendorf et al. 2016), RTKs that function in cell migration and proliferation (Christensen et al. 2012), and TGF-β receptors (TGF-βRs) that are important in heart development (Koefoed et al. 2014). The latter two classes of receptors will be described in more detail below. Further, the membrane of primary cilia has been associ-

ated with receptors for extracellular matrix (ECM) proteins (McGlashan et al. 2006; Seeger-Nukpezah and Golemis 2012), purinergic receptors (Masyuk et al. 2008), as well as multiple types of ion channels of the TRP family, which similar to ciliary receptors for ECM proteins (McGlashan et al. 2006; Seeger-Nukpezah and Golemis 2012) were proposed to function in flow sensing and/or mechanosensation (Phua et al. 2015). Finally, Notch receptors, which are activated by the δ-like and Jagged families of transmembrane ligands expressed in trans on neighboring cells (Hori et al. 2013), were shown to localize to primary cilia to control skin development (Ezratty et al. 2011).

The multiplicity of ciliary receptor systems opens up a whole realm of possibilities for the primary cilium to coordinate the cross talking between different signaling pathways, which in a concerted action balances the biological output. To illustrate this, primary cilia are associated with numerous pathways that are well known to form complex signaling networks. These include mitogen-activated protein kinase (MAPK), phosphoinositide 3-kinase (PI3K)-AKT, Hippo, nuclear factor κ light-chain enhancer of activated B cells (NF-κB), and mammalian target of rapamycin (mTOR) pathways (Boehlke et al. 2010; Christensen et al. 2012; Wann et al. 2014; Hansen et al. 2015; Umberger and Caspary 2015). In addition, primary cilia were proposed to play a critical role in coordinating the balanced regulation of WNT/β-catenin versus WNT/PCP (planar cell polarity) pathways (Lienkamp et al. 2012; Oh and Katsanis 2013; Veland et al. 2013; Li et al. 2015; Saito et al. 2015), which extensively cross talk with SHH, TGF-β, RTK, MAPK, Hippo, and Notch signaling (Bernascone and Martin-Belmonte 2013; Zhang et al. 2014; Borggrefe et al. 2016; Zhang et al. 2016).

The sensory capacity of primary cilia evidently relies on regulation of ciliary formation and maintenance as well as dynamic localization of receptors and their downstream signaling components along the cilium–centrosome axis. Although much of the existing work on ciliary signaling has focused on discrete pathways and isolated cellular processes, there is an increasing interest in understanding how these pathways cooperate in the context of primary cilia, for instance, how HH signaling is modulated by Notch signaling (Kong et al. 2015; Stasiulewicz et al. 2015). Recent advances in proteome-wide analyses of ciliary protein networks (Mick et al. 2015) may prove helpful in profiling the mechanisms that set up and translate ciliary pathway interactions and how perturbation of these mechanisms causes disease.

PRIMARY CILIA AND RTK SIGNALING

Receptor tyrosine kinases comprise a large family of cell-surface receptors that play key roles in regulating cell proliferation, differentiation, survival, metabolism, migration, and cell-cycle control. The human genome encodes 58 known RTKs that are grouped into 20 distinct subfamilies, which all share a similar structural organization comprising an extracellular ligand-binding domain, a single transmembrane helix, a cytoplasmic region harboring the protein tyrosine kinase domain, as well as additional carboxy-terminal and juxta-membrane regulatory regions. On ligand binding, RTKs undergo phosphorylation at intracellular tyrosine residues, which triggers recruitment and binding of signaling proteins at these sites followed by activation of downstream signaling pathways (Lemmon and Schlessinger 2010). Interestingly, different RTKs generally signal via the same downstream signaling pathways (e.g., the MAPK, PI3K-AKT, and phospholipase Cγ [PLCγ]) pathways, yet produce distinct outcomes in many different cell types, most likely because of differences in signal timing, magnitude, and duration (Vasudevan et al. 2015). Ligand binding usually also induces receptor dimerization although some RTKs form dimers or higher-order oligomers even in the absence of ligand. Many RTKs are able to form both homodimeric as well as heterodimeric species with different ligand-binding and signaling properties, thereby expanding their functional repertoire. Moreover, RTKs cross talk extensively with each other as well as with other types of signaling receptors, thereby contributing to the formation of highly complex signaling networks (Lemmon and Schlessinger 2010). Given their importance in regulating a range of basic cellular processes, it is not surprising that mutations in genes coding for specific RTKs have been linked to numerous diseases, including many cancers, and the structure and function of many RTKs have, therefore, been intensely investigated over the years (for recent reviews, see Lemmon and Schlessinger 2010; Fantauzzo and Soriano 2015; McDonell et al. 2015). Here we will limit our discussion to RTKs that have been found to localize to primary cilia and/or that have been functionally coupled to this organelle.

Primary Cilia and Regulation of PDGFRαα Signaling

One of the first RTKs that was shown to localize to primary cilia is platelet-derived growth factor receptor α (PDGFRα) (Fig. 2) (Schneider et al. 2005), which is broadly required during embryogenesis (e.g., during development of the central nervous system, neural crest, and neural crest mesenchyme-derived structures, such as the cardiac outflow tract, the thymus, and skeletal components of the facial region). Consequently, targeted disruption of *Pdgfra* in the mouse is embryonic lethal and results in embryos with a range of developmental phenotypes (Soriano 1997; Andrae et al. 2008; Fantauzzo and Soriano 2015). Similar abnormalities (e.g., isolated cleft palate) have been observed in human patients harboring *PDGFRA* mutations (Rattanasopha et al. 2012). Moreover, dysregulation of PDGFRα signaling has been linked to various cancers and fibrotic diseases in humans (Andrae et al. 2008; Corless et al. 2011). In addition to PDGFRα, the mammalian PDGF signaling network includes a related RTK, PDGFRβ, as well as four ligands, PDGFA−D, that are known to bind as dimers to homodimeric PDGFR species with different affinities. For example, PDGF-AA specifically binds and activates homodimeric PDGFRαα, whereas PDGF-DD solely activates PDGFRββ (Andrae et al. 2008; Fantauzzo and Soriano 2015). Although a heterodimeric species of PDGFRαβ is known to exist (Rupp et al.

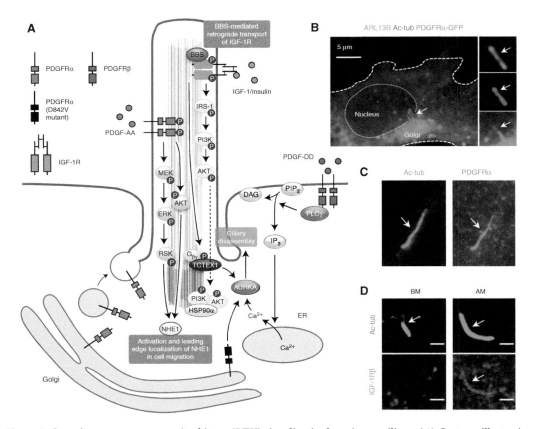

Figure 2. Overview on receptor tyrosine kinase (RTK) signaling in the primary cilium. (*A*) Cartoon illustrating signaling pathways regulated by platelet-derived growth factor receptor (PDGFR)αα and insulin-like growth factor (IGF)-1R in the primary cilium as well as extraciliary RTK signaling in ciliary disassembly. (*B*) Fluorescence microscopy analysis on the localization of green fluorescent protein (GFP)-PDGFRα (green) to the Golgi and the primary cilium in retinal pigment epithelium (RPE) cells. Cilia (arrows) were costained with anti-ARL13B (red) and antiacetylated α-tubulin (Ac-tub, blue). (Panel from Nielsen et al. 2015; reprinted, with permission, from Company of Biologists.) (*C*) Fluorescence microscopy analysis on the localization of endogenous PDGFRα (green) to the primary cilium (Ac-tub, red) in mouse embryonic fibroblasts. (Panel from Schneider et al. 2005; reprinted, with permission, from Elsevier.) (*D*) Fluorescence microscopy analysis on ciliary length and localization of endogenous IGF-1Rβ (red) to primary cilia (Ac-tub, green) in human mesenchymal stem cells (hMSCs) cultured in basal media (BM), adipogenic media (AM), and basal medium (BM), and after 5 days. Scale bar, 10 μm. (Panel from Dalbay et al. 2015; reprinted under the terms of the Creative Commons Attribution License.)

1994), its specificity and function remains unclear (Andrae et al. 2008; Gerhardt et al. 2013). Following ligand binding and receptor dimerization the PDGFR dimers undergo autophosphorylation to activate downstream signaling via the MAPK, PI3K-AKT, and PLCγ pathways (Andrae et al. 2008).

Using specific antibodies and expression of GFP-tagged PDGFRα, Schneider and col-

leagues first showed that PDGFRα localizes to the primary cilium of mouse NIH3T3 cells and mouse embryonic fibroblasts (MEFs) (Fig. 1C), whereas PDGFRβ is largely absent from this compartment and predominantly distributed to patches on the plasma membrane (Schneider et al. 2005, 2010). In line with this result, analysis of PDGFRα mRNA and protein levels showed that PDGFRα is highly upregulated during serum deprivation in cultured cells, concomitantly with formation of the primary cilium (Lih et al. 1996; Schneider et al. 2005). Subsequent studies have confirmed cilia-specific localization of PDGFRα in a range of additional cell types, including rat astrocytes and neuroblasts (Danilov et al. 2009), mouse heart ventricular cells (Gerhardt et al. 2013), human embryonic stem cells (Awan et al. 2010), ovarian surface epithelial cells (Egeberg et al. 2012), and mouse osteoblasts (Noda et al. 2016). Of note, in some ciliated cell types, such as rat oligodendrocytes (Falcon-Urrutia et al. 2015) and mouse heart atrial cells (Gerhardt et al. 2013), PDGFRα seems to be conspicuously absent from the organelle, whereas in retinal pigment epithelial (RPE) cells, which are commonly used in ciliary studies, PDGFRα seems hardly to be expressed at all (Lei et al. 2011; Nielsen et al. 2015). Nevertheless, when GFP-tagged PDGFRα was expressed in RPE cells, the fusion protein was found to localize to the primary cilium (Fig. 2B) (Nielsen et al. 2015), suggesting that these cells indeed contain the machinery for targeting the receptor to this organelle. Moreover, careful examination of newly divided, cultured NIH3T3 cells revealed that PDGFRα localizes asynchronously to cilia of two sister cells, with the receptor preferentially accumulating in the cilium emanating from the cell with the oldest mother centriole (Anderson and Stearns 2009). It is possible that such age-dependent recruitment of PDGFRα to the cilium also operates in vivo, which may explain why the receptor is absent from cilia in certain cell types such as rat oligodendrocytes (Falcon-Urrutia et al. 2015) and mouse heart atrial cells (Gerhardt et al. 2013). The molecular determinants that confer cell-type- and/or age-dependent ciliary targeting of PDGFRα, as well as the mechanisms regulating its cilium- and/or cell-type-dependent expression are interesting avenues for future research.

Although studies using antibody staining of endogenous receptor or heterologous expression of GFP-tagged receptor have now confirmed localization of PDGFRα to the primary cilium in a variety of cell types (see above), cell-based assays in a range of mutant cells with specific ciliary assembly defects have substantiated a requirement for primary cilia in regulating PDGFRα-mediated signaling. First, in MEFs that contain a hypomorphic mutation in the gene encoding IFT-B complex protein IFT88 ($Tg737^{orpk}$) (Murcia et al. 2000; Pazour et al. 2000), PDGFRαα-mediated signaling via the MEK1/2-ERK1/2 and PI3K-AKT pathways is ablated (Fig. 2A) (Schneider et al. 2005, 2010), and this is associated with impaired directional cell migration and wound healing by a mechanism involving the Na^+/H^+ exchanger NHE1 (Schneider et al. 2009; Clement et al. 2013b; for review, see Christensen et al. 2012). Mutations affecting other components of the IFT machinery also lead to impaired PDGFRαα signaling in MEFs in a manner that depends on the nature of the mutation and how it affects IFT and ciliary structure (Umberger and Caspary 2015). For example, mutant cells with defects in anterograde IFT components generally display low PDGFRα levels, whereas retrograde IFT mutants display normal levels of the receptor. Nevertheless, both classes of mutant cells fail to respond properly to PDGF-AA stimulation, presumably caused by dysregulated mTORC1 signaling and PP2A activity, which affect PDGFRα expression levels and PDGF-AA mediated signaling (Umberger and Caspary 2015). Furthermore, in heart ventricles from $Ftm^{-/-}$ mutant mice embryos, which lack the ciliary TZ protein RPGRIP1L, ciliary enrichment of PDGFRα and expression of the downstream target gene of PDGFRα signaling, $Hif1\alpha$, are reduced (Gerhardt et al. 2013). These studies highlight the importance of cilia integrity for regulating PDGFRαα signaling at both the cellular as well as organ level, and are in line with the known roles of PDGFRα during development (Soriano 1997; Andrae et al. 2008; Fantauzzo and Soriano 2015). In addition, specific kinase-activating mutations in PDGFRα that are associated with increased PLCγ signaling and development of gastrointestinal stromal tumors (GISTs), such as the PDGFRα D842V mutation (Olson and Soriano 2009; Corless et al. 2011; Bahlawane et al. 2015), may lead to cilia loss by triggering calmodulin and Aurora A kinase-dependent cilia disassembly (Fig. 2A) (Pugacheva and Golemis 2006; Plotnikova et al. 2012; Nielsen et al. 2015). However, it remains to be determined whether cilia loss contributes to disease progression in GIST patients with the PDGFRα D842V mutation, although precursor cells of GIST are known to possess primary cilia (Castiella et al. 2013).

Primary Cilia and Regulation of Insulin and IGF-1 Signaling

In their inactive state the insulin and insulin-like growth factor (IGF-1) receptors (IR and IGF-1R) are unlike most RTKs covalent dimers composed of two extracellular α subunits and two transmembrane β subunits containing the tyrosine kinase domains (Hubbard 2013). The IR further comes in two isoforms, IR-A and IR-B, which can form receptor hybrids with IGF-1R that have different affinities for insulin and insulin-like growth factors (IGF-1 and IGF-2) (Belfiore et al. 2009). On ligand binding the dimers alter receptor conformation, which allows the catalytic domains to become activated followed by autophosphorylation at tyrosine residues beyond the catalytic site for recruitment of adaptor and effector proteins in signal transduction, including insulin receptor substrate 1 (IRS-1), which activates the PI3K/AKT pathway (Hubbard 2013; Cabail et al. 2015). In their activated forms, receptors of the insulin family play an essential role in the control of diverse cellular and physiological processes, including cell-cycle control, cell survival, programmed cell death, cell migration, cell differentiation, as well as metabolism (Taniguchi et al. 2006; Belfiore et al. 2009; Cohen and LeRoith 2012).

A link between primary cilia and IGF-1 signaling was first shown in 3T3-L1 preadipocytes, where cilia-mediated IGF-1R signaling was found to be important for induction of preadipocyte differentiation to adipocytes. Specifically, when ciliogenesis was ablated by knockdown of IFT88 or the anterograde IFT motor subunit Kif3a, the ability of the cells to respond to insulin was reduced; this was accompanied by reduced phosphorylation of IGF-1R and AKT at the base of the primary cilium (Fig. 2A), as well as decreased cellular expression of adipocyte transcription factors C/EBPα and PPARγ. Furthermore, a fraction of the cellular pool of IGF-1R was detected in primary cilia of the differentiating preadipocytes, and it was proposed that the receptors localized in the cilium are more sensitive to insulin stimulation than those present in the plasma membrane (Zhu et al. 2009). Similarly, in human mesenchymal stem cells (hMSCs), adipogenic differentiation was shown to involve cilia elongation and recruitment of IGF-1Rβ onto cilia (Fig. 2D), which was accompanied by increased nuclear accumulation of the early adipogenesis marker PPARγ (Dalbay et al. 2015). Interestingly, downstream transmission of ciliary IGF-1R signaling appears to rely on formation of a signaling platform at the ciliary base, composed of a complex of HSP90α, IRS-1, and AKT (Fig. 2A) (Wang et al. 2015). Ciliary compartmentalization of IGF-1/insulin signaling components may furthermore involve active intraciliary transport by the BBSome and IFT system (Fig. 2A), because a recent proteomics analysis indicated that IGF-1R accumulates dramatically in photoreceptor outer segments from *Bbs17/Lztfl1* mutant mice (Datta et al. 2015), in which association of the BBSome with the retrograde IFT machinery is impaired (Seo et al. 2011; Lechtreck 2015). Furthermore, studies using additional *Bbs* mutant animal models as well as BBS patient cells have confirmed a role for BBS proteins in regulating trafficking and function IGF-1/insulin signaling components, specifically the IR-B isoform (Gerdes et al. 2014; Starks et al. 2015), which in turn is important for controlling whole body insulin action and glucose metabolism. However, the precise mechanisms by which BBS proteins control insulin signaling and metabolism are likely to be complex, and may involve cilium-dependent as well as cilium-independent functions of BBS proteins (Marion et al. 2012a; Gerdes et al. 2014; Starks et al. 2015). Finally, in addition to regulating cilia/BBS-dependent adipocyte differentiation and metabolism, IGF-1/insulin signaling components have also been implicated in ciliary disassembly and cell-cycle progression. Specifically, activation of IGF-1R localized on the cilia of mouse fibroblasts and RPE cells was found to accelerate cilia resorption and G_1-S phase progression via a noncanonical Gβγ signaling pathway culminating in recruitment of phospho(T94)Tctex-1 to the ciliary TZ (Fig. 2A) (Yeh et al. 2013). This pathway also seems to operate in neuronal progenitor cells such as radial glia where TZ localized phospho(T94)Tctex-1 may control cell fate choice by activating the ciliary disassembly factors HDAC6 and Aurora A kinase (Li et al. 2011; Yeh et al. 2013). Interestingly, patients with mutations in *IGF-1* or *IGF-1R* were reported to suffer from microcephaly and mental retardation (Walenkamp et al. 2005; Walenkamp and Wit 2006), which may be caused by defects in cilia disassembly of neural progenitor cells (Gabriel et al. 2016). Of note, additional RTKs may cooperate with IGF-1R to promote cilia disassembly in a cell- or tissue-dependent manner, because in some cell types (e.g., mouse fibroblasts) significant IGF-1-induced ciliary disassembly was observed only in combination with other RTK ligands such as PDGF-AA (Jacoby et al. 2009; Nielsen et al. 2015). Moreover, because activation of nonciliary RTKs such as PDGFRβ induces robust ciliary disassembly, at least in some cell types (Nielsen et al. 2015), the signaling network(s) involved in control of ciliary disassembly and cell-cycle control is likely to be quite complex.

Additional RTKs Associated with Primary Cilia

Although PDGFRα and insulin/IGF-1 receptors are the most studied RTKs from a ciliary perspective, a number of other RTKs have also been linked to primary cilia in various ways (reviewed in Christensen et al. 2012). For example, epidermal growth factor (EGF) receptors were reported to localize to primary cilia in kidney epithelial cells (Ma et al. 2005), astrocytes, and neuroblasts (Danilov et al. 2009), as well as in airway smooth muscle cells where EGF signaling may contribute to mechanosensation and directed cell migration in cooperation with integrins and polycystins 1 and 2 (Wu et al. 2009). Polycystin signaling may also involve interaction with another RTK, c-Met, which was

shown in mouse kidney cells to rely on polycystin 1 for appropriate ubiquitylation and signaling down-regulation in response to stimulation with the c-Met ligand, hepatocyte growth factor (HGF) (Qin et al. 2010). Interestingly, another member of the HGF receptor subfamily of RTKs, recepteur d'origine nantais (RON), was shown to localize to motile cilia in human airway epithelial cells where it may participate in regulation of ciliary beat frequency (Manzanares et al. 2007). Functional links to cilia have also been reported for fibroblast growth factor receptors (FGFRs), which seem to regulate ciliogenesis by promoting transcription of ciliogenic genes (Christensen et al. 2012). Further, FGFRs were reported to localize to motile cilia in the airways of the rhesus monkey (Evans et al. 2002) as well as to the basal body of primary cilia in mouse neural progenitor cells (Garcia-Gonzalez et al. 2016). Finally, angiopoietin receptors Tie1 and Tie2 were shown to localize to cilia of the murine female reproductive organs, including primary cilia of the surface epithelium of the ovary, bursa, and extraovarian rete ducts, as well as to motile cilia of the oviduct (Teilmann and Christensen 2005).

In summary, a growing number of RTKs have now been shown to localize to primary (and/or motile) cilia in a variety of cell types and tissues where they control important cellular and physiological processes in a context-dependent fashion. Ciliary RTKs are likely to cross talk extensively with each other, as well as with other ciliary signaling systems such as SHH signaling (Parathath et al. 2008; McGowan and McCoy 2013) to ensure balanced signaling outputs.

PRIMARY CILIA AND TGF-β SIGNALING

The superfamily of TGF-β signaling provides one of the most fascinating systems of cellular communication, in which the cellular response greatly relies on the cellular context. This means that the effects of the same ligand can be quite different depending on the cell type and the conditions. The superfamily comprises more than 30 different ligand types of the TGF-β–activin–Nodal and bone morphogenetic protein (BMP) subfamilies, which act in paracrine or autocrine manners to activate receptor serine/threonine kinases of types I and II (TGF-βRI/II and BMP-RI/II, respectively) (Massague 2012). Canonical signaling is propagated through the activation of SMAD transcription factors (R-SMADs) and a plethora of so-called noncanonical pathways, which together with other receptor systems form extensive circuits of cross talking. In canonical signaling, the type I receptor is responsible for phosphorylation of R-SMADs, which comprise SMAD2/3 in TGF-β−activin−Nodal signaling and SMAD1/5/8 in BMP signaling. The R-SMADs then interact with SMAD4 for nuclear translocation as well as with different DNA-binding transcription factors, which are activated by other receptor pathways, such as in RTK, HH, Hippo, and WNT signaling (Guo and Wang 2009; Varelas and Wrana 2012; Zhang et al. 2016). Signaling includes the expression of SMAD7, which antagonizes TGF-β signaling through multiple mechanisms in the cytoplasm and in the nucleus, including SMURF1/2-mediated ubiquitination and degradation of TGF-βRI and R-SMADs (Yan and Chen 2011). SMAD7 also suppresses WNT/β-catenin signaling (Han et al. 2006), illustrating another layer of cross talking based on feedback inhibition.

Likewise, both type I and II receptors are able to regulate a series of non-R-SMAD pathways that impinge on multiple cellular and physiological processes. As an example, TGF-βRI-mediated activation of TGF-β-activated kinase 1 (TAK1) plays a critical role in activation of MAPKs, RhoA/Cdc42 GTPases, and NF-κB signaling, which defines a critical step in balancing diverse signaling networks (Yan and Chen 2011). In NF-κB signaling, TAK1 and TAK1 bindings proteins (TAB1/2) mediate the phosphorylation of the IKKα/β/λ complex, which in turn phosphorylates inhibitor of NF-κB (I-κB), leading to its proteasomal degradation followed by nuclear translocation of NF-κB and activation of NF-κB target genes (Kim and Choi 2012). Further, TGF-βRII directly phosphorylates the cell polarity protein PAR6, which controls axon formation in neocortical neurons (Yi et al. 2010) as well as cell polarity processes,

which when aberrantly regulated are associated with both tumor promoting and premetastatic effects (Heldin et al. 2012). Indeed, the extensive cross talking between different pathways from both canonical and noncanonical signaling pathways are responsible for the balanced and cell-context-dependent expression of thousands of different target genes to control a myriad of cellular processes (Ranganathan et al. 2007).

A critical step in the balancing of diverse pathways in TGF-β/BMP signaling includes the internalization of receptors by CME (Balogh et al. 2013; Ehrlich 2016). A well-described example includes activation of SMAD2/3 signaling, in which internalization of activated receptors into early endosomes (EEs) greatly enhances the phosphorylation of the R-SMADs. Here, SARA (SMAD anchor for receptor activation) binds to the PtdIns3P-enriched membrane of the endosomes via its FYVE zinc finger domain to enable the association between TGF-βRI and the R-SMADs (Sorkin and von-Zastrow 2009). Conversely, caveolae-mediated endocytosis has been proposed to enhance noncanonical signaling through MAPKs and PI3K-AKT (Zuo and Chen 2009). However, this view has been challenged by the recent findings that clathrin-coated and caveolae vesicles may fuse to form multifunctional compartments in TGF-β signaling (He et al. 2015). Because the CiPo is a site for extensive CME, this has led to the suggestion that the primary cilium may play a role in coordination of R-SMAD signaling (Clement et al. 2013a) as well as other pathways, which rely on receptor internalization, including HH (Pal et al. 2016) and tumor necrosis factor α (TNF-α) signaling (Rattner et al. 2010). In the following, we will give an overview on the association between TGF-β signaling and primary cilia and how this coupling is linked to the regulation of cellular and physiological processes.

Coupling Primary Cilia to TGF-β Signaling

Ciliopathies comprise a series of syndromic disorders, which extensively overlaps with phenotypes associated with aberrant TGF-β/BMP signaling. Prominent examples include structural heart defects associated with congenital heart disease (CHD) (Koefoed et al. 2014), suggesting that cilia may play a critical role in heart development through the coordination of TGF-β/BMP signaling. To some extent, this is true for nodal cilia, which are motile units that create a leftward flow of fluid across the embryonic node for defining lateral asymmetry (Hirokawa et al. 2012). In this scenario, the flow creates a gradient of Nodal ligands that accumulate at the left side of the node to activate SMAD2/3 signaling, which regulates specified gene expression profiles for asymmetric morphogenesis (Shiratori and Hamada 2014). Consequently, defects in nodal cilia cause heterotaxy and isolated CHD in humans that that arise from abnormal looping and remodeling of the heart tube into a multichambered organ (Chen et al. 2010). Further, primary cilia are present throughout the embryonic heart, and, in some cases, these cilia are expressed and distributed in a spatiotemporal manner during the developmental stages of the heart development and morphogenesis. Therefore, cardiac primary cilia may also contribute to cellular events regulated by TGF-β/BMP signaling morphogenetic events during heart development.

In support of a function of cardiac cilia in TGF-β signaling during heart development, the formation of primary cilia was reported to suppress SMAD2/3 signaling in endothelial cells thereby preventing shear stress-mediated EndoMT, which is required to populate the endocardial cushions for proper development into fibrous valves (Egorova et al. 2011). Although these studies implicated a suppressive function of primary cilia in canonical TGF-β signaling, other studies have shown that primary cilia in conjunction with CME at the CiPo are required to maintain an operative level of TGF-β signaling in fibroblasts as well as in stem cells undergoing in vitro cardiomyogenesis (Clement et al. 2013a). In fibroblasts, different components of the TGF-β signaling machinery, including TGF-βRI, TGF-βRII, SMAD2/3, SMAD4, and SMAD7 are present at the cilia–centrosome axis as evidenced by immunolocalization with specific antibodies (Fig. 3), and upon

TGF-β1 stimulation the receptors accumulate at the CiPo for CME-mediated activation of SMAD2/3 (Fig. 3A–C). Similarly, TGF-β receptors localize to primary cilia of human embryonic stem cells (Vestergaard et al. 2016), and TGF-β-mediated in vitro differentiation of mouse stem cells into cardiomyocytes is linked to a dramatic buildup of these receptors at the ciliary base where SMAD2/3 is activated. Stimulation with TGF-β1 ligand in fibroblasts also leads to increased activation of the ERK1/2 (Fig. 3A) at the ciliary base, but this seems to be independent of CME, because phosphorylation of ERK1/2 is not affected by inhibition of clathrin-mediated processes. In support of a function of primary cilia in TGF-β signaling, proteomics of primary cilia in IMCD3 cells based on proximity labeling identified SMAD2/3 and SMAD4 as well as several proteins in noncanonical signaling, such as TAB1, ERK1/2, AKT, and Rho GTPases (Mick et al. 2015).

Interestingly, $Tg737^{orpk}$ mutant fibroblasts, which contain a hypormorphic mutation in the IFT88 gene, show reduced TGF-β1-mediated SMAD2/3 signaling at the base of stunted primary cilia (Clement et al. 2013a). Although this observation is in line with the idea that cilia formation is required for proper TGF-β signaling, further studies showed that mutant cells display reduced CME at the ciliary base region (Clement et al. 2013a). This may indicate that IFT88 plays a role in setting up the CiPo for endocytic events that control TGF-β signaling. In support of this idea, IFT88 was shown to be essential for organization, orientation, and function of the flagellar pocket in *Trypanosoma brucei* (Absalon et al. 2008), which roots the flagellum and defines a unique microdomain for exocytosis and receptor-meditated endocytosis at the posterior end of this organism (Field and Carrington 2009). Indeed, cells of *T. brucei* subjected RNAi-mediated depletion of IFT88 display perturbations of vesicular trafficking, including a drastic reduction in endocytosis (Absalon et al. 2008).

Finally, TGF-β signaling has been linked to length control and TZ function in both primary and motile cilia. As an example, inhibition of

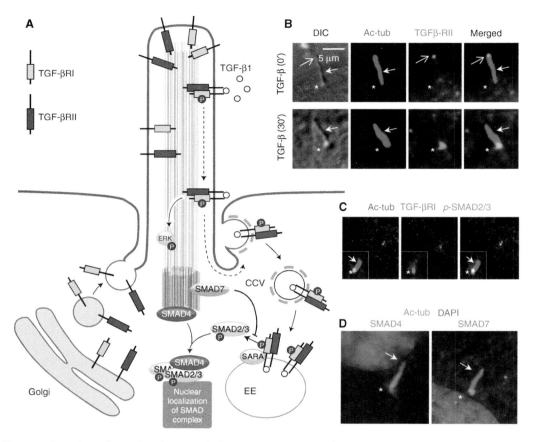

Figure 3. Overview of transforming growth factor β (TGF-β) signaling in the primary cilium. (*A*) Cartoon illustrating TGF-β signaling at the primary cilium. (*B*) Fluorescence microcopy analysis on the localization of endogenous TGF-βRII (green) at the primary cilium before (0′) and 30 min after TGF-β1 stimulation (30′). Primary cilia are shown with differential interference contrast microcopy (DIC) and stained with antiacetylated α-tubulin (Ac-tub, blue, closed arrows). The ciliary base region is marked with asterisks and the ciliary tip with open arrows. (*C*) Fluorescence microcopy analysis on the accumulation of TGF-βRI (green) and phospho-SMAD2/3 (*p*-SMAD2/3, red) at the ciliary base region after 30 min of TGF-β1 stimulation. (*D*) Fluorescence microscopy analysis of localization of SMAD4 (green, *left* panel) and SMAD7 (green, *right* panel) to the ciliary base region (asterisks). The cilium (Ac-tub, red) is marked with closed arrows, and nuclei were stained with DAPI (blue). All images were obtained from cultures of human foreskin fibroblasts. (From Clement et al. 2013a; reprinted, with permission, from Elsevier.)

SMAD2 signaling reduces the length of motile cilia at the gastrocoel roof plate, at the *Xenopus* left–right (LR) organizer, and at the neural tube and the epidermis, and blockage of TGF-β signaling was suggested to impair the structure and/or function of the ciliary TZ as evidenced by lack of the TZ protein B9D1/MSKR-1 in cilia emerging from the epidermis (Tozser et al. 2015). In chondrocytic cells, TGF-β signaling was reported to suppress the levels of Ift88

mRNA stability, leading to a reduced average length and number of primary cilia (Kawasaki et al. 2015), and in the disorganized growth plate of *Smad1/5^CKO* mutant mice, the orientation of projecting primary cilia is disturbed (Ascenzi et al. 2011). These results suggest that TGF-β signaling plays multiple roles in cilia, and future studies should focus on the mechanisms by which the CiPo is organized and regulated to control endo- and exocytic events in ciliary signaling, and how TGF-β signaling itself impacts on ciliary length and TZ organization. Further, it will be important to address the mechanisms that control the movement of TGF-β receptors into and out of the cilium proper and how this impinges on the balanced regulation of diverse signaling pathways to control cellular and physiological processes during development and in tissue homeostasis.

CONCLUDING REMARKS

Primary cilia play a critical role in the coordination of multiple cellular signaling pathways that control diverse cellular and physiological processes, and when aberrantly regulated may be the cause of syndromic disorders and disease. Here we have presented an overview of our current understanding of the connection between primary cilia and regulation of RTK and TGF-β signaling in mammalian cells, which offer a platform from which to understand complex signaling machineries during development and in function of tissues and organs in the adult. Clearly, many questions still need to be resolved; especially in terms of the molecular mechanisms of ciliary targeting and dynamic trafficking of RTKs and TGF-β receptors into and out of the cilium proper, and how the spatiotemporal regulation of these events contributes to the balanced regulation of signaling networks. Furthermore, it will be of interest to investigate how endocytic events as well as recycling of receptors to the cilium may contribute to these networks.

ACKNOWLEDGMENTS

We apologize to our colleagues whose work we could not cite because of space limitations. This work is supported by grants from the Danish Council for Independent Research (1331-00254; 6108-004578), the Novo Nordisk Foundation (NNF15OC0016886), and the University of Copenhagen Excellence Programme for Interdisciplinary Research (2016 Funds). S.K.M. and J.B.M. are partially supported by PhD fellowships from the Department of Biology, University of Copenhagen.

REFERENCES

Absalon S, Blisnick T, Bonhivers M, Kohl L, Cayet N, Toutirais G, Buisson J, Robinson D, Bastin P. 2008. Flagellum elongation is required for correct structure, orientation and function of the flagellar pocket in *Trypanosoma brucei. J Cell Sci* 121: 3704–3716.

Al JA, Lemaitre AI, Delgehyr N, Faucourt M, Spassky N, Meunier A. 2014. Centriole amplification by mother and daughter centrioles differs in multiciliated cells. *Nature* 516: 104–107.

Anderson CT, Stearns T. 2009. Centriole age underlies asynchronous primary cilium growth in mammalian cells. *Curr Biol* 19: 1498–1502.

Andrae J, Gallini R, Betsholtz C. 2008. Role of platelet-derived growth factors in physiology and medicine. *Genes Dev* 22: 1276–1312.

Ascenzi MG, Blanco C, Drayer I, Kim H, Wilson R, Retting KN, Lyons KM, Mohler G. 2011. Effect of localization, length and orientation of chondrocytic primary cilium on murine growth plate organization. *J Theor Biol* 285: 147–155.

Awan A, Oliveri RS, Jensen PL, Christensen ST, Andersen CY. 2010. Immunoflourescence and mRNA analysis of human embryonic stem cells (hESCs) grown under feeder-free conditions. *Methods Mol Biol* 584: 195–210.

Bahlawane C, Eulenfeld R, Wiesinger MY, Wang J, Muller A, Girod A, Nazarov PV, Felsch K, Vallar L, Sauter T, et al. 2015. Constitutive activation of oncogenic PDGFRα-mutant proteins occurring in GIST patients induces receptor mislocalisation and alters PDGFRα signalling characteristics. *Cell Commun Signal* 13: 21.

Balogh P, Katz S, Kiss AL. 2013. The role of endocytic pathways in TGF-β signaling. *Pathol Oncol Res* 19: 141–148.

Bangs FK, Schrode N, Hadjantonakis AK, Anderson KV. 2015. Lineage specificity of primary cilia in the mouse embryo. *Nat Cell Biol* 17: 113–122.

Bauss K, Knapp B, Jores P, Roepman R, Kremer H, Wijk EV, Marker T, Wolfrum U. 2014. Phosphorylation of the Usher syndrome 1G protein SANS controls Magi2-mediated endocytosis. *Hum Mol Genet* 23: 3923–3942.

Belfiore A, Frasca F, Pandini G, Sciacca L, Vigneri R. 2009. Insulin receptor isoforms and insulin receptor/insulin-like growth factor receptor hybrids in physiology and disease. *Endocr Rev* 30: 586–623.

Benmerah A. 2013. The ciliary pocket. *Curr Opin Cell Biol* 25: 78–84.

Bernascone I, Martin-Belmonte F. 2013. Crossroads of Wnt and Hippo in epithelial tissues. *Trends Cell Biol* 23: 380–389.

Bhogaraju S, Engel BD, Lorentzen E. 2013. Intraflagellar transport complex structure and cargo interactions. *Cilia* 2: 10.

Blitzer AL, Panagis L, Gusella GL, Danias J, Mlodzik M, Iomini C. 2011. Primary cilia dynamics instruct tissue patterning and repair of corneal endothelium. *Proc Natl Acad Sci* 108: 2819–2824.

Boehlke C, Kotsis F, Patel V, Braeg S, Voelker H, Bredt S, Beyer T, Janusch H, Hamann C, Godel M, et al. 2010.

Primary cilia regulate mTORC1 activity and cell size through Lkb1. *Nat Cell Biol* **12:** 1115–1122.

Borggrefe T, Lauth M, Zwijsen A, Huylebroeck D, Oswald F, Giaimo BD. 2016. The Notch intracellular domain integrates signals from Wnt, Hedgehog, TGF-β/BMP and hypoxia pathways. *Biochim Biophys Acta* **1863:** 303–313.

Cabail MZ, Li S, Lemmon E, Bowen ME, Hubbard SR, Miller WT. 2015. The insulin and IGF1 receptor kinase domains are functional dimers in the activated state. *Nat Commun* **6:** 6406.

Castiella T, Munoz G, Luesma MJ, Santander S, Soriano M, Junquera C. 2013. Primary cilia in gastric gastrointestinal stromal tumours (GISTs): An ultrastructural study. *J Cell Mol Med* **17:** 844–853.

Chen CM, Norris D, Bhattacharya S. 2010. Transcriptional control of left–right patterning in cardiac development. *Pediatr Cardiol* **31:** 371–377.

Christensen ST, Clement CA, Satir P, Pedersen LB. 2012. Primary cilia and coordination of receptor tyrosine kinase (RTK) signalling. *J Pathol* **226:** 172–184.

Clement CA, Ajbro KD, Koefoed K, Vestergaard ML, Veland IR, Henriques de Jesus MP, Pedersen LB, Benmerah A, Andersen CY, Larsen LA, et al. 2013a. TGF-β signaling is associated with endocytosis at the pocket region of the primary cilium. *Cell Rep* **3:** 1806–1814.

Clement DL, Mally S, Stock C, Lethan M, Satir P, Schwab A, Pedersen SF, Christensen ST. 2013b. PDGFRα signaling in the primary cilium regulates NHE1-dependent fibroblast migration via coordinated differential activity of MEK1/2-ERK1/2-p90RSK and AKT signaling pathways. *J Cell Sci* **126:** 953–965.

Cohen DH, LeRoith D. 2012. Obesity, type 2 diabetes, and cancer: The insulin and IGF connection. *Endocr Relat Cancer* **19:** F27–F45.

Corless CL, Barnett CM, Heinrich MC. 2011. Gastrointestinal stromal tumours: Origin and molecular oncology. *Nat Rev Cancer* **11:** 865–878.

Dalbay MT, Thorpe SD, Connelly JT, Chapple JP, Knight MM. 2015. Adipogenic differentiation of hMSCs is mediated by recruitment of IGF-1R onto the primary cilium associated with cilia elongation. *Stem Cells* **33:** 1952–1961.

Danilov AI, Gomes-Leal W, Ahlenius H, Kokaia Z, Carlemalm E, Lindvall O. 2009. Ultrastructural and antigenic properties of neural stem cells and their progeny in adult rat subventricular zone. *Glia* **57:** 136–152.

Datta P, Allamargot C, Hudson JS, Andersen EK, Bhattarai S, Drack AV, Sheffield VC, Seo S. 2015. Accumulation of non-outer segment proteins in the outer segment underlies photoreceptor degeneration in Bardet–Biedl syndrome. *Proc Natl Acad Sci* **112:** E4400–E4409.

Dwyer ND, Adler CE, Crump JG, L'Etoile ND, Bargmann CI. 2001. Polarized dendritic transport and the AP-1 mu1 clathrin adaptor UNC-101 localize odorant receptors to olfactory cilia. *Neuron* **31:** 277–287.

Egeberg DL, Lethan M, Manguso R, Schneider L, Awan A, Jorgensen TS, Byskov AG, Pedersen LB, Christensen ST. 2012. Primary cilia and aberrant cell signaling in epithelial ovarian cancer. *Cilia* **1:** 15.

Egorova AD, Khedoe PP, Goumans MJ, Yoder BK, Nauli SM, ten DP, Poelmann RE, Hierck BP. 2011. Lack of primary cilia primes shear-induced endothelial-to-mesenchymal transition. *Circ Res* **108:** 1093–1101.

Eguether T, San Agustin JT, Keady BT, Jonassen JA, Liang Y, Francis R, Tobita K, Johnson CA, Abdelhamed ZA, Lo CW, et al. 2014. IFT27 links the BBSome to IFT for maintenance of the ciliary signaling compartment. *Dev Cell* **31:** 279–290.

Ehrlich M. 2016. Endocytosis and trafficking of BMP receptors: Regulatory mechanisms for fine-tuning the signaling response in different cellular contexts. *Cytokine Growth Factor Rev* **27:** 35–42.

Evans MJ, Fanucchi MV, Van Winkle LS, Baker GL, Murphy AE, Nishio SJ, Sannes PL, Plopper CG. 2002. Fibroblast growth factor-2 during postnatal development of the tracheal basement membrane zone. *Am J Physiol Lung Cell Mol Physiol* **283:** L1263–L1270.

Ezratty EJ, Stokes N, Chai S, Shah AS, Williams SE, Fuchs E. 2011. A role for the primary cilium in Notch signaling and epidermal differentiation during skin development. *Cell* **145:** 1129–1141.

Falcon-Urrutia P, Carrasco CM, Lois P, Palma V, Roth AD. 2015. Shh signaling through the primary cilium modulates rat oligodendrocyte differentiation. *PLoS ONE* **10:** e0133567.

Fantauzzo KA, Soriano P. 2015. Receptor tyrosine kinase signaling: Regulating neural crest development one phosphate at a time. *Curr Top Dev Biol* **111:** 135–182.

Field MC, Carrington M. 2009. The trypanosome flagellar pocket. *Nat Rev Microbiol* **7:** 775–786.

Gabriel E, Wason A, Ramani A, Gooi LM, Keller P, Pozniakovsky A, Poser I, Noack F, Telugu NS, Calegari F, et al. 2016. CPAP promotes timely cilium disassembly to maintain neural progenitor pool. *EMBO J* **35:** 803–819.

Garcia-Gonzalez D, Murcia-Belmonte V, Esteban PF, Ortega F, Diaz D, Sanchez-Vera I, Lebron-Galan R, Escobar-Castanondo L, Martinez-Millan L, Weruaga E, et al. 2016. Anosmin-1 over-expression increases adult neurogenesis in the subventricular zone and neuroblast migration to the olfactory bulb. *Brain Struct Funct* **221:** 239–260.

Garcia-Gonzalo FR, Corbit KC, Sirerol-Piquer MS, Ramaswami G, Otto EA, Noriega TR, Seol AD, Robinson JF, Bennett CL, Josifova DJ, et al. 2011. A transition zone complex regulates mammalian ciliogenesis and ciliary membrane composition. *Nat Genet* **43:** 776–784.

Gerdes JM, Christou-Savina S, Xiong Y, Moede T, Moruzzi N, Karlsson-Edlund P, Leibiger B, Leibiger IB, Ostenson CG, Beales PL, et al. 2014. Ciliary dysfunction impairs β-cell insulin secretion and promotes development of type 2 diabetes in rodents. *Nat Commun* **5:** 5308.

Gerhardt C, Lier JM, Kuschel S, Ruther U. 2013. The ciliary protein Ftm is required for ventricular wall and septal development. *PLoS ONE* **8:** e57545.

Gilula NB, Satir P. 1972. The ciliary necklace. A ciliary membrane specialization. *J Cell Biol* **53:** 494–509.

Guo X, Wang XF. 2009. Signaling cross-talk between TGF-β/BMP and other pathways. *Cell Res* **19:** 71–88.

Han G, Li AG, Liang YY, Owens P, He W, Lu S, Yoshimatsu Y, Wang D, Ten DP, Lin X, et al. 2006. Smad7-induced β-catenin degradation alters epidermal appendage development. *Dev Cell* **11:** 301–312.

Hansen CG, Moroishi T, Guan KL. 2015. YAP and TAZ: A nexus for Hippo signaling and beyond. *Trends Cell Biol* **25:** 499–513.

Haycraft CJ, Banizs B, Aydin-Son Y, Zhang Q, Michaud EJ, Yoder BK. 2005. Gli2 and Gli3 localize to cilia and require the intraflagellar transport protein polaris for processing and function. *PLoS Genet* **1:** e53.

He K, Yan X, Li N, Dang S, Xu L, Zhao B, Li Z, Lv Z, Fang X, Zhang Y, et al. 2015. Internalization of the TGF-β type I receptor into caveolin-1 and EEA1 double-positive early endosomes. *Cell Res* **25:** 738–752.

Heldin CH, Vanlandewijck M, Moustakas A. 2012. Regulation of EMT by TGF-β in cancer. *FEBS Lett* **586:** 1959–1970.

Hildebrandt F, Benzing T, Katsanis N. 2011. Ciliopathies. *N Engl J Med* **364:** 1533–1543.

Hilgendorf KI, Johnson CT, Jackson PK. 2016. The primary cilium as a cellular receiver: Organizing ciliary GPCR signaling. *Curr Opin Cell Biol* **39:** 84–92.

Hirokawa N, Tanaka Y, Okada Y. 2012. Cilia, KIF3 molecular motor and nodal flow. *Curr Opin Cell Biol* **24:** 31–39.

Hori K, Sen A, Artavanis-Tsakonas S. 2013. Notch signaling at a glance. *J Cell Sci* **126:** 2135–2140.

Hu J, Wittekind SG, Barr MM. 2007. STAM and Hrs down-regulate ciliary TRP receptors. *Mol Biol Cell* **18:** 3277–3289.

Hubbard SR. 2013. The insulin receptor: Both a prototypical and atypical receptor tyrosine kinase. *Cold Spring Harb Perspect Biol* **5:** a008946.

Jacoby M, Cox JJ, Gayral S, Hampshire DJ, Ayub M, Blockmans M, Pernot E, Kisseleva MV, Compere P, Schiffmann SN, et al. 2009. INPP5E mutations cause primary cilium signaling defects, ciliary instability and ciliopathies in human and mouse. *Nat Genet* **41:** 1027–1031.

Jensen VL, Li C, Bowie RV, Clarke L, Mohan S, Blacque OE, Leroux MR. 2015. Formation of the transition zone by Mks5/Rpgrip1L establishes a ciliary zone of exclusion (CIZE) that compartmentalises ciliary signalling proteins and controls PIP2 ciliary abundance. *EMBO J* **34:** 2537–2556.

Kaplan OI, Doroquez DB, Cevik S, Bowie RV, Clarke L, Sanders AA, Kida K, Rappoport JZ, Sengupta P, Blacque OE. 2012. Endocytosis genes facilitate protein and membrane transport in *C. elegans* sensory cilia. *Curr Biol* **22:** 451–460.

Kawasaki M, Ezura Y, Hayata T, Notomi T, Izu Y, Noda M. 2015. TGF-β suppresses Ift88 expression in chondrocytic ATDC5 cells. *J Cell Physiol* **230:** 2788–2795.

Keady BT, Samtani R, Tobita K, Tsuchya M, San Agustin JT, Follit JA, Jonassen JA, Subramanian R, Lo CW, Pazour GJ. 2012. IFT25 links the signal-dependent movement of Hedgehog components to intraflagellar transport. *Dev Cell* **22:** 940–951.

Kim SI, Choi ME. 2012. TGF-β-activated kinase-1: New insights into the mechanism of TGF-β signaling and kidney disease. *Kidney Res Clin Pract* **31:** 94–105.

Kirouac DC, Saez-Rodriguez J, Swantek J, Burke JM, Lauffenburger DA, Sorger PK. 2012. Creating and analyzing pathway and protein interaction compendia for modelling signal transduction networks. *BMC Syst Biol* **6:** 29.

Kobayashi T, Dynlacht BD. 2011. Regulating the transition from centriole to basal body. *J Cell Biol* **193:** 435–444.

Koefoed K, Veland IR, Pedersen LB, Larsen LA, Christensen ST. 2014. Cilia and coordination of signaling networks during heart development. *Organogenesis* **10:** 108–125.

Kong JH, Yang L, Dessaud E, Chuang K, Moore DM, Rohatgi R, Briscoe J, Novitch BG. 2015. Notch activity modulates the responsiveness of neural progenitors to sonic hedgehog signaling. *Dev Cell* **33:** 373–387.

Lage K, Mollgard K, Greenway S, Wakimoto H, Gorham JM, Workman CT, Bendsen E, Hansen NT, Rigina O, Roque FS, et al. 2010. Dissecting spatio-temporal protein networks driving human heart development and related disorders. *Mol Syst Biol* **6:** 381.

Lechtreck KF. 2015. IFT-cargo interactions and protein transport in cilia. *Trends Biochem Sci* **40:** 765–778.

Lei H, Rheaume MA, Velez G, Mukai S, Kazlauskas A. 2011. Expression of PDGFRα is a determinant of the PVR potential of ARPE19 cells. *Invest Ophthalmol Vis Sci* **52:** 5016–5021.

Lemmon MA, Schlessinger J. 2010. Cell signaling by receptor tyrosine kinases. *Cell* **141:** 1117–1134.

Li A, Saito M, Chuang JZ, Tseng YY, Dedesma C, Tomizawa K, Kaitsuka T, Sung CH. 2011. Ciliary transition zone activation of phosphorylated Tctex-1 controls ciliary resorption, S-phase entry and fate of neural progenitors. *Nat Cell Biol* **13:** 402–411.

Li BI, Matteson PG, Ababon MF, Nato AQ Jr, Lin Y, Nanda V, Matise TC, Millonig JH. 2015. The orphan GPCR, Gpr161, regulates the retinoic acid and canonical Wnt pathways during neurulation. *Dev Biol* **402:** 17–31.

Lienkamp S, Ganner A, Walz G. 2012. Inversin, Wnt signaling and primary cilia. *Differentiation* **83:** S49–S55.

Lih CJ, Cohen SN, Wang C, Lin-Chao S. 1996. The platelet-derived growth factor α-receptor is encoded by a growth-arrest-specific (*gas*) gene. *Proc Natl Acad Sci* **93:** 4617–4622.

Loktev AV, Zhang Q, Beck JS, Searby CC, Scheetz TE, Bazan JF, Slusarski DC, Sheffield VC, Jackson PK, Nachury MV. 2008. A BBSome subunit links ciliogenesis, microtubule stability, and acetylation. *Dev Cell* **15:** 854–865.

Ma R, Li WP, Rundle D, Kong J, Akbarali HI, Tsiokas L. 2005. PKD2 functions as an epidermal growth factor-activated plasma membrane channel. *Mol Cell Biol* **25:** 8285–8298.

Manzanares D, Monzon ME, Savani RC, Salathe M. 2007. Apical oxidative hyaluronan degradation stimulates airway ciliary beating via RHAMM and RON. *Am J Respir Cell Mol Biol* **37:** 160–168.

Marion V, Mockel A, De MC, Obringer C, Claussmann A, Simon A, Messaddeq N, Durand M, Dupuis L, Loeffler JP, et al. 2012a. BBS-induced ciliary defect enhances adipogenesis, causing paradoxical higher-insulin sensitivity, glucose usage, and decreased inflammatory response. *Cell Metab* **16:** 363–377.

Marion V, Stutzmann F, Gerard M, De MC, Schaefer E, Claussmann A, Helle S, Delague V, Souied E, Barrey C, et al. 2012b. Exome sequencing identifies mutations in LZTFL1, a BBSome and smoothened trafficking regulator, in a family with Bardet–Biedl syndrome with situs inversus and insertional polydactyly. *J Med Genet* **49:** 317–321.

Massague J. 2012. TGF-β signalling in context. *Nat Rev Mol Cell Biol* **13:** 616–630.

Masyuk AI, Gradilone SA, Banales JM, Huang BQ, Masyuk TV, Lee SO, Splinter PL, Stroope AJ, LaRusso NF. 2008. Cholangiocyte primary cilia are chemosensory organelles that detect biliary nucleotides via P2Y12 purinergic receptors. *Am J Physiol Gastrointest Liver Physiol* **295:** G725–G734.

McDonell LM, Kernohan KD, Boycott KM, Sawyer SL. 2015. Receptor tyrosine kinase mutations in developmental syndromes and cancer: Two sides of the same coin. *Hum Mol Genet* **24:** R60–R66.

McGlashan SR, Jensen CG, Poole CA. 2006. Localization of extracellular matrix receptors on the chondrocyte primary cilium. *J Histochem Cytochem* **54:** 1005–1014.

McGowan SE, McCoy DM. 2013. Platelet-derived growth factor-A and sonic hedgehog signaling direct lung fibroblast precursors during alveolar septal formation. *Am J Physiol Lung Cell Mol Physiol* **305:** L229–L239.

Mick DU, Rodrigues RB, Leib RD, Adams CM, Chien AS, Gygi SP, Nachury MV. 2015. Proteomics of primary cilia by proximity labeling. *Dev Cell* **35:** 497–512.

Mourão A, Christensen ST, Lorentzen E. 2016. The intraflagellar transport machinery in ciliary signaling. *Curr Opin Struct Biol* **41:** 98–108.

Mukhopadhyay S, Rohatgi R. 2014. G-protein-coupled receptors, Hedgehog signaling and primary cilia. *Semin Cell Dev Biol* **33:** 63–72.

Murcia NS, Richards WG, Yoder BK, Mucenski ML, Dunlap JR, Woychik RP. 2000. The Oak Ridge Polycystic Kidney (orpk) disease gene is required for left–right axis determination. *Development* **127:** 2347–2355.

Nachury MV, Loktev AV, Zhang Q, Westlake CJ, Peranen J, Merdes A, Slusarski DC, Scheller RH, Bazan JF, Sheffield VC, et al. 2007. A core complex of BBS proteins cooperates with the GTPase Rab8 to promote ciliary membrane biogenesis. *Cell* **129:** 1201–1213.

Nielsen BS, Malinda RR, Schmid FM, Pedersen SF, Christensen ST, Pedersen LB. 2015. PDGFRβ and oncogenic mutant PDGFRα D842V promote disassembly of primary cilia through a PLCγ- and AURKA-dependent mechanism. *J Cell Sci* **128:** 3543–3549.

Noda K, Kitami M, Kitami K, Kaku M, Komatsu Y. 2016. Canonical and noncanonical intraflagellar transport regulates craniofacial skeletal development. *Proc Natl Acad Sci* **113:** E2589–E2597.

Norris DP, Grimes DT. 2012. Mouse models of ciliopathies: The state of the art. *Dis Model Mech* **5:** 299–312.

Oh EC, Katsanis N. 2013. Context-dependent regulation of Wnt signaling through the primary cilium. *J Am Soc Nephrol* **24:** 10–18.

Olson LE, Soriano P. 2009. Increased PDGFRα activation disrupts connective tissue development and drives systemic fibrosis. *Dev Cell* **16:** 303–313.

Paintrand M, Moudjou M, Delacroix H, Bornens M. 1992. Centrosome organization and centriole architecture: their sensitivity to divalent cations. *J Struct Biol* **108:** 107–128.

Pal K, Hwang SH, Somatilaka B, Badgandi H, Jackson PK, DeFea K, Mukhopadhyay S. 2016. Smoothened determines β-arrestin-mediated removal of the G protein-coupled receptor Gpr161 from the primary cilium. *J Cell Biol* **212:** 861–875.

Parathath SR, Mainwaring LA, Fernandez L, Campbell DO, Kenney AM. 2008. Insulin receptor substrate 1 is an effector of sonic hedgehog mitogenic signaling in cerebellar neural precursors. *Development* **135:** 3291–3300.

Pazour GJ, Dickert BL, Vucica Y, Seeley ES, Rosenbaum JL, Witman GB, Cole DG. 2000. Chlamydomonas IFT88 and its mouse homologue, polycystic kidney disease gene tg737, are required for assembly of cilia and flagella. *J Cell Biol* **151:** 709–718.

Pedersen LB, Veland IR, Schroder JM, Christensen ST. 2008. Assembly of primary cilia. *Dev Dyn* **237:** 1993–2006.

Pedersen LB, Mogensen JB, Christensen ST. 2016. Endocytic control of cellular signaling at the primary cilium. *Trends Biochem Sci* doi: 10.1016/j.tibs.2016.06.002.

Phua SC, Lin YC, Inoue T. 2015. An intelligent nano-antenna: Primary cilium harnesses TRP channels to decode polymodal stimuli. *Cell Calcium* **58:** 415–422.

Plotnikova OV, Nikonova AS, Loskutov YV, Kozyulina PY, Pugacheva EN, Golemis EA. 2012. Calmodulin activation of Aurora-A kinase (AURKA) is required during ciliary disassembly and in mitosis. *Mol Biol Cell* **23:** 2658–2670.

Pugacheva EN, Golemis EA. 2006. HEF1-Aurora A interactions: Points of dialog between the cell cycle and cell attachment signaling networks. *Cell Cycle* **5:** 384–391.

Qin S, Taglienti M, Nauli SM, Contrino L, Takakura A, Zhou J, Kreidberg JA. 2010. Failure to ubiquitinate c-Met leads to hyperactivation of mTOR signaling in a mouse model of autosomal dominant polycystic kidney disease. *J Clin Invest* **120:** 3617–3628.

Ranganathan P, Agrawal A, Bhushan R, Chavalmane AK, Kalathur RK, Takahashi T, Kondaiah P. 2007. Expression profiling of genes regulated by TGF-β: differential regulation in normal and tumour cells. *BMC Genomics* **8:** 98.

Rattanasopha S, Tongkobpetch S, Srichomthong C, Siriwan P, Suphapeetiporn K, Shotelersuk V. 2012. PDGFRa mutations in humans with isolated cleft palate. *Eur J Hum Genet* **20:** 1058–1062.

Rattner JB, Sciore P, Ou Y, van der Hoorn FA, Lo IK. 2010. Primary cilia in fibroblast-like type B synoviocytes lie within a cilium pit: A site of endocytosis. *Histol Histopathol* **25:** 865–875.

Reiter JF, Blacque OE, Leroux MR. 2012. The base of the cilium: Roles for transition fibres and the transition zone in ciliary formation, maintenance and compartmentalization. *EMBO Rep* **13:** 608–618.

Roepman R, Wolfrum U. 2007. Protein networks and complexes in photoreceptor cilia. *Subcell Biochem* **43:** 209–235.

Rupp E, Siegbahn A, Ronnstrand L, Wernstedt C, Claesson-Welsh L, Heldin CH. 1994. A unique autophosphorylation site in the platelet-derived growth factor α receptor from a heterodimeric receptor complex. *Eur J Biochem* **225:** 29–41.

Saito S, Tampe B, Muller GA, Zeisberg M. 2015. Primary cilia modulate balance of canonical and noncanonical

Wnt signaling responses in the injured kidney. *Fibrogenesis Tissue Repair* **8**: 6.

Satir P, Christensen ST. 2007. Overview of structure and function of mammalian cilia. *Annu Rev Physiol* **69**: 377–400.

Satir P, Pedersen LB, Christensen ST. 2010. The primary cilium at a glance. *J Cell Sci* **123**: 499–503.

Schaefer E, Lauer J, Durand M, Pelletier V, Obringer C, Claussmann A, Braun JJ, Redin C, Mathis C, Muller J, et al. 2014. Mesoaxial polydactyly is a major feature in Bardet–Biedl syndrome patients with LZTFL1 (BBS17) mutations. *Clin Genet* **85**: 476–481.

Schneider L, Clement CA, Teilmann SC, Pazour GJ, Hoffmann EK, Satir P, Christensen ST. 2005. PDGFRαα signaling is regulated through the primary cilium in fibroblasts. *Curr Biol* **15**: 1861–1866.

Schneider L, Stock CM, Dieterich P, Jensen BH, Pedersen LB, Satir P, Schwab A, Christensen ST, Pedersen SF. 2009. The Na$^+$/H$^+$ exchanger NHE1 is required for directional migration stimulated via PDGFR-α in the primary cilium. *J Cell Biol* **185**: 163–176.

Schneider L, Cammer M, Lehman J, Nielsen SK, Guerra CF, Veland IR, Stock C, Hoffmann EK, Yoder BK, Schwab A, et al. 2010. Directional cell migration and chemotaxis in wound healing response to PDGF-AA are coordinated by the primary cilium in fibroblasts. *Cell Physiol Biochem* **25**: 279–292.

Schou KB, Pedersen LB, Christensen ST. 2015. Ins and outs of GPCR signaling in primary cilia. *EMBO Rep* **16**: 1099–1113.

Seeger-Nukpezah T, Golemis EA. 2012. The extracellular matrix and ciliary signaling. *Curr Opin Cell Biol* **24**: 652–661.

Seo S, Zhang Q, Bugge K, Breslow DK, Searby CC, Nachury MV, Sheffield VC. 2011. A novel protein LZTFL1 regulates ciliary trafficking of the BBSome and Smoothened. *PLoS Genet* **7**: e1002358.

Shah AS, Ben-Shahar Y, Moninger TO, Kline JN, Welsh MJ. 2009. Motile cilia of human airway epithelia are chemosensory. *Science* **325**: 1131–1134.

Shiratori H, Hamada H. 2014. TGF-β signaling in establishing left-right asymmetry. *Semin Cell Dev Biol* **32**: 80–84.

Soriano P. 1997. The PDGF α receptor is required for neural crest cell development and for normal patterning of the somites. *Development* **124**: 2691–2700.

Sorkin A, von-Zastrow M. 2009. Endocytosis and signalling: Intertwining molecular networks. *Nat Rev Mol Cell Biol* **10**: 609–622.

Starks RD, Beyer AM, Guo DF, Boland L, Zhang Q, Sheffield VC, Rahmouni K. 2015. Regulation of insulin receptor trafficking by Bardet–Biedl syndrome proteins. *PLoS Genet* **11**: e1005311.

Stasiulewicz M, Gray SD, Mastromina I, Silva JC, Bjorklund M, Seymour PA, Booth D, Thompson C, Green RJ, Hall EA, et al. 2015. A conserved role for Notch signaling in priming the cellular response to Shh through ciliary localisation of the key Shh transducer Smo. *Development* **142**: 2291–2303.

Szymanska K, Johnson CA. 2012. The transition zone: An essential functional compartment of cilia. *Cilia* **1**: 10.

Takao D, Verhey KJ. 2015. Gated entry into the ciliary compartment. *Cell Mol Life Sci* **73**: 119–127.

Taniguchi CM, Emanuelli B, Kahn CR. 2006. Critical nodes in signalling pathways: Insights into insulin action. *Nat Rev Mol Cell Biol* **7**: 85–96.

Teilmann SC, Christensen ST. 2005. Localization of the angiopoietin receptors Tie-1 and Tie-2 on the primary cilia in the female reproductive organs. *Cell Biol Int* **29**: 340–346.

Teilmann SC, Byskov AG, Pedersen PA, Wheatley DN, Pazour GJ, Christensen ST. 2005. Localization of transient receptor potential ion channels in primary and motile cilia of the female murine reproductive organs. *Mol Reprod Dev* **71**: 444–452.

Teilmann SC, Clement CA, Thorup J, Byskov AG, Christensen ST. 2006. Expression and localization of the progesterone receptor in mouse and human reproductive organs. *J Endocrinol* **191**: 525–535.

Tozser J, Earwood R, Kato A, Brown J, Tanaka K, Didier R, Megraw TL, Blum M, Kato Y. 2015. TGF-β signaling regulates the differentiation of motile cilia. *Cell Rep* **11**: 1000–1007.

Umberger NL, Caspary T. 2015. Ciliary transport regulates PDGF-AA/αα signaling via elevated mammalian target of rapamycin signaling and diminished PP2A activity. *Mol Biol Cell* **26**: 350–358.

Valente EM, Rosti RO, Gibbs E, Gleeson JG. 2014. Primary cilia in neurodevelopmental disorders. *Nat Rev Neurol* **10**: 27–36.

Varelas X, Wrana JL. 2012. Coordinating developmental signaling: Novel roles for the Hippo pathway. *Trends Cell Biol* **22**: 88–96.

Vasudevan HN, Mazot P, He F, Soriano P. 2015. Receptor tyrosine kinases modulate distinct transcriptional programs by differential usage of intracellular pathways. *eeLife* **4**: e07186.

Veland IR, Montjean R, Eley L, Pedersen LB, Schwab A, Goodship J, Kristiansen K, Pedersen SF, Saunier S, Christensen ST. 2013. Inversin/nephrocystin-2 is required for fibroblast polarity and directional cell migration. *PLoS ONE* **8**: e60193.

Vestergaard ML, Awan A, Warzecha CB, Christensen ST, Andersen CY. 2016. Immunofluorescence microscopy and mRNA analysis of human embryonic stem cells (hESCs) including primary cilia associated signaling pathways. *Methods Mol Biol* **1307**: 123–140.

Vieira OV, Gaus K, Verkade P, Fullekrug J, Vaz WL, Simons K. 2006. FAPP2, cilium formation, and compartmentalization of the apical membrane in polarized Madin–Darby canine kidney (MDCK) cells. *Proc Natl Acad Sci* **103**: 18556–18561.

Walenkamp MJ, Wit JM. 2006. Genetic disorders in the growth hormone–insulin-like growth factor-I axis. *Horm Res* **66**: 221–230.

Walenkamp MJ, Karperien M, Pereira AM, Hilhorst-Hofstee Y, van DJ, Chen JW, Mohan S, Denley A, Forbes B, van Duyvenvoorde HA, et al. 2005. Homozygous and heterozygous expression of a novel insulin-like growth factor-I mutation. *J Clin Endocrinol Metab* **90**: 2855–2864.

Wang H, Zou X, Wei Z, Wu Y, Li R, Zeng R, Chen Z, Liao K. 2015. Hsp90α forms a stable complex at the cilium neck for the interaction of signalling molecules in IGF-1 receptor signalling. *J Cell Sci* **128:** 100–108.

Wann AK, Chapple JP, Knight MM. 2014. The primary cilium influences interleukin-1β-induced NFκB signalling by regulating IKK activity. *Cell Signal* **26:** 1735–1742.

Waters AM, Beales PL. 2011. Ciliopathies: An expanding disease spectrum. *Pediatr Nephrol* **26:** 1039–1056.

Wu J, Du H, Wang X, Mei C, Sieck GC, Qian Q. 2009. Characterization of primary cilia in human airway smooth muscle cells. *Chest* **136:** 561–570.

Yan X, Chen YG. 2011. Smad7: Not only a regulator, but also a cross-talk mediator of TGF-β signalling. *Biochem J* **434:** 1–10.

Yeh C, Li A, Chuang JZ, Saito M, Caceres A, Sung CH. 2013. IGF-1 activates a cilium-localized noncanonical Gβγ signaling pathway that regulates cell-cycle progression. *Dev Cell* **26:** 358–368.

Yi JJ, Barnes AP, Hand R, Polleux F, Ehlers MD. 2010. TGF-β signaling specifies axons during brain development. *Cell* **142:** 144–157.

Zhang Y, Pizzute T, Pei M. 2014. A review of crosstalk between MAPK and Wnt signals and its impact on cartilage regeneration. *Cell Tissue Res* **358:** 633–649.

Zhang J, Tian XJ, Xing J. 2016. Signal transduction pathways of EMT Induced by TGF-β, SHH, and WNT and their crosstalks. *J Clin Med* **5:** E41.

Zhu D, Shi S, Wang H, Liao K. 2009. Growth arrest induces primary-cilium formation and sensitizes IGF-1-receptor signaling during differentiation induction of 3T3-L1 preadipocytes. *J Cell Sci* **122:** 2760–2768.

Zuo W, Chen YG. 2009. Specific activation of mitogen-activated protein kinase by transforming growth factor-β receptors in lipid rafts is required for epithelial cell plasticity. *Mol Biol Cell* **20:** 1020–1029.

Primary Cilia and Mammalian Hedgehog Signaling

Fiona Bangs and Kathryn V. Anderson

Developmental Biology Program, Sloan Kettering Institute, Memorial Sloan Kettering Cancer Center, New York, New York 10065

Correspondence: k-anderson@ski.mskcc.org

It has been a decade since it was discovered that primary cilia have an essential role in Hedgehog (Hh) signaling in mammals. This discovery came from screens in the mouse that identified a set of genes that are required for both normal Hh signaling and for the formation of primary cilia. Since then, dozens of mouse mutations have been identified that disrupt cilia in a variety of ways and have complex effects on Hedgehog signaling. Here, we summarize the genetic and developmental studies used to deduce how Hedgehog signal transduction is linked to cilia and the complex effects that perturbation of cilia structure can have on Hh signaling. We conclude by describing the current status of our understanding of the cell-type-specific regulation of ciliogenesis and how that determines the ability of cells to respond to Hedgehog ligands.

Hedgehog (Hh) is one of a handful of signaling pathways that is used repeatedly for intercellular communication in development. Hh is critical for the development of nearly every organ in mammals, as well as in homeostasis and regeneration, and Hh signaling is disrupted in several types of cancer. Unlike other core developmental signaling pathways, vertebrate Hh signaling is completely dependent on a highly specialized organelle, the primary cilium.

The *Hh* gene was discovered in *Drosophila* based on the striking phenotype of fly larvae that lack Hh—the mutants do not develop the segmented anterior-to-posterior body plan and have ectopic denticles resembling a Hedgehog (Nüsslein-Volhard and Wieschaus 1980). Mutations in other genes with related phenotypes defined a signaling pathway in which Hh is the ligand that acts through the membrane receptor Patched (PTCH1) and the seven transmembrane spanning protein Smoothened (SMO) to control the activity of a transcription factor Cubitus interruptus (Ci) (Forbes et al. 1993; Ingham 1998). In the mid-1990s, embryological experiments in chicks and targeted mutations in the mouse homologs of the *Drosophila* Hh pathway genes showed that the Hh signaling pathway is also critical for the development of many tissues and organs in vertebrates. Both the core players and their pathway relationships are conserved (Fig. 1) (Ingham and McMahon 2001). Because of this evolutionary conservation, it was completely unexpected when phenotype-based genetic screens in the mouse identified a set of proteins required for the formation of primary cilia that were also required

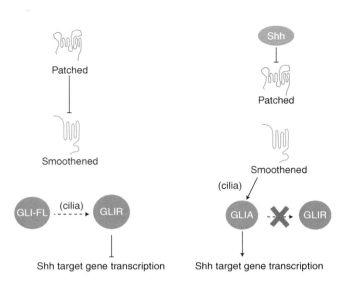

Figure 1. The backbone of the Hedgehog (Hh) signal transduction pathway. The core of the Hh signaling pathway is conserved between *Drosophila* and vertebrates. In the absence of ligand, the Hh receptor Patched (PTCH1) keeps the pathway off by inhibiting the activity of the seven transmembrane-domain protein Smoothened (SMO). When SMO is inactive, the GLI/Ci (glioblastoma/Cubitus interruptus) transcription factors are proteolytically processed to make a transcriptional repressor that binds to Hh target genes and blocks their transcription. Binding of Hh to PTCH1 inhibits its activity, relieving the repression of SMO, which promotes conversion of full-length GLI/Ci into a transcriptional activator. In vertebrates, cilia are required for the production of both GLI-repressor and GLI-activator.

for mammalian (but not *Drosophila*) Hh signaling. The relationship between primary cilia and mammalian Hh signaling has turned out to be fascinating, complex, and directly relevant for specific human diseases, the ciliopathies.

Here, we describe the nature of primary cilia Hh-dependent patterning during mammalian development, with a primary focus on proteins that have been shown to affect Hh signaling in vivo in genetic studies. The combination of genetics, developmental biology, and cell biology has provided overwhelming evidence that mammalian Hh signal transduction absolutely depends on the primary cilium, defined the features of the cilium that are important for signaling, and characterized how different changes in cilia structure have distinct effects on Hh signaling. From these studies, the view emerges that Hh signaling and the primary cilium have coevolved, concentrating critical signaling components in the very small volume at the tip of the cilium to allow efficient responses to low levels of ligand.

HEDGEHOG SIGNALING IN MAMMALIAN DEVELOPMENT

There are three mammalian Hh proteins, Sonic (SHH), Indian (IHH), and Desert (DHH). SHH and IHH have important, and sometimes overlapping, functions in many tissues. SHH has particularly striking roles in specification of cell types in the nervous system and in patterning of the limbs, whereas IHH has critical roles in skeletal development. Many studies have connected both SHH and IHH signaling to the primary cilium (see below). DHH appears to be restricted to the gonads and may also depend on primary cilia for its activity (Nygaard et al. 2015).

Drosophila Hedgehog has important roles in the patterning of adult organs, such as the eye, wing, and leg (Burke and Basler 1997). In these contexts, Hh acts as a short-range inducer of cell fate, inducing target gene expression in a range of 2−3 cell diameters away from the source of the signal. *Drosophila* Hh can activate expression of secreted proteins, such as Dpp (Bmp4) that can move further in tissues; as a result, *Drosophila* Hh has a profound role in the patterning of entire organs.

In contrast to the short-range signaling of *Drosophila*, the vertebrate Hh proteins can act over a field of cells that are many cell diameters away from the source of the protein. In neural patterning, SHH is first expressed in the notochord, which lies below the neural tube, and acts directly on cells in the ventral half of the neural tube to specify neural progenitors (Fig. 2) (Briscoe and Ericson 2001), and it appears that cells at least 20 cell diameters away from the source respond to Shh. In the limb, Shh is expressed in a classical organizer, the zone of polarizing activity (ZPA), and influences the pattern across most of the developing limb, again at a distance from its source (Ahn and Joyner 2004; Harfe et al. 2004). Indian hedgehog also acts at a distance to control proliferation and differentiation of chondrocytes (Long et al. 2001; Mak et al. 2008). Thus, one significant difference between Hh signaling in vertebrates and *Drosophila* appears to be that Hh ligands act at greater distances from their source in vertebrates; we hypothesize that long-range signaling requires a sensitized signal transduction system dependent on the primary cilium.

INTRAFLAGELLAR TRANSPORT IS REQUIRED FOR SHH SIGNAL TRANSDUCTION IN THE MOUSE EMBRYO

In the developing mouse embryo, perhaps the best-studied function of Shh is to direct patterning of neural progenitors, in which molecular markers make it possible to distinguish six different cell types that are specified by different levels of Shh activity (Briscoe and Ericson 2001; Dessaud et al. 2008). SHH made by the notochord induces formation of the floor plate and then specifies five additional neural cell types— V3 interneuron progenitors are specified adjacent to the floor plate by prolonged high levels of Shh signaling; motor neurons are determined by lower levels of Shh activity; and more dorsal interneurons are specified by yet lower levels of Shh. The final activity of the pathway depends on both ligand concentration and duration of exposure to ligand, and final cell fate depends on a regulatory network set into motion by SHH (Dessaud et al. 2007; Balaskas et al. 2012). In the absence of SHH or its positive effector SMO, ventral neural cell types fail to be specified, whereas loss of the negative regulator PTCH1 leads to the formation of ventral cell types by all neural progenitors (Fig. 2).

The seminal discovery that linked the primary cilium to Hedgehog signaling in the mouse came from the characterization of chemically induced mutations, such as the *wimple* mutation that lacked Shh-dependent ventral neural cell types and affected genes that encoded intraflagellar transport (IFT) proteins (Huangfu et al. 2003). IFT is the process by which the cilium is assembled and maintained. It is powered by two conserved, dedicated microtubule motors—anterograde transport from the base to the tip depends on a plus-end-directed heterotrimeric kinesin-2 complex, KIF3a, KIF3b, and KIF associated protein 3 (KAP3) (Cole 1999) and retrograde transport from the tip to the base requires the minus-end-directed cytoplasmic dynein2 motor made up of a heavy chain, an intermediate chain, light intermediate chain, and several light chains (Hou and Witman 2015). These motors are coupled to the cargos by large electron-dense IFT trains, made up of two protein complexes, IFT-A and IFT-B (Pigino et al. 2009; Behal and Cole 2013).

Map-based cloning showed that the *wimple* mutation disrupted the gene encoding IFT172, an IFT-B complex protein, resulting in a complete lack of primary cilia. The same screen identified a partial loss-of-function allele of another IFTB complex protein, IFT88, which partially disrupted cilia assembly and showed a milder disruption of neural patterning. Targeted null alleles of *Ift88* and *Kif3a* also caused the complete absence of cilia accompanied by a loss of Shh-dependent ventral neural cell types. Although *wimple* mutant embryos lacked ventral neural cell types, the mutants still expressed SHH, suggesting that without a cilium these mutants are unable to respond to Shh signaling (Huangfu et al. 2003). It was subsequently discovered that null mutations in other IFT-B

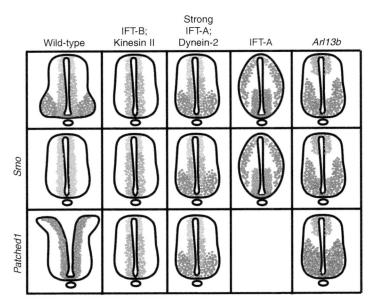

Figure 2. Mutations in cilia genes alter Hedgehog (Hh)-dependent neural patterning. Schematics of the spatial distribution of neural cell types in the developing lumbar neural tube in mutants with abnormal cilia. Dorsal up, ventral down; the notochord is the small oval below the neural tube. SHH ligand released from the notochord and floorplate patterns the dorsoventral axis of the neural tube. The highest levels of Shh signaling specify floorplate cells (pink). Dorsal to the floorplate are V3 neural progenitors, which also require high levels of Shh signaling (magenta). Motor neurons (green) require Shh but are specified at lower concentrations of ligand. V2, V1, and V0 interneurons (orange) require even lower levels of Shh activity. Dorsal progenitors (yellow) are specified by default, and high levels of Shh signaling inhibit the specification of dorsal progenitors. *Smo* mutant embryos fail to specify ventral neural subtypes and all neural progenitors follow a dorsal fate. In *Patched1* mutant embryos, all neural progenitors follow a ventral fate and express markers of the floorplate and V3 neural progenitors. Mouse embryos that lack a core intraflagellar transport (IFT-B) protein or kinesin-II lack primary cilia and therefore are unable to respond to Hh signaling; these mutants lack floorplate, V3 interneurons, and most motor neurons, V0-V2 interneurons and dorsal cell types extend ventrally, but V0-V2 interneurons are specified. *Dync2h1* mutants, which lack the heavy chain of the dynein retrograde IFT motor and strong loss of IFT-A mutant embryos, have very short, bulged primary cilia and cannot transduce Hh signals efficiently; these mutants lack floor plate V3 interneurons and some motor neurons. In combination with Shh, Smo, or Patched1, double mutant embryos resemble single dynein or strong IFT-A mutants indicating that like IFT-B these components of the cilium assembly machinery are required downstream from SMO and PTCH1. Surprisingly, IFT-A mutant embryos that have a milder effect on cilia morphology (e.g., with cilia of near-normal length and bulged tips) have the opposite effect on neural patterning, with expanded ventral neural cell types. Cilia in $Arl13b^{hnn/hnn}$ mutant embryos have structural defects in the microtubule axoneme, and the mutants fail to specify both the most ventral and most dorsal neural cell types. Double mutants that lack both a cilia component and *Smo* or *Patched1* have phenotypes similar to the cilia single mutants, indicating that the cilia machinery is required downstream from *Ptch1* and *Smo*.

genes (IFT52, IFT54, and IFT57) also block cilia formation and Shh-dependent specification of ventral neural cell types (Table 1), providing strong evidence that the primary cilium is required for a cell to receive Shh signals.

Double-mutant analysis defined the step in the Hh signaling pathway that requires IFT. PTCH1, the Hh receptor, is a negative regulator of Hh signaling. $Ptch1^{-/-}$ mutant embryos die at E9.0 and express markers of ventral neural cell types (floor plate and V3 interneuron progenitors) throughout the neural plate, even far from the source of SHH (Fig. 2). In contrast, double mutants that

Table 1. Subtle changes in cilia structure can decrease or increase Hh pathway activity, suggesting that the cilium provides more than just a permissive context for Hh signaling

Mouse gene	Function in cilia	Mouse embryo Hh pathway null phenotype (neural and limb)	Human phenotype
Kif3a	Anterograde IFT motor	Loss of SHH-dependent ventral neural cell types; polydactyly; midgestation lethality (Huangfu et al. 2003)	
Dync2h1	Retrograde IFT motor	Partial loss of ventral neural cell types; polydactyly; midgestation lethality (Huangfu and Anderson 2005; May et al. 2005)	Short rib dysplasia, with or without polydactyly (SRP) (Dagoneau et al. 2009; Merrill et al. 2009)
Dync2li1 (D2lic)	Dynein light intermediate chain		SRP (Taylor et al. 2015), Jeune syndrome (JATD) (Kessler et al. 2015)
Wdr34	Dynein light chain		SRP (Huber et al. 2013)
Tctex1d2	Light chain, IFT dynein		Jeune syndrome
Ift25 (Hspb11)	IFT-B	Perinatal lethality; weak loss of ventral neural cell types; partially penetrant polydactyly (Keady et al. 2012)	
Ift27	IFT-B	Perinatal lethality; weak loss of ventral neural cell types; partially penetrant polydactyly (Eguether et al. 2014)	
Ift38 (Cluap1)	IFT-B	Midgestation lethality; abnormal neural patterning (Botilde et al. 2013)	
Ift52	IFT-B	Loss of SHH-dependent ventral neural cell types; midgestation lethality; polydactyly (Liu et al. 2005)	
Ift54 (Traf3ip1)	IFT-B	Loss of SHH-dependent ventral neural cell types; midgestation lethality; polydactyly (Berbari et al. 2011)	Senior−Løken syndrome (Bizet et al. 2015)
Ift57 (Hippi)	IFT-B	Loss of SHH-dependent ventral neural cell types; midgestation lethality; polydactyly (Houde et al. 2006)	
Ift80 (WDR56)	IFT-B	Hypomorphic allele only: short long bones; polydactyly; late gestation lethality (Rix et al. 2011)	SRP, JATD
IftT88 (Polaris; Tg737)	IFT-B	Loss of SHH-dependent ventral neural cell types; midgestation lethality; polydactyly (Huangfu et al. 2003)	
Ift172	IFT-B	Loss of SHH-dependent ventral neural cell types; midgestation lethality (Huangfu et al. 2003)	Jeune (Halbritter et al. 2013)
Ttc26 (hop)	IFT-B	Partial loss of function allele only; abnormal neural patterning (Swiderski et al. 2014)	
Ift121 (Wdr35)	IFT-A	Polydactyly (Mill et al. 2011)	SRP (Mill et al. 2011); Ellis−van Creveld (EVC) syndrome (Caparrós-Martín et al. 2015)

Continued

Table 1. *Continued*

Mouse gene	Function in cilia	Mouse embryo Hh pathway null phenotype (neural and limb)	Human phenotype
Ift122	IFT-A	Expansion of SHH-dependent ventral neural cell types; midgestation lethality; polydactyly (Qin et al. 2011)	Sensenbrenner syndrome (Walczak-Sztulpa et al. 2010)
Ift139 (Ttc21b/ Thm1/ alien)	IFT-A	Expansion of SHH-dependent ventral neural cell types; late gestation lethality; polydactyly (Tran et al. 2008)	NPHP12, JATD/SRP (Davis et al. 2011)
Ift140	IFT-A	Polydactyly, neural tube closure defects, (Miller et al. 2013)	SRTD9 (Perrault et al. 2012)
Ift144 (Wdr19)	IFT-A	Weak allele: GOF neural; polydactyly. Strong allele: LOF (Liem et al. 2012) (Wicking)	Various: Senior–Løken, Sensenbrenner, Jeune, NPHP, RP (Bredrup et al. 2011; Fehrenbach et al. 2014)
Tulp3	IFT-A-associated	Expansion of SHH-dependent ventral neural cell types; midgestation lethality; polydactyly (Norman et al. 2009; Patterson et al. 2009)	
Gpr161	GPCR	Expansion of SHH-dependent ventral neural cell types; midgestation lethality (Mukhopadhyay et al. 2013)	
C2cd3 (chick talpid2)	Distal appendage protein	Loss of SHH-dependent ventral neural cell types; midgestation lethality; polydactyly (Hoover et al. 2008; Ye et al. 2014)	Skeletal dysplasia (Cortés et al. 2016)
Ofd1	Centriole and centriolar satellites	Loss of SHH-dependent ventral neural cell types; midgestation lethality (Ferrante et al. 2006)	Oral-facial-digital syndrome (Ferrante et al. 2001)
Ttbk2	Kinase required for cilia initiation; associates with distal appendages	Loss of SHH-dependent ventral neural cell types; midgestation lethality (Goetz et al. 2012)	Spinocerebellar ataxia type 11 (Houlden et al. 2007)
Talpid3	Associated with distal appendages and PCM	Loss of SHH-dependent ventral neural cell types; midgestation lethality; polydactyly (Bangs et al. 2011)	Ciliopathy spectrum (Alby et al. 2015); Joubert syndrome (Stephen et al. 2015)
Inpp5e	Cilia membrane composition	Bielas et al. 2009; Jacoby et al. 2009	Joubert syndrome (JBTS1) (Bielas et al. 2009); MORM syndrome (Jacoby et al. 2009)
Rab23	Not clear	GOF neural; polydactyly (Eggenschwiler et al. 2001)	Carpenter syndrome (Jenkins et al. 2007)
Kif7	Kinesin-4 protein; controls dynamics of axonemal microtubules	Mild expansion of SHH-dependent ventral neural cell types; polydactyly, defective diaphragm (Cheung et al. 2009; Endoh-Yamagami et al. 2009; Liem et al. 2009; Coles and Ackerman 2013; He et al. 2014)	Joubert syndrome; acrocallosal syndrome

Continued

 Cite this article as *Cold Spring Harb Perspect Biol* doi: 10.1101/cshperspect.a028175

Table 1. *Continued*

Mouse gene	Function in cilia	Mouse embryo Hh pathway null phenotype (neural and limb)	Human phenotype
Tectonic (*Tctn1*)	Transition zone protein	Abnormal patterning of SHH-dependent ventral neural cell types (Reiter and Skarnes 2006)	Joubert (Garcia-Gonzalo et al. 2011)
Mks1	Transition zone protein	Abnormal patterning of SHH-dependent ventral neural cell types (Weatherbee et al. 2009)	Meckel syndrome (Kyttälä et al. 2006)
Fuzzy	CPLANE complex	Loss of SHH-dependent ventral neural cell types; late gestation lethality; polydactyly (Heydeck et al. 2009)	
Inturned	CPLANE complex	Loss of SHH-dependent ventral neural cell types; late gestation lethality; polydactyly (Zeng et al. 2010)	
Jbts17	CPLANE complex	Skeletal and heart defects; no change in neural patterning detected (Damerla et al. 2015)	Joubert syndrome (Srour et al. 2012)
Rsg1	Small GTPase, CPLANE-associated	Loss of SHH-dependent ventral neural cell types; late gestation lethality; polydactyly (S Agbu and KV Anderson, in prep.)	
Evc1/Evc2	Ciliary membrane protein	EVC-like (Ruiz-Perez et al. 2007; Caparrós-Martín et al. 2013)	Ellis–van Creveld syndrome (Ruiz-Perez et al. 2003)
Arl13b	Ciliary membrane GTPase	Loss of both the most dorsal and most ventral neural cell types (Caspary et al. 2007)	Joubert syndrome (Cantagrel et al. 2008)
Sas-4 (*p53*) (*Cpap*/ *Cenpj*)	Centriole duplication	Loss of SHH-dependent ventral neural cell types; midgestation lethality (Bazzi and Anderson 2014)	Microcephaly (Bond et al. 2005)
DZip1/ *Iguana*	Cilia base	Loss of SHH-dependent ventral neural cell types; midgestation lethality (Wang et al. 2013)	

References noted in table are for specific mouse and human proteins and phenotypes. This is not a complete list of references, and includes only the earliest mouse references, and is not a complete list of phenotypes.

IFT, intraflagellar transport; GPCR, G-protein-coupled receptor; PCM, pericentriolar material; CPLANE, ciliogenesis and planar polarity effector; SHH, Sonic Hedgehog; GOF, gain of function; LOF, loss of function; MORM, mental retardation, truncal obesity, retinal dystrophy, and micropenis.

lack both *Ptch1* and *Ift172* lack all ventral neural cell types, like the *Ift172* single mutants (Fig. 2) (Huangfu et al. 2003). Thus, the ectopic activation of the Hh pathway caused by loss of *Ptch1* depends on IFT172, arguing that IFT is required at a step in the signal transduction pathway downstream from PTCH1. Similar double mutant analysis showed that IFT172, IFT88, and KIF3A act downstream from both PTCH1 and SMO and upstream of the GLI proteins, at the heart of the signal transduction cascade (Fig. 1).

CENTRIOLE AND BASAL BODY PROTEINS ARE ALSO REQUIRED FOR THE RESPONSE TO Hh

In addition to IFT proteins, other proteins that build cilia are also required for the ability of cells to respond to Hh family ligands (Table 1). The doublet microtubules of the cilium are templated by the triplet microtubules of the mother centriole. Embryos that lack centrioles because of the lack of proteins required for centriole duplication, such as SAS4 (CPAP/CENPJ), can go through normal segregation at mitosis but fail to assemble cilia. These embryos can survive to midgestation (if a cell-cycle checkpoint is blocked by removal of p53) but Shh-dependent ventral neural cell types are not specified (Bazzi and Anderson 2014). The distal appendages of the mother centriole are required for docking on the plasma membrane (Tanos et al. 2013). The distal appendage protein C2cd3 is required for the centriole to dock on the ciliary vesicle (Ye et al. 2014) and *C2cd3* null mutant embryos lack cilia and lack Hh-dependent ventral neural cell types (Hoover et al. 2008). Talpid3 and OFD1 are both centrosome-associated proteins required for ciliogenesis and for the response to SHH in the neural tube (Ferrante et al. 2006; Bangs et al. 2011). Tau tubulin kinase 2 (TTBK2) is required for cilia initiation and mutants lack all cilia and Hh-dependent ventral neural cell types (Goetz et al. 2012). Embryos that lack CP110, a protein that localizes to the distal ends of mother centrioles, also have fewer cilia and mild defects in Hh signaling (Yadav et al. 2016).

The CPLANE (ciliogenesis and planar polarity effector) protein complex is located at the basal body and includes downstream mediators of planar cell polarity in *Drosophila*, *Fuzzy*, *Inturned*, and *Fritz* (*Wdpcp*), and *Jbts17*, a gene responsible for Joubert syndrome in humans (Damerla et al. 2015; Toriyama et al. 2016). The CPLANE complex is required for efficient ciliogenesis as it functions to recruit IFT-A proteins to the cilium RSG1, a small GTPase that appears to be a peripheral member of the CPLANE complex. *Rsg1* mutant mice have fewer cilia of normal morphology, suggesting *Rsg1* has a specific role in cilia initiation (S Agbu and KV Anderson, in prep.). These mutants survive longer than IFT-B mutants, yet still display Hh signaling defects. Thus, diverse classes of proteins are required to build cilia and ensure efficient ciliogenesis. Loss of any type of protein required for normal cilia formation in the mouse prevents or attenuates cells from responding to Shh. Thus, it is the primary cilium, rather than another activity of the IFT machinery, which is required for mammalian cells to respond to Hh ligands.

CILIA ARE REQUIRED FOR RESPONSES TO Hh FAMILY LIGANDS IN ALL VERTEBRATE TISSUES

Conditional genetic deletion of *Ift88* and/or *Kif3a* has been used to test the role of cilia in Hh signaling in other mouse tissues. In every context examined, removal of cilia blocks the response to both SHH and IHH (Table 2). The defects observed include shortened bones of the limb, craniofacial defects, and loss of neural stem and progenitor cells.

Cilia are also required for Hh signaling in other vertebrates. As in the mouse, absence of cilia in the chick blocks the ability of cells to receive Hh signals (Yin et al. 2009). In zebrafish, loss of cilia leads to milder defects in Hh-dependent patterning, in part because of the maternal contribution of cilia proteins or mRNAs to the zebrafish embryo, which masks the phenotype in homozygous embryos, and in part because of the different roles of the zebrafish GLI proteins in Hh responses (Huang and Schier 2009; Tay et al. 2010; Ben et al. 2011).

In humans, it is not possible to directly examine the requirement of cilia for Hh signaling during development, as mutations that completely disrupt IFT genes would presumably result in lethality in the first half of gestation, as in mice, and would not be detected. However, many mutations in human cilia genes do not completely ablate cilia, allowing survival to the end of gestation with comparatively mild Hh pathway defects (Table 1). For example, mutations in the cytoplasmic dynein-2 complex cause short ribs (narrow chest)—short-rib dysplasia, with or without polydactyly (SRP) or Jeune asphyxiating thoracic dystrophy (JATD) (Dagoneau et al. 2009; Merrill et al. 2009).

Further evidence that cilia are important for Hh signaling in humans comes from their impact on two human tumor types, medulloblastoma and basal cell carcinoma, which are both caused by gain of Hh signaling. In mouse medulloblastoma or basal cell carcinoma triggered by activated SMO, conditional removal of cilia blocks tumor formation, consistent with the requirement for cilia at a step downstream from SMO. The same tumors can also be caused by

Table 2. In every context examined, removal of cilia blocks the response to both SHH and IHH

Tissue	Conditional allele	Cre	Phenotype
Cranial neural crest	*Kif3a*	*Wnt1*	Craniofacial defects (Liu et al. 2014)
Endochondral bone (Ihh-dependent)	*Ift88*	*Prx1*	Short bones (Haycraft et al. 2007)
Postnatal cartilage	*Kif3a*	*Col2a1-Cre*	Craniofacial defects (Koyama et al. 2007)
Neural stem cells	*Kif3a*	*hGFAP*	Loss of hippocampal stem cells (Han et al. 2008)
Hippocampal stem cells	*Ift20*	*mGFAP*	Reduction hippocampal amplifying progenitors (Amador-Arjona et al. 2011)
Postnatal B1 SVZ neural stem cells	*Kif3a*	*Adeno-Cre (injected)*	Decreased proliferation of neural stem cells (Tong et al. 2014)
Cerebellar granule cell precursors	*Kif3a*	*hGFAP*	Loss of cerebellar granule cell progenitors (Chizhikov et al. 2007; Spassky et al. 2008)
Basal cell carcinoma	*Kif3a; Ift88*	*Keratin14*	Removal of cilia inhibited tumors induced by activated Smoothened. Removal of cilia accelerated tumors induced by activated Gli2 (Wong et al. 2009)
Medulloblastoma	*Kif3a*	*hGFAP*	Removal of cilia inhibited tumors induced by activated Smoothened. Removal of cilia accelerated tumors induced by activated Gli2 (Han et al. 2009)

SHH, Sonic Hedgehog; IHH, Indian Hedgehog; SVZ, subventricular zone.

activation of GLI2; in these cases, removal of cilia makes the tumors more aggressive because GLI3 repressor is no longer generated resulting in a greater net GLI activation.

TRAFFICKING OF PATHWAY PROTEINS WITHIN THE CILIUM IS REQUIRED FOR Hh SIGNALING

Immunolocalization experiments showed that all the proteins required for transduction of Hh signals are enriched in primary cilia and change their distribution in response to ligand (Fig. 3). The first demonstration of cilia localization came for the membrane protein SMO, which accumulates in the cilium within an hour following stimulation of the Hh pathway (Corbit et al. 2005; Rohatgi et al. 2007). As total SMO protein levels are unaltered, accumulation of SMO in the cilium is a consequence of translocation of a ready-made pool of SMO and not from newly synthesized protein (Rohatgi et al. 2007). A common hydrophobic and basic residue motif following the seventh transmembrane domain at the carboxyl terminus of SMO is required for localization of SMO to the cilium in the presence of SHH (Händel

et al. 1999; Brailov et al. 2000; Dwyer et al. 2001; Corbit et al. 2005; Aanstad et al. 2009). Single-molecule imaging indicates that once in the cilium SMO moves by diffusion within the membrane, rather than by IFT (Milenkovic et al. 2015). Localization of SMO to the cilium is, however, not sufficient for signaling, as SMO accumulates in primary cilia that lack Dync2h1, the heavy change of the retrograde dynein IFT motor, but *Dync2h1* mutants lack ventral neural cell types (Huangfu and Anderson 2005; May et al. 2005).

Even more remarkable, all three transcription factors of the GLI family, the mediators of Hh-regulated transcription, are enriched at cilia tips in the absence of Hh ligand and become further enriched in response to pathway activation (Haycraft et al. 2005). Only full-length GLI proteins localize to cilia, whereas the proteolytically processed repressor forms that lack the carboxy-terminal half of the protein do not (Wen et al. 2010; Santos and Reiter 2014). Deletion analysis identified a 330-amino-acid (out of the 1544 in the full-length protein) central region of GLI2 is required for cilia targeting. SUFU, a key negative regulator of vertebrate Hh signaling, binds to GLI preventing its acti-

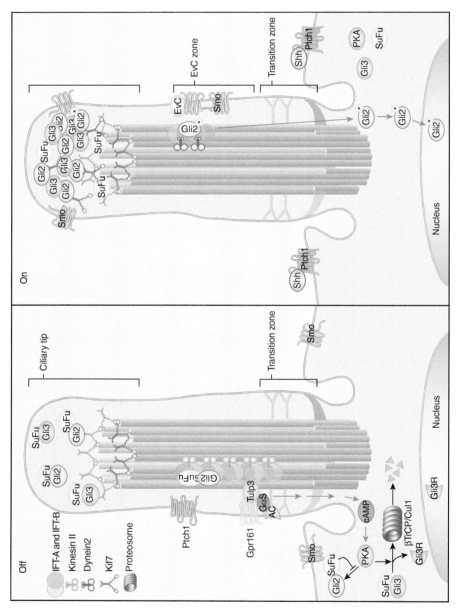

Figure 3. Hedgehog (Hh) signal transduction in the primary cilium. The primary cilium has several spatially distinct regions that promote normal Hh signal transduction. Entry into the cilium is gated at the transition zone; the EvC zone facilitates pathway activation in some cell types; the cilia tip compartment, defined by KIF7, is the site of enrichment and activation of the GLI/SUFU complex; and proteolytic processing to make GLI repressors that may occur at the base where protein kinase A (PKA) is localized and leads to formation of Gli3R. (*Legend continues on following page.*)

vation and is also highly enriched at cilia tips (Haycraft et al. 2005).

PTCH1 has 12 transmembrane domains and the region carboxy-terminal to the last transmembrane domain is required for cilia localization and signaling (Kim et al. 2015). PTCH1 is localized to cilia in the absence of ligand and binding of SHH to PTCH1 triggers its removal from the cilium, although removal of PTCH1 from the cilium is not required for Hh pathway activation. During the same interval after exposure to ligand when PTCH moves out of the cilium, SMO translocates into the cilium (Rohatgi et al. 2007).

CILIA ARE REQUIRED FOR BOTH KEEPING THE PATHWAY OFF AND TURNING THE PATHWAY ON

In mammals, the GLI family of transcription factors, GLI1, GLI2, and GLI3 implement the transcriptional responses to Hh family ligands (Bai et al. 2004). In the absence of ligand, proteolytically processed forms of GLI proteins repress expression of Hh target genes (primarily GLI3, with a minor role for GLI2). In the presence of ligand, processing of the repressor forms is blocked and, instead, full-length GLI proteins are converted (by an unknown mechanism) to GLI activators, principally mediated by GLI2. The formation of both GLI repressor and activator depend on cilia (Huangfu and Anderson 2005; Liu et al. 2005; May et al. 2005).

Proteolytic processing of the GLI proteins depends on phosphorylation by cAMP-dependent PKA, a key evolutionarily conserved component of the Hh signal transduction pathway (Wang et al. 2014). In contrast to *Drosophila*, mammalian PKA does not act only by promoting formation of GLI repressor, it is also required to prevent inappropriate activation of GLI2 (Tuson et al. 2011). In addition to the phosphorylation sites on GLI2 required for proteolytic processing, phosphorylation of two additional PKA sites of GLI2 and GLI3 prevents formation of the fully activated form of the transcription factors (Niewiadomski et al. 2014).

Both the catalytic and regulator subunits of PKA are highly enriched at the base of the cilium (Barzi et al. 2010; Tuson et al. 2011). Mutant mouse embryos that lack both genes that encode catalytic subunits (*PKA nulls*) show a very strong activation of the Hh pathway, in which all cells in the neural tube acquire the most ventral fates (Tuson et al. 2011). Compound mutants that lack both PKA catalytic subunits and also lack cilia have a neural phenotype similar to cilia mutants, showing that the action of PKA depends on events that take place in cilia. Although both catalytic and regulatory subunits of PKA are highly enriched at the base of the cilium, recent proximity ligation experiments provide biochemical evidence that regulatory subunits of PKA localize to the cilium (Mick et al. 2015), suggesting that sensing of cAMP by PKA may take place inside cilia.

PKA is activated by cAMP, and several adenylyl cyclases that produce cAMP, including AC3, AC5, and AC6, are enriched in cilia (Bishop et al. 2007; Mick et al. 2015; Vuolo et al. 2015). These, in turn, are regulated by G-protein-coupled receptors (GPCRs), including

Figure 3. (*Continued*) In the absence of Hh signal, PTCH1 and Gpr161, both negative regulators of the pathway, are present in the cilia membrane. Gpr161 trafficking into the cilium depends on TULP3 and intraflagellar transport (IFT)-A; GPR161 appears to activate Gαs, which activates adenylyl cyclase, which increases the levels of cAMP, thereby activating PKA. Activated PKA phosphorylates sites on GLI3 that promote partial proteolysis by βTrCP/Cul1 and the proteasome, generating Gli3 repressor (Gli3R), which moves to the nucleus and represses expression of Hh target genes. GLI2 and GLI3 are trafficked to the tip of the cilium in the absence of ligand in a complex with SuFu, and processing of GLI3 depends on it having transited the cilium. Binding of Hh to PTCH1 triggers its removal from the cilium, allowing Smoothened (SMO) to translocate into the cilium where it can activate downstream signaling. Gpr161 also exits the cilium after exposure to ligand. Binding of EvC to SMO near the base of the cilium promotes SMO activity. In the presence of ligand, the GLI/SUFU complex accumulates to high levels at the tip of the cilium, where dissociation of the complex allows formation of the activator form of GLI2.

Gpr161, which activates ACIII and thereby increases cAMP levels (Mukhopadhyay et al. 2013). Like PTCH1, Gpr161 is present in the cilia membrane in the absence of ligand and moves out of the cilium in response to ligand. This movement is regulated by TULP3, a member of the vertebrate tubby-like family of proteins, which links GPCRs, including Gpr161, to IFT-A, which transports them into the cilium (Nishina et al. 1998; Mukhopadhyay et al. 2010).

Small G proteins, including Gαs, mediate the actions of GPCRs. In mouse embryos that lack Gαs (Gnas), the Hh pathway is ectopically activated with a phenotype similar to that of *PKA* or *Ptch1* mutant embryos, indicating that Gαs negatively regulates the Hh pathway (Regard et al. 2013). Thus, the Hh pathway is kept off in the absence of ligand by Gpr161, which is localized to the cilium by TULP3 and activated by Gαs. Gpr161 subsequently activates adenylyl cyclase, leading to increased cAMP and increased PKA activity, which promotes GLI repressor formation and prevents GLI activation (Fig. 3) (Mukhopadhyay and Rohatgi 2014).

CHANGES IN AXONEMAL STRUCTURE OR CILIA MEMBRANE COMPOSITION CAN INCREASE OR DECREASE Hh PATHWAY ACTIVITY

The requirement for primary cilia for Hh signaling and the localization of pathway proteins to the cilium would be consistent with the cilium acting as a scaffold for Hh signal transduction or a site for concentration of Hh pathway proteins. However, subtle changes in cilia structure can decrease or increase Hh pathway activity, suggesting that the cilium provides more than just a permissive context for Hh signaling. (References to specific mouse and human proteins and phenotypes are in Table 1.)

Embryos carrying mutations in the gene that encodes the heavy chain of the retrograde motor cytoplasmic dynein 2 (*Dync2h1*) assemble cilia; however, the axoneme is bulged because of accumulation of IFT complexes and other proteins trapped in the cilium as a consequence of defective retrograde transport. Like IFT-B or *Kif3a* mutant embryos, *Dync2h1* mu-

tants lack floor plate and V3 interneuron progenitors indicating these cells cannot respond to high levels of Shh signaling. However, unlike IFT-B or *Kif3a* mutants, motor neurons are specified in the caudal spinal cord of *Dync2h1* mutants in which they are intermingled with V2 interneurons (Fig. 2) (Huangfu and Anderson 2005; May et al. 2005). Thus, low levels of Hh signaling can still be transduced in the bulged *Dync2h1* cilia.

Although null mutations in most IFT-B components block cilia formation, abnormal cilia form in most mutants that lack one of the six proteins of the IFT-A complex. Mutations that strongly disrupt the mouse IFT-A proteins IFT122 or IFT139 cause bulged cilia, similar to mutants that lack Dynein2. However, unlike mutants that disrupt Dynein2, the IFT-A mutants show a gain-of-activity of the Hh pathway in the neural tube, with expanded domains of Shh-dependent ventral neuron cell types (e.g., motor neurons) (Fig. 2) (Tran et al. 2008; Qin et al. 2011). This gain-of-function for Hh pathway activity is independent of ligand, suggesting that IFT-A plays a specific role in keeping the pathway off in the absence of ligand. A similar phenotype is seen in embryos that lack TULP3 (Mukhopadhyay et al. 2010).

The effects of mutations in IFT-A genes on Hh signaling are complex. A weak allele of *Ift144* (WDR19) shows bulged cilia and ventralization of the neural tube similar to that seen in IFT122 or 139 (Table 1). However, a stronger allele that produces only very short bloated cilia with a highly disrupted axoneme causes a loss of Hh signaling in the neural tube. Double mutants that carry both the weak allele of *Ift144* and an allele of *Ift122* show the short bulged cilia and loss-of-function Hh neural phenotype, suggesting that some IFT-A proteins have overlapping functions and that a complete lack of IFT-A complex activity leads to the formation of very short cilia and a loss of Hh activity (Liem et al. 2012). Human mutations in IFT-A genes cause a spectrum of phenotypes, ranging in severity from retinitis pigmentosum in the adult to SRP associated with perinatal lethality (Table 1).

Even small changes in IFT function can have subtle effects on Hh signaling activity. IFT25 and IFT27 form an IFT-B subcomplex. In contrast to other IFT-B proteins, mutants that lack either of these proteins make primary cilia that have apparently normal structure. The mouse mutants survive to birth, when they show mild Hh pathway phenotypes and defects in trafficking of GLI2, SMO, and PTCH1. IFT27 appears to be important for loading of retrograde IFT cargo in trypanosomes (Huet et al. 2014), and given the evolutionary conservation of IFT proteins, it is likely that the Hh defects in mouse *Ift25* and *Ift27* mutants are the result of relatively subtle changes in global IFT. This indicates that Hh signaling is sensitive to even small changes in IFT function.

ARL13B is an ARF-family GTPase that is a component of the ciliary membrane that is required for normal cilia structure—in the absence of ARL13B, cilia are short and the B-tubules of the axoneme are frequently open. The neural tube of *Arl13b* mouse mutants lacks both the most ventral and the most dorsal cell types, indicating that when ciliary structure is disrupted in this way it is not possible to achieve the highest level of Hh signaling or to prevent ectopic low-level activation of the pathway.

It has been supposed that cilia length affects Hh signaling; however, it appears not to be the case. Cilia on *Rfx3* mutant cells are about half the normal length (Bonnafe et al. 2004) and embryos that overexpress ARL13B have cilia that are 50% longer than wild-type (Bangs et al. 2015), but both genotypes are viable and apparently have a normal pattern of Hh-dependent neural cell types.

The ciliary membrane has a distinct lipid composition; the inositol phosphate PI(4)P is enriched in the entire ciliary membrane and PI(4,5)P_2 is enriched at the base of the cilium. Inpp5e, a ciliary phosphoinositide 5-phosphatase, is required to maintain this distribution and is one of the genes mutated in Joubert syndrome (Chávez et al. 2015; Garcia-Gonzalo et al. 2015). In *Inpp5e* mutant mice, ciliary PI(4,5)P_2 levels are elevated, causing disrupted Hh signaling, apparently a result of inappropriate accumulation of TULP3, which binds to these phosphoinositides. *Inpp5e* mutant mice survive to birth when they show polydactyly, suggesting

that they have a significant, but relatively mild disruption of Hh signal transduction.

KIF7 ORGANIZES THE CILIA TIP COMPARTMENT

One rationale for why Hh signaling requires the primary cilium is that it concentrates components of the Hh signal transduction cascade in a small volume, promoting their interactions by mass action. It has been calculated that concentrating the entire pool of a protein into the cilium would increase its concentration by two to three orders of magnitude (Nachury 2014). However, certain Hh components, GLI, SuFu, and KIF7, are strongly enriched at the cilia tip. Full-length Gli proteins form complexes with SuFu, and data indicate that separation of SuFu and GLI proteins is critical for GLI activation (Humke et al. 2010); it is likely that this separation occurs or is regulated by events that happen at the tip of the cilium. KIF7 also binds directly to both GLI2 and GLI3 (Cheung et al. 2009; Endoh-Yamagami et al. 2009).

The tip compartment of the primary cilium is organized, at least in part, by KIF7, an evolutionarily conserved core component of the Hh pathway (He et al. 2014). KIF7 and its *Drosophila* homolog Cos2 have both positive and negative roles in Hh signal transduction. KIF7, a member of the Kinesin-4 family, binds microtubule plus-ends at the tip of the cilium to organize microtubule architecture at the tip (He et al. 2014). In the absence of KIF7, axonemal microtubules have variable lengths, and GLI and SuFu are localized in multiple puncta along the ciliary axoneme that appear to correspond to ectopic cilia tip compartments. The mislocalization of the GLI/SUFU complex causes inappropriate, ligand-independent GLI activation and the mild ectopic activation of the pathway seen in *Kif7*$^{-/-}$ mutant embryos.

CELL-TYPE-SPECIFIC DIFFERENCES IN CILIA COMPOSITION CAN MODULATE Hh SIGNALING

Most proteins that have a role in Hh signaling in the cilium act in all cell types. Two exceptions are EVC and EVC2, transmembrane proteins responsible for Ellis–van Creveld syndrome, a distinctive ciliopathy associated with short limbs, a narrow chest (short ribs), and polydactyly (Ruiz-Perez et al. 2003). Mouse *Evc* is highly expressed in differentiating chondrocytes and all the cartilaginous components of the skeleton. *Evc* mutant mice survive to birth, and some can survive to adulthood; like the affected people, the mice have short bones, short ribs, and abnormal teeth, although the response to Shh in the neural tube appears to be nearly normal (Caparrós-Martín et al. 2013). Specific partial loss-of-function mutations that disrupt the IFT-A protein IFT121/WDR35 also cause EVC syndrome, and both EVC proteins and SMO fail to localize to *Ift121*/*Wdr35* mutant MEF cilia (Caparrós-Martín et al. 2015). As other IFT-A proteins are also required for the recruitment of transmembrane proteins, including SMO, to cilia (Liem et al. 2012), it appears IFT-A has a general role in recruiting membrane proteins, including the EVC proteins, to the cilium, and chondrocytes appear to be particularly sensitive to the loss of IFT-A.

The Tectonic complex localizes to the transition zones and appears to have complex tissue-specific functions in ciliogenesis (Table 1) (see Vaisse et al. 2016). Mutant fibroblasts assemble cilia but cells in the mutant embryonic node and mesenchymal cells adjacent to the neural tube fail to form cilia. As mutants show both defects in neural patterning and polydactyly, the tissue specificity of the Tectonic complex merits further study.

LINEAGE DETERMINES WHICH CELLS HAVE PRIMARY CILIA IN THE MOUSE EMBRYO

In cell culture, primary cilia are present in the G_0 phase of the cell cycle and can be induced by serum removal (Plotnikova et al. 2009). In contrast, many proliferating cells in the mouse embryo and adult organs in vivo have a primary cilium. Scattered reports indicate that some specific cell types in the whole animal, such as acinar cells of the adult pancreas (Aughsteen 2001), lack primary cilia. Transgenic mice that carry fluorescent markers for basal bodies and cilia have been used to systematically identify ciliated and unciliated cells (Bangs et al. 2015). No cilia are present on cells of the preimplantation embryo; cilia first appear on cells of the epiblast lineage (the embryo proper) shortly after cavitation (E6.0) and all cells derived from the epiblast (ectoderm, mesoderm, and definitive endoderm) are ciliated in the midgestation embryo. In contrast, cells of extraembryonic lineages, the visceral endoderm and trophectoderm, have centrioles but lack primary cilia. Cells of both the trophectoderm lineage and from the embryonic mesoderm contribute to the placenta and cells of the visceral endoderm and mesoderm contribute to the yolk sac. These lineages lack cilia until at least E14.5 (Bangs et al. 2015).

Stem cell lines that correspond to the different embryonic lineages (Ralston and Rossant 2005) recapitulate the cilia status of the lineages in the embryo—all nondividing epiblast stem cells (EpiSCs) (which recapitulate the status of the E6.5 epiblast) are ciliated, whereas extraembryonic endoderm stem cells (XEN cells) and trophectoderm stem cells (TS cells) lack cilia (Bangs et al. 2015). For all three stem cell types, the presence or absence of primary cilia on embryo-derived stem cells is independent of the presence of serum.

The absence of primary cilia on cells of extraembryonic origin implies that these cells cannot respond to Hh ligands. The first time that Hh signaling is active in the mouse embryo is at the beginning of gastrulation, when Ihh expressed in the extraembryonic visceral endoderm signals to the adjacent epiblast derived extraembryonic mesoderm, which is ciliated. This is necessary to promote blood island formation (Fig. 4) (Farrington et al. 1997; Dyer et al. 2001). In the placenta, the only cells that respond to Hh signaling are embryo-derived ciliated cells that surround the fetal blood vessels (Fig. 4) (Jiang and Herman 2006). It is interesting to speculate that normal development of the placenta and yolk sac depends on preventing extraembryonic lineages from responding to Hh signals and this is achieved by blocking ciliogenesis in these cells.

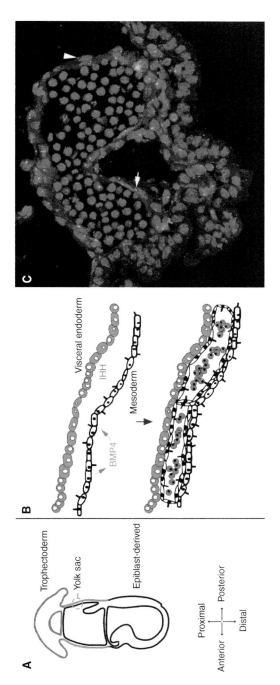

Figure 4. Lineage-dependent cilia formation in the early mouse embryo. (*A*) Three lineages defined in the preimplantation embryo persist in the postimplantation embryo (E8.0 embryo shown). Two extraembryonic lineages, the trophectoderm (blue) and visceral endoderm (red) contribute to the placenta and yolk sac respectively; these two lineages lack primary cilia. Nearly all nonmitotic cells of the third lineage, the epiblast (black), are ciliated; this lineage gives rise to almost all cells of the embryo proper, as well as to mesoderm-derived components of the placenta and yolk sac. (*B*) Paracrine Hh signaling in the yolk sac. *Upper* drawing represents the two layers of the E8.0 yolk sac (lassoed in panel *A*). The unciliated visceral endoderm layer produces IHH, but does not activate Hh target genes. The adjacent ciliated extraembryonic mesoderm responds to the IHH produced by the visceral endoderm and produces BMP4, which leads to formation of the blood islands (*lower* panel). (Image based on data in Baron 2001.) (*C*) A blood vessel in the E14.5 yolk sac from an embryo expressing a Centrin2-GFP, a marker for the basal body, and ARL13b-mCherry, a marker of the cilia membrane. The lineage relationships are preserved from earlier development—the visceral endoderm (cell layer at the top of the image) has GFP$^+$ centrioles, but lacks cilia (arrowhead), whereas the extraembryonic mesoderm that surrounds the vessel filled with round blood cells is ciliated (arrow). (Image from Bangs et al. 2015; modified, with permission, from the authors.)

These cases of Hh ligand made by extraembryonic epithelia signaling to adjacent Hh-responsive ciliated cells reflect one of the common modes of Hh signaling in other organs. Paracrine signaling from ligand-producing cells to adjacent Hh responsive cells has been seen in a number of organs, including the digestive tract and the prostate (Yu et al. 2009; Mao et al. 2010). Thus, it will be interesting to see whether tissue-specific absence of primary cilia enforces paracrine, rather than autocrine Hh signaling.

CONCLUDING REMARKS

Vertebrate Hedgehog signal transduction, which is essential for the development and maintenance of most organs, takes place in the primary cilium. The responses to Hh ligands are exquisitely sensitive to alterations in cilia structure—disruptions of many cilia components reduce the response to Hh ligands, but disruption of specific cilia proteins can also cause inappropriate activity of the pathway, and some proteins can be altered in different ways that enhance or interfere with Hh signal transduction. Dozens, and probably hundreds, of proteins are required to build the primary cilium; because of the many essential roles of Hh signaling, mutations in cilia genes collectively have a broad impact on human health.

The central effectors of Hh signaling, GLI proteins and the negative regulator SUFU, are regulated in the tiny compartment at the tip of the primary cilium. A single core conserved component of the Hh pathway, the kinesin-family protein KIF7, has an essential role in organizing the cilia tip. The dual roles of KIF7 in Hh signal transduction and primary cilia structure argue for an ancient intertwining of Hh signal transduction with the structure of the primary cilium.

The distribution of primary cilia determines the ability of tissues to respond to Hh ligands, but the tissue distribution of primary cilia and the mechanisms that regulate cilia assembly and disassembly in vivo are poorly understood. A deeper understanding of the mechanisms that control cilia formation in vivo is likely to provide the foundation for therapeutic interventions in both ciliopathies and cancer.

ACKNOWLEDGMENTS

We thank Meg Distinti for valuable assistance with the manuscript. Work in the Anderson laboratory in this area is supported by the National Institutes of Health (NIH) Grants R37 HD03455 and R01 NS044385 to K.V.A. and the MSKCC Cancer Center Support Grant (P30 CA008748).

REFERENCES

*Reference is also in this collection.

Aanstad P, Santos N, Corbit KC, Scherz PJ, Trinh le A, Salvenmoser W, Huisken J, Reiter JF, Stainier DY. 2009. The extracellular domain of Smoothened regulates ciliary localization and is required for high-level Hh signaling. Curr Biol 19: 1034–1039.

Ahn S, Joyner AL. 2004. Dynamic changes in the response of cells to positive Hedgehog signaling during mouse limb patterning. Cell 118: 505–516.

Alby C, Piquand K, Huber C, Megarbané A, Ichkou A, Legendre M, Pelluard F, Encha-Ravazi F, Abi-Tayeh G, Bessières B, et al. 2015. Mutations in KIAA0586 cause lethal ciliopathies ranging from a hydrolethalus phenotype to short-rib polydactyly syndrome. Am J Hum Genet 97: 311–318.

Amador-Arjona A, Elliott J, Miller A, Ginbey A, Pazour GJ, Enikolopov G, Roberts AJ, Terskikh AV. 2011. Primary cilia regulate proliferation of amplifying progenitors in adult hippocampus: Implications for learning and memory. J Neurosci 31: 9933–9944.

Aughsteen AA. 2001. The ultrastructure of primary cilia in the endocrine and excretory duct cells of the pancreas of mice and rats. Eur J Morphol 39: 277–283.

Bai CB, Stephen D, Joyner AL. 2004. All mouse ventral spinal cord patterning by hedgehog is Gli dependent and involves an activator function of Gli3. Dev Cell 6: 103–115.

Balaskas N, Ribeiro A, Panovska J, Dessaud E, Sasai N, Page KM, Briscoe J, Ribes V. 2012. Gene regulatory logic for reading the Sonic Hedgehog signaling gradient in the vertebrate neural tube. Cell 148: 273–284.

Bangs F, Antonio N, Thongnuek P, Welten M, Davey MG, Briscoe J, Tickle C. 2011. Generation of mice with functional inactivation of talpid3, a gene first identified in chicken. Development 138: 3261–72.

Bangs FK, Schrode N, Hadjantonakis AK, Anderson KV. 2015. Lineage specificity of primary cilia in the mouse embryo. Nat Cell Biol 17: 113–122.

Baron MH. 2001. Molecular regulation of embryonic hematopoiesis and vascular development: A novel pathway. J Hematother Stem Cell Res 10: 587–594.

Barzi M, Berenguer J, Menendez A, Alvarez-Rodriguez R, Pons S. 2010. Sonic-Hedgehog-mediated proliferation requires the localization of PKA to the cilium base. *J Cell Sci* **123:** 62–69.

Bazzi H, Anderson KV. 2014. Acentriolar mitosis activates a p53-dependent apoptosis pathway in the mouse embryo. *Proc Natl Acad Sci* **111:** E1491–E500.

Behal RH, Cole DG. 2013. Analysis of interactions between intraflagellar transport proteins. *Methods Enzymol* **524:** 171–94.

Ben J, Elworthy S, Ng AS, van Eeden F, Ingham PW. 2011. Targeted mutation of the talpid3 gene in zebrafish reveals its conserved requirement for ciliogenesis and Hedgehog signalling across the vertebrates. *Development* **138:** 4969–4978.

Berbari NF, Kin NW, Sharma N, Michaud EJ, Kesterson RA, Yoder BK. 2011. Mutations in Traf3ip1 reveal defects in ciliogenesis, embryonic development, and altered cell size regulation. *Dev Biol* **360:** 66–76.

Bielas SL, Silhavy JL, Brancati F, Kisseleva MV, Al-Gazali L, Sztriha L, Bayoumi RA, Zaki MS, Abdel-Aleem A, Rosti RO, et al. 2009. Mutations in *INPP5E*, encoding inositol polyphosphate-5-phosphatase E, link phosphatidyl inositol signaling to the ciliopathies. *Nat Genet* **41:** 1032–1036.

Bishop GA, Berbari NF, Lewis J, Mykytyn K. 2007. Type III adenylyl cyclase localizes to primary cilia throughout the adult mouse brain. *J Comp Neurol* **505:** 562–571.

Bizet AA, Becker-Heck A, Ryan R, Weber K, Filhol E, Krug P, Halbritter J, Delous M, Lasbennes MC, Linghu B, et al. 2015. Mutations in TRAF3IP1/IFT54 reveal a new role for IFT proteins in microtubule stabilization. *Nat Commun* **6:** 8666.

Bond J, Roberts E, Springell K, Lizarraga SB, Scott S, Higgins J, Hampshire DJ, Morrison EE, Leal GF, Silva EO, et al. 2005. A centrosomal mechanism involving CDK5RAP2 and CENPJ controls brain size. *Nat Genet* **37:** 353–355.

Bonnafe E, Touka M, AitLounis A, Baas D, Barras E, Ucla C, Moreau A, Flamant F, Dubruille R, Couble P, et al. 2004. The transcription factor RFX3 directs nodal cilium development and left–right asymmetry specification. *Mol Cell Biol* **24:** 4417–4427.

Botilde Y, Yoshiba S, Shinohara K, Hasegawa T, Nishimura H, Shiratori H, Hamada H. 2013. Cluap1 localizes preferentially to the base and tip of cilia and is required for ciliogenesis in the mouse embryo. *Dev Biol* **381:** 203–212.

Brailov I, Bancila M, Brisorgueil MJ, Miquel MC, Hamon M, Vergé D. 2000. Localization of 5-HT$_6$ receptors at the plasma membrane of neuronal cilia in the rat brain. *Brain Res* **872:** 271–275.

Bredrup C, Saunier S, Oud MM, Fiskerstrand T, Hoischen A, Brackman D, Leh SM, Midtbø M, Filhol E, Bole-Feysot C, et al. 2011. Ciliopathies with skeletal anomalies and renal insufficiency due to mutations in the IFT-A gene *WDR19. Am J Hum Genet* **89:** 634–643.

Briscoe J, Ericson J. 2001 Specification of neuronal fates in the ventral neural tube. *Curr Opin Neurobiol* **11:** 43–49.

Burke R, Basler K. 1997. Hedgehog signaling in *Drosophila* eye and limb development—Conserved machinery, divergent roles? *Curr Opin Neurobiol* **7:** 55–61.

Cantagrel V, Silhavy JL, Bielas SL, Swistun D, Marsh SE, Bertrand JY, Audollent S, Attié-Bitach T, Holden KR,

Dobyns WB, et al. 2008. Mutations in the cilia gene *ARL13B* lead to the classical form of Joubert syndrome. *Am J Hum Genet* **83:** 170–179.

Caparrós-Martín JA, Valencia M, Reytor E, Pacheco M, Fernandez M, Perez-Aytes A, Gean E, Lapunzina P, Peters H, Goodship JA, et al. 2013. The ciliary Evc/Evc2 complex interacts with Smo and controls Hedgehog pathway activity in chondrocytes by regulating Sufu/Gli3 dissociation and Gli3 trafficking in primary cilia. *Hum Mol Genet* **22:** 124–139.

Caparrós-Martín JA, De Luca A, Cartault F, Aglan M, Temtamy S, Otaify GA, Mehrez M, Valencia M, Vázquez L, Alessandri JL, et al. 2015. Specific variants in *WDR35* cause a distinctive form of Ellis–van Creveld syndrome by disrupting the recruitment of the EvC complex and SMO into the cilium. *Hum Mol Genet* **24:** 4126–4137.

Caspary T, Larkins CE, Anderson KV. 2007. The graded response to Sonic Hedgehog depends on cilia architecture. *Dev Cell* **12:** 767–778.

Chávez M, Ena S, Van Sande J, de Kerchove d'Exaerde A, Schurmans S, Schiffmann SN. 2015. Modulation of ciliary phosphoinositide content regulates trafficking and Sonic Hedgehog signaling output. *Dev Cell* **34:** 338–350.

Cheung HO, Zhang X, Ribeiro A, Mo R, Makino S, Puviindran V, Law KK, Briscoe J, Hui CC. 2009. The kinesin protein Kif7 is a critical regulator of Gli transcription factors in mammalian hedgehog signaling. *Sci Signal* **2:** ra29.

Chizhikov VV, Davenport J, Zhang Q, Shih EK, Cabello OA, Fuchs JL, Yoder BK, Millen KJ. 2007. Cilia proteins control cerebellar morphogenesis by promoting expansion of the granule progenitor pool. *J Neurosci* **27:** 9780–9789.

Cole DG. 1999. Kinesin-II, coming and going. *J Cell Biol* **147:** 463–466.

Coles GL, Ackerman KG. 2013. Kif7 is required for the patterning and differentiation of the diaphragm in a model of syndromic congenital diaphragmatic hernia. *Proc Natl Acad Sci* **110:** E1898–E905.

Corbit KC, Aanstad P, Singla V, Norman AR, Stainier DY, Reiter JF. 2005. Vertebrate smoothened functions at the primary cilium. *Nature* **437:** 1018–1021.

Cortés CR, McInerney-Leo AM, Vogel I, Rondón Galeano MC, Leo PJ, Harris JE, Anderson LK, Keith PA, Brown MA, Ramsing M, et al. 2016. Mutations in human *C2CD3* cause skeletal dysplasia and provide new insights into phenotypic and cellular consequences of altered C2CD3 function. *Sci Rep* **6:** 24083.

Dagoneau N, Goulet M, Geneviève D, Sznajer Y, Martinovic J, Smithson S, Huber C, Baujat G, Flori E, Tecco L, et al. 2009. *DYNC2H1* mutations cause asphyxiating thoracic dystrophy and short rib-polydactyly syndrome, type III. *Am J Hum Genet* **84:** 706–711.

Damerla RR, Cui C, Gabriel GC, Liu X, Craige B, Gibbs BC, Francis R, Li Y, Chatterjee B, San Agustin JT, et al. 2015. Novel *Jbts17* mutant mouse model of Joubert syndrome with cilia transition zone defects and cerebellar and other ciliopathy related anomalies. *Hum Mol Genet* **24:** 3994–4005.

Davis EE, Zhang Q, Liu Q, Diplas BH, Davey LM, Hartley J, Stoetzel C, Szymanska K, Ramaswami G, Logan CV, et al. 2011. *TTC21B* contributes both causal and modifying alleles across the ciliopathy spectrum. *Nat Genet* **43:** 189–196.

Dessaud E, Yang LL, Hill K, Cox B, Ulloa F, Ribeiro A, Mynett A, Novitch BG, Briscoe J. 2007. Interpretation of the Ssonic Hedgehog morphogen gradient by a temporal adaptation mechanism. *Nature* **450:** 717–720.

Dessaud E, McMahon AP, Briscoe J. 2008. Pattern formation in the vertebrate neural tube: A sonic hedgehog morphogen-regulated transcriptional network. *Development* **135:** 2489–503.

Dwyer ND, Adler CE, Crump JG, L'Etoile ND, Bargmann CI. 2001. Polarized dendritic transport and the AP-1 μ1 clathrin adaptor UNC-101 localize odorant receptors to olfactory cilia. *Neuron* **31:** 277–287.

Dyer MA, Farrington SM, Mohn D, Munday JR, Baron MH. 2001. Indian hedgehog activates hematopoiesis and vasculogenesis and can respecify prospective neurectodermal cell fate in the mouse embryo. *Development* **128:** 1717–1730.

Eggenschwiler JT, Espinoza E, Anderson KV. 2001. Rab23 is an essential negative regulator of the mouse Sonic hedgehog signalling pathway. *Nature* **412:** 194–198.

Eguether T, San Agustin JT, Keady BT, Jonassen JA, Liang Y, Francis R, Tobita K, Johnson CA, Abdelhamed ZA, Lo CW, et al. 2014. IFT27 links the BBSome to IFT for maintenance of the ciliary signaling compartment. *Dev Cell* **31:** 279–290.

Endoh-Yamagami S, Evangelista M, Wilson D, Wen X, Theunissen JW, Phamluong K, Davis M, Scales SJ, Solloway MJ, de Sauvage FJ, et al. 2009. The mammalian Cos2 homolog Kif7 plays an essential role in modulating Hh signal transduction during development. *Curr Biol* **19:** 1320–1326.

Farrington SM, Belaoussoff M, Baron MH. 1997. *Winged-helix*, *Hedgehog* and *Bmp* genes are differentially expressed in distinct cell layers of the murine yolk sac. *Mech Dev* **62:** 197–211.

Fehrenbach H, Decker C, Eisenberger T, Frank V, Hampel T, Walden U, Amann KU, Krüger-Stollfuß I, Bolz HJ, Häffner K, et al. 2014. Mutations in *WDR19* encoding the intraflagellar transport component IFT144 cause a broad spectrum of ciliopathies. *Pediatr Nephrol* **29:** 1451–1456.

Ferrante MI, Giorgio G, Feather SA, Bulfone A, Wright V, Ghiani M, Selicorni A, Gammaro L, Scolari F, Woolf AS, et al. 2001. Identification of the gene for oral–facial–digital type I syndrome. *Am J Hum Genet* **68:** 569–576.

Ferrante MI, Zullo A, Barra A, Bimonte S, Messaddeq N, Studer M, Dollé P, Franco B. 2006. Oral–facial–digital type I protein is required for primary cilia formation and left–right axis specification. *Nat Genet* **38:** 112–117.

Forbes AJ, Nakano Y, Taylor AM, Ingham PW. 1993. Genetic analysis of hedgehog signalling in the *Drosophila* embryo. *Dev Suppl* **1993:** 115–124.

Garcia-Gonzalo FR, Corbit KC, Sirerol-Piquer MS, Ramaswami G, Otto EA, Noriega TR, Seol AD, Robinson JF, Bennett CL, Josifova DJ, et al. 2011. A transition zone complex regulates mammalian ciliogenesis and ciliary membrane composition. *Nat Genet* **43:** 776–784.

Garcia-Gonzalo FR, Phua SC, Roberson EC, Garcia G III, Abedin M, Schurmans S, Inoue T, Reiter JF. 2015. Phosphoinositides regulate ciliary protein trafficking to modulate Hedgehog signaling. *Dev Cell* **34:** 400–409.

Goetz SC, Liem KF Jr, Anderson KV. 2012. The spinocerebellar ataxia-associated gene *Tau tubulin kinase 2* controls the initiation of ciliogenesis. *Cell* **151:** 847–858.

Halbritter J, Bizet AA, Schmidts M, Porath JD, Braun DA, Gee HY, McInerney-Leo AM, Krug P, Filhol E, Davis EE, et al. 2013. Defects in the IFT-B component IFT172 cause Jeune and Mainzer–Saldino syndromes in humans. *Am J Hum Genet* **93:** 915–925.

Han YG, Spassky N, Romaguera-Ros M, Garcia-Verdugo JM, Aguilar A, Schneider-Maunoury S, Alvarez-Buylla A. 2008. Hedgehog signaling and primary cilia are required for the formation of adult neural stem cells. *Nat Neurosci* **11:** 277–284.

Han YG, Kim HJ, Dlugosz AA, Ellison DW, Gilbertson RJ, Alvarez-Buylla A. 2009. Dual and opposing roles of primary cilia in medulloblastoma development. *Nat Med* **15:** 1062–1065.

Händel M, Schulz S, Stanarius A, Schreff M, Erdtmann-Vourliotis M, Schmidt H, Wolf G, Höllt V. 1999. Selective targeting of somatostatin receptor 3 to neuronal cilia. *Neuroscience* **89:** 909–926.

Harfe B, Scherz PJ, Nissim S, Tian H, McMahon AP, Tabin CJ. 2004. Evidence for an expansion-based temporal Shh gradient in specifying vertebrate digit identities. *Cell* **118:** 517–528.

Haycraft CJ, Banizs B, Aydin-Son Y, Zhang Q, Michaud EJ, Yoder BK. 2005. Gli2 and Gli3 localize to cilia and require the intraflagellar transport protein polaris for processing and function. *PLoS Genet* **1:** e53.

Haycraft CJ, Zhang Q, Song B, Jackson WS, Detloff PJ, Serra R, Yoder BK. 2007. Intraflagellar transport is essential for endochondral bone formation. *Development* **134:** 307–316.

He M, Subramanian R, Bangs F, Omelchenko T, Liem KF Jr, Kapoor TM, Anderson KV. 2014. The kinesin-4 protein Kif7 regulates mammalian Hedgehog signalling by organizing the cilium tip compartment. *Nat Cell Biol* **16:** 663–672.

Heydeck W, Zeng H, Liu A. 2009. Planar cell polarity effector gene *Fuzzy* regulates cilia formation and Hedgehog signal transduction in mouse. *Dev Dyn* **238:** 3035–3042.

Hoover AN, Wynkoop A, Zeng H, Jia J, Niswander LA, Liu A. 2008. C2cd3 is required for cilia formation and Hedgehog signaling in mouse. *Development* **135:** 4049–4058.

Hou Y, Witman GB. 2015. Dynein and intraflagellar transport. *Exp Cell Res* **334:** 26–34.

Houde C, Dickinson RJ, Houtzager VM, Cullum R, Montpetit R, Metzler M, Simpson EM, Roy S, Hayden MR, Hoodless PA, et al. 2006. Hippi is essential for node cilia assembly and Sonic hedgehog signaling. *Dev Biol* **300:** 523–533.

Houlden H, Johnson J, Gardner-Thorpe C, Lashley T, Hernandez D, Worth P, Singleton AB, Hilton DA, Holton J, Revesz T, et al. 2007. Mutations in *TTBK2*, encoding a kinase implicated in tau phosphorylation, segregate with spinocerebellar ataxia type 11. *Nat Genet* **39:** 1434–1436.

Huang P, Schier AF. 2009. Dampened Hedgehog signaling but normal Wnt signaling in zebrafish without cilia. *Development* **136:** 3089–3098.

Huangfu D, Anderson KV. 2005. Cilia and Hedgehog responsiveness in the mouse. *Proc Natl Acad Sci* **102**: 11325–11330.

Huangfu D, Liu A, Rakeman AS, Murcia NS, Niswander L, Anderson KV. 2003. Hedgehog signalling in the mouse requires intraflagellar transport proteins. *Nature* **426**: 83–87.

Huber C, Wu S, Kim AS, Sigaudy S, Sarukhanov A, Serre V, Baujat G, Le Quan Sang KH, Rimoin DL, Cohn DH, et al. 2013. *WDR34* mutations that cause short-rib polydactyly syndrome type III/severe asphyxiating thoracic dysplasia reveal a role for the NF-κB pathway in cilia. *Am J Hum Genet* **93**: 926–931.

Huet D, Blisnick T, Perrot S, Bastin P. 2014. The GTPase IFT27 is involved in both anterograde and retrograde intraflagellar transport. *eLife* **3**: e02419.

Humke EW, Dorn KV, Milenkovic L, Scott MP, Rohatgi R. 2010. The output of Hedgehog signaling is controlled by the dynamic association between Suppressor of Fused and the Gli proteins. *Genes Dev* **24**: 670–682.

Ingham PW. 1998. Transducing Hedgehog: The story so far. *EMBO J* **17**: 3505–3511.

Ingham PW, McMahon AP. 2001. Hedgehog signaling in animal development: Paradigms and principles. *Genes Dev* **15**: 3059–3087.

Jacoby M, Cox JJ, Gayral S, Hampshire DJ, Ayub M, Blockmans M, Pernot E, Kisseleva MV, Compère P, Schiffmann SN, et al. 2009. *INPP5E* mutations cause primary cilium signaling defects, ciliary instability and ciliopathies in human and mouse. *Nat Genet* **41**: 1027–1031.

Jenkins D, Seelow D, Jehee FS, Perlyn CA, Alonso LG, Bueno DF, Donnai D, Josifova D, Mathijssen IM, Morton JE, et al. 2007. RAB23 mutations in Carpenter syndrome imply an unexpected role for hedgehog signaling in cranial-suture development and obesity. *Am J Hum Genet* **80**: 1162–1170.

Jiang F, Herman GE. 2006. Analysis of *Nsdhl*-deficient embryos reveals a role for Hedgehog signaling in early placental development. *Hum Mol Genet* **15**: 3293–3305.

Keady BT, Samtani R, Tobita K, Tsuchya M, San Agustin JT, Follit JA, Jonassen JA, Subramanian R, Lo CW, Pazour GJ. 2012. IFT25 links the signal-dependent movement of Hedgehog components to intraflagellar transport. *Dev Cell* **22**: 940–951.

Kessler K, Wunderlich I, Uebe S, Falk NS, Gießl A, Brandstätter JH, Popp B, Klinger P, Ekici AB, Sticht H, et al. 2015. *DYNC2LI1* mutations broaden the clinical spectrum of dynein-2 defects. *Sci Rep* **5**: 11649.

Kim J, Hsia EY, Brigui A, Plessis A, Beachy PA, Zheng X. 2015. The role of ciliary trafficking in Hedgehog receptor signaling. *Sci Signal* **8**: ra55.

Koyama E, Young B, Nagayama M, Shibukawa Y, Enomoto-Iwamoto M, Iwamoto M, Maeda Y, Lanske B, Song B, Serra R, et al. 2007. Conditional *Kif3a* ablation causes abnormal hedgehog signaling topography, growth plate dysfunction, and excessive bone and cartilage formation during mouse skeletogenesis. *Development* **134**: 2159–2169.

Kyttälä M, Tallila J, Salonen R, Kopra O, Kohlschmidt N, Paavola-Sakki P, Peltonen L, Kestilä M. 2006. *MKS1*, encoding a component of the flagellar apparatus basal body proteome, is mutated in Meckel syndrome. *Nat Genet* **38**: 155–157.

Liem KF Jr, He M, Ocbina PJ, Anderson KV. 2009. Mouse Kif7/Costal2 is a cilia-associated protein that regulates Sonic hedgehog signaling. *Proc Natl Acad Sci* **106**: 13377–13382.

Liem KF Jr, Ashe A, He M, Satir P, Moran J, Beier D, Wicking C, Anderson KV. 2012. The IFT-A complex regulates Shh signaling through cilia structure and membrane protein trafficking. *J Cell Biol* **197**: 789–800.

Liu A, Wang B, Niswander LA. 2005. Mouse intraflagellar transport proteins regulate both the activator and repressor functions of Gli transcription factors. *Development* **132**: 3103–3111.

Liu B, Chen S, Johnson C, Helms JA. 2014. A ciliopathy with hydrocephalus, isolated craniosynostosis, hypertelorism, and clefting caused by deletion of Kif3a. *Reprod Toxicol* **48**: 88–97.

Long F, Zhang XM, Karp S, Yang Y, McMahon AP. 2001. Genetic manipulation of hedgehog signaling in the endochondral skeleton reveals a direct role in the regulation of chondrocyte proliferation. *Development* **128**: 5099–5108.

Mak KK, Kronenberg HM, Chuang PT, Mackem S, Yang Y. 2008. Indian hedgehog signals independently of PTHrP to promote chondrocyte hypertrophy. *Development* **135**: 1947–1956.

Mao J, Kim BM, Rajurkar M, Shivdasani RA, McMahon AP. 2010. Hedgehog signaling controls mesenchymal growth in the developing mammalian digestive tract. *Development* **137**: 1721–1729.

May SR, Ashique AM, Karlen M, Wang B, Shen Y, Zarbalis K, Reiter J, Ericson J, Peterson AS. 2005. Loss of the retrograde motor for IFT disrupts localization of Smo to cilia and prevents the expression of both activator and repressor functions of Gli. *Dev Biol* **287**: 378–389.

Merrill AE, Merriman B, Farrington-Rock C, Camacho N, Sebald ET, Funari VA, Schibler MJ, Firestein MH, Cohn ZA, Priore MA, et al. 2009. Ciliary abnormalities due to defects in the retrograde transport protein DYNC2H1 in short-rib polydactyly syndrome. *Am J Hum Genet* **84**: 542–549.

Mick DU, Rodrigues RB, Leib RD, Adams CM, Chien AS, Gygi SP, Nachury MV. 2015. Proteomics of primary cilia by proximity labeling. *Dev Cell* **35**: 497–512.

Milenkovic L, Weiss LE, Yoon J, Roth TL, Su YS, Sahl SJ, Scott MP, Moerner WE. 2015. Single-molecule imaging of Hedgehog pathway protein Smoothened in primary cilia reveals binding events regulated by Patched1. *Proc Natl Acad Sci* **112**: 8320–8325.

Mill P, Lockhart PJ, Fitzpatrick E, Mountford HS, Hall EA, Reijns MA, Keighren M, Bahlo M, Bromhead CJ, Budd P, et al. 2011. Human and mouse mutations in *WDR35* cause short-rib polydactyly syndromes due to abnormal ciliogenesis. *Am J Hum Genet* **88**: 508–515.

Miller KA, Ah-Cann CJ, Welfare MF, Tan TY, Pope K, Caruana G, Freckmann ML, Savarirayan R, Bertram JF, Dobbie MS, et al. 2013. *Cauli*: A mouse strain with an Ift140 mutation that results in a skeletal ciliopathy modelling Jeune syndrome. *PLoS Genet* **9**: e1003746.

Mukhopadhyay S, Rohatgi R. 2014. G-protein-coupled receptors, Hedgehog signaling and primary cilia. *Semin Cell Dev Biol* **33**: 63–72.

Mukhopadhyay S, Wen X, Chih B, Nelson CD, Lane WS, Scales SJ, Jackson PK. 2010. TULP3 bridges the IFT-A complex and membrane phosphoinositides to promote trafficking of G-protein-coupled receptors into primary cilia. *Genes Dev* **24**: 2180–2193.

Mukhopadhyay S, Wen X, Ratti N, Loktev A, Rangell L, Scales SJ, Jackson PK. 2013. The ciliary G-protein-coupled receptor Gpr161 negatively regulates the Sonic hedgehog pathway via cAMP signaling. *Cell* **152**: 210–223.

Nachury MV. 2014. How do cilia organize signalling cascades? *Philos Trans R Soc Lond B Biol Sci* **369**: 20130465.

Niewiadomski P, Kong JH, Ahrends R, Ma Y, Humke EW, Khan S, Teruel MN, Novitch BG, Rohatgi R. 2014. Gli protein activity is controlled by multisite phosphorylation in vertebrate Hedgehog signaling. *Cell Rep* **6**: 168–181.

Nishina PM, North MA, Ikeda A, Yan Y, Naggert JK. 1998. Molecular characterization of a novel tubby gene family member, *TULP3*, in mouse and humans. *Genomics* **54**: 215–220.

Norman RX, Ko HW, Huang V, Eun CM, Abler LL, Zhang Z, Sun X, Eggenschwiler JT. 2009. Tubby-like protein 3 (TULP3) regulates patterning in the mouse embryo through inhibition of Hedgehog signaling. *Hum Mol Genet* **18**: 1740–1754.

Nüsslein-Volhard C, Wieschaus E. 1980. Mutations affecting segment number and polarity in *Drosophila*. *Nature* **287**: 795–801.

Nygaard MB, Almstrup K, Lindbæk L, Christensen ST, Svingen T. 2015. Cell context–specific expression of primary cilia in the human testis and ciliary coordination of Hedgehog signalling in mouse Leydig cells. *Sci Rep* **5**: 10364.

Patterson VL, Damrau C, Paudyal A, Reeve B, Grimes DT, Stewart ME, Williams DJ, Siggers P, Greenfield A, Murdoch JN. 2009. Mouse *hitchhiker* mutants have spina bifida, dorso-ventral patterning defects and polydactyly: Identification of Tulp3 as a novel negative regulator of the Sonic hedgehog pathway. *Hum Mol Genet* **18**: 1719–1739.

Perrault I, Saunier S, Hanein S, Filhol E, Bizet AA, Collins F, Salih MA, Gerber S, Delphin N, Bigot K, et al. 2012. Mainzer–Saldino syndrome is a ciliopathy caused by *IFT140* mutations. *Am J Hum Genet* **90**: 864–870.

Pigino G, Geimer S, Lanzavecchia S, Paccagnini E, Cantele F, Diener DR, Rosenbaum JL, Lupetti P. 2009. Electron-tomographic analysis of intraflagellar transport particle trains in situ. *J Cell Biol* **187**: 135–148.

Plotnikova OV, Pugacheva EN, Golemis EA. 2009. Primary cilia and the cell cycle. *Methods Cell Biol* **94**: 137–160.

Qin J, Lin Y, Norman RX, Ko HW, Eggenschwiler JT. 2011. Intraflagellar transport protein 122 antagonizes Sonic Hedgehog signaling and controls ciliary localization of pathway components. *Proc Natl Acad Sci* **108**: 1456–1461.

Ralston A, Rossant J. 2005. Genetic regulation of stem cell origins in the mouse embryo. *Clin Genet* **68**: 106–112.

Regard JB, Malhotra D, Gvozdenovic-Jeremic J, Josey M, Chen M, Weinstein LS, Lu J, Shore EM, Kaplan FS, Yang Y. 2013. Activation of Hedgehog signaling by loss of GNAS causes heterotopic ossification. *Nat Med* **19**: 1505–1512.

Reiter JF, Skarnes WC. 2006. Tectonic, a novel regulator of the Hedgehog pathway required for both activation and inhibition. *Genes Dev* **20**: 22–27.

Rix S, Calmont A, Scambler PJ, Beales PL. 2011. An *Ift80* mouse model of short rib polydactyly syndromes shows defects in hedgehog signalling without loss or malformation of cilia. *Hum Mol Genet* **20**: 1306–1314.

Rohatgi R, Milenkovic L, Scott MP. 2007. Patched1 regulates hedgehog signaling at the primary cilium. *Science* **317**: 372–376.

Ruiz-Perez VL, Tompson SW, Blair HJ, Espinoza-Valdez C, Lapunzina P, Silva EO, Hamel B, Gibbs JL, Young ID, Wright MJ, et al. 2003. Mutations in two nonhomologous genes in a head-to-head configuration cause Ellis-van Creveld syndrome. *Am J Hum Genet* **72**: 728–732.

Ruiz-Perez VL, Blair HJ, Rodriguez-Andres ME, Blanco MJ, Wilson A, Liu YN, Miles C, Peters H, Goodship JA. 2007. Evc is a positive mediator of Ihh-regulated bone growth that localises at the base of chondrocyte cilia. *Development* **134**: 2903–2912.

Santos N, Reiter JF. 2014. A central region of Gli2 regulates its localization to the primary cilium and transcriptional activity. *J Cell Sci* **127**: 1500–1510.

Spassky N, Han YG, Aguilar A, Strehl L, Besse L, Laclef C, Ros MR, Garcia-Verdugo JM, Alvarez-Buylla A. 2008. Primary cilia are required for cerebellar development and Shh-dependent expansion of progenitor pool. *Dev Biol* **317**: 246–259.

Srour M, Schwartzentruber J, Hamdan FF, Ospina LH, Patry L, Labuda D, Massicotte C, Dobrzeniecka S, Capo-Chichi JM, Papillon-Cavanagh S, et al. 2012. Mutations in *C5ORF42* cause Joubert syndrome in the French Canadian population. *Am J Hum Genet* **90**: 693–700.

Stephen LA, Tawamie H, Davis GM, Tebbe L, Nürnberg P, Nürnberg G, Thiele H, Thoenes M, Boltshauser E, Uebe S, et al. 2015. TALPID3 controls centrosome and cell polarity and the human ortholog *KIAA0586* is mutated in Joubert syndrome (*JBTS23*). *eLife* **4**: e08077.

Swiderski RE, Nakano Y, Mullins RF, Seo S, Bánfi B. 2014. A mutation in the mouse *ttc26* gene leads to impaired hedgehog signaling. *PLoS Genet* **10**: e1004689.

Tanos BE, Yang HJ, Soni R, Wang WJ, Macaluso FP, Asara JM, Tsou MF. 2013. Centriole distal appendages promote membrane docking, leading to cilia initiation. *Genes Dev* **27**: 163–168.

Tay SY, Yu X, Wong KN, Panse P, Ng CP, Roy S. 2010. The iguana/DZIP1 protein is a novel component of the ciliogenic pathway essential for axonemal biogenesis. *Dev Dyn* **239**: 527–534.

Taylor SP, Dantas TJ, Duran I, Wu S, Lachman RS, University of Washington Center for Mendelian Genomics Consortium, Nelson SF, Cohn DH, Vallee RB, Krakow D. 2015. Mutations in *DYNC2LI1* disrupt cilia function and cause short rib polydactyly syndrome. *Nat Commun* **6**: 7092.

Tong CK, Han YG, Shah JK, Obernier K, Guinto CD, Alvarez-Buylla A. 2014. Primary cilia are required in a unique

Cite this article as *Cold Spring Harb Perspect Biol* doi: 10.1101/cshperspect.a028175

subpopulation of neural progenitors. *Proc Natl Acad Sci* **111**: 12438–12443.

Toriyama M, Lee C, Taylor SP, Duran I, Cohn DH, Bruel AL, Tabler JM, Drew K, Kelly MR, Kim S, et al. 2016. The ciliopathy-associated CPLANE proteins direct basal body recruitment of intraflagellar transport machinery. *Nat Genet* **48**: 648–656.

Tran PV, Haycraft CJ, Besschetnova TY, Turbe-Doan A, Stottmann RW, Herron BJ, Chesebro AL, Qiu H, Scherz PJ, Shah JV, et al. 2008. THM1 negatively modulates mouse sonic hedgehog signal transduction and affects retrograde intraflagellar transport in cilia. *Nat Genet* **40**: 403–410.

Tuson M, He M, Anderson KV. 2011. Protein kinase A acts at the basal body of the primary cilium to prevent Gli2 activation and ventralization of the mouse neural tube. *Development* 2011 **138**: 4921–4930.

* Vaisse C, Reiter JF, Berbari NF. 2016. Cilia and obesity. *Cold Spring Harb Perspect Biol* doi: 10.1101/cshperspect.a028217.

Vuolo L, Herrera A, Torroba B, Menendez A, Pons S. 2015. Ciliary adenylyl cyclases control the Hedgehog pathway. *J Cell Sci* **128**: 2928–2937.

Walczak-Sztulpa J, Eggenschwiler J, Osborn D, Brown DA, Emma F, Klingenberg C, Hennekam RC, Torre G, Garshasbi M, Tzschach A, et al. 2010. Cranioectodermal dysplasia, Sensenbrenner syndrome, is a ciliopathy caused by mutations in the *IFT122* gene. *Am J Hum Genet* **86**: 949–956.

Wang C, Low WC, Liu A, Wang B. 2013. Centrosomal protein DZIP1 regulates Hedgehog signaling by promoting cytoplasmic retention of transcription factor GLI3 and affecting ciliogenesis. *J Biol Chem* **288**: 29518–29529.

Wang H, Kane AW, Lee C, Ahn S. 2014. Gli3 repressor controls cell fates and cell adhesion for proper establishment of neurogenic niche. *Cell Rep* **8**: 1093–1104.

Weatherbee SD, Niswander LA, Anderson KV. 2009. A mouse model for Meckel syndrome reveals Mks1 is required for ciliogenesis and Hedgehog signaling. *Hum Mol Genet* **18**: 4565–4575.

Wen X, Lai CK, Evangelista M, Hongo JA, de Sauvage FJ, Scales SJ. 2010. Kinetics of hedgehog-dependent full-length Gli3 accumulation in primary cilia and subsequent degradation. *Mol Cell Biol* **30**: 1910–1922.

Wong SY, Seol AD, So PL, Ermilov AN, Bichakjian CK, Epstein EH Jr, Dlugosz AA, Reiter JF. 2009. Primary cilia can both mediate and suppress Hedgehog pathway–dependent tumorigenesis. *Nat Med* **15**: 1055–1061.

Yadav SP, Sharma NK, Liu C, Dong L, Li T, Swaroop A. 2016. Centrosomal protein CP110 controls maturation of the mother centriole during cilia biogenesis. *Development* **143**: 1491–1501.

Ye X, Zeng H, Ning G, Reiter JF, Liu A. 2014. C2cd3 is critical for centriolar distal appendage assembly and ciliary vesicle docking in mammals. *Proc Natl Acad Sci* **111**: 2164–2169.

Yin Y, Bangs F, Paton IR, Prescott A, James J, Davey MG, Whitley P, Genikhovich G, Technau U, Burt DW, et al. 2009. The *Talpid3* gene (*KIAA0586*) encodes a centrosomal protein that is essential for primary cilia formation. *Development* **136**: 655–664.

Yu M, Gipp J, Yoon JW, Iannaccone P, Walterhouse D, Bushman W. 2009. Sonic Hedgehog–responsive genes in the fetal prostate. *J Biol Chem* **284**: 5620–5629.

Zeng H, Hoover AN, Liu A. 2010. PCP effector gene *Inturned* is an important regulator of cilia formation and embryonic development in mammals. *Dev Biol* **339**: 418–428.

G-Protein-Coupled Receptor Signaling in Cilia

Kirk Mykytyn[1,2] and Candice Askwith[2,3]

[1]Department of Biological Chemistry and Pharmacology, The Ohio State University, Ohio 43210

[2]Neuroscience Research Institute, The Ohio State University, Ohio 43210

[3]Department of Neuroscience, The Ohio State University, Ohio 43210

Correspondence: mykytyn.1@osu.edu

G-protein-coupled receptors (GPCRs) are the largest and most versatile family of signaling receptors in humans. They respond to diverse external signals, such as photons, proteins, peptides, chemicals, hormones, lipids, and sugars, and mediate a myriad of functions in the human body. Signaling through GPCRs can be optimized by enriching receptors and downstream effectors in discrete cellular domains. Many GPCRs have been found to be selectively targeted to cilia on numerous mammalian cell types. Moreover, investigations into the pathophysiology of human ciliopathies have implicated GPCR ciliary signaling in a number of developmental and cellular pathways. Thus, cilia are now appreciated as an increasingly important nexus for GPCR signaling. Yet, we are just beginning to understand the precise signaling pathways mediated by most ciliary GPCRs and how they impact cellular function and mammalian physiology.

It is estimated that the human genome encodes approximately 950 G-protein-coupled receptors (GPCRs), of which 500 correspond to odorant or taste receptors (Takeda et al. 2002). Approximately 150 of the remaining 450 GPCRs have no known natural ligand and so are referred to as orphan GPCRs (Tang et al. 2012). GPCRs represent the largest group of therapeutic drug targets, with more than a third of all drugs acting on GPCRs (Rask-Andersen et al. 2011). Because of the functional diversity of GPCRs, there is little conservation of amino acid sequence across the GPCR superfamily. Yet, all GPCRs share a common structure: an extracellular amino terminus, seven transmembrane domains, and an intracellular carboxyl terminus.

In canonical GPCR signaling at the plasma membrane, agonist binding to a receptor causes a change in receptor conformation and results in activation of heterotrimeric GTP-binding proteins (G proteins) (Fig. 1A) (Shenoy and Lefkowitz 2011). G proteins consist of three associated protein subunits: α, β, and γ. G proteins are classified based on the nature of their α-subunits and there are 16 known α-subunits that are functionally categorized into four subfamilies: $G\alpha_s$, $G\alpha_i$, $G\alpha_q$, and $G\alpha_{12}$. When inactive, the α-subunit is bound to GDP and a $\beta\gamma$-complex to form a trimeric protein complex (Fig. 1A). On agonist binding, the receptor facilitates GDP release, GTP binding to the α-subunit, and dissociation of the α-subunit from the $\beta\gamma$-

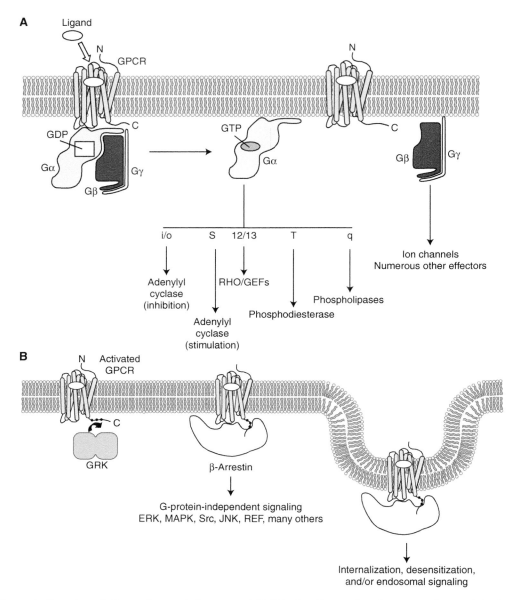

Figure 1. Overview of G-protein-coupled receptor (GPCR) signaling at the plasma membrane. (*A*) Ligand binding to a GPCR facilitates GDP release from the G-protein α-subunit and stimulates GTP binding to the α-subunit, which leads to dissociation of the α-subunit from the βγ-complex. Both the α-subunit and βγ-complex can then regulate various intracellular effectors. (*B*) Activated GPCRs are phosphorylated at specific sites on their intracellular domains predominantly by G-protein-coupled receptor kinases (GRKs). Phosphorylated receptors are targets for the recruitment of β-arrestins, which prevent further G-protein activation and mediate internalization of receptors by promoting clathrin-mediated endocytosis. β-Arrestins bind to numerous intracellular signaling proteins and can act as signal transducers independently of G-protein coupling. In some cases, GPCR signaling can be sustained or enhanced on endocytosis.

complex (Fig. 1A). Both the α-subunit and βγ-complex can then regulate various intracellular effectors (e.g., adenylyl cyclases by the α-subunit and potassium channels by the βγ-complex).

Activated GPCRs are then phosphorylated at specific sites on their intracellular domains predominantly by G-protein-coupled receptor kinases (GRKs) (Fig. 1B) (Marchese et al. 2008), but also other kinases such as protein kinase A (PKA) or protein kinase C (PKC) (Kelly et al. 2008). Once phosphorylated, the receptors become targets for the recruitment and binding of scaffolding proteins, termed β-arrestins, which prevent further G-protein activation and mediate internalization of receptors by promoting clathrin-mediated endocytosis (Fig. 1B) (Shenoy and Lefkowitz 2011). This process is known as homologous desensitization. Although internalization of GPCRs is generally associated with a decrease in signaling, in some cases receptor signaling can be sustained or enhanced on endocytosis (Sorkin and von Zastrow 2009; McMahon and Boucrot 2011). In addition, β-arrestins bind to numerous intracellular signaling proteins, including Src, ERK1/2, p38, and PI3K, and can act as signal transducers independent of G-protein coupling (DeFea 2011; Shukla et al. 2011). There are two β-arrestin isoforms (1 and 2) that are expressed ubiquitously and regulate most GPCRs.

There is an ever-expanding list of GPCRs that are enriched in cilia on a variety of cell types (Table 1; Fig. 2). Numerous GPCR effector molecules have also been localized to cilia (Fig. 2) (Hilgendorf et al. 2016), suggesting that cilia mediate signaling of a diverse set of GPCRs. Importantly, ciliopathies are associated with alterations in GPCR signaling. In this review, we will focus on mammalian cilia-mediated GPCR signaling transduction pathways. We will begin with a brief account of the well-described signaling pathways mediated by the prototypical ciliary GPCRs, odorant receptors, and opsins. Then we will discuss more recently described examples of GPCR ciliary signaling with a focus on the potential functional impacts of cilia on GPCR signaling.

OLFACTORY RECEPTOR SIGNALING

Ciliopathies can be associated with deficits in olfaction (Kulaga et al. 2004; Iannaccone et al. 2005; McEwen et al. 2007). Mammalian olfaction is mediated by olfactory sensory neurons (OSNs) that project from the olfactory bulb in the brain to the olfactory epithelium located in the nasal cavity (Fig. 3A). OSNs are bipolar neurons with a single axon that projects distally to the olfactory bulb and a single dendrite that projects apically to the olfactory epithelium. At the apical end of the OSN, the dendritic tip is enlarged to form a dendritic knob from which 10 to 30 nonmotile 9+2 cilia project (Menco 1980, 1997). These olfactory cilia range from 50 to 60 μm in length and extend into the olfactory mucus where they are directly exposed to odorants (Jenkins et al. 2009). There are two important consequences of the ciliary structure. First, the presence of numerous cilia increases the surface area that is exposed to the external environment by about 40 times and enhances our ability to detect odorants (Menco 1992). Second, the small diameter of the distal ends of these cilia (~0.1 μm) leads to a large ratio of membrane surface area to cytoplasmic volume (Menco 1980), allowing a small signal to generate a large effect.

The molecular elements required for olfactory transduction are concentrated within the ciliary compartment (Fig. 3B). Olfaction begins with binding of an odorant to an olfactory receptor (OR) on the ciliary membrane. In rodents, each OSN predominantly expresses one of approximately 1000 ORs (Ressler et al. 1993; Vassar et al. 1993), which triggers the activation of the heterotrimeric stimulatory G protein comprising $G\alpha_{olf}$, β_1, and γ_{13} (Jones and Reed 1989; Kerr et al. 2008; Li et al. 2013). $G\alpha_{olf}$ then activates type 3 adenylyl cyclase (AC3), which increases cAMP levels within the cilium (Bakalyar and Reed 1990). The cAMP then binds to and activates cyclic-nucleotide-gated (CNG) channels on the ciliary membrane, allowing the entry of calcium ions and depolarizing the membrane potential. Increased ciliary Ca^{2+} levels leads to activation and opening of Ca^{2+}-gated chloride channels,

Table 1. Nonodorant/nonvisual G-protein-coupled receptors (GPCRs) with confirmed ciliary localization

GPCR	Cell type	References
β2-adrenergic receptor (β2AR)	Neurons	Yao et al. 2015
Bile acid receptor (TGR5)	Cholangiocytes	Keitel et al. 2010; Masyuk et al. 2013
Bitter taste receptors (T2R)	Airway epithelial cells	Shah et al. 2009
Dopamine receptor 1 (D1)	Neurons	Domire et al. 2011
Dopamine receptor 5 (D5)	Vascular endothelial cells, renal epithelial cells	Abdul-Majeed and Nauli 2011; Jin et al. 2014b
Galanin receptor 3 (GALR3)	Neurons	Loktev and Jackson 2013
GPR83	Neurons	Loktev and Jackson 2013
GPR161	Neurons, mouse embryonic fibroblasts	Mukhopadhyay et al. 2013
GPR175	Mouse embryonic fibroblasts	Singh et al. 2015
Kisspeptin receptor 1 (KISS1R)	Neurons	Koemeter-Cox et al. 2014
Melanin-concentrating hormone receptor 1 (MCHR1)	Neurons	Berbari et al. 2008
Muscarinic acetylcholine receptor 3 (M3R)	Olfactory sensory neurons	Jiang et al. 2015
Neuropeptide Y receptor 2 (NPY2R)	Neurons	Loktev and Jackson 2013
Neuropeptide Y receptor 5 (NPY5R)	Neurons	Loktev and Jackson 2013
Prolactin-releasing hormone receptor (PRLHR)	Glial cells	Omori et al. 2015
Prostaglandin E receptor 4 (EP4)	Human retinal pigment epithelial cells	Jin et al. 2014a
Pyroglutamylated RFamide peptide receptor (QRFPR)	Neurons	Loktev and Jackson 2013
Serotonin receptor 6 (HTR6)	Neurons	Brailov et al. 2000
Smoothened (SMO)	Fibroblasts, nodal cells	
Somatostatin receptor 3 (SSTR3)	Neurons	Handel et al. 1999
Trace amine-associated receptor 1 (TAAR1)	Thyroid epithelial cells	Szumska et al. 2015
Vasopressin receptor 2 (V2R)	Renal epithelial cells	Raychowdhury et al. 2009

causing an efflux of Cl^- ions that augments depolarization of the neuron (Kleene 1993; Lowe and Gold 1993; Stephan et al. 2009), which eventually initiates an action potential that propagates along the axon to the olfactory bulb.

β-Arrestin 2 is recruited to activated ORs to mediate desensitization and receptor internalization at the dendritic knob (Dawson et al. 1993; Mashukova et al. 2006). Interestingly, a nonodorant GPCR has been found to localize to olfactory cilia and modulate OR signaling in mice. Specifically, activation of type 3 muscarinic acetylcholine receptor (M3-R) on the ciliary membrane inhibits the recruitment of β-arrestin 2 to ORs, thereby potentiating odor-induced signaling (Fig. 3C) (Jiang et al. 2015). The olfactory epithelium is innervated by nerve endings that release acetylcholine (Baraniuk and Merck 2009). Thus, acetylcholine release may enhance the sensitivity of OR signaling via M3-Rs.

In summary, olfactory cilia possess several critical attributes that optimize OR signaling: (1) They extend into the olfactory epithelium where ORs are exposed to odorants; (2) they allow for enrichment and concentration of the molecular components of the olfactory transduction cascade, thereby optimizing signaling; and (3) they are present in large numbers and

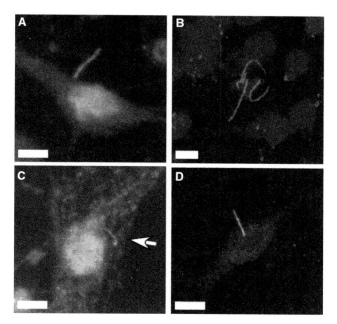

Figure 2. Examples of G-protein-coupled receptors (GPCRs) and effectors that are enriched in primary cilia. (*A*) Image of a day 7 mouse hippocampal neuron immunolabeled with an antibody to somatostatin receptor subtype 3 (SSTR3) showing an SSTR3-positive cilium projecting from the cell body. (*B*) Adult mouse brain section corresponding to the medial hypothalamus immunolabeled with an antibody to kisspeptin receptor 1 (KISS1R). Note the presence of multiple KISS1R-positive cilia. (*C*) Image of a day 7 mouse hippocampal neuron treated with somatostatin and immunolabeled with an antibody to β-arrestin. Arrow indicates β-arrestin ciliary localization. (*D*) Image of a day 7 mouse hippocampal neuron immunolabeled with an antibody to type 3 adenylyl cyclase (AC3) showing an AC3-positive cilium projecting from the cell body. Scale bars, 5 μm.

have a large surface-to-volume ratio, which increases sensitivity to odorants.

OPSIN SIGNALING

Vision is initiated when photons are absorbed by the rod and cone photoreceptors in the retina (Arshavsky and Burns 2012). Photoreceptors are highly polarized neurons with a distal end that is comprised of the light-sensing outer segment and a proximal end that synapses on downstream neurons. The outer segment is a highly modified primary cilium packed with membrane disks containing light-sensitive GPCRs and downstream signaling effectors. Upon light activation in rods, rhodopsin activates the G protein transducin, which then stimulates its effector, cGMP phosphodiesterase. This leads to a reduction in intracellular cGMP levels and causes cGMP-gated channels to close, thereby hyperpolarizing the cell and generating a transient photoresponse within milliseconds. Rhodopsin kinase then phosphorylates rhodopsin, which leads to visual arrestin binding and a block in transducin activation, thereby terminating the signal. In response to sustained bright light, there is a massive redistribution of phototransduction proteins that involves transducin exiting the rod outer segment and visual arrestin accumulating in the outer segment. This adaptive mechanism plays a role in setting photoreceptor sensitivity and may protect rods from the adverse effects of persistent light exposure (Arshavsky and Burns 2012). Thus, the ciliary outer segment facilitates signaling in response to light and provides a protective mechanism by allowing the physical separation

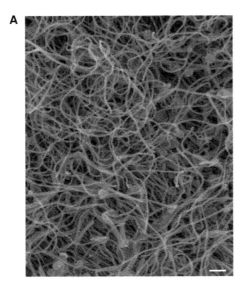

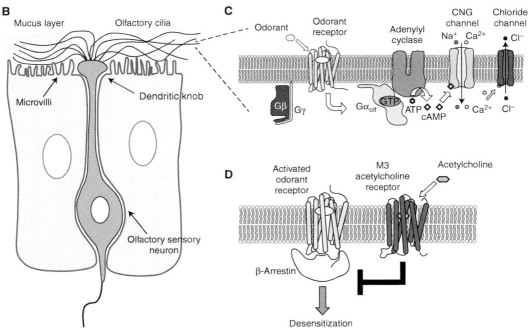

Figure 3. Overview of odorant receptor signaling in olfactory sensory neurons. (*A*) Scanning electron microscopy image of the surface of the mouse olfactory epithelium. Scale bar, 1 μm (courtesy of Jeff Martens). (*B*) Schematic of a single olfactory sensory neuron with cilia projecting into the olfactory epithelium. (*C*) Odorant activation of olfactory G-protein-coupled receptors (GPCRs) triggers the activation of the stimulatory G protein $G\alpha_{olf}$, which then activates type 3 adenylyl cyclase (AC3) and increases cAMP levels within the cilium. The cAMP binds to and activates cyclic-nucleotide-gated (CNG) channels on the ciliary membrane, leading to an increase in Ca^{2+} levels, subsequent activation of Ca^{2+}-gated chloride channels, and depolarization of the neuron. (*D*) β-Arrestin binding to activated odorant receptors mediates desensitization. The type 3 muscarinic (M3) acetylcholine receptor can inhibit the recruitment of β-arrestin to odorant receptors, thereby potentiating odor-induced signaling.

Cite this article as *Cold Spring Harb Perspect Biol* doi: 10.1101/cshperspect.a028183

of components of the phototransduction cascade.

GPCR MODULATION OF HEDGEHOG SIGNALING

Hedgehog (Hh) signaling, which plays an essential role in mammalian development, requires the presence of primary cilia (Huangfu et al. 2003). Briefly, in the absence of Hh ligand, the 12-transmembrane Hh receptor patched (Ptch1) is enriched on the ciliary membrane and the GPCR Smoothened (Smo) is excluded from the cilium (Rohatgi et al. 2007). In this "OFF" state, cAMP-dependent PKA functions at the base of the cilium to phosphorylate members of the Gli family of transcription factors, which promotes the formation of truncated Gli repressors and inhibits transcription of Hh target genes (Sasaki et al. 1999; Pan et al. 2006; Tempe et al. 2006; Tuson et al. 2011). Repression of Hh signaling is further enforced by the ciliary GPCR Gpr161. Gpr161 has constitutive activity and couples to $G\alpha_s$ to increase cellular cAMP levels, thereby increasing activation of PKA (Mukhopadhyay et al. 2013). It is hypothesized that Gpr161 establishes a basal cAMP gradient within the cilium that is important for proper regulation of Hh signaling. Indeed, disruption of Gpr161 in mouse is embryonic lethal and causes increased Hh signaling in the neural tube (Mukhopadhyay et al. 2013).

In the presence of Hh ligand, Ptch1 and Gpr161 leave the cilium, allowing Smo to enter the cilium, activate Gli transcription factors, and initiate signaling (Corbit et al. 2005; Rohatgi et al. 2007; Mukhopadhyay et al. 2013). Recently, another orphan GPCR, Gpr175, has been shown to localize to cilia in response to Hh treatment and enhances Hh signaling in several mammalian cell lines (Singh et al. 2015). Specifically, Gpr175 interacts with ciliary $G\alpha_i$, which leads to a lowering of cAMP levels and an inhibition of PKA activity and Gli repressor formation (Singh et al. 2015). Depletion of Gpr175 in cell lines has a relatively modest effect on signaling ($\sim$50%), suggesting that it plays a regulatory role rather than an essential role in Hh signaling. Indeed, as opposed to

Gpr161 knockout mice, Gpr175 knockout mice are viable and do not have any developmental defects (Singh et al. 2015). The effect of Gpr175 on Hh signaling is dependent on Smo activity, suggesting it acts on $G\alpha_i$ downstream from Smo to modulate PKA activity. Although it does not play an essential role in Hh signaling, it may enhance signaling in certain contexts (Singh et al. 2015).

CILIARY GPCR SIGNALING IN THE RENAL SYSTEM

A link between renal cilia dysfunction and cystic disease is well established (Cramer and Guay-Woodford 2015). Primary cilia on renal epithelial cells are generally regarded as mechanosensors that illicit Ca^{2+} signals in response to fluid flow (Fig. 4A). However, there is increasing evidence that renal cilia also mediate GPCR signaling. For example, the type 2 vasopressin receptor (V2R), which regulates Na^+ and water reabsorption in the mammalian nephron, localizes to cilia on renal epithelial cells (Raychowdhury et al. 2009). In response to vasopressin, ciliary V2R functionally couples with adenylyl cyclase to increase local cAMP concentrations and activate a cation-selective channel (Fig. 4B) (Raychowdhury et al. 2009). These data suggest the presence of a GPCR-mediated cAMP-dependent second-messenger signaling mechanism in renal cilia that regulates intraciliary Ca^{2+} signals. This signaling, in turn, may modulate different cellular processes, including cell proliferation, ciliary microtubule stability, and/or the ciliary membrane resting potential. Interestingly, renal cAMP levels are increased in numerous animal models of polycystic kidney disease (PKD) (Torres and Harris 2014) and treatment with V2R antagonists inhibits cyst formation (Gattone et al. 2003; Torres et al. 2004). Yet, it is unclear how much V2R signaling within the cilium contributes to these effects.

Dopaminergic signaling in the kidney plays an important role in controlling renal sodium excretion and blood pressure (Carey 2013). Recently, dopamine receptor type 5 (D5) has been localized to cilia on renal epithelial cells (Jin

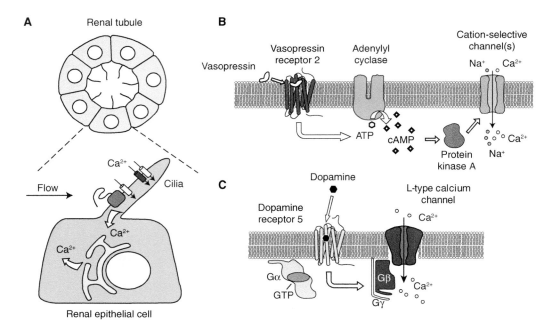

Figure 4. Overview of ciliary signaling in renal cilia. (*A*) Cross section of a renal tubule showing primary cilia projecting into the lumen of the tubule (*top*). Schematic of flow-induced Ca^{2+} signaling (*bottom*). (*B*) Vasopressin binding to vasopressin receptor 2 (V2R) on the ciliary membrane activates adenylyl cyclase. The increase in local cAMP concentrations activates a cation-selective channel, possibly through protein kinase A, thereby regulating intraciliary Ca^{2+} signals. (*C*) Agonist binding to dopamine receptor 5 on the ciliary membrane results in $Ca_V1.2$ channel activation, possibly through the action of dissociated $G\beta\gamma$, which increases intraciliary Ca^{2+} levels.

et al. 2014b; Upadhyay et al. 2014). Evidence for D5-mediated signaling on cilia comes from studies looking at calcium signaling in the ciliary compartment. Specifically, Jin et al. (2014b) used a ciliary-targeted calcium sensor to show that treatment of renal epithelial cells with the D5 agonist fenoldopam causes an increase in calcium levels in the cilium that precedes an increase in calcium levels in the cytosol of the cell. This calcium signal is dependent on the $Ca_V1.2$ L-type calcium channel, which is localized in the cilium (Jin et al. 2014b). With regard to a functional consequence, fenoldopam treatment also causes an actin-mediated increase in cilia length and increased calcium signaling in response to fluid flow (Upadhyay et al. 2014). Taken together, these results suggest that agonist binding to D5 on the ciliary membrane results in $Ca_V1.2$ channel activation, possibly through the action of dissociated $G\beta\gamma$, which increases intraciliary Ca^{2+} levels (Fig. 4C) (Atkinson et al.

2015). This increased Ca^{2+} concentration subsequently leads to cilia elongation and confers greater sensitivity to fluid-shear stress. Interestingly, the most frequent target found in a chemical screen of pathways involved in flagellar length control in the unicellular green alga *Chlamydomonas* was the family of dopamine binding GPCRs (Avasthi et al. 2012). Thus, ciliary dopaminergic signaling may be an evolutionarily conserved mechanism for regulating ciliary length, which may then impact sensitivity to signals.

GPCR SIGNALING IN CHOLANGIOCYTE CILIA

Cholangiocytes are ciliated epithelial cells that line bile ducts and are responsible for bile acid transport and bicarbonate secretion (Tabibian et al. 2013). Cholangiocyte primary cilia have been determined to be mechano-, chemo-, and

Cite this article as *Cold Spring Harb Perspect Biol* doi: 10.1101/cshperspect.a028183

osmosensory organelles that regulate cholangiocyte proliferation (Masyuk et al. 2006, 2008a,b; Gradilone et al. 2007). The importance of these cilia is highlighted by the fact that polycystic liver disease, which is characterized by the development of fluid-filled hepatic cysts arising from cholangiocytes, is associated with ciliopathies (Masyuk et al. 2015). In cholangiocytes, bile acid signaling is transmitted through TGR5, a GPCR that is localized to the apical plasma membrane, subapical compartment, and cilium (Keitel et al. 2009, 2010; Keitel and Haussinger 2011; Keitel and Haussinger 2012; Masyuk et al. 2013). TGR5 is coupled to $G\alpha_s$ and activation of TGR5 by bile acids causes an increase in intracellular cAMP levels (Maruyama et al. 2002; Kawamata et al. 2003). Downstream effectors of GPCR signaling, including adenylyl cyclase, PKA, and the exchange protein directly activated by cAMP 2 (EPAC-2), have also been localized to cholangiocyte cilia (Masyuk et al. 2006, 2008b). In addition, $G\alpha_i$ localizes to the base of cilia on ciliated cholangiocytes. Taken together, these data suggest that cholangiocyte cilia mediate bile acid signaling through TGR5.

Intriguingly, the presence or absence of cilia on cultured cholangiocytes determines the impact of TGR5 agonists (Masyuk et al. 2013). Specifically, agonist treatment of nonciliated cholangiocytes increases colocalization of TGR5 with $G\alpha_s$ and results in increased cAMP signaling, inhibition of ERK signaling, and increased cellular proliferation. Agonist treatment of ciliated cells, on the other hand, results in increased colocalization of TGR5 with $G\alpha_i$ and decreased cAMP signaling, activation of ERK signaling, and decreased cellular proliferation. Together, these results suggest that TGR5 is functionally coupled to $G\alpha_s$ on the plasma membrane and stimulates cellular proliferation in response to bile acid signaling, but is functionally coupled to $G\alpha_i$ in the cilium to prevent cellular proliferation in response to bile acid signaling. Thus, cholangiocyte cilia provide a compartment for TGR5 to functionally couple with different effectors and provide an alternative signal to TGR5 signaling on the plasma membrane.

CILIARY GPCR SIGNALING ON CENTRAL NEURONS

Most, if not all, adult neurons in the mammalian brain possess a primary cilium (Handel et al. 1999; Fuchs and Schwark 2004; Bishop et al. 2007). Numerous GPCRs are selectively enriched in neuronal cilia (Table 1). Seminal studies using mouse knockout models have provided compelling evidence for cilia-dependent GPCR signaling in the brain. For example, mice lacking cilia on specific neuronal subpopulations in the brain manifest prominent phenotypes, such as obesity and learning and memory deficits (Davenport et al. 2007; Berbari et al. 2013, 2014). Moreover, mice lacking ciliary GPCRs or ciliary-enriched downstream effectors of GPCR signaling display similar phenotypes (Wang et al. 2009, 2011; Einstein et al. 2010). Together, these results suggest that neuronal cilia provide a unique platform for GPCRs to signal in response to factors in the extracellular milieu. Recent studies have begun to elucidate these signaling pathways and how they impact neuronal function.

Neuropeptide Y (NPY) is one of the most abundant neuropeptides in the mammalian brain and plays an important role in regulating food intake and energy expenditure (Herzog 2003). Recently, two of the NPY receptor subtypes, NPY2R and NPY5R, were found to be enriched in neuronal cilia in mice (Loktev and Jackson 2013). Interestingly, genetically modified mice that are unable to transport NPY2R into neuronal cilia are obese and do not respond to administration of the anorexigenic ligand PYY3-36 (Loktev and Jackson 2013), suggesting that NPY2R ciliary localization is important for ligand-dependent signaling in vivo. In support of this model, quantification of cAMP signaling on RPE cells expressing NPY2R revealed that ligand treatment produced a more pronounced inhibition of cAMP signaling in cells with a cilium (Fig. 5A) (Loktev and Jackson 2013). Thus, cilia localization seemingly enhances NPY2R signaling and may provide a more robust signal to control food intake.

Somatostatin is a widely distributed neurotransmitter and modulator of neural activity

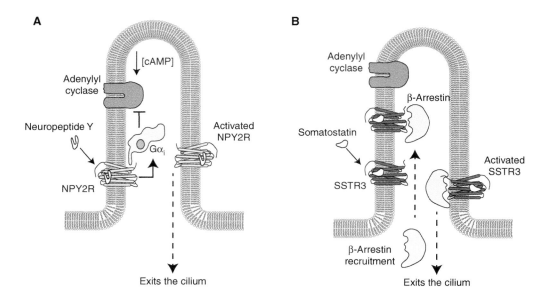

Figure 5. Overview of G-protein-coupled receptor (GPCR) signaling on neuronal cilia. (*A*) Ligand binding to neuropeptide Y receptor 2 (NPY2R) on the ciliary membrane may activate Gα_i and inhibit adenylyl cyclase, thereby leading to a reduction in cAMP levels. Ligand treatment also leads to a reduction in NPY2R ciliary localization, suggesting that activated receptor exits the cilium. (*B*) Somatostatin treatment stimulates endogenous β-arrestin recruitment into somatostatin receptor subtype 3 (SSTR3)-positive cilia. Somatostatin treatment also causes a β-arrestin-dependent decrease in SSTR3 ciliary localization, suggesting that β-arrestin mediates SSTR3 ciliary export.

that can affect many physiological processes, including motor activity and cognitive function (Patel 1999; Barnett 2003; Olias et al. 2004; Viollet et al. 2008). Somatostatin receptor subtype 3 (SSTR3) colocalizes with AC3 in cilia throughout the mouse brain (Handel et al. 1999). Interestingly, mice lacking SSTR3, AC3, or cilia in the hippocampus show similar deficits in learning and memory (Wang et al. 2009; Einstein et al. 2010; Berbari et al. 2014). Together, these results suggest that SSTR3 signals on cilia and this signaling is required for proper learning and memory. In support of this model, it was recently shown that somatostatin treatment stimulates endogenous β-arrestin recruitment into SSTR3-positive cilia on hippocampal neurons (Green et al. 2016). This recruitment is reminiscent of β-arrestin translocation to activated GPCRs on the plasma membrane. Furthermore, expressing SSTR3-containing mutations that prevent agonist binding or phosphorylation blocks β-arrestin recruitment into cilia. These results suggest that agonist binds to SSTR3 on the ciliary membrane and leads to phosphorylation of the receptor, which facilitates β-arrestin ciliary localization (Fig. 5B).

Another finding from this study was that somatostatin treatment causes a rapid β-arrestin-dependent decrease in the ciliary localization of endogenous SSTR3 (Green et al. 2016). These findings suggest a model whereby activation of SSTR3 on the ciliary membrane stimulates β-arrestin recruitment, which binds to the receptor and mediates export of SSTR3 from the cilium (Fig. 5B). There are several potential functional consequences of β-arrestin ciliary recruitment, including (1) SSTR3 desensitization, (2) potentiation of SSTR3 signaling through internalization, and/or (3) β-arrestin-mediated ciliary signaling. Yet, additional studies are required to determine the functional consequences of β-arrestin recruitment into cilia and whether this is a ubiquitous mechanism for modulating ciliary GPCR signaling on neurons.

The kisspeptin receptor (KISS1R), which regulates the onset of puberty and adult reproductive function, has recently been found to be enriched in cilia on mouse gonadotropin-releasing hormone (GnRH) neurons (Koemeter-Cox et al. 2014). Intriguingly, GnRH neurons in adult animals possess multiple KISS1R cilia and the percentage of multiciliated GnRH neurons increases during postnatal development and correlates with sexual maturation. Disruption of GnRH cilia leads to a significant reduction in kisspeptin-mediated GnRH neuronal activity (Koemeter-Cox et al. 2014), suggesting that cilia enhance KISS1R signaling. More recently, the β2-adrenergic receptor (β2AR) has been shown to be localized to neuronal cilia in the mouse hippocampus (Yao et al. 2015). β2AR is activated by noradrenalin and plays a role in hippocampal synaptic plasticity (Hagena et al. 2016). Interestingly, β2AR colocalizes in neuronal cilia with the nonselective cation channel polycystic kidney disease 2-like 1 (Pkd2l1). However, β2AR ciliary localization is disrupted in mice lacking Pkd2l1, suggesting that Pkd2l1 is required for β2AR ciliary localization (Yao et al. 2015). Pkd2l1 mice have decreased cAMP levels in the brain and increased susceptibility to pentylenetetrazol-induced seizures (Yao et al. 2015). As β2AR is coupled to $G\alpha_s$, a potential model is that ciliary localization of β2AR and Pkd2l1 form a ciliary complex that enhances cAMP production, which inhibits neuronal excitability. Further studies are required to directly test whether KISS1R and β2AR signal within cilia.

GPCR SIGNALING IN MOTILE CILIA

The primary function of motile cilia on human airway epithelia is to move mucus out of the lung and their disruption results in airway disease (Fliegauf et al. 2007). Interestingly, several members of the bitter taste receptor (T2R) family have been localized to cilia on human airway epithelia (Shah et al. 2009; Lee et al. 2012). Downstream effectors of the T2R signal transduction pathway include the G-protein α-gustducin and the enzyme phospholipase C-β2 (PLC-β2) (Devillier et al. 2015). In ciliated airway epithelial cells, α-gustducin localizes to cilia and PLC-β2 localizes to the apical portion of the cell below the cilia (Shah et al. 2009). Application of bitter compounds causes an increase in intracellular calcium concentrations only in ciliated cells, which further results in a ~25% increase in ciliary beat frequency (Shah et al. 2009). Thus, T2R ciliary localization may facilitate sensing of noxious compounds and generation of a signal that leads to an increase in ciliary activity to eliminate the substance.

POTENTIAL FUNCTIONAL CONSEQUENCES OF CILIA ON GPCR SIGNALING

A key question is how ciliary localization impacts GPCR signaling. The examples of ciliary GPCR signaling discussed above highlight several general functional consequences cilia can confer on GPCR signaling. First, ciliary localization can enhance GPCR signaling. This may be due to several reasons. Cilia may extend toward the origin of a signal, thereby increasing sensitivity. The cilium may enhance signaling by allowing more efficient coupling of the receptor and its effectors. Note that these two functions are not mutually exclusive. Another possibility is GPCRs generate a unique signal when they are activated on the ciliary membrane versus the plasma membrane. This may be the result of coupling to distinct effectors in the cilium, as in cholangiocyte cilia. As β-arrestins can function as signal transducers, the finding that endogenous β-arrestin is recruited into cilia on somatostatin treatment is particularly provocative. This could be analogous to biased agonism, whereby a ligand preferentially triggers G-protein- or β-arrestin-mediated signaling pathways (Rajagopal et al. 2010). Perhaps activation of a receptor on the ciliary membrane activates β-arrestin-mediated signaling but not G-protein signaling, or vice versa. Another potential mechanism for generating a unique signal is by facilitating heteromerization of different GPCRs, which can alter ligand binding, G-protein coupling, and/or desensitization and internalization. There is evidence that SSTR3 and melanin-concentrating hormone receptor

1 heteromerize in cilia in multiple mouse brain regions (Green et al. 2012). Alternatively, ciliary localization may act as an insulator to prevent GPCR cross regulation (Marley et al. 2013). It is possible that some or all of these mechanisms are used in ciliary GPCR signaling and the precise effects vary between cell types or even between different cells in the same tissue.

CONCLUDING REMARKS

Given the prevalence of GPCRs as drug targets, understanding ciliary GPCR signaling will likely have important ramifications for therapeutic development. The remaining overarching challenges in the ciliary GPCR field are to identify the complete complement of ciliary GPCRs, define the signaling pathways mediated by ciliary GPCRs and determine how these signaling pathways impact cellular function. Meeting these challenges will require the development of new tools to visualize and/or modulate ciliary GPCR signaling as well as assays to determine how these signals impact cellular function. A better understanding of GPCR signaling will likely yield new therapeutic strategies to target-specific aspects of GPCR function and lend important insight into the consequences of cilia loss and disruption for human health.

ACKNOWLEDGMENTS

This work is supported by research project Grant R21 MH107021 from the National Institutes of Health/National Institute of Mental Health (NIH/NIMH) to K.M.

REFERENCES

Abdul-Majeed S, Nauli SM. 2011. Dopamine receptor type 5 in the primary cilia has dual chemo- and mechano-sensory roles. *Hypertension* **58:** 325–331.

Arshavsky VY, Burns ME. 2012. Photoreceptor signaling: Supporting vision across a wide range of light intensities. *J Biol Chem* **287:** 1620–1626.

Atkinson KF, Kathem SH, Jin X, Muntean BS, Abou-Alaiwi WA, Nauli AM, Nauli SM. 2015. Dopaminergic signaling within the primary cilia in the renovascular system. *Front Physiol* **6:** 103.

Avasthi P, Marley A, Lin H, Gregori-Puigjane E, Shoichet BK, von Zastrow M, Marshall WF. 2012. A chemical screen identifies class a G-protein coupled receptors as regulators of cilia. *ACS Chem Biol* **7:** 911–919.

Bakalyar HA, Reed RR. 1990. Identification of a specialized adenylyl cyclase that may mediate odorant detection. *Science* **250:** 1403–1406.

Baraniuk JN, Merck SJ. 2009. New concepts of neural regulation in human nasal mucosa. *Acta Clin Croat* **48:** 65–73.

Barnett P. 2003. Somatostatin and somatostatin receptor physiology. *Endocrine* **20:** 255–264.

Berbari NF, Johnson AD, Lewis JS, Askwith CC, Mykytyn K. 2008. Identification of ciliary localization sequences within the third intracellular loop of G protein-coupled receptors. *Mol Biol Cell* **19:** 1540–1547.

Berbari NF, Pasek RC, Malarkey EB, Yazdi SM, McNair AD, Lewis WR, Nagy TR, Kesterson RA, Yoder BK. 2013. Leptin resistance is a secondary consequence of the obesity in ciliopathy mutant mice. *Proc Natl Acad Sci* **110:** 7796–7801.

Berbari NF, Malarkey EB, Yazdi SM, McNair AD, Kippe JM, Croyle MJ, Kraft TW, Yoder BK. 2014. Hippocampal and cortical primary cilia are required for aversive memory in mice. *PLoS ONE* **9:** e106576.

Bishop GA, Berbari NF, Lewis JS, Mykytyn K. 2007. Type III adenylyl cyclase localizes to primary cilia throughout the adult mouse brain. *J Comp Neurol* **505:** 562–571.

Brailov I, Bancila M, Brisorgueil MJ, Miquel MC, Hamon M, Verge D. 2000. Localization of 5-HT(6) receptors at the plasma membrane of neuronal cilia in the rat brain. *Brain Res* **872:** 271–275.

Carey RM. 2013. The intrarenal renin-angiotensin and dopaminergic systems: Control of renal sodium excretion and blood pressure. *Hypertension* **61:** 673–680.

Corbit KC, Aanstad P, Singla V, Norman AR, Stainier DY, Reiter JF. 2005. Vertebrate Smoothened functions at the primary cilium. *Nature* **437:** 1018–1021.

Cramer MT, Guay-Woodford LM. 2015. Cystic kidney disease: A primer. *Adv Chronic Kidney Dis* **22:** 297–305.

Davenport JR, Watts AJ, Roper VC, Croyle MJ, van Groen T, Wyss JM, Nagy TR, Kesterson RA, Yoder BK. 2007. Disruption of intraflagellar transport in adult mice leads to obesity and slow-onset cystic kidney disease. *Curr Biol* **17:** 1586–1594.

Dawson TM, Arriza JL, Jaworsky DE, Borisy FF, Attramadal H, Lefkowitz RJ, Ronnett GV. 1993. β-Adrenergic receptor kinase-2 and β-arrestin-2 as mediators of odorant-induced desensitization. *Science* **259:** 825–829.

DeFea KA. 2011. β-Arrestins as regulators of signal termination and transduction: How do they determine what to scaffold? *Cell Signal* **23:** 621–629.

Devillier P, Naline E, Grassin-Delyle S. 2015. The pharmacology of bitter taste receptors and their role in human airways. *Pharmacol Ther* **155:** 11–21.

Domire JS, Green JA, Lee KG, Johnson AD, Askwith CC, Mykytyn K. 2011. Dopamine receptor 1 localizes to neuronal cilia in a dynamic process that requires the Bardet-Biedl syndrome proteins. *Cell Mol Life Sci* **68:** 2951–2960.

Einstein EB, Patterson CA, Hon BJ, Regan KA, Reddi J, Melnikoff DE, Mateer MJ, Schulz S, Johnson BN, Tallent MK. 2010. Somatostatin signaling in neuronal cilia is critical for object recognition memory. *J Neurosci* **30:** 4306–4314.

Fliegauf M, Benzing T, Omran H. 2007. When cilia go bad: Cilia defects and ciliopathies. *Nat Rev Mol Cell Biol* **8:** 880–893.

Fuchs JL, Schwark HD. 2004. Neuronal primary cilia: A review. *Cell Biol Int* **28:** 111–118.

Gattone VH II, Wang X, Harris PC, Torres VE. 2003. Inhibition of renal cystic disease development and progression by a vasopressin V2 receptor antagonist. *Nat Med* **9:** 1323–1326.

Gradilone SA, Masyuk AI, Splinter PL, Banales JM, Huang BQ, Tietz PS, Masyuk TV, Larusso NF. 2007. Cholangiocyte cilia express TRPV4 and detect changes in luminal tonicity inducing bicarbonate secretion. *Proc Natl Acad Sci* **104:** 19138–19143.

Green JA, Gu C, Mykytyn K. 2012. Heteromerization of ciliary G protein-coupled receptors in the mouse brain. *PLoS ONE* **7:** e46304.

Green JA, Schmid CL, Bley E, Monsma PC, Brown A, Bohn LM, Mykytyn K. 2016. Recruitment of β-arrestin into neuronal cilia modulates somatostatin receptor subtype 3 ciliary localization. *Mol Cell Biol* **36:** 223–235.

Hagena H, Hansen N, Manahan-Vaughan D. 2016. β-Adrenergic control of hippocampal function: Subserving the choreography of synaptic information storage and memory. *Cereb Cortex* **26:** 1349–1364.

Handel M, Schulz S, Stanarius A, Schreff M, Erdtmann-Vourliotis M, Schmidt H, Wolf G, Hollt V. 1999. Selective targeting of somatostatin receptor 3 to neuronal cilia. *Neuroscience* **89:** 909–926.

Herzog H. 2003. Neuropeptide Y and energy homeostasis: Insights from Y receptor knockout models. *Eur J Pharmacol* **480:** 21–29.

Hilgendorf KI, Johnson CT, Jackson PK. 2016. The primary cilium as a cellular receiver: Organizing ciliary GPCR signaling. *Curr Opin Cell Biol* **39:** 84–92.

Huangfu D, Liu A, Rakeman AS, Murcia NS, Niswander L, Anderson KV. 2003. Hedgehog signalling in the mouse requires intraflagellar transport proteins. *Nature* **426:** 83–87.

Iannaccone A, Mykytyn K, Persico AM, Searby CC, Baldi A, Jablonski MM, Sheffield VC. 2005. Clinical evidence of decreased olfaction in Bardet–Biedl syndrome caused by a deletion in the *BBS4* Gene. *Am J Med Genet A* **132A:** 343–346.

Jenkins PM, McEwen DP, Martens JR. 2009. Olfactory cilia: Linking sensory cilia function and human disease. *Chem Senses* **34:** 451–464.

Jiang Y, Li YR, Tian H, Ma M, Matsunami H. 2015. Muscarinic acetylcholine receptor M3 modulates odorant receptor activity via inhibition of β-arrestin-2 recruitment. *Nat Commun* **6:** 6448.

Jin D, Ni TT, Sun J, Wan H, Amack JD, Yu G, Fleming J, Chiang C, Li W, Papierniak A, et al. 2014a. Prostaglandin signalling regulates ciliogenesis by modulating intraflagellar transport. *Nat Cell Biol* **16:** 841–851.

Jin X, Mohieldin AM, Muntean BS, Green JA, Shah JV, Mykytyn K, Nauli SM. 2014b. Cilioplasm is a cellular compartment for calcium signaling in response to

mechanical and chemical stimuli. *Cell Mol Life Sci* **71:** 2165–2178.

Jones DT, Reed RR. 1989. Golf: An olfactory neuron specific-G protein involved in odorant signal transduction. *Science* **244:** 790–795.

Kawamata Y, Fujii R, Hosoya M, Harada M, Yoshida H, Miwa M, Fukusumi S, Habata Y, Itoh T, Shintani Y, et al. 2003. A G protein-coupled receptor responsive to bile acids. *J Biol Chem* **278:** 9435–9440.

Keitel V, Haussinger D. 2011. TGR5 in the biliary tree. *Dig Dis* **29:** 45–47.

Keitel V, Haussinger D. 2012. Perspective: TGR5 (Gpbar-1) in liver physiology and disease. *Clin Res Hepatol Gastroenterol* **36:** 412–419.

Keitel V, Cupisti K, Ullmer C, Knoefel WT, Kubitz R, Haussinger D. 2009. The membrane-bound bile acid receptor TGR5 is localized in the epithelium of human gallbladders. *Hepatology* **50:** 861–870.

Keitel V, Ullmer C, Haussinger D. 2010. The membrane-bound bile acid receptor TGR5 (Gpbar-1) is localized in the primary cilium of cholangiocytes. *Biol Chem* **391:** 785–789.

Kelly E, Bailey CP, Henderson G. 2008. Agonist-selective mechanisms of GPCR desensitization. *Br J Pharmacol* **153:** S379–388.

Kerr DS, Von Dannecker LE, Davalos M, Michaloski JS, Malnic B. 2008. Ric-8B interacts with Gαolf and Gγ13 and co-localizes with Gαolf, Gβ1 and Gγ13 in the cilia of olfactory sensory neurons. *Mol Cell Neurosci* **38:** 341–348.

Kleene SJ. 1993. Origin of the chloride current in olfactory transduction. *Neuron* **11:** 123–132.

Koemeter-Cox AI, Sherwood TW, Green JA, Steiner RA, Berbari NF, Yoder BK, Kauffman AS, Monsma PC, Brown A, Askwith CC, et al. 2014. Primary cilia enhance kisspeptin receptor signaling on gonadotropin-releasing hormone neurons. *Proc Natl Acad Sci* **111:** 10335–10340.

Kulaga HM, Leitch CC, Eichers ER, Badano JL, Lesemann A, Hoskins BE, Lupski JR, Beales PL, Reed RR, Katsanis N. 2004. Loss of BBS proteins causes anosmia in humans and defects in olfactory cilia structure and function in the mouse. *Nat Genet* **36:** 994–998.

Lee RJ, Xiong G, Kofonow JM, Chen B, Lysenko A, Jiang P, Abraham V, Doghramji L, Adappa ND, Palmer JN, et al. 2012. T2R38 taste receptor polymorphisms underlie susceptibility to upper respiratory infection. *J Clin Invest* **122:** 4145–4159.

Li F, Ponissery-Saidu S, Yee KK, Wang H, Chen ML, Iguchi N, Zhang G, Jiang P, Reisert J, Huang L. 2013. Heterotrimeric G protein subunit Gγ13 is critical to olfaction. *J Neurosci* **33:** 7975–7984.

Loktev AV, Jackson PK. 2013. Neuropeptide Y family receptors traffic via the Bardet-Biedl syndrome pathway to signal in neuronal primary cilia. *Cell Rep* **5:** 1316–1329.

Lowe G, Gold GH. 1993. Nonlinear amplification by calcium-dependent chloride channels in olfactory receptor cells. *Nature* **366:** 283–286.

Marchese A, Paing MM, Temple BR, Trejo J. 2008. G protein-coupled receptor sorting to endosomes and lysosomes. *Ann Rev Pharmacol Toxicol* **48:** 601–629.

Marley A, Choy RW, von Zastrow M. 2013. GPR88 reveals a discrete function of primary cilia as selective insulators of GPCR cross-talk. *PLoS ONE* **8:** e70857.

Maruyama T, Miyamoto Y, Nakamura T, Tamai Y, Okada H, Sugiyama E, Itadani H, Tanaka K. 2002. Identification of membrane-type receptor for bile acids (M-BAR). *Biochem Biophys Res Commun* **298:** 714–719.

Mashukova A, Spehr M, Hatt H, Neuhaus EM. 2006. β-Arrestin2-mediated internalization of mammalian odorant receptors. *J Neurosci* **26:** 9902–9912.

Masyuk AI, Masyuk TV, Splinter PL, Huang BQ, Stroope AJ, LaRusso NF. 2006. Cholangiocyte cilia detect changes in luminal fluid flow and transmit them into intracellular Ca^{2+} and cAMP signaling. *Gastroenterology* **131:** 911–920.

Masyuk AI, Gradilone SA, Banales JM, Huang BQ, Masyuk TV, Lee SO, Splinter PL, Stroope AJ, Larusso NF. 2008a. Cholangiocyte primary cilia are chemosensory organelles that detect biliary nucleotides via P2Y12 purinergic receptors. *Am J Physiol Gastrointest Liver Physiol* **295:** G725–G734.

Masyuk AI, Masyuk TV, LaRusso NF. 2008b. Cholangiocyte primary cilia in liver health and disease. *Dev Dyn* **237:** 2007–2012.

Masyuk AI, Huang BQ, Radtke BN, Gajdos GB, Splinter PL, Masyuk TV, Gradilone SA, LaRusso NF. 2013. Ciliary subcellular localization of TGR5 determines the cholangiocyte functional response to bile acid signaling. *Am J Physiol Gastrointest Liver Physiol* **304:** G1013–G1024.

Masyuk TV, Masyuk AI, LaRusso NF. 2015. TGR5 in the cholangiociliopathies. *Dig Dis* **33:** 420–425.

McEwen DP, Koenekoop RK, Khanna H, Jenkins PM, Lopez I, Swaroop A, Martens JR. 2007. Hypomorphic CEP290/NPHP6 mutations result in anosmia caused by the selective loss of G proteins in cilia of olfactory sensory neurons. *Proc Natl Acad Sci* **104:** 15917–15922.

McMahon HT, Boucrot E. 2011. Molecular mechanism and physiological functions of clathrin-mediated endocytosis. *Nat Rev Mol Cell Biol* **12:** 517–533.

Menco BP. 1980. Qualitative and quantitative freeze-fracture studies on olfactory and nasal respiratory epithelial surfaces of frog, ox, rat, and dog. III: Tight-junctions. *Cell Tissue Res* **211:** 361–373.

Menco B. 1992. Ultrastructural studies on membrane, cytoskeletal, mucous, and protective compartments in olfaction. *Microsc Res Tech* **22:** 215–224.

Menco BP. 1997. Ultrastructural aspects of olfactory signaling. *Chem Senses* **22:** 295–311.

Mukhopadhyay S, Wen X, Ratti N, Loktev A, Rangell L, Scales SJ, Jackson PK. 2013. The ciliary G-protein-coupled receptor Gpr161 negatively regulates the Sonic Hedgehog pathway via cAMP signaling. *Cell* **152:** 210–223.

Olias G, Viollet C, Kusserow H, Epelbaum J, Meyerhof W. 2004. Regulation and function of somatostatin receptors. *J Neurochem* **89:** 1057–1091.

Omori Y, Chaya T, Yoshida S, Irie S, Tsujii T, Furukawa T. 2015. Identification of G protein-coupled receptors (GPCRs) in primary cilia and their possible involvement in body weight control. *PLoS ONE* **10:** e0128422.

Pan Y, Bai CB, Joyner AL, Wang B. 2006. Sonic Hedgehog signaling regulates Gli2 transcriptional activity by suppressing its processing and degradation. *Mol Cell Biol* **26:** 3365–3377.

Patel YC. 1999. Somatostatin and its receptor family. *Front Neuroendocrinol* **20:** 157–198.

Rajagopal S, Rajagopal K, Lefkowitz RJ. 2010. Teaching old receptors new tricks: Biasing seven-transmembrane receptors. *Nat Rev Drug Discov* **9:** 373–386.

Rask-Andersen M, Almen MS, Schioth HB. 2011. Trends in the exploitation of novel drug targets. *Nat Rev Drug Discov* **10:** 579–590.

Raychowdhury MK, Ramos AJ, Zhang P, McLaughin M, Dai XQ, Chen XZ, Montalbetti N, Del Rocio Cantero M, Ausiello DA, Cantiello HF. 2009. Vasopressin receptor-mediated functional signaling pathway in primary cilia of renal epithelial cells. *Am J Physiol Renal Physiol* **296:** F87–97.

Ressler KJ, Sullivan SL, Buck LB. 1993. A zonal organization of odorant receptor gene expression in the olfactory epithelium. *Cell* **73:** 597–609.

Rohatgi R, Milenkovic L, Scott MP. 2007. Patched1 regulates hedgehog signaling at the primary cilium. *Science* **317:** 372–376.

Sasaki H, Nishizaki Y, Hui C, Nakafuku M, Kondoh H. 1999. Regulation of Gli2 and Gli3 activities by an amino-terminal repression domain: Implication of Gli2 and Gli3 as primary mediators of Shh signaling. *Development* **126:** 3915–3924.

Shah AS, Ben-Shahar Y, Moninger TO, Kline JN, Welsh MJ. 2009. Motile cilia of human airway epithelia are chemosensory. *Science* **325:** 1131–1134.

Shenoy SK, Lefkowitz RJ. 2011. β-Arrestin-mediated receptor trafficking and signal transduction. *Trends Pharmacol Sci* **32:** 521–533.

Shukla AK, Xiao K, Lefkowitz RJ. 2011. Emerging paradigms of β-arrestin-dependent seven transmembrane receptor signaling. *Trends Biochem Sci* **36:** 457–469.

Singh J, Wen X, Scales SJ. 2015. The orphan G protein-coupled receptor Gpr175 (Tpra40) enhances Hedgehog signaling by modulating cAMP levels. *J Biol Chem* **290:** 29663–29675.

Sorkin A, von Zastrow M. 2009. Endocytosis and signalling: Intertwining molecular networks. *Nat Rev Mol Cell Biol* **10:** 609–622.

Stephan AB, Shum EY, Hirsh S, Cygnar KD, Reisert J, Zhao H. 2009. ANO2 is the cilial calcium-activated chloride channel that may mediate olfactory amplification. *Proc Natl Acad Sci* **106:** 11776–11781.

Szumska J, Qatato M, Rehders M, Fuhrer D, Biebermann H, Grandy DK, Kohrle J, Brix K. 2015. Trace amine-associated receptor 1 localization at the apical plasma membrane domain of fisher rat thyroid epithelial cells is confined to cilia. *Eur Thyroid J* **4:** 30–41.

Tabibian JH, Masyuk AI, Masyuk TV, O'Hara SP, LaRusso NF. 2013. Physiology of cholangiocytes. *Compr Physiol* **3:** 541–565.

Cite this article as *Cold Spring Harb Perspect Biol* doi: 10.1101/cshperspect.a028183

Takeda S, Kadowaki S, Haga T, Takaesu H, Mitaku S. 2002. Identification of G protein-coupled receptor genes from the human genome sequence. *FEBS Lett* **520:** 97–101.

Tang XL, Wang Y, Li DL, Luo J, Liu MY. 2012. Orphan G protein-coupled receptors (GPCRs): Biological functions and potential drug targets. *Acta Pharmacol Sin* **33:** 363–371.

Tempe D, Casas M, Karaz S, Blanchet-Tournier MF, Concordet JP. 2006. Multisite protein kinase A and glycogen synthase kinase 3β phosphorylation leads to Gli3 ubiquitination by SCF$^{\beta TrCP}$. *Mol Cell Biol* **26:** 4316–4326.

Torres VE, Harris PC. 2014. Strategies targeting cAMP signaling in the treatment of polycystic kidney disease. *J Am Soc Nephrol* **25:** 18–32.

Torres VE, Wang X, Qian Q, Somlo S, Harris PC, Gattone VH II. 2004. Effective treatment of an orthologous model of autosomal dominant polycystic kidney disease. *Nat Med* **10:** 363–364.

Tuson M, He M, Anderson KV. 2011. Protein kinase A acts at the basal body of the primary cilium to prevent Gli2 activation and ventralization of the mouse neural tube. *Development* **138:** 4921–4930.

Upadhyay VS, Muntean BS, Kathem SH, Hwang JJ, Aboualaiwi WA, Nauli SM. 2014. Roles of dopamine receptor on chemosensory and mechanosensory primary cilia in renal epithelial cells. *Front Physiol* **5:** 72.

Vassar R, Ngai J, Axel R. 1993. Spatial segregation of odorant receptor expression in the mammalian olfactory epithelium. *Cell* **74:** 309–318.

Viollet C, Lepousez G, Loudes C, Videau C, Simon A, Epelbaum J. 2008. Somatostatinergic systems in brain: Networks and functions. *Mol Cell Endocrinol* **286:** 75–87.

Wang Z, Li V, Chan GC, Phan T, Nudelman AS, Xia Z, Storm DR. 2009. Adult type 3 adenylyl cyclase-deficient mice are obese. *PLoS ONE* **4:** e6979.

Wang Z, Phan T, Storm DR. 2011. The type 3 adenylyl cyclase is required for novel object learning and extinction of contextual memory: Role of cAMP signaling in primary cilia. *J Neurosci* **31:** 5557–5561.

Yao G, Luo C, Harvey M, Wu M, Schreiber TH, Du Y, Basora N, Su X, Contreras D, Zhou J. 2015. Disruption of polycystin-L causes hippocampal and thalamocortical hyperexcitability. *Hum Mol Genet* **25:** 448–458.

Multiciliated Cells in Animals

Alice Meunier[1] and Juliette Azimzadeh[2]

[1]Institut de Biologie de l'Ecole Normale Supérieure, Institut National de la Santé et de la Recherche Médicale U1024, Centre National de la Recherche Scientifique UMR8197, 75005 Paris, France

[2]Institut Jacques Monod, Centre National de la Recherche Scientifique UMR7592, Université Paris-Diderot, 75013 Paris, France

Correspondence: ameunier@biologie.ens.fr; juliette.azimzadeh@ijm.fr

Many animal cells assemble single cilia involved in motile and/or sensory functions. In contrast, multiciliated cells (MCCs) assemble up to 300 motile cilia that beat in a coordinate fashion to generate a directional fluid flow. In the human airways, the brain, and the oviduct, MCCs allow mucus clearance, cerebrospinal fluid circulation, and egg transportation, respectively. Impairment of MCC function leads to chronic respiratory infections and increased risks of hydrocephalus and female infertility. MCC differentiation during development or repair involves the activation of a regulatory cascade triggered by the inhibition of Notch activity in MCC progenitors. The downstream events include the simultaneous assembly of a large number of basal bodies (BBs)—from which cilia are nucleated—in the cytoplasm of the differentiating MCCs, their migration and docking at the plasma membrane associated to an important remodeling of the actin cytoskeleton, and the assembly and polarization of motile cilia. The direction of ciliary beating is coordinated both within cells and at the tissue level by a combination of planar polarity cues affecting BB position and hydrodynamic forces that are both generated and sensed by the cilia. Herein, we review the mechanisms controlling the specification and differentiation of MCCs and BB assembly and organization at the apical surface, as well as ciliary assembly and coordination in MCCs.

Multiciliated cells (MCCs) form hundreds of motile cilia that beat in a coordinated fashion to generate a fluid flow or displace particles and cells. In the respiratory tract, motile cilia are required for the clearance of mucus that traps inhaled particles and pathogens. MCCs also drive the flow of cerebrospinal fluid in the brain ventricles and allow egg transportation along the oviduct. As a consequence, genetic diseases that disrupt cilia-generated fluid flow, such as primary ciliary dyskinesia, result in chronic recurrent respiratory infections, increased risks of hydrocephalus, and female in-

fertility (Lee 2011, 2013; Boon et al. 2014; Popatia et al. 2014; Wallmeier et al. 2014). Outside vertebrates, MCCs are involved in generating fluid flows or driving the locomotion of a variety of animal species (Gibbons 1961; Tyler 1981; Tamm and Tamm 1988). Cilia in MCCs are assembled from basal bodies (BBs), which are ninefold symmetrical microtubule (MT)-based structures related to the centrioles found within the centrosome. In most mammalian cell types, the older centriole, called the mother centriole, can convert to a BB and template the assembly of a primary cilium, which is typically a nonmotile cilium involved in sensory functions (Goetz and Anderson 2010). In contrast, MCCs simultaneously assemble up to several hundred motile cilia while undergoing terminal differentiation. These cilia are formed following assembly a large number of BBs in the cytoplasm and their migration to the apical membrane (Sorokin 1968; Steinman 1968; Anderson and Brenner 1971; Dirksen 1971; Sandoz and Boisvieux-Ulrich 1976; Tyler 1981).

MCC cilia usually show a planar beating pattern, with a fast power stroke and a slower recovery stroke occurring in the same plane. To generate a directional fluid flow, cilia must beat in the right orientation with respect to the polarity of the whole tissue. The direction of ciliary beat depends on the position of BBs within the plane of the apical membrane, or BB rotational polarity. This polarity is evidenced by the presence of BB appendages that align with the axis of ciliary beating: the basal foot and the striated ciliary rootlet (Gibbons 1961; Sorokin 1968; Dirksen 1971; Anderson 1972; Reed et al. 1984). These appendages anchor cytoskeletal networks linking the BBs to each other and to the apical cell junctions. Polarity proteins localize to the centrioles and to the cell junctions to connect BB polarity to the planar polarity of the epithelium.

We will first give an overview of the phylogenetic distribution of MCCs in animals, and then we will review the current knowledge on MCC specification and differentiation, BB assembly, docking, and positioning at the apical membrane, ciliogenesis, cellular and tissue-level polarization of MCCs, and the regulation of MCC function.

MULTICILIATED EPITHELIA ACROSS ANIMALS

In mammals, MCCs are present in the respiratory tract, the brain ventricles, the oviduct, and the efferent ducts (Brightman and Palay 1963; Sorokin 1968; Anderson and Brenner 1971; Dirksen 1971; Danielian et al. 2016). MCCs are also found transiently in kidney tubules and in the esophagus during fetal development (Katz and Morgan 1984; Menard 1995). In Xen-

opus tadpoles, MCCs are found in the skin, the trachea, and the digestive tract (Steinman 1968; Werner and Mitchell 2012; Walentek et al. 2015). In addition, MCCs are present in the pronephros in zebrafish and Xenopus to facilitate urine flow (Vize et al. 2003; Kramer-Zucker et al. 2005; Liu et al. 2007). Cell types with multiple nonmotile cilia also exist, such as olfactory neurons in vertebrates (Cuschieri and Bannister 1975; Zielinski and Hara 1988; Ying et al. 2014). Outside of vertebrates, MCCs with similar ultrastructural features, in particular BBs decorated by a basal foot, are found within a group of protostome animals called Lophotrochozoa. Within this group, MCCs are particularly well described in mollusks and flatworms (Gibbons 1961; Rieger 1981; Reed et al. 1984; Basquin et al. 2015). In flatworms, epidermal MCCs are required for locomotion, either by ciliary gliding along solid substrates or swimming through water (Rompolas et al. 2013; Basquin et al. 2015). MCCs are also present in protonephridia, the branched tubules that form the excretory system of flatworms and other Lophotrochozoa (Rink et al. 2011; Thi-Kim Vu et al. 2015). MCCs are absent in arthropods and nematodes (Giribet and Ribera 1998). Outside bilaterians, MCCs are found in ctenophores (Tamm and Tamm 1988) and in sponges; although, in the latter case, each BB is associated to an accessory BB (i.e., like in a centrosome), suggesting that distinct mechanisms are used for BB assembly in these cells (Boury-Esnault et al. 1999). It is worth mentioning that MCCs are also found outside of animals in phyla as diverse as plants, ciliates, or amoebozoa (Tamm et al. 1975; Mikrjukov and Mylnikov 1998; Hodges et al. 2010). This distribution is most likely the result of convergent evolution, however, because these phyla originate from distinct mono- or biflagellated unicellular ancestors (Adl et al. 2012; Azimzadeh 2014).

MCC SPECIFICATION AND DIFFERENTIATION

Notch Inhibition

The very first steps of the MCC differentiation process are particularly well described in the mammalian respiratory system. The adult airways are lined by a pseudostratified epithelium containing MCCs, mucus-secreting cells, and basal cells. During development, they all derive from p63$^+$ progenitors (Rock and Hogan 2011). Notch sets the balance between ciliated versus secretory lineage. In the absence of Notch signaling, the MCC population expands at the expense of the secretory lineage (Guseh et al. 2009; Tsao et al. 2009; Morimoto et al. 2010). The inverse happens after transgenic Notch activation in the embryonic airways (Guseh et al. 2009). The same dynamics exists between MCCs and transporting cells in the zebrafish pronephros (Liu et al. 2007). Notch inhibition also drives MCC specification in mouse oviducts and brain ventricles (Kessler et al. 2015; Kyrousi et al. 2015). Recently, bone morphogenetic protein (BMP) signaling has been shown to interact with the Notch pathway to control mucociliary differentiation of vertebrate epithelia (Cibois et al. 2015).

In the adult, ciliated cells can differentiate from both basal progenitors and secretory cells in mouse airways (Rawlins et al. 2009; Lafkas et al. 2015; Pardo-Saganta et al. 2015b; Watson et al. 2015). Electron microscopy (EM) suggests the same events of (trans-)differentiation from secretory-to-ciliated cells in human and quail oviducts under hormonal control (Sandoz and Boisvieux-Ulrich 1976; Sandoz et al. 1976; Hagiwara 1995). Secretory cells are thus proposed to act as transit-amplifying cells by self-renewing and giving rise to terminally differentiated postmitotic MCCs. Depending on the context, the mechanism by which Notch proceeds to segregate secretory and ciliated lineages seems to differ. Classical Notch-mediated lateral inhibition has been proposed, during development (Tsao et al. 2009; Morimoto et al. 2010; Gomi et al. 2015; Mori et al. 2015) or repair (Pardo-Saganta et al. 2015a), to lead to the mosaic distribution of MCC among secretory cells in mouse airways (Fig. 1). In the zebrafish pronephros and the *Xenopus* skin, Notch-lateral inhibition is also reported to drive the mosaic differentiation pattern of MCC versus transporting or secretory epithelial cells (Deblandre et al. 1999; Liu et al. 2007; Ma and Jiang 2007). In addition, microRNAs, such as miR449, are specifically expressed in MCC progenitors in *Xenopus* skin and human airway developing epithelium, where they repress the Notch pathway by targeting Notch1 and its ligand DLL1 (Marcet et al. 2011). Finally, in the adult mouse homeostatic airways, a recent study proposes that basal cells continuously supply their daughter secretory cells with a Notch ligand to avoid their terminal (trans-)differentiation into MCC (Lafkas et al. 2015; Pardo-Saganta et al. 2015b).

Master Regulators of the MCC Differentiation Program

Two master regulators of MCC fate differentiation downstream from Notch inhibition have been discovered recently: GEMC1 (geminin coiled-coil containing protein 1) and IDAS

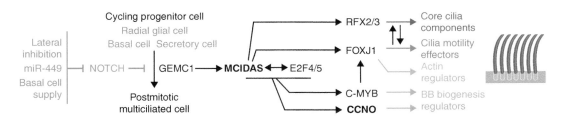

Figure 1. Regulation of multiciliated cell (MCC) differentiation in vertebrates. MCC differentiation is triggered by Notch inhibition in radial glial cells of the ependyma, and in basal cells or secretory cells (trans-differentiation) in mucociliary epithelia. Notch inhibition triggers the activation of a regulatory cascade that involves the geminin-related proteins GEMC1 and MCIDAS, the transcription factors E2F4/5, RFX2/3, FOXJ1, and C-MYB, and the cyclin-like protein CCNO. This cascade triggers cell-cycle exit, basal body (BB) amplification, cytoskeleton remodeling, and ciliogenesis (see text for details).

(a cousin of the Gemini in ancient Greek my-thology). As their name suggests, both have coiled-coil domains related to geminin, a pro-tein involved in cell-cycle progression and in the balance between proliferation and differentia-tion (Kroll 2007). GEMC1 was primarily found to promote initiation of DNA replication in vertebrate cells (Balestrini et al. 2010) and was recently found to be necessary for MCC post-mitotic commitment in zebrafish pronephros, *Xenopus* skin, and mouse brain MCCs (Kyrousi et al. 2015; Zhou et al. 2015). IDAS was first found to bind to geminin and to regulate its function in DNA replication licensing in S phase (Pefani et al. 2011). It was later found to be required for MCC differentiation in *Xenopus* skin/kidney and mammalian airway/brain ep-ithelia, and therefore renamed multicilin or MCIDAS (Stubbs et al. 2012; Kyrousi et al. 2015). Both GEMC1 and MCIDAS are reported to be sufficient to trigger early-onset differenti-ation of MCC progenitors in the mouse brain (Kyrousi et al. 2015). In addition, they are able to drive ectopic differentiation of MCCs in *Xen-opus* skin/kidney and mouse airway epithelia (Stubbs et al. 2012; Zhou et al. 2015). GEMC1 and MCIDAS are therefore sufficient to trigger the full developmental program for secretory-to-ciliated cell conversion, which is consistent with the finding that this (*trans-*)differentiation process is Notch-dependent (Lafkas et al. 2015; Pardo-Saganta et al. 2015b). Hierarchically, MCIDAS lies downstream from GEMC1, al-though the molecular aspect of this control is unknown (Fig. 1). Interestingly, geminin binds to MCIDAS and represses MCC formation in the *Xenopus* skin (Ma et al. 2014). Because gem-inin is silenced in cells exiting the cell cycle (Kroll 2007), these three geminin-family pro-teins may constitute the central network for triggering postmitotic MCC transcriptional program.

Downstream Effectors of the MCC Differentiation Program

MCIDAS lacks a DNA-binding domain. In *Xen-opus*, it binds to the cell-cycle-repressing tran-scription factors E2F4/5 (Sadasivam and De-Caprio 2013; Ma et al. 2014). Expression of a dominant-negative form of E2F4 in the *Xenopus* skin blocks BB amplification and multicilia-tion in presumptive MCCs. RNA-seq analysis after coexpression of an inducible form of MCIDAS and of the dominant-negative form of E2F4 suggests that the MCIDAS/E2F4 as-sociation is required for the up-regulation of BB components and ciliogenic transcription factors (Fig. 1) (Ma et al. 2014). In addition, mouse mutants for E2F4 and/or 5 fail to form MCCs in the airways and in the efferent ducts (Danielian et al. 2007, 2016). Together with the role of E2F4/5 in cell-cycle exit (Sadasivam and DeCaprio 2013), MCIDAS/E2F4 interaction is proposed to induce the MCC terminal differen-tiation program. *Mcidas* mutations are found in patients with congenital mucociliary clearance disorder with reduced generation of multiple motile cilia (RGMC) (Boon et al. 2014). Inter-estingly, MCIDAS/E2F association is impaired by certain of mutations found in patients (Ma et al. 2014).

One transcription factor proposed to medi-ate BB amplification MCIDAS/E2F4 complex is C-MYB (Fig. 1) (Stubbs et al. 2012; Ma et al. 2014). C-MYB normally promotes S-phase entry and enhances cycling in a variety of pro-genitor cells (Ramsay and Gonda 2008). It is, however, up-regulated in zebrafish, *Xenopus,* and mouse brain/tracheal presumptive MCC progenitors (Tan et al. 2013; Wang et al. 2013; Pardo-Saganta et al. 2015a) where it controls the expression of key BB biogenesis regulators. Depletion of C-MYB leads to a block or a delay in BB amplification in MCCs from zebrafish, *Xenopus*, and mouse tracheal cells (Tan et al. 2013; Wang et al. 2013). Its effect seems, how-ever, to be redundant with at least one other factor because MCCs can partially recover mul-ticiliogenesis after *cmyb* deletion in the mouse trachea (Tan et al. 2013). Downstream from the MCIDAS/E2F complex also lies CCNO (Fig. 1) (Stubbs et al. 2012; Ma et al. 2014; Wallmeier et al. 2014), a cyclin-like protein first shown to be involved in oocyte meiosis resumption and apoptosis (Roig et al. 2009; Ma et al. 2013), and then shown to be expressed in MCC progenitors in the *Xenopus* skin and in the mouse brain,

Cite this article as *Cold Spring Harb Perspect Biol* doi: 10.1101/cshperspect.a028233

airways, and oviducts (Stubbs et al. 2012; Funk et al. 2015; Amirav et al. 2016). Although its partners and molecular function remain to be determined, CCNO mutations or depletion lead to defects comparable to the C-MYB phenotype: defects in BB amplification, docking, and ciliogenesis (Wallmeier et al. 2014; Funk et al. 2015). In humans, mutations in CCNO have been identified as a cause of RGMC and associated with hydrocephalus and female infertility (Wallmeier et al. 2014; Amirav et al. 2016). Interestingly, human *mcidas*, *ccno*, and miR449 host gene *cdc20b* are physically colocalized in a genomic region of chromosome 5 and are syntenic with those of *Xenopus* and mouse (Marcet et al. 2011; Stubbs et al. 2012; Wallmeier et al. 2014).

In parallel to BB amplification, multiple motile cilia biogenesis is controlled by the deployment of the RFX (regulatory factor X)/ FOXJ1 (forkhead box J1) ciliogenic network (Fig. 1). The current view is that the RFX family of transcription factors is involved in primary and motile cilia formation by directing the expression of core components of all types of cilia (Choksi et al. 2014). The FOXJ1 family has emerged as an additional factor important for the generation of motile cilia by directing the expression of genes required for BB docking, axoneme outgrowth, and motility (Chen et al. 1998; Brody et al. 2000). This view is, however, progressing in MCC epithelia because RFX2 and 3 are also shown to direct the expression of core components of the cilia motility machinery, and to affect cilia beating in *Xenopus* and mouse (El Zein et al. 2009; Chung et al. 2012, 2014). Interestingly, FoxJ1 and RFX2/3 cross-regulate their expression during vertebrate MCC formation (Yu et al. 2008; El Zein et al. 2009; Didon et al. 2013). In the human airways, RFX3 can enhance FOXJ1-dependent transcription and immunoprecipitation experiments provide evidence for RFX2/3–FOXJ1 interactions, suggesting a role of RFX3 as a cofactor of FOXJ1 for motile ciliogenesis (Didon et al. 2013). Altogether, this raises the possibility that RFX and FOXJ1 form a transcriptional complex in MCC in which RFX factors regulate core cilia genes and cooperate with FOXJ1 to regulate motility genes. Both RFX2 and FOXJ1 expressions are up-regulated following inducible MCIDAS expression in the *Xenopus* skin, suggesting a control by MCIDAS of the deployment of this ciliogenic program (Ma et al. 2014). C-MYB was also reported to drive the expression of FOXJ1 in addition to BB biogenesis regulators, leading to the identification of C-MYB as a core regulator of both early and late steps of multiciliogenesis. In this line, it was shown to trigger the complete multiciliogenesis program in the *Xenopus* skin, but not in the mouse trachea (Tan et al. 2013).

BB ASSEMBLY

Dynamics

Like most other cell types, MCC progenitors contain a centrosome with two centrioles—a mother and a daughter—which serves as the main MT-organizing center. During the course of their differentiation, MCCs must then produce from 30 up to 300 additional centrioles/ BBs. BB amplification has been extensively described by EM in quail, chick, *Xenopus*, and mammalian MCCs (Sorokin 1968; Steinman 1968; Kalnins and Porter 1969; Dirksen 1971; Sandoz and Boisvieux-Ulrich 1976). Two pathways have been shown to contribute to the final BB population: a "centriolar pathway," where two to six BBs are formed orthogonally to the two preexisting centrosomal centrioles in a manner akin to centriole duplication, and a "deuterosome pathway," where tens of BBs are seen growing on electron-dense spherical structures called deuterosomes. This second pathway evokes the de novo generation of centrioles from blepharoplasts observed in plant male gametes (Mizukami and Gall 1966; Hepler 1976). Some investigators, however, provide evidence supporting the hypothesis that the clusters of centrioles growing on the deuterosomes are primarily nucleated from the centrosomal centrioles (Kalnins and Porter 1969).

Live imaging recently helped to clarify the very first events of BB amplification and disclosed the origin of deuterosomes in mouse brain MCCs (Fig. 2) (Al Jord et al. 2014).

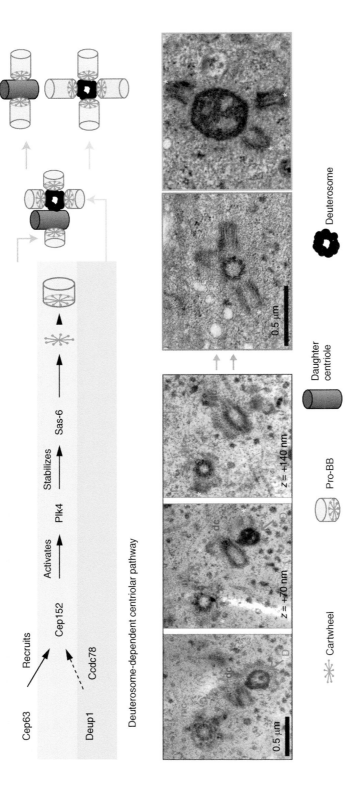

Figure 2. Basal body (BB) amplification in mouse multiciliated cells (MCCs). (*Upper* panel) BB assembly occurs both through a deuterosome-dependent and a deuterosome-independent centriolar pathway. The deuterosome-independent pathway is initiated by the recruitment of CEP63 at centrosomal mother and daughter centrioles (for clarity, only the daughter centriole is represented). The deuterosome-dependent pathway is initiated by the recruitment of DEUP1 at the daughter centriole. Both pathways involve the same downstream molecular components for BB assembly, including CEP152, PLK4, and SAS-6 (Al Jord et al. 2014). Deuterosome-dependent assembly also involves the CCDC78 protein in *Xenopus* (Klos Dehring et al. 2013). (*Left lower* panel) Serial section electron micrographs showing pro–BB assembly in mouse ependymal MCCs. mc, mother centriole; dc, daughter centriole; D, deuterosome. Asterisks mark pro–BBs assembling near the centrosomal centrioles or around the deuterosome. (*Right lower* panel) Electron micrographs showing elongated pro–BBs (asterisks) around a centrosomal centriole (*left*) or a deuterosome (*right*).

Time-lapse sequences show that deuterosomes together with the pro-BBs they support arise from the proximal part of the daughter centriole. Multiple rounds of deuterosome formation provide the cell with more than a hundred pro-BBs. Electron and superresolution microscopies indicate that the short pro-BBs remain latent during this phase of amplification (Kalnins and Porter 1969; Al Jord et al. 2014). Once the last deuterosome has been produced, pro-BBs then elongate simultaneously from all the deuterosomes and from the centrosomal centrioles (Zhao et al. 2013; Al Jord et al. 2014). This explains the coexistence of both "deuterosome" and "centriolar" pathways during MCC differentiation, which are seemingly the outcome of the same process originating at the centrosome. Immunostaining experiments suggest that the same dynamics of deuterosome formation and pro-BB amplification occurs in mouse tracheal MCCs (Al Jord et al. 2014).

Molecular Control

Insights into the molecular control of deuterosome formation and BB amplification in MCCs come from the process of centriole duplication in cycling cells. Centriole duplication is tightly controlled to avoid supernumerary centrosomes responsible for chromosome segregation defects during mitosis (Firat-Karalar and Stearns 2014). Briefly, at the G_1/S transition, centrosomal proteins 63 (CEP63) and centrosomal protein 152 (CEP152) form a complex at the proximal end of each centriole, initiating centriole biogenesis (Fig. 2) (Sir et al. 2011; Brown et al. 2013; Lukinavicius et al. 2013). CEP152 is then thought to recruit and/or activate polo-like kinase 4 (PLK4), which stabilizes spindle-assembly 6 homolog (SAS-6) to form the cartwheel structure that templates procentriole assembly in S phase. The level of centriole biogenesis initiators is regulated throughout the cell cycle to avoid the formation of supernumerary centrioles (Firat-Karalar and Stearns 2014).

As mentioned in the previous paragraph, MCC differentiation involves a massive up-regulation of BB biogenesis regulators. Under the control of MCIDAS/E2F complex and involving C-MYB transcription factor, the expression levels of CEP63, CEP152, and PLK4, among others, is increased at the onset of differentiation (Hoh et al. 2012; Klos Dehring et al. 2013; Wang et al. 2013; Zhao et al. 2013; Ma et al. 2014; Funk et al. 2015). In addition, a paralog of CEP63, CCDC67 (also called deuterosome protein 1, DEUP1) is specifically expressed in vertebrate multiciliated tissues (Zhao et al. 2013; Ma et al. 2014). Although BB amplification through the centriolar pathway seems to be driven by a duplication-like process, downstream from CEP63-CEP152, the deuterosome pathway is driven by DEUP1 (Fig. 2). DEUP1 depletion blocks deuterosome formation and BB amplification in Xenopus skin and mouse tracheal MCCs. Its overexpression in bacteria leads to the formation of spherical structures suggesting self-oligomerization properties. In some transformed cycling cells, DEUP1 can drive the formation of deuterosome-like structures capable of procentriole nucleation (Zhao et al. 2013). It thus seems to be a core structural and functional component of deuterosomes. Consistent with live imaging and EM experiments, DEUP1 accumulates on the proximal part of the daughter centriole during deuterosome formation in mouse brain MCCs (Al Jord et al. 2014). DEUP1 shares 37% of homology with CEP63 and has retained its ability to associate with centrosomal centrioles and to interact with CEP152. Knockdown experiments in mouse tracheal cells and Xenopus skin suggest that the recruitment of CEP152 by DEUP1 and the subsequent activation of PLK4, enable pro-BB nucleation from deuterosomal structures (Klos Dehring et al. 2013; Zhao et al. 2013). The current view is therefore that in MCCs, CEP152 is recruited by CEP63 or its paralog DEUP1 to drive the centriolar and the deuterosome pathway, respectively, via a unique molecular cascade highjacked from the centriole duplication process (Fig. 2). Although the BB duplication players downstream from SAS-6 are also up-regulated during vertebrate MCC differentiation, their involvement in centriole amplification remains to be determined (Hoh et al. 2012; Klos

Dehring et al. 2013; Wang et al. 2013; Zhao et al. 2013; Ma et al. 2014; Funk et al. 2015).

The centriole-associated protein CCDC78 has been shown to associate with deuterosomes and to be required for CEP152 recruitment and deuterosome-mediated pro-BB formation in mouse tracheal and *Xenopus* skin MCCs (Klos Dehring et al. 2013). The CCDC78 and DEUP1 relationship has not been tested yet. CCNO, a cyclin-like protein, is also involved in deuterosome-mediated BB amplification. Although CCNO partners and function are unknown, patients with mutations in *ccno* have MCCs with reduced number of cilia (Wallmeier et al. 2014). In particular, *ccno* mutant mice tracheal cells display larger deuterosomes with fewer pro-BBs suggesting a role for CCNO in the early onset of amplification (Funk et al. 2015).

Evolutionary Perspective

DEUP1 appears to be a key component for BB amplification in MCCs. Although CEP63 enables centriole duplication in cycling cells or mild amplification in MCCs, turning on DEUP1 expression allows the switch to massive BB production. Phylogenetic tree analyses suggest that DEUP1 was duplicated and diverged from CEP63 during vertebrate evolution after ray-finned fish specification (Zhao et al. 2013). *Deup1* is thus absent in zebrafish, which specify their MCCs under the same transcriptional cascade as the one of mammalian species and *Xenopus* (see previous paragraph) (Wang et al. 2013; Zhou et al. 2015). The availability of the single CEP63-mediated centriolar pathway may explain why zebrafish MCCs, observed in the olfactory placode and the pronephric ducts, are reported to have less than 16 cilia (Hansen and Zeiske 1993; Kramer-Zucker et al. 2005). Supporting this hypothesis, mammalian olfactory neurons, which harbor a dozen cilia, are proposed to produce their centrioles through a deuterosome-independent centriolar pathway (Cuschieri and Bannister 1975). Different situations exist in flatworms. Although the marine flatworm *Macrostomum lignano* has centrosomes and amplifies centrioles through a deuterosome-centriolar pathway (Tyler 1981),

the freshwater planarian *Schmidtea mediterranea* lacks centrosomes and thus amplifies centrioles de novo (Azimzadeh et al. 2012). In both cases, *gemc1*, *mcidas*, and *deup1* genes are absent, suggesting that centrioles can be massively amplified through centriolar and de novo pathways independently from this specific transcriptional cascade in animal cells. Conserved centriole duplication players are nevertheless required in *S. mediterranea* MCCs showing that de novo and centriole/deuterosome-derived amplification use unique upstream regulators but common downstream effectors in animals (Azimzadeh et al. 2012).

BB DOCKING AND CILIOGENESIS

Role of the Actin Cytoskeleton in BB Migration and Docking to the Apical Membrane

Following assembly and simultaneous release from either deuterosomes or centrosomal centrioles, BBs migrate and dock en masse at the apical membrane of MCCs (Ioannou et al. 2013; Al Jord et al. 2014). Pioneering work on differentiating MCCs from quail oviduct revealed that BB migration and docking depend on actin–myosin (Klotz et al. 1986; Lemullois et al. 1988; Boisvieux-Ulrich et al. 1990). A dense actin meshwork is present at the apical cortex in MCCs and myosin is also enriched in this area (Reverdin et al. 1975; Reed et al. 1984; Klotz et al. 1986; Lemullois et al. 1988; Sandoz et al. 1988; Park et al. 2006, 2008; Pan et al. 2007; Werner et al. 2011). The formation of the apical actin network is temporally linked to BB docking and involves conserved actin regulators, such as the RhoA and Rac1GTPases, the membrane-cytoskeletal linker ezrin, the cross-linking protein filamin A, the nucleator dishevelled-associated activator of morphogenesis 1 (DAAM1), or the F-actin capping protein CapZ-interacting protein (CapZIP) (Gomperts et al. 2004; Pan et al. 2007; Park et al. 2008; Chevalier et al. 2015; Epting et al. 2015; Miyatake et al. 2015; Yasunaga et al. 2015). The expression of several of these actin regulators is triggered by the regulatory network that controls MCC differentiation (Fig. 1) (Gomperts et al. 2004; Pan et al. 2007; Chevalier et al. 2015; Miyatake et al. 2015). Apical actin assembly and BB docking in MCCs are orchestrated by the planar cell polarity (PCP) pathway. Originally identified in the *Drosophila* wing, the PCP pathway is a conserved signaling pathway that governs a range of developmental processes (reviewed in Bayly and Axelrod 2011; Goodrich and Strutt 2011; Devenport 2014). Core PCP components include the intracellular proteins dishevelled (Dvl), diversin (diego in *Drosophila*), prickle (Pk), the transmembrane proteins frizzled (Fz), Van Gogh-like (Vangl) (strabismus or Van Gogh in *Drosophila*), cadherin, EGF-like, laminin G-like, seven-pass, and G-type receptor (Celsr) (flamingo in *Drosophila*). The control of BB docking by the PCP pathway

is an evolutionarily conserved feature because depletion of PCP components inhibit this process both in vertebrates and planarian MCCs (Park et al. 2006, 2008; Tissir et al. 2010; Almuedo-Castillo et al. 2011). Depleting the core PCP component Dvl or the PCP effectors inturned and fuzzy in *Xenopus* MCCs decreases the density of the apical actin meshwork and strongly affects ciliogenesis (Park et al. 2006, 2008). RhoA, which is present throughout the apical cortex and is essential to BB docking, is activated (i.e., converted to its GTP-bound form) specifically at BBs in a Dvl-dependent manner (Pan et al. 2007; Park et al. 2008). Recruitment of both Dvl and RhoA at BBs is dependent on inturned, which is itself recruited by nephrocystin 4 (NPHP4), a component of the

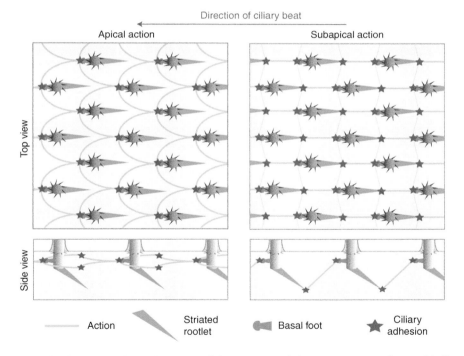

Figure 3. A model illustrating the architecture of the actin cytoskeleton in *Xenopus* skin multiciliated cells (MCCs). The drawings represent the basal bodies (BBs) and either the actin apical or subapical pools in mature MCCs. This organization of the subapical network is proposed in Antoniades et al. (2014) and the model of apical actin organization is based on immunofluorescence data from the same study. Both actin pools are connected to the BBs by ciliary adhesions located in the vicinity of the basal feet (note that the exact localization—near or at the basal feet—remains to be determined) (Antoniades et al. 2014). In addition, actin bridges forming the subapical pool are connected to rootlet tips both along the direction of ciliary beat and perpendicular to this direction (Werner et al. 2011; Antoniades et al. 2014). For clarity, only the actin bridges connecting the BBs along the direction of ciliary beat are represented in the side view of the subapical actin network.

ciliary transition zone that also binds to the striated rootlet in MCCs (Park et al. 2008; Yasunaga et al. 2015). In addition, inturned recruits the PCP effector DAAM1, and Dvl facilitates the activation of CapZIP by the kinase extracellular signal-regulated kinase 7 (ERK7) (Miyatake et al. 2015; Yasunaga et al. 2015). Another potential PCP effector affecting the actin cytoskeleton is the recently characterized flattop (FLTP) protein, which is required for BB docking in mouse MCCs (Gegg et al. 2014). Taken together, these results support that BB docking and the organization of apical actin are, in large part, orchestrated at the BBs by the local, PCP-dependent recruitment of actin regulators.

Studies in *Xenopus* reveal that the apical actin cap in mature MCCs is composed of two distinct but interconnected actin pools (Fig. 3). The apical pool is a very stable meshwork-like network that surrounds the BBs (Werner et al. 2011; Ioannou et al. 2013; Antoniades et al. 2014). The apical actin pool starts accumulating in the form of actin foci during early steps of MCC intercalation and becomes organized after BBs have docked (Ioannou et al. 2013). The subapical pool forms just below the apical pool ($\sim$0.5 μm below the apical surface) and consists of short actin bridges that connect neighboring BBs (Werner et al. 2011; Antoniades et al. 2014). During intercalation, migrating BBs are clustered together and connected to the cell periphery by actin cables. Disruption of this internal actin network by partially depleting the nucleotide-binding protein 1 (Nubp1) ATPase correlates with defective BB migration. Following BB docking, the internal actin network is replaced, or possibly remodeled, into the subapical network (Ioannou et al. 2013). Remarkably, both the internal actin network and subapical actin bridges are connected to the BBs via a protein complex known to regulate focal adhesions, which link the actin cytoskeleton to the extracellular matrix. The focal adhesion components vinculin, paxilin, talin, and focal adhesion kinase (FAK) localize at the BBs, forming anchor sites called "ciliary adhesions" by analogy to focal adhesions. In agreement with this idea, BBs in intercalating MCCs from FAK morphants are more dispersed

that in control cells and many BBs fail to associate to actin cables, which correlates with a defect in BB migration and docking (Antoniades et al. 2014).

BB Docking and Ciliary Assembly

As initially hypothesized by Sergei Sorokin, mother centriole docking at the onset of primary ciliogenesis requires the binding of a ciliary vesicle at the distal centriole and subsequent fusion of the ciliary vesicle with the plasma membrane (Sorokin 1962; and reviews by Rohatgi and Snell 2010; Ishikawa and Marshall 2011; Reiter et al. 2012; Wei et al. 2015). The ciliary vesicle is formed after recruitment of smaller vesicles to mother centriole distal appendages and fusion of these vesicles under the control of membrane-remodeling proteins and the Rab11/Rab8 cascade of GTPases (Lu et al. 2015). Rab8 and its activator Rabin8 interact with the distal appendage component CEP164 (Schmidt et al. 2012). BBs with docked vesicles have also been observed in vertebrate MCCs (Sorokin 1968; Dirksen 1971; Lemullois et al. 1988; Boisvieux-Ulrich et al. 1989). Recent work has provided further evidence that BB docking in MCCs indeed involves a ciliary vesicle. In *Xenopus*, BBs with tightly associated vesicles are seen during differentiation of control MCCs but not in Dvl morphants. Sec8, a component of a vesicle-trafficking complex called the exocyst, is present at BBs in control cells but not following Dvl depletion. This suggests that the PCP pathway controls the recruitment of Sec8-positive vesicles to the BBs to allow BB docking and ciliary membrane extension (Park et al. 2008). In agreement with this, the exocyst is required for primary ciliogenesis (Feng et al. 2012; Seixas et al. 2016). In mouse airway MCCs, ciliary vesicle docking further requires the BB component Chibby, which binds to CEP164 to promote the recruitment of Rabin8 and Rab8 (Burke et al. 2014). Primary cilium assembly also involves recruitment by CEP164 of tau tubulin kinase (TTBK2), which promotes the removal of the distal centriole component CP110 and the recruitment of the intraflagellar transport (IFT) machinery (Spektor et al. 2007;

Cite this article as *Cold Spring Harb Perspect Biol* doi: 10.1101/cshperspect.a028233

Goetz et al. 2012; Cajanek and Nigg 2014). CP110 is a distal centriole/BB that negatively regulates ciliogenesis by preventing growth of the axoneme (reviewed in Tsang and Dynlacht 2013). During differentiation of mouse and *Xenopus* MCCs, CP110 expression is repressed by the miR-34/449 microRNA. Inhibition of miR-34/449 induces a defect in ciliogenesis that can be rescued by codepleting CP110, indeed supporting that CP110 function is conserved in MCCs (Song et al. 2014).

Like for most ciliated cells, cilia assembly and maintenance in MCCs relies on the IFT machinery, a bidirectional transport system that moves ciliary components along axoneme microtubules. In vertebrates, mutations in IFT components that affect the assembly of primary cilia also usually affect the assembly of motile cilia (reviewed in Ishikawa and Marshall 2011; Sung and Leroux 2013; Lechtreck 2015). In addition, mutations that specifically affect ciliary motility are among the most frequent causes of primary ciliary dyskinesia (Lee 2011).

CILIARY BEAT ORIENTATION

BB Organization

In *Xenopus* MCCs, the subapical pool of actin is essential to BB proper spacing and orientation during MCC differentiation (Werner et al. 2011; Ioannou et al. 2013; Antoniades et al. 2014; Epting et al. 2015; Yasunaga et al. 2015). The subapical pool consists of actin bridges connecting the BBs along the axis of ciliary beating so that each BB is connected to the tip of the ciliary rootlet of the neighboring BB facing the direction of beating (Fig. 3) (Werner et al. 2011; Antoniades et al. 2014). Focal adhesion components are present at each end of these actin bridges, forming an apical ciliary adhesion in the vicinity of the basal foot and a subapical ciliary adhesion at the tip of the rootlet. A second set of actin bridges at a right angle with respect to the first set connects BBs laterally, likely also via the apical and subapical ciliary adhesions (Antoniades et al. 2014). In conditions that specifically disrupt the subapical actin pool but leave the apical network unaffected,

such as low doses of the actin polymerization inhibitor cytochalasin D or partial inhibition of actin regulators, BBs dock but show defects in spacing (Werner et al. 2011; Ioannou et al. 2013; Antoniades et al. 2014; Epting et al. 2015; Yasunaga et al. 2015). These results support that subapical actin bridges are required for spacing BBs both along the axis of cilia orientation and perpendicular to this axis. MT depolymerization has no effect on BB spacing in *Xenopus* MCCs, but it does affect spacing in mouse tracheal cells, and so does inhibition of MT anchoring at the BBs, suggesting that some aspects of this process might be different between these species (Werner et al. 2011; Kunimoto et al. 2012; Vladar et al. 2012).

Disruption of the subapical actin network also impacts centriole rotational polarity (Werner et al. 2011; Ioannou et al. 2013; Yasunaga et al. 2015). Treating differentiating *Xenopus* MCCs with low doses of cytochalasin D results in a strong increase in BB circular standard deviation (CSD) across the cell surface, although BB orientation remains locally coordinated. In contrast, MT depolymerization disrupts the coordination of rotational polarity even between neighboring BBs, supporting that actin and MTs control different aspects of BB rotational polarity (Werner et al. 2011). In mouse tracheal cells, tyrosinated MTs connect BBs to the cell junction in a polarized fashion, and transmission electron microscopy (TEM) analysis reveals the presence of MTs connecting the basal foot of peripheral BBs to the apical junctions (Vladar et al. 2012). Basal foot assembly requires a protein called ODF2/cenexin (Kunimoto et al. 2012), which is also a centrosome component essential for the assembly of mother centriole appendages (Ishikawa et al. 2005; Tateishi et al. 2013). Inactivation of *Odf2* in mouse leads to centriole and ciliary beat disorientation in MCCs, in agreement with the MT drug experiments in the *Xenopus* skin (Werner et al. 2011; Kunimoto et al. 2012). The microtubule nucleator γ-tubulin localizes at the basal foot cap, which suggests that MTs are assembled from the basal foot (Hagiwara et al. 2000; Clare et al. 2014). Actin filaments emanating from the basal foot cap have also been observed in various species, which is in line with the observation that focal adhesion components localize in the vicinity of the basal foot in *Xenopus* (Fig. 3) (Reed et al. 1984; Sandoz et al. 1988; Chailley et al. 1989; Antoniades et al. 2014). In addition, connecting the basal foot to the cytoskeleton involves ζ-tubulin in *Xenopus* MCCs. ζ-tubulin localizes at the tip of the basal foot and is required for BB spacing and orientation. Interestingly, depletion of ζ-tubulin does not affect binding of MTs to the basal feet, but it perturbs the assembly of the apical and subapical actin networks, which further suggests that basal feet connect BBs to the actin cytoskeleton (Turk et al. 2015). Mature MCCs in the quail oviduct contain a well-developed cytokeratin network that binds to desmosomes and to the striated rootlets, suggesting that intermediate filaments could also participate in BB organization in these cells (Lemullois et al. 1987).

MCC Polarization by the PCP Pathway

The polarization of multiciliated epithelia involves not only the coordination of BB rotational polarity within cells but also between cells.

Both intracellular and tissue-level polarities are governed by the PCP pathway in vertebrate MCCs. Establishment of PCP at the tissue-level involves the asymmetric partitioning of PCP components at the cell cortex and intercellular communication to coordinate polarity between neighboring cells. In mouse ependymal MCCs, Vangl2 localizes transiently at the posterior side of the apical membrane (upstream of flow) during MCC differentiation (Guirao et al. 2010; Boutin et al. 2014). In tracheal cells, Vangl1/2 and Pk2 also localize upstream of flow at the posterior side, whereas Dvl1/3 and Fzd3/6 are at the opposite side (Vladar et al. 2012). In the *Xenopus* skin, a subset of PCP components displays an increasingly asymmetric distribution during MCC differentiation. Intriguingly, this distribution is reversed compared with that of mouse MCCs: Vangl1 and Pk2 localize downstream from flow, whereas Dvl1 and Fzd6 localize upstream of flow (Fig. 4) (Butler and Wallingford 2015).

In *Drosophila*, PCP components act noncell-autonomously to establish tissue-level polarity and this property is conserved in vertebrate MCCs. In *Xenopus*, expression of Vangl2

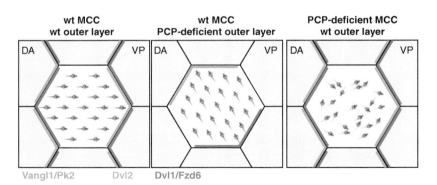

Figure 4. Multiciliated cell (MCC) polarization by the planar cell polarity (PCP) pathway in the *Xenopus* skin. Combined action of the PCP pathway and hydrodynamic forces align the BBs in a dorsoanterior (DA) to ventroposterior (VP) direction in wild-type (wt) *Xenopus* embryos. PCP components are asymmetrically distributed at the apical junctions, with a Dvl1/Fzd6 domain located upstream of flow and a Vangl1/Pk2 domain located downstream from flow (Butler and Wallingford 2015). When wt MCCs intercalate within a PCP-deficient outer epidermal layer in grafting experiments, intracellular basal body (BB) orientation is coordinated but improperly aligned with tissue polarity. When PCP-deficient MCCs intercalate within a wt outer layer, both intracellular and tissue-level polarity are lost (Mitchell et al. 2009). Note that in mouse ependymal and tracheal MCCs, the orientation of the Dvl/Fzd and Vangl/Pk domains with respect to the direction of ciliary beat is reversed (Vladar et al. 2012; Boutin et al. 2014).

and Fzd3 in the outer epidermal layer is required for BB polarization in intercalating MCCs. When wild-type MCCs intercalate within a PCP-deficient outer layer, BB polarity is homogeneous within MCCs, but overall polarity with respect to the anterior–posterior (AP) axis is perturbed (Fig. 4) (Mitchell et al. 2009). In addition, the PCP pathway is required cell-autonomously to control BB docking and orientation in mouse and *Xenopus* (Park et al. 2008; Mitchell et al. 2009; Guirao et al. 2010; Vladar et al. 2012). Perturbation of Dvl function in differentiating *Xenopus* MCCs results in low coordination of BB orientation within cells even when they intercalate within a wild-type outer layer (Fig. 4) (Mitchell et al. 2009). In the mouse oviduct, Celsr1 depletion affects rotational polarity both within cells and between cells, although BBs remain locally coordinated with neighboring BBs (Shi et al. 2014). In ependymal MCCs, cilia are not present across the apical surface but clustered in a ciliary patch located toward the anterior end of the apical surface, a property called translational polarity (Mirzadeh et al. 2010). The PCP pathway controls both BB rotational and translational polarities by coordinating between adjacent cells the direction of BB orientation and that of BB cluster displacement—but not the displacement itself (Hirota et al. 2010; Mirzadeh et al. 2010; Tissir et al. 2010; Boutin et al. 2014; Ohata et al. 2014). Remarkably, translational polarity and rotational polarity involve partially distinct PCP components. Coordinating the direction of displacement of the BB cluster between cells requires Celsr1, whereas organization of BBs within a cluster requires Celsr2 and Celsr3 (Boutin et al. 2014). Translational polarity is first established in ependymal cell progenitors (radial glial cells [RGCs]) in which the primary cilium becomes off-centered in the anterior direction. The presence of a primary cilium in RGCs is essential for proper asymmetric localization of PCP components and localization of the BB cluster in ependymal cells (Mirzadeh et al. 2010). Remarkably, this function of the primary cilium depends on the mechanosensory proteins polycystic kidney disease (PKD) 1 and 2, which localize to the primary cilium in

RGCs. Primary cilia possibly sense a weak preexisting flow generated by the posterior secretion and anterior reabsorption of CSF and respond by polarizing in the direction of the flow (Ohata et al. 2015).

How does the PCP pathway control BB polarization in MCCs? The core PCP component Dvl2 localizes at the BBs to trigger RhoA activation and BB docking in the *Xenopus* skin (Park et al. 2008). Following BB docking and ciliogenesis, Dvl2 localizes at the ciliary rootlet in *Xenopus* and mouse MCCs (Park et al. 2008; Hirota et al. 2010; Vladar et al. 2012). Expression of the Dvl2 dominant-negative construct Xdd1 in the *Xenopus* skin disrupts BB orientation without affecting docking, supporting that these two aspects of Dvl2 function can be uncoupled (Park et al. 2008). Accumulation of Dvl2 at the base of cilia following inactivation of the anaphase-promoting complex, which targets Dvl2 for degradation, results in BB disorientation in the *Xenopus* skin, suggesting that BB polarization requires a tight control of Dvl2 levels at the BBs (Ganner et al. 2009). Besides Dvl2, diversin is present at the BBs in mouse airway and *Xenopus* skin MCCs (Yasunaga et al. 2011; Vladar et al. 2012), and Pk2 associates to BBs in mature *Xenopus* skin MCCs (Butler and Wallingford 2015). PCP effectors affecting the actin cytoskeleton such as inturned, DAAM1, and RhoA-GTP localize at the BBs, and following mild perturbations of these effectors BBs dock but fail to orient (Park et al. 2006, 2008; Yasunaga et al. 2015). Together, these data support that the PCP pathway locally controls actin assembly or remodeling to position the BBs. In addition, the establishment of tissue-level polarity requires coordinating BB orientation with the asymmetric PCP domains at the apical junctions. In mouse tracheal cells, MTs link the BBs to the apical junction corresponding to the Fzd3/6-Dvl1/3 domain (Vladar et al. 2012). In mature ependymal cells, MTs connects the BB cluster to the apical junction opposite to the Vangl2 domain, which corresponds to a presumptive Fzd3 domain (Boutin et al. 2014). PCP domains at the apical junctions most likely influence the polarity of the apical actin networks as well, particularly in *Xenopus* MCCs in which actin perturbation strongly impacts tissue-level polarity (Werner et al. 2011).

What acts upstream of the PCP pathway? A recent study found that mechanical strain imposed during gastrulation on *Xenopus* ectodermal cells, which gives rise to the larval epidermis, instructs planar polarity (Chien et al. 2015). During gastrulation, the ectoderm is strained along the AP axis by pulling forces generated by involution of the mesoderm. As a result, cells elongate and align apical MTs along the axis of the strain. Polarization of apical MTs then allows activation of PCP complexes specifically at apical junctions that are perpendicular to the axis of the strain. No clear asymmetric localization of PCP components is observed at this stage, but FRAP analysis shows that Fzd3-GFP is stabilized on perpendicular junctions during gastrulation or under externally imposed strain (Chien et al. 2015). In differentiating MTECs as in other planar polarized tissues, apical MTs oriented in the axis of planar polarization are required for asymmetric localization of core PCP components (Shimada et al. 2006; Harumoto et al. 2010; Vladar et al. 2012). MTs are also required for asymmetric partitioning of PCP complexes during polarization of ependymal cells (Boutin et al. 2014). Thus, alignment of apical MTs is likely a conserved feature of PCP establishment in MCCs. Whether signaling cues only or a combination of signaling and mechanical cues acts upstream of the PCP pathway in other multiciliated epithelia remains to be determined.

Hydrodynamic Forces

Studies in *Xenopus* and mouse have shown that hydrodynamic forces play a major role in BB polarization by favoring the coordinated beating of neighboring cilia (Guirao and Joanny 2007; Mitchell et al. 2007; Guirao et al. 2010). In *Xenopus* MCCs, PCP signaling cues first induce a bias in the orientation of cilia beating (Park et al. 2008; Mitchell et al. 2009). This bias initiates a weak flow that establishes a positive-feedback loop in which cilia respond to the flow by refining their orientation until they are well aligned. Ciliary motility is essential for the re-

finement of BB orientation (Mitchell et al. 2007). BBs fail to polarize in response to hydrodynamic forces following perturbation of either MT or actin networks, suggesting that the cytoskeleton is actively remodeled during this process (Werner et al. 2011). Unlike *Xenopus*, BBs dock in random orientation in mouse ependymal cells. In vitro differentiated ependymal cells submitted to a continuous shear flow align their cilia in the direction of the flow, and this response requires Vangl2. This suggests a model whereby ependymal cilia align in response to hydrodynamic forces initially generated by a weak preexisting CSF flow, in a PCP-dependent manner (Guirao et al. 2010). One possible model is that the PCP is required for remodeling the cytoskeleton to allow reorientation of the BBs. Ciliary motility is essential for this response as rotational polarity is altered in ependymal cells of the *Hydin* and *Kintoun* (*Ktu*) mutants, which assemble cilia with abnormal motility or fully immotile cilia, respectively (Lechtreck et al. 2008; Matsuo et al. 2013). Interestingly, ciliary motility is not required for the refinement of BB orientation in the trachea, because $Ktu^{-/-}$ mice display properly aligned BBs in this tissue (Matsuo et al. 2013). Like primary cilia, motile cilia in MCCs have sensory functions (Shah et al. 2009; Jain et al. 2012) and a proposed hypothesis is that MCC cilia could sense a bidirectional flow produced in the trachea through fetal breathing movements at the time when refinement occurs (Shah et al. 2009; Jain et al. 2012; Matsuo et al. 2013). Importantly, cilia realign in response to flow only during a developmental window during which MCCs polarize. Past this time, mature MCCs appear to lock their ciliary orientation (Boisvieux-Ulrich and Sandoz 1991; Mitchell et al. 2007; Guirao et al. 2010). The protein basal body orientation factor 1 (Bbof1), which localizes to the ciliary rootlet in *Xenopus* MCCs, is possibly involved in locking ciliary orientation, although the corresponding mechanism remains unclear (Chien et al. 2013).

Regulation of Ciliary Beating

Computational models support that the formation of metachronal waves in mature MCCs results from the local coordination of ciliary beat by hydrodynamic coupling (Guirao and Joanny 2007; Guirao et al. 2010; Elgeti and Gompper 2013). Coordination possibly involves a feedback mechanism that regulates the activity of axonemal dyneins in response to hydrodynamic cues (Rompolas et al. 2010). Formation of metachronal waves also requires the coordination of ciliary beat by the cytoskeleton. Disruption of the subapical actin bridges in already polarized *Xenopus* MCCs alters the ciliary beat coordination, suggesting that these bridges serve as a physical lever to facilitate propagation of the metachronal wave (Werner et al. 2011). In addition, hormonal or neurotransmitter control appears to affect ciliary beat in MCCs. In planarians, ablation of serotonergic neurons disrupts metachrony in the ventral epidermis without affecting the beating pattern of individual cilia (Currie and Pearson 2013). Serotonin-secreting cells are also present in the *Xenopus* skin, and serotonin regulates ciliary beat frequency in this tissue (Walentek et al. 2014). Ciliary beat frequency can also be regulated in response to external stimuli. In the mammalian airways, receptors to bitter compounds localize on motile cilia. In the presence of bitter compounds, intracellular calcium and ciliary beat frequency increase, thus providing a cell-autonomous mechanism to eliminate toxic compounds (Shah et al. 2009).

CONCLUDING REMARKS

Remarkable progress has been made during the past decade in our understanding of developmental and cellular mechanisms underlying MCC differentiation and function. In particular, the identification of the regulatory cascade controlling the specification and differentiation of MCCs in vertebrates opens important avenues in terms of both fundamental knowledge and biomedical applications. The ability to differentiate human MCCs in vitro from induced pluripotent stem cells constitutes a promising therapeutic approach for treating respiratory diseases known to affect the differentiation and function of MCCs, such as primary ciliary dyskinesia, cystic fibrosis, and severe asthma (Thomas et al. 2010; Firth et al. 2014; Hoegger et al. 2014).

There are many questions that remain to be addressed. Several of the identified regulators of MCC differentiation and BB assembly (MCIDAS, GEMC1, DEUP1) result from gene duplications that occurred in the vertebrate lineage. Did this regulatory network evolve to control the differentiation of mosaic multiciliated epithelia composed by distinct but highly related ciliated and mucus-secreting cells? How exactly do these regulators interact with each other to control MCC differentiation? Known deuterosome components include DEUP1, which by itself can form spherical structures resembling deuterosomes, and CCDC78 (Klos Dehring et al. 2013; Zhao et al. 2013). Are there other deuterosome components? How is deuterosome assembly regulated? In mouse MCCs, deuterosomes are assembled at the daughter centriole of the centrosome (Al Jord et al. 2014). Is this a conserved mechanism or are there differences between species? BB number in MCCs can vary up to 10-fold depending on the tissue. How are these numbers controlled? The involvement of the PCP pathway and the cytoskeleton in coordinating BB migration, docking, and rotational polarity is now well established, but what are the precise mechanisms by which the PCP pathway controls the organization of actin and MTs? Hydrodynamic forces are required for achieving the proper polarization of ciliary beating. What are the respective impacts of ciliary motility and sensory functions in this process? With the current pace of discovery, we expect that many new findings will extend our understanding of MCC biology in a near future.

ACKNOWLEDGMENTS

The authors thank Andrea Aguilar, Camille Boutin, Bénédicte Durand, Laurent Kodjabachian, Brice Marcet, and Diego Revinski for useful comments on the manuscript. Our work is funded by Agence Nationale de la Recherche and La Ligue contre le Cancer.

REFERENCES

Adl SM, Simpson AG, Lane CE, Lukes J, Bass D, Bowser SS, Brown MW, Burki F, Dunthorn M, Hampl V, et al. 2012. The revised classification of eukaryotes. *J Eukaryot Microbiol* **59:** 429–493.

Al Jord A, Lemaitre AI, Delgehyr N, Faucourt, Spassky N, Meunier A. 2014. Centriole amplification by mother and daughter centrioles differs in multiciliated cells. *Nature* **516:** 104–107.

Almuedo-Castillo M, Salo E, Adell T. 2011. Dishevelled is essential for neural connectivity and planar cell polarity in planarians. *Proc Natl Acad Sci* **108:** 2813–2818.

Amirav I, Wallmeier J, Loges NT, Menchen T, Pennekamp P, Mussaffi H, Abitbul R, Avital A, Bentur L, Dougherty GW, et al. 2016. Systematic analysis of *CCNO* variants in a defined population: Implications for clinical phenotype and differential diagnosis. *Hum Mutat* **37:** 396–405.

Anderson RG. 1972. The three-dimensional structure of the basal body from the rhesus monkey oviduct. *J Cell Biol* **54:** 246–265.

Anderson RG, Brenner RM. 1971. The formation of basal bodies (centrioles) in the Rhesus monkey oviduct. *J Cell Biol* **50:** 10–34.

Antoniades I, Stylianou P, Skourides PA. 2014. Making the connection: Ciliary adhesion complexes anchor basal bodies to the actin cytoskeleton. *Dev Cell* **28:** 70–80.

Azimzadeh J. 2014. Exploring the evolutionary history of centrosomes. *Philos Trans R Soc Lond B Biol Sci* **369:** 20130453.

Azimzadeh J, Wong ML, Downhour DM, Sanchez Alvarado A, Marshall WF. 2012. Centrosome loss in the evolution of planarians. *Science* **335:** 461–463.

Balestrini A, Cosentino C, Errico A, Garner E, Costanzo V. 2010. GEMC1 is a TopBP1-interacting protein required for chromosomal DNA replication. *Nat Cell Biol* **12:** 484–491.

Basquin C, Orfila AM, Azimzadeh J. 2015. The planarian *Schmidtea mediterranea* as a model for studying motile cilia and multiciliated cells. *Methods Cell Biol* **127:** 243–262.

Bayly R, Axelrod JD. 2011. Pointing in the right direction: New developments in the field of planar cell polarity. *Nat Rev Genet* **12:** 385–391.

Boisvieux-Ulrich E, Sandoz D. 1991. Determination of ciliary polarity precedes differentiation in the epithelial cells of quail oviduct. *Biol Cell* **72:** 3–14.

Boisvieux-Ulrich E, Laine MC, Sandoz D. 1989. In vitro effects of taxol on ciliogenesis in quail oviduct. *J Cell Sci* **92:** 9–20.

Boisvieux-Ulrich E, Laine MC, Sandoz D. 1990. Cytochalasin D inhibits basal body migration and ciliary elongation in quail oviduct epithelium. *Cell Tissue Res* **259:** 443–454.

Boon M, Wallmeier J, Ma L, Loges NT, Jaspers M, Olbrich H, Dougherty GW, Raidt J, Werner C, Amirav I, et al. 2014. *MCIDAS* mutations result in a mucociliary clearance disorder with reduced generation of multiple motile cilia. *Nat Commun* **5:** 4418.

Boury-Esnault N, Efremova S, Bézac C, Vacelet J. 1999. Reproduction of a hexactinellid sponge: First description of gastrulation by cellular delamination in the Porifera. *Invertebr Reprod Dev* **35:** 187–201.

Boutin C, Labedan P, Dimidschstein J, Richard F, Cremer H, Andre P, Yang Y, Montcouquiol M, Goffinet AM, Tissir F. 2014. A dual role for planar cell polarity genes in ciliated cells. *Proc Natl Acad Sci* **111:** E3129–E3138.

Brightman MW, Palay SL. 1963. The fine structure of ependyma in the brain of the rat. *J Cell Biol* **19:** 415–439.

Brody SL, Yan XH, Wuerffel MK, Song SK, Shapiro SD. 2000. Ciliogenesis and left–right axis defects in forkhead factor HFH-4-null mice. *Am J Respir Cell Mol Biol* **23:** 45–51.

Brown NJ, Marjanovic M, Luders J, Stracker TH, Costanzo V. 2013. Cep63 and cep152 cooperate to ensure centriole duplication. *PLoS ONE* **8:** e69986.

Burke MC, Li FQ, Cyge B, Arashiro T, Brechbuhl HM, Chen X, Siller SS, Weiss MA, O'Connell CB, Love D, et al. 2014. Chibby promotes ciliary vesicle formation and basal body docking during airway cell differentiation. *J Cell Biol* **207:** 123–137.

Butler MT, Wallingford JB. 2015. Control of vertebrate core planar cell polarity protein localization and dynamics by Prickle 2. *Development* **142:** 3429–3439.

Cajanek L, Nigg EA. 2014. Cep164 triggers ciliogenesis by recruiting Tau tubulin kinase 2 to the mother centriole. *Proc Natl Acad Sci* **111:** E2841–E2850.

Chailley B, Nicolas G, Laine MC. 1989. Organization of actin microfilaments in the apical border of oviduct ciliated cells. *Biol Cell* **67:** 81–90.

Chen J, Knowles HJ, Hebert JL, Hackett BP. 1998. Mutation of the mouse hepatocyte nuclear factor/forkhead homologue 4 gene results in an absence of cilia and random left–right asymmetry. *J Clin Invest* **102:** 1077–1082.

Chevalier B, Adamiok A, Mercey O, Revinski DR, Zaragosi LE, Pasini A, Kodjabachian L, Barbry P, Marcet B. 2015. miR-34/449 control apical actin network formation during multiciliogenesis through small GTPase pathways. *Nat Commun* **6:** 8386.

Chien YH, Werner ME, Stubbs J, Joens MS, Li J, Chien S, Fitzpatrick JA, Mitchell BJ, Kintner C. 2013. Bbof1 is required to maintain cilia orientation. *Development* **140:** 3468–3477.

Chien YH, Keller R, Kintner C, Shook DR. 2015. Mechanical strain determines the axis of planar polarity in ciliated epithelia. *Curr Biol* **25:** 2774–2784.

Choksi SP, Lauter G, Swoboda P, Roy S. 2014. Switching on cilia: Transcriptional networks regulating ciliogenesis. *Development* **141:** 1427–1441.

Chung MI, Peyrot SM, LeBoeuf S, Park TJ, McGary KL, Marcotte EM, Wallingford JB. 2012. RFX2 is broadly required for ciliogenesis during vertebrate development. *Dev Biol* **363:** 155–165.

Chung MI, Kwon T, Tu F, Brooks ER, Gupta R, Meyer M, Baker JC, Marcotte EM, Wallingford JB. 2014. Coordinated genomic control of ciliogenesis and cell movement by RFX2. *eLife* **3:** e01439.

Cibois M, Luxardi G, Chevalier B, Thome V, Mercey O, Zaragosi, Barbry P, Pasini A, Marcet B, Kodjabachian L. 2015. BMP signalling controls the construction of vertebrate mucociliary epithelia. *Development* **142:** 2352–2363.

Clare DK, Magescas J, Piolot T, Dumoux M, Vesque C, Pichard E, Dang T, Duvauchelle B, Poirier F, Delacour D. 2014. Basal foot MTOC organizes pillar MTs required for coordination of beating cilia. *Nat Commun* **5:** 4888.

Currie KW, Pearson BJ. 2013. Transcription factors *lhx1/5-1* and pitx are required for the maintenance and regeneration of serotonergic neurons in planarians. *Development* **140:** 3577–3588.

Cuschieri A, Bannister LH. 1975. The development of the olfactory mucosa in the mouse: Electron microscopy. *J Anat* **119:** 471–498.

Danielian PS, Bender Kim CF, Caron AM, Vasile E, Bronson RT, Lees JA. 2007. *E2f4* is required for normal development of the airway epithelium. *Dev Biol* **305:** 564–576.

Danielian PS, Hess RA, Lees JA. 2016. *E2f4* and *E2f5* are essential for the development of the male reproductive system. *Cell Cycle* **15:** 250–260.

Deblandre GA, Wettstein DA, Koyano-Nakagawa N, Kintner C. 1999. A two-step mechanism generates the spacing pattern of the ciliated cells in the skin of *Xenopus* embryos. *Development* **126:** 4715–4728.

Devenport D. 2014. The cell biology of planar cell polarity. *J Cell Biol* **207:** 171–179.

Didon L, Zwick RK, Chao IW, Walters MS, Wang R, Hackett NR, Crystal RG. 2013. RFX3 modulation of FOXJ1 regulation of cilia genes in the human airway epithelium. *Respir Res* **14:** 70.

Dirksen ER. 1971. Centriole morphogenesis in developing ciliated epithelium of the mouse oviduct. *J Cell Biol* **51:** 286–302.

Elgeti J, Gompper G. 2013. Emergence of metachronal waves in cilia arrays. *Proc Natl Acad Sci* **110:** 4470–4475.

El Zein L, Ait-Lounis A, Morle L, Thomas J, Chhin B, Spassky N, Reith W, Durand B. 2009. RFX3 governs growth and beating efficiency of motile cilia in mouse and controls the expression of genes involved in human ciliopathies. *J Cell Sci* **122:** 3180–3189.

Epting D, Slanchev K, Boehlke C, Hoff S, Loges NT, Yasunaga T, Indorf L, Nestel S, Lienkamp SS, Omran H, et al. 2015. The Rac1 regulator ELMO controls basal body migration and docking in multiciliated cells through interaction with Ezrin. *Development* **142:** 174–184.

Feng S, Knodler A, Ren J, Zhang J, Zhang X, Hong Y, Huang S, Peranen J, Guo W. 2012. A Rab8 guanine exchange factor-effector interaction network regulates primary ciliogenesis. *J Biol Chem* **287:** 15602–15609.

Firat-Karalar EN, Stearns T. 2014. The centriole duplication cycle. *Philos Trans R Soc Lond B Biol Sci* **369:** 20130460.

Firth AL, Dargitz CT, Qualls SJ, Menon T, Wright R, Singer O, Gage FH, Khanna A, Verma IM. 2014. Generation of multiciliated cells in functional airway epithelia from human induced pluripotent stem cells. *Proc Natl Acad Sci* **111:** E1723–E1730.

Funk MC, Bera AN, Menchen T, Kuales G, Thriene K, Lienkamp SS, Dengjel J, Omran H, Frank M, Arnold SJ. 2015. *Cyclin O (Ccno)* functions during deuterosome-mediated centriole amplification of multiciliated cells. *EMBO J* **34:** 1078–1089.

Ganner A, Lienkamp S, Schafer T, Romaker D, Wegierski T, Park TJ, Spreitzer S, Simons M, Gloy J, Kim E, et al. 2009.

Regulation of ciliary polarity by the APC/C. *Proc Natl Acad Sci* **106:** 17799–17804.

Gegg M, Bottcher A, Burtscher I, Hasenoeder S, Van Campenhout C, Aichler M, Walch A, Grant SG, Lickert H. 2014. Flattop regulates basal body docking and positioning in mono- and multiciliated cells. *eLife* **3:** e03842.

Gibbons IR. 1961. The relationship between the fine structure and direction of beat in gill cilia of a lamellibranch mollusc. *J Biophys Biochem Cytol* **11:** 179–205.

Giribet G, Ribera C. 1998. The position of arthropods in the animal kingdom: A search for a reliable outgroup for internal arthropod phylogeny. *Mol Phylogenet Evol* **9:** 481–488.

Goetz SC, Anderson KV. 2010. The primary cilium: A signalling centre during vertebrate development. *Nat Rev Genet* **11:** 331–344.

Goetz SC, Liem KF Jr, Anderson KV. 2012. The spinocerebellar ataxia-associated gene Tau tubulin kinase 2 controls the initiation of ciliogenesis. *Cell* **151:** 847–858.

Gomi K, Arbelaez V, Crystal RG, Walters MS. 2015. Activation of NOTCH1 or NOTCH3 signaling skews human airway basal cell differentiation toward a secretory pathway. *PLoS ONE* **10:** e0116507.

Gomperts BN, Gong-Cooper X, Hackett BP. 2004. Foxj1 regulates basal body anchoring to the cytoskeleton of ciliated pulmonary epithelial cells. *J Cell Sci* **117:** 1329–1337.

Goodrich LV, Strutt D. 2011. Principles of planar polarity in animal development. *Development* **138:** 1877–1892.

Guirao B, Joanny JF. 2007. Spontaneous creation of macroscopic flow and metachronal waves in an array of cilia. *Biophys J* **92:** 1900–1917.

Guirao B, Meunier A, Mortaud S, Aguilar A, Corsi JM, Strehl L, Hirota Y, Desoeuvre A, Boutin C, Han YG, et al. 2010. Coupling between hydrodynamic forces and planar cell polarity orients mammalian motile cilia. *Nat Cell Biol* **12:** 341–350.

Guseh JS, Bores SA, Stanger BZ, Zhou Q, Anderson WJ, Melton DA, Rajagopal J. 2009. Notch signaling promotes airway mucous metaplasia and inhibits alveolar development. *Development* **136:** 1751–1759.

Hagiwara H. 1995. Electron microscopic studies of ciliogenesis and ciliary abnormalities in human oviduct epithelium. *Ital J Anat Embryol* **100:** 451–459.

Hagiwara H, Kano A, Aoki T, Ohwada N, Takata K. 2000. Localization of γ-tubulin to the basal foot associated with the basal body extending a cilium. *Histochem J* **32:** 669–671.

Hansen A, Zeiske E. 1993. Development of the olfactory organ in the zebrafish, *Brachydanio rerio*. *J Comp Neurol* **333:** 289–300.

Harumoto T, Ito M, Shimada Y, Kobayashi TJ, Ueda HR, Lu B, Uemura T. 2010. Atypical cadherins Dachsous and Fat control dynamics of noncentrosomal microtubules in planar cell polarity. *Dev Cell* **19:** 389–401.

Hepler PK. 1976. The blepharoplast of Marsilea: Its de novo formation and spindle association. *J Cell Sci* **21:** 361–390.

Hirota Y, Meunier A, Huang S, Shimozawa T, Yamada O, Kida YS, Inoue M, Ito T, Kato H, Sakaguchi M, et al. 2010. Planar polarity of multiciliated ependymal cells involves the anterior migration of basal bodies regulated by non-muscle myosin II. *Development* **137**: 3037–3046.

Hodges ME, Scheumann N, Wickstead B, Langdale JA, Gull K. 2010. Reconstructing the evolutionary history of the centriole from protein components. *J Cell Sci* **123**: 1407–1413.

Hoegger MJ, Fischer AJ, McMenimen JD, Ostedgaard LS, Tucker AJ, Awadalla MA, Moninger TO, Michalski AS, Hoffman EA, Zabner J, et al. 2014. Impaired mucus detachment disrupts mucociliary transport in a piglet model of cystic fibrosis. *Science* **345**: 818–822.

Hoh RA, Stowe TR, Turk E, Stearns T. 2012. Transcriptional program of ciliated epithelial cells reveals new cilium and centrosome components and links to human disease. *PLoS ONE* **7**: e52166.

Ioannou A, Santama N, Skourides PA. 2013. *Xenopus laevis* nucleotide binding protein 1 (xNubp1) is important for convergent extension movements and controls ciliogenesis via regulation of the actin cytoskeleton. *Dev Biol* **380**: 243–258.

Ishikawa H, Marshall WF. 2011. Ciliogenesis: Building the cell's antenna. *Nat Rev Mol Cell Biol* **12**: 222–234.

Ishikawa H, Kubo A, Tsukita S. 2005. Odf2-deficient mother centrioles lack distal/subdistal appendages and the ability to generate primary cilia. *Nat Cell Biol* **7**: 517–524.

Jain R, Javidan-Nejad C, Alexander-Brett J, Horani A, Cabellon MC, Walter MJ, Brody SL. 2012. Sensory functions of motile cilia and implication for bronchiectasis. *Front Biosci (Schol Ed)* **4**: 1088–1098.

Kalnins VI, Porter KR. 1969. Centriole replication during ciliogenesis in the chick tracheal epithelium. *Z Zellforsch Mikrosk Anat* **100**: 1–30.

Katz SM, Morgan JJ. 1984. Cilia in the human kidney. *Ultrastruct Pathol* **6**: 285–294.

Kessler M, Hoffmann K, Brinkmann V, Thieck O, Jackisch S, Toelle B, Berger H, Mollenkopf HJ, Mangler M, Sehouli J, et al. 2015. The Notch and Wnt pathways regulate stemness and differentiation in human fallopian tube organoids. *Nat Commun* **6**: 8989.

Klos Dehring DA, Vladar EK, Werner ME, Mitchell JW, Hwang P, Mitchell BJ. 2013. Deuterosome-mediated centriole biogenesis. *Dev Cell* **27**: 103–112.

Klotz C, Bordes N, Laine MC, Sandoz D, Bornens M. 1986. Myosin at the apical pole of ciliated epithelial cells as revealed by a monoclonal antibody. *J Cell Biol* **103**: 613–619.

Kramer-Zucker AG, Olale F, Haycraft CJ, Yoder BK, Schier AF, Drummond IA. 2005. Cilia-driven fluid flow in the zebrafish pronephros, brain and Kupffer's vesicle is required for normal organogenesis. *Development* **132**: 1907–1921.

Kroll KL. 2007. Geminin in embryonic development: Coordinating transcription and the cell cycle during differentiation. *Front Biosci* **12**: 1395–1409.

Kunimoto K, Yamazaki Y, Nishida T, Shinohara K, Ishikawa H, Hasegawa T, Okanoue T, Hamada H, Noda T, Tamura A, et al. 2012. Coordinated ciliary beating requires Odf2-mediated polarization of basal bodies via basal feet. *Cell* **148**: 189–200.

Kyrousi C, Arbi M, Pilz GA, Pefani DE, Lalioti ME, Ninkovic J, Gotz M, Lygerou Z, Taraviras S. 2015. Mcidas and

GemC1 are key regulators for the generation of multiciliated ependymal cells in the adult neurogenic niche. *Development* **142**: 3661–3674.

Lafkas D, Shelton A, Chiu C, de Leon Boenig G, Chen Y, Stawicki SS, Siltanen C, Reichelt M, Zhou M, Wu X, et al. 2015. Therapeutic antibodies reveal Notch control of transdifferentiation in the adult lung. *Nature* **528**: 127–131.

Lechtreck KF. 2015. IFT–cargo interactions and protein transport in cilia. *Trends Biochem Sci* **40**: 765–778.

Lechtreck KF, Delmotte P, Robinson ML, Sanderson MJ, Witman GB. 2008. Mutations in *Hydin* impair ciliary motility in mice. *J Cell Biol* **180**: 633–643.

Lee L. 2011. Mechanisms of mammalian ciliary motility: Insights from primary ciliary dyskinesia genetics. *Gene* **473**: 57–66.

Lee L. 2013. Riding the wave of ependymal cilia: Genetic susceptibility to hydrocephalus in primary ciliary dyskinesia. *J Neurosci Res* **91**: 1117–1132.

Lemullois M, Gounon P, Sandoz D. 1987. Relationships between cytokeratin filaments and centriolar derivatives during ciliogenesis in the quail oviduct. *Biol Cell* **61**: 39–49.

Lemullois M, Boisvieux-Ulrich E, Laine MC, Chailley B, Sandoz D. 1988. Development and functions of the cytoskeleton during ciliogenesis in metazoa. *Biol Cell* **63**: 195–208.

Liu Y, Pathak N, Kramer-Zucker A, Drummond IA. 2007. Notch signaling controls the differentiation of transporting epithelia and multiciliated cells in the zebrafish pronephros. *Development* **134**: 1111–1122.

Lu Q, Insinna C, Ott C, Stauffer J, Pintado PA, Rahajeng J, Baxa U, Walia V, Cuenca A, Hwang YS, et al. 2015. Early steps in primary cilium assembly require EHD1/EHD3-dependent ciliary vesicle formation. *Nat Cell Biol* **17**: 228–240.

Lukinavicius G, Lavogina D, Orpinell M, Umezawa K, Reymond L, Garin N, Gonczy P, Johnsson K. 2013. Selective chemical crosslinking reveals a Cep57-Cep63-Cep152 centrosomal complex. *Curr Biol* **23**: 265–270.

Ma M, Jiang YJ. 2007. Jagged2a-notch signaling mediates cell fate choice in the zebrafish pronephric duct. *PLoS Genet* **3**: e18.

Ma JY, Ou-Yang YC, Luo YB, Wang ZB, Hou Y, Han ZM, Liu Z, Schatten H, Sun QY. 2013. Cyclin O regulates germinal vesicle breakdown in mouse oocytes. *Biol Reprod* **88**: 110.

Ma L, Quigley I, Omran H, Kintner C. 2014. Multicilin drives centriole biogenesis via E2f proteins. *Genes Dev* **28**: 1461–1471.

Marcet B, Chevalier B, Luxardi G, Coraux C, Zaragosi LE, Cibois M, Robbe-Sermesant K, Jolly T, Cardinaud B, Moreilhon C, et al. 2011. Control of vertebrate multiciliogenesis by miR-449 through direct repression of the Delta/Notch pathway. *Nat Cell Biol* **13**: 693–699.

Matsuo M, Shimada A, Koshida S, Saga Y, Takeda H. 2013. The establishment of rotational polarity in the airway and ependymal cilia: Analysis with a novel cilium motility mutant mouse. *Am J Physiol Lung Cell Mol Physiol* **304**: L736–L745.

Menard D. 1995. Morphological studies of the developing human esophageal epithelium. *Microsc Res Tech* **31:** 215–225.

Mikrjukov K, Mylnikov A. 1998. The fine structure of a carnivorous multiflagellar protist, *Multicilia marina* Cienkowski, 1881 (flagellata incertae sedis). *Eur J Protistol* **34:** 391–401.

Mirzadeh Z, Han YG, Soriano-Navarro M, Garcia-Verdugo JM, Alvarez-Buylla A. 2010. Cilia organize ependymal planar polarity. *J Neurosci* **30:** 2600–2610.

Mitchell B, Jacobs R, Li J, Chien S, Kintner C. 2007. A positive feedback mechanism governs the polarity and motion of motile cilia. *Nature* **447:** 97–101.

Mitchell B, Stubbs JL, Huisman F, Taborek P, Yu C, Kintner C. 2009. The PCP pathway instructs the planar orientation of ciliated cells in the *Xenopus* larval skin. *Curr Biol* **19:** 924–929.

Miyatake K, Kusakabe M, Takahashi C, Nishida E. 2015. ERK7 regulates ciliogenesis by phosphorylating the actin regulator CapZIP in cooperation with Dishevelled. *Nat Commun* **6:** 6666.

Mizukami I, Gall J. 1966. Centriole replication. II: Sperm formation in the fern, *Marsilea*, and the cycad, *Zamia*. *J Cell Biol* **29:** 97–111.

Mori M, Mahoney JE, Stupnikov MR, Paez-Cortez JR, Szymaniak AD, Varelas X, Herrick DB, Schwob J, Zhang H, Cardoso WV. 2015. Notch3-Jagged signaling controls the pool of undifferentiated airway progenitors. *Development* **142:** 258–267.

Morimoto M, Liu Z, Cheng HT, Winters N, Bader D, Kopan R. 2010. Canonical Notch signaling in the developing lung is required for determination of arterial smooth muscle cells and selection of Clara versus ciliated cell fate. *J Cell Sci* **123:** 213–224.

Ohata S, Nakatani J, Herranz-Perez V, Cheng J, Belinson H, Inubushi T, Snider WD, Garcia-Verdugo JM, Wynshaw-Boris A, Alvarez-Buylla A. 2014. Loss of Dishevelleds disrupts planar polarity in ependymal motile cilia and results in hydrocephalus. *Neuron* **83:** 558–571.

Ohata S, Herranz-Perez V, Nakatani J, Boletta A, Garcia-Verdugo JM, Alvarez-Buylla A. 2015. Mechanosensory genes *Pkd1* and *Pkd2* contribute to the planar polarization of brain ventricular epithelium. *J Neurosci* **35:** 11153–11168.

Pan J, You Y, Huang T, Brody SL. 2007. RhoA-mediated apical actin enrichment is required for ciliogenesis and promoted by Foxj1. *J Cell Sci* **120:** 1868–1876.

Pardo-Saganta A, Law BM, Tata PR, Villoria J, Saez B, Mou H, Zhao R, Rajagopal J. 2015a. Injury induces direct lineage segregation of functionally distinct airway basal stem/progenitor cell subpopulations. *Cell Stem Cell* **16:** 184–197.

Pardo-Saganta A, Tata PR, Law BM, Saez B, Chow R, Prabhu M, Gridley T, Rajagopal J. 2015b. Parent stem cells can serve as niches for their daughter cells. *Nature* **523:** 597–601.

Park TJ, Haigo SL, Wallingford JB. 2006. Ciliogenesis defects in embryos lacking inturned or fuzzy function are associated with failure of planar cell polarity and Hedgehog signaling. *Nat Genet* **38:** 303–311.

Park TJ, Mitchell BJ, Abitua PB, Kintner C, Wallingford JB. 2008. Dishevelled controls apical docking and planar polarization of basal bodies in ciliated epithelial cells. *Nat Genet* **40:** 871–879.

Pefani DE, Dimaki M, Spella M, Karantzelis N, Mitsiki E, Kyrousi C, Symeonidou IE, Perrakis A, Taraviras S, Lygerou Z. 2011. Idas, a novel phylogenetically conserved geminin-related protein, binds to geminin and is required for cell cycle progression. *J Biol Chem* **286:** 23234–23246.

Popatia R, Haver K, Casey A. 2014. Primary ciliary dyskinesia: An update on new diagnostic modalities and review of the literature. *Pediatr Allergy Immunol Pulmonol* **27:** 51–59.

Ramsay RG, Gonda TJ. 2008. MYB function in normal and cancer cells. *Nat Rev Cancer* **8:** 523–534.

Rawlins EL, Okubo T, Xue Y, Brass DM, Auten RL, Hasegawa H, Wang F, Hogan BL. 2009. The role of Scgb1a1$^+$ Clara cells in the long-term maintenance and repair of lung airway, but not alveolar, epithelium. *Cell Stem Cell* **4:** 525–534.

Reed W, Avolio J, Satir P. 1984. The cytoskeleton of the apical border of the lateral cells of freshwater mussel gill: Structural integration of microtubule and actin filament-based organelles. *J Cell Sci* **68:** 1–33.

Reiter JF, Blacque OE, Leroux MR. 2012. The base of the cilium: Roles for transition fibres and the transition zone in ciliary formation, maintenance and compartmentalization. *EMBO Rep* **13:** 608–618.

Reverdin N, Gabbiani G, Kapanci Y. 1975. Actin in tracheobronchial ciliated epithelial cells. *Experientia* **31:** 1348–1350.

Rieger RM. 1981. Morphology of the Turbellaria at the ultrastructural level. *Hydrobiologia* **84:** 213–229.

Rink JC, Vu HT, Sanchez Alvarado A. 2011. The maintenance and regeneration of the planarian excretory system are regulated by EGFR signaling. *Development* **138:** 3769–3780.

Rock JR, Hogan BL. 2011. Epithelial progenitor cells in lung development, maintenance, repair, and disease. *Annu Rev Cell Dev Biol* **27:** 493–512.

Rohatgi R, Snell WJ. 2010. The ciliary membrane. *Curr Opin Cell Biol* **22:** 541–546.

Roig MB, Roset R, Ortet L, Balsiger NA, Anfosso A, Cabellos L, Garrido M, Alameda F, Brady HJ, Gil-Gomez G. 2009. Identification of a novel cyclin required for the intrinsic apoptosis pathway in lymphoid cells. *Cell Death Differ* **16:** 230–243.

Rompolas P, Patel-King RS, King SM. 2010. An outer arm dynein conformational switch is required for metachronal synchrony of motile cilia in planaria. *Mol Biol Cell* **21:** 3669–3679.

Rompolas P, Azimzadeh J, Marshall WF, King SM. 2013. Analysis of ciliary assembly and function in planaria. *Method Enzymol* **525:** 245–264.

Sadasivam S, DeCaprio JA. 2013. The DREAM complex: Master coordinator of cell cycle-dependent gene expression. *Nat Rev Cancer* **13:** 585–595.

Sandoz D, Boisvieux-Ulrich E. 1976. Ciliogenesis in the mucous cells of the quail oviduct. I: Ultrastructural study in the laying quail. *J Cell Biol* **71:** 449–459.

Sandoz D, Boisvieux-Ulrich E, Laugier C, Brard E. 1976. Ciliogenesis in the mucous cells of the quail oviduct. II: Hormonal control. *J Cell Biol* **71:** 460–471.

Sandoz D, Chailley B, Boisvieux-Ulrich E, Lemullois M, Laine MC, Bautista-Harris G. 1988. Organization and functions of cytoskeleton in metazoan ciliated cells. *Biol Cell* **63:** 183–193.

Schmidt KN, Kuhns S, Neuner A, Hub B, Zentgraf H, Pereira G. 2012. Cep164 mediates vesicular docking to the mother centriole during early steps of ciliogenesis. *J Cell Biol* **199:** 1083–1101.

Seixas C, Choi SY, Polgar N, Umberger NL, East MP, Zuo X, Moreiras H, Ghossoub R, Benmerah A, Kahn RA, et al. 2016. Arl13b and the exocyst interact synergistically in ciliogenesis. *Mol Biol Cell* **27:** 308–320.

Shah AS, Ben-Shahar Y, Moninger TP, Kline JN, Welsh MJ. 2009. Motile cilia of human airway epithelia are chemosensory. *Science* **325:** 1131–1134.

Shi D, Komatsu K, Hirao M, Toyooka Y, Koyama H, Tissir F, Goffinet M, Uemura T, Fujimori T. 2014. Celsr1 is required for the generation of polarity at multiple levels of the mouse oviduct. *Development* **141:** 4558–4568.

Shimada Y, Yonemura S, Ohkura H, Strutt D, Uemura T. 2006. Polarized transport of Frizzled along the planar microtubule arrays in *Drosophila* wing epithelium. *Dev Cell* **10:** 209–222.

Sir JH, Barr AR, Nicholas AK, Carvalho OP, Khurshid M, Sossick A, Reichelt S, D'Santos C, Woods CG, Gergely F. 2011. A primary microcephaly protein complex forms a ring around parental centrioles. *Nat Genet* **43:** 1147–1153.

Song R, Walentek P, Sponer N, Klimke A, Lee JS, Dixon G, Harland R, Wan Y, Lishko P, Lize M, et al. 2014. miR-34/449 miRNAs are required for motile ciliogenesis by repressing *cp110*. *Nature* **510:** 115–120.

Sorokin S. 1962. Centrioles and the formation of rudimentary cilia by fibroblasts and smooth muscle cells. *J Cell Biol* **15:** 363–377.

Sorokin SP. 1968. Reconstructions of centriole formation and ciliogenesis in mammalian lungs. *J Cell Sci* **3:** 207–230.

Spektor A, Tsang WY, Khoo D, Dynlacht BD. 2007. Cep97 and CP110 suppress a cilia assembly program. *Cell* **130:** 678–690.

Steinman RM. 1968. An electron microscopic study of ciliogenesis in developing epidermis and trachea in the embryo of *Xenopus laevis*. *Am J Anat* **122:** 19–55.

Stubbs JL, Vladar EK, Axelrod JD, Kintner C. 2012. Multicilin promotes centriole assembly and ciliogenesis during multiciliate cell differentiation. *Nat Cell Biol* **14:** 140–147.

Sung CH, Leroux MR. 2013. The roles of evolutionarily conserved functional modules in cilia-related trafficking. *Nat Cell Biol* **15:** 1387–1397.

Tamm S, Tamm SL. 1988. Development of macrociliary cells in Beroe. I: Actin bundles and centriole migration. *J Cell Sci* **89:** 67–80.

Tamm SL, Sonneborn TM, Dippell RV. 1975. The role of cortical orientation in the control of the direction of ciliary beat in paramecium. *J Cell Biol* **64:** 98–112.

Tan FE, Vladar EK, Ma L, Fuentealba LC, Hoh R, Espinoza FH, Axelrod JD, Alvarez-Buylla A, Stearns T, Kintner C, et al. 2013. Myb promotes centriole amplification and later steps of the multiciliogenesis program. *Development* **140:** 4277–4286.

Tateishi K, Yamazaki Y, Nishida T, Watanabe S, Kunimoto K, Ishikawa H, Tsukita S. 2013. Two appendages homologous between basal bodies and centrioles are formed using distinct Odf2 domains. *J Cell Biol* **203:** 417–425.

Thi-Kim Vu H, Rink JC, McKinney SA, McClain M, Lakshmanaperumal N, Alexander R, Sanchez Alvarado A. 2015. Stem cells and fluid flow drive cyst formation in an invertebrate excretory organ. *eLife* **4:** e07405.

Thomas B, Rutman A, Hirst RA, Haldar P, Wardlaw AJ, Bankart J, Brightling CE, O'Callaghan C. 2010. Ciliary dysfunction and ultrastructural abnormalities are features of severe asthma. *J Allergy Clin Immunol* **126:** 722–729.e2.

Tissir F, Qu Y, Montcouquiol M, Zhou L, Komatsu K, Shi D, Fujimori T, Labeau J, Tyteca D, Courtoy P, et al. 2010. Lack of cadherins Celsr2 and Celsr3 impairs ependymal ciliogenesis, leading to fatal hydrocephalus. *Nat Neurosci* **13:** 700–707.

Tsang WY, Dynlacht BD. 2013. CP110 and its network of partners coordinately regulate cilia assembly. *Cilia* **2:** 9.

Tsao PN, Vasconcelos M, Izvolsky KI, Qian J, Lu J, Cardoso WV. 2009. Notch signaling controls the balance of ciliated and secretory cell fates in developing airways. *Development* **136:** 2297–2307.

Turk E, Wills AA, Kwon T, Sedzinski J, Wallingford JB, Stearns T. 2015. Zeta-tubulin is a member of a conserved tubulin module and is a component of the centriolar basal foot in multiciliated cells. *Curr Biol* **25:** 2177–2183.

Tyler S. 1981. Development of cilia in embryos of the turbellarian Macrostomum. *Hydrobiologia* **84:** 231–239.

Vize PD, Carroll TJ, Wallingford JB. 2003. Induction, development, and physiology of the pronephric tubules. In *The kidney: From normal development to congenital disease* (ed. Vize PD, Woolf AS, Bard JBL), pp. 19–50. Academic, San Diego.

Vladar EK, Bayly RD, Sangoram AM, Scott MP, Axelrod JD. 2012. Microtubules enable the planar cell polarity of airway cilia. *Curr Biol* **22:** 2203–2212.

Walentek P, Bogusch S, Thumberger T, Vick P, Dubaissi E, Beyer T, Blum M, Schweickert A. 2014. A novel serotonin-secreting cell type regulates ciliary motility in the mucociliary epidermis of *Xenopus* tadpoles. *Development* **141:** 1526–1533.

Walentek P, Hagenlocher C, Beyer T, Muller C, Feistel K, Schweickert A, Harland RM, Blum M. 2015. ATP4 and ciliation in the neuroectoderm and endoderm of *Xenopus* embryos and tadpoles. *Data Brief* **4:** 22–31.

Wallmeier J, Al-Mutairi DA, Chen CT, Loges NT, Pennekamp P, Menchen T, Ma L, Shamseldin HE, Olbrich H, Dougherty GW, et al. 2014. Mutations in CCNO result in congenital mucociliary clearance disorder with reduced generation of multiple motile cilia. *Nat Genet* **46:** 646–651.

Wang L, Fu C, Fan H, Du T, Dong M, Chen Y, Jin Y, Zhou Y, Deng M, Gu A, et al. 2013. miR-34b regulates multiciliogenesis during organ formation in zebrafish. *Development* **140:** 2755–2764.

Cite this article as *Cold Spring Harb Perspect Biol* doi: 10.1101/cshperspect.a028233

Watson JK, Rulands S, Wilkinson AC, Wuidart A, Ousset M, Van Keymeulen A, Gottgens B, Blanpain C, Simons BD, Rawlins EL. 2015. Clonal dynamics reveal two distinct populations of basal cells in slow-turnover airway epithelium. *Cell Rep* **12**: 90–101.

Wei Q, Ling K, Hu J. 2015. The essential roles of transition fibers in the context of cilia. *Curr Opin Cell Biol* **35**: 98–105.

Werner ME, Mitchell BJ. 2012. Understanding ciliated epithelia: The power of *Xenopus*. *Genesis* **50**: 176–185.

Werner ME, Hwang P, Huisman F, Taborek P, Yu CC, Mitchell BJ. 2011. Actin and microtubules drive differential aspects of planar cell polarity in multiciliated cells. *J Cell Biol* **195**: 19–26.

Yasunaga T, Itoh K, Sokol SY. 2011. Regulation of basal body and ciliary functions by diversin. *Mech Dev* **128**: 376–386.

Yasunaga T, Hoff S, Schell C, Helmstadter M, Kretz O, Kuechlin S, Yakulov TA, Engel C, Muller B, Bensch R, et al. 2015. The polarity protein inturned links NPHP4 to Daam1 to control the subapical actin network in multiciliated cells. *J Cell Biol* **211**: 963–973.

Ying G, Avasthi P, Irwin M, Gerstner CD, Frederick JM, Lucero MT, Baehr W. 2014. Centrin 2 is required for mouse olfactory ciliary trafficking and development of ependymal cilia planar polarity. *J Neurosci* **34**: 6377–6388.

Yu X, Ng CP, Habacher H, Roy S. 2008. Foxj1 transcription factors are master regulators of the motile ciliogenic program. *Nat Genet* **40**: 1445–1453.

Zhao H, Zhu L, Zhu Y, Cao J, Li S, Huang Q, Xu T, Huang X, Yan X, Zhu X. 2013. The Cep63 paralogue Deup1 enables massive de novo centriole biogenesis for vertebrate multiciliogenesis. *Nat Cell Biol* **15**: 1434–1444.

Zhou F, Narasimhan V, Shboul M, Chong YL, Reversade B, Roy S. 2015. Gmnc is a master regulator of the multiciliated cell differentiation program. *Curr Biol* **25**: 3267–3273.

Zielinski B, Hara TJ. 1988. Morphological and physiological development of olfactory receptor cells in rainbow trout (*Salmo gairdneri*) embryos. *J Comp Neurol* **271**: 300–311.

Cilia in Left–Right Symmetry Breaking

Kyosuke Shinohara[1] and Hiroshi Hamada[2]

[1]Department of Biotechnology and Life Science, Tokyo University of Agriculture and Technology, Tokyo 184-8588, Japan

[2]Developmental Genetics Group, Graduate School of Frontier Biosciences, Osaka University, Osaka 565-0871, Japan

Correspondence: k_shino@cc.tuat.ac.jp; hamada@fbs.osaka-u.ac.jp

Visceral organs of vertebrates show left–right (L–R) asymmetry with regard to their position and morphology. Cilia play essential role in generating L–R asymmetry. A number of genes required for L–R asymmetry have now been identified in vertebrates, including human, many of which contribute to the formation and motility of cilia. In the mouse embryo, breaking of L–R symmetry occurs in the ventral node, where two types of cilia (motile and immotile) are present. Motile cilia are located at the central region of the node, and generate a leftward fluid flow. These motile cilia at the node are unique in that they rotate in the clockwise direction, unlike other immotile cilia such as airway cilia that show planar beating. The second type of cilia essential for L–R asymmetry is immotile cilia that are peripherally located immotile cilia. They sense a flow-dependent signal, which is either chemical or mechanical in nature. Although Ca^{2+} signaling is implicated in flow sensing, the precise mechanism remains unknown.

Visceral organs in all vertebrates are left–right (L–R) asymmetric with regards to their position, pattern, and size. Cilia, both motile and immotile in the mouse, play essential roles in L–R symmetry breaking. Four steps are required to generate such L–R asymmetric morphology in most of vertebrates including the mouse (Fig. 1): (1) Symmetry breaking by a leftward fluid flow (nodal flow) generated by the rotational movement of primary cilia at the node; (2) transmission of an asymmetric signal produced in or around the node (most likely Nodal activity) to the lateral plate mesoderm (LPM); (3) asymmetric expression of *Nodal* and the gene for its feedback inhibitor Lefty2 in the left LPM; and (4) situs-specifc organo-genesis as a result of asymmetric expression of *Pitx2*, which encodes a transcription factor induced by Nodal signaling.

CILIA-DRIVEN FLUID FLOW DETERMINES LEFT–RIGHT ASYMMETRY IN THE MOUSE EMBRYO

L–R asymmetry of the mouse embryo is determined in the node cavity at 8 days after fertilization (E8.0 mouse embryo). In 1994, Sulik et al. reported the node cell harbors single cilium in the E8.0 mouse embryo, but the role of this cilium had remained obscure (Fig. 2). In 1998, Nonaka et al. discovered that the cilium is motile and generates unidirectional fluid flow

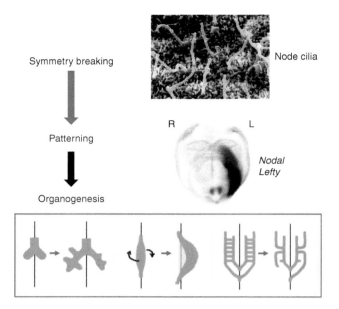

Figure 1. Three step for L−R asymmetry. Three steps that contribute to the generation of L−R asymmetry are shown: (1) symmetry breaking, (2) differential patterning of the lateral plate mesoderm (LPM), and (3) asymmetric organogenesis. Scanning microscopic view of node cilia is shown for the symmetry-breaking step, whereas left-sided expression of *Nodal* and *Lefty* in the lateral plate is shown for the differential patterning step. For asymmetric organogenesis, three different mechanisms that can give rise to asymmetric anatomical structures are illustrated: differential branching, directional looping, and one-sided regression.

(Nonaka et al. 1998; Tabin and Vogan 2003; Hirokawa et al. 2006). This leftward flow is generated by the rotational movement of cilia, which are solitary motile structures that contain microtubules with dynein arms (Hirokawa et al. 2006). In general, there are two types of motile cilia. The 9+2 type cilia exist in the airway, brain, and oviduct, containing nine peripheral doublet microtubules with dynein arms, one pair of single microtubules at the center of the axoneme (central-pair), and radial spokes that bridge the peripheral microtubules to the central microtubules (Fig. 2). These cilia show planar beating and generate directional fluid flow. In contrast, the 9+0 type motile cilia present in the node cavity of the mouse embryo possess nine doublet microtubules with dynein arms but lack the central microtubules and radial spokes (Nonaka et al. 1998; Takeda et al. 1999; Hirokawa et al. 2006; Shinohara et al. 2015; Odate et al. 2016). The driving force of ciliary bending is generated by sliding of dynein arms between the peripheral doublet microtubules of the axoneme (Summers et al. 1971; Fox et al. 1987). Motile cilia harbor two types of dynein arm complexes: outer and inner dynein arms. Dynein arm consists of heavy chain, intermediate chain, light chain of dynein, and accessory proteins (Inaba 2007). The outer dynein arm contains two or three types of heavy chains, whereas inner dynein arm consists of seven types of heavy chains in *Chlamydomonas* (Bui et al. 2008; Yagi et al. 2009). In transmission electron micrographs (TEMs) described previously, mouse node cilia seem to contain outer dynein arms with two heads, but the inner dynein arm was not identified (Takeda et al. 1999; Caspary et al. 2007; Shinohara et al. 2015).

Allelic mouse mutations *inversus viscerum* (*iv*) caused loss of motility of node cilia and left−right patterning defect in mice (Supp et al. 1997, 1999). Positional cloning identified its causative gene *Lrd*, which had a single amino acid substitution in the *iv* mutant allele at a conserved position of Lrd protein. Lrd/Dnah11 is thought to be a heavy-chain protein of outer

Cite this article as *Cold Spring Harb Perspect Biol* doi: 10.1101/cshperspect.a028282

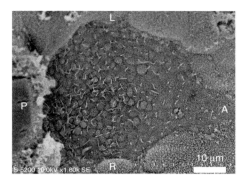

Figure 2. Node cilia in the E8.0 mouse embryo. Cells harbor one single cilium inside the node cavity in the E8.0 mouse embryo. The length of cilium is between 2 to 6 μm. Scale bar, 10 μm. A, Anterior; P, posterior; L, left; R, right.

dynein arm of motile cilia (Horani et al. 2014). Furthermore, mutation in DNAH5 (Dnah5), another putative heavy-chain protein of outer dynein arm, also causes left–right patterning defects in humans and mice (Ibanez-Tallon et al. 2002; Olbrich et al. 2002; Hornef et al. 2006). Although the role of inner dynein arm remains obscure, it seems that outer dynein arm plays a critical role in motility of mouse node cilia.

The direction of the flow is determined by a combination of two features of node cilia: their posterior tilt and clockwise rotation. Clockwise rotation of posteriorly tilted cilia thus generates a leftward effective stroke and rightward recovery stroke near the surface of the cell (Cartwright et al. 2004; Okada et al. 2005; Nonaka et al. 2005). Rotational movement of the cilia would generate slow rightward fluid flow near the cell surface and fast leftward flow in the middle of the node cavity (Cartwright et al. 2007). The precise mechanism of the tilt of node cilia is mentioned in the latter part of this review.

COMBINATION OF PLANAR CELL POLARITY AND ROTATIONAL MOVEMENT OF CILIA INDUCE LEFTWARD FLUID FLOW

Planar Cell Polarity Governs Posteriorly Tilted Rotational Axis of the Mouse Node Cilia

According to hydrodynamics, if the node cilia stand vertically to the apical surface of the node cells, they would generate the vertical flow inside the node cavity. However, the rotational axis of node cilia is actually tilted toward the posterior direction of the E8.0 mouse embryo (Nonaka et al. 2005; Okada et al. 2005; Hashimoto and Hamada 2010; Hashimoto et al. 2010; Song et al. 2010). Clockwise rotation of posteriorly tilted cilia generates a leftward fluid flow by effective stroke far from the surface of the cell and rightward recovery stroke near the surface (Cartwright et al. 2004). To create the tilted axis, the basal body position of node cilia has to be tightly regulated, because the apical surface of the node cell is curved/spherical (not flat) (Hashimoto and Hamada 2010).

How is the position of basal body determined? Recent studies have shown that planar cell polarity (PCP) signaling positions the basal body at the posterior side of the node cells (Hashimoto et al. 2010; Song et al. 2010). PCP is a mechanism that governs the cell orientation within an epithelial sheet in multicellular organisms (Lawrence et al. 2007). The large majority of PCP signaling components and pathways have been discovered in *Drosophila* and vertebrates. It is known that two pathways mainly control PCP signaling. One system consists of atypical Cadherin Dachsous, Fat, and Fourjointed. The other system consists of transmembrane proteins Frizzled, Celsr, Vangl, and membrane proteins Dishevelled and Prickle. In both systems, the core proteins show asymmetric localization at the apical membrane of the epithelium through a feedback loop (Lawrence et al. 2007). In E8.0 mouse embryo, it is suggested that Dishevelled (Hashimoto et al. 2010), Prickle (Antic et al. 2010), and Vangl (Antic et al. 2010; Song et al. 2010) are required for planar polarization of node cilia. A mouse genetics-based mosaic assay reveals that Dishevelled protein is localized to the posterior side of the apical membrane of node cells (Hashimoto et al. 2010). Vangl and Prickle seem to be localized to the anterior side (Fig. 3) (Antic et al. 2010). Vangl is critical also for left–right determination in zebrafish and *Xenopus* embryos (Antic et al. 2010; Borovina et al. 2010), suggesting that the role of PCP signaling in basal body positioning is evolutionarily conserved among vertebrates.

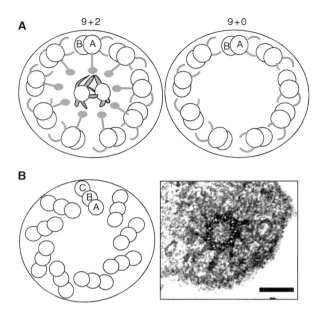

Figure 3. Ultrastructure of mouse motile cilia. (*A*) Two types of motile cilia in mammals. 9+2 type cilia have one central pair of microtubule and radial spokes at the center of the axoneme, whereas 9+0 type cilia do not contain any central structure. (*B*) Node cilia basal body. Scale bar, 200 nm. A, A-tubule; B, B-tubule; C, C-tubule.

Clockwise Rotation of the Mouse Node Cilia

Node cilia show the clockwise rotation, which is another reason why they can generate the leftward flow. However, how node cilia are able to sustain stable clockwise rotation has remained unknown. One can address this question into two essential questions: (1) Why do they rotate instead of beating planarly? and (2) Why do they rotate into the clockwise direction?

To address the first question, we should examine the role of regular arrangement of doublet microtubule found in mouse node cilia. A structural data-driven computer simulation of mouse node cilia suggests that the regular circular arrangement of doublet microtubules is essential for the stable rotational movement of node cilia (Shinohara et al. 2015). Furthermore, this work reported that absence of radial spoke is also critical for rotational movement of node cilia. To explore the role of radial spoke in mice, the investigators generated mice that lack the *Rsph4a* gene, an ortholog of Rsp4, which encodes the head of the radial spoke in *Chlamydomonas* (Shinohara et al. 2015). In *Rsph4a*

knockout mice, airway cilia showed clockwise rotation as well as mouse node cilia, suggesting that the radial spokes translate ciliary motion pattern from the clockwise rotation to the planar beating. Although the mechanism of switching of ciliary motion pattern remains unknown, this work proposes the mouse node cilia need to lose the radial spoke to acquire rotational movement. In the mouse embryo, the dorsoventral (DV) axis and the anteroposterior (AP) axis of the embryo are already established until E8.0. Described above, the node cells recognize the AP axis and move the basal body to the posterior side. Thus, the mouse embryo acquires the LR axis based on the AP axis (Hashimoto and Hamada 2010). The planar beating of ciliary motion makes unidirectional fluid flow in the airway, brain, and oviduct of mice (Fliegauf et al. 2007). If the node cilia harbored the planar beating, they have made fluid flow along the AP axis. To translate AP polarity to LR polarity, the clockwise rotation rather than the planar beating of ciliary motion pattern is indispensable and we speculate that this is the reason why the mouse node cilia have lost radial spokes during evolution. A very recent study reports the presence of minor population of cells in the mouse node containing cilia of 9+2 structure (Odate et al. 2016). It is tempting to propose that the mouse embryo might still be evolving toward the loss of 9+2 arrangement.

To address the second question, we should consider the origin of microtubule polarity in the cilia node. Cilia harbor doublet microtubules that consist of complete A-tubules and incomplete B-tubules (Fig. 4). Previous studies revealed that the outer dynein arms are always attached to the A-tubules via dynein-docking complex in *Chlamydomonas* flagella (Takada et al. 1994; Koutoulis et al. 1997). Thus, the clockwise rotation of node cilia is probably originated from the orientation of A-B-tubules and sliding direction of outer dynein arms. During ciliogenesis, it is thought that formation of the doublet microtubules occurred by the template of the basal body. Basal bodies contain nine pairs of triplet microtubules that consist of A-, B-, and C-tubules (basal bodies of mouse node cells in Fig. 4) (Li et al. 2012). Because the basal body originally acquires the arrangement of A-B-C-tubule in order, A-tubules locate at the most central position, whereas C-tubules locate at the most peripheral side of the basal body. Thus, the order of each tubule is originated from the order of triplet tubules found in the basal body. To understand how the triplets acquire this order, we should further examine and determine the fine structure of basal bodies found in motile cilia and in the mouse node cilia.

ACTION OF THE FLUID FLOW FOR L–R SYMMETRY BREAKING

Immotile Cilia Act as Flow Sensor

In addition to motile cilia that generate the flow, there are immotile cilia at the edge of the node (Yoshiba et al. 2012; Yoshiba and Hamada 2014). Those cells with immotile cilia are often called crown cells, and express three signaling molecules required for correct L–R patterning, Nodal, Gdf1, and Cerl2. Why are they immotile? Presumably these immotile cilia are immotile because they lack outer dynein arms like other immotile cilia, but this has not been addressed experimentally.

Genetic evidence has shown that the immotile cilia located at the periphery of the node are required to sense the flow (Fig. 5) (Yoshiba et al. 2012). Thus, $Kif3a^{-/-}$ mouse embryo that lacks all the cilia are unable to respond to the flow. However, when $Kif3a$ expression was rescued in crown cells by the crown-cell-specific enhancer, such embryos now contain immotile cilia and can respond to flow (strictly speaking, these results still leave a possibility that motile cilia act as flow sensors, because cilia rescued by the crown-cell-specific enhancer are mainly immotile but most likely contain a small number of motile cilia as well).

It is still unknown what exactly immotile cilia sense. They may sense a chemical molecule(s) that is transported toward the left side by the nodal flow and functions as the left-side determinant. Such a molecule should fulfill the

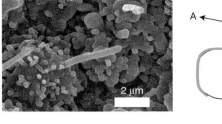
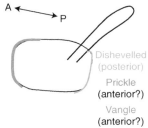

Figure 4. Planar cell polarity (PCP) of node cells. The PCP core proteins are localized at the apical membrane of the node cells: Dishevelled is localized at the posterior side of the apical membrane and controls positioning of the node cilia basal body. Scale bar, 2 μm.

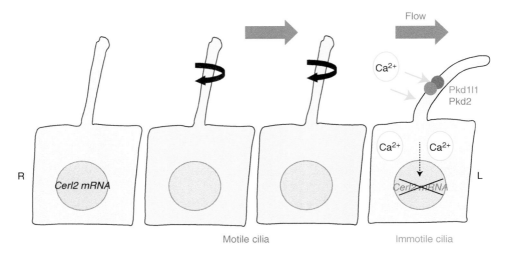

Figure 5. Immotile cilia sense the flow. Two types of cilia found in the node are shown: motile cilia located in the central region of the node generate the flow, while immotile cilia located peripherally (pink) sense the flow. Flow sensing involves a ciliary localized Pkd1l1–Pkd2 complex with Ca^{2+} channel activity. Flow-mediated signals lead to degradation of *Cerl2* mRNA. In this model, an immotile cilium on the *left* side is bent in response to the flow. However, this has not been confirmed in vivo.

following criteria: (1) It is an extracellular molecule that can be transported by the flow, (2) its absence would lead to the loss of left-sided *Nodal* expression in the lateral plate, and (3) its bilateral/ubiquitous presence would induce bilateral *Nodal* expression in the lateral plate. Up to now, there is no candidate molecule that satisfies all of these criteria. Alternatively, immotile cilia may sense the mechanical force produced by the flow. It has been suggested that some immotile cilia can sense mechanical forces. While this may represent a more likely mechanism, it is difficult to envisage how immotile cilia on both sides sense mechanical forces differently. Currently, there has been no direct evidence either supporting or discarding either possibility.

Involvement of Ca^{2+}

Although it is not clear what exactly the immotile cilia of crown cells sense, many lines of evidence suggest the involvement of Ca^{2+} in flow sensing. Most convincing evidence is the essential role of Pkd2 and Pkd1l1, a Ca^{2+} channel complex, in L–R symmetry breaking at the node (Pennekamp et al. 2002; Field et al. 2011;

Kamura et al. 2011). Pkd2 is required in crown cells, those cells with immotile cilia, acting upstream of *Cerl2*, a target gene of the fluid flow. In particular, Pkd2 protein is localized at the immotile cilia node and its ciliary localization is indeed required for correct L–R establishment (Yoshiba et al. 2012). Furthermore, treatment of mouse embryos with various Ca^{2+} signaling inhibitors perturbs L–R asymmetry at the node (Yoshiba et al. 2012). Inhibition of IP3 signaling in *Xenopus* embryos is known to randomize the L–R pattern (Hatayama et al. 2011).

When Ca^{2+} influx was directly examined with Fluo4- or Ca^{2+}-sensing fluorescent protein, L–R asymmetry in Ca^{2+} oscillation was indeed detected at the node of the mouse embryo (Takao et al. 2013) and at the Kupffer's vesicle of zebrafish embryo (Yuan et al. 2015).

Asymmetry in *Cerl2* mRNA as the Readout of the Flow Sensing

Crown cells respond to the fluid flow when their immotile cilia sense it. The most immediate gene that responds to the nodal flow is *Cerl2* (Schweickert et al. 2010; Shinohara et al. 2012). *Cerl2*, encoding a Nodal antagonist, is asymmetrically (R>L) expressed in crown cells at the node. In the mouse embryo, expression of *Cerl2* is initially symmetric, but it becomes R>L as the nodal flow increases its velocity, with expression on the left side being down-regulated (Kawasumi et al. 2011; Shinohara et al. 2012). L–R asymmetry in the level of *Cerl2* mRNA is determined not at the transcriptional level but at a posttranscriptional level (Nakamura et al. 2012). Thus, *Cerl2* mRNA in crown cells on the left side undergoes degradation via its 3' untranslated region. Preferential decay of *Cerl2* mRNA on the left is initiated by the leftward flow and further enhanced by the *Wnt–Cerl2* interlinked feedback loops, in which Wnt3 up-regulates *Wnt3* expression and promotes *Cerl2* mRNA decay, whereas Cerl2 promotes Wnt degradation.

L–R asymmetry of *Cerl2* mRNA at the node is the earliest molecular L–R asymmetry at least in the mouse embryo, and plays an essential role in symmetry breaking (Marques et al. 2004). Cerl2 protein inhibits Nodal activity, most likely by interacting with Nodal protein. The level of *Nodal* mRNA (and also Nodal protein) is bilaterally symmetric, but the R>L expression of *Cerl2* renders Nodal activity in crown cells higher on the left side (Kawasumi et al. 2011). R<L pattern of Nodal activity will eventually be transmitted to the lateral plate and induces left-sided *Nodal* expression there (Shiratori and Hamada 2006; Yoshiba and Hamada 2014).

PERSPECTIVES AND REMAINING QUESTIONS

We have reviewed the generation and sensing of fluid flow by cilia during L–R determination in the mouse embryo. The node cilia make leftward fluid flow driven by the axonemal dynein arm complex. PCP and motion pattern of cilia determine the direction of fluid flow via the positioning of the basal body and the clockwise rotation of cilia. Although previous studies provide a lot of knowledge on the mechanism of L–R determination, several essential questions still remain to be solved: (1) How cilia sense fluid flow during L–R determination: do motile cilia act as mechanosensors or chemosensors?

(2) How the node cilia rotate in the clockwise direction? (3) What is the positional cue that polarizes node cells along the anteroposterior axis? We expect that advanced technologies, including CRISPR-mediated genetics, live imaging, and electron microscopy, will address these questions in the future.

ACKNOWLEDGMENTS

We thank T. Nishida and T. Hasegawa for their help in obtaining SEMs and TEMs, and members of the Hiroshi Hamada Laboratory for fruitful discussions.

REFERENCES

Antic D, Stubbs JL, Suyama K, Kintner C, Scott MP, Axelrod JD. 2010. Planar cell polarity enables posterior localization of nodal cilia and left–right axis determination during mouse and *Xenopus* embryogenesis. *PLoS ONE* **5:** e899.

Borovina A, Superina S, Voskas D, Ciruna B. 2010. Vangl2 directs the posterior tilting and asymmetric localization of motile primary cilia. *Nat Cell Biol* **12:** 407–412.

Bui KH, Sakakibara H, Movassagh T, Oiwa K, Ishikawa T. 2008. Molecular architecture of inner dynein arms in situ in *Chlamydomonas reinhardtii* flagella. *J Cell Biol* **183:** 923–932.

Cartwright JHE, Piro O, Tuval I. 2004. Fluid-dynamical basis of the embryonic development of left–right asymmetry in vertebrates. *Proc Natl Acad Sci* **101:** 7234–7239.

Cartwright JHE, Piro N, Piro O, Tuval I. 2007. Embryonic nodal flow and the dynamics of nodal vesicular parcels. *J R Soc Interface* **4:** 49–55.

Caspary T, Larkins CE, Anderson KV. 2007. The graded response to Sonic Hedgehog depends on cilia architecture. *Dev Cell* **12:** 767–778.

Field S, Riley KL, Grimes DT, Hilton H, Simon M, Powles-Glover N, Siggers P, Bogani D, Greenfield A, Norris DP. 2011. Pkd1l1 establishes left–right asymmetry and physically interacts with Pkd2. *Development* **138:** 1131–1142.

Fliegauf M, Benzing T, Omran H. 2007. When cilia go bad: Cilia defects and ciliopathies. *Nat Rev Mol Cell Biol* **8:** 880–893.

Fox LA, Sale WS. 1987. Direction of force generated by the inner row of dynein arms on flagellar microtubules. *J Cell Biol* **105:** 1781–1787.

Hashimoto M, Hamada H. 2010. Translation of anterior–posterior polarity into left–right polarity in the mouse embryo. *Curr Opin Genet Dev* **20:** 433–437.

Hashimoto M, Shinohara K, Wang J, Ikeuchi S, Yoshiba S, Meno C, Nonaka S, Takada S, Hatta K, Wynshaw-Boris A, et al. 2010. Planar polarization of node cells determines the rotational axis of the node cilia. *Nat Cell Biol* **12**: 170–176.

Hatayama M, Mikoshiba K, Aruga J. 2011. IP3 signaling is required for cilia formation and left–right body axis determination in *Xenopus* embryos. *Biochem Biophys Res Commun* **410**: 520–524.

Hirokawa N, Tanaka Y, Okada Y, Takeda S. 2006. Nodal flow and the generation of left–right asymmetry. *Cell* **125**: 33–45.

Horani A, Brody SL, Ferkol TW. 2014. Picking up speed: Advances in the genetics of primary ciliary dyskinesia. *Pediatr Res* **75**: 158–164.

Hornef N, Olbrich H, Horvath J, Zariwala MA, Fliegauf M, Loges NT, Wildhaber J, Noone PG, Kennedy M, Antonarakis SE, et al. DNAH5 mutations are a common cause of primary ciliary dyskinesia with outer dynein arm defects. *Am J Resp Crit Med* **174**: 120–126.

Ibanez-Tallon I, Gorokhova S, Heintz N. 2002. Loss of function of axonemal dynein Mdnah5 causes primary ciliary dyskinesia and hydrocephalus. *Hum Mol Genet* **11**: 715–721.

Inaba K. 2007. Molecular basis of sperm flagellar axonemes. Structural and evolutionary aspects. *Ann NY Acad Sci* **1101**: 506–526.

Kamura K, Kobayashi D, Uehara Y, Koshida S, Iijima N, Kudo A, Yokoyama T, Takeda H. 2011. Pkd1l1 complexes with Pkd2 on motile cilia and functions to establish the left–right axis. *Development* **138**: 1121–1129.

Kawasumi A, Nakamura T, Iwai N, Yashiro K, Saijoh Y, Belo JA, Shiratori H, Hamada H. 2011. Left–right asymmetry in the level of active Nodal protein produced in the node is translated into left–right asymmetry in the lateral plate of mouse embryos. *Dev Biol* **353**: 321–330.

Koutoulis A, Pazour GJ, Wilkerson CG, Inaba K, Sheng H, Takada S, Witman GB. 1997. The *Chlamydomonas reinhardtii ODA3* gene encodes a protein of the outer dynein arm docking complex. *J Cell Biol* **137**: 1069–1080.

Lawrence PA, Struhl G, Casal J. 2007. Planar cell polarity: One or two pathways? *Nat Rev Genet* **8**: 555–563

Li S, Fernandez JJ, Marshall WF, Agard DA. 2012. Three-dimensional structure of basal body triplet revealed by electron cryo-tomography. *EMBO J* **31**: 552–562.

Marques S, Borges AC, Silva AC, Freitas S, Cordenonsi M, Belo JA. 2004. The activity of the Nodal antagonist Cerl-2 in the mouse node is required for correct L/R body axis. *Genes Dev* **18**: 2342–2347.

Nakamura T, Saito D, Kawasumi A, Shinohara K, Asai Y, Takaoka K, Dong F, Takamatsu A, Belo JA, Mochizuki A, et al. 2012. Fluid flow and interlinked feedback loops establish left–right asymmetric decay of *Cerl2* mRNA. *Nat Commun* **3**: 1322.

Nonaka S, Tanaka Y, Okada Y, Takeda S, Harada A, Kanai Y, Kido M, Hirokawa N. 1998. Randomization of left–right asymmetry due to loss of nodal cilia generating leftward flow of extraembryonic fluid in mice lacking KIF3B motor protein. *Cell* **95**: 829–837.

Nonaka S, Yoshiba S, Watanabe D, Ikeuchi S, Goto T, Marshall WF, Hamada H. 2005. De novo formation of left–right asymmetry by posterior tilt of nodal cilia. *PLoS Biol* **3**: e268.

Odate T, Takeda S, Narita K, Kawahara T. 2016. 9+0 and 9+2 cilia are randomly dispersed in the mouse node. *Microscopy* **65**: 119–126.

Okada Y, Takeda S, Tanaka Y, Belmonte JC, Hirokawa N. 2005. Mechanism of nodal flow: A conserved symmetry breaking event in left–right axis determination. *Cell* **121**: 633–644.

Olbrich H, Haffner K, Kispert A, Volkel A, Volz A, Sasmaz G, Reinhardt R, Hennig S, Lehrach H, Konietzko N, et al. 2002. Mutations in *DNAH5* cause primary ciliary dyskinesia and randomization of left–right and asymmetry. *Nat Genet* **30**: 143–144.

Pennekamp P, Karcher C, Fischer A, Schweickert A, Skryabin B, Horst J, Blum M, Dworniczak B. 2002. The ion channel polycystin-2 is required for left–right axis determination in mice. *Curr Biol* **12**: 938–943.

Schweickert A, Vick P, Getwan M, Weber T, Schneider I, Eberhardt M, Beyer T, Pachur A, Blum M. 2010. The nodal inhibitor Coco is a critical target of leftward flow in *Xenopus*. *Curr Biol* **20**: 738–743.

Shinohara K, Kawasumi A, Takamatsu A, Yoshiba S, Botilde Y, Motoyama N, Reith W, Durand B, Shiratori H, Hamada H. 2012. Two rotating cilia in the node cavity are sufficient to break left–right symmetry in the mouse embryo. *Nat Commun* **3**: 622.

Shinohara K, Chen D, Nishida T, Misaki K, Yonemura S, Hamada H. 2015. Absence of radial spokes in mouse node cilia is required for rotational movement but confers ultrastructural instability as a trade-off. *Dev Cell* **35**: 236–246.

Shiratori H, Hamada H. 2006. The left–right axis in the mouse: From origin to morphology. *Development* **133**: 2095–2104.

Song H, Hu J, Chen W, Elliott G, Andre P, Gao B, Yang Y. 2010. Planar cell polarity breaks bilateral symmetry by controlling ciliary positioning. *Nature* **466**: 378–382.

Sulik K, Dehart DB, Inagaki T, Carson JL, Vrablic T, Gesteland K, Schoenwolf GC. 1994. Morphogenesis of the murine node and notochordal plate. *Dev Dyn* **201**: 260–278.

Summers KE, Gibbons IR. 1971. Adenosine triphosphate-induced sliding of tubules in Trypsin-treated flagella of sea-urchin sperm. *Proc Natl Acad Sci* **68**: 3092–3096.

Supp DM, Witte DP, Potter SS, Brueckner M. 1997. Mutation of an axonemal dynein affects left–right asymmetry in inversus viscerum mice. *Nature* **389**: 963–966.

Supp DM, Brueckner M, Kuehn MR, Witte DP, Lowe LA, McGrath J, Corrales J, Potter SS. 1999. Targeted deletion of the ATP binding domain of left–right dynein confirms its role in specifying development of left–right asymmetries. *Development* **126**: 5495–5504.

Tabin CJ, Vogan KJ. 2003. A two-cilia model for vertebrate left–right axis specification. *Gene Dev* **17**: 1–6.

Takada S, Kamiya R. 1994. Functional reconstitution of *Chlamydomonas* outer dynein arms from α-β and γ subunits: Requirement of a third factor. *J Cell Biol* **126**: 737–745.

Takao D, Nemoto T, Abe T, Kiyonari H, Kajiura-Kobayashi H, Shiratori H, Nonaka S. 2013. Asymmetric distribution of dynamic calcium signals in the node of mouse embryo during left–right axis formation. *Dev Biol* **376:** 23–30.

Takeda S, Yonekawa Y, Tanaka Y, Okada Y, Nonaka S, Hirokawa N. 1999. Left–right asymmetry and kinesin superfamily protein KIF3A: New insights in determination of laterality and mesoderm induction by *kif3*$^{-/-}$ mice analysis. *J Cell Biol* **145:** 825–836.

Yagi A, Uematsu K, Liu Z, Kamiya R. 2009. Identification of dyneins that localize exclusively to the proximal portion of *Chlamydomonas* flagella. *J Cell Sci* **122:** 1306–1314.

Yoshiba S, Hamada H. 2014. Roles of cilia, fluid flow, and Ca^{2+} signaling in breaking of left-right symmetry. *Trends Genet* **30:** 10–17.

Yoshiba S, Shiratori H, Kuo IY, Kawasumi A, Shinohara K, Nonaka S, Asai Y, Sasaki G, Belo JA, Sasaki H, et al. 2012. Cilia at the node of mouse embryos sense fluid flow for left–right determination via Pkd2. *Science* **338:** 226–231.

Yuan SL, Zhao L, Brueckner M, Sun ZX. 2015. Intraciliary calcium oscillations initiate vertebrate left–right asymmetry. *Curr Biol* **25:** 556–567.

Ciliopathies

Daniela A. Braun and Friedhelm Hildebrandt

Division of Nephrology, Harvard Medical School, Boston Children's Hospital,
Boston, Massachusetts 02115

Correspondence: Friedhelm.hildebrandt@childrens.harvard.edu

Nephronophthisis-related ciliopathies (NPHP-RC) are a group of inherited diseases that affect genes encoding proteins that localize to primary cilia or centrosomes. With few exceptions, ciliopathies are inherited in an autosomal recessive manner, and affected individuals manifest early during childhood or adolescence. NPHP-RC are genetically very heterogeneous, and, currently, mutations in more than 90 genes have been described as single-gene causes. The phenotypes of NPHP-RC are very diverse, and include cystic-fibrotic kidney disease, brain developmental defects, retinal degeneration, skeletal deformities, facial dimorphism, and, in some cases, laterality defects, and congenital heart disease. Mutations in the same gene can give rise to diverse phenotypes depending on the mutated allele. At the same time, there is broad phenotypic overlap between different monogenic genes. The identification of monogenic causes of ciliopathies has furthered the understanding of molecular mechanism and cellular pathways involved in the pathogenesis.

Cilia are organelles that are present on the apical surface of almost every cell type in various tissues and organs. They are involved in a variety of cellular functions such as planar cell polarity, cell-cycle regulation and mechanosensation. Furthermore, cilia integrate multiple signaling pathways that are of critical importance for vertebrate development and organ differentiation. Hence, ciliary dysfunction gives rise to a wide spectrum of human disease phenotypes involving various organ systems. Cilia can be categorized as motile cilia and immotile cilia, also known as primary cilia. Motile cilia are present on respiratory epithelial cells, ependymal cells of cerebrospinal fluid spaces, sperm cells, and cells of the embryonic node during development. Dysfunction of motile cilia results in the human phenotype of primary ciliary dyskinesia and Kartagener syndrome (OMIM #244400), which is characterized by impaired mucociliary clearance resulting in recurrent infections of the upper respiratory tract, recurrent pneumonia, and progressive destruction of functional respiratory tissue. Additional disease symptoms include heterotaxy, congenital heart disease, asplenia, infertility, and *situs inversus* in 50% of patients (Leigh et al. 2009). Disruption of primary cilia has first been linked to autosomal-recessive and autosomal-dominant polycystic kidney diseases, which are discussed in more detail in Ma et al. (2016). In contrast to polycystic kidney disease that typically presents with enlarged kidneys and massive cysts, the second group of cilia-related cystic kidney diseases, the group of nephronophthisis-related ciliopathies (NPHP-RC), rather presents with shrunken or normal-sized fibrotic kidneys and small cysts at the corticomedullary junction. The renal phenotype of NPHP frequently occurs in a syndromic manner and is accompanied by anomalies in other organ systems, specifically retinal degeneration, cerebellar vermis hypoplasia, hepatic fibrosis, skeletal anomalies, ectodermal dysplasia, brain malformations, and neurological impairment. Frequently, a broad phenotypic spectrum is caused by mutations in the same monogenic gene on an allelic basis (Table 1; Fig. 1).

NEPHRONOPHTHISIS AND RELATED DISORDERS

Nephronophthisis (NPHP) is an inherited disease that represents one of the most frequent monogenic causes of end-stage renal failure in children and young adults. NPHP is inherited in an autosomal-recessive manner, and currently mutations in up to 90 genes have been identified as disease causing. Depending on the composition of the examined cohort, mutations in these known monogenic genes account for up to 63% of cases (Braun et al. 2015). The first genetic cause of NPHP, the gene NPHP1, which encodes the protein nephrocystin was described in 1997 (Hildebrandt et al. 1997a; Saunier et al. 1997). It later became apparent that homozygous deletions in the NPHP1 gene are the most frequent cause of NPHP and that mutations in this gene alone account for 20%–25% of all cases (Halbritter et al. 2013b). Each of the subsequently identified monogenic genes accounts only for a small fraction of affected individuals (Halbritter et al. 2013b). The initial clinical presentation of NPHP is typically mild, and an increase in serum-creatinine, as an indicator of impaired renal function, is typically not noted before an average age of 9 years (Gretz 1989). Clinical symptoms are polyuria with secondary enuresis, polydipsia with regular fluid intake at night, anemia, and growth retardation (Kleinknecht 1989). Based on the age of onset, NPHP is further categorized into infantile (Gagnadoux et al. 1989), juvenile (Hildebrandt et al. 1992), and adolescent (Omran et al. 2000) NPHP, in which the median onset

of clinical symptoms is at an age of 1, 13, and 15 years, respectively. In contrast to other renal diseases, patients usually do not develop arterial hypertension until the renal function is severely impaired. The clinical diagnosis of NPHP is typically based on renal ultrasound presentation with normal or small-sized kidneys, increased renal echogenicity, and loss of corticomedullary differentiation. Renal cysts in NPHP are not a necessary diagnostic criterion, and, if present, are rare, remain small, and are typically located in the corticomedullary junction region. This is in contrast to autosomal-dominant or -recessive polycystic kidney disease in which cysts are large and arise from all regions of the kidney. Hallmarks of NPHP in renal histology are thickening and disintegration of the tubular basement membrane, atrophy of renal tubular structure, and tubulointerstitial fibrosis. The prevalence of NPHP does not show regional clustering or gender predisposition (Kleinknecht 1989). Patients with NPHP have been described worldwide and in all ethnicities (Kleinknecht 1989). The incidence of NPHP has been reported as nine patients/8.3 million in the United States (Potter et al. 1980) or as one in 50,000 live births in Canada (Waldherr et al. 1982; Pistor et al. 1985). A North American study assessing causes of end-stage renal failure in pediatric patients estimated that NPHP-related ciliopathies account for ~5% of all cases (Avner 1994; Warady et al. 1997).

Because of shared features in renal histology, namely, cysts that primarily arise from the corticomedullary region and tubulointerstitial fibrosis, NPHP was previously grouped with medullary–cystic kidney disease (MCKD) under the term "NPHP–MCKD complex." However, they are now considered distinct disease entities because of differences in mode of inheritance, age of onset, extrarenal involvement, and molecular disease cause. MCKD is an autosomal-dominant disease that presents with adult-onset chronic kidney disease and renal salt wasting. In contrast to NPHP, end-stage renal failure typically occurs around the sixth decade of life (Wolf et al. 2004). MCKD does not present as a syndromic disease; the only described extrarenal associations are hyperuricemia and

Table 1. Phenotypic spectrum of 92 monogenic genes of NPHP-RC

Nephronopthisis (NPHP), Senior–Loken syndrome (SLS), Joubert syndrome (JBTS), Meckel–Gruber syndrome (MKS)

OMIM disease	Gene symbol (first description)	Alias	Mode of inheritance	Human disease phenotype								Subcellular localization
				Kidney	Eye	Brain	Liver	Skeletal anomalies	Laterality defects	Congenital heart disease	Other/ syndromic	
NPHP1/ SLS1/ JBTS4	*NPHP1* (Hildebrandt et al. 1997a; Saunier et al. 1997)	Nephrocystin	AR	NPHP, juvenile	RP, OMA	CVH, rare						TZ
NPHP2	*INVS* (Otto et al. 2003b)	Inversin	AR	NPHP, infantile	RP		LF, BDP		SI		PH, OH	TZ
NPHP3/ MKS7/ RHPD1	*NPHP3* (Olbrich et al. 2003)	NPHP3	AR	NPHP	RP		LF, LC		SI	CHD	MPD, MKS	TZ
NPHP4/ SLS4	*NPHP4* (Mollet et al. 2002; Otto et al. 2002)	NPHP4, nephroretinin	AR	NPHP	RP, OMA		LF, BDP		HTX	CHD		BB, CA
SLS5	*IQCB1* (Otto et al. 2005)	NPHP5, IQCB1	AR	NPHP	RP (all cases)							TZ, BB
NPHP6/ SLS6/ JBTS5/ MKS4/ BBS14	*CEP290* (Sayer et al. 2006; Valente et al. 2006)	CEP290, NPHP6, BBS14	AR	NPHP	RP	CVH, CBD	LF, BDP	PD		VSD, rare	MKS, BBS	BB
NPHP7	*GLIS2* (Attanasio et al. 2007)	NPHP7, GLIS2	AR	NPHP								CA
NPHP8/ JBTS7/ MKS5/ COACH	*RPGRIP1L* (Delous et al. 2007)	NPHP8, RPGRIP1L	AR	NPHP	RP, OMA, CB	CVH, CBD	LF, BDP	PD			MKS, BBS	TZ
NPHP9	*NEK8* (Otto et al. 2008)	NPHP9, NEK8	AR	NPHP		ID	LF			CHD	MKS	TZ
SLS7/BBS16	*SDCCAG8* (Otto et al. 2010)	SDCCAG8, NPHP10	AR	NPHP	RP						BBS	BB

Continued

Table 1. *Continued*

OMIM disease	Gene symbol (first description)	Alias	Mode of inheritance	Human disease phenotype								Subcellular localization
				Kidney	Eye	Brain	Liver	Skeletal anomalies	Laterality defects	Congenital heart disease	Other/ syndromic	
NPHP11/ JBTS6/ MKS3/ COACH	*TMEM67* (Smith et al. 2006)	Meckelin, TMEM67	AR	NPHP	RP, OMA, CB	CVH	LF, BDP	PD			MKS	TZ
NPHP12/ JBTS11/ SRTD4	*TTC21B* (Davis et al. 2011)	TTC21B, NPHP12	AR	NPHP		CVH		JATD			MKS	CA
NPHP13/ SLS8/ CED4/ SRTD5	*WDR19* (Bredrup et al. 2011)	WDR19, IFT144	AR	NPHP	RP		LF	JATD, CED			CED	CA
NPHP14/ JBTS19	*ZNF423* (Chaki et al. 2012)	NPHP14, ZNF423	AR/AD	NPHP	RP	CVH			SI			
NPHP15	*CEP164* (Chaki et al. 2012)	NPHP15, CEP164	AR	NPHP	RP	CVH, ID	LF	PD			OB	BB
NPHP16	*ANKS6* (Hoff et al. 2013)	NPHP16, ANKS6	AR	NPHP			LF		SI	CHD		TZ
NPHP17/ SRTD10	*IFT172* (Halbritter et al. 2013a)	IFT172	AR	NPHP	RP	CVH, ID	LF	SRTD, PD, CSE		VSD	OB	CA
NPHP18	*CEP83* (Failler et al. 2014)	CEP83, CCDC41	AR	NPHP	RP	ID, HC	LF				OB	BB
NPHP19	*DCDC2* (Schueler et al. 2015)	DCDC2	AR	NPHP			LF					CA
IFT81	*IFT81* (Perrault et al. 2015)	IFT81	AR	NPHP	RP	CVH		PD				CA
SLS9	*IFT54* (Bizet et al. 2015)	TRAF3IP1	AR	NPHP	RP	ID		BD			OB	CA, BB
JBTS1	*INPP5E* (Bielas et al. 2009)	INPP5E	AR	NPHP, rare	RP, OMA, CB	CVH, CBD, ID	LF	PD, rare			MORM	CA

Cite this article as *Cold Spring Harb Perspect Biol* doi: 10.1101/cshperspect.a028191

Continued

JBTS2/MKS2	*TMEM216* (Valente et al. 2010)	AR	NPHP	RP, OMA, CB	CVH, ID, CBD, HC		PD	MKS	TZ
JBTS3	*AHI1*, Jouberin (Ferland et al. 2004)	AR	NPHP, rare	RP, OMA	CVH, ID				
JBTS8	*ARL13B* (Cantagrel et al. 2008)	AR			CVH, ID				CA
JBTS9/MKS6	*CC2D2A* (Gorden et al. 2008)	AR		RP	CVH, ID	LF		MKS	TZ
JBTS12	*KIF7* (Putoux et al. 2011)	AR			CBD, ID, CVH	VSD	PD, FD	ACS, BBS, HLS	CA
JBTS13	*TECT1* (Garcia-Gonzalo et al. 2011)	AR			CVH		Abnormal limbs		TZ
JBTS14	*TMEM237* (Huang et al. 2011)	AR	NPHP	OpA, CB, Nys	CVH, HC				TZ
JBTS15	*CEP41* (Lee et al. 2012a)	AR		OMA, RP	CVH, ID		PD		BB
JBTS16	*TMEM138* (Lee et al. 2012b)	AR	NPHP, rare	OMA, CB	CVH		PD		TZ
JBTS17/OFD6	*C5ORF42* (Srour et al. 2012b)	AR	RC	OMA, RP	CVH, ID		PD, FD		–
JBTS18/OFD4	*TCTN3*, C10orf61 (Thomas 2012)	AR	RC	OMA	CVH, OE, ID	LF, BDP	PD, FD, Abnormal limbs		TZ
JBTS20/OFD3/MKS11	*TMEM231* (Srour et al. 2012a)	AR	RC	OMA, RP	CVH, ID, OE		FD, PD	MKS	BB
JBTS21	*CSPP1* (Akizu et al. 2014)	AR	NPHP	OMA, RP	CVH, CBD, ID	LF	JATD	SD, PH	BB
JBTS22	*PDE6D* (Thomas et al. 2014)	AR	DK	CB, MO, RP	CVH, CBD, ID		FD, PD		BB
JBTS23/SRTD14	*KIAA0586*, TALPID3 (Alby et al. 2015; Bachmann-Gagescu et al. 2015)	AR		OMA, CB	CVH, HC, CBD	ASD	PD, FD, SRTD	HLS, PH	BB

Table 1. *Continued*

OMIM disease	Gene symbol (first description)	Alias	Mode of inheritance	Human disease phenotype								Subcellular localization
				Kidney	Eye	Brain	Liver	Skeletal anomalies	Laterality defects	Congenital heart disease	Other/syndromic	
JBTS24/MKS8	TCTN2 (Sang et al. 2011)	TECTONIC2	AR	DK	Nys, MO	CVH, ID, CBD	LF; BDP	PD			MKS	TZ
JBTS25	CEP104 (Srour et al. 2015)	CEP104	AR	NPHP	OMA, RP	CVH, ID, CBD		Abnormal limbs, PD				BB, ciliary tip
JBTS26	KIAA0556 (Sanders et al. 2015)	KIAA0556, KATNIP	AR			CVH, ID		FD				Ciliary base
MKS9	B9D1 (Hopp et al. 2011)	B9D1	AR	RC, DK		OE		Abnormal limbs				TZ
MKS10	B9D2 (Dowdle et al. 2011)	B9D2	AR	RC		OE, AE	DPM, LF	PD				BB, TZ
MKS12	KIF14 (Filges et al. 2014)	KIF14	AR	DK, RA		MiC, CVH, CBD		Abnormal limbs, FD				CA
n/a	TMEM107 (Shaheen et al. 2015a; Lambacher et al. 2016; Shylo et al. 2016)			DK	RP	CVH, ID, OE		FD, PD			JBTS, OFD, MKS	TZ
Bardet–Biedl syndrome (BBS)												
BBS1	BBS1 (Mykytyn et al. 2002)	BBS1	AR/DR	RC, DK	RP	ID	LF, rare	PD, BD, FD			OB	BBSome
BBS2	BBS2 (Katsanis et al. 2001a; Nishimura et al. 2001)	BBS2	AR	RC, DK	RP	ID	LF, rare	PD, BD, FD			OB	BBSome
BBS3	ARL6 (Fan et al. 2004)	ARL6/BBS3	AR/DR	RC, DK	RP	ID	LF, rare	PD, BD, FD			OB	BBSome
BBS4	BBS4 (Katsanis et al. 2002)	BBS4	AR	RC, DK	RP	ID	LF, rare	PD, BD, FD			OB	BBSome
BBS5	BBS5 (Li et al. 2004)	BBS5	AR	RC, DK	RP	ID	LF, rare	PD, BD, FD			OB	BBSome

Cite this article as *Cold Spring Harb Perspect Biol* doi: 10.1101/cshperspect.a028191

Locus	Gene (reference)	Gene										
BBS6/MKKS	MKKS (Katsanis et al. 2000; Slavotinek et al. 2000)	BBS6/MKKS	AR	RC, CAKUT	RP	ID	LF, rare	PD, BD, FD, congenital dislocation of hips		CHD	MKKS, OB, HL	BBSome
BBS7	BBS7 (Badano et al. 2003)	BBS7	AR	RC, DK	RP	ID	LF, rare	PD, BD, FD			OB	BBSome
BBS8	TTC8 (Ansley et al. 2003)	BBS8/TTC8	AR	RC, DK	RP	ID	LF, rare	PD, BD, FD			OB	BBSome
BBS9	PTHB1 (Nishimura et al. 2005)	PTHB1/BBS9	AR	RC, DK	RP	ID	LF, rare	PD, BD, FD			OB	BBSome
BBS10	BBS10 (Stoetzel et al. 2006)	BBS10/C12orf58	AR	RC, DK	RP	ID	LF, rare	PD, BD, FD			OB	BBSome
BBS11	TRIM32 (Chiang et al. 2006)	BBS11/TRIM32	AR	RC, DK	RP	ID	LF, rare	PD, BD, FD			LGMD2H, OB	-
BBS12	BBS12 (Stoetzel et al. 2007)	C4orf24	AR	RC, DK	RP	ID	LF, rare	PD, BD, FD			OB	BBSome
BBS13/MKS1	MKS1 (Kyttala et al. 2006; Leitch et al. 2008)	MKS1/BBS13	AR	RC, DK, RA	RP, CB, MO	ID, MiC, OE, AA	LE, BDP	PD, BD, FD		CHD	OB, MKS, PH	BBSome
BBS15	WDPCP (Kim et al. 2010)	WDPCP/BBS15	AR	RC, DK	RP	ID	LF, rare	PD, BD, FD		CHD	OB	BBSome
BBS17	LZTFL1 (Marion et al. 2012)	LZTFL1/BBS17	AR	RC, DK	RP	ID	LF, rare	PD, BD, FD	SI, HTX		OB	BBSome
BBS18	BBIP1 (Scheidecker et al. 2014)	BBIP1/BBS18	AR	RC, DK	RP	ID	LF, rare	PD, BD, FD			OB	BBSome
BBS19	IFT27 (Aldahmesh et al. 2014)	IFT27/RABL4	AR	RC, DK	RP	ID	LF, rare	PD, BD, FD			OB	BBSome
Skeletal ciliopathies: oral-facial-digital syndrome (OFD), cranioectodermal dysplasia (CED), short-rib thoracic dysplasia (SRTD)												
OFD1/JBTS10/SGBS2	OFD1 (Ferrante et al. 2001; Budny et al. 2006)	OFD1/CXorf5	XD	RC	RP	ID, CBD, SZ, CVH	LC, LF	FD, OA, PD, BD		CHD	PC	BB
OFD5	DDX59 (Shamseldin et al. 2013)	DDX59	AR	DK, rare		ID, CBD		PD, FD, OA		CHD, rare		-

Continued

Table 1. *Continued*

OMIM disease	Gene symbol (first description)	Alias	Mode of inheritance	Human disease phenotype								
				Kidney	Eye	Brain	Liver	Skeletal anomalies	Laterality defects	Congenital heart disease	Other/ syndromic	Subcellular localization
OFD9 (VUS)	*SCLT1/TBC1D32* (Adly et al. 2014)	TBC1D32/ C6orf170, SCLT1	AR		CB, MO	MiC, CBD, CVH		PD, FD, OA		CHD		-
OFD14	*C2CD3* (Thauvin-Robinet et al. 2014)	C2CD3	AR		RP	ID, CBD, CVH		PD, FD, OA				BB
CED1	*IFT122* (Walczak-Sztulpa et al. 2010)		AR	NPHP	RP, Nys		LF, LC	FD, OA, SRTD, BD		CHD		CA
CED2/ SRTD7	*WDR35* (Gilissen et al. 2010; Mill et al. 2011)	WDR35/IFT121	AR	NPHP	RP		LF, BDP, LC	FD, OA, SRTD, BD, PD			PH, PC	CA
CED3	*IFT43* (Arts et al. 2011)	IFT43	AR	NPHP		LF		FD, OA, BD, PD, SRTD				CA
SRTD2	*IFT80* (Beales et al. 2007)	IFT80	AR					PD, BD, SRTD, TA				CA
SRTD3	*DYN2H1* (Dagoneau et al. 2009)	DYNC2H1	AR/DR	CK, DK	RP, rare	CBD, rare	LF, rare	PD, BD, SRTD, FD		CHD, rare	JATD	CA
SRTD6	*NEK1* (Thiel et al. 2011)	NEK1	AR/DR	CK		HC, rare	LF, rare	PD, SRTD, FD, OA, Dw		CHD, rare	HF, PH	Ncl
SRTD8	*WDR60* (McInerney-Leo et al. 2013)	WDR60	AR	CK		MaC,	DPM, LF	PD, SRTD		VSD	PH, PF	BB
SRTD9 / MZSDS	*IFT140* (Perrault et al. 2012; Schmidts et al. 2013b)	IFT140	AR	NPHP	RP, Nys	MiC, ID, DD	LF	SRTD, CSE, craniosynostosis				CA

Disease	Gene (reference)	Protein	Inheritance	Kidney	Eye	Other	Skeletal	Neuro	Heart	Other	Cilia
SRTD11	WDR34 (Huber et al. 2013; Schmidts et al. 2013c)	WDR34	AR				BD, PD, SRTD, TA			PH	CA
SRTD13	CEP120 (Shaheen et al. 2015b)	CEP120, CCDC100	AR	NPHP		CVH	SRTD, PD, FD, OA			PH	BB
EVC	EVC (Ruiz-Perez et al. 2000)						PD, SRTD, CED, Dw, OA	ID, DWM	CHD		CA + BB
EVC	EVC2 (Galdzicka et al. 2002)						PD, SRTD, CED, Dw, OA	ID, DWM	CHD		CA + BB
–	IFT57 (Thevenon et al. 2016)	IFT57	AR		VL, Nys		FD, PD				CA
–	IFT52 (Girisha et al. 2016)	IFT52	AR				PD, BD, FD, SRTD				CA
Alstrom and Usher syndrome											
ALMS	ALMS1 (Hearn et al. 2002)	ALMS1	AR	NPHP, CAKUT	RP, Nys		Skeletal anomalies	LF	DCM	SD, OB, HT, HU, Dm	BB
USH1B	MYO7A (Weil et al. 1995)	MYO7A	AR		RP, VL	DD				SD	
USH1C	USH1C (Verpy et al. 2000)	USH1C/ HARMONIN	AR		RP, VL					SD	
USH1D	CDH23 (Bolz et al. 2001)	CDH23	AR/DR		RP, VL					SD	
USH1E	USH1E (Chaib et al. 1997)	USH1E	AR		RP, VL					SD	
USH1F	PCDH15 (Ahmed et al. 2001)	PCDH15	AR		RP, VL					SD	
USH1G	SANS (Weil et al. 2003)	SANS	AR		RP, VL					SD	
USH1J/ DFNB48	CIB2 (Riazuddin et al. 2012)		AR		RP, VL					SD	
USH2A/ RP39	USH2A (Eudy et al. 1998)	USHERIN	AR		RP, VL					SD	
USH2C	GPR98 (Weston et al. 2004) PDZD7, DR (Ebermann et al. 2010)	GPR98	AR/DR		RP, VL					SD	
USH2D/ DFNB31	WHRN (Ebermann et al. 2007)	WHRN	AR		RP, VL					SD	

Continued

Table 1. *Continued*

OMIM disease	Gene symbol (first description)	Alias	Mode of inheritance	Human disease phenotype								
				Kidney	Eye	Brain	Liver	Skeletal anomalies	Laterality defects	Congenital heart disease	Other/ syndromic	Subcellular localization
USH3A	*CLRN1* (Joensuu et al. 2001)	CLRN1	AR		RP, VL						SD	
USH3B	*HARS* (Puffenberger et al. 2012)	HARS	AR		RP, VL						SD	

Columns represent disease name, gene symbol, encoded protein/aliases, human disease phenotype, and subcellular localization. The first description of each gene is cited in the second column. Diseases are grouped as (1) nephronophthisis (NPHP), Senior–Løken syndrome (SLS), Joubert syndrome (JBTS), Meckel–Gruber syndrome (MKS); (2) Bardet–Biedl syndrome (BBS); and (3) skeletal ciliopathies, oral-facial-digital syndrome (OFD), cranioectodermal dysplasia (CED), and short-rib thoracic dysplasia (SRTD). Grouping is in accordance with Figures 1 and 2.

ACS, acrocallosal syndrome; AE, anencephaly; AR, autosomal recessive; BB, basal body; BD, brachydactyly; BDP, bile duct proliferation; CA, ciliary axoneme; CAKUT, congenital anomalies of the kidney and urinary tract; CB, coloboma; CBD, congenital brain defects; CHD, congenital heart defect; COACH, COACH (cerebellar vermis defect, oligophrenia, ataxia, coloboma, hepatic fibrosis) syndrome; CSE, cone-shaped epiphyses; CVH, cerebellar vermis hypoplasia; DCM, dilated cardiomyopathy; DD, developmental delay; DK, dysplastic kidneys; Dm, diabetes mellitus; DPM, ductal plate malformations; DR, digenic recessive; Dw, dwarfism; DWM, Dandy–Walker malformation; EVC, Ellis–van Creveld syndrome; FD, facial dysmorphism; HC, hydrocephalus; HF, hydrops fetalis; HLS, hydrolethalus syndrome; HT, hypertension; HTX, heterotaxia; HU, hyperuricemia; ID, intellectual disability; IATD, Jeune asphyxiating thoracic dystrophy; LC, liver cysts; LF, liver fibrosis; MaC, macrocephaly; MiC, microcephaly; MKKS, McKusick–Kaufman syndrome; MO, microphthalmia, MORM, MORM (mental retardation, truncal obesity, retinal dystrophy, and micropenis) syndrome; MPD, multiorgan polycystic disease; Ncl, nuclear; Nys, nystagmus; OA, oral anomalies; OB, obesity; OE, occipital omphalocele; OH, oligohydramnios; OMA, oculomotor apraxia; OMIM, Online Mendelian Inheritance in Man, Johns Hopkins University; OpA, optic atrophy; PD, polydactyly; PF, pancreatic fibrosis; PC, pancreatic cysts; PH, pulmonary hypoplasia; RA, renal agenesis; RC, renal cysts; Rf, renal failure; RHPD1, renal-hepatic-pancreatic dysplasia, type 1; RP, retinitis pigmentosa; SD, sensorineural deafness; SGBS2, SI, *situs inversus*; SZ, seizures; Simpson–Golabi–Behmel syndrome, type 2; TA, trident acetabulum; TZ, transition zone; VL, vision loss; VSD, ventricular septal defect; VUS, variant of unknown significance.

Cite this article as *Cold Spring Harb Perspect Biol* doi: 10.1101/cshperspect.a028191

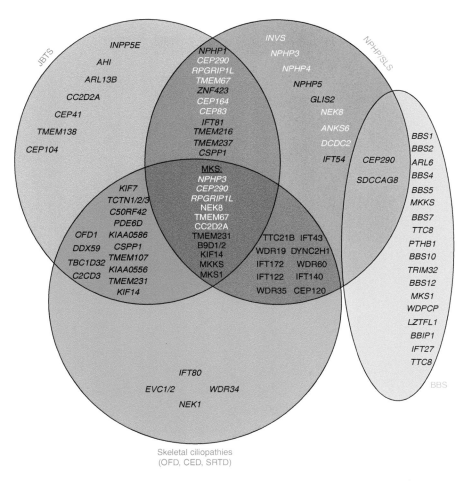

Figure 1. Monogenic genes of nephronophthisis-related ciliopathies (NPHP-RC) cause distinct but widely overlapping phenotypes. Monogenic genes of NPHP-RC are categorized into four major phenotypes, namely, JBTS (blue, Joubert syndrome [JBTS]: congenital brain malformations, cerebellar vermis hypoplasia, and intellectual disability), NPHP/SLS (red, nephronophthisis/Senior–Løken syndrome [SLS]: nephronophthisis [NPHP], retinal degeneration, coloboma), BBS (yellow, Bardet–Biedl syndrome [BBS]: obesity, intellectual disability, retinal degeneration, cystic kidney disease, polydactyly, hypogonadisms), and skeletal ciliopathies (green, oral-facial-digital syndrome [OFD], cranioectodermal dysplasia (CED), short-rib thoracic dysplasia [SRTD]). As shown in a Venn diagram, numerous genes can give rise to overlapping phenotypes if mutated. Meckel–Gruber syndrome (MKS), the most severe clinical manifestation of NPHP-RC, can be caused by mutations in 10 monogenic genes. With the exception of the genes *B9D1* and *B9D2* that have not been described in association with other phenotypes, MKS is caused by mutations in monogenic genes of nephronophthisis, Joubert syndrome, and Bardet–Biedl syndrome on an allelic basis. White text indicates genes in which liver involvement has been reported.

gout. By now, two monogenic causes of MCDK have been described, namely, the genes *MUC1* (MCKD, type 1; OMIM #174000) and *UMOD* (MCKD, type 2; OMIM #603860). Interestingly, the UMOD protein localizes to renal primary ciliary and colocalizes with other ciliary proteins such as nephrocystin (Zaucke et al. 2010). Besides MCKD2, mutations of *UMOD* also give rise to the renal diseases glomerulocystic kidney disease with hyperuricemia and isosthenuria (OMIM #609886) and familial juvenile hyperuricemic nephropathy (OMIM #162000).

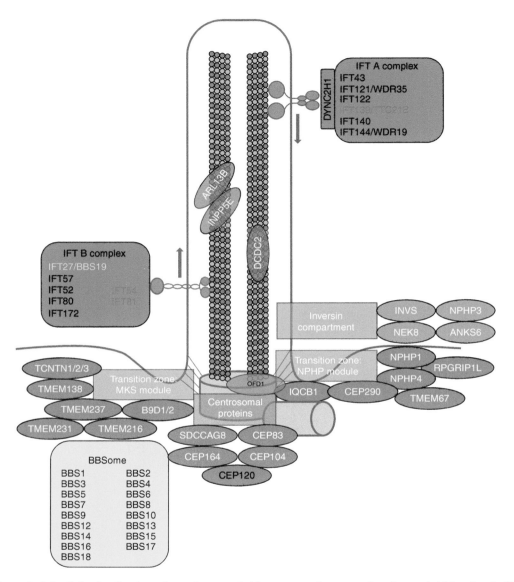

Figure 2. Subcellular localization of proteins encoded by monogenic genes of nephronophthisis-related cili-opathies (NPHP-RC). Subcellular localization of proteins encoded by monogenic genes of NPHP-RC is de-picted. Proteins are color-coded based on their respective disease group as shown in Table 1 (red/orange/pink: nephronophthisis [NPHP], Senior–Løken syndrome [SLS], Joubert syndrome [JBTS], Meckel–Gruber syn-drome [MKS]; yellow: Bardet–Biedl syndrome [BBS]; green: skeletal ciliopathies). It becomes apparent that disease groups cluster to distinct subcellular localizations. IFT, Intraflagellar transport.

RENAL PHENOTYPE OF NPHP

Clinical Presentation of NPHP

The clinical presentation of NPHP was first de-scribed in 1951 by Guido Fanconi who intro-duced the term "nephronophthisis," which translates as disappearance of nephrons to de-scribe the renal histology of affected children (Fanconi et al. 1951). With the exception of infantile NPHP, caused by mutations in the gene *IVSN* (NPHP type 2), clinical symptoms of NPHP typically start at an average age of 4–6 years (Kleinknecht 1989). The major pathology in this early phase is an impaired ability of the kidney to concentrate urine and retain water. This results in reduced urine osmolality and explains the hallmark symptoms of NPHP, namely, polyuria and polydipsia, which are present in ~80% of affected patients (Kleinknecht 1989). Secondary enuresis and regular fluid intake at night are characteristic features of NPHP and should be specifically investigated when taking the patient's history. In contrast to other pediatric kidney diseases edema, hypertension, and recurrent infections of the urinary tract are uncommon in patients with NPHP. Later in the course of disease progression, anemia and growth retardation occur. It has been noticed that, when adjusted to the degree of renal failure, both symptoms are pronounced in NPHP as compared to other renal diseases (Ala-Mello et al. 1996). With further deterioration of renal function, the clinical presentation is dominated by classical symptoms of end-stage renal failure, such as fatigue, weakness, pruritus, pallor, electrolyte imbalance, and fluid overload. If the diagnosis is missed at this stage of disease progression, patients are at risk for cardiac arrhythmia and sudden death because of electrolyte imbalances, particularly hyperkalemia.

Therapy and Prognosis of NPHP

Despite intensive research, a curative or targeted therapy for NPHP is currently not available. Treatment of NPHP is therefore centered on correcting the secondary consequences of chronic kidney disease, such electrolyte imbalances, fluid overload, anemia, renal osteodystrophy, secondary hyperparathyroidism, and growth retardation. Psychosocial counseling and intense patient/family education are an integral part of successful treatment strategies as these patients have to cope with a chronic condition that requires intense livelong therapy. Once end-stage renal disease (ESRD) has occurred, renal replacement therapy with either dialysis or renal transplantation is the only remaining treatment option. Recurrence of NPHP after renal transplant has never been reported (Steel et al. 1980).

Across all subtypes of NPHP, chronic renal failure typically develops within the first three decades of life (Hildebrandt et al. 1997b; Haider et al. 1998; Omran et al. 2000). Onset of ESRD later than 30 years of age is uncommon in all types of NPHP. Therefore, the clinical diagnosis should be reevaluated and other differential diagnoses should be considered if renal failure has not occurred by age 30 years. Interestingly, a high concordance in the rate of deterioration of renal function has been noticed between monozygotic twins, suggesting that the individual genotype determines the disease course and the development of ESRD (Mongeau and Worthen 1967; Makker et al. 1973).

Renal Pathology in NPHP

Macroscopic Pathology

The renal pathology of NPHP differs considerably from other cystic kidney diseases such as autosomal-dominant and autosomal-recessive polycystic kidney disease (ARPKD/ADPKD) (Waldherr et al. 1982). The kidney size is typically normal or slightly reduced. Renal echogenicity is characteristically enhanced as a result of renal fibrosis. Cysts are present in ~70% of affected individuals, but are not a prerequisite for the diagnosis of NPHP. Furthermore, progressive intestinal fibrosis rather than cyst development represents the primary pathogenic mechanism driving disease progression (Bernstein and Gardner 1992). If cysts are present, they are typically small in size, arise from the corticomedullary border region of the kidney, (Sherman et al. 1971) and develop late in the course of disease (Sworn and Eisinger 1972). In contrast to this classical renal phenotype, mutations in some monogenic genes of NPHP, notably the genes *INVS* (Otto et al. 2003b) and *ANKS6* (Hoff et al. 2013) (NPHP types 2 and 16), can cause a phenotype that resembles ARPKD and is characterized by enlarged, multicystic kidneys (Igarashi and Somlo 2002). Cysts in other organs have not been described in isolated NPHP. However, severe forms of renal ciliopathies can present as multicystic, dysplastic developmental phenotypes affecting all organ systems. In NPHP, there is always bilateral renal involvement.

Microscopic Pathology

Histologic hallmark findings of NPHP are summarized in a histological triad consisting of (a) disintegration and irregular thickening of the tubular basement membrane (TBM), (b) interstitial inflammation with round-cell infiltration and consecutive fibrosis, (c) tubular atrophy and cyst formation at the corticomedullary junction (Zollinger et al. 1980; Waldherr et al. 1982). Cyst formation in NPHP is thought to be secondary to degeneration and atrophy of kidney tissue rather than being an independent, primary process triggering disease progression. Early histological changes in NPHP are most prominent in distal tubules and, in contrast to other monogenic kidney disease, cells of the renal glomerulus are characteristically unaffected in NPHP. Transmission electron microscopy (TEM) in NPHP reveals characteristic changes of the tubular membrane, namely, thickening, attenuation, splitting, and granular disintegration. The presence of these alterations is discontinuous, and shows wide variability between different sections of an affected kidney. In end-stage NPHP, the histology is dominated by severe, diffuse sclerotic tubulointerstitial nephropathy.

Diagnostics and Differential Diagnosis of NPHP

Molecular genetic analysis represents the only method to diagnose NPHP-RC with certainty and thus provide patients and families with an unequivocal diagnosis. Because of an increasing number of potentially causative monogenic genes as well as advances in next-generation sequencing, whole-exome sequencing has mostly replaced targeted-sequencing panels in the diagnosis of NPHP-RC (Halbritter et al. 2012, 2013b; Gee et al. 2014). By applying this method, a causative single-gene mutation can be detected in up to 60% of cases depending on the composition of the cohort (Braun et al. 2015). However, if no mutation is detected, the diagnosis of NPHP is not excluded. Impor-

tantly, genetic testing should always be combined with thorough phenotyping as well as genetic counseling. Renal ultrasound imaging represents a very useful tool for the early diagnosis of otherwise asymptomatic renal ciliopathies. Characteristic findings include increased echogenicity, loss of corticomedullary differentiation and corticomedullary cysts (Blowey et al. 1996; Aguilera et al. 1997; Ala-Mello et al. 1998). Renal ultrasonography can help to distinguish NPHP from other inherited kidney diseases such as congenital anomalies of the kidney and urinary tract (CAKUT) or nephrocalcinosis. In some cases, magnetic resonance imaging (MRI) can provide additional insides. Other than molecular genetic testing, there is no laboratory test that specifically detects NPHP. However, serum chemistry analysis is very useful to monitor progressive renal impairment, as well as complications of chronic kidney disease. Unaffected siblings should be carefully examined and closely monitored to allow early diagnosis. However, it is generally not recommended to perform genetic testing in unaffected siblings to avoid stigmatization (Ross et al. 2013). In contrast to other monogenic kidney diseases, proteinuria, hematuria, or recurrent infections are characteristically absent in NPHP. Lack of hypertension, differences in kidney size, and distinct localization of cysts, can help to distinguish NPHP from other forms of cystic kidney disease. However, especially in late stages differential diagnosis can be challenging.

EXTRARENAL MANIFESTATIONS OF NPHP-RELATED CILIOPATHIES

Renal ciliopathies are frequently associated with additional clinical symptoms that affect other organ systems than the kidney. Typical extrarenal manifestations of renal ciliopathies include retina, central nervous system (CNS), liver, and bones. The association of NPHP with retinal degeneration (Senior–Løken syndrome) is most frequent and occurs in ~10% of all patients with NPHP (Løken et al. 1961; Senior et al. 1961). Furthermore, NPHP can be accompanied by oculomotor apraxia (Cogan syndrome) (Betz et al. 2000), by cerebellar vermis hypoplasia and retinal coloboma (Joubert syndrome) (Saraiva and Baraitser 1992; Valente et al. 2006), by liver fibrosis (Boichis et al. 1973), by ectodermal dysplasia (Sensenbrenner syndrome, oral-facial-digital syndrome type I) (Bredrup et al. 2011; Fehrenbach et al. 2014), and in rare cases by congenital heart defects (Otto et al. 2003a; Rajagopalan et al. 2016). Skeletal malformations associated with renal ciliopathies include thoracic dystrophy (Jeune syndrome) (Halbritter et al. 2013a; Huber et al. 2013; Schmidts et al. 2013a), poly- and brachydactyly, and cone-shaped epiphysis (Mainzer-Saldino syndrome) (Perrault et al. 2012). In addition, patients with infantile NPHP can show heterotaxy, *situs inversus* and potentially lung involvement (Otto et al. 2003a). Because extrarenal symptoms can be discrete but result in severe complications, screening of affected patients is generally recommended. Therefore, in addition to thorough genotyping, all patients should undergo ophthalmoscopy to exclude retinal involvement, as well as liver ultrasound examination and laboratory test for liver function. Characteristic extrarenal phenotypes for monogenic genes of human ciliopathies are summarized in Table 1. As shown in Figure 1, there is broad phenotypic overlap between different monogenic disease genes. Advances in next-generation sequencing have drastically accelerated the identification of novel single-gene causes of human ciliopathies over the last years. With increasing numbers of identified disease genes, it has become apparent that the proteins encoded by genes responsible for different subtypes of ciliopathies cluster in protein complexes that localize to distinct subcellular localizations (Fig. 2).

Retinal Involvement (Senior–Løken Syndrome)

Senior–Løken syndrome (SLS) or renal–retinal dysplasia describes the syndromic association of NPHP with retinal involvement. This association was first reported by Løken et al. (1961), Senior et al. (1961), and Fairley et al. (1963). The retinal pathology in SLS has been described as retinitis pigmentosa, tapetoretinal degenera-

tion, and retinal aplasia, which likely represent a phenotypic spectrum ranging from dysplastic, developmental, to degenerative defects (Saraux et al. 1970). Similarly, early- and late-onset forms of SLS can be distinguished based on age of onset and clinical presentation. The early onset type (a form of Leber congenital amaurosis) typically manifests at birth with blindness, reduced or absent response in electroretinography, and nystagmus. If not present at birth, disease symptoms develop rapidly within the first 2 years of life (Medhioub et al. 1994). The late-onset form typically develops slowly, starting with night blindness and progressive vision loss during school age. The clinical diagnosis is based on electroretinography (ERG) that shows a constant and complete extinction of response to stimuli reflecting retinal degeneration, as well as characteristic findings on funduscopy. Typically, ophthalmoscopic alterations are present in all SLS patients by the age of 10 years. Specific findings in ophthalmoscopy in these patients include retinitis pigmentosa, characterized by increased retinal pigmentation, attenuation of retinal vessels, atrophy of retinal pigment epithelium, pallor of the optic disc, and taptoretinal degeneration. Additional eye symptoms include atrophy of the optic nerve, nystagmus, choroidal coloboma, and ametropia (myopia, hyperopia, amblyopia, and strabismus) (Adams et al. 2007). Retinitis pigmentosa is the most common extrarenal manifestation of NPHP-related ciliopathies. It occurs in ∼10% of affected individuals and affects predominantly patients with causative mutations in the genes *IQCB1/NPHP5* (Otto et al. 2005) and *CEP290/NPHP6* (Valente et al. 2006). Remarkably, to date, all described cases of mutations of *IQCB1/NPHP5* showed retinal involvement (Otto et al. 2005). However, it can be present in virtually all types of NPHP. Interestingly, the renal phenotype of patients with and without retinal involvement is indistinguishable.

Oculomotor Apraxia (Cogan Syndrome)

Oculomotor apraxia, type Cogan, is a disorder of the CNS that affects the coordination of voluntary eye movements. Affected individuals are unable to execute purposeful horizontal eye movement to focus on or follow an object of interest. The clinical presentation is dominated by jerking head movements or compensatory head turning for side vision, as well as optico-kinetic nystagmus. This symptom has been described in patients with mutations in the genes *NPHP1* (Betz et al. 2000) and *NPHP4* (Mollet et al. 2002). Frequently, the defect is transient and improves after the first years of life. In contrast to retinitis pigmentosa, in which the defect lies in the eye itself, the underlying defect in Cogan syndrome affects central nervous regions responsible for motor control of eye movements, such as the nuclei of the abducens or oculomotoric nerves as well as supranuclear control regions.

Cerebellar Vermis Hypoplasia (Joubert Syndrome)

Joubert syndrome is a developmental disorder that manifests with a complex phenotype and involves various organ systems. Most prominent are developmental defects of the CNS, particularly the brain stem region and the cerebellum (Romani et al. 2013). The hallmark symptom of Joubert syndrome on radiology imaging of the CNS, namely, the so-called molar tooth sign, is explained by agenesis or hypoplasia of the cerebellar vermis. Furthermore, occipital meningo- or meningomyocele can be present in some cases. Clinically affected patients present with various degrees of psychomotor retardation, intellectual disability, and disrupted motor coordination causing hypotonia, ataxia, and oculomotor apraxia. Severe cases can manifest with defects of neonatal breathing regulation and newborn tachypnea. Outside the CNS, Joubert syndrome can involve facial dysmorphism, retinal dysplasia, coloboma of the optic nerve, and, less frequently, polydactyly, renal cysts, and hepatic fibrosis. Genetically, Joubert syndrome is remarkably heterogeneous and, by now, 27 monogenic genes have been identified as causative when mutated. Among the first monogenic genes to be described in patients with Joubert syndrome were the genes *NPHP1* (Parisi et al. 2004) and

AHI (Dixon-Salazar et al. 2004), which encode for a protein called Jouberin. Jouberin interacts with NPHP1, NPHP3, and NPHP4 and, interestingly, Jouberin as well as other proteins that are altered in Joubert syndrome localizes to a defined region of primary cilia, the so-called transition zone (Fig. 2) (Omran 2010). With increasing numbers of molecularly diagnosed patients, specific genotype–phenotype correlations have been observed for specific genes (Hildebrandt et al. 2011). For example, mutations in *NPHP1/JBTS4* typically cause isolated nephronophthisis with Joubert syndrome (Parisi et al. 2004), whereas mutations in *CEP290/JBTS5* (Sayer et al. 2006; Valente et al. 2006) and *RPGRIP1L/JBTS7* (Arts et al. 2007; Delous et al. 2007) frequently show retinal–renal syndrome, although mutations of *TMEM67/JBTS6* characteristically involve liver fibrosis (Table 1; Fig. 1) (Otto et al. 2009).

Renal Ciliopathies with Liver Involvement

In patients with mutations in the genes *NPHP3* (Olbrich et al. 2003), *TMEM67* (Otto et al. 2009), *ANKS6* (Hoff et al. 2013), and most recently *DCDC2* (Schueler et al. 2015), renal ciliopathies can be associated with liver involvement. The liver phenotype in these patients is characterized by hepatomegaly, periportal liver fibrosis, ductal plate malformations, and bile duct proliferation. The phenotype of renal–hepatic ciliopathies is characteristically different from congenital liver fibrosis, which is dominated by dysplasia and agenesis of bile ducts. The association of cerebellar vermis hypoplasia, oligophrenia, congenital ataxia, ocular coloboma, and hepatic fibrosis is known as COACH syndrome and is most frequently caused by mutations of *TMEM67* (NPHP11/JBTS6/MKS3) (Brancati et al. 2009) or in rare cases by mutations of *CC2D2A* or *RPGRIP1L* (Doherty et al. 2010). In patients with infantile nephronophthisis (caused by mutations of *NPHP2*), a transient elevation of liver enzymes without histopathologic changes or disturbed liver architecture has been described (Haider et al. 1998). Interestingly, in some cases, different phenotypes can be caused by mutations of the same gene on an allelic basis. Among others, this is true for mutations of the gene *ANKS6* in which protein-truncating nonsense mutations results in a severe, syndromic phenotype that involves liver fibrosis, whereas hypomorphic mutations results in isolated nephronophthisis only (Hoff et al. 2013).

Skeletal Phenotypes and Polydactyly

Short-rib thoracic dysplasia with and without polydactyly (SRTD) describes a group of ciliopathies with skeletal involvement, including poly-/brachydactyly, shortening of the long bones, cone-shaped epiphysis, and malformations of the ribs and thoracic cage. The most severe form, which is known as Jeune syndrome or asphyxiating thoracic dystrophy (ATD), is characterized by severely shortened ribs resulting in respiratory distress after birth (Jeune et al. 1955; Donaldson et al. 1985). Other diseases summarized in this group are Ellis–van Creveld syndrome (chondroectodermal dysplasia, OMIM #225500) (Moudgil et al. 1998) and Mainzer–Saldino syndrome (Mortellaro et al. 2010). The phenotypes observed in cranioectodermal dysplasia (Sensenbrenner syndrome, OMIM #218330) partially overlap with those of SRTD (Gilissen et al. 2010). Interestingly, many of the proteins encoded by genes that are mutated in patients with skeletal ciliopathies play a role in ciliary transport, such as the components of the intraflagellar transport (IFT) modules or dynein motor components (Fig. 2) (Perrault et al. 2012; Halbritter et al. 2013a; Schmidts et al. 2013a). It has been postulated that disruption of sonic hedgehog signaling, a pathway that is particularly important for bone morphogenesis and that depends on intact primary cilia for signal transduction, represents the molecular pathogenic link between ciliary dysfunction and skeletal anomalies (Huangfu et al. 2003; Qin et al. 2011).

Laterality Defects and Congenital Heart Disease

Although laterality defects such as *situs inversus* and *situs solitus* are frequently observed in pa-

tients with defects of motile cilia, they have only been described in very limited subgroups of patients with NPHP-RC. The most prominent example for this combination of phenotypes is represented by patients with mutations in the gene *NPHP2/inversin*, in which the association between infantile nephronophthisis and *situs inversus* was first described in 2003 (Otto et al. 2003a). Similar to what had been observed in the *inv/inv* knockout mouse (Mochizuki et al. 1998; Morgan et al. 1998), patients also showed congenital heart defects, such as heterotaxy, ventricular septum defects, and vascular malformations (Otto et al. 2003a). Comparably, knockdown of the ortholog of the *inversin* gene in zebrafish results in randomization of heart looping (Otto et al. 2003a). With growing numbers of detected mutations, there is increasing evidence that also proteins encoded by other monogenic genes of NPHP-RC play a role in the determination of left−right asymmetry. For example, nonsense mutations in the gene *NPHP3* cause a severe developmental phenotype resulting in embryonic lethality, which among other features also includes *situs inversus* as well as congenital heart defects (Bergmann et al. 2008). Similarly, *situs inversus* and congenital heart defects have recently been described in patients with mutations of *ANKS6* (Hoff et al. 2013) and *NEK8* (Frank et al. 2013). Interestingly, the encoded proteins of all four genes interact and colocalize to a specific compartment of primary cilia, the so-called transition zone (Czarnecki et al. 2015). This finding underlines that specific "proteins modules" may give rise to distinct clinical phenotypes.

Other Syndromic Ciliopathies

Meckel−Gruber syndrome (MKS, OMIM #249000) represents the most severe clinical manifestation of human ciliopathies. It is an autosomal-recessive developmental disorder that can affect various organ systems and shows vast phenotypic pleiotropy. Frequently, the combination of severe developmental defects results in embryonic or perinatal lethality. Characteristic symptoms include malformations of the CNS, such as hydrocephalus, microcephaly, occipital encephalocele, and, most severely, complete anencephaly. In addition, most patient show cystic–dysplastic kidney disease, liver involvement, congenital heart defects, dysmorphic features, and skeletal malformation, predominantly polydactyly. MKS presents with oligohydramnios caused by impaired kidney function, and can be diagnosed on perinatal ultrasound. With the exception of the genes *B9D1* (Hopp et al. 2011) and *B9D2* (Dowdle et al. 2011) that have not been described in association with other phenotypes, MKS is caused by mutations in monogenic genes of nephronophthisis, Joubert syndrome, and Bardet–Biedl syndrome on an allelic basis. Typically, truncating mutations give rise to the severe developmental phenotype of MKS, whereas hypomorphic mutations result in limited, organ specific disease that is degenerative rather than developmental (Hildebrandt et al. 2011).

Oral-facial-digital syndrome, type 1 (OFD1, OMIM #311200) is an X-linked dominant ciliopathy with prominent dysmorphic and external features. Because the disease is embryonic lethal in males, all surviving affected patients are female. The syndrome is characterized by oral anomalies (i.e., defective dentation, cleft lip and palate, lobulated tongue), facial dysmorphism, and polydactyly, as well as other malformations of fingers and toes. About 50% of affected individuals show structural defects of the CNS, including microcephaly, defective gyration, agenesis of the corpus callosum, and arachnoidal cysts. Seizures and variable degrees of intellectual disability can be present (Macca and Franco 2009). In contrast to other subtypes of oral-facial-digital syndrome, adult-onset cystic kidney disease has only been described in patients with OFD type 1 in which it is present in ∼50% of cases. Furthermore, fibrocystic liver disease with elevated liver enzymes can be present (Chetty-John et al. 2010). The syndrome is caused by mutations in the gene *OFD1*, which on an allelic basis can also give rise to Joubert syndrome (JBTS10, OMIM #300804) as well as Simpson–Golabi–Behmel syndrome type 2 (GBS2, OMIM #300209). Interestingly, the gene product of *OFD1* localizes to the centrosome and ciliary basal body (Chetty-John et al. 2010).

Bardet–Biedl syndrome (BBS, OMIM #209900) is a complex, syndromic ciliopathy that affects numerous organ systems. Classical symptoms include intellectual disability and behavioral anomalies, as well obesity, cystic kidney disease, retinitis pigmentosa, hypogonadism, and polydactyly. BBS is genetically very heterogeneous and by now, mutations in 19 monogenic genes have been described as causative. Although BBS is typically inherited in an autosomal-recessive manner, recent publications have suggested that digenic inheritance may also be possible (Katsanis et al. 2001b; Fan et al. 2004; Katsanis 2004). Thus far, BBS is the only ciliopathy for which a mode of inheritance different from strictly recessive (autosomal or X-linked) has been described. Interestingly, the protein products encoded by known monogenic genes of BBS interact with each other to form a tight protein complex, the so-called BBSome, which carries important functions for ciliary trafficking (Loktev et al. 2008; Jin and Nachury 2009; Zhang et al. 2013).

Alstrom syndrome (OMIM #203800) is an autosomal-recessive disorder that involves multiple organ systems and resembles BBS in some aspects of its clinical presentation. Prominent symptoms include progressive sensineural deafness and vision loss caused by cone-rod degeneration, as well dilated cardiomyopathy with congestive heart failure. Affected patients show childhood-onset obesity and can have endocrine disorders, predominantly diabetes mellitus type 2. In some cases, kidney involvement with progressive renal failure, recurrent pulmonary infections, and hepatomegaly have been described. To date, the gene *ALMS1* (Collin et al. 2002; Hearn et al. 2002) is the only described single-gene cause of Alstrom syndrome.

Usher syndrome (OMIM #276901) is a ciliopathy that is characterized by sensorineural deafness and progressive vision loss caused by retinal degeneration. Based on severity and age of onset, Usher syndrome is further classified in subtypes 1, 2, and 3. In the most severe type 1, children are predominantly born deaf and progressive vision loss begins in childhood. The syndrome is genetically heterogeneous and, by now, mutations in 14 genes have been identified as causative in humans (Table 1). Although the mode of inheritance is classically autosomal-recessive, digenic inheritance has been described for some genes.

MOLECULAR GENETICS OF NEPHRONOPHTHISIS-RELATED CILIOPATHIES

Identification of monogenic causes of NPHP-RC has furthered the understanding of its pathogenesis. The observation that the protein products of most NPHP-RC genes localizes to primary cilia, basal bodies, or centrosomes has led to the term ciliopathies to classify this group of diseases (Hildebrandt et al. 2011). Mutational analysis allows an etiologic classification of different subtypes of NPHP-RC, and enables physicians to provide patients and families with an unequivocal diagnosis. For the identification of the first single-gene causes of NPHP, namely, *NPHP1-GLIS2/NPHP7*, researchers relied on linkage analysis and positional cloning. Because of recent progress in next-generation sequencing, gene discovery has exponentially increased and by now, in up to 63% of affected individuals, a causative single gene mutation can be identified (Halbritter et al. 2012, 2013b; Braun et al. 2015). The combination of whole-exome sequencing with homozygosity mapping in consanguineous kindred, or with linkage analysis in familial cases, has proven to be a powerful tool for the identification of novel ciliopathy genes (Otto et al. 2010; Chaki et al. 2012; Schueler et al. 2015). By now, 19 classical NPHP genes and more than 90 additional ciliopathy genes have been identified as mutated in affected individuals. With the exception of the gene *NPHP1*, which explains about 25% of all cases with NPHP, mutations in other subsequently identified genes each account for only small percentages of affected individuals.

Nephronophthisis Type 1 (*NPHP1*)

In 1997, using total genome search and linkage analysis, a gene locus for juvenile nephronophthisis was mapped to chromosome 2q12-q13 (Antignac et al. 1993; Hildebrandt et al. 1993).

Molecular cloning subsequently allowed the identification of the gene *NPHP1* within that region (Hildebrandt et al. 1997a; Saunier et al. 2000). Future studies showed that about 25% of all patients with NPHP harbor causative mutations in this gene. About 85% of the mutations in *NPHP1* are large homozygous deletions of the complete *NPHP1* gene (Konrad et al. 1996; Saunier et al. 2000; Hildebrandt et al. 2001). The high degree of genomic rearrangements is caused by two large inverted duplications that flank the *NPHP1* locus on both sites (Saunier et al. 2000). The *NPHP1* locus was later described as one of 24 regions within the human genome in which sequence duplications results in hot spots of genomic instability that subsequently give rise to human disease (Bailey et al. 2002). In patients with mutations of *NPHP*, end-stage renal disease is reached at a median age of 13 years. About 10% show a combined phenotype of retinitis pigmentosa and NPHP (Caridi et al. 2000). The occurrence of oculomotor apraxia (Betz et al. 2000) and cerebellar vermis hypoplasia (Parisi et al. 2004) in patients with *NPHP1* mutations has been described but is very rare. Interestingly, there is no recognizable correlation between genotype and phenotype.

EVOLUTIONARY CONSERVATION AND MODEL ORGANISMS OF NPHP-RC

Numerous ciliary proteins and their functions are highly conserved throughout evolution. Most strikingly, the ciliary intraflagellar transport system is conserved down to the green algae *chlamydomonas rheinhardii*, and this model system has proven very useful for the identification of novel proteins and mechanisms that are part of this machinery (Pedersen et al. 2006; Engel et al. 2012; Taschner et al. 2016). Orthologs of ciliary proteins are expressed in sensory neurons of the nematode *Caenorhabditis elegans*. Expression of green fluorescent protein (GFP)-tagged fusion proteins containing the promotor regions of the orthologs of the human ciliopathy genes *NPHP1* and *NPHP4*, showed that these genes are specifically expressed in ciliated sensory neurons in the head and tail regions (Wolf et al. 2005). Interestingly, on RNAi knockdown of *nph-1* and *nph-4* male animals showed abnormal mating behavior (Wolf et al. 2005), which was similar to the phenotype observed on knockdown of the orthologs of *PKD1* and *PKD2* (Barr et al. 2001). Mutations in *nphp-8*, the *C. elegans* ortholog of *RPGRIP1L*, resulted in reduced ciliary length and functional impairment of ciliated neuron (abnormal dye filling and reduced chemotaxis) (Liu et al. 2011). *C. elegans* ciliated neurons are a unique tool for in vivo microscopy, and, as one example, this system has helped to unravel the evolutionary conserved role of the BBSome for the assembly of intraflagellar transport particles (Wei et al. 2012). Furthermore, the strong evolutionary conservation of ciliary genes allows, using the *C. elegans* system, to study newly identified human disease genes, and test for pathogenicity of specific allelic mutations in these genes (Roberson et al. 2015). Genetic mouse and zebrafish models, as well as *Xenopus tropicalis* are established vertebrate organisms to model and study gene defects that give rise to human ciliopathies. The *inv/inv* mouse, in which the murine ortholog of *NPHP2* is mutated, shows phenotypic features compatible with the one described in affected patients, in particular, cystic kidney disease, disrupted left–right determination, congenital heart defects, and liver involvement (Phillips et al. 2004). Similarly, the *pcy* mouse, in which the ortholog of human *NPHP3* is disrupted, shows tubulointerstitial renal fibrosis and cysts. The genetic zebrafish model *scorpion*[hi459] (Duldulao et al. 2009) carries a zygotic null allele for the ortholog of the human ciliopathy gene *ALR13b*. Mutant zebrafish show a characteristic ciliopathy phenotype including body axis curvature and pronephic renal cysts (Duldulao et al. 2009). Furthermore, a canine model of NPHP (Finco et al. 1970, 1977; Finco 1976) and a chicken with a mutation of *Talpid3* (Izpisua-Belmonte et al. 1992; Davey et al. 2006), the ortholog of the human Joubert gene *KIAA0586* (Alby et al. 2015; Bachmann-Gagescu et al. 2015; Malicdan et al. 2015; Roosing et al. 2015), have been described.

CONCLUDING REMARKS

Identification of monogenic causes of human ciliopathies has furthered the understanding of underlying pathogenic mechanisms, and has implicated numerous cellular pathways in pathogenesis and disease progression. To date, no treatment for affected children is available, and further research will be needed to better understand the molecular causes of disease, and thereby identify future therapeutic strategies and potential drug targets.

REFERENCES

*Reference is also in this collection.

Adams NA, Awadein A, Toma HS. 2007. The retinal ciliopathies. *Ophthalmic Genet* **28:** 113–125.

Adly N, Alhashem A, Ammari A, Alkuraya FS. 2014. Ciliary genes TBC1D32/C6orf170 and SCLT1 are mutated in patients with OFD type IX. *Hum Mutat* **35:** 36–40.

Aguilera A, Rivera M, Gallego N, Nogueira J, Ortuno J. 1997. Sonographic appearance of the juvenile nephronophthisis-cystic renal medulla complex. *Nephrol Dial Transplant* **12:** 625–626.

Ahmed ZM, Riazuddin S, Bernstein SL, Ahmed Z, Khan S, Griffith AJ, Morell RJ, Friedman TB, Wilcox ER. 2001. Mutations of the protocadherin gene PCDH15 cause Usher syndrome type 1F. *Am J Hum Genet* **69:** 25–34.

Akizu N, Silhavy JL, Rosti RO, Scott E, Fenstermaker AG, Schroth J, Zaki MS, Sanchez H, Gupta N, Kabra M, et al. 2014. Mutations in CSPP1 lead to classical Joubert syndrome. *Am J Hum Genet* **94:** 80–86.

Ala-Mello S, Kivivuori SM, Ronnholm KA, Koskimies O, Siimes MA. 1996. Mechanism underlying early anaemia in children with familial juvenile nephronophthisis. *Pediatr Nephrol* **10:** 578–581.

Ala-Mello S, Jaaskelainen J, Koskimies O. 1998. Familial juvenile nephronophthisis. An ultrasonographic follow-up of seven patients. *Acta Radiol* **39:** 84–89.

Alby C, Piquand K, Huber C, Megarbane A, Ichkou A, Legendre M, Pelluard F, Encha-Ravazi F, Abi-Tayeh G, Bessieres B, et al. 2015. Mutations in KIAA0586 cause lethal ciliopathies ranging from a hydrolethalus phenotype to short-rib polydactyly syndrome. *Am J Hum Genet* **97:** 311–318.

Aldahmesh MA, Li Y, Alhashem A, Anazi S, Alkuraya H, Hashem M, Awaji AA, Sogaty S, Alkharashi A, Alzahrani S, et al. 2014. IFT27, encoding a small GTPase component of IFT particles, is mutated in a consanguineous family with Bardet–Biedl syndrome. *Hum Mol Genet* **23:** 3307–3315.

Ansley SJ, Badano JL, Blacque OE, Hill J, Hoskins BE, Leitch CC, Kim JC, Ross AJ, Eichers ER, Teslovich TM, et al. 2003. Basal body dysfunction is a likely cause of pleiotropic Bardet–Biedl syndrome. *Nature* **425:** 628–633.

Cite this article as *Cold Spring Harb Perspect Biol* doi: 10.1101/cshperspect.a028191

Antignac C, Arduy CH, Beckmann JS, Benessy F, Gros F, Medhioub M, Hildebrandt F, Dufier JL, Kleinknecht C, Broyer M, et al. 1993. A gene for familial juvenile nephronophthisis (recessive medullary cystic kidney disease) maps to chromosome 2p. *Nat Genet* **3:** 342–345.

Arts HH, Doherty D, van Beersum SE, Parisi MA, Letteboer SJ, Gorden NT, Peters TA, Marker T, Voesenek K, Kartono A, et al. 2007. Mutations in the gene encoding the basal body protein RPGRIP1L, a nephrocystin-4 interactor, cause Joubert syndrome. *Nat Genet* **39:** 882–888.

Arts HH, Bongers EM, Mans DA, van Beersum SE, Oud MM, Bolat E, Spruijt L, Cornelissen EA, Schuurs-Hoeijmakers JH, de Leeuw N, et al. 2011. C14ORF179 encoding IFT43 is mutated in Sensenbrenner syndrome. *J Med Genet* **48:** 390–395.

Attanasio M, Uhlenhaut NH, Sousa VH, O'Toole JF, Otto E, Anlag K, Klugmann C, Treier AC, Helou J, Sayer JA, et al. 2007. Loss of GLIS2 causes nephronophthisis in humans and mice by increased apoptosis and fibrosis. *Nat Genet* **39:** 1018–1024.

Avner ED. 1994. Medullary cystic disease and medullary sponge kidney. In *Primer on kidney diseases* (ed. Greenberg A). Academic, Boston.

Bachmann-Gagescu R, Phelps IG, Dempsey JC, Sharma VA, Ishak GE, Boyle EA, Wilson M, Marques Lourenco C, Arslan M, Shendure J, et al. 2015. KIAA0586 is mutated in Joubert syndrome. *Hum Mutat* **36:** 831–835.

Badano JL, Ansley SJ, Leitch CC, Lewis RA, Lupski JR, Katsanis N. 2003. Identification of a novel Bardet–Biedl syndrome protein, BBS7, that shares structural features with BBS1 and BBS2. *Am J Hum Genet* **72:** 650–658.

Bailey JA, Gu Z, Clark RA, Reinert K, Samonte RV, Schwartz S, Adams MD, Myers EW, Li PW, Eichler EE. 2002. Recent segmental duplications in the human genome. *Science* **297:** 1003–1007.

Barr MM, DeModena J, Braun D, Nguyen CQ, Hall DH, Sternberg PW. 2001. The *Caenorhabditis elegans* autosomal dominant polycystic kidney disease gene homologs lov-1 and pkd-2 act in the same pathway. *Curr Biol* **11:** 1341–1346.

Beales PL, Bland E, Tobin JL, Bacchelli C, Tuysuz B, Hill J, Rix S, Pearson CG, Kai M, Hartley J, et al. 2007. IFT80, which encodes a conserved intraflagellar transport protein, is mutated in Jeune asphyxiating thoracic dystrophy. *Nat Genet* **39:** 727–729.

Bergmann C, Fliegauf M, Bruchle NO, Frank V, Olbrich H, Kirschner J, Schermer B, Schmedding I, Kispert A, Kranzlin B, et al. 2008. Loss of nephrocystin-3 function can cause embryonic lethality, Meckel–Gruber-like syndrome, *situs inversus*, and renal-hepatic-pancreatic dysplasia. *Am J Hum Genet* **82:** 959–970.

Bernstein J, Gardner KD. 1992. Hereditary tubulo interstitial nephritis. In *Oxford textbook of clinical nephrology* (ed. Cotran RS, Brenner BM, Stein JH). Oxford University Press, Oxford.

Betz R, Rensing C, Otto E, Mincheva A, Zehnder D, Lichter P, Hildebrandt F. 2000. Children with ocular motor apraxia type Cogan carry deletions in the gene (*NPHP1*) for juvenile nephronophthisis. *J Pediatr* **136:** 828–831.

Bielas SL, Silhavy JL, Brancati F, Kisseleva MV, Al-Gazali L, Sztriha L, Bayoumi RA, Zaki MS, Abdel-Aleem A, Rosti RO, et al. 2009. Mutations in INPP5E, encoding inositol polyphosphate-5-phosphatase E, link phosphatidyl inositol signaling to the ciliopathies. *Nat Genet* **41:** 1032–1036.

Bizet AA, Becker-Heck A, Ryan R, Weber K, Filhol E, Krug P, Halbritter J, Delous M, Lasbennes MC, Linghu B, et al. 2015. Mutations in TRAF3IP1/IFT54 reveal a new role for IFT proteins in microtubule stabilization. *Nat Commun* **6:** 8666.

Blowey DL, Querfeld U, Geary D, Warady BA, Alon U. 1996. Ultrasound findings in juvenile nephronophthisis. *Pediatr Nephrol* **10:** 22–24.

Boichis H, Passwell J, David R, Miller H. 1973. Congenital hepatic fibrosis and nephronophthisis. A family study. *Q J Med* **42:** 221–233.

Bolz H, von Brederlow B, Ramirez A, Bryda EC, Kutsche K, Nothwang HG, Seeliger M, del CSCM, Vila MC, Molina OP, et al. 2001. Mutation of CDH23, encoding a new member of the cadherin gene family, causes Usher syndrome type 1D. *Nat Genet* **27:** 108–112.

Brancati F, Iannicelli M, Travaglini L, Mazzotta A, Bertini E, Boltshauser E, D'Arrigo S, Emma F, Fazzi E, Gallizzi R, et al. 2009. MKS3/TMEM67 mutations are a major cause of COACH syndrome, a Joubert syndrome related disorder with liver involvement. *Hum Mutat* **30:** E432–E442.

Braun DA, Schueler M, Halbritter J, Gee HY, Porath JD, Lawson JA, Airik R, Shril S, Allen SJ, Stein D, et al. 2015. Whole exome sequencing identifies causative mutations in the majority of consanguineous or familial cases with childhood-onset increased renal echogenicity. *Kidney Int* **89:** 468–475.

Bredrup C, Saunier S, Oud MM, Fiskerstrand T, Hoischen A, Brackman D, Leh SM, Midtbo M, Filhol E, Bole-Feysot C, et al. 2011. Ciliopathies with skeletal anomalies and renal insufficiency due to mutations in the IFT-A gene WDR19. *Am J Hum Genet* **89:** 634–643.

Budny B, Chen W, Omran H, Fliegauf M, Tzschach A, Wisniewska M, Jensen LR, Raynaud M, Shoichet SA, Badura M, et al. 2006. A novel X-linked recessive mental retardation syndrome comprising macrocephaly and ciliary dysfunction is allelic to oral-facial-digital type I syndrome. *Hum Genet* **120:** 171–178.

Cantagrel V, Silhavy JL, Bielas SL, Swistun D, Marsh SE, Bertrand JY, Audollent S, Attie-Bitach T, Holden KR, Dobyns WB, et al. 2008. Mutations in the cilia gene ARL13B lead to the classical form of Joubert syndrome. *Am J Hum Genet* **83:** 170–179.

Caridi G, Dagnino M, Gusmano R, Ginevri F, Murer L, Ghio L, Piaggio G, Ciardi MR, Perfumo F, Ghiggeri GM. 2000. Clinical and molecular heterogeneity of juvenile nephronophthisis in Italy: Insights from molecular screening. *Am J Kidney Dis* **35:** 44–51.

Chaib H, Kaplan J, Gerber S, Vincent C, Ayadi H, Slim R, Munnich A, Weissenbach J, Petit C. 1997. A newly identified locus for Usher syndrome type I, USH1E, maps to chromosome 21q21. *Hum Mol Genet* **6:** 27–31.

Chaki M, Airik R, Ghosh AK, Giles RH, Chen R, Slaats GG, Wang H, Hurd TW, Zhou W, Cluckey A, et al. 2012. Exome capture reveals ZNF423 and CEP164 mutations, linking renal ciliopathies to DNA damage response signaling. *Cell* **150:** 533–548.

Chetty-John S, Piwnica-Worms K, Bryant J, Bernardini I, Fischer RE, Heller T, Gahl WA, Gunay-Aygun M. 2010. Fibrocystic disease of liver and pancreas; under-recognized features of the X-linked ciliopathy oral-facial-digital syndrome type 1 (OFD I). *Am J Med Genet Part A* **152A:** 2640–2645.

Chiang AP, Beck JS, Yen HJ, Tayeh MK, Scheetz TE, Swiderski RE, Nishimura DY, Braun TA, Kim KY, Huang J, et al. 2006. Homozygosity mapping with SNP arrays identifies TRIM32, an E3 ubiquitin ligase, as a Bardet–Biedl syndrome gene (BBS11). *Proc Natl Acad Sci* **103:** 6287–6292.

Collin GB, Marshall JD, Ikeda A, So WV, Russell-Eggitt I, Maffei P, Beck S, Boerkoel CF, Sicolo N, Martin M, et al. 2002. Mutations in ALMS1 cause obesity, type 2 diabetes and neurosensory degeneration in Alstrom syndrome. *Nat Genet* **31:** 74–78.

Czarnecki PG, Gabriel GC, Manning DK, Sergeev M, Lemke K, Klena NT, Liu X, Chen Y, Li Y, San Agustin JT, et al. 2015. ANKS6 is the critical activator of NEK8 kinase in embryonic situs determination and organ patterning. *Nat Commun* **6:** 6023.

Dagoneau N, Goulet M, Genevieve D, Sznajer Y, Martinovic J, Smithson S, Huber C, Baujat G, Flori E, Tecco L, et al. 2009. DYNC2H1 mutations cause asphyxiating thoracic dystrophy and short rib-polydactyly syndrome, type III. *Am J Hum Genet* **84:** 706–711.

Davey MG, Paton IR, Yin Y, Schmidt M, Bangs FK, Morrice DR, Smith TG, Buxton P, Stamataki D, Tanaka M, et al. 2006. The chicken talpid3 gene encodes a novel protein essential for Hedgehog signaling. *Genes Dev* **20:** 1365–1377.

Davis EE, Zhang Q, Liu Q, Diplas BH, Davey LM, Hartley J, Stoetzel C, Szymanska K, Ramaswami G, Logan CV, et al. 2011. TTC21B contributes both causal and modifying alleles across the ciliopathy spectrum. *Nat Genet* **43:** 189–196.

Delous M, Baala L, Salomon R, Laclef C, Vierkotten J, Tory K, Golzio C, Lacoste T, Besse L, Ozilou C, et al. 2007. The ciliary gene RPGRIP1L is mutated in cerebello-oculo-renal syndrome (Joubert syndrome type B) and Meckel syndrome. *Nat Genet* **39:** 875–881.

Dixon-Salazar T, Silhavy JL, Marsh SE, Louie CM, Scott LC, Gururaj A, Al-Gazali L, Al-Tawari AA, Kayserili H, Sztriha L, et al. 2004. Mutations in the AHI1 gene, encoding jouberin, cause Joubert syndrome with cortical polymicrogyria. *Am J Hum Genet* **75:** 979–987.

Doherty D, Parisi MA, Finn LS, Gunay-Aygun M, Al-Mateen M, Bates D, Clericuzio C, Demir H, Dorschner M, van Essen AJ, et al. 2010. Mutations in 3 genes (MKS3, CC2D2A and RPGRIP1L) cause COACH syndrome (Joubert syndrome with congenital hepatic fibrosis). *J Med Genet* **47:** 8–21.

Donaldson MD, Warner AA, Trompeter RS, Haycock GB, Chantler C. 1985. Familial juvenile nephronophthisis, Jeune's syndrome, and associated disorders. *Arch Dis Child* **60:** 426–434.

Dowdle WE, Robinson JF, Kneist A, Sirerol-Piquer MS, Frints SG, Corbit KC, Zaghloul NA, van Lijnschoten G, Mulders L, Verver DE, et al. 2011. Disruption of a ciliary B9 protein complex causes Meckel syndrome. *Am J Hum Genet* **89:** 94–110.

Duldulao NA, Lee S, Sun Z. 2009. Cilia localization is essential for in vivo functions of the Joubert syndrome protein Arl13b/Scorpion. *Development* **136:** 4033–4042.

Ebermann I, Scholl HP, Charbel Issa P, Becirovic E, Lamprecht J, Jurklies B, Millan JM, Aller E, Mitter D, Bolz H. 2007. A novel gene for Usher syndrome type 2: Mutations in the long isoform of whirlin are associated with retinitis pigmentosa and sensorineural hearing loss. *Hum Genet* **121:** 203–211.

Ebermann I, Phillips JB, Liebau MC, Koenekoop RK, Schermer B, Lopez I, Schafer E, Roux AF, Dafinger C, Bernd A, et al. 2010. PDZD7 is a modifier of retinal disease and a contributor to digenic Usher syndrome. *J Clin Invest* **120:** 1812–1823.

Engel BD, Ishikawa H, Wemmer KA, Geimer S, Wakabayashi K, Hirono M, Craige B, Pazour GJ, Witman GB, Kamiya R, et al. 2012. The role of retrograde intraflagellar transport in flagellar assembly, maintenance, and function. *J Cell Biol* **199:** 151–167.

Eudy JD, Weston MD, Yao S, Hoover DM, Rehm HL, Ma-Edmonds M, Yan D, Ahmad I, Cheng JJ, Ayuso C, et al. 1998. Mutation of a gene encoding a protein with extracellular matrix motifs in Usher syndrome type IIa. *Science* **280:** 1753–1757.

Failler M, Gee HY, Krug P, Joo K, Halbritter J, Belkacem L, Filhol E, Porath JD, Braun DA, Schueler M, et al. 2014. Mutations of CEP83 cause infantile nephronophthisis and intellectual disability. *Am J Hum Genet* **94:** 905–914.

Fairley KF, Leighton PW, Kincaid-Smith P. 1963. Familial visual defects associated with polycystic kidney and medullary sponge kidney. *Br Med J* **1:** 1060–1063.

Fan Y, Esmail MA, Ansley SJ, Blacque OE, Boroevich K, Ross AJ, Moore SJ, Badano JL, May-Simera H, Compton DS, et al. 2004. Mutations in a member of the Ras superfamily of small GTP-binding proteins causes Bardet–Biedl syndrome. *Nat Genet* **36:** 989–993.

Fanconi G, Hanhart E, Albertini A, Uhlinger E, Dolivo G, Prader A. 1951. Die familiäre juvenile nephronophthise. *Helv Paediatr Acta* **6:** 1–49.

Fehrenbach H, Decker C, Eisenberger T, Frank V, Hampel T, Walden U, Amann KU, Kruger-Stollfuss I, Bolz HJ, Haffner K, et al. 2014. Mutations in WDR19 encoding the intraflagellar transport component IFT144 cause a broad spectrum of ciliopathies. *Pediatr Nephrol* **29:** 1451–1456.

Ferland RJ, Eyaid W, Collura RV, Tully LD, Hill RS, Al-Nouri D, Al-Rumayyan A, Topcu M, Gascon G, Bodell A, et al. 2004. Abnormal cerebellar development and axonal decussation due to mutations in AHI1 in Joubert syndrome. *Nat Genet* **36:** 1008–1013.

Ferrante MI, Giorgio G, Feather SA, Bulfone A, Wright V, Ghiani M, Selicorni A, Gammaro L, Scolari F, Woolf AS, et al. 2001. Identification of the gene for oral-facial-digital type I syndrome. *Am J Hum Genet* **68:** 569–576.

Filges I, Nosova E, Bruder E, Tercanli S, Townsend K, Gibson WT, Rothlisberger B, Heinimann K, Hall JG, Gregory-Evans CY, et al. 2014. Exome sequencing identifies mutations in KIF14 as a novel cause of an autosomal recessive lethal fetal ciliopathy phenotype. *Clin Genet* **86:** 220–228.

Finco DR. 1976. Familial renal disease in Norwegian Elkhound dogs: Physiologic and biochemical examinations. *Am J Vet Res* **37:** 87–91.

Cite this article as *Cold Spring Harb Perspect Biol* doi: 10.1101/cshperspect.a028191

Finco DR, Kurtz HJ, Low DG, Perman V. 1970. Familial renal disease in Norwegian Elkhound dogs. *J Am Vet Med Assoc* **156:** 747–760.

Finco DR, Duncan JD, Crowell WA, Hulsey ML. 1977. Familial renal disease in Norwegian Elkhound dogs: Morphologic examinations. *Am J Vet Res* **38:** 941–947.

Frank V, Habbig S, Bartram MP, Eisenberger T, Veenstra-Knol HE, Decker C, Boorsma RA, Gobel H, Nurnberg G, Griessmann A, et al. 2013. Mutations in NEK8 link multiple organ dysplasia with altered Hippo signalling and increased c-MYC expression. *Hum mol genet* **22:** 2177–2185.

Gagnadoux MF, Bacri JL, Broyer M, Habib R. 1989. Infantile chronic tubulo-interstitial nephritis with cortical microcysts: Variant of nephronophthisis or new disease entity? *Pediatr Nephrol* **3:** 50–55.

Galdzicka M, Patnala S, Hirshman MG, Cai JF, Nitowsky H, Egeland JA, Ginns EI. 2002. A new gene, EVC2, is mutated in Ellis–van Creveld syndrome. *Mol Genet Metab* **77:** 291–295.

Garcia-Gonzalo FR, Corbit KC, Sirerol-Piquer MS, Ramaswami G, Otto EA, Noriega TR, Seol AD, Robinson JF, Bennett CL, Josifova DJ, et al. 2011. A transition zone complex regulates mammalian ciliogenesis and ciliary membrane composition. *Nat Genet* **43:** 776–784.

Gee HY, Otto EA, Hurd TW, Ashraf S, Chaki M, Cluckey A, Vega-Warner V, Saisawat P, Diaz KA, Fang H, et al. 2014. Whole-exome resequencing distinguishes cystic kidney diseases from phenocopies in renal ciliopathies. *Kidney Int* **85:** 880–887.

Gilissen C, Arts HH, Hoischen A, Spruijt L, Mans DA, Arts P, van Lier B, Steehouwer M, van Reeuwijk J, Kant SG, et al. 2010. Exome sequencing identifies WDR35 variants involved in Sensenbrenner syndrome. *Am J Hum Genet* **87:** 418–423.

Girisha KM, Shukla A, Trujillano D, Bhavani GS, Hebbar M, Kadavigere R, Rolfs A. 2016. A homozygous nonsense variant in IFT52 is associated with a human skeletal ciliopathy. *Clin Genet* doi: 10.1111.cge.12762.

Gorden NT, Arts HH, Parisi MA, Coene KL, Letteboer SJ, van Beersum SE, Mans DA, Hikida A, Eckert M, Knutzen D, et al. 2008. CC2D2A is mutated in Joubert syndrome and interacts with the ciliopathy-associated basal body protein CEP290. *Am J Hum Genet* **83:** 559–571.

Gretz N. 1989. Rate of deterioration of renal function in juvenile nephronophthisis. *Pediatr Nephrol* **3:** 56–60.

Haider NB, Carmi R, Shalev H, Sheffield VC, Landau D. 1998. A Bedouin kindred with infantile nephronophthisis demonstrates linkage to chromosome 9 by homozygosity mapping. *Am J Hum Genet* **63:** 1404–1410.

Halbritter J, Diaz K, Chaki M, Porath JD, Tarrier B, Fu C, Innis JL, Allen SJ, Lyons RH, Stefanidis CJ, et al. 2012. High-throughput mutation analysis in patients with a nephronophthisis-associated ciliopathy applying multiplexed barcoded array-based PCR amplification and next-generation sequencing. *J Med Genet* **49:** 756–767.

Halbritter J, Bizet AA, Schmidts M, Porath JD, Braun DA, Gee HY, McInerney-Leo AM, Krug P, Filhol E, Davis EE, et al. 2013a. Defects in the IFT-B component IFT172 cause Jeune and Mainzer-Saldino syndromes in humans. *Am J Hum Genet* **93:** 915–925.

Halbritter J, Porath JD, Diaz KA, Braun DA, Kohl S, Chaki M, Allen SJ, Soliman NA, Hildebrandt F, Otto EA. 2013b. Identification of 99 novel mutations in a worldwide cohort of 1,056 patients with a nephronophthisis-related ciliopathy. *Hum Genet* **132:** 865–884.

Hearn T, Renforth GL, Spalluto C, Hanley NA, Piper K, Brickwood S, White C, Connolly V, Taylor JF, Russell-Eggitt I, et al. 2002. Mutation of ALMS1, a large gene with a tandem repeat encoding 47 amino acids, causes Alstrom syndrome. *Nat Genet* **31:** 79–83.

Hildebrandt F, Otto E, Rensing C, Nothwang HG, Vollmer M, Adolphs J, Hanusch H, Brandis M. 1997a. A novel gene encoding an SH3 domain protein is mutated in nephronophthisis type 1. *Nat Genet* **17:** 149–153.

Hildebrandt F, Strahm B, Nothwang HG, Gretz N, Schnieders B, Singh-Sawhney I, Kutt R, Vollmer M, Brandis M. 1997b. Molecular genetic identification of families with juvenile nephronophthisis type 1: Rate of progression to renal failure. APN Study Group. Arbeitsgemeinschaft fur Padiatrische Nephrologie. *Kidney Int* **51:** 261–269.

Hildebrandt F, Rensing C, Betz R, Sommer U, Birnbaum S, Imm A, Omran H, Leipoldt M, Otto E. 2001. Establishing an algorithm for molecular genetic diagnostics in 127 families with juvenile nephronophthisis. *Kidney Int* **59:** 434–445.

Hildebrandt F, Benzing T, Katsanis N. 2011. Ciliopathies. *N Engl J Med* **364:** 1533–1543.

Hoff S, Halbritter J, Epting D, Frank V, Nguyen TM, van Reeuwijk J, Boehlke C, Schell C, Yasunaga T, Helmstadter M, et al. 2013. ANKS6 is a central component of a nephronophthisis module linking NEK8 to INVS and NPHP3. *Nat Genet* **45:** 951–956.

Hopp K, Heyer CM, Hommerding CJ, Henke SA, Sundsbak JL, Patel S, Patel P, Consugar MB, Czarnecki PG, Gliem TJ, et al. 2011. B9D1 is revealed as a novel Meckel syndrome (MKS) gene by targeted exon-enriched next-generation sequencing and deletion analysis. *Hum Mol Genet* **20:** 2524–2534.

Huang L, Szymanska K, Jensen VL, Janecke AR, Innes AM, Davis EE, Frosk P, Li C, Willer JR, Chodirker BN, et al. 2011. TMEM237 is mutated in individuals with a Joubert syndrome related disorder and expands the role of the TMEM family at the ciliary transition zone. *Am J Hum Genet* **89:** 713–730.

Huangfu D, Liu A, Rakeman AS, Murcia NS, Niswander L, Anderson KV. 2003. Hedgehog signalling in the mouse requires intraflagellar transport proteins. *Nature* **426:** 83–87.

Huber C, Wu S, Kim AS, Sigaudy S, Sarukhanov A, Serre V, Baujat G, Le Quan Sang KH, Rimoin DL, Cohn DH, et al. 2013. WDR34 mutations that cause short-rib polydactyly syndrome type III/severe asphyxiating thoracic dysplasia reveal a role for the NF-κB pathway in cilia. *Am J Hum Genet* **93:** 926–931.

Igarashi P, Somlo S. 2002. Genetics and pathogenesis of polycystic kidney disease. *J Am Soc Nephrol* **13:** 2384–2398.

Izpisua-Belmonte JC, Ede DA, Tickle C, Duboule D. 1992. The mis-expression of posterior Hox-4 genes in talpid (ta3) mutant wings correlates with the absence of anteroposterior polarity. *Development* **114:** 959–963.

Jeune M, Beraud C, Carron R. 1955. Dystrophie thoracique asphyxiante de caractere familial [Asphyxiating thoracic dystrophy with familial characteristics]. *Arch Fr Pediatr* **12:** 886–891.

Jin H, Nachury MV. 2009. The BBSome. *Curr Biol* **19:** R472–473.

Joensuu T, Hamalainen R, Yuan B, Johnson C, Tegelberg S, Gasparini P, Zelante L, Pirvola U, Pakarinen L, Lehesjoki AE, et al. 2001. Mutations in a novel gene with transmembrane domains underlie Usher syndrome type 3. *Am J Hum Genet* **69:** 673–684.

Katsanis N. 2004. The oligogenic properties of Bardet–Biedl syndrome. *Hum Mol Genet* **13:** R65–71.

Katsanis N, Beales PL, Woods MO, Lewis RA, Green JS, Parfrey PS, Ansley SJ, Davidson WS, Lupski JR. 2000. Mutations in MKKS cause obesity, retinal dystrophy and renal malformations associated with Bardet–Biedl syndrome. *Nat Genet* **26:** 67–70.

Katsanis N, Ansley SJ, Badano JL, Eichers ER, Lewis RA, Hoskins BE, Scambler PJ, Davidson WS, Beales PL, Lupski JR. 2001a. Triallelic inheritance in Bardet–Biedl syndrome, a Mendelian recessive disorder. *Science* **293:** 2256–2259.

Katsanis N, Lupski JR, Beales PL. 2001b. Exploring the molecular basis of Bardet–Biedl syndrome. *Hum Mol Genet* **10:** 2293–2299.

Katsanis N, Eichers ER, Ansley SJ, Lewis RA, Kayserili H, Hoskins BE, Scambler PJ, Beales PL, Lupski JR. 2002. BBS4 is a minor contributor to Bardet–Biedl syndrome and may also participate in triallelic inheritance. *Am J Hum Genet* **71:** 22–29.

Kim SK, Shindo A, Park TJ, Oh EC, Ghosh S, Gray RS, Lewis RA, Johnson CA, Attie-Bittach T, Katsanis N, et al. 2010. Planar cell polarity acts through septins to control collective cell movement and ciliogenesis. *Science* **329:** 1337–1340.

Kleinknecht C. 1989. The inheritance of nephronophthisis. In *Inheritance of kidney and urinary tract diseases* (ed. Spitzer A, Avner ED), p. 464. Kluwer, Boston.

Konrad M, Saunier S, Heidet L, Silbermann F, Benessy F, Calado J, Le Paslier D, Broyer M, Gubler MC, Antignac C. 1996. Large homozygous deletions of the 2q13 region are a major cause of juvenile nephronophthisis. *Hum Mol Genet* **5:** 367–371.

Kyttala M, Tallila J, Salonen R, Kopra O, Kohlschmidt N, Paavola-Sakki P, Peltonen L, Kestila M. 2006. MKS1, encoding a component of the flagellar apparatus basal body proteome, is mutated in Meckel syndrome. *Nat Genet* **38:** 155–157.

Lambacher NJ, Bruel AL, van Dam TJ, Szymanska K, Slaats GG, Kuhns S, McManus GJ, Kennedy JE, Gaff K, Wu KM, et al. 2016. TMEM107 recruits ciliopathy proteins to subdomains of the ciliary transition zone and causes Joubert syndrome. *Nat Cell Biol* **18:** 122–131.

Lee JE, Silhavy JL, Zaki MS, Schroth J, Bielas SL, Marsh SE, Olvera J, Brancati F, Iannicelli M, Ikegami K, et al. 2012a. CEP41 is mutated in Joubert syndrome and is required for tubulin glutamylation at the cilium. *Nat Genet* **44:** 193–199.

Lee JH, Silhavy JL, Lee JE, Al-Gazali L, Thomas S, Davis EE, Bielas SL, Hill KJ, Iannicelli M, Brancati F, et al. 2012b.

Evolutionarily assembled *cis*-regulatory module at a human ciliopathy locus. *Science* **335:** 966–969.

Leigh MW, Pittman JE, Carson JL, Ferkol TW, Dell SD, Davis SD, Knowles MR, Zariwala MA. 2009. Clinical and genetic aspects of primary ciliary dyskinesia/Kartagener syndrome. *Genet Med* **11:** 473–487.

Leitch CC, Zaghloul NA, Davis EE, Stoetzel C, Diaz-Font A, Rix S, Alfadhel M, Lewis RA, Eyaid W, Banin E, et al. 2008. Hypomorphic mutations in syndromic encephalocele genes are associated with Bardet–Biedl syndrome. *Nat Genet* **40:** 443–448.

Li JB, Gerdes JM, Haycraft CJ, Fan Y, Teslovich TM, May-Simera H, Li H, Blacque OE, Li L, Leitch CC, et al. 2004. Comparative genomics identifies a flagellar and basal body proteome that includes the BBS5 human disease gene. *Cell* **117:** 541–552.

Liu L, Zhang M, Xia Z, Xu P, Chen L, Xu T. 2011. *Caenorhabditis elegans* ciliary protein NPHP-8, the homologue of human RPGRIP1L, is required for ciliogenesis and chemosensation. *Biochem Biophys Res Commun* **410:** 626–631.

Løken AC, Hanssen O, Halvorsen S, Jolster NJ. 1961. Hereditary renal dysplasia and blindness. *Acta Paediatr* **50:** 177–184.

Loktev AV, Zhang Q, Beck JS, Searby CC, Scheetz TE, Bazan JF, Slusarski DC, Sheffield VC, Jackson PK, Nachury MV. 2008. A BBSome subunit links ciliogenesis, microtubule stability, and acetylation. *Dev Cell* **15:** 854–865.

* Ma M, Gallagher A-R, Somlo S. 2016. Ciliary mechanisms of cyst formation in polycystic kidney disease. *Cold Spring Harb Perspect Biol* doi: 10.1101/cshperspect.a028209.

Macca M, Franco B. 2009. The molecular basis of oral-facial-digital syndrome, type 1. *Am J Med Genet C Semin Med Genet* **151C:** 318–325.

Makker SP, Grupe WE, Perrin E. 1973. Identical progression of juvenile hereditary nephronophthisis in monozygotic twins. *J Pediatr* **82:** 773–779.

Malicdan MC, Vilboux T, Stephen J, Maglic D, Mian L, Konzman D, Guo J, Yildirimli D, Bryant J, Fischer R, et al. 2015. Mutations in human homologue of chicken talpid3 gene (KIAA0586) cause a hybrid ciliopathy with overlapping features of Jeune and Joubert syndromes. *J Med Genet* **52:** 830–839.

Marion V, Stutzmann F, Gerard M, De Melo C, Schaefer E, Claussmann A, Helle S, Delague V, Souied E, Barrey C, et al. 2012. Exome sequencing identifies mutations in LZTFL1, a BBSome and smoothened trafficking regulator, in a family with Bardet–Biedl syndrome with *situs inversus* and insertional polydactyly. *J Med Genet* **49:** 317–321.

McInerney-Leo AM, Schmidts M, Cortes CR, Leo PJ, Gener B, Courtney AD, Gardiner B, Harris JA, Lu Y, Marshall M, et al. 2013. Short-rib polydactyly and Jeune syndromes are caused by mutations in WDR60. *Am J Hum Genet* **93:** 515–523.

Medhioub M, Cherif D, Benessy F, Silbermann F, Gubler MC, Le Paslier D, Cohen D, Weissenbach J, Beckmann J, Antignac C. 1994. Refined mapping of a gene (NPH1) causing familial juvenile nephronophthisis and evidence for genetic heterogeneity. *Genomics* **22:** 296–301.

Mill P, Lockhart PJ, Fitzpatrick E, Mountford HS, Hall EA, Reijns MA, Keighren M, Bahlo M, Bromhead CJ, Budd P, et al. 2011. Human and mouse mutations in WDR35 cause short-rib polydactyly syndromes due to abnormal ciliogenesis. *Am J Hum Genet* **88:** 508–515.

Mochizuki T, Saijoh Y, Tsuchiya K, Shirayoshi Y, Takai S, Taya C, Yonekawa H, Yamada K, Nihei H, Nakatsuji N, et al. 1998. Cloning of inv, a gene that controls left/right asymmetry and kidney development. *Nature* **395:** 177–181.

Mollet G, Salomon R, Gribouval O, Silbermann F, Bacq D, Landthaler G, Milford D, Nayir A, Rizzoni G, Antignac C, et al. 2002. The gene mutated in juvenile nephronophthisis type 4 encodes a novel protein that interacts with nephrocystin. *Nat Genet* **32:** 300–305.

Mongeau JG, Worthen HG. 1967. Nephronophthisis and medullary cystic disease. *Am J Med* **43:** 345–355.

Morgan D, Turnpenny L, Goodship J, Dai W, Majumder K, Matthews L, Gardner A, Schuster G, Vien L, Harrison W, et al. 1998. Inversin, a novel gene in the vertebrate left-right axis pathway, is partially deleted in the inv mouse. *Nat Genet* **20:** 149–156.

Mortellaro C, Bello L, Pucci A, Lucchina AG, Migliario M. 2010. Saldino-Mainzer syndrome: Nephronophthisis, retinitis pigmentosa, and cone-shaped epiphyses. *J Craniofac Surg* **21:** 1554–1556.

Moudgil A, Bagga A, Kamil ES, Rimoin DL, Lachman RS, Cohen AH, Jordan SC. 1998. Nephronophthisis associated with Ellis–van Creveld syndrome. *Pediatr Nephrol* **12:** 20–22.

Mykytyn K, Nishimura DY, Searby CC, Shastri M, Yen HJ, Beck JS, Braun T, Streb LM, Cornier AS, Cox GF, et al. 2002. Identification of the gene (BBS1) most commonly involved in Bardet–Biedl syndrome, a complex human obesity syndrome. *Nat Genet* **31:** 435–438.

Nishimura DY, Searby CC, Carmi R, Elbedour K, Van Maldergem L, Fulton AB, Lam BL, Powell BR, Swiderski RE, Bugge KE, et al. 2001. Positional cloning of a novel gene on chromosome 16q causing Bardet–Biedl syndrome (BBS2). *Hum Mol Genet* **10:** 865–874.

Nishimura DY, Swiderski RE, Searby CC, Berg EM, Ferguson AL, Hennekam R, Merin S, Weleber RG, Biesecker LG, Stone EM, et al. 2005. Comparative genomics and gene expression analysis identifies BBS9, a new Bardet–Biedl syndrome gene. *Am J Hum Genet* **77:** 1021–1033.

Olbrich H, Fliegauf M, Hoefele J, Kispert A, Otto E, Volz A, Wolf MT, Sasmaz G, Trauer U, Reinhardt R, et al. 2003. Mutations in a novel gene, *NPHP3*, cause adolescent nephronophthisis, tapeto-retinal degeneration and hepatic fibrosis. *Nat Genet* **34:** 455–459.

Omran H. 2010. NPHP proteins: Gatekeepers of the ciliary compartment. *J Cell Biol* **190:** 715–717.

Omran H, Fernandez C, Jung M, Haffner K, Fargier B, Villaquiran A, Waldherr R, Gretz N, Brandis M, Ruschendorf F, et al. 2000. Identification of a new gene locus for adolescent nephronophthisis, on chromosome 3q22 in a large Venezuelan pedigree. *Am J Hum Genet* **66:** 118–127.

Otto E, Hoefele J, Ruf R, Mueller AM, Hiller KS, Wolf MT, Schuermann MJ, Becker A, Birkenhager R, Sudbrak R, et al. 2002. A gene mutated in nephronophthisis and retinitis pigmentosa encodes a novel protein, nephroretinin, conserved in evolution. *Am J Hum Genet* **71:** 1167–1171.

Otto EA, Schermer B, Obara T, O'Toole JF, Hiller KS, Mueller AM, Ruf RG, Hoefele J, Beekmann F, Landau D, et al. 2003a. Mutations in INVS encoding inversin cause nephronophthisis type 2, linking renal cystic disease to the function of primary cilia and left–right axis determination. *Nat Genet* **34:** 413–420.

Otto EA, Schermer B, Obara T, O'Toole JF, Hiller KS, Mueller AM, Ruf RG, Hoefele J, Beekmann F, Landau D, et al. 2003b. Mutations in INVS encoding inversin cause nephronophthisis type 2, linking renal cystic disease to the function of primary cilia and left–right axis determination. *Nat Genet* **34:** 413–420.

Otto E, Loeys B, Khanna H, Hellemans J, Sudbrak R, Fan S, Muerb U, O'Toole JF, Helou J, Attanasio M, et al. 2005. A novel ciliary IQ domain protein, NPHP5, is mutated in Senior–Løken syndrome (nephronophthisis with retinitis pigmentosa), and interacts with RPGR and calmodulin. *Nat Genet* **37:** 282–288.

Otto EA, Trapp ML, Schultheiss UT, Helou J, Quarmby LM, Hildebrandt F. 2008. NEK8 mutations affect ciliary and centrosomal localization and may cause nephronophthisis. *J Am Soc Nephrol* **19:** 587–592.

Otto EA, Tory K, Attanasio M, Zhou W, Chaki M, Paruchuri Y, Wise EL, Wolf MT, Utsch B, Becker C, et al. 2009. Hypomorphic mutations in meckelin (MKS3/ TMEM67) cause nephronophthisis with liver fibrosis (NPHP11). *J Med Genet* **46:** 663–670.

Otto EA, Hurd TW, Airik R, Chaki M, Zhou W, Stoetzel C, Patil SB, Levy S, Ghosh AK, Murga-Zamalloa CA, et al. 2010. Candidate exome capture identifies mutation of SDCCAG8 as the cause of a retinal-renal ciliopathy. *Nat Genet* **42:** 840–850.

Parisi MA, Bennett CL, Eckert ML, Dobyns WB, Gleeson JG, Shaw DW, McDonald R, Eddy A, Chance PF, Glass IA. 2004. The NPHP1 gene deletion associated with juvenile nephronophthisis is present in a subset of individuals with Joubert syndrome. *Am J Hum Genet* **75:** 82–91.

Pedersen LB, Geimer S, Rosenbaum JL. 2006. Dissecting the molecular mechanisms of intraflagellar transport in chlamydomonas. *Curr Biol* **16:** 450–459.

Perrault I, Saunier S, Hanein S, Filhol E, Bizet AA, Collins F, Salih MA, Gerber S, Delphin N, Bigot K, et al. 2012. Mainzer–Saldino syndrome is a ciliopathy caused by IFT140 mutations. *Am J Hum Genet* **90:** 864–870.

Perrault I, Halbritter J, Porath JD, Gerard X, Braun DA, Gee HY, Fathy HM, Saunier S, Cormier-Daire V, Thomas S, et al. 2015. IFT81, encoding an IFT-B core protein, as a very rare cause of a ciliopathy phenotype. *J Med Genet* **52:** 657–665.

Phillips CL, Miller KJ, Filson AJ, Nurnberger J, Clendenon JL, Cook GW, Dunn KW, Overbeek PA, Gattone VH II, Bacallao RL. 2004. Renal cysts of *inv/inv* mice resemble early infantile nephronophthisis. *J Am Soc Nephrol* **15:** 1744–1755.

Pistor K, Schärer K, Olbing H, Tamminen-Möbius T. 1985. Children with chronic renal failure in the Federal Republic of Germany. II: Primary renal diseases, age and intervals from early renal failure to renal death. *Clin Nephrol* **23:** 278–284.

Potter DE, Holliday MA, Piel CF, Feduska NJ, Belzer FO, Salvatierra O Jr. 1980. Treatment of end-stage renal disease in children: A 15-year experience. *Kidney Int* **18**: 103–109.

Puffenberger EG, Jinks RN, Sougnez C, Cibulskis K, Willert RA, Achilly NP, Cassidy RP, Fiorentini CJ, Heiken KF, Lawrence JJ, et al. 2012. Genetic mapping and exome sequencing identify variants associated with five novel diseases. *PloS ONE* **7**: e28936.

Putoux A, Thomas S, Coene KL, Davis EE, Alanay Y, Ogur G, Uz E, Buzas D, Gomes C, Patrier S, et al. 2011. KIF7 mutations cause fetal hydrolethalus and acrocallosal syndromes. *Nat Genet* **43**: 601–606.

Qin J, Lin Y, Norman RX, Ko HW, Eggenschwiler JT. 2011. Intraflagellar transport protein 122 antagonizes Sonic Hedgehog signaling and controls ciliary localization of pathway components. *Proc Natl Acad Sci* **108**: 1456–1461.

Rajagopalan R, Grochowski CM, Gilbert MA, Falsey AM, Coleman K, Romero R, Loomes KM, Piccoli DA, Devoto M, Spinner NB. 2016. Compound heterozygous mutations in NEK8 in siblings with end-stage renal disease with hepatic and cardiac anomalies. *Am J Hum Genet Part A* **170**: 750–753.

Riazuddin S, Belyantseva IA, Giese AP, Lee K, Indzhykulian AA, Nandamuri SP, Yousaf R, Sinha GP, Lee S, Terrell D, et al. 2012. Alterations of the CIB2 calcium- and integrin-binding protein cause Usher syndrome type 1J and non-syndromic deafness DFNB48. *Nat Genet* **44**: 1265–1271.

Roberson EC, Dowdle WE, Ozanturk A, Garcia-Gonzalo FR, Li C, Halbritter J, Elkhartoufi N, Porath JD, Cope H, Ashley-Koch A, et al. 2015. TMEM231, mutated in orofaciodigital and Meckel syndromes, organizes the ciliary transition zone. *J Cell Biol* **209**: 129–142.

Romani M, Micalizzi A, Valente EM. 2013. Joubert syndrome: Congenital cerebellar ataxia with the molar tooth. *Lancet Neurol* **12**: 894–905.

Roosing S, Hofree M, Kim S, Scott E, Copeland B, Romani M, Silhavy JL, Rosti RO, Schroth J, Mazza T, et al. 2015. Functional genome-wide siRNA screen identifies KIAA0586 as mutated in Joubert syndrome. *eLife* **4**: e06602.

Ross LF, Saal HM, David KL, Anderson RR. 2013. Technical report: Ethical and policy issues in genetic testing and screening of children. *Genet Med* **15**: 234–245.

Ruiz-Perez VL, Ide SE, Strom TM, Lorenz B, Wilson D, Woods K, King L, Francomano C, Freisinger P, Spranger S, et al. 2000. Mutations in a new gene in Ellis–van Creveld syndrome and Weyers acrodental dysostosis. *Nat Genet* **24**: 283–286.

Sanders AA, de Vrieze E, Alazami AM, Alzahrani F, Malarkey EB, Sorusch N, Tebbe L, Kuhns S, van Dam TJ, Alhashem A, et al. 2015. KIAA0556 is a novel ciliary basal body component mutated in Joubert syndrome. *Genome Biol* **16**: 293.

Sang L, Miller JJ, Corbit KC, Giles RH, Brauer MJ, Otto EA, Baye LM, Wen X, Scales SJ, Kwong M, et al. 2011. Mapping the NPHP-JBTS-MKS protein network reveals ciliopathy disease genes and pathways. *Cell* **145**: 513–528.

Saraiva JM, Baraitser M. 1992. Joubert syndrome: A review. *Am J Med Genet* **43**: 726–731.

Saraux H, Dhermy P, Fontaine JL, Boulesteix J, Lasfargue G, Grenet P, N'Ghiem M, Laplane R. 1970. La dégénérescence rétino-tubulaire de Senior et Løken [Senior–Løken retino-tubular degeneration]. *Arch Ophtalmol Rev Gen Ophtalmol* **30**: 683–696.

Saunier S, Calado J, Heilig R, Silbermann F, Benessy F, Morin G, Konrad M, Broyer M, Gubler MC, Weissenbach J, et al. 1997. A novel gene that encodes a protein with a putative src homology 3 domain is a candidate gene for familial juvenile nephronophthisis. *Hum Mol Genet* **6**: 2317–2323.

Saunier S, Calado J, Benessy F, Silbermann F, Heilig R, Weissenbach J, Antignac C. 2000. Characterization of the NPHP1 locus: Mutational mechanism involved in deletions in familial juvenile nephronophthisis. *Am J Hum Genet* **66**: 778–789.

Sayer JA, Otto EA, O'Toole JF, Nurnberg G, Kennedy MA, Becker C, Hennies HC, Helou J, Attanasio M, Fausett BV, et al. 2006. The centrosomal protein nephrocystin-6 is mutated in Joubert syndrome and activates transcription factor ATF4. *Nat Genet* **38**: 674–681.

Scheidecker S, Etard C, Pierce NW, Geoffroy V, Schaefer E, Muller J, Chennen K, Flori E, Pelletier V, Poch O, et al. 2014. Exome sequencing of Bardet–Biedl syndrome patient identifies a null mutation in the BBSome subunit BBIP1 (BBS18). *J Med Genet* **51**: 132–136.

Schmidts M, Arts HH, Bongers EM, Yap Z, Oud MM, Antony D, Duijkers L, Emes RD, Stalker J, Yntema JB, et al. 2013a. Exome sequencing identifies DYNC2H1 mutations as a common cause of asphyxiating thoracic dystrophy (Jeune syndrome) without major polydactyly, renal or retinal involvement. *J Med Genet* **50**: 309–323.

Schmidts M, Frank V, Eisenberger T, Al Turki S, Bizet AA, Antony D, Rix S, Decker C, Bachmann N, Bald M, et al. 2013b. Combined NGS approaches identify mutations in the intraflagellar transport gene IFT140 in skeletal ciliopathies with early progressive kidney disease. *Hum Mutat* **34**: 714–724.

Schmidts M, Vodopiutz J, Christou-Savina S, Cortes CR, McInerney-Leo AM, Emes RD, Arts HH, Tuysuz B, D'Silva J, Leo PJ, et al. 2013c. Mutations in the gene encoding IFT dynein complex component WDR34 cause Jeune asphyxiating thoracic dystrophy. *Am J Hum Genet* **93**: 932–944.

Schueler M, Braun DA, Chandrasekar G, Gee HY, Klasson TD, Halbritter J, Bieder A, Porath JD, Airik R, Zhou W, et al. 2015. DCDC2 mutations cause a renal-hepatic ciliopathy by disrupting Wnt signaling. *Am J Hum Genet* **96**: 81–92.

Senior B, Friedmann AI, Braudo JL. 1961. Juvenile familial nephropathy with tapetoretinal degeneration: A new oculorenal dystrophy. *Am J Ophthalmol* **52**: 625–633.

Shaheen R, Almoisheer A, Faqeih E, Babay Z, Monies D, Tassan N, Abouelhoda M, Kurdi W, Al Mardawi E, Khalil MM, et al. 2015a. Identification of a novel MKS locus defined by TMEM107 mutation. *Hum Mol Genet* **24**: 5211–5218.

Shaheen R, Schmidts M, Faqeih E, Hashem A, Lausch E, Holder I, Superti-Furga A, Mitchison HM, Almoisheer A, Alamro R, et al. 2015b. A founder CEP120 mutation in Jeune asphyxiating thoracic dystrophy expands the role of centriolar proteins in skeletal ciliopathies. *Hum Mol Genet* **24**: 1410–1419.

Shamseldin HE, Rajab A, Alhashem A, Shaheen R, Al-Shidi T, Alamro R, Al Harassi S, Alkuraya FS. 2013. Mutations in DDX59 implicate RNA helicase in the pathogenesis of orofaciodigital syndrome. *Am J Hum Genet* **93:** 555–560.

Sherman FE, Studnicki FM, Fetterman G. 1971. Renal lesions of familial juvenile nephronophthisis examined by microdissection. *Am J Clin Pathol* **55:** 391–400.

Shylo NA, Christopher KJ, Iglesias A, Daluiski A, Weatherbee SD. 2016. TMEM107 is a critical regulator of ciliary protein composition and is mutated in orofaciodigital syndrome. *Hum Mutat* **37:** 155–159.

Slavotinek AM, Stone EM, Mykytyn K, Heckenlively JR, Green JS, Heon E, Musarella MA, Parfrey PS, Sheffield VC, Biesecker LG. 2000. Mutations in MKKS cause Bardet–Biedl syndrome. *Nat Genet* **26:** 15–16.

Smith UM, Consugar M, Tee LJ, McKee BM, Maina EN, Whelan S, Morgan NV, Goranson E, Gissen P, Lilliquist S, et al. 2006. The transmembrane protein meckelin (MKS3) is mutated in Meckel–Gruber syndrome and the wpk rat. *Nat Genet* **38:** 191–196.

Srour M, Hamdan FF, Schwartzentruber JA, Patry L, Ospina LH, Shevell MI, Desilets V, Dobrzeniecka S, Mathonnet G, Lemyre E, et al. 2012a. Mutations in TMEM231 cause Joubert syndrome in French Canadians. *J Med Genet* **49:** 636–641.

Srour M, Schwartzentruber J, Hamdan FF, Ospina LH, Patry L, Labuda D, Massicotte C, Dobrzeniecka S, Capo-Chichi JM, Papillon-Cavanagh S, et al. 2012b. Mutations in C5ORF42 cause Joubert syndrome in the French Canadian population. *Am J Hum Genet* **90:** 693–700.

Srour M, Hamdan FF, McKnight D, Davis E, Mandel H, Schwartzentruber J, Martin B, Patry L, Nassif C, Dionne-Laporte A, et al. 2015. Joubert syndrome in French Canadians and identification of mutations in CEP104. *Am J Hum Genet* **97:** 744–753.

Steel BT, Lirenman DS, Battie CW. 1980. Nephronophthisis. *Am J Med* **68:** 531–538.

Stoetzel C, Laurier V, Davis EE, Muller J, Rix S, Badano JL, Leitch CC, Salem N, Chouery E, Corbani S, et al. 2006. BBS10 encodes a vertebrate-specific chaperonin-like protein and is a major BBS locus. *Nat Genet* **38:** 521–524.

Stoetzel C, Muller J, Laurier V, Davis EE, Zaghloul NA, Vicaire S, Jacquelin C, Plewniak F, Leitch CC, Sarda P, et al. 2007. Identification of a novel BBS gene (BBS12) highlights the major role of a vertebrate-specific branch of chaperonin-related proteins in Bardet–Biedl syndrome. *Am J Hum Genet* **80:** 1–11.

Sworn MJ, Eisinger AJ. 1972. Medullary cystic disease and juvenile nephronophthisis in separate members of the same family. *Arch Dis Child* **47:** 278.

Taschner M, Weber K, Mourao A, Vetter M, Awasthi M, Stiegler M, Bhogaraju S, Lorentzen E. 2016. Intraflagellar transport proteins 172, 80, 57, 54, 38, and 20 form a stable tubulin-binding IFT-B2 complex. *EMBO J* **35:** 773–790.

Thauvin-Robinet C, Lee JS, Lopez E, Herranz-Perez V, Shida T, Franco B, Jego L, Ye F, Pasquier L, Loget P, et al. 2014. The oral-facial-digital syndrome gene C2CD3 encodes a positive regulator of centriole elongation. *Nat Genet* **46:** 905–911.

Thevenon J, Duplomb L, Phadke S, Eguether T, Saunier A, Avila M, Carmignac V, Bruel AL, St-Onge J, Duffourd Y, et al. 2016. Autosomal recessive IFT57 hypomorphic mutation cause ciliary transport defect in unclassified oral-facial-digital syndrome with short stature and brachymesophalangia. *Clin Genet* doi: 10.111/cge.12785.

Thiel C, Kessler K, Giessl A, Dimmler A, Shalev SA, von der Haar S, Zenker M, Zahnleiter D, Stoss H, Beinder E, et al. 2011. NEK1 mutations cause short-rib polydactyly syndrome type Majewski. *Am J Hum Genet* **88:** 106–114.

Thomas S, Legendre M, Saunier S, Bessieres B, Alby C, Bonniere M, Toutain A, Loeuillet L, Szymanska K, Jossic F, et al. 2012. TCTN3 mutations cause Mohr-Majewski syndrome. *Am J Hum Genet* **91:** 372–378.

Thomas S, Wright KJ, Le Corre S, Micalizzi A, Romani M, Abhyankar A, Saada J, Perrault I, Amiel J, Litzler J, et al. 2014. A homozygous PDE6D mutation in Joubert syndrome impairs targeting of farnesylated INPP5E protein to the primary cilium. *Hum Mutat* **35:** 137–146.

Valente EM, Silhavy JL, Brancati F, Barrano G, Krishnaswami SR, Castori M, Lancaster MA, Boltshauser E, Boccone L, Al-Gazali L, et al. 2006. Mutations in CEP290, which encodes a centrosomal protein, cause pleiotropic forms of Joubert syndrome. *Nat Genet* **38:** 623–625.

Valente EM, Logan CV, Mougou-Zerelli S, Lee JH, Silhavy JL, Brancati F, Iannicelli M, Travaglini L, Romani S, Illi B, et al. 2010. Mutations in TMEM216 perturb ciliogenesis and cause Joubert, Meckel and related syndromes. *Nat Genet* **42:** 619–625.

Verpy E, Leibovici M, Zwaenepoel I, Liu XZ, Gal A, Salem N, Mansour A, Blanchard S, Kobayashi I, Keats BJ, et al. 2000. A defect in harmonin, a PDZ domain-containing protein expressed in the inner ear sensory hair cells, underlies Usher syndrome type 1C. *Nat Genet* **26:** 51–55.

Walczak-Sztulpa J, Eggenschwiler J, Osborn D, Brown DA, Emma F, Klingenberg C, Hennekam RC, Torre G, Garshasbi M, Tzschach A, et al. 2010. Cranioectodermal dysplasia, Sensenbrenner syndrome, is a ciliopathy caused by mutations in the IFT122 gene. *Am J Hum Genet* **86:** 949–956.

Waldherr R, Lennert T, Weber HP, Fodisch HJ, Scharer K. 1982. The nephronophthisis complex. A clinicopathologic study in children. *Virchows Arch A Pathol Anat Histol* **394:** 235–254.

Warady BA, Hébert D, Sullivan EK, Alexander SR, Tejani A. 1997. Renal transplantation, chronic dialysis, and chronic renal insufficiency in children and adolescents. The 1995 Annual Report of the North American Pediatric Renal Transplant Cooperative Study. *Pediatr Nephrol* **11:** 49–64.

Wei Q, Zhang Y, Li Y, Zhang Q, Ling K, Hu J. 2012. The BBSome controls IFT assembly and turnaround in cilia. *Nat Cell Biol* **14:** 950–957.

Weil D, Blanchard S, Kaplan J, Guilford P, Gibson F, Walsh J, Mburu P, Varela A, Levilliers J, Weston MD, et al. 1995. Defective myosin VIIA gene responsible for Usher syndrome type 1B. *Nature* **374:** 60–61.

Weil D, El-Amraoui A, Masmoudi S, Mustapha M, Kikkawa Y, Laine S, Delmaghani S, Adato A, Nadifi S, Zina ZB, et al. 2003. Usher syndrome type I G (USH1G) is caused by mutations in the gene encoding SANS, a protein that associates with the USH1C protein, harmonin. *Hum Mol Genet* **12:** 463–471.

Weston MD, Luijendijk MW, Humphrey KD, Moller C, Kimberling WJ. 2004. Mutations in the VLGR1 gene implicate G-protein signaling in the pathogenesis of Usher syndrome type II. *Am J Hum Genet* **74:** 357–366.

Wolf MT, van Vlem B, Hennies HC, Zalewski I, Karle SM, Puetz M, Panther F, Otto E, Fuchshuber A, Lameire N, et al. 2004. Telomeric refinement of the MCKD1 locus on chromosome 1q21. *Kidney Int* **66:** 580–585.

Wolf MT, Lee J, Panther F, Otto EA, Guan KL, Hildebrandt F. 2005. Expression and phenotype analysis of the nephrocystin-1 and nephrocystin-4 homologs in *Caenorhabditis elegans*. *J Am Soc Nephrol* **16:** 676–687.

Zaucke F, Boehnlein JM, Steffens S, Polishchuk RS, Rampoldi L, Fischer A, Pasch A, Boehm CW, Baasner A, Attanasio M, et al. 2010. Uromodulin is expressed in renal primary cilia and UMOD mutations result in decreased ciliary uromodulin expression. *Hum Mol Genet* **19:** 1985–1997.

Zhang Q, Nishimura D, Vogel T, Shao J, Swiderski R, Yin T, Searby C, Carter CS, Kim G, Bugge K, et al. 2013. BBS7 is required for BBSome formation and its absence in mice results in Bardet–Biedl syndrome phenotypes and selective abnormalities in membrane protein trafficking. *J Cell Sci* **126:** 2372–2380.

Zollinger HU, Mihatsch MJ, Edefonti A, Gaboardi F, Imbasciati E, Lennert T. 1980. Nephronophthisis (medullary cystic disease of the kidney). A study using electron microscopy, immunofluorescence, and a review of the morphological findings. *Helv Paediatr Acta* **35:** 509–530.

Cite this article as *Cold Spring Harb Perspect Biol* doi: 10.1101/cshperspect.a028191

Ciliary Mechanisms of Cyst Formation in Polycystic Kidney Disease

Ming Ma,[1] Anna-Rachel Gallagher,[1] and Stefan Somlo[1,2]

[1]Department of Internal Medicine, Yale University School of Medicine, New Haven, Connecticut 06520-8029

[2]Department of Genetics, Yale University School of Medicine, New Haven, Connecticut 06520-8029

Correspondence: stefan.somlo@yale.edu

Autosomal-dominant polycystic kidney disease (ADPKD) is a disease of defective tissue homeostasis resulting in active remodeling of nephrons and bile ducts to form fluid-filled sacs called cysts. The causal genes *PKD1* and *PKD2* encode transmembrane proteins polycystin 1 (PC1) and polycystin 2 (PC2), respectively. Together, the polycystins localize to the solitary primary cilium that protrudes from the apical surface of most kidney tubule cells and is thought to function as a privileged compartment that the cell uses for signal integration of sensory inputs. It has been proposed that PC1 and PC2 form a receptor-channel complex that detects external stimuli and transmit a local calcium-mediated signal, which may control a multitude of cellular processes by an as-yet unknown mechanism. Genetic studies using mouse models of cilia and polycystin dysfunction have shown that polycystins regulate an unknown cilia-dependent signal that is normally part of the homeostatic maintenance of nephron structure. ADPKD ensues when this pathway is dysregulated by absence of polycystins from intact cilia, but disruption of cilia also disrupts this signaling mechanism and ameliorates ADPKD even in the absence of polycystins. Understanding the role of cilia and ciliary signaling in ADPKD is challenging, but success will provide saltatory advances in our understanding of how tubule structure is maintained in healthy kidneys and how disruption of polycystin or cilia function leads to the pathological tissue remodeling process underlying ADPKD.

Autosomal dominant polycystic kidney disease (ADPKD) is one of the most common human monogenic diseases with an occurrence of ~1:1000 live births and more than 12 million people worldwide meeting radiographic diagnostic criteria for the disease. ADPKD patients develop cysts in the kidney that typically become clinically apparent by the fourth decade of life and increase in number and size over time, culminating in end-stage renal disease (ESRD) in ~50% of affected individuals in the sixth decade. Extrarenal cystic manifestations of ADPKD commonly include liver cysts, which arise from bile ducts. In the United States, 4%–5% of the prevalent ESRD population has ADPKD, which represents a significant burden to the patients, their families, and the health-care system. There are currently no approved therapies specifically targeting ADPKD in the United States.

The causal genes for ADPKD are *PKD1* and *PKD2*, encoding polycystin 1 (PC1) and polycystin 2 (PC2), respectively (The European Polycystic Kidney Disease Consortium 1994; The International Polycystic Kidney Disease Consortium 1995; Mochizuki et al. 1996). PC1 and PC2 are multimembrane-spanning proteins that interact with each other through their respective carboxy-terminal coiled-coil domains (Qian et al. 1997). Mature forms of both proteins localize to the primary cilium, a minute membrane-encased microtubule-based structure that protrudes from the apical surface of kidney tubule epithelial cells (Pazour et al. 2002; Yoder et al. 2002) that is used by cells as a sensory organelle for signal integration. Mutations in *PKD1* account for ~78% of ADPKD cases, mutations in *PKD2* account for the ~15% of cases, and the remaining 7%–8% of cases have no mutation detected in either gene (Audrezet et al. 2012; Heyer et al. 2016). PC1 is a 4302-amino-acid protein consisting of a large ~3000-amino-acid extracellular domain, 11 transmembrane domains, and an intracellular carboxyl terminus. PC1 undergoes autoproteolytic G-protein-coupled receptor proteolytic site (GPS) cleavage in the endoplasmic reticulum (ER) (Qian et al. 2002; Yu et al. 2007), which requires the structural integrity of the GPCR autoproteolysis-inducing (GAIN) domain (Arac et al. 2012; Tesmer 2012). This generates an amino-terminal fragment (PC1–NTF) and a carboxy-terminal fragment (PC1–CTF) that remains associated and they traffic together to the primary cilium (Cai et al. 2014; Kim et al. 2014; Gainullin et al. 2015). Additional subcellular locations have been reported for PC1 at the apical membrane and the lateral membrane, as well as in desmosomes (Ibraghimov-Beskrovnaya et al. 1997; Scheffers et al. 2000; Chapin et al. 2010).

PC2 has been described as a nonselective calcium permeable cation channel belonging to the transient receptor potential (TRP) channel family (Koulen et al. 2002), although direct measurement of its wild-type channel properties has not yet been achieved (DeCaen et al. 2013; Delling et al. 2013). PC2 has six transmembrane domains with a cytoplasmic NH_2-and carboxyl termini. The carboxyl terminus contains a coiled-coil domain that facilitates its interaction with PC1 (Tsiokas et al. 1997) and an EF-hand for calcium binding (Mochizuki et al. 1996; Celic et al. 2008). The amino terminus contains an RVxP motif that is important for targeting PC2 to the cilium (Geng et al. 2006). The carboxyl terminus contains a putative interaction domain that retains PC2 in the ER (Cai et al. 1999), and interaction with PC1 is required for full-length PC2 to leave the ER and reach the cilium (Kim et al. 2014; Gainullin et al. 2015). PC2 is serine/threonine phosphorylated in the carboxyl terminus and the phosphorylation changes the calcium activation properties of the channel measured in lipid bilayers and the trafficking of PC2 in cultured cells (Cai et al. 2004; Kottgen et al. 2005; Plotnikova et al. 2011; Streets et al. 2013). The in vivo physiological role of PC2 phosphorylation, including its potential role in the pathogenesis of ADPKD, remains to be determined.

ADPKD is typically transmitted as a dominant trait by loss-of-function alleles in *PKD1* or *PKD2*. Cyst initiation in ADPKD generally occurs by somatic "second-hit" mutations that inactivate the normal copy of the respective disease gene rendering it recessive at the cellular level (Qian et al. 1996; Watnick et al. 1998; Wu et al. 1998; Pei et al. 1999). Because the second hits are stochastic and can occur throughout life and throughout the kidney, cysts are focal and varied in size. This may, in part, account for the substantial intrafamilial clinical variation in ADPKD families, although genetic modifiers may also contribute to clinical variation. Recessive inheritance of hypomorphic *PKD1* alleles has been reported with clinical presentation of more homogeneous polycystic kidneys (Rossetti et al. 2009). Polycystic kidney disease progresses more rapidly in *PKD1* patients than *PKD2* patients. Individual cysts grow exponentially (Grantham et al. 2006). The rate of cyst growth in *PKD1* and *PKD2* are similar and the difference in severity is largely accounted for by the presence of more cysts early in life of *PKD1* compared with *PKD2* patients (Harris et al. 2006). This implicates cyst initiation, rather than cyst expansion, as the main determinant

Cite this article as *Cold Spring Harb Perspect Biol* doi: 10.1101/cshperspect.a028209

of disease progression rate (Grantham et al. 2008). Because the *PKD1* coding sequence is four times as long as that of *PKD2*, it is expected to sustain more second hits contributing to increased cyst initiation events.

Rather than being an all-or-none phenomenon, cyst formation has been shown to be a continuum that is dependent on the degree of functional impairment of PC1, which serves as the rate limiting step in the PC1–PC2 complex (Fedeles et al. 2011, 2014). The functional dosage of PC1 (and PC2) in human polycystic diseases can be modulated by several factors. The most extreme level is total loss of function, which occurs when both the germline PKD gene mutation and the second hit mutation result in complete loss of functional polycystin protein. A subset of germline nonsynonymous amino-acid substitution mutations in PC1 have reduced but not absent function (i.e., hypomorphic alleles), and these result in a milder course for ADPKD (Hopp et al. 2012; Cornec-Le Gall et al. 2013; Heyer et al. 2016). These hypomorphic alleles typically produce intact PC1 protein that may be relatively deficient in one of several putative functions. Changes that may lead to reduced function include relative decrease in the steady-state expression of PC1, defects in cleavage at the GPS, or impaired trafficking to cilia (Fedeles et al. 2011; Hopp et al. 2012; Cai et al. 2014). These features may not be entirely independent. For example, a recent study showed that mutations in the GAIN domain that abrogate GPS cleavage also impair cilia trafficking both in vivo and in vitro (Cai et al. 2014). Mutations in amino-terminal domains that do not affect GPS cleavage can nonetheless impair cilia trafficking, whereas mutations in the extracellular IgG-like PKD domains allow both cleavage and trafficking to cilia yet still result in a loss of function (Cai et al. 2014).

Other factors have also been shown to affect cyst growth. Cysts generated in utero grow extraordinarily fast as evidenced by monitoring cyst volume of individual cysts using magnetic resonance for more than 3 years (Grantham et al. 2010). Similarly, mouse genetic models of ADPKD show that cyst growth rates are much faster when Pkd gene inactivation occurs during kidney development than when it occurs in adult kidneys (Piontek et al. 2007). Different cell types and nephron segments in the kidney also have different cyst growth potential. Collecting duct cysts develop much faster and predominate over cysts originating in other nephron segments (Wu et al. 1998; Shibazaki et al. 2008; Fedeles et al. 2011). Although most models showing this rely on Cre recombinase–mediated gene inactivation, predominance of collecting duct cysts was also observed in $Pkd2^{ws25/-}$ mice (Wu et al. 1998). This model has an unstable allele that undergoes stochastic, rather than enzymatic, somatic inactivation, which eliminates any bias introduced by activity properties of the latter. Evidence suggests that collecting duct cysts also predominate in human ADPKD (Torres et al. 2012).

A UNIFYING THEORY OF CILIARY DYSFUNCTION UNDERLIES CYST FORMATION?

Cilia structural and compositional abnormalities underlie a wide range of recessive fibrocystic genetic diseases including nephronophthisis, Joubert syndrome, Meckel–Gruber syndrome, and Bardet–Biedl syndrome (Gerdes et al. 2009; Hildebrandt et al. 2011). These diseases are collectively termed "ciliopathies" and have pleiotropic manifestations in many organs, including fibrosis and cyst formation in the kidney and liver. It is important to note that all of the ciliopathies are recessively inherited. The primary cilium is a pivotal organelle for the pathogenesis of cystic kidney diseases. Almost all eukaryotic cells have a primary cilium. The cilium is connected with transition zone and basal body, which functionally separate the cilia compartment and its overlying membrane from the rest of the cell body and apical membrane of epithelial cells. Signaling molecules express on the ciliary membrane and mediate sensory signal inputs to detect and transmit stimuli from outside the cell (Menco et al. 1997; Corbit et al. 2005; Rohatgi et al. 2007). The protein products of many ciliopathy disease genes localize to the cilia–basal body complex (Hildebrandt et al. 2009; Jin et al. 2010; Garcia-Gonzalo et al.

2011; Sang et al. 2011). Recessive fibrocystic kidney phenotypes can be produced by genetic models that result in total loss of cilia (Lin et al. 2003; Weatherbee et al. 2009), in structural abnormalities of cilia (Bielas et al. 2009; Cui et al. 2011), and disturbance of membrane protein composition in cilia (Nishimura et al. 2004; Eichers et al. 2006; Garcia-Gonzalo et al. 2011). These data led to a unifying hypothesis that the mechanism of cyst formation in different genetic diseases is a result of cilia dysfunction, including abnormalities in cilia structure, composition, and signaling (Watnick and Germino 2003; Hildebrandt and Otto 2005; Hildebrandt and Zhou 2007). In this formulation, the detailed molecular mechanisms for individual diseases were lacking, and it was hard to know whether the exact function of cilia for each disease was independent of the others or had a more complex relationship.

Although cysts share some similarities across all polycystic kidney and liver diseases, the histopathological features and cellular events associated with the different diseases are not identical. The ADPKD cysts are focal and originate from all segments of the nephron. The autosomal recessive polycystic kidney disease cysts are fusiform and originate exclusively in the collecting duct. Kidney cysts in nephronophthisis are typically located in the corticomedullary junctional region and are associated with extensive fibrosis at earlier stages (Hildebrandt and Zhou 2007). Epithelial cell proliferation is associated with cyst formation in mouse models of ADPKD, whereas apoptosis is not apparent in cyst-lining cells (Shibazaki et al. 2008; Ma et al. 2013). In kidney cysts of *pcy* and *jck* mice, the respective genetic models of *Nphp3* and *Nphp9*, extensive apoptosis is observed in addition to epithelial cell proliferation (Omori et al. 2006; Smith et al. 2006). The different cellular events and histopathology associated with the different cystic diseases imply that distinct molecular mechanisms may underlie cyst formation in different forms of disease. Furthermore, heterozygous carrier adults for the ciliopathy disorders are typically asymptomatic throughout life. Presumably, second hits also affect the normal alleles in the adult heterozygous ciliopathy gene carriers, but they do not result in a discernible clinical phenotype during adult life. Additionally, conditional genetic removal of cilia in the kidney tubules of rodents result in a cystic phenotype in both early-onset and adult models (Lin et al. 2003; Davenport et al. 2007), but cysts progress at a much slower pace than in *Pkd1* conditional mutants models in which only a single cilia component protein is removed (Piontek et al. 2007; Shibazaki et al. 2008). ADPKD may be related to ciliopathy disorders by virtue of the ciliary location of the affected gene products, but it is genetically and mechanistically distinct from ciliopathies. Together, these and other findings suggested that the mechanisms of ADPKD and ciliopathies may be distinct.

CILIA STIMULATE CYST GROWTH IN ADPKD MODELS

Experimental support for a distinct mechanistic relationship between polycystins and cilia came from studies showing that removing cilia in mouse models of *Pkd1* or *Pkd2* inactivation suppresses cyst growth (Ma et al. 2013). Conditional mouse models in which cilia and either *Pkd1 or Pkd2* were inactivated in the same population of cells in kidney tubules and liver bile ducts showed that removal of cilia suppresses cyst growth in both early and adult-inactivation models in both organs (Ma et al. 2013). Cilia-dependent cyst growth was applicable to all nephron segments, including proximal tubule, medullary thick ascending loop of Henle, and collecting duct. Loss of polycystins results in slow, but elevated, proliferation rates compared with wild-type kidney tubule epithelia. This increased proliferation is a marker for disease progression, but is likely not sufficient to generate cysts as a forced expression of *Cux1* in cilia (*Ift88*) mutant kidney does not increase cysts (Sharma et al. 2013). The increased proliferation in ADPKD models is suppressed in cilia−polycystin double mutants (Ma et al. 2013). Transgenic overexpression of PC1 in vivo in a cilia mutant did not have an impact on cystic burden, suggesting that the cysts resulting from cilia removal were independent of PC1 function

(Ma et al. 2013; Wills et al. 2014). Finally, the severity of the polycystic kidney disease in the cilia–polycystin double knockouts is directly correlated to the length of time between the initial disappearance of polycystin protein and the subsequent involution of preformed cilia (Ma et al. 2013).

Together, these findings provide genetic evidence for an unidentified ciliary signal(s) termed cilia-dependent cyst activation (CDCA), which is dependent on intact cilia for activity and is normally inhibited by the functioning of polycystins (Fig. 1) (Ma et al. 2013). CDCA is presumably derepressed in a regulated manner by the normal physiological function of polycystins, although the nature of this physiological role is unknown. Following somatic second hit inactivation of polycystins, CDCA becomes constitutively and pathologically derepressed if intact cilia persist, giving rise to ADPKD (Fig. 1). The aggregate data indicate that cilia-dependent cyst growth is a universal mechanism of cyst formation in ADPKD because of mutations in *Pkd1* and *Pkd2*. This suggests that it is essential when testing therapeutic compounds for ADPKD to use orthologous models and to reasonably expect that an effective preclinical intervention should work in both early and late inactivation models, *Pkd1* and *Pkd2* models, all nephron segments, and the bile duct as well.

WHAT DOES CDCA DO?

What may the physiological role of CDCA be? One hypothesis may be that it is a dynamic regulator of nephron tubule lumen diameter and of nephron tubule epithelial cell morphology (Fig. 1). Under normal conditions, PC1–PC2-regulated CDCA achieves a dynamic balance to modulate tubule lumen diameter and epithelial-lining cell shape under different physiological conditions including, perhaps, altered luminal flow rate or ligand binding. PC1 is the rate limiting component for CDCA and likely encodes a graded signal whose functional activity is dependent on the quantitative nature of the signal input or PC1 dosage (Fedeles et al. 2011, 2014; Hopp et al. 2012). When PC1–PC2

signaling is attenuated below a certain threshold by mutation (Fedeles et al. 2011; Hopp et al. 2012), inhibition of CDCA is chronically weakened. Under normal conditions, polycystin-dependent loss of inhibition is reversible when the signal to reduce polycystin function is reversed. In the case of chronically weakened of absent polycystin function, the otherwise reversible process becomes indolent and persistent and tubules begin to remodel the surrounding basement membranes and parenchyma with attendant cell proliferation and change in morphology eventually causing cysts to form.

MAMMALIAN HEDGEHOG SIGNALING AND THE POTENTIAL COMPLEXITY OF CDCA

The biggest challenge in understanding the cellular molecular mechanisms of ADPKD is to discover the specific local polycystin-regulated signaling processes that occur in cilia. One paradigm for the operation of CDCA–polycystin signaling in cilia may be found in the mammalian Hedgehog (Hh) signaling pathway, in which cilia play an essential, but complex, role. The molecular components of Hh signaling were originally uncovered from a genetic screen for segment polarity mutants in *Drosophila melanogaster* larvae (Nusslein-Volhard and Wieschaus 1980). Forward genetic screens in the mouse identified the primary cilium to be essential for the mammalian Hh pathway particularly in neural tubule patterning (Huangfu et al. 2003). Almost all known conserved Hh pathway components, including Patched1 (Ptch) (Rohatgi et al. 2007), Smoothened (Corbit et al. 2005), Gli2, Gli3 (Huangfu and Anderson 2005), suppressor of fused (Sufu) (Haycraft et al. 2005), and Kif7 (He et al. 2014), localize to the cilia. In the absence of Hh ligand, Ptch is expressed on primary cilia and the signaling is shut off. When Hh ligand binds to ciliary Ptch, Ptch exits the cilia, allowing Smoothened to translocate into the cilia and transduce downstream signaling (Corbit et al. 2005; Rohatgi et al. 2007). Cilia are also required for generating the functional forms of downstream Gli proteins, including generating the truncated func-

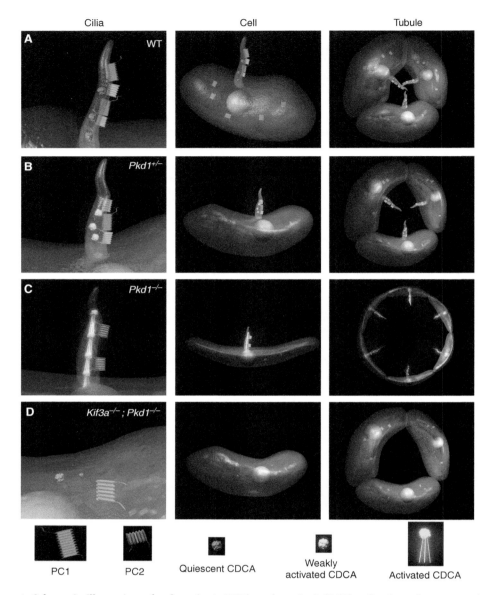

Figure 1. Schematic illustration of polycystin 1 (PC1)–polycystin 2 (PC2)–cilia-dependent cyst activation (CDCA) signaling. (*A*) The PC1–PC2 complex is expressed in cilia and maintains CDCA in a physiologically regulated quiescent state with epithelial cells retaining a columnar shape and the tubule with normal lumen diameter. Normal physiological input would adjust CDCA to physiological needs of cell shape, lumen morphology, cell transport, metabolic properties, etc. WT, Wild type. (*B*) Reduced PC1 dosage in heterozygous cells of autosomal-dominant polycystic kidney disease (ADPKD) patients may lead to weak constitutive activation of CDCA and a modest steady-state change in cell shape (less columnar) and lumen diameter (increased). (*C*) Loss of PC1 in the presence of intact cilia, the condition for cyst initiation in ADPKD, leads to inexorable activation of CDCA, profound changes in cells to a more squamoid shape, low-level proliferation, active remodeling of surrounding kidney parenchyma, and growth of cysts. Note that the images of cells and tubules (cysts) in panel *C* are at a much lower illustrative "magnification" than *A*, *B*, or *D*. (*D*) Loss of cilia in the absence of PC1 markedly reduces the activation of CDCA and maintains cell shape and tubule lumen diameter in a more normal range.

tional repressor form of Gli3R (Haycraft et al. 2005; Huangfu and Anderson 2005). Primary cilia have been shown to both mediate and suppress Hh pathway-dependent tumorigenesis (Han et al. 2009; Wong et al. 2009). The primary cilium does not participate in the signaling per se; rather, it is a signal integration compartment that positively and negatively regulates the signaling events that are transduced (Goetz and Anderson 2010; Ocbina et al. 2011). To draw the analogy to Hh signaling, the normal physiological function of PC1 and PC2 in cilia is an inhibitory signal, like that of Ptch. Ptch mutants result in unchecked activation of the Hh pathway (Goodrich et al. 1997), whereas in Ptch-cilia double mutants, this activation is attenuated (Huangfu et al. 2003). This has parallels to what is observed in polycystin mutants and polycystin–cilia double mutant. Given the complexity of the components of Hh downstream from Ptch, one may begin to get an appreciation of the potential nuances of CDCA. A critical step toward understanding CDCA will be the improved discovery of the other proteins present in cilia (Mukhopadhyay et al. 2013; Mick et al. 2015).

CYST FORMATION IN OTHER CILIOPATHIES IN RELATION TO PC1–PC2–CDCA BALANCE

The cystic phenotypes in nephronophthisis, Joubert syndrome, Meckel–Gruber syndrome, and Bardet–Biedl syndrome might be the result of imbalance of CDCA–PC1–PC2 signaling pathway in cilia. The gene products involved for the recessive fibrocystic disorders in part regulate structural architecture and composition of cilia and centrosomes through control of trafficking of proteins in and out of these compartments (Garcia-Gonzalo et al. 2011). The animal models for the recessive ciliopathies develop less severe cystic disease than ADPKD models (Nishimura et al. 2004; Eichers et al. 2006; Omori et al. 2006; Davenport et al. 2007; Weatherbee et al. 2009). Such an outcome may be the result of compromised polycystin-mediated inhibitory function resulting in modest activation of CDCA as opposed to unchecked

activation of CDCA resulting in more aggressive cyst growth in ADPKD models (Piontek et al. 2007; Shibazaki et al. 2008). The less severe kidney disease observed in the recessive ciliopathy models than that in ADPKD mouse models may be analogous to the less severe neural tube dorsoventral patterning defect seen in the cilia structural mutants than in Ptch mutants (Goetz and Anderson 2010).

Polycystins and ciliary CDCA components are likely the cargo and targets of the transition zone sorting and gating machinery subsumed by some of the recessive ciliopathy gene products. It has been shown that some mutations in ciliary gating machinery components result in defective trafficking of ciliary membrane proteins, including Smoothened and PC2 (Garcia-Gonzalo et al. 2011). In this context, polycystins and CDCA might not traffic sufficiently to cilia and thus generate a slow forming cystic phenotype. Interestingly PC1 is reported to interact with BBS1, BBS4, BBS5, and BBS8 and ciliary PC1 expression is reduced in BBS1 knockdown cells and BBS3 mutant cells (Su et al. 2014). These findings support the possibility that relative PC1 deficiency may underlie a component of the cystic kidney phenotype in Bardet–Biedl syndrome. Based on this hypothesis, it would be expected the overexpression of physiologically active PC1 would be able to ameliorate the kidney cystic phenotypes in some of the recessive ciliopathy mutants in a manner analogous to what has been shown for isolated polycystic liver diseases (Fedeles et al. 2011).

CILIARY SIGNALING PATHWAYS AS CANDIDATES FOR CDCA

The cilium is essential for the signal transduction of a number of signaling pathways, any of which may be candidate pathways for CDCA. There also exist a plethora of proteins that localize to the cilium, but whose roles in cilia signal transduction are not fully understood. CDCA may well be among these unknown pathways. Here, we briefly discuss the known cilia associated signaling pathways and comment on their roles as candidates for CDCA and ADPKD.

Hedgehog Signaling

Recent studies have described genetic interaction of the Hh signaling pathway with cystic kidney diseases models. These findings raise the question of whether Smoothened could be a component of CDCA and could Hh signaling modulate the CDCA–PC1–PC2 to control ADPKD? Hh pathway transcriptional target genes are elevated in *Thm1* (IFT complex A) and *jck* (*Nphp9*, *Nek8*) cystic kidneys and in kidneys of an early inactivation model of *Pkd1* (Tran et al. 2014). Reduced Gli2 activity slowed cyst growth in the *Thm1* mutant kidney. In *Pkd2l1* mutant cilia, where ciliary calcium signaling is impaired as had been hypothesized (but not shown) for PC1–PC2 mutants, Hh agonist- mediated expression of Gli1 is reduced and cilia tip localization of Gli2 is diminished (Delling et al. 2013). Examination of the *Pkd2l1* mutant mice shows low penetrance of a gut looping chirality defect, a process dependent on Hh signaling (Zhang et al. 2001; Delling et al. 2013). Therefore, ciliary polycystin paralogs and some recessive ciliopathy genes modulate ciliary Hh signaling. Nonetheless, the contribution of Hh signaling to cyst formation in ADPKD remains an open question that is worth resolving, especially in adult-onset ADPKD models. If inactivation of Hh pathway suppresses cyst growth in adult-onset ADPKD models, this would suggest that Hh signaling promotes cyst progression making it a candidate for CDCA.

cAMP

Elevated cAMP levels have been associated with ADPKD, ARPKD (*pck*), and NPHP (*pcy*) rodent models and vasopressin receptor 2 antagonist administration slows cyst growth in these models (Gattone et al. 2003; Torres et al. 2004). Moreover, a competitive vasopressin receptor 2 antagonist has undergone human clinical trials with some reduction in kidney volume in ADPKD patients (Torres et al. 2012), although the drug has not been approved for use in ADPKD in the United States. Studies have shown that PC2 and phosphodiesterase 4C are components of a ciliary A-kinase-anchoring protein complex that is disrupted in the absence of *Pkd2* (Choi et al. 2011). Interestingly, conditional inactivation of the cilia-expressed adenylyl cyclase 6 (AC6) concomitantly with inactivation of Pkd1 in the collecting duct reduces cystic burden in an early-onset model of ADPKD (Rees et al. 2014). These findings suggest that further studies are needed to evaluate whether AC6 and cAMP inhibition are also effective in other nephron segments, in adult-onset models and in other organs like liver—features that would suggest they may be central to CDCA in ADPKD. Even if that is not the case, because a large portion of cysts in human ADPKD are derived from the collecting duct in which cAMP activation has been observed, blocking this pathway genetically or pharmacologically remains an attractive target (Torres et al. 2004, 2012; Rees et al. 2014).

Ciliary Lkb1 and mTOR Signaling

Mammalian target of rapamycin (mTOR) signaling integrates the nutrient, oxygen, and energy state of the cells and controls cell growth, proliferation, and survival. The Lkb1–AMPK branch of this pathway negatively regulates mTOR activity. Lkb1 localizes to the primary cilium and regulates mTORC1 activity and cell size in response to flow (Boehlke et al. 2010). In this model, inactivation of PC2 reduced calcium transients in cells but had no effect on cilia-dependent mTORC1 activation. Ciliary Lkb1-mediated mTORC1 signaling is independent of PC2 signaling and Lkb1 is less likely to be part of the CDCA pathway (Boehlke et al. 2010). mTOR signaling is also unlikely to be candidate of CDCA because loss of PC1 is not sufficient to activate mTOR. Activation of this pathway is mosaic in cyst-lining epithelia that are completely devoid of *Pkd1* (Shillingford et al. 2006, 2010). In another study, mTOR activation was not observed in an adult-onset ADPKD model at both precystic and cystic stage when CDCA was active (Ma et al. 2013). The mosaic activation of mTOR signaling in cyst-lining cells, however, could serve as a booster for cyst growth in ADPKD. Pharmacological inhibition of the mTOR pathway ameliorates cystic burden in

ADPKD mouse models (Shillingford et al. 2010) but has yet to prove beneficial in ADPKD patients perhaps because of limitations in delivering the needed dose to the kidney (Serra et al. 2010; Torres et al. 2010; Walz et al. 2010).

Ciliary PDGFRαα and MAPK/ERK Signaling

PDGFRαα (PdgfrAA) is expressed in the primary cilia of fibroblast cells and on treatment with PDGF ligand, phosphorylation of PdgfrAA, and phospho-MEK1/2 is detected in cilia (Schneider et al. 2005). It has been shown that the activation of PdgfrAA is reduced in Ift88 mutant Tg737 fibroblast cells that have short and stumpy cilia compared with that in control cells, showing the integrity of cilia is required for efficient activation of the pathway (Schneider et al. 2005). The PdgfrAA pathway in animal models of ADPKD has not been directly evaluated for a role in cyst initiation or progression. MAPK/ERK signaling is not likely to be CDCA as its activation is mainly restricted to the distal nephron and at later stages of cyst formation in ADPKD models (Shibazaki et al. 2008; Ma et al. 2013). Moreover, attenuation of the MAPK/ERK pathway pharmacologically has no impact on cyst progression in a developmental model of ADPKD, although the effects of longer-term therapy could not be evaluated (Shibazaki et al. 2008). MAPK/ERK activation is found in cyst lining cells of *pcy* (*Nphp3*) mice and blocking activation of the pathway with an inhibitor attenuates cyst progression in these animals (Omori et al. 2006). Thus MAPK/ERK signaling acts as a driver for cyst progression in some ciliopathy models model, but not ADPKD models.

Extracellular Matrix and Integrin Signaling

Multiple genetic interactions between polycystin signaling and extracellular matrix (ECM) components have been shown in a number of organ systems including the kidney. Knockdown of *Pkd1a/b* and *Pkd2* in zebrafish results in collagen deposition in the notochord and generates the curly tail "up" phenotype (Mangos et al. 2010), a commonly observed characteristic in cilia defective mutants (Sun et al. 2004). Another ECM protein, fibrillin, encoded by *FBN1*, which is the causal gene in Marfan syndrome, has been shown to genetically interact with *Pkd1*. *Pkd1* and *Fbn1* compound heterozygote mice exacerbate vascular complications observed in *Fbn1* heterozygotes resulting in increased TGF-β signaling. Inactivation of *Pkd1* in the vascular smooth muscle cells is sufficient to generate vascular complications associated with elevated TGF-β signaling (Liu et al. 2014). More direct evidence of ECM signaling involvement in the pathogenesis of ADPKD came from the recent finding that inactivation of β1-integrin suppresses cyst formation in a developmental model of ADPKD (Lee et al. 2014). In vitro knockdown of β1-integrin reduces cyst formation by *Pkd1* knockdown cells in a matrix−gel model (Subramanian et al. 2012). Interestingly, multiple integrin receptors are expressed in primary cilia of MDCK cells (Praetorius et al. 2004) and chondrocytes (McGlashan et al. 2007) and β1-integrin has been localized to cilia in the rat kidney. Ciliary β1-integrin has been shown in MDCK cells to potentiate fibronectin-induced calcium signaling (Praetorius et al. 2004). Considering that the removal of β1-integrin ameliorates the cystic phenotype in the early inactivation model of *Pkd1* (Lee et al. 2014), further studies on the role of β1-integrin in adult-onset models of ADPKD would be reasonable. If confirmed, investigation of the ciliary β1-integrin as a candidate for CDCA may be indicated.

THE MECHANOSENSOR HYPOTHESIS FOR POLYCYSTINS IN CILIA

PC1 has been hypothesized to have receptor-type function on cilia. It has long been proposed to be a flow sensor (Nauli et al. 2003) although recent studies have called that into question (Delling et al. 2016). More recently, it has been proposed to be a Wnt ligand coreceptor (Kim et al. 2016). A common downstream effect as part of these hypotheses is polycystin-regulated calcium entry through the PC1−PC2 complex. The recent development of direct channel measurements by patch clamping of cilia and of genetically encoded calcium sensors targeted to cilia have begun to allow more direct measurement and visualization of intraciliary calcium transients under different physiological conditions (DeCaen et al. 2013; Delling et al. 2013; Su et al. 2013; Jin et al. 2014). Studies directly measuring channel activity on the surface of cilia have been unable to detect the PC1–PC2 channel activity (DeCaen et al. 2013) and have instead argued that ciliary calcium entry could not be the cellular signaling event of polycystin action emerging from cilia (Delling et al. 2013; Delling et al. 2016). Specifically, primary cilia of human retinal pigment epithelial cells and other cells are specialized calcium signaling organelles whose resting calcium concentration is up to sevenfold higher than cytosolic calcium (DeCaen et al. 2013). The intracilia calcium appears to be regulated by Pkd1l2–Pkd2l1, and not the anticipated PC1–PC2 complex (Delling et al. 2013). Ciliary Pkd1l2–Pkd2l1 is essential for efficient activation of Hh signaling via modulating the expression of Gli1 and Gli2 localization at the ciliary tip. *Pkd2l1* mutant mice display low penetrance of gut looping chirality defect, a process dependent on Hh signaling during development (Delling et al. 2013). Applying physiologically relevant levels of mechanical force to bend or deflect cilia resulted in no alteration of calcium transients in cilia of the kidney tubule and other cell types (Delling et al. 2016). Laminar flow applied to ciliated cells generates shear stress and initiates calcium transients in the cytoplasm, which then propagate to the cilia (Delling et al. 2016), not vice versa. Delling et al. (2016) propose that the mechanosensation function of cilia, if it exists, is through mediators other than whole cell calcium transients. It is possible that PC2 exerts its function via allosteric transfer of calcium ion from PC1–PC2 to other signaling molecules, like CDCA components in the local context of the cilium, but massive cellular calcium transients are not required.

Polycystins, particularly PC2, function in other tissues aside from the kidney and liver. One example is in the vertebrate embryonic node where the left–right axis is established during early embryonic development (Nonaka

et al. 1998; Essner et al. 2002; Pennekamp et al. 2002; McGrath et al. 2003). In mammals, the motile cilia at the center of the node propel fluid to generate leftward flow, and the crown cells at the periphery of the node express PC2 on nonmotile cilia that may sense this flow (McGrath et al. 2003; Yoshiba et al. 2012). Loss of PC2 results in abnormal left–right axis formation (Pennekamp et al. 2002). Reexpression of PC2 specifically in the crown cells in *Pkd2* mutant nodes is sufficient to initiate lateralized *Nodal* and *Pitx2* expression, and ciliary localization of PC2 is required for this process (Yoshiba et al. 2012). *Pkd1l1*, a homolog of *Pkd1*, may act as the partner with PC2 in nodal cilia. Together, they play a role in establishing the left–right axis, a function that is conserved in vertebrates (Vogel et al. 2010; Field et al. 2011; Kamura et al. 2011). In cell culture systems, artificial flow can deflect cilia and trigger intraciliary calcium transients (Su et al. 2013). Live imaging of calcium transients in zebrafish Kupffer's vesicle (KV), the zebrafish counterpart of mouse embryonic node, shows that calcium transients propagate from cilia to the cytoplasm of the crown cells on the left side of the KV and that the calcium transients oscillate over time followed by lateralized gene expression (Yuan et al. 2015). These calcium transients are highly reduced in *Pkd2* knockdown embryos or when motile cilia are paralyzed. Introduction of cilia-targeted genetically encoded calcium chelators dampen ciliary calcium transients and generate left–right defective embryos. The investigators conclude that intraciliary calcium mediated by PC2 is essential to initiate left–right axis determination in zebrafish (Yuan et al. 2015).

A number of salient differences in the experimental systems used could in part explain the disparate ciliary calcium kinetics in these studies (DeCaen et al. 2013; Delling et al. 2013; Su et al. 2013; Jin et al. 2014; Yuan et al. 2015). The differences in combinatorial pairing of ciliary PC1 and PC2 or their respective homologs might generate different calcium responses that are further specialized by cell and tissue type to meet specific physiologic functional requirements. The differences in channel activity and transients between mouse epithelial cells and fish KV may be a result of physiological differences between stationary primary cilia in the mammalian cells and the active motile cilia in KV in zebrafish (Becker-Heck et al. 2011). Furthermore, the milieu in cell culture experiments and the fluid in the KV have different compositions of chemical and bioactive molecules. The differences in observed activity may also stem from incomplete knowledge of biological and physiological regulators of the respective cilia channels. Nonetheless, the genetically encoded calcium indicators and chelators and the technical advances in patch clamping offer novel physiological tools to measure the in vivo intraciliary calcium transients in kidney and liver and relate PC1–PC2 function to physiological properties of cilia. These advances will help dissect the function of the intracilia calcium and polycystins across a spectrum of biological systems in the future.

HOW DOES PC1 REGULATE CDCA?

Given the unique structural features of PC1 and the importance of GPS cleavage to its function (Qian et al. 2002; Yu et al. 2007; Cai et al. 2014), it is worth considering hypotheses based on the functional inferences from other GPS cleaved proteins. Adhesion G-protein-coupled receptors (aGPCR) are a class of GPCRs that are characterized by a large extracellular domain (ECD), seven transmembrane spans (7TM), and a GAIN/GPS domain located near the junction of the ECD and the first transmembrane domain. GAIN/GPS domain containing aGPCRs undergo autoproteolytic cleavage to generate ECD and 7TM that associate with each other noncovalently (Liebscher et al. 2013), in a manner analogous to what has been proposed for PC1. ECD dissociation from the 7TM is thought to result in activation of disparate downstream G protein signaling by GPR56 and GPR110, implicating that the association of ECD with 7TM is autoinhibitory (Stoveken et al. 2015). The ECD displacement unmasks the amino-terminal stalk regions of 7TM, which serve as autoligands essential for G protein signaling (Stoveken et al. 2015). A missense variant (p.C492Y) in the GAIN domain of another aGPCR, ADGRE2 (EMR2), weakens the ECD and 7TM autoinhibitory interaction and sensitizes mast cells to vibration-induced degranulation in a familial form vibratory urticaria (Boyden et al. 2016).

The autoinhibitory action of ECD on 7TM in aGPCR could have implications for how PC1–NTF may regulate PC1–CTF and how their dynamic interactions regulate CDCA to control target phenotypes such as lumen diameter or kidney epithelial cell shape (Fig. 2). Strong association of PC1–NTF with PC1–CTF suppresses the activation of CDCA. Either ligand binding or mechanical stimuli such as luminal flow may trigger stimulus dosage-dependent displacement of PC1–NTF from PC1–CTF, which could unmask the amino-terminal stalk region of PC1–CTF. This stalk protein could act as an autoactivating ligand to relieve PC1–CTF inhibition of CDCA with consequent downstream activation of CDCA-dependent changes in the tubule epithelial cells. The PC1 signal can be turned off either by reassociation of PC1–NTF with PC1–CTF or, if the signal deactivation has a longer time course, by removal of the unbound PC1–CTF from cilia and its replacement with intact PC1–NTF/CTF complex. The latter model would seem more likely in cilia projecting into the urinary space of the nephron lumen. In the case of ADPKD, when there is little or no PC1 present in the cilia, the net effect of this absence is the same as having PC1–CTF present but dissociated from PC1–NTF. It is the intact PC1–NTF/CTF complex that keeps CDCA in check. The identification of CDCA and the study of its interaction with PC1 and PC2 may explain how tubule lumen diameter or cell shape or differentiation state is dynamically fine-tuned to meet the demand of various physiological conditions and how unchecked CDCA activation leads to the indolent, but profound tissue remodeling processes that underlie ADPKD.

CONCLUSION

Polycystic kidney diseases encompass a broad spectrum of genetic diseases. Cysts can develop via nonidentical and nonoverlapping cellular

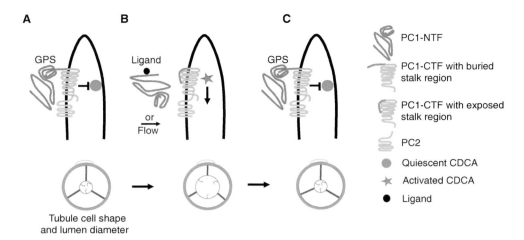

Figure 2. Model of PC1 regulation after GPS cleavage. (*A*) Following G-protein-coupled receptor proteolytic site (GPS) cleavage, PC1−NTF and PC1−CTF remain associated with each other and traffic to cilia where the PC1−PC2 complex maintains cilia-dependent cyst activation (CDCA) in a regulated quiescent state. (*B*) In the presence of a stimulus, either ligand binding or mechanical input such as flow, PC1−NTF is displaced from PC1−CTF. This exposes the buried stalk region of PC1−CTF (red), which may serve as an autoligand that releases PC1−CTF inhibition of CDCA. Transient activation of CDCA leads to alterations in cell shape and lumen diameter (*bottom* panels). (*C*) Either reassociation of PC1−NTF with PC1−CTF or retrieval of unbound PC1−CTF from cilia and its replacement with PC1−NTF/CTF−PC2 complex returns CDCA to a reduced state of activity and the tubules to a baseline morphology.

events, which implies different molecular mechanisms for the pathogenesis of each group or class of disease entity. The pathogenetic mechanism of each disease requires focused investigation. In the case of ADPKD caused by mutations in *PKD1* or *PKD2*, the protective role of loss of cilia suggests that the PC1−PC2 complex serves to regulate a ciliary signaling pathway(s) (CDCA) that is necessary to drive cyst growth when either polycystin is inactivated. This serves as a strong proof of concept that targeting the components of CDCA should slow cyst growth. The challenge remains in discovering CDCA, whose components most likely localize to and function in cilia. The advances in generation and use of animal models, the development of novel biological tools such as cilia-targeted genetically encoded calcium sensors and chelators and enhanced cilia proteomics, and the improved understanding of the biochemical properties of polycystins and related proteins offer exciting prospects for novel discoveries for the next phase of understanding and eventually treating ADPKD.

ACKNOWLEDGMENTS

We apologize to the community that many papers could not be cited because of space limitations. We thank Zhaoning Liu for artwork. This work is supported by grants to S.S. from the National Institutes of Health (NIH) (R01 DK54053 and R01 DK100592) and the Lillian Goldman Charitable Trust. M.M. and A.-R.G. are supported by Polycystic Kidney Disease Foundation Research Grants-in-Aid (51ga13 and 46ga13).

REFERENCES

Arac D, Boucard AA, Bolliger MF, Nguyen J, Soltis SM, Sudhof TC, Brunger AT. 2012. A novel evolutionarily conserved domain of cell-adhesion GPCRs mediates autoproteolysis. *EMBO J* **31:** 1364–1378.

Audrezet MP, Cornec-Le Gall E, Chen JM, Redon S, Quere I, Creff J, Benech C, Maestri S, Le Meur Y, Ferec C. 2012. Autosomal dominant polycystic kidney disease: Comprehensive mutation analysis of *PKD1* and *PKD2* in 700 unrelated patients. *Hum Mutat* **33:** 1239–1250.

Becker-Heck A, Zohn IE, Okabe N, Pollock A, Lenhart KB, Sullivan-Brown J, McSheene J, Loges NT, Olbrich H,

Haeffner K, et al. 2011. The coiled-coil domain containing protein CCDC40 is essential for motile cilia function and left–right axis formation. *Nat Genet* **43:** 79–84.

Bielas SL, Silhavy JL, Brancati F, Kisseleva MV, Al-Gazali L, Sztriha L, Bayoumi RA, Zaki MS, Abdel-Aleem A, Rosti RO, et al. 2009. Mutations in *INPP5E*, encoding inositol polyphosphate-5-phosphatase E, link phosphatidyl inositol signaling to the ciliopathies. *Nat Genet* **41:** 1032–1036.

Boehlke C, Kotsis F, Patel V, Braeg S, Voelker H, Bredt S, Beyer T, Janusch H, Hamann C, Godel M, et al. 2010. Primary cilia regulate mTORC1 activity and cell size through Lkb1. *Nat Cell Biol* **12:** 1115–1122.

Boyden SE, Desai A, Cruse G, Young ML, Bolan HC, Scott LM, Eisch AR, Long RD, Lee CC, Satorius CL, et al. 2016. Vibratory urticaria associated with a missense variant in *ADGRE2*. *N Engl J Med* **374:** 656–663.

Cai Y, Maeda Y, Cedzich A, Torres VE, Wu G, Hayashi T, Mochizuki T, Park JH, Witzgall R, Somlo S. 1999. Identification and characterization of polycystin-2, the *PKD2* gene product. *J Biol Chem* **274:** 28557–28565.

Cai Y, Anyatonwu G, Okuhara D, Lee KB, Yu Z, Onoe T, Mei CL, Qian Q, Geng L, Wiztgall R, et al. 2004. Calcium dependence of polycystin-2 channel activity is modulated by phosphorylation at Ser812. *J Biol Chem* **279:** 19987–19995.

Cai Y, Fedeles SV, Dong K, Anyatonwu G, Onoe T, Mitobe M, Gao JD, Okuhara D, Tian X, Gallagher AR, et al. 2014. Altered trafficking and stability of polycystins underlie polycystic kidney disease. *J Clin Invest* **124:** 5129–5144.

Celic A, Petri ET, Demeler B, Ehrlich BE, Boggon TJ. 2008. Domain mapping of the polycystin-2 C-terminal tail using de novo molecular modeling and biophysical analysis. *J Biol Chem* **283:** 28305–28312.

Chapin HC, Rajendran V, Caplan MJ. 2010. Polycystin-1 surface localization is stimulated by polycystin-2 and cleavage at the G protein-coupled receptor proteolytic site. *Mol Biol Cell* **21:** 4338–4348.

Choi YH, Suzuki A, Hajarnis S, Ma Z, Chapin HC, Caplan MJ, Pontoglio M, Somlo S, Igarashi P. 2011. Polycystin-2 and phosphodiesterase 4C are components of a ciliary A-kinase anchoring protein complex that is disrupted in cystic kidney diseases. *Proc Natl Acad Sci* **108:** 10679–10684.

Corbit KC, Aanstad P, Singla V, Norman AR, Stainier DY, Reiter JF. 2005. Vertebrate smoothened functions at the primary cilium. *Nature* **437:** 1018–1021.

Cornec-Le Gall E, Audrezet MP, Chen JM, Hourmant M, Morin MP, Perrichot R, Charasse C, Whebe B, Renaudineau E, Jousset P, et al. 2013. Type of *PKD1* mutation influences renal outcome in ADPKD. *J Am Soc Nephrol* **24:** 1006–1013.

Cui C, Chatterjee B, Francis D, Yu Q, SanAgustin JT, Francis R, Tansey T, Henry C, Wang B, Lemley B, et al. 2011. Disruption of Mks1 localization to the mother centriole causes cilia defects and developmental malformations in Meckel–Gruber syndrome. *Dis Model Mech* **4:** 43–56.

Davenport JR, Watts AJ, Roper VC, Croyle MJ, van Groen T, Wyss JM, Nagy TR, Kesterson RA, Yoder BK. 2007. Disruption of intraflagellar transport in adult mice leads to obesity and slow-onset cystic kidney disease. *Curr Biol* **17:** 1586–1594.

DeCaen PG, Delling M, Vien TN, Clapham DE. 2013. Direct recording and molecular identification of the calcium channel of primary cilia. *Nature* **504:** 315–318.

Delling M, DeCaen PG, Doerner JF, Febvay S, Clapham DE. 2013. Primary cilia are specialized calcium signalling organelles. *Nature* **504:** 311–314.

Delling M, Indzhykulian AA, Liu X, Li Y, Xie T, Corey DP, Clapham DE. 2016. Primary cilia are not calcium-responsive mechanosensors. *Nature* **531:** 656–660.

Eichers ER, Abd-El-Barr MM, Paylor R, Lewis RA, Bi W, Lin X, Meehan TP, Stockton DW, Wu SM, Lindsay E, et al. 2006. Phenotypic characterization of *Bbs4* null mice reveals age-dependent penetrance and variable expressivity. *Hum Genet* **120:** 211–226.

Essner JJ, Vogan KJ, Wagner MK, Tabin CJ, Yost HJ, Brueckner M. 2002. Conserved function for embryonic nodal cilia. *Nature* **418:** 37–38.

Fedeles SV, Tian X, Gallagher AR, Mitobe M, Nishio S, Lee SH, Cai Y, Geng L, Crews CM, Somlo S. 2011. A genetic interaction network of five genes for human polycystic kidney and liver diseases defines polycystin-1 as the central determinant of cyst formation. *Nat Genet* **43:** 639–647.

Fedeles SV, Gallagher AR, Somlo S. 2014. Polycystin-1: A master regulator of intersecting cystic pathways. *Trends Mol Med* **20:** 251–260.

Field S, Riley KL, Grimes DT, Hilton H, Simon M, Powles-Glover N, Siggers P, Bogani D, Greenfield A, Norris DP. 2011. Pkd1l1 establishes left–right asymmetry and physically interacts with *Pkd2*. *Development* **138:** 1131–1142.

Gainullin VG, Hopp K, Ward CJ, Hommerding CJ, Harris PC. 2015. Polycystin-1 maturation requires polycystin-2 in a dose-dependent manner. *J Clin Invest* **125:** 607–620.

Garcia-Gonzalo FR, Corbit KC, Sirerol-Piquer MS, Ramaswami G, Otto EA, Noriega TR, Seol AD, Robinson JF, Bennett CL, Josifova DJ, et al. 2011. A transition zone complex regulates mammalian ciliogenesis and ciliary membrane composition. *Nat Genet* **43:** 776–784.

Gattone VH II, Wang X, Harris PC, Torres VE. 2003. Inhibition of renal cystic disease development and progression by a vasopressin V2 receptor antagonist. *Nat Med* **9:** 1323–1326.

Geng L, Okuhara D, Yu Z, Tian X, Cai Y, Shibazaki S, Somlo S. 2006. Polycystin-2 traffics to cilia independently of polycystin-1 by using an N-terminal RVxP motif. *J Cell Sci* **119:** 1383–1395.

Gerdes JM, Davis EE, Katsanis N. 2009. The vertebrate primary cilium in development, homeostasis, and disease. *Cell* **137:** 32–45.

Goetz SC, Anderson KV. 2010. The primary cilium: A signalling centre during vertebrate development. *Nat Rev Genet* **11:** 331–344.

Goodrich LV, Milenkovic L, Higgins KM, Scott MP. 1997. Altered neural cell fates and medulloblastoma in mouse patched mutants. *Science* **277:** 1109–1113.

Grantham JJ, Torres VE, Chapman AB, Guay-Woodford LM, Bae KT, King BF Jr, Wetzel LH, Baumgarten DA, Kenney PJ, Harris PC, et al. 2006. Volume progression in polycystic kidney disease. *N Engl J Med* **354:** 2122–2130.

Grantham JJ, Cook LT, Torres VE, Bost JE, Chapman AB, Harris PC, Guay-Woodford LM, Bae KT. 2008. Determinants of renal volume in autosomal-dominant polycystic kidney disease. *Kidney Int* **73:** 108–116.

Grantham JJ, Cook LT, Wetzel LH, Cadnapaphornchai MA, Bae KT. 2010. Evidence of extraordinary growth in the progressive enlargement of renal cysts. *Clin J Am Soc Nephrol* **5:** 889–896.

Han YG, Kim HJ, Dlugosz AA, Ellison DW, Gilbertson RJ, Alvarez-Buylla A. 2009. Dual and opposing roles of primary cilia in medulloblastoma development. *Nat Med* **15:** 1062–1065.

Harris PC, Bae KT, Rossetti S, Torres VE, Grantham JJ, Chapman AB, Guay-Woodford LM, King BF, Wetzel LH, Baumgarten DA, et al. 2006. Cyst number but not the rate of cystic growth is associated with the mutated gene in autosomal dominant polycystic kidney disease. *J Am Soc Nephrol* **17:** 3013–3019.

Haycraft CJ, Banizs B, Aydin-Son Y, Zhang Q, Michaud EJ, Yoder BK. 2005. Gli2 and Gli3 localize to cilia and require the intraflagellar transport protein polaris for processing and function. *PLoS Genet* **1:** e53.

He M, Subramanian R, Bangs F, Omelchenko T, Liem KF Jr, Kapoor TM, Anderson KV. 2014. The kinesin-4 protein Kif7 regulates mammalian Hedgehog signalling by organizing the cilium tip compartment. *Nat Cell Biol* **16:** 663–672.

Heyer CM, Sundsbak JL, Abebe KZ, Chapman AB, Torres VE, Grantham JJ, Bae KT, Schrier RW, Perrone RD, Braun WE, et al. 2016. Predicted mutation strength of nontruncating *PKD1* mutations aids genotype–phenotype correlations in autosomal dominant polycystic kidney disease. *J Am Soc Nephrol* doi: 10.1681/ASN.2015050583.

Hildebrandt F, Otto E. 2005. Cilia and centrosomes: A unifying pathogenic concept for cystic kidney disease? *Nat Rev Genet* **6:** 928–940.

Hildebrandt F, Zhou W. 2007. Nephronophthisis-associated ciliopathies. *J Am Soc Nephrol* **18:** 1855–1871.

Hildebrandt F, Attanasio M, Otto E. 2009. Nephronophthisis: Disease mechanisms of a ciliopathy. *J Am Soc Nephrol* **20:** 23–35.

Hildebrandt F, Benzing T, Katsanis N. 2011. Ciliopathies. *N Engl J Med* **364:** 1533–1543.

Hopp K, Ward CJ, Hommerding CJ, Nasr SH, Tuan HF, Gainullin VG, Rossetti S, Torres VE, Harris PC. 2012. Functional polycystin-1 dosage governs autosomal dominant polycystic kidney disease severity. *J Clin Invest* **122:** 4257–4273.

Huangfu D, Anderson KV. 2005. Cilia and Hedgehog responsiveness in the mouse. *Proc Natl Acad Sci* **102:** 11325–11330.

Huangfu D, Liu A, Rakeman AS, Murcia NS, Niswander L, Anderson KV. 2003. Hedgehog signalling in the mouse requires intraflagellar transport proteins. *Nature* **426:** 83–87.

Ibraghimov-Beskrovnaya O, Dackowski WR, Foggensteiner L, Coleman N, Thiru S, Petry LR, Burn TC, Connors TD, Van Raay T, Bradley J, et al. 1997. Polycystin: In vitro synthesis, in vivo tissue expression, and subcellular localization identifies a large membrane-associated protein. *Proc Natl Acad Sci* **94:** 6397–6402.

Jin H, White SR, Shida T, Schulz S, Aguiar M, Gygi SP, Bazan JF, Nachury MV. 2010. The conserved Bardet–Biedl syndrome proteins assemble a coat that traffics membrane proteins to cilia. *Cell* **141:** 1208–1219.

Jin X, Mohieldin AM, Muntean BS, Green JA, Shah JV, Mykytyn K, Nauli SM. 2014. Cilioplasm is a cellular compartment for calcium signaling in response to mechanical and chemical stimuli. *Cell Mol Life Sci* **71:** 2165–2178.

Kamura K, Kobayashi D, Uehara Y, Koshida S, Iijima N, Kudo A, Yokoyama T, Takeda H. 2011. Pkd1l1 complexes with Pkd2 on motile cilia and functions to establish the left–right axis. *Development* **138:** 1121–1129.

Kim H, Xu H, Yao Q, Li W, Huang Q, Outeda P, Cebotaru V, Chiaravalli M, Boletta A, Piontek K, et al. 2014. Ciliary membrane proteins traffic through the Golgi via a Rabep1/GGA1/Arl3-dependent mechanism. *Nat Commun* **5:** 5482.

Kim S, Nie H, Nesin V, Tran U, Outeda P, Bai CX, Keeling J, Maskey D, Watnick T, Wessely O, et al. 2016. The polycystin complex mediates Wnt/Ca^{2+} signalling. *Nat Cell Biol* **18:** 752–764.

Kottgen M, Benzing T, Simmen T, Tauber R, Buchholz B, Feliciangeli S, Huber TB, Schermer B, Kramer-Zucker A, Hopker K, et al. 2005. Trafficking of TRPP2 by PACS proteins represents a novel mechanism of ion channel regulation. *EMBO J* **24:** 705–716.

Koulen P, Cai Y, Geng L, Maeda Y, Nishimura S, Witzgall R, Ehrlich BE, Somlo S. 2002. Polycystin-2 is an intracellular calcium release channel. *Nat Cell Biol* **4:** 191–197.

Lee K, Boctor S, Barisoni LM, Gusella GL. 2014. Inactivation of integrin-β1 prevents the development of polycystic kidney disease after the loss of polycystin-1. *J Am Soc Nephrol* **26:** 888–895.

Liebscher I, Schoneberg T, Promel S. 2013. Progress in demystification of adhesion G protein–coupled receptors. *Biol Chem* **394:** 937–950.

Lin F, Hiesberger T, Cordes K, Sinclair AM, Goldstein LS, Somlo S, Igarashi P. 2003. Kidney-specific inactivation of the KIF3A subunit of kinesin-II inhibits renal ciliogenesis and produces polycystic kidney disease. *Proc Natl Acad Sci* **100:** 5286–5291.

Liu D, Wang CJ, Judge DP, Halushka MK, Ni J, Habashi JP, Moslehi J, Bedja D, Gabrielson KL, Xu H, et al. 2014. A Pkd1–Fbn1 genetic interaction implicates TGF-β signaling in the pathogenesis of vascular complications in autosomal dominant polycystic kidney disease. *J Am Soc Nephrol* **25:** 81–91.

Ma M, Tian X, Igarashi P, Pazour GJ, Somlo S. 2013. Loss of cilia suppresses cyst growth in genetic models of autosomal dominant polycystic kidney disease. *Nat Genet* **45:** 1004–1012.

Mangos S, Lam PY, Zhao A, Liu Y, Mudumana S, Vasilyev A, Liu A, Drummond IA. 2010. The ADPKD genes *pkd1a/b* and *pkd2* regulate extracellular matrix formation. *Dis Model Mech* **3:** 354–365.

McGlashan SR, Haycraft CJ, Jensen CG, Yoder BK, Poole CA. 2007. Articular cartilage and growth plate defects are associated with chondrocyte cytoskeletal abnormalities in Tg737^{orpk} mice lacking the primary cilia protein polaris. *Matrix Biol* **26:** 234–246.

McGrath J, Somlo S, Makova S, Tian X, Brueckner M. 2003. Two populations of node monocilia initiate left–right asymmetry in the mouse. *Cell* **114:** 61–73.

Menco BP, Cunningham AM, Qasba P, Levy N, Reed RR. 1997. Putative odour receptors localize in cilia of olfactory receptor cells in rat and mouse: A freeze-substitution ultrastructural study. *J Neurocytol* **26:** 691–706.

Mick DU, Rodrigues RB, Leib RD, Adams CM, Chien AS, Gygi SP, Nachury MV. 2015. Proteomics of primary cilia by proximity labeling. *Dev Cell* **35:** 497–512.

Mochizuki T, Wu G, Hayashi T, Xenophontos SL, Veldhuisen B, Saris JJ, Reynolds DM, Cai Y, Gabow PA, Pierides A, et al. 1996. *PKD2*, a gene for polycystic kidney disease that encodes an integral membrane protein. *Science* **272:** 1339–1342.

Mukhopadhyay S, Wen X, Ratti N, Loktev A, Rangell L, Scales SJ, Jackson PK. 2013. The ciliary G-protein-coupled receptor Gpr161 negatively regulates the Sonic hedgehog pathway via cAMP signaling. *Cell* **152:** 210–223.

Nauli SM, Alenghat FJ, Luo Y, Williams E, Vassilev P, Li X, Elia AE, Lu W, Brown EM, Quinn SJ, et al. 2003. Polycystins 1 and 2 mediate mechanosensation in the primary cilium of kidney cells. *Nat Genet* **33:** 129–137.

Nishimura DY, Fath M, Mullins RF, Searby C, Andrews M, Davis R, Andorf JL, Mykytyn K, Swiderski RE, Yang B, et al. 2004. *Bbs2*-null mice have neurosensory deficits, a defect in social dominance, and retinopathy associated with mislocalization of rhodopsin. *Proc Natl Acad Sci* **101:** 16588–16593.

Nonaka S, Tanaka Y, Okada Y, Takeda S, Harada A, Kanai Y, Kido M, Hirokawa N. 1998. Randomization of left–right asymmetry due to loss of nodal cilia generating leftward flow of extraembryonic fluid in mice lacking KIF3B motor protein. *Cell* **95:** 829–837.

Nusslein-Volhard C, Wieschaus E. 1980. Mutations affecting segment number and polarity in *Drosophila*. *Nature* **287:** 795–801.

Ocbina PJ, Eggenschwiler JT, Moskowitz I, Anderson KV. 2011. Complex interactions between genes controlling trafficking in primary cilia. *Nat Genet* **43:** 547–553.

Omori S, Hida M, Fujita H, Takahashi H, Tanimura S, Kohno M, Awazu M. 2006. Extracellular signal-regulated kinase inhibition slows disease progression in mice with polycystic kidney disease. *J Am Soc Nephrol* **17:** 1604–1614.

Pazour GJ, San Agustin JT, Follit JA, Rosenbaum JL, Witman GB. 2002. Polycystin-2 localizes to kidney cilia and the ciliary level is elevated in *orpk* mice with polycystic kidney disease. *Curr Biol* **12:** R378–380.

Pei Y, Watnick T, He N, Wang K, Liang Y, Parfrey P, Germino G, St George-Hyslop P. 1999. Somatic PKD2 mutations in individual kidney and liver cysts support a "two-hit" model of cystogenesis in type 2 autosomal dominant polycystic kidney disease. *J Am Soc Nephrol* **10:** 1524–1529.

Pennekamp P, Karcher C, Fischer A, Schweickert A, Skryabin B, Horst J, Blum M, Dworniczak B. 2002. The ion channel polycystin-2 is required for left–right axis determination in mice. *Curr Biol* **12:** 938–943.

Piontek K, Menezes LF, Garcia-Gonzalez MA, Huso DL, Germino GG. 2007. A critical developmental switch defines the kinetics of kidney cyst formation after loss of *Pkd1*. *Nat Med* **13:** 1490–1495.

Plotnikova OV, Pugacheva EN, Golemis EA. 2011. Aurora A kinase activity influences calcium signaling in kidney cells. *J Cell Biol* **193:** 1021–1032.

Praetorius HA, Praetorius J, Nielsen S, Frokiaer J, Spring KR. 2004. β1-integrins in the primary cilium of MDCK cells potentiate fibronectin-induced Ca^{2+} signaling. *Am J Physiol Renal Physiol* **287:** F969–978.

Qian F, Watnick TJ, Onuchic LF, Germino GG. 1996. The molecular basis of focal cyst formation in human autosomal dominant polycystic kidney disease type I. *Cell* **87:** 979–987.

Qian F, Germino FJ, Cai Y, Zhang X, Somlo S, Germino GG. 1997. PKD1 interacts with PKD2 through a probable coiled-coil domain. *Nat Genet* **16:** 179–183.

Qian F, Boletta A, Bhunia AK, Xu H, Liu L, Ahrabi AK, Watnick TJ, Zhou F, Germino GG. 2002. Cleavage of polycystin-1 requires the receptor for egg jelly domain and is disrupted by human autosomal-dominant polycystic kidney disease 1–associated mutations. *Proc Natl Acad Sci* **99:** 16981–16986.

Rees S, Kittikulsuth W, Roos K, Strait KA, Van Hoek A, Kohan DE. 2014. Adenylyl cyclase 6 deficiency ameliorates polycystic kidney disease. *J Am Soc Nephrol* **25:** 232–237.

Rohatgi R, Milenkovic L, Scott MP. 2007. Patched1 regulates hedgehog signaling at the primary cilium. *Science* **317:** 372–376.

Rossetti S, Kubly VJ, Consugar MB, Hopp K, Roy S, Horsley SW, Chauveau D, Rees L, Barratt TM, van't Hoff WG, et al. 2009. Incompletely penetrant *PKD1* alleles suggest a role for gene dosage in cyst initiation in polycystic kidney disease. *Kidney Int* **75:** 848–855.

Sang L, Miller JJ, Corbit KC, Giles RH, Brauer MJ, Otto EA, Baye LM, Wen X, Scales SJ, Kwong M, et al. 2011. Mapping the NPHP–JBTS–MKS protein network reveals ciliopathy disease genes and pathways. *Cell* **145:** 513–528.

Scheffers MS, van der Bent P, Prins F, Spruit L, Breuning MH, Litvinov SV, de Heer E, Peters DJ. 2000. Polycystin-1, the product of the polycystic kidney disease 1 gene, co-localizes with desmosomes in MDCK cells. *Hum Mol Genet* **9:** 2743–2750.

Schneider L, Clement CA, Teilmann SC, Pazour GJ, Hoffmann EK, Satir P, Christensen ST. 2005. PDGFRαα signaling is regulated through the primary cilium in fibroblasts. *Curr Biol* **15:** 1861–1866.

Serra AL, Poster D, Kistler AD, Krauer F, Raina S, Young J, Rentsch KM, Spanaus KS, Senn O, Kristanto P, et al. 2010. Sirolimus and kidney growth in autosomal dominant polycystic kidney disease. *N Engl J Med* **363:** 820–829.

Sharma N, Malarkey EB, Berbari NF, O'Connor AK, Vanden Heuvel GB, Mrug M, Yoder BK. 2013. Proximal tubule proliferation is insufficient to induce rapid cyst formation after cilia disruption. *J Am Soc Nephrol* **24:** 456–464.

Shibazaki S, Yu Z, Nishio S, Tian X, Thomson RB, Mitobe M, Louvi A, Velazquez H, Ishibe S, Cantley LG, et al. 2008. Cyst formation and activation of the extracellular regulated kinase pathway after kidney specific inactivation of *Pkd1*. *Hum Mol Genet* **17:** 1505–1516.

Shillingford JM, Murcia NS, Larson CH, Low SH, Hedgepeth R, Brown N, Flask CA, Novick AC, Goldfarb DA, Kramer-Zucker A, et al. 2006. The mTOR pathway is regulated by polycystin-1, and its inhibition reverses renal cystogenesis in polycystic kidney disease. *Proc Natl Acad Sci* **103:** 5466–5471.

Shillingford JM, Piontek KB, Germino GG, Weimbs T. 2010. Rapamycin ameliorates PKD resulting from conditional inactivation of *Pkd1*. *J Am Soc Nephrol* **21:** 489–497.

Smith LA, Bukanov NO, Husson H, Russo RJ, Barry TC, Taylor AL, Beier DR, Ibraghimov-Beskrovnaya O. 2006. Development of polycystic kidney disease in juvenile cystic kidney mice: Insights into pathogenesis, ciliary abnormalities, and common features with human disease. *J Am Soc Nephrol* **17:** 2821–2831.

Stoveken HM, Hajduczok AG, Xu L, Tall GG. 2015. Adhesion G protein-coupled receptors are activated by exposure of a cryptic tethered agonist. *Proc Natl Acad Sci* **112:** 6194–6199.

Streets AJ, Wessely O, Peters DJ, Ong AC. 2013. Hyperphosphorylation of polycystin-2 at a critical residue in disease reveals an essential role for polycystin-1-regulated dephosphorylation. *Hum Mol Genet* **22:** 1924–1939.

Su S, Phua SC, DeRose R, Chiba S, Narita K, Kalugin PN, Katada T, Kontani K, Takeda S, Inoue T. 2013. Genetically encoded calcium indicator illuminates calcium dynamics in primary cilia. *Nat Methods* **10:** 1105–1107.

Su X, Driscoll K, Yao G, Raed A, Wu M, Beales PL, Zhou J. 2014. Bardet–Biedl syndrome proteins 1 and 3 regulate the ciliary trafficking of polycystic kidney disease 1 protein. *Hum Mol Genet* **23:** 5441–5451.

Subramanian B, Ko WC, Yadav V, DesRochers TM, Perrone RD, Zhou J, Kaplan DL. 2012. The regulation of cystogenesis in a tissue engineered kidney disease system by abnormal matrix interactions. *Biomaterials* **33:** 8383–8394.

Sun Z, Amsterdam A, Pazour GJ, Cole DG, Miller MS, Hopkins N. 2004. A genetic screen in zebrafish identifies cilia genes as a principal cause of cystic kidney. *Development* **131:** 4085–4093.

Tesmer JJ. 2012. A GAIN in understanding autoproteolytic G protein-coupled receptors and polycystic kidney disease proteins. *EMBO J* **31:** 1334–1335.

The European Polycystic Kidney Disease Consortium. 1994. The polycystic kidney disease 1 gene encodes a 14 kb transcript and lies within a duplicated region on chromosome 16. *Cell* **78:** 725.

The International Polycystic Kidney Disease Consortium. 1995. Polycystic kidney disease: The complete structure of the *PKD1* gene and its protein. *Cell* **81:** 289–298.

Torres VE, Wang X, Qian Q, Somlo S, Harris PC, Gattone VH II. 2004. Effective treatment of an orthologous model of autosomal dominant polycystic kidney disease. *Nat Med* **10:** 363–364.

Torres VE, Boletta A, Chapman A, Gattone V, Pei Y, Qian Q, Wallace DP, Weimbs T, Wuthrich RP. 2010. Prospects for mTOR inhibitor use in patients with polycystic kidney disease and hamartomatous diseases. *Clin J Am Soc Nephrol* **5:** 1312–1329.

Torres VE, Chapman AB, Devuyst O, Gansevoort RT, Grantham JJ, Higashihara E, Perrone RD, Krasa HB, Ouyang J, Czerwiec FS, et al. 2012. Tolvaptan in patients with autosomal dominant polycystic kidney disease. *N Engl J Med* **367:** 2407–2418.

Tran PV, Talbott GC, Turbe-Doan A, Jacobs DT, Schonfeld MP, Silva LM, Chatterjee A, Prysak M, Allard BA, Beier DR. 2014. Downregulating hedgehog signaling reduces renal cystogenic potential of mouse models. *J Am Soc Nephrol* **25:** 2201–2212.

Tsiokas L, Kim E, Arnould T, Sukhatme VP, Walz G. 1997. Homo- and heterodimeric interactions between the gene products of *PKD1* and *PKD2*. *Proc Natl Acad Sci* **94:** 6965–6970.

Vogel P, Read R, Hansen GM, Freay LC, Zambrowicz BP, Sands AT. 2010. Situs inversus in *Dpcd/Poll*$^{-/-}$, *Nme7*$^{-/-}$, and *Pkd1l1*$^{-/-}$ mice. *Veterinary Pathol* **47:** 120–131.

Walz G, Budde K, Mannaa M, Nurnberger J, Wanner C, Sommerer C, Kunzendorf U, Banas B, Horl WH, Obermuller N, et al. 2010. Everolimus in patients with autosomal dominant polycystic kidney disease. *N Engl J Med* **363:** 830–840.

Watnick T, Germino G. 2003. From cilia to cyst. *Nat Genet* **34:** 355–356.

Watnick TJ, Torres VE, Gandolph MA, Qian F, Onuchic LF, Klinger KW, Landes G, Germino GG. 1998. Somatic mutation in individual liver cysts supports a two-hit model of cystogenesis in autosomal dominant polycystic kidney disease. *Mol Cell* **2:** 247–251.

Weatherbee SD, Niswander LA, Anderson KV. 2009. A mouse model for Meckel syndrome reveals Mks1 is required for ciliogenesis and Hedgehog signaling. *Hum Mol Genet* **18:** 4565–4575.

Wills ES, Roepman R, Drenth JP. 2014. Polycystic liver disease: Ductal plate malformation and the primary cilium. *Trends Mol Med* **20:** 261–270.

Wong SY, Seol AD, So PL, Ermilov AN, Bichakjian CK, Epstein EH Jr, Dlugosz AA, Reiter JF. 2009. Primary cilia can both mediate and suppress Hedgehog pathway-dependent tumorigenesis. *Nat Med* **15:** 1055–1061.

Wu G, D'Agati V, Cai Y, Markowitz G, Park JH, Reynolds DM, Maeda Y, Le TC, Hou H Jr, Kucherlapati R, et al. 1998. Somatic inactivation of *Pkd2* results in polycystic kidney disease. *Cell* **93:** 177–188.

Yoder BK, Hou X, Guay-Woodford LM. 2002. The polycystic kidney disease proteins, polycystin-1, polycystin-2, polaris, and cystin, are co-localized in renal cilia. *J Am Soc Nephrol* **13:** 2508–2516.

Yoshiba S, Shiratori H, Kuo IY, Kawasumi A, Shinohara K, Nonaka S, Asai Y, Sasaki G, Belo JA, Sasaki H, et al. 2012. Cilia at the node of mouse embryos sense fluid flow for left–right determination via Pkd2. *Science* **338:** 226–231.

Yu S, Hackmann K, Gao J, He X, Piontek K, Garcia-Gonzalez MA, Menezes LF, Xu H, Germino GG, Zuo J, et al. 2007. Essential role of cleavage of Polycystin-1 at G protein-coupled receptor proteolytic site for kidney tubular structure. *Proc Natl Acad Sci* **104:** 18688–18693.

Yuan S, Zhao L, Brueckner M, Sun Z. 2015. Intraciliary calcium oscillations initiate vertebrate left–right asymmetry. *Curr Biol* **25:** 556–567.

Zhang XM, Ramalho-Santos M, McMahon AP. 2001. Smoothened mutants reveal redundant roles for Shh and Ihh signaling including regulation of L/R asymmetry by the mouse node. *Cell* **105:** 781–792.

Cilia and Obesity

Christian Vaisse,[1,4] Jeremy F. Reiter,[2,4] and Nicolas F. Berbari[3,4]

[1]Diabetes Center and Department of Medicine, University of California San Francisco, San Francisco, California 94143

[2]Department of Biochemistry and Biophysics, Cardiovascular Research Institute, University of California, San Francisco, San Francisco, California 94158

[3]Department of Biology, Indiana University-Purdue University Indianapolis, Indianapolis, Indiana 46202

Correspondence: nberbari@iupui.edu

The ciliopathies Bardet–Biedl syndrome and Alström syndrome cause obesity. How ciliary dysfunction leads to obesity has remained mysterious, partly because of a lack of understanding of the physiological roles of primary cilia in the organs and pathways involved in the regulation of metabolism and energy homeostasis. Historically, the study of rare monogenetic disorders that present with obesity has informed our molecular understanding of the mechanisms involved in nonsyndromic forms of obesity. Here, we present a framework, based on genetic studies in mice and humans, of the molecular and cellular pathways underlying long-term regulation of energy homeostasis. We focus on recent progress linking these pathways to the function of the primary cilia with a particular emphasis on the roles of neuronal primary cilia in the regulation of satiety.

The primary cilium has emerged as a clinically important organelle with ciliary dysfunction underlying several human syndromes collectively called the ciliopathies. Ciliopathies are associated with diverse phenotypes affecting nearly every tissue and organ system (see Braun and Hildebrandt 2016). Ciliopathies such as Bardet–Biedl syndrome (BBS) and Alström syndrome (ALMS) present with pediatric obesity as a clinical feature. Although an understanding of how ciliary defects cause certain ciliopathy-associated phenotypes, such as limb and neural tube defects, is emerging (see Bangs and Anderson 2016), how disrupted ciliary function leads to obesity remains poorly understood.

Here, we describe some neuroendocrine signaling pathways involved in the control of energy homeostasis in mammals, defects in which cause obesity in humans. We then describe how the use of mouse genetic models of ciliopathies has yielded interesting and sometimes contradictory results about the mechanisms through which primary cilia regulate body weight.

In particular, we examine a potential role for cilia in the leptin–melanocortin signaling axis. We also discuss the *FTO* locus and a neighboring gene involved in ciliary function, *RPGRIP1L*, as influencing obesity through effects on leptin signaling. We close with a discussion on the emerging literature on the potential of adipocyte or preadipocyte cilia to impact me-

---

[4]These authors contributed equally to this work.

Cite this article as *Cold Spring Harb Perspect Biol* doi: 10.1101/cshperspect.a028217

tabolism and obesity. It is clear that the primary cilium, a long-neglected organelle, plays important roles in controlling mammalian energy homeostasis. Future studies on the roles of cilia in energy homeostasis may reveal therapeutic targets for this common disease.

HUMAN OBESITY AND CILIOPATHIES

Pathophysiology of Obesity

Obesity is an increase in energy stored as fat in sufficient magnitude to result in adverse health consequences, such as diabetes and cardiovascular disease. Mechanistically, obesity is caused by long-term caloric intake in excess of energy expenditure. The control of food intake and energy expenditure is accomplished through afferent signals that sense the energy status of the individual, integration in the brain including the hypothalamus, and efferent signals including those determining the intensity of hunger. One common misconception is that this physiological system is dedicated to the prevention of obesity. Instead, this system's essential role is the prevention of starvation and ensuring adequate energy intake to meet the energy requirements of basal metabolism, physical activity, growth, and reproduction.

Genetic Predisposition to Human Obesity

Environmental influences, such as diet and exercise, interact with genetic factors to influence the onset and progression of weight gain. Genetic studies, including twin studies, have revealed that genetic variation accounts for 40%–70% of weight variation, and that the heritability of obesity increases with its severity (Allison et al. 1996; O'Rahilly and Farooqi 2000). Both common genetic variants with small effects and rare genetic variants of larger effect contribute to the predisposition to obesity. The common genetic variants identified by genome-wide association studies (GWAS) have small, clinically insignificant effects individually. Most of these variants are located outside gene-coding regions, making it difficult to identify how they act. A case in point is the variant

most strongly associated with obesity in multiple populations, a polymorphism present within an intron of the *FTO* gene (Frayling et al. 2007). One *FTO*-associated variant predisposes to obesity with an odds ratio of 1.3 to 1.7, which translates into a 2–3 kg weight gain. Even after extensive research efforts, the mechanism behind *FTO*-associated increases in weight remains unclear. We discuss below how the *FTO*-associated SNP may impact a neighboring gene, *RPGRIP1L*, encoding a critical component of the ciliary transition zone.

On the other end of the genetic effect spectrum from GWAS-identified single-nucleotide polymorphisms (SNPs), are single-gene defects that cause severe obesity, both in humans and mice. Analysis of these single genes has provided valuable insights into how energy stores are regulated in response to variable access to nutrition and demands for energy expenditure. Examples of single genes in which mutations cause human obesity include *LEPTIN* (*LEP*), its receptor (*LEPR*), *PROOPIOMELANOCORTIN* (*POMC*), *MELANOCORTIN RECEPTOR 4* (*MC4R*), and *SIM1*, all of which are components of the leptin–melanocortin system.

The Leptin–Melanocortin Pathway Regulates Energy Homeostasis

One major afferent signal allowing the brain to sense the level of energy stores is the hormone leptin (Zhang et al. 1994). This cytokine-like 167-amino-acid protein is released by adipocytes in proportion to fat mass. Decreasing leptin levels inform the brain of diminishing fat storage resulting from a negative energy balance, and compensatory effects on appetite and energy expenditure that can replenish the stores and reestablish energy balance. Leptin's action is mediated by the leptin receptor (LEPR), a single-transmembrane domain member of the class I cytokine receptor family. On binding to leptin, LEPR homodimerizes, leading to phosphorylation of JAK and STAT3. On phosphorylation, STAT3 dimerizes, translocates to the nucleus, and activates transcription of LEPTIN target genes (for a review on leptin signaling, see Myers et al. 2010).

Cite this article as *Cold Spring Harb Perspect Biol* doi: 10.1101/cshperspect.a028217

The LEPTIN-responsive isoform of LEPR is expressed mainly in the hypothalamus, a brain region that interprets and integrates the peripheral signals that communicate energy balance. Within the hypothalamus, LEPTIN differentially affects the activity of two adjacent groups of neurons in the arcuate nucleus (ARC). LEPTIN inhibits the orexigenic agouti-related peptide (AgRP) and neuropeptide Y (NPY)–producing neurons, and activates the anorexigenic proopiomelanocortin (POMC)-producing neurons. In these latter neurons, POMC is cleaved by the proteases proconvertase-1 and proconvertase-2 to generate the anorexigenic neuropeptide α-melanocyte-stimulating hormone (α-MSH).

Interactions between these neuronal populations within the ARC allow for cross talk and modulation of the neuronal output. The development of sophisticated inducible conditional alleles and reporters in mouse models has begun to reveal the complex regulation of subpopulations of neurons within the ARC (for recent reviews on ARC-mediated control of food intake, see Begg and Woods 2013; Mountjoy 2015).

AgRP/NPY- and POMC-producing neurons in the ARC appear to respond directly to circulating signaling factors such as LEPTIN. These ARC neurons send axonal projections to second-order neurons in other regions of the hypothalamus, such as the paraventricular nucleus (PVN) and the lateral hypothalamic area (LHA), as well as to the hindbrain. Both α-MSH and AgRP act on a common G-protein-coupled receptor, melanocortin 4 receptor (MC4R), expressed by a subset of PVN neurons. α-MSH binds to and activates MC4R, whereas AgRP inhibits MC4R activity. To date, mutations in *MC4R* are the most common monogenic cause of severe human obesity, accounting for ∼2.5% of cases. Similar to the POMC and AgRP neurons of the ARC, new tools have begun to reveal diversity among MC4R-expressing neurons and signaling mechanisms (Garfield et al. 2015; Ghamari-Langroudi et al. 2015). MC4R signaling and downstream neuronal circuitry remains an attractive target for therapeutics to treat obesity.

Human patients and mouse models mutant for components of the leptin–melanocortin pathway have helped reveal the physiological functions of this signaling system. For example, rare humans with complete LEPTIN deficiency show behaviors and physiological signs of starvation despite being extremely obese. LEPTIN replacement abolishes the hyperphagia, leading to normalization of weight, showing the necessary role for LEPTIN in regulating human energy homeostasis (Farooqi et al. 1999).

Although LEPTIN was initially thought to be a potential treatment for common obesity, all obese humans and animal models develop a resistance to LEPTIN's anorectic actions, despite having elevated adipose and thus increased circulating serum leptin levels. The molecular mechanisms associated with LEPTIN resistance are an active area of research and include altered transport of leptin across the blood–brain barrier, hypothalamic inflammation and ER stress, and diminished hypothalamic LEPTIN signaling (for a recent review of leptin resistance, see Aragones et al. 2016).

Ciliopathies Associated with Obesity

Ciliopathies for which obesity is an integral component of the clinical presentation include BBS (OMIM #209900) and ALMS (OMIM #203800). In addition to obesity, BBS is characterized by retinal degeneration, postaxial polydactyly, and kidney cysts. Other associated findings include anosmia, mental retardation, hepatic fibrosis, and type 2 diabetes mellitus (Forsythe and Beales 2013). Obesity in patients with BBS ranges from mild to severe, and is reversible with caloric restriction and exercise (Beales et al. 1999). BBS is rare and genetically heterogeneous. To date, mutations in 21 genes, *BBS1–21*, have been identified that contribute to the development of the phenotype (see Braun and Hildebrandt 2016). Many BBS gene products form a large protein complex termed the BBSome (Nachury et al. 2007). The BBSome functions in the localization of select transmembrane proteins to, and removal from, the cilium (Berbari et al. 2008; Jin et al. 2010;

Domire et al. 2011; Koemeter-Cox et al. 2014; Liew et al. 2014; Mourao et al. 2014).

Unlike the genetic heterogeneity underlying BBS, ALMS is associated with mutations in a single gene, *ALMS1*. Together with retinal degeneration and hearing loss, early-onset obesity is also one of the hallmarks. In addition, ALMS is associated with cardiomyopathy, liver and kidney dysfunction, and delayed puberty (for a review on ALMS, see Girard and Petrovsky 2011). Like BBS, the pathogenesis of ALMS has been linked to dysfunction of the primary cilium. The ALMS1 protein localizes to the centrosome and ciliary basal body and likely has a role in the formation or maintenance of primary cilia (Hearn et al. 2002; Li et al. 2007; Knorz et al. 2010). The human mutations that cause ALMS truncate ALMS1. These truncated proteins are able to support ciliogenesis, but may alter ciliary function or long-term maintenance, leading to the development of ALMS.

LESSONS FROM MOUSE MODELS OF CILIA-ASSOCIATED OBESITY

Mouse models of monogenic forms of obesity, which recapitulate the human phenotypes, have been invaluable for elucidating molecular mechanisms underlying the pathogenesis of obesity. Similarly, a better understanding of the role of the primary cilia in obesity and the relationship between cilia function and the leptin–melanocortin pathway has been driven by the study of mice carrying mutations affecting *Alms1* or BBS-associated genes, as well as by mice carrying conditional mutations affecting the formation and maintenance of primary cilia.

BBS and ALMS Ciliopathy Mouse Models Are Obese

As in humans, mice mutant for many of the BBS-associated genes show obesity (Table 1) (Mykytyn et al. 2004; Nishimura et al. 2004; Davis et al. 2007). Mouse mutations that model ALMS include a gene-trap ($Alms1^{Gt(XH152)Byg}$) and a spontaneous mutation (*fat aussie, foz*) (Collin et al. 2005; Arsov et al. 2006). Unlike Bbs mutant mice, *Alms1* mutant mice are born at a normal weight, similar to the clinical observations of ALMS in humans. Hyperphagia and obesity accompanied by hyperinsulinemia and type 2 diabetes occur postnatally in *Alms1* mutant mice.

Although Bbs and *Alms1* gene mutations cause obesity and Bbs and Alms1 proteins are implicated in the function of the primary cilia, these observations do not show that the associated obesity is caused by alteration in cilia function, or that cilia are essential for the regulation of energy homeostasis.

Conditional Disruption of Primary Cilia Causes Obesity in Mice

Intraflagellar transport (IFT) is the process of protein transport within cilia critical for both their formation and maintenance. Conditional deletion of genes essential for IFT, such as *Ift88* or *Kif3a*, is useful for delineating when and where cilia are required for a specific phenotype. Inducing organism-wide loss of cilia in adult mice causes hyperphagia and subsequent obesity (Davenport et al. 2007). Restricting mice lacking cilia to the diet of control animals prevented obesity, indicating that cilia restrict weight gain by inhibiting the consumption of food rather than by affecting metabolism or locomotor activity. To test where cilia function to restrict food consumption, *Ift88* or *Kif3a* were removed exclusively in neurons using synapsin1-Cre (Davenport et al. 2007). As with organism-wide loss of cilia, removing cilia specifically in neurons causes obesity, strongly implicating neuronal cilia in the regulation of appetite and satiety.

The Neurons Involved in Obesity-Associated Ciliopathy

As discussed above, the hypothalamus regulates appetite through neurons that make POMC or AgRP. Importantly, hypothalamic neurons each possess a single primary cilium, although the precise functions of these cilia are largely unknown. To address the role of cilia in these hypothalamic neurons, *Ift88* or *Kif3a* were conditionally removed in POMC- or AgRP-expressing neurons (Xu et al. 2005). By 6 weeks

Table 1. Cilia-associated mouse models of obesity

Gene	Allele (MGI)	Type of allele	Mouse phenotypes	References
Adcy3	*Adcy3[tm1Drs]*	Knockout	Obesity, anosmia	Wang et al. 2009
Adcy3	*Adcy3[Jll]*	ENU gain-of-function	Resistant to diet-induced obesity, less adipose	Pitman et al. 2014
Alms1	*Alms1[Gt(XH152)Byg]*	Genetrap	Obesity, retinopathy, male infertility, late-onset hearing loss	Collin et al. 2005
Alms1	*Alms1[foz]*	Spontaneous	Obesity, male infertility, late-onset hearing loss	Arsov et al. 2006
Bbs1	*Bbs1[tm1Vcs]*	Knockin	Obesity, retinopathy, male infertility, ventriculomegaly	Davis et al. 2007
Bbs2	*Bbs2[tm1Vcs]*	Knockout	Obesity, retinopathy, renal cysts, male infertility, anosmia, social submissiveness, ventriculomegaly	Nishimura et al. 2004; Davis et al. 2007
Bbs3/ Arl6[a]	*Arl6[tm2Vcs]*	Knockout	Increased fat mass, retinopathy, male infertility, hydrocephalus, elevated blood pressure	Zhang et al. 2011
Bbs4[b]	*Bbs4[tm1Vcs]*	Knockout	Obesity, retinopathy, male infertility, social submissiveness, ventriculomegaly	Mykytyn et al. 2004; Nishimura et al. 2004; Davis et al. 2007
Bbs4[b]	*Bbs4[Gt1Nk]*	Genetrap	Obesity (sex-dependent penetrance and severity), retinopathy, social submissiveness, increased anxiety	Eichers et al. 2006
Bbs6/ Mkks	*Mkks[tm1Vcs]*	Knockout	Obesity, retinopathy, male infertility, anosmia, elevated blood pressure, social submissiveness, ventriculomegaly	Fath et al. 2005; Davis et al. 2007
Bbs7	*Bbs7[tm1Vcs]*	Knockout	Obesity, male infertility, ventriculomegaly	Zhang et al. 2013
Bbs8/ Ttc8	*Ttc8[tm1Reed]*	Knockout	Obesity, anosmia, retinal degeneration, renal tubule anomalies	Tadenev et al. 2011
Bbs11/ Trim32	*Trim32[Gt(BGA355)Byg]*	Genetrap	Increased body weight, muscular myopathy	Kudryashova et al. 2009
Bbs12	*Bbs12[tm1.1Vmar]*	Knockout	Obesity, retinal degeneration	Marion et al. 2012
Ift88	*Ift88[tm1.1Bky]*	Conditional knockout	Obesity, renal cysts, hepatic cysts	Davenport et al. 2007
Kif3a	*Kif3a[tm1Gsn]*	Conditional knockout	Obesity, renal cysts, hepatic cysts	Davenport et al. 2007
Rpgrip1l	*Rpgrip1l[tm1a(EUCOMM)Wtsi]*	Knockout	Obesity	Stratigopoulos et al. 2014

[a]*Bbs3* mutant mice do not become obese, but do display increased fat mass. Likewise, *Bbs11* mutant mice have not been reported to be obese, but do display a significant and persistent increase in body weight starting at 2 months of age.

[b]Two different *Bbs4[−/−]* knockout mouse lines have been independently generated and different penetrance and severity of obesity have been reported for each.

of age, mice lacking cilia on POMC-expressing neurons weighed significantly more than control mice, and continued to become obese during adulthood (Davenport et al. 2007). Mice lacking cilia on AgRP-expressing neurons did not show increased weight (NF Berbari and BK Yoder, unpubl.). In addition to obesity, mice lacking cilia on POMC-expressing neurons displayed increased levels of leptin, fasting serum glucose, and insulin. These increases were only present in obese mutants, not those kept at the control weight by pair feeding, indicating that these changes were secondary to obesity.

Potential Lepr Involvement in Ciliopathy-Associated Obesity

Although this work indicates that hypothalamic cilia restrain feeding, it does not reveal how they do so (Davenport et al. 2007). A suggestion of a molecular mechanism, Bbs1, a component of the BBSome, directly binds to Lepr and may participate in Lepr trafficking (Seo et al. 2009). Like mice lacking cilia on POMC-expressing neurons, ad libitum−fed Bbs2, Bbs4, and Bbs6 mutant mice show elevated leptin levels (Rahmouni et al. 2008; Seo et al. 2009). Importantly, these Bbs mutants fail to reduce food intake in response to injection of leptin, raising the possibility that a diminished response to leptin contributes to obesity in BBS (Rahmouni et al. 2008).

Nearly all obese mice and humans show elevated levels of circulating leptin, but this leptin is insufficient to suppress appetite, a phenomenon known as leptin resistance (Maffei et al. 1995; Considine et al. 1996). Thus, leptin resistance can either be a cause or a consequence of obesity. Interestingly, when caloric restriction was used to normalize leptin levels in Bbs mutant mice they still failed to respond to leptin with diminished food intake (Seo et al. 2009). The investigators concluded that leptin resistance was the primary deficit initiating hyperphagia and obesity in Bbs mice, but did not take into account the food anticipatory behavior that is observed on calorie restriction in mice. A growing literature reports that maintaining

calorie restriction in rodents can have prolonged effects on meal structure and circadian rhythm (for a review on anticipatory feeding behavior, see Mistlberger 2009).

If both body composition and anticipatory feeding behavior are controlled for, adult mice lacking Ift88, or Bbs mutant mice, before the onset of obesity, display unaltered responses to leptin injected intraperitoneally (Berbari et al. 2013). Lepr activity was similar between these ciliopathy models and control mice, and other phenotypes associated with leptin and LEPR mutations, such as changes in thermoregulation and locomotor activity, are not present. These data suggest that cilia are not directly involved in leptin signaling (Berbari et al. 2013).

Very recent work by Guo et al. (2016) report a cilium-independent function of the BBSome. It appears to be required for trafficking of Lepr to the plasma membrane, and in Bbs mutants the obesity appears to be primarily a result of deficits in leptin sensitivity. This is in contrast to their findings with conditional loss of IFT88 in which they report leptin resistance on increases in adiposity. This work begins to show that cilia mutant mouse models may display obesity through different and independent mechanisms (Fig. 1). This raises the exciting potential for cilia-mediated signaling and obesity to be broadly relevant beyond the rare ciliopathies.

Leptin-Independent Mechanisms by Which Neuronal Cilia May Affect Energy Homeostasis

Mutations affecting the mouse orthologs of BBS-associated genes, including Bbs2, Bbs3, and Bbs4, disrupt the localization of at least some GPCRs to their cilia, including melanin-concentrating hormone receptor 1 (Mchr1) (Berbari et al. 2008; Zhang et al. 2011).

Defective Mchr1 signaling is an attractive candidate for mediating obesity in Bbs mutants, as Mchr1 regulates feeding behavior. Either pharmacological or genetic activation of the Mchr1 pathway is associated with hyperphagia, whereas repression is associated with anorectic behavior (Qu et al. 1996; Shimada et al. 1998; Ludwig et al. 2001; Borowsky et al. 2002; Chen

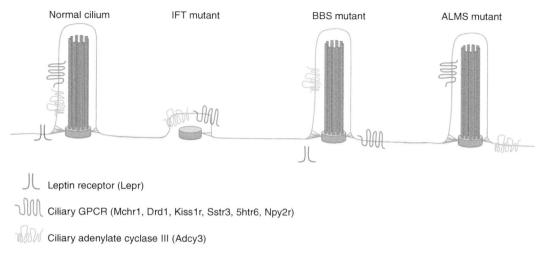

Leptin receptor (Lepr)

Ciliary GPCR (Mchr1, Drd1, Kiss1r, Sstr3, 5htr6, Npy2r)

Ciliary adenylate cyclase III (Adcy3)

Figure 1. Model of how ciliary signaling may contribute to energy homeostasis. Certain G-protein-coupled receptors (GPCRs) and signaling machinery, such as Sstr3, Mchr1, Drd1, Kiss1r, Htr6, and Adcy3, localize to the cilia membrane of neurons in specific brain regions. Intraflagellar transport (IFT) mutation results in loss of the cilium. In mouse models of Bardet–Biedl syndrome (BBS), both ciliary GPCRs and membrane-associated Lepr localization are perturbed. In Alström syndrome (ALMS) mouse models, GPCRs remain at cilia, but Adcy3 no longer localizes appropriately to cilia. Taken together, these models convey the complexity of the cilium as a signaling center and indicate that there are differing requirements for membrane protein localization to cilia.

et al. 2002). Thus, for Mchr1 signaling to underlie obesity in Bbs mutants, the failure of Mchr1 to reach the cilium would have to be associated with ectopic activation of Mchr1 signaling. This is possible, although inactivation through sequestration in the cilium has not been previously described for GPCRs.

Apart from Mchr1, there is a growing list of other GPCRs that can preferentially localize to neuronal cilia. Some of these ciliary GPCRs include somatostatin receptor 3 (Sstr3), serotonin receptor 6 (5HT6), dopamine receptor 1 (Drd1), neuropeptide Y receptor 2 (Npy2r), and kisspeptin receptor 1 (Kiss1r) (Hamon et al. 1999; Handel et al. 1999; Brailov et al. 2000; Schulz et al. 2000; Marley and von Zastrow 2010; Domire et al. 2011; Loktev and Jackson 2013; Koemeter-Cox et al. 2014). While the functional significance of the localization of these receptors to cilia remains unclear, it is possible that they affect appetite, satiety, or metabolism, especially when given that somatostatin, serotonin, dopamine, neuropeptide Y, and kisspeptin, have all been implicated in either reward, feeding behaviors, metabolism, or glucose handling

(Vijayan and McCann 1977; Pollock and Rowland 1981; Aponte et al. 1984; Salamone et al. 1990; Tolson et al. 2014).

Interestingly, $Alms1^{foz/foz}$ mice display a progressive loss of ciliary adenylyl cyclase III (Adcy3) in their hypothalamus, but no changes in Sstr3 or Mchr1 localization to cilia (Bishop et al. 2007; Heydet et al. 2013). Loss of Adcy3 function can itself cause obesity in mice through alterations in activity, hyperphagia, and leptin resistance (Wang et al. 2009). Conversely, a gain-of-function mutation in Adcy3 confers protection from diet-induced obesity (Pitman et al. 2014). Future studies aimed at dissecting the roles of cilia GPCRs and their associated signaling proteins will shed light on cilia-associated changes in feeding behaviors, and other behaviors that neuronal cilia may modulate.

RELATIONSHIP OF THE TRANSITION ZONE PROTEIN RPGRIP1L TO OBESITY

Several common SNPs associated with obesity occur within the first intron of the Fat Mass and

Obesity-Associated (*FTO*) gene (Dina et al. 2007; Frayling et al. 2007; Scuteri et al. 2007; Meyre et al. 2010). While studies have suggested mechanisms by which Fto can affect body weight, SNPs can alter regulatory elements that control the expression of distant genes, raising the possibility that *FTO*-associated SNPs may alter the expression of other genes. Chromatin conformation capture (3C) and circular chromosome conformation capture followed by high-throughput sequencing (4C-seq) revealed that, in addition to the *Fto* promoter, the region of the obesity-associated SNPs also interacts with the promoter of *Irx3*, a distant gene encoding a homeobox transcription factor (Smemo et al. 2014). At least in the human cerebellum, the presence of weight-associated SNPs and *Irx3* expression, but not *Fto* expression, were correlated (Smemo et al. 2014). Thus, SNPs within the *FTO* gene may affect the brain expression of *IRX3* to affect human body weight.

In addition to *IRX3*, *FTO* is nearby to *retinitis pigmentosa GTPase regulator-interacting protein-1 like* (*RPGRIP1L*). *RPGRIP1L* encodes a key transition zone component involved in the localization of many other transition zone proteins (Liu et al. 2011; Williams et al. 2011). Like *Fto*, fasting lowers expression of *Rpgrip1l* in the mouse hypothalamus (Stratigopoulos et al. 2011), suggesting that transcriptional regulation of *Rpgrip1l* may control energy homeostasis. Homozygous mutations of mouse *Rpgrip1l* cause embryonic phenotypes consistent with a severe disruption in ciliogenesis, complicating the analysis of how Rpgrip1l may impact energy metabolism (Delous et al. 2007). However, heterozygous mice express half the normal levels of Rpgrip1l protein (and wild-type levels of *Fto* and *Irx3*), and are hyperphagic and fatter than wild-type controls (Stratigopoulos et al. 2014). The number of Adcy3-positive cilia in the ARC of *Rpgrip1l* heterozygotes is modestly reduced, suggesting that reduced expression of Rpgrip1l may alter neuronal cilia, increasing feeding (Stratigopoulos et al. 2014). One of the *FTO*-associated SNPs overlaps with a binding site for the p110 isoform of CUX1, a long-range transcriptional regulator that can affect *RPGRIP1L* expression in vitro (Stratigopoulos et al. 2011;

Vadnais et al. 2013). Thus, diminished expression of this key transition zone component may alter ciliogenesis to account for how the *FTO*-associated SNPs affect human weight. Future studies looking at conditional changes in Rpgrip1l expression in development and the adult animal will help to elucidate the potential role for Rpgrip1l in energy homeostasis.

POTENTIAL ROLES FOR PERIPHERAL CILIA IN METABOLIC REGULATION AND OBESITY

Functions for cilia in the central control of energy metabolism do not preclude additional roles for cilia relevant to obesity in peripheral tissues, including adipocytes. Like many mesenchymal cell types, preadipocytes can be ciliated (Marion et al. 2009; Zhu et al. 2009; Dalbay et al. 2015). Interestingly, preadipocytes are transiently ciliated during the transition from proliferation to terminal differentiation, leading to mature adipocytes that lack cilia (Marion et al. 2009). Interfering with ciliary function by knocking down *BBS10* or *BBS12* in human preadipocytes increased PPARγ nuclear localization, a marker of adipogenesis (Marion et al. 2009). In contrast, knocking down *Ift88* in preadipocytes decreased PPARγ levels and nuclear localization, and inhibited fat droplet formation, a marker of mature adipocytes (Zhu et al. 2009; Dalbay et al. 2015). Although it is not clear how these different manipulations of BBS-associated and ciliogenic genes relate to each other, one possibility is that cilia promote adipogenesis in a way that is restrained by the activity of BBS proteins.

One candidate for mediating the effects of cilia on adipogenesis is Hedgehog (Hh) signaling (see Bangs and Anderson 2016). Sustained activation of Hh signaling can suppress adipogenesis, and down-regulation of Hh signaling is concomitant with terminal differentiation of adipocytes (Cousin et al. 2006; Suh et al. 2006; Fontaine et al. 2008; James et al. 2010). The Hh pathway mediator Smoothened (Smo) localizes to preadipocyte cilia, suggesting that active ciliary Hh signaling may restrain premature differentiation into mature adipocytes (Marion et al. 2009), although there are differing data on whether Smo inhibition is sufficient to promote adipocyte differentiation (Suh et al. 2006; Fontaine et al. 2008; James et al. 2010). In most cell types, BBS proteins have minor roles in Hh signaling, but it will be interesting to determine whether down-regulation of Hh signaling in preadipocytes may account for the increased adipogenesis caused by BBS protein loss-of-function.

The decision of preadipocytes to terminally differentiate may depend on ciliary signaling beyond Hh signaling. One pathway that may oppose antiadipogenic Hh signaling may be insulin-like growth factor 1 receptor (IGF1R) signaling. IGF1R, a receptor tyrosine kinase, is an important activator of both the expansion and differentiation of 3T3-L1 preadipocytes (Qiu et al. 2001; Xu and Liao 2004). IGF1R phosphorylates its adapter protein, insulin receptor substrate 1 (IRS1), to indirectly activate AKT1, a serine/threonine protein kinase involved in adipocyte metabolism (Fischer-Posovszky et al. 2012). Activated IGF1R localizes to 3T3-L1 cilia, and activated IRS1 and AKT1 to the basal body (Zhu et al. 2009), suggesting an important role for cilia in sensitizing preadipocytes to the prodifferentiative effects of insulin. Consistent with this hypothesis, inhibiting ciliogenesis by knockdown of either *Ift88* or *Kif3a* inhibits the activation of IGF1R or AKT1 in 3T3-L1 preadipocytes the same (Zhu et al. 2009).

In addition to effects on the differentiation of preadipocyte into mature adipocytes, ciliary signaling may impact adipogenesis through regulation of cellular metabolism. Stimulating 3T3-L1 adipocytes with Sonic Hedgehog ligand promoted aerobic glycolysis on a time scale inconsistent with a transcriptional effect, suggesting that Hh signals can impact metabolism through Gli-independent "noncanonical" mechanisms (Teperino et al. 2012).

Could disruption of a nontranscriptional, noncanonical ciliary Hh signaling pathway decrease aerobic glycolysis and thereby at least partially account for how ciliary defects cause obesity? SAG and cyclopamine, two small molecules that promote ciliary localization of Smo but have opposite effects on the Gli-dependent Hh transcriptional program, both promote glucose uptake in 3T3-L1 cells (Teperino et al. 2012). Moreover, cyclopamine increases glucose uptake by mouse brown adipose tissue and muscle in vivo and increases core body temperature by ∼1°C, consistent with increased thermogenesis. It will be interesting to determine whether genetic removal of Smo or cilia from brown adipose tissue and muscle blocks these effects, confirming that cyclopamine acts through a ciliary Hh pathway to activate thermogenesis.

One way that ciliary Smo may affect metabolism is through AMP-activated protein kinase (AMPK), a critical regulator of cellular energy homeostasis. Pharmacological stimulation of Smo activated AMPK in a way that was dependent on *Ift88* and *Kif3a*, two genes required for ciliogenesis (Teperino et al. 2012). AMPK and its upstream regulator, LKB1, have been identified as proteins that can, at least partially, localize to cilia (Boehlke et al. 2010; Mick et al. 2015), further suggesting that Smo may regulate AMPK at the cilium. However, small molecule antagonists of Smo that block its localization to cilia also induced AMPK activation, indicating that the cilium may not be crucial for Hh-mediated modulation of AMPK (Teperino et al. 2012). In contrast to Hh pathway stimulation, AMPK reduces glucose uptake and restricts aerobic glycolysis, at least in some cell types (Faubert et al. 2013), suggesting that the effects of Hh signaling on metabolism may be at least partly mediated through an AMPK-independent mechanism. Although neuronal cilia clearly have important roles in satiety control, alterations in ciliary signaling in preadipocytes, adipocytes, or muscle may contribute to ciliopathy-associated obesity, a possibility that remains to be examined in ciliopathy models.

CONCLUSIONS

Important questions about the pathogenesis of ciliopathy-associated obesity concern some of the following: which tissues are involved, which signaling pathways are involved, and whether the obesity arises from developmental or physiological changes. Conditional genetic ablation of the primary cilia in mice has established that neuronal primary cilia suppress feeding behavior, suggesting that some form of ciliary signaling promotes satiety. A clear hypothesis is that human ciliopathy mutations could alter the localization and/or function of receptor(s) involved in energy homeostasis at the primary cilia. While LEPR is an excellent candidate for such a role, it may not be one of the direct culprits. Moreover, it is not yet clear whether the obesity caused by these mutations results from a developmental or postdevelopmental alteration of cilia function, with the possibility that it may be both. Of interest will be the systematic conditional removal of both ciliogenic and BBS-associated genes in adult hypothalamic neuronal populations implicated in feeding behavior to elucidate which neurons possess the cilia that function in appetite and satiety.

REFERENCES

*Reference is also in this collection.

Allison DB, Kaprio J, Korkeila M, Koskenvuo M, Neale MC, Hayakawa K. 1996. The heritability of body mass index among an international sample of monozygotic twins reared apart. *Int J Obes Relat Metab Disord* **20:** 501–506.

Aponte G, Leung P, Gross D, Yamada T. 1984. Effects of somatostatin on food intake in rats. *Life Sci* **35:** 741–746.

Aragones G, Ardid-Ruiz A, Ibars M, Suarez M, Blade C. 2016. Modulation of leptin resistance by food compounds. *Mol Nutr Food Res* **60:** 1789–1803.

Arsov T, Silva DG, O'Bryan MK, Sainsbury A, Lee NJ, Kennedy C, Manji SS, Nelms K, Liu C, Vinuesa CG, et al. 2006. Fat aussie—A new Alström syndrome mouse showing a critical role for ALMS1 in obesity, diabetes, and spermatogenesis. *Mol Endocrinol* **20:** 1610–1622.

* Bangs F, Anderson KV. 2016. Primary cilia and mammalian Hedgehog signaling. *Cold Spring Harb Perspect Biol* doi: 10.1101/cshperspect.a028175.

Beales PL, Elcioglu N, Woolf AS, Parker D, Flinter FA. 1999. New criteria for improved diagnosis of Bardet–Biedl syndrome: Results of a population survey. *J Med Genet* **36:** 437–446.

Begg DP, Woods SC. 2013. The endocrinology of food intake. *Nat Rev Endocrinol* **9:** 584–597.

Berbari NF, Lewis JS, Bishop GA, Askwith CC, Mykytyn K. 2008. Bardet–Biedl syndrome proteins are required for the localization of G protein-coupled receptors to primary cilia. *Proc Natl Acad Sci* **105:** 4242–4246.

Berbari NF, Pasek RC, Malarkey EB, Yazdi SM, McNair AD, Lewis WR, Nagy TR, Kesterson RA, Yoder BK. 2013. Leptin resistance is a secondary consequence of the obesity in ciliopathy mutant mice. *Proc Natl Acad Sci* **110:** 7796–7801.

Bishop GA, Berbari NF, Lewis J, Mykytyn K. 2007. Type III adenylyl cyclase localizes to primary cilia throughout the adult mouse brain. *J Comp Neurol* **505:** 562–571.

Boehlke C, Kotsis F, Patel V, Braeg S, Voelker H, Bredt S, Beyer T, Janusch H, Hamann C, Godel M, et al. 2010. Primary cilia regulate mTORC1 activity and cell size through Lkb1. *Nat Cell Biol* **12:** 1115–1122.

Borowsky B, Durkin MM, Ogozalek K, Marzabadi MR, DeLeon J, Lagu B, Heurich R, Lichtblau H, Shaposhnik Z, Daniewska I, et al. 2002. Antidepressant, anxiolytic and anorectic effects of a melanin-concentrating hormone-1 receptor antagonist. *Nat Med* **8:** 825–830.

Brailov I, Bancila M, Brisorgueil MJ, Miquel MC, Hamon M, Verge D. 2000. Localization of 5-HT(6) receptors at the plasma membrane of neuronal cilia in the rat brain. *Brain Res* **872:** 271–275.

* Braun DA, Hildebrandt F. 2016. Ciliopathies. *Cold Spring Harb Perspect Biol* doi: 10.1101/cshperspect.a028191.

Chen Y, Hu C, Hsu CK, Zhang Q, Bi C, Asnicar M, Hsiung HM, Fox N, Slieker LJ, Yang DD, et al. 2002. Targeted disruption of the melanin-concentrating hormone receptor-1 results in hyperphagia and resistance to diet-induced obesity. *Endocrinology* **143:** 2469–2477.

Collin GB, Cyr E, Bronson R, Marshall JD, Gifford EJ, Hicks W, Murray SA, Zheng QY, Smith RS, Nishina PM, et al. 2005. *Alms1*-disrupted mice recapitulate human Alström syndrome. *Hum Mol Genet* **14:** 2323–2333.

Considine RV, Sinha MK, Heiman ML, Kriauciunas A, Stephens TW, Nyce MR, Ohannesian JP, Marco CC, McKee LJ, Bauer TL, et al. 1996. Serum immunoreactive-leptin concentrations in normal-weight and obese humans. *N Engl J Med* **334:** 292–295.

Cousin W, Dani C, Peraldi P. 2006. Inhibition of the anti-adipogenic Hedgehog signaling pathway by cyclopamine does not trigger adipocyte differentiation. *Biochem Biophys Res Commun* **349:** 799–803.

Dalbay MT, Thorpe SD, Connelly JT, Chapple JP, Knight MM. 2015. Adipogenic differentiation of hMSCs is mediated by recruitment of IGF-1r onto the primary cilium associated with cilia elongation. *Stem Cells* **33:** 1952–1961.

Davenport JR, Watts AJ, Roper VC, Croyle MJ, van Groen T, Wyss JM, Nagy TR, Kesterson RA, Yoder BK. 2007. Disruption of intraflagellar transport in adult mice leads to obesity and slow-onset cystic kidney disease. *Curr Biol* **17:** 1586–1594.

Davis RE, Swiderski RE, Rahmouni K, Nishimura DY, Mullins RF, Agassandian K, Philp AR, Searby CC, Andrews MP, Thompson S, et al. 2007. A knockin mouse model of the Bardet–Biedl syndrome 1 M390R mutation has cilia defects, ventriculomegaly, retinopathy, and obesity. *Proc Natl Acad Sci* **104:** 19422–19427.

Delous M, Baala L, Salomon R, Laclef C, Vierkotten J, Tory K, Golzio C, Lacoste T, Besse L, Ozilou C, et al. 2007. The ciliary gene RPGRIP1L is mutated in cerebello-oculo-renal syndrome (Joubert syndrome type B) and Meckel syndrome. *Nat Genet* **39:** 875–881.

Dina C, Meyre D, Gallina S, Durand E, Korner A, Jacobson P, Carlsson LM, Kiess W, Vatin V, Lecoeur C, et al. 2007. Variation in *FTO* contributes to childhood obesity and severe adult obesity. *Nat Genet* **39:** 724–726.

Domire JS, Green JA, Lee KG, Johnson AD, Askwith CC, Mykytyn K. 2011. Dopamine receptor 1 localizes to neuronal cilia in a dynamic process that requires the Bardet–Biedl syndrome proteins. *Cell Mol Life Sci* **68:** 2951–2960.

Eichers ER, Abd-El-Barr MM, Paylor R, Lewis RA, Bi W, Lin X, Meehan TP, Stockton DW, Wu SM, Lindsay E, et al. 2006. Phenotypic characterization of *Bbs4* null mice reveals age-dependent penetrance and variable expressivity. *Hum Genet* **120:** 211–226.

Farooqi IS, Jebb SA, Langmack G, Lawrence E, Cheetham CH, Prentice AM, Hughes IA, McCamish MA, O'Rahilly S. 1999. Effects of recombinant leptin therapy in a child with congenital leptin deficiency. *N Engl J Med* **341:** 879–884.

Fath MA, Mullins RF, Searby C, Nishimura DY, Wei J, Rahmouni K, Davis RE, Tayeh MK, Andrews M, Yang B, et al. 2005. Mkks-null mice have a phenotype resembling Bardet–Biedl syndrome. *Hum Mol Genet* **14:** 1109–1118.

Faubert B, Boily G, Izreig S, Griss T, Samborska B, Dong Z, Dupuy F, Chambers C, Fuerth BJ, Viollet B, et al. 2013. AMPK is a negative regulator of the Warburg effect and suppresses tumor growth in vivo. *Cell Metab* **17:** 113–124.

Fischer-Posovszky P, Tews D, Horenburg S, Debatin KM, Wabitsch M. 2012. Differential function of Akt1 and Akt2 in human adipocytes. *Mol Cell Endocrinol* **358:** 135–143.

Fontaine C, Cousin W, Plaisant M, Dani C, Peraldi P. 2008. Hedgehog signaling alters adipocyte maturation of human mesenchymal stem cells. *Stem Cells* **26:** 1037–1046.

Forsythe E, Beales PL. 2013. Bardet–Biedl syndrome. *Eur J Hum Genet* **21:** 8–13.

Frayling TM, Timpson NJ, Weedon MN, Zeggini E, Freathy RM, Lindgren CM, Perry JR, Elliott KS, Lango H, Rayner NW, et al. 2007. A common variant in the *FTO* gene is associated with body mass index and predisposes to childhood and adult obesity. *Science* **316:** 889–894.

Garfield AS, Li C, Madara JC, Shah BP, Webber E, Steger JS, Campbell JN, Gavrilova O, Lee CE, Olson DP, et al. 2015. A neural basis for melanocortin-4 receptor-regulated appetite. *Nat Neurosci* **18:** 863–871.

Ghamari-Langroudi M, Digby GJ, Sebag JA, Millhauser GL, Palomino R, Matthews R, Gillyard T, Panaro BL, Tough IR, Cox HM, et al. 2015. G-protein-independent coupling of MC4R to Kir7.1 in hypothalamic neurons. *Nature* **520:** 94–98.

Girard D, Petrovsky N. 2011. Alström syndrome: Insights into the pathogenesis of metabolic disorders. *Nat Rev Endocrinol* **7:** 77–88.

Guo DF, Cui H, Zhang Q, Morgan DA, Thedens DR, Nishimura D, Grobe JL, Sheffield VC, Rahmouni K. 2016. The BBSome controls energy homeostasis by mediating the transport of the leptin receptor to the plasma membrane. *PLoS Genet* **12:** e1005890.

Hamon M, Doucet E, Lefevre K, Miquel MC, Lanfumey L, Insausti R, Frechilla D, Del Rio J, Verge D. 1999. Antibodies and antisense oligonucleotide for probing the distribution and putative functions of central 5-HT6 receptors. *Neuropsychopharmacology* **21:** 68S–76S.

Handel M, Schulz S, Stanarius A, Schreff M, Erdtmann-Vourliotis M, Schmidt H, Wolf G, Hollt V. 1999. Selective targeting of somatostatin receptor 3 to neuronal cilia. *Neuroscience* **89:** 909–926.

Hearn T, Renforth GL, Spalluto C, Hanley NA, Piper K, Brickwood S, White C, Connolly V, Taylor JF, Russell-Eggitt I, et al. 2002. Mutation of *ALMS1*, a large gene with a tandem repeat encoding 47 amino acids, causes Alström syndrome. *Nat Genet* **31:** 79–83.

Heydet D, Chen LX, Larter CZ, Inglis C, Silverman MA, Farrell GC, Leroux MR. 2013. A truncating mutation of Alms1 reduces the number of hypothalamic neuronal cilia in obese mice. *Dev Neurobiol* **73:** 1–13.

James AW, Leucht P, Levi B, Carre AL, Xu Y, Helms JA, Longaker MT. 2010. Sonic Hedgehog influences the balance of osteogenesis and adipogenesis in mouse adipose-derived stromal cells. *Tissue Eng Part A* **16:** 2605–2616.

Jin H, White SR, Shida T, Schulz S, Aguiar M, Gygi SP, Bazan JF, Nachury MV. 2010. The conserved Bardet–Biedl syndrome proteins assemble a coat that traffics membrane proteins to cilia. *Cell* **141:** 1208–1219.

Knorz VJ, Spalluto C, Lessard M, Purvis TL, Adigun FF, Collin GB, Hanley NA, Wilson DI, Hearn T. 2010. Centriolar association of ALMS1 and likely centrosomal functions of the ALMS motif-containing proteins C10orf90 and KIAA1731. *Mol Biol Cell* **21:** 3617–3629.

Koemeter-Cox AI, Sherwood TW, Green JA, Steiner RA, Berbari NF, Yoder BK, Kauffman AS, Monsma PC, Brown A, Askwith CC, et al. 2014. Primary cilia enhance kisspeptin receptor signaling on gonadotropin-releasing hormone neurons. *Proc Natl Acad Sci* **111:** 10335–10340.

Kudryashova E, Wu J, Havton LA, Spencer MJ. 2009. Deficiency of the E3 ubiquitin ligase TRIM32 in mice leads to a myopathy with a neurogenic component. *Hum Mol Genet* **18:** 1353–1367.

Li G, Vega R, Nelms K, Gekakis N, Goodnow C, McNamara P, Wu H, Hong NA, Glynne R. 2007. A role for Alström syndrome protein, alms1, in kidney ciliogenesis and cellular quiescence. *PLoS Genet* **3:** e8.

Liew GM, Ye F, Nager AR, Murphy JP, Lee JS, Aguiar M, Breslow DK, Gygi SP, Nachury MV. 2014. The intraflagellar transport protein IFT27 promotes BBSome exit from cilia through the GTPase ARL6/BBS3. *Dev Cell* **31:** 265–278.

Liu L, Zhang M, Xia Z, Xu P, Chen L, Xu T. 2011. *Caenorhabditis elegans* ciliary protein NPHP-8, the homologue of human RPGRIP1L, is required for ciliogenesis and chemosensation. *Biochem Biophys Res Commun* **410:** 626–631.

Loktev AV, Jackson PK. 2013. Neuropeptide Y family receptors traffic via the Bardet–Biedl syndrome pathway to signal in neuronal primary cilia. *Cell Rep* **5:** 1316–1329.

Ludwig DS, Tritos NA, Mastaitis JW, Kulkarni R, Kokkotou E, Elmquist J, Lowell B, Flier JS, Maratos-Flier E. 2001. Melanin-concentrating hormone overexpression in transgenic mice leads to obesity and insulin resistance. *J Clin Invest* **107:** 379–386.

Maffei M, Halaas J, Ravussin E, Pratley RE, Lee GH, Zhang Y, Fei H, Kim S, Lallone R, Ranganathan S, et al. 1995. Leptin levels in human and rodent: Measurement of plasma leptin and *ob* RNA in obese and weight-reduced subjects. *Nat Med* **1:** 1155–1161.

Marion V, Stoetzel C, Schlicht D, Messaddeq N, Koch M, Flori E, Danse JM, Mandel JL, Dollfus H. 2009. Transient ciliogenesis involving Bardet–Biedl syndrome proteins is a fundamental characteristic of adipogenic differentiation. *Proc Natl Acad Sci* **106:** 1820–1825.

Marion V, Mockel A, De Melo C, Obringer C, Claussmann A, Simon A, Messaddeq N, Durand M, Dupuis L, Loeffler JP, et al. 2012. BBS-induced ciliary defect enhances adipogenesis, causing paradoxical higher-insulin sensitivity, glucose usage, and decreased inflammatory response. *Cell Metab* **16:** 363–377.

Marley A, von Zastrow M. 2010. DISC1 regulates primary cilia that display specific dopamine receptors. *PLoS ONE* **5:** e10902.

Meyre D, Proulx K, Kawagoe-Takaki H, Vatin V, Gutierrez-Aguilar R, Lyon D, Ma M, Choquet H, Horber F, Van Hul W, et al. 2010. Prevalence of loss-of-function *FTO* mutations in lean and obese individuals. *Diabetes* **59:** 311–318.

Mick DU, Rodrigues RB, Leib RD, Adams CM, Chien AS, Gygi SP, Nachury MV. 2015. Proteomics of primary cilia by proximity labeling. *Dev Cell* **35:** 497–512.

Mistlberger RE. 2009. Food-anticipatory circadian rhythms: Concepts and methods. *Eur J Neurosci* **30:** 1718–1729.

Mountjoy KG. 2015. Pro-Opiomelanocortin (POMC) neurones, POMC-derived peptides, melanocortin receptors and obesity: How understanding of this system has changed over the last decade. *J Neuroendocrinol* **27:** 406–418.

Mourao A, Nager AR, Nachury MV, Lorentzen E. 2014. Structural basis for membrane targeting of the BBSome by ARL6. *Nat Struct Mol Biol* **21:** 1035–1041.

Myers MG Jr, Leibel RL, Seeley RJ, Schwartz MW. 2010. Obesity and leptin resistance: Distinguishing cause from effect. *Trends Endocrinol Metab* **21:** 643–651.

Mykytyn K, Mullins RF, Andrews M, Chiang AP, Swiderski RE, Yang B, Braun T, Casavant T, Stone EM, Sheffield VC. 2004. Bardet–Biedl syndrome type 4 (BBS4)-null mice implicate Bbs4 in flagella formation but not global cilia assembly. *Proc Natl Acad Sci* **101:** 8664–8669.

Nachury MV, Loktev AV, Zhang Q, Westlake CJ, Peranen J, Merdes A, Slusarski DC, Scheller RH, Bazan JF, Sheffield VC, et al. 2007. A core complex of BBS proteins cooperates with the GTPase Rab8 to promote ciliary membrane biogenesis. *Cell* **129:** 1201–1213.

Nishimura DY, Fath M, Mullins RF, Searby C, Andrews M, Davis R, Andorf JL, Mykytyn K, Swiderski RE, Yang B, et al. 2004. *Bbs2*-null mice have neurosensory deficits, a defect in social dominance, and retinopathy associated with mislocalization of rhodopsin. *Proc Natl Acad Sci* **101:** 16588–16593.

O'Rahilly S, Farooqi IS. 2000. The genetics of obesity in humans. In *Endotext* (ed. De Groot LJ, et al.). MDText, South Dartmouth, MA.

Pitman JL, Wheeler MC, Lloyd DJ, Walker JR, Glynne RJ, Gekakis N. 2014. A gain-of-function mutation in adenylate cyclase 3 protects mice from diet-induced obesity. *PLoS ONE* **9:** e110226.

Pollock JD, Rowland N. 1981. Peripherally administered serotonin decreases food intake in rats. *Pharmacol Biochem Behav* **15:** 179–183.

Qiu Z, Wei Y, Chen N, Jiang M, Wu J, Liao K. 2001. DNA synthesis and mitotic clonal expansion is not a required step for 3T3-L1 preadipocyte differentiation into adipocytes. *J Biol Chem* **276:** 11988–11995.

Qu D, Ludwig DS, Gammeltoft S, Piper M, Pelleymounter MA, Cullen MJ, Mathes WF, Przypek R, Kanarek R, Maratos-Flier E. 1996. A role for melanin-concentrating hormone in the central regulation of feeding behaviour. *Nature* **380:** 243–247.

Rahmouni K, Fath MA, Seo S, Thedens DR, Berry CJ, Weiss R, Nishimura DY, Sheffield VC. 2008. Leptin resistance contributes to obesity and hypertension in mouse models of Bardet–Biedl syndrome. *J Clin Invest* **118:** 1458–1467.

Salamone JD, Zigmond MJ, Stricker EM. 1990. Characterization of the impaired feeding behavior in rats given haloperidol or dopamine-depleting brain lesions. *Neuroscience* **39:** 17–24.

Schulz S, Handel M, Schreff M, Schmidt H, Hollt V. 2000. Localization of five somatostatin receptors in the rat central nervous system using subtype-specific antibodies. *J Physiol Paris* **94:** 259–264.

Scuteri A, Sanna S, Chen WM, Uda M, Albai G, Strait J, Najjar S, Nagaraja R, Orru M, Usala G, et al. 2007. Genome-wide association scan shows genetic variants in the *FTO* gene are associated with obesity-related traits. *PLoS Genet* **3:** e115.

Seo S, Guo DF, Bugge K, Morgan DA, Rahmouni K, Sheffield VC. 2009. Requirement of Bardet–Biedl syndrome proteins for leptin receptor signaling. *Hum Mol Genet* **18:** 1323–1331.

Shimada M, Tritos NA, Lowell BB, Flier JS, Maratos-Flier E. 1998. Mice lacking melanin-concentrating hormone are hypophagic and lean. *Nature* **396:** 670–674.

Smemo S, Tena JJ, Kim KH, Gamazon ER, Sakabe NJ, Gomez-Marin C, Aneas I, Credidio FL, Sobreira DR, Wasserman NF, et al. 2014. Obesity-associated variants within in *FTO* form long-range functional connections with IRX3. *Nature* **507:** 371–375.

Stratigopoulos G, LeDuc CA, Cremona ML, Chung WK, Leibel RL. 2011. Cut-like homeobox 1 (CUX1) regulates expression of the fat mass and obesity-associated and retinitis pigmentosa GTPase regulator-interacting protein-1-like (RPGRIP1L) genes and coordinates leptin receptor signaling. *J Biol Chem* **286:** 2155–2170.

Stratigopoulos G, Martin Carli JF, O'Day DR, Wang L, Leduc CA, Lanzano P, Chung WK, Rosenbaum M, Egli D, Doherty DA, et al. 2014. Hypomorphism for *RPGRIP1L*, a ciliary gene vicinal to the FTO locus, causes increased adiposity in mice. *Cell Metab* **19:** 767–779.

Suh JM, Gao X, McKay J, McKay R, Salo Z, Graff JM. 2006. Hedgehog signaling plays a conserved role in inhibiting fat formation. *Cell Metab* **3:** 25–34.

Tadenev AL, Kulaga HM, May-Simera HL, Kelley MW, Katsanis N, Reed RR. 2011. Loss of Bardet–Biedl syndrome protein-8 (BBS8) perturbs olfactory function, protein localization, and axon targeting. *Proc Natl Acad Sci* **108:** 10320–10325.

Teperino R, Amann S, Bayer M, McGee SL, Loipetzberger A, Connor T, Jaeger C, Kammerer B, Winter L, Wiche G, et al. 2012. Hedgehog partial agonism drives Warburg-like metabolism in muscle and brown fat. *Cell* **151:** 414–426.

Tolson KP, Garcia C, Yen S, Simonds S, Stefanidis A, Lawrence A, Smith JT, Kauffman AS. 2014. Impaired kisspeptin signaling decreases metabolism and promotes glucose intolerance and obesity. *J Clin Invest* **124:** 3075–3079.

Vadnais C, Awan AA, Harada R, Clermont PL, Leduy L, Berube G, Nepveu A. 2013. Long-range transcriptional regulation by the p110 CUX1 homeodomain protein on the ENCODE array. *BMC Genomics* **14:** 258.

Vijayan E, McCann SM. 1977. Suppression of feeding and drinking activity in rats following intraventricular injection of thyrotropin releasing hormone (TRH). *Endocrinology* **100:** 1727–1730.

Wang Z, Li V, Chan GC, Phan T, Nudelman AS, Xia Z, Storm DR. 2009. Adult type 3 adenylyl cyclase-deficient mice are obese. *PLoS ONE* **4:** e6979.

Williams CL, Li C, Kida K, Inglis PN, Mohan S, Semenec L, Bialas NJ, Stupay RM, Chen N, Blacque OE, et al. 2011. MKS and NPHP modules cooperate to establish basal body/transition zone membrane associations and ciliary gate function during ciliogenesis. *J Cell Biol* **192:** 1023–1041.

Xu J, Liao K. 2004. Protein kinase B/AKT 1 plays a pivotal role in insulin-like growth factor-1 receptor signaling induced 3T3-L1 adipocyte differentiation. *J Biol Chem* **279:** 35914–35922.

Xu AW, Kaelin CB, Takeda K, Akira S, Schwartz MW, Barsh GS. 2005. PI3K integrates the action of insulin and leptin on hypothalamic neurons. *J Clin Invest* **115:** 951–958.

Zhang Y, Proenca R, Maffei M, Barone M, Leopold L, Friedman JM. 1994. Positional cloning of the mouse obese gene and its human homologue. *Nature* **372:** 425–432.

Zhang Q, Nishimura D, Seo S, Vogel T, Morgan DA, Searby C, Bugge K, Stone EM, Rahmouni K, Sheffield VC. 2011. Bardet–Biedl syndrome 3 (Bbs3) knockout mouse model reveals common BBS-associated phenotypes and Bbs3 unique phenotypes. *Proc Natl Acad Sci* **108:** 20678–20683.

Zhang Q, Nishimura D, Vogel T, Shao J, Swiderski R, Yin T, Searby C, Carter CS, Kim G, Bugge K, et al. 2013. BBS7 is required for BBSome formation and its absence in mice results in Bardet–Biedl syndrome phenotypes and selective abnormalities in membrane protein trafficking. *J Cell Sci* **126:** 2372–2380.

Zhu D, Shi S, Wang H, Liao K. 2009. Growth arrest induces primary-cilium formation and sensitizes IGF-1-receptor signaling during differentiation induction of 3T3-L1 preadipocytes. *J Cell Sci* **122:** 2760–2768.

Sperm Sensory Signaling

Dagmar Wachten,[1] Jan F. Jikeli,[1] and U. Benjamin Kaupp[2]

[1]Minerva Max Planck Research Group, Molecular Physiology, Center of Advanced European Studies and Research (caesar), 53175 Bonn, Germany

[2]Department Molecular Sensory Systems, Center of Advanced European Studies and Research (caesar), 53175 Bonn, Germany

Correspondence: dagmar.wachten@caesar.de; u.b.kaupp@caesar.de

Fertilization is exceptionally complex and, depending on the species, happens in entirely different environments. External fertilizers in aquatic habitats, like marine invertebrates or fish, release their gametes into the seawater or freshwater, whereas sperm from most internal fertilizers like mammals cross the female genital tract to make their way to the egg. Various chemical and physical cues guide sperm to the egg. Quite generally, these cues enable signaling pathways that ultimately evoke a cellular Ca^{2+} response that modulates the waveform of the flagellar beat and, hence, the swimming path. To cope with the panoply of challenges to reach and fertilize the egg, sperm from different species have developed their own unique repertoire of signaling molecules and mechanisms. Here, we review the differences and commonalities for sperm sensory signaling in marine invertebrates (sea urchin), fish (zebrafish), and mammals (mouse, human).

Sperm carry a motile hair-like protrusion—called flagellum—that extends from the head. The flagellum serves both as sensory antenna and propelling motor. Sperm flagella and motile cilia share a similar $9 + 2$ axoneme structure. The sperm cell is propelled by bending waves traveling down the flagellum. For steering, sperm modulate the asymmetry of the flagellar waveform. Sperm motility has been mostly studied in two dimensions (2D) at the glass/water interface of shallow observation chambers (Böhmer et al. 2005; Alvarez et al. 2012). When the flagellum beats symmetrical with respect to the long axis of the ellipsoid-shaped head, sperm move on a straight path; when the flagellum beats more asymmetrical, the swimming path is curved. If unrestricted, sperm from several species swim on helical paths or chiral ribbons (Crenshaw 1990, 1993a,b; Crenshaw and Edelstein-Keshet 1993; Corkidi et al. 2008; Su et al. 2012; Jikeli et al. 2015).

Various chemical and physical cues guide sperm to the egg. Quite generally, these cues enable signaling pathways that ultimately evoke a cellular Ca^{2+} response that in turn modulates the waveform of the flagellar beat and hence the swimming path (Böhmer et al. 2005; Wood et al. 2005; Shiba et al. 2008). Four different mechanisms have been identified that guide sperm to the egg: chemotaxis, haptotaxis, thermotaxis, and rheotaxis (Box 1).

Chemotaxis has been firmly established in marine invertebrates, notably in sperm from the sea urchin *Arbacia punctulata* (Fig. 1A). The

---

BOX 1

Chemotaxis refers to directed movement of cells in a gradient of a chemical substance. When sperm probe a chemical gradient along a circular or helical path, the spatial gradient is translated into a temporal stimulus pattern, which elicits periodic steering responses. This stimulus pattern is composed of a fast periodic component, which results from the helical or circular swimming path, and a mean stimulus component that increases or decreases when swimming up or down the gradient, respectively. During 3D navigation, the torsion of the helical 3D path is in phase ($\varphi = 0°$) and the path curvature is in antiphase ($\varphi = 180°$) with respect to the fast periodic stimulus component; the helix continuously bends to align with the attractant gradient ("on response"). If sperm happen to swim down the gradient, the mean stimulus level decreases and sperm respond with a strong correcting turn toward higher attractant concentrations ("off response") (Jikeli et al. 2015).

Rheotaxis refers to the directed movement of cells against a fluid flow. It is a mechanical sense because cells register a gradient of flow velocities. Because freely swimming sperm rotate around the longitudinal axis of the flagellum, mouse sperm are exposed to different flow velocities along their length.

Haptotaxis refers to the directed movement of cells on a surface that is covered with tethered chemoattractant molecules that form a chemical density gradient. Haptotaxis could occur on the surface of fish eggs and the epithelial layer of the fallopian tube.

Thermotaxis refers to directed movement in a temperature gradient. In one concept, sperm "tumble" by hyperactivation followed by straighter and faster swimming periods up the temperature gradient, similar to the tumble-and-run mechanism of bacteria during chemotaxis.

---

corresponding chemoattractant and the sequence of signaling events have been identified. The signaling pathway endows *Arbacia* sperm with exquisite sensitivity: they can register binding of a single chemoattractant molecule and transduce this event into a cellular Ca^{2+} response. Moreover, the navigation strategy in 2D and 3D chemoattractant gradients has been deciphered (Böhmer et al. 2005; Alvarez et al. 2012; Jikeli et al. 2015). In a 2D gradient, sperm swim on looping trajectories; in a 3D gradient, the swimming helix bends to align with the gradient (Jikeli et al. 2015).

Sperm of teleost fish are not actively attracted to the egg from afar (Yanagimachi et al. 2013). Instead, male fish deposit sperm directly onto the egg surface. Hydrodynamic interactions provide a theoretical framework that explains why sperm accumulate at interfaces (Rothschild 1963; Winet et al. 1984; Cosson et al. 2003), follow boundaries in microchannels, or for that matter stay at the egg surface (Elgeti and Gompper 2009; Lauga and Powers 2009; Elgeti et al. 2010; Denissenko et al. 2012). For fertilization, fish sperm must search for the narrow entrance to a cone-shaped funnel in the

egg coat—the micropyle—that provides access to the egg membrane for fusion (Fig. 1B) (Yanagimachi et al. 2013). Sperm reach the micropyle probably by haptic interactions with tethered molecules that line the egg surface and the opening or interior of the micropyle. Thus, fish sperm motility might be governed by specific hydrodynamic and haptic interactions with the egg surface and the micropyle.

In mammals, chemotaxis, rheotaxis, and thermotaxis have been proposed as guiding mechanisms (Bretherton and Rothschild 1961; Bahat et al. 2003, 2012; Eisenbach and Giojalas 2006; Miki and Clapham 2013; Kantsler et al. 2014; Boryshpolets et al. 2015; Bukatin et al. 2015; Perez-Cerezales et al. 2015; Zhang et al. 2016). Neither one of these mechanisms nor their respective contribution to sperm guidance across the long and narrow oviduct is well understood (Fig. 1C). This ignorance is a result of the technical difficulties to emulate the complex native conditions encountered by mammalian sperm during fertilization. On their journey to the egg, mammalian sperm undergo two processes—capacitation and hyperactivation—necessary to acquire the potential

Cite this article as *Cold Spring Harb Perspect Biol* doi: 10.1101/cshperspect.a028225

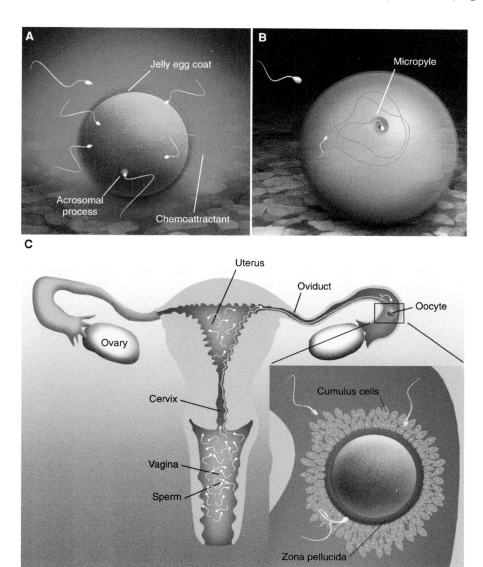

Figure 1. Fertilization in marine invertebrates, fish, and mammals. (*A*) In many marine invertebrates, sperm are attracted by chemotaxis (blue gradient); sperm can elicit the acrosomal process and fertilize the egg at any site of the egg surface. (*B*) Fish sperm are probably not attracted from afar by chemoattractants. Instead, sperm navigate on the egg surface to a small opening, the micropyle. The mechanisms and molecules underlying directed movement on the egg surface are not known. (*C*) Mammalian sperm are guided by several mechanisms across the female genital tract. The specifics and underlying molecules are discussed.

to fertilize the egg (Suarez et al. 1993; Suarez 2008). At any given time, only a small fraction of sperm is capacitated or hyperactive; therefore, mammalian sperm populations are heterogeneous. Moreover, mammalian sperm are spatially constrained in the convoluted female reproductive tract and interact with the ciliated epithelial layer that lines the oviduct. Notwithstanding this complexity, major advances have been achieved identifying the signaling molecules and delineating the signaling events that encode Ca^{2+} responses.

In this review, we discuss what is known about signaling pathways that control sperm sensory signaling in sea urchin, zebrafish, and in mammals.

SIGNALING IN SEA URCHIN SPERM AND OTHER MARINE INVERTEBRATES

Overview

Oocytes from the sea urchin *A. punctulata* release a short, species-specific chemoattractant peptide (resact, 14 amino acids in length). The chemoattractant binds to a receptor guanylate cyclase (GC) and thereby stimulates the synthesis of cGMP (Fig. 2A). In turn, cGMP opens K^+-selective cyclic nucleotide-gated (CNGK) channels. The ensuing hyperpolarization—the resting potential V_{rest} is ~ -50 mV—activates two other signaling components: a voltage-gated sodium/proton exchanger (sNHE) and a hyperpolarization-activated, cyclic nucleotide-gated (HCN) channel. sNHE activity causes a small intracellular alkalization that shifts the voltage dependence of the sperm-specific Ca^{2+} channel (CatSper, cation channel of sperm) to more negative values and thereby "primes" CatSper to open during the return of V_m to resting value. This return is initiated by the HCN channel that carries an Na^+ inward current. Recovery from the Ca^{2+} response is accomplished by a sodium/calcium/potassium exchanger (NCKX) and a plasma membrane Ca^{2+}-ATPase (PMCA) that restore Ca^{2+} levels, and a phosphodiesterase (PDE) that breaks down cGMP.

How common is this chemotactic signaling pathway? Sperm of the sea urchin *Strongylocentrotus purpuratus* harbor a similar cGMP-signaling pathway; however, in shallow recording chambers frequently used for the study of chemotactic behavior under the microscope, *S. purpuratus* sperm do not display chemotaxis (Guerrero et al. 2010). Presumably, the geometric conditions under which 2D chemotaxis can be observed are severely restricted for *S. purpuratus* sperm. Sperm of the sea star *Asterias amurensis* also use a peptide as chemoattractant, a GC as chemoreceptor (Nishigaki et al. 1996; Matsumoto et al. 2003), and binding of the chemoattractant elicits a rapid transient increase of cGMP that evokes a Ca^{2+} response (Matsumoto et al. 2003). *Arbacia* and *Strongylocentrotus* diverged 200 million years ago, whereas the split between asteroids (seastars) and echinoids (sea urchins) occurred approximately 500 million years ago. Thus, the cGMP-signaling pathway has been conserved for at least 500 million years in many marine invertebrates across several phyla. In the following, the signaling components are discussed in more detail.

THE CHEMORECEPTOR GUANYLATE CYCLASE

The chemoreceptor GC is composed of three functional domains (Potter 2011): an extracellular domain binds the chemoattractant peptide resact, an intracellular catalytic domain synthesizes cGMP, and a single transmembrane domain connects the binding and catalytic domain and relays the binding event to the cell interior (Fig. 2A). The oligomeric structure of the GC is not known. Mammalian orthologs exist either as dimers or trimers (Wilson and Chinkers 1995; Yu et al. 1999; Vaandrager 2002; Ogawa et al. 2004). The overall chemotactic sensitivity relies on a high efficacy to capture the chemoattractant; the efficacy is maximal if every molecule that hits the flagellum binds to a receptor and activates it. The receptor density and affinity determine the capture efficacy. The flagellum harbors about 300,000 GC copies at a density of 9500 GC molecules/μm^2 (Pichlo et al. 2014). In fact, the GC rivals with rhodopsin in photoreceptors ($\sim 26,000-45,000$ rhodopsin molecules/μm^2) as one of the most densely packed membrane receptors (Fotiadis et al. 2003; Gunkel et al. 2015). At very low receptor occupancy, the binding affinity of the GC is in the picomolar range ($K_{1/2} = 90$ pM). Thus, high capture efficacy is achieved by combining an extraordinary high GC density and ligand affinity. At higher occupancy, the GC affinity is lower ($K_{1/2} = 0.65$ nM). In fact, chemoattractant binding spans six orders of magnitude; the broad operational range might involve negative cooperation among GC subunits, a negative cellular feedback, or a combination of both. The affinity adjustment ensures that, at high chemo-

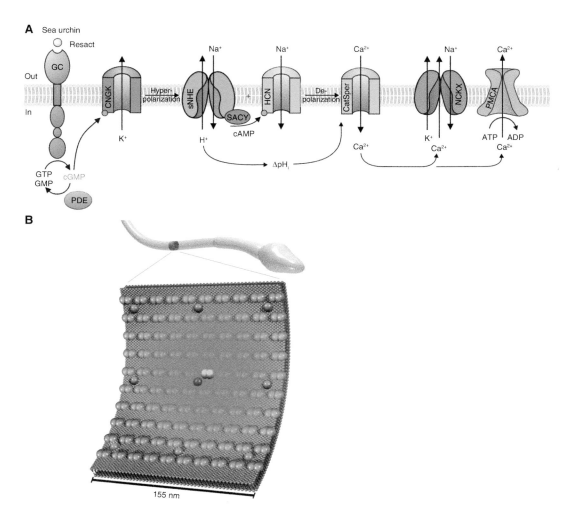

Figure 2. Signaling pathway in sea urchin sperm. (*A*) The guanylate cyclase (GC) serves as the receptor for the chemoattractant resact. Resact binding activates the GC, resulting in cGMP synthesis. cGMP activates the K^+-selective cyclic nucleotide-gated channel (CNGK). Opening of CNGK hyperpolarizes the cell and activates a hyperpolarization-activated and cyclic nucleotide-gated (HCN) channel and a sperm-specific voltage-dependent Na^+/H^+ exchanger (sNHE). Opening of HCN channels restores the resting potential, whereas activation of sNHE increases the intracellular pH. Both events activate the principal Ca^{2+} channel CatSper, leading to a Ca^{2+} influx. The Ca^{2+} levels are restored by Ca^{2+} extrusion through a $Na^+/Ca^{2+}/K^+$ exchanger (NCKX) and a Ca^{2+} ATPase PMCA, whereas cGMP is hydrolyzed by a phosphodiesterase (PDE). On resact binding, sea urchin sperm not only synthesize cGMP, but also cAMP, probably through activation of the soluble adenylate cyclase (SACY), which, at least in mouse sperm, forms a complex with sNHE. One known target for cAMP is the HCN channel, whose voltage-dependent opening is modulated by cAMP. (*B*) Schematic distribution of GC receptors and CNGK channel on the flagellum. The GC (gray) and CNGK (green) densities are drawn to scale (ratio GC dimer/CNGK is 10:1). The cGMP gradient is depicted in shades of magenta.

attractant concentrations prevailing near the egg, vacant receptors are still available.

The turnover number of active GC (GC*) is 72 cGMP molecules/GC*/second (Pichlo et al. 2014). GC* activity ceases within 150 ms probably by autodephosphorylation: at rest, six conserved serine residues carry phosphate groups that are removed on chemoattractant binding (Pichlo et al. 2014). GC* inactivation by multistage autodephosphorylation might allow for precise lifetime control, which could produce uniform Ca^{2+} responses and thereby reduce "molecule noise," which limits the precision of gradient sensing. A precedent for such a mechanism is the visual pigment rhodopsin: stepwise inactivation by two phosphorylation reactions and "capping" of phosphorylated rhodopsin by arrestin, a stop protein, has been proposed to control its lifetime and thereby reduce photon noise (Whitlock and Lamb 1999; Mendez et al. 2000; Doan et al. 2006). However, uniform single-photon responses also involve other mechanisms (Bisegna et al. 2008; Caruso et al. 2011; Gross et al. 2012a,b).

THE CYCLIC NUCLEOTIDE-GATED K^+ CHANNEL

The first electrical event in sea urchin sperm is a chemoattractant-induced hyperpolarization mediated by a cyclic nucleotide-gated K^+ channel (CNGK). The CNGK is unique compared with classic CNG channels (Bönigk et al. 2009; Cukkemane et al. 2011). Like voltage-activated Ca_v and Na_v channels, the large pore-forming polypeptide consists of four homologous repeats; each repeat carries the prototypical GYGD pore motif of K^+ channels and a cyclic nucleotide-binding domain (CNBD). The CNGK channel can respond to small changes in cGMP, because it is exquisitely sensitive to cGMP ($K_{1/2} = 20$ nM) and is activated in a noncooperative fashion. Disabling each of the four CNBDs through mutagenesis revealed that binding of a single cGMP molecule to the third repeat is necessary and sufficient to activate the channel (Bönigk et al. 2009). The other three CNBD domains either do not bind cGMP or fail to gate the channel pore. Thus, CNGK has

developed a noncooperative mechanism of activation that is different from the activation of CNG channels in photoreceptors and olfactory neurons that require the cooperative binding of several ligands (Biskup et al. 2007) and that operate in the micromolar rather than the nanomolar range of cAMP or cGMP concentrations (Kaupp and Seifert 2002).

A SPERM-SPECIFIC SODIUM/PROTON EXCHANGER (sNHE)

Perhaps one of the most enigmatic signaling events is a sodium/proton exchange. As little as we know, in sea urchin sperm, exchange of Na^+ against H^+ is electroneutral and its activity is controlled by voltage (Lee 1984a,b, 1985; Lee and Garbers 1986). The exchange is thought to be catalyzed by members of a sperm-specific subfamily of solute carriers (SLC9C1 or sNHE) that are structurally unique. They share with other NHEs a membrane-spanning exchange domain that features at least 12 transmembrane segments (Wang et al. 2003). Unlike any other SLC family members, sNHE harbors a voltage-sensor domain (VSD) similar to voltage-gated K^+-, Na^+-, and Ca^{2+} channels. Moreover, sNHE carry a CNBD similar to those in CNG and HCN channels. The presence of a VSD and a CNBD domain is both intriguing and presumably meaningful. However, sNHE molecules have not been functionally expressed. Therefore, the functions of these domains are unknown, even more so as some channels that harbor a VSD are not gated by voltage (e.g., CNG channels) and some CNG channels are not gated by cyclic nucleotides (Brelidze et al. 2013; Carlson et al. 2013; Haitin et al. 2013; Fechner et al. 2015). Considering the exquisite pH sensitivity of several signaling molecules, pH_i homeostasis by sodium/proton exchange is an important research area for future work.

HCN CHANNEL

Two HCN channel isoforms, SpHCN1 and SpHCN2, have been identified in S. purpuratus (Gauss et al. 1998; Galindo et al. 2005). SpHCN1 (originally called SpIH) (Gauss et al. 1998) has been functionally characterized in a

 Cite this article as *Cold Spring Harb Perspect Biol* doi: 10.1101/cshperspect.a028225

mammalian cell line bathed in normal Ringer solution. The ionic strength and composition of seawater and Ringer solution are entirely different. Moreover, whether the two orthologs form heteromers is not known. Thus, the physiological properties of the native HCN channel might be distinctively different from those of heterologously expressed SpHCN1. Notwithstanding, SpHCN1 shares several basic properties with mammalian HCN channels in neurons and the heart. HCN channels become activated when the membrane is hyperpolarized (i.e., at V_m more negative than ~0 mV); their activity is enhanced by cAMP; and their permeability is three- to fourfold larger for K^+ than for Na^+ ions. Therefore, under physiological conditions ($V_m > -30$ mV; high Na^+ outside, high K^+ inside), HCN channels carry a depolarizing inward Na^+ current.

HCN channels may serve multiple functions in sea urchin sperm. First, they probably contribute to the unusually low V_{rest} together with another ion channel with different ion selectivity (Fig. 3). There is a precedent in rod photoreceptors, which hyperpolarize on light stimulation. Two channels set V_{rest} (-40 mV): a K^+ channel in the inner segment (reversal potential $V_{rev} = -75$ mV) and a nonselective

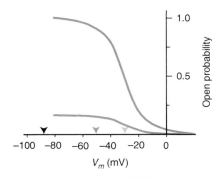

Figure 3. Membrane potential. The resting potential V_{rest} of ~-50 mV (green arrowhead) is probably determined by the hyperpolarization and cyclic nucleotide-gated (HCN) channels ($V_{rev} \sim -30$ mV in Ringer solution, orange arrowhead) and the K^+-selective cyclic nucleotide-gated (CNGK) channel (V_{rev} 9~-0 mV, black arrowhead). The $V_{1/2}$ of HCN opening is similar with and without cAMP, whereas open probability P_o is much larger in the presence of cAMP (red) than without (blue).

CNG channel in the outer segment ($V_{rev} = 0$ mV). In sperm, the HCN channel ($V_{rev} = -30$ mV in Ringer solution) and the CNGK channel ($V_{rev} = -90$ mV) or another K^+ channel could set V_{rest} at ~-50 mV (Fig. 3). In fact, a fraction of SpHCN1 channels is constitutively open at rest (Gauss et al. 1998).

Second, HCN channels carry an inward Na^+ current and thus counter the hyperpolarization. In retinal rods, HCN channels also repolarize the cell quickly in bright light and prevent V_m from reaching the K^+ equilibrium potential E_K. HCN channels in sperm are likely to serve a similar function: they initiate the rapid recovery from stimulation and allow sperm to encode a wide range of chemoattractant concentrations.

Third, sperm HCN channels are set apart from their mammalian cousins by two unique properties. SpHCN1, after a hyperpolarizing voltage step, first activates and then inactivates. cAMP removes the inactivation and consequently enhances the open probability P_o (Fig. 3). In contrast, mammalian HCN channels do not inactivate and cAMP shifts the $P_o - V_m$ relation to less negative potentials without affecting much the maximal current. Because of the exquisite cAMP sensitivity ($K_{1/2} = 0.75$ µM) and the large effect of cAMP on P_o, modulation of sperm HCN channels by cAMP might control the sensitivity of sperm, that is, contribute to adaptation.

Finally, HCN channels are often referred to as pacemakers, because they control rhythmic electrical activity in neurons and cardiac myocytes. HCN channels may serve a similar function in sperm. The periodic swimming path temporally organizes the stimulus pattern that is perceived by sperm. The ensuing periodic stimulation pattern entrains Ca^{2+} oscillations (Böhmer et al. 2005). Future studies need to examine whether HCN channels are important for pacing Ca^{2+} oscillations in sperm.

CatSper CHANNEL

CatSper is one of the most complex voltage-gated ion channels. Like in mammalian sperm, the CatSper channel in *A. punctulata* comprises four homologous α subunits (CatSper 1–4)

and at least three auxiliary subunits: CatSper β, CatSper γ, and CatSper δ (Seifert et al. 2015). At rest, CatSper is closed and is activated by a "switch-like" mechanism that involves two steps. The alkalization by sNHE shifts the voltage dependence to more negative values; the pH dependence of this cooperative shift is exceptionally steep (Hill coefficient of about 11) (Seifert et al. 2015). During recovery from hyperpolarization, CatSper opens (Fig. 2A). The high cooperativity permits CatSper to transduce the minute elementary changes in pH_i and V_m into a Ca^{2+} response.

ADENYLATE CYCLASE AND cAMP — IN SEARCH OF A FUNCTION

In contrast to cGMP, much less is known about the regulation and function of cAMP. Stimulation with resact evokes a rise of cAMP that is delayed with respect to the increase of cGMP (Kaupp et al. 2003). The synthesis of cAMP, presumably by a soluble adenylate cyclase (SACY) (Nomura et al. 2005), is enhanced under hyperpolarizing conditions (Beltrán et al. 1996) and alkaline pH_i (Cook and Babcock 1993), but is insensitive to Ca^{2+} (Nomura et al. 2005). Each of these properties is at odds with those of other SACYs, which are activated by bicarbonate and Ca^{2+} (Chen et al. 2000; Jaiswal and Conti 2003; Steegborn et al. 2005; Kleinboelting et al. 2014). Either the SACY in sperm of sea urchin mammals are entirely different, or a different membrane-spanning AC is involved in sea urchins. Protein kinase A (PKA) in mammalian sperm represents a sperm-specific isoform that is different from that in somatic cells (Nolan et al. 2004; Burton and McKnight 2007).

MOLECULES FOR RECOVERY

Although the signaling events that excite sea urchin sperm have been delineated in great detail, much less is known about recovery and adaptation. For recovery from chemoattractant stimulation cGMP, cAMP, pH_i, $[Ca^{2+}]_i$, and V_m must return to baseline levels. The elevated cGMP level is probably lowered by a cGMP-specific PDE 5 (Su and Vacquier 2006). The molecules that reestablish resting pH_i are not known. Ca^{2+} levels return to baseline owing to the activity of a sodium/calcium/potassium exchanger (NCKX) (Su and Vacquier 2002), and a Ca^{2+}-ATPase in the plasma membrane (PMCA) (Gunaratne et al. 2006). How these molecules are regulated is not known (Fig. 2A).

SINGLE-MOLECULE SENSITIVITY BY THE NUMBERS

Several mechanisms have been proposed that explain ultrasensitivity in cells. These concepts largely originate from the study of signaling in photoreceptors, olfactory neurons, and bacteria. Mechanisms include (1) lattices of highly cooperative chemoreceptors in bacteria (Maddock and Shapiro 1993; Bray et al. 1998; Duke and Bray 1999; Gestwicki and Kiessling 2002; Sourjik and Berg 2004); (2) high multistage gain provided by a cascade of enzymatic reactions in photoreceptors (Pugh and Lamb 2000; Yau and Hardie 2009; Kaupp 2010); (3) local signaling by supramolecular complexes (transducisome or signalosome) (Huber et al. 1996; Tsunoda et al. 1997; Scott and Zuker 1998); and (4) restricted diffusion of chemical messengers in confined cellular subcompartments (Rich et al. 2000, 2001).

As will become clear, sperm use different mechanisms to achieve single-molecule sensitivity. First, sperm do not rely on arrays of receptor clusters that display positive cooperativity, although the GC may adopt a supramolecular organization that, however, serves other functions.

Second, amplification at the receptor level is orders of magnitude lower in sperm compared with rod photoreceptors. Rods also entertain a cGMP-signaling pathway that endows these cells with single-photon sensitivity. Capture of a photon initiates the hydrolysis of 2000 to 72,000 cGMP molecules, depending on the species, by two-stage amplification (Pugh and Lamb 2000; Burns and Pugh 2010; Arshavsky and Burns 2014). For sperm, a lower and upper limit of the number of cGMP molecules involved in single-molecule events has been estimated by two different methods. During its life-

Cite this article as *Cold Spring Harb Perspect Biol* doi: 10.1101/cshperspect.a028225

time, a GC* synthesizes ∼11 cGMP molecules (Pichlo et al. 2014). Using fluorescent caged cGMP, the number of cGMP molecules required for a single-molecule response was estimated to be about 47 cGMP molecules (Bönigk et al. 2009). Both estimates are fraught with uncertainties. The turnover rate of cGMP synthesis was determined at relatively high chemoattractant concentrations and has been linearly extrapolated to the single-molecule regime. However, the catalytic turnover might depend on the occupation level of the GC. For example, if the GC exists as a dimer and if each subunit binds a resact molecule (stoichiometry 2:2), the turnover of single- or double-occupied receptors might be different. Finally, it is assumed that the concentration of membrane-permeant caged cGMP has completely equilibrated across the membrane. Apart from these uncertainties, cGMP amplification in rods is about 600-fold larger than in sperm. Nonetheless, rod and sperm operate at a similar level of sensitivity: binding of a molecule to sperm or capture of a photon by rods each evokes a ΔV_m of 1–2 mV (Pugh and Lamb 2000; Strünker et al. 2006). Taking into account a volume ratio $V_{rod}/V_{flagellum}$ of 80/2 fl = 40:1, the change in cGMP concentration per unit volume is only 15× larger in rods compared with sperm.

Third, it has been argued that second-messenger concentrations steeply decay from the site of synthesis (Rich et al. 2000, 2001), which calls either for a signaling complex between receptors and downstream targets or for restricted diffusion in subcellular compartments channeling the messenger to the target. In order for a GC–CNGK complex to be effective, receptor (GC) and target (CNGK channel) ought to be present stoichiometrically, whereas, in fact, the GC is ∼24-fold more abundant than the CNGK (Pichlo et al. 2014). Thus, a "transducisome" between GC and CNGK is unlikely to contribute to single-molecule sensitivity of the sea urchin sperm.

The impact of cGMP diffusion from a point source at the membrane in a small cylindrical compartment like the flagellum was assessed (Pichlo et al. 2014). Within ∼15 µs, the cGMP concentration equilibrates across the fla-

gellar diameter and, within ∼100 ms, it equilibrates along the length of the flagellum. In a 155-nm-long segment of the flagellum, neglecting cGMP binding to high-affinity buffers or hydrolysis by PDEs, the cGMP transiently increases to micromolar concentrations that saturate any nearby CNGK channels ($K_{1/2} =$ 26 nM) (Bönigk et al. 2009). For a regular arrangement of ∼100 GC dimers in a 155 × 155-nm membrane patch, an active GC* would be girded by nine CNGK channels at a distance not farther than ∼100 nm (Fig. 2B); these next neighbors would be first served with cGMP molecules. Finally, the input resistance of sperm ($\gtrsim 5$ GΩ) (Navarro et al. 2007; Zeng et al. 2013) is at least five-fold higher than that of mammalian rod photoreceptors (1–2 GΩ) (Schneeweis and Schnapf 1995); thus, by changing the open probability of only a few CNGK channels, sperm can produce a single-molecule response.

In conclusion, there is no need to invoke high amplification, signaling complexes, or restricted diffusion to account for single-molecule sensitivity of sperm. The exquisite cGMP sensitivity of CNGK channels, the minuscule flagellar volume, and the high input resistance are key. A word of caution: all of these conclusions have been inferred from population measurements. Future work requires to study single-molecule sensitivity in single sperm cells. Notwithstanding, the hallmarks of this signaling mechanism might provide a blueprint for chemical sensing in small compartments such as olfactory cilia, insect antennae, or even synaptic boutons.

SIGNALING IN FISH

Navigation of fish sperm and the underlying signaling pathways must be arguably different from those of marine invertebrates and mammals. First, teleost fish lack CatSper channels (Cai and Clapham 2008). However, activation of sperm motility requires Ca^{2+} influx (Billard 1986; Takai and Morisawa 1995; Alavi and Cosson 2006; Cosson et al. 2008; Morisawa 2008). Motility is activated by hyper- or hypoosmotic shock after spawning of sperm into seawater or freshwater, respectively (Krasznai et al. 2000; Vines et al. 2002; Alavi and Cosson 2006; Cherr et al. 2008; Morisawa 2008). Therefore, Ca^{2+} influx in fish sperm involves other Ca^{2+} channels or Ca^{2+} release from intracellular stores.

Second, the ionic milieu seriously constrains ion channel function. Sperm of freshwater fish, marine invertebrates, and mammals are facing entirely different ionic milieus. K^+ and Na^+ concentrations in freshwater are extremely low (70 μM and 200 μM, respectively) compared with the orders-of-magnitude higher concentrations in seawater or the oviduct (Alavi and Cosson 2006; Hugentobler et al. 2007). Furthermore, $[Ca^{2+}]$ in seawater is high (10 mM), whereas in freshwater it is low (<1 mM). The low salt concentrations in freshwater probably require ion channels that are differently designed. In fact, none of the ion channels controlling electrical and Ca^{2+} signaling of fish sperm had been known until recently.

Unexpectedly, although the principal targets of hyperpolarization—the sNHE and CatSper—are absent in fish, orthologs of the sperm CNGK channel are present in various fish genomes (Fig. 4) (Fechner et al. 2015). Surprisingly, the CNGK channel of zebrafish differs from its sea urchin cousin: it is activated by alkalinization, but not by cyclic nucleotides; moreover, the channel is localized in the head rather the flagellum; finally, the sea urchin CNGK is blocked by intracellular Na^+, whereas the block of the zebrafish CNGK is much weaker. Future studies need to identify the molecules that are located upstream and downstream of CNGK in the signaling pathways in fish.

MAMMALIAN SPERM—DIFFERENCES AND COMMONALITIES

Mammalian sperm navigate across the female genital tract to reach the egg. During this transit, sperm undergo capacitation and hyperactivation. Capacitation is a complex, ill-defined maturation process. Hyperactivation is initiated during capacitation and is characterized by a whip-like beat of the flagellum, which is essential to penetrate the egg's vestments—the zona pellucida and cumulus cells (Suarez et al. 1993; Suarez 2008).

During their journey from the epididymis via the female genital tract to the egg, sperm experience an ever-changing environment, including large differences in pH, ionic milieu, viscosity, and epithelial surfaces. To adapt to these changes, mammalian sperm have developed species-specific signaling mechanisms to reach and fertilize the egg. These differences are also reflected by large variations in sperm size and shape. Rodent sperm have a sickle-shaped head and a fairly long flagellum, whereas primate sperm feature an oval-shaped head and a shorter flagellum (Miller et al. 2015). In the following, differences and commonalities of sensory signaling between mouse and human sperm are described.

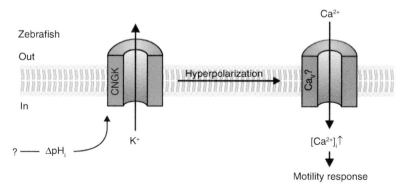

Figure 4. Signaling pathway in zebrafish. Only the K^+-selective cyclic nucleotide-gated (CNGK) channel has been identified. The other components, in particular the Ca^{2+} channel, are not known. A change in pH_i that is produced by unknown mechanisms upstream of CNGK causes hyperpolarization and, in turn, Ca^{2+} influx.

Cite this article as *Cold Spring Harb Perspect Biol* doi: 10.1101/cshperspect.a028225

CatSper IS KEY FOR Ca^{2+} SIGNALING

Ca^{2+} is also key for mammalian sperm navigation and fertilization: Ca^{2+} controls the flagellar beat pattern and, thus, sperm motility, navigation, rheotaxis, and hyperactivation. Ca^{2+} is also required for the acrosome reaction, which is needed to penetrate the egg's vestments and for capacitation.

Many different Ca^{2+} channels have been proposed to control sperm function (Darszon et al. 2011). Although some voltage-dependent Ca_v channels might be present during spermatogenesis, CatSper has been identified as the principal Ca^{2+} channel in mature mammalian sperm (Fig. 5) (Lishko et al. 2012). CatSper forms a heteromeric channel complex made up of at least seven different subunits. Catsper1−4 (α-subunits) form the pore, whereas CatSperβ, γ, and δ represent auxiliary subunits that are associated with the pore-forming complex (Ren et al. 2001; Carlson et al. 2003; Liu et al. 2007; Qi et al. 2007; Wang et al. 2009; Chung et al. 2011). Although basal motility is not affected by targeted disruption of CatSper subunits in mice, hyperactivation is abolished and capacitation is impaired (Carlson et al. 2003; Quill et al. 2003; Chung et al. 2014). Finally, rheotaxis is abolished in CatSper null sperm (Miki and Clapham 2013). Whether other potential navigation strategies like thermotaxis or chemotaxis also rely on CatSper is not known. However, a double knockout of CatSper and KSper conclusively shows that mouse sperm harbor no voltage and pH-dependent ion channels other than CatSper and KSper (Zeng et al. 2013). In line with the findings from mouse sperm, mutations in the human CatSper result in male infertility (Avidan et al. 2003; Zhang et al. 2007; Avenarius et al. 2009; Smith et al. 2013; Jaiswal et al. 2014).

CatSper serves as a platform along the flagellum to create signaling domains. CatSper channels form a quadrilateral arrangement in three dimensions that organizes structurally distinct Ca^{2+} signaling domains (Chung et al. 2014). Loss of any one of the CatSper channel subunits destroys the organization of these signaling domains and hyperactivated motility.

Furthermore, these Ca^{2+} domains also organize the spatiotemporal pattern of sperm capacitation (Chung et al. 2014).

The CatSper channel is a polymodal sensor that registers changes in membrane voltage, pH_i, and the concentration of various ligands. Mouse CatSper is less voltage-dependent than human CatSper, but shows a higher pH sensitivity (Kirichok et al. 2006; Lishko and Kirichok 2010; Lishko et al. 2011). At physiological pH_i (7.4), the $V_{1/2}$ for human and mouse CatSper is $+85$ mV and $+11$ mV, respectively (Lishko and Kirichok 2010; Lishko et al. 2011), indicating that at a resting membrane potential of $V_{rest} = -30$ mV, mouse CatSper would be partially open, whereas the human CatSper would be mostly closed.

In human sperm, the female sex hormone progesterone and prostaglandins in the oviductal fluid activate CatSper and cause a Ca^{2+} influx (Lishko et al. 2011; Strünker et al. 2011; Brenker et al. 2012). Progesterone has been proposed to act as a chemoattractant, controlling sperm navigation and fertilization (Teves et al. 2006; Oren-Benaroya et al. 2008). The CatSper activation occurs almost instantaneously and does not involve G-protein-coupled receptors (GPCRs) or G-proteins (Lishko et al. 2011; Strünker et al. 2011; Brenker et al. 2012), suggesting that CatSper is gated either directly by progesterone or a closely associated receptor. Recently, the orphan enzyme α/β hydrolase domain-containing protein 2 (ABHD2) has been identified as the progesterone receptor in human sperm (Fig. 5) (Miller et al. 2016). On progesterone binding, ABHD2 cleaves the endocannabinoids 1- and 2-arachidonoylglycerol (AGs) into free glycerol and arachidonic acid (Miller et al. 2016). AGs inhibit the CatSper current $I_{CatSper}$; however, hydrolysis of AGs by ABHD2 relieves their inhibition (Miller et al. 2016). Mouse sperm are insensitive to stimulation with progesterone. The species specificity of progesterone seems to be based on the difference in lipid homeostasis and localization of ABHD2 (Miller et al. 2016). In mouse sperm, ABHD2 is localized to the acrosome rather than the sperm flagellum (in contrast to human sperm). Thus, ABHD2 does not colocalize

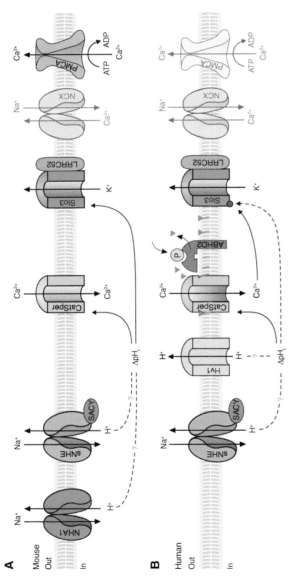

Figure 5. Comparison of signaling pathways in mouse and human sperm. (*A*) Mouse sperm. The principal Ca²⁺ channel is CatSper. CatSper opening is regulated by changes in intracellular pH (ΔpH$_i$) and changes in membrane potential. The membrane potential is controlled by the pH-dependent Slo3 K⁺ channel. Its auxiliary subunit LRRC52 controls the pH- and voltage-dependent opening of Slo3. The activation of Slo3 during Ca²⁺ signaling is ill defined. Prominent candidates for controlling the intracellular pH are two Na⁺/H⁺ exchangers, the sperm-specific sNHE and NHA1. sNHE is localized in a protein complex with the soluble adenylate cyclase (SACY). However, the role of sNHE and NHA1 in controlling the intracellular pH has yet to be confirmed. The recovery of the Ca²⁺ homeostasis after CatSper opening is not well understood. In mouse sperm, the Ca²⁺-ATPase PMCA4 extrudes Ca²⁺ and controls sperm hyperactivation. The role of a Na⁺/Ca²⁺ exchanger (NCX) has yet to be confirmed. (*B*) Human sperm. As in mouse sperm, the principal Ca²⁺ channel is CatSper, which is also regulated by changes in pH$_i$ and membrane voltage. Furthermore, CatSper is activated by binding of progesterone (P) to the lipid hydrolase ABHD2, which hydrolyzes the endocannabinoid 2-arachidonoylglycerol (2AG) to arachidonic acid and glycerol. This relieves CatSper inhibition by 2AG and opens the channel. In human sperm, the principal K⁺ channel is also Slo3; it is regulated by Ca²⁺ and to a lesser extent by changes in pH$_i$. Presumably, Slo3 is placed downstream from CatSper on the recovery branch of Ca²⁺ signaling. Human sperm contain an H⁺ channel (H$_V$1), which carries an outward rectifying H⁺ current. H$_V$1 and sNHE are the candidates to control pH$_i$ in human sperm. The recovery of the Ca²⁺ response is also thought to be regulated by members of the NCX and PMCA family. However, their molecular identity is yet to be confirmed. Pale icons indicate molecules whose identity is yet to be confirmed; dotted lines present signaling pathways with weak experimental evidence; question marks indicate hypothetical signaling pathways that have not yet been confirmed experimentally.

with CatSper. In addition, mouse sperm lose AGs during their transit through the epididymis; thus, CatSper is no longer inhibited by AGs and does not require progesterone-dependent hydrolysis of AGs for activation (Miller et al. 2016). In contrast, human sperm maintain high AG levels and require stimulation with progesterone for activation (Miller et al. 2016).

In addition to steroids and prostaglandins, other chemicals as diverse as odorants, menthol, and analogs of cyclic nucleotides, as well as various endocrine-disrupting chemicals (EDCs), can also activate CatSper (Brenker et al. 2012; Tavares et al. 2013; Schiffer et al. 2014). EDCs are omnipresent in food, household, and personal care products and have been linked to decreasing fertility rates in the Western world (Bergman et al. 2013). Whether these chemicals activate CatSper via ABHD2 or another mechanism is not known.

K^+ CHANNELS AND THE CONTROL OF THE MEMBRANE POTENTIAL

The principal K^+ channel in mouse sperm is Slo3, a member of the Slo family of ion channels (Schreiber et al. 1998; Santi et al. 2010; Zeng et al. 2011). Deletion of Slo3 severely impairs male fertility (Santi et al. 2010; Zeng et al. 2011). The Slo3 channel in mouse sperm is activated at $pH_i > 6.0$ and membrane potentials >0 mV (pH 7); it carries a hyperpolarizing outward current (Navarro et al. 2007; Zeng et al. 2011). Apart from the pore-forming subunit encoded by the *Slo3* gene, the channel complex also contains the auxiliary subunits LRRC52 and LRRC26 (leucine-rich repeat-containing proteins). LRRC52 regulates the pH- and voltage-dependent opening of the channel complex (Yang et al. 2011; Zeng et al. 2015). The development of hyperactivated motility during capacitation of mouse sperm is dependent on cytosolic alkalization followed by an increase of $[Ca^{2+}]_i$ (Suarez 2008). Slo3 mediates the hyperpolarization on alkalization (Zeng et al. 2011) and recent evidence suggests that hyperpolarization indirectly activates CatSper by promoting a rise of pH_i through a voltage-dependent mechanism (Chavez et al. 2014). Together, Slo3 and CatSper are the sole

ion channels in mouse sperm that regulate membrane potential and Ca^{2+} influx in response to alkalization (Zeng et al. 2013).

In contrast to mouse sperm, the K^+ current in human sperm is only weakly pH-dependent, but strongly regulated by Ca^{2+} (Brenker et al. 2014). Consequently, the membrane potential in human sperm is sensitive to changes in Ca^{2+} and less so to pH_i. Furthermore, the K^+ current is inhibited by progesterone (Mannowetz et al. 2013; Brenker et al. 2014). Efforts to identify the molecules underlying the K^+ channel in human sperm provided mixed results. Initially, it was thought that Slo1, the Ca^{2+}-regulated member of the Slo family carries the K^+ current in human sperm (Mannowetz et al. 2013). However, overwhelming evidence now shows that Slo1 is absent in human sperm and that a Ca^{2+}-regulated Slo3 channel carries K^+ currents (Brenker et al. 2014). In particular, the properties of heterologously expressed human Slo3 match those of native K^+ currents, including block by quinidine and clofilium, inhibition by progesterone, modest pH sensitivity but large Ca^{2+}-sensitivity, and single-channel conductance (70 pS) (Brenker et al. 2014). Thus, human sperm switched the ligand selectivity of Slo3 from pH_i to Ca^{2+} rather than to adopt a new Slo isoform. Activation of the Ca^{2+}-sensitive Slo3 might curtail Ca^{2+} influx via CatSper, suggesting that Slo3 in human sperm is placed downstream of CatSper on the recovery branch of Ca^{2+} signaling, whereas in mouse sperm, Slo3 at alkaline pH could hyperpolarize the cell and would open CatSper through a voltage-dependent alkalization.

Ca^{2+} CLEARANCE MECHANISMS

To restore Ca^{2+} levels, Ca^{2+} needs to be extruded from the cytoplasm or stored in intracellular organelles. In mouse sperm, the plasma membrane Ca^{2+}-ATPase PMCA4 seems to be important to maintain Ca^{2+} homeostasis (Okunade et al. 2004; Schuh et al. 2004). PMCA4 null male mice are infertile and are unable to undergo hyperactivation. PMCA pumps are the fastest Ca^{2+} extrusion mechanisms in sperm. Na^+/Ca^{2+} exchanger (NCX) and mitochondrial

Ca^{2+} uniporter are slower (Wennemuth et al. 2003a). However, the underlying molecules in mouse sperm are ill defined. In principal, similar Ca^{2+} clearance mechanisms also exist in human sperm. Future studies need to investigate the recovery branch of Ca^{2+} signaling in more detail.

REGULATION OF pH_i—ENIGMATIC IN MOUSE, BUT NOT SO MUCH IN HUMAN SPERM

Changes in pH_i control capacitation, hyperactivation, motility, and many signaling molecules are pH-dependent. Several molecules have been proposed to control pH_i, including the voltage-gated proton-specific channel H_v1 (Lishko et al. 2010), different members of the protein family of solute carriers (SLC; sodium–proton exchangers and bicarbonate transporters), and carbonic anhydrases (Nishigaki et al. 2014).

In human sperm, the proton channel H_v1 carries a large outwardly rectifying H^+ current (Lishko et al. 2010). The channel features unique characteristics: it is activated by membrane depolarization, regulated by pH_i and pH_o, and inhibited by zinc (Ramsey et al. 2006; Sasaki et al. 2006). These characteristics may be particularly important during the transit through the female genital tract. When sperm are ejaculated, they are mixed with the seminal plasma that contains millimolar concentrations of zinc, rendering the channel inactive (Lishko and Kirichok 2010). During the transit through the female genital tract, zinc is chelated by proteins, suggesting that this causes a gradual activation of the channel (Lishko and Kirichok 2010). H_v1 is also thought be activated by capacitation (Lishko et al. 2010). However, the underlying mechanism is still enigmatic.

In contrast to human sperm, mouse sperm do not contain an outwardly rectifying H^+ current and H_v1 null male mice do not show a fertility defect (Miller et al. 2015). How the pH_i is regulated in mouse sperm is not known. The most likely candidate has been an Na^+/H^+ exchanger specific for sperm (sNHE). Indeed, mice lacking sNHE are infertile because the sperm are immotile (Wang et al. 2003). However, sNHE is found in a complex with the SACY, the predominant source for cAMP in mammalian sperm (Wang et al. 2007). In fact, the loss of sNHE results in a concomitant loss of SACY. Thus, it is difficult to disentangle the physiological function of sNHE and the SACY. Cyclic AMP is essential for sperm development, motility, and maturation in the female genital tract (Visconti et al. 1995; Wennemuth et al. 2003b; Krähling et al. 2013). Mice lacking SACY are infertile and the sperm are immotile (Esposito et al. 2004). The motility defect in sNHE-null mice can be rescued by application of membrane permeable cAMP analogs or by optogenetic stimulation of cAMP production (Jansen et al. 2015), showing that the motility defect and thereby the infertile phenotype is solely due to the loss of SACY, but not sNHE (Wang et al. 2003). Recently, another NHE, NHA1, has been shown to control sperm function (Liu et al. 2010; Chen et al. 2016). However, information about its role in controlling pH_i in mouse sperm is missing. Surprisingly, NHA1-knockout mice also seem to display attenuated cAMP signaling and impaired sperm motility. Future studies need to reveal whether NHA1 regulates pH_i in mouse sperm.

SPERM COMPARTMENTALIZATION AND SUPRAMOLECULAR STRUCTURES

Mammalian sperm show different levels of compartmentalization. The two major compartments are the head and the flagellum. The flagellum itself is separated into the midpiece, principal piece, and the endpiece, where signaling molecules and second messenger dynamics are spatially organized.

As described above, the CatSper complex creates structurally distinct Ca^{2+}-signaling domains along the principal piece of mouse sperm (Chung et al. 2014). In human sperm, CatSper is also localized in the principal piece, and progesterone stimulation first elicits a Ca^{2+} increase in the principal piece and midpiece, which then propagates to the head (Servin-Vences et al. 2012). Furthermore, it has been shown that cAMP signaling pathways are compartmentalized in the head and flagellum (Wertheimer et al. 2013) and that cAMP dynamics are also organized in distinct domains along the flagellum (Mukherjee et al. 2016). However, the molecular mechanisms underlying the compartmentalization of cAMP dynamics have not been identified yet. The combination of genetically encoded biosensors with optogenetics holds great promise to map the dynamics of cAMP signaling in live cells in precise spatiotemporal and quantitative terms.

OUTLOOK

Although many molecules in the flagellum of mammalian sperm have been identified that are involved in electrical and Ca^{2+} signaling, we do not know which sensory modality they serve: rheotaxis, chemotaxis, haptotaxis, or thermotaxis? In fact, other molecules, like rhodopsin, G-proteins, phospholipase C, TRPC3 channels, and IP_3-gated channels have been implied in thermosensation of human sperm (Perez-Cerezales et al. 2015). Moreover, yet other molecules have been suggested for chemotaxis of human sperm in a progesterone gradient (Teves et al. 2009). Future work is required to unequivocally assign signaling pathways to behavioral responses during directed movement.

ACKNOWLEDGMENTS

We thank Dr. R. Pascal for preparing the figures and H. Krause for preparing the manuscript.

REFERENCES

Alavi SM, Cosson J. 2006. Sperm motility in fishes. II: Effects of ions and osmolality: A review. *Cell Biol Int* **30:** 1–14.

Alvarez L, Dai L, Friedrich BM, Kashikar ND, Gregor I, Pascal R, Kaupp UB. 2012. The rate of change in Ca^{2+} concentration controls sperm chemotaxis. *J Cell Biol* **196:** 653–663.

Arshavsky VY, Burns ME. 2014. Current understanding of signal amplification in phototransduction. *Cell Logist* **4:** e29390.

Avenarius MR, Hildebrand MS, Zhang Y, Meyer NC, Smith LL, Kahrizi K, Najmabadi H, Smith RJ. 2009. Human male infertility caused by mutations in the CATSPER1 channel protein. *Am J Hum Genet* **84:** 505–510.

Avidan N, Tamary H, Dgany O, Cattan D, Pariente A, Thulliez M, Borot N, Moati L, Barthelme A, Shalmon L, et al.

2003. CATSPER2, a human autosomal nonsyndromic male infertility gene. *Eur J Hum Genet* **11:** 497–502.

Bahat A, Tur-Kaspa I, Gakamsky A, Giojalas LC, Breitbart H, Eisenbach M. 2003. Thermotaxis of mammalian sperm cells: A potential navigation mechanism in the female genital tract. *Nat Med* **9:** 149–150.

Bahat A, Caplan SR, Eisenbach M. 2012. Thermotaxis of human sperm cells in extraordinarily shallow temperature gradients over a wide range. *PLoS ONE* **7:** e41915.

Beltrán C, Zapata O, Darszon A. 1996. Membrane potential regulates sea urchin sperm adenylylcyclase. *Biochemistry* **35:** 7591–7598.

Bergman A, Heindel JJ, Kasten T, Kidd KA, Jobling S, Neira M, Zoeller RT, Becher G, Bjerregaard P, Bornman R, et al. 2013. The impact of endocrine disruption: A consensus statement on the state of the science. *Environ Health Perspect* **121:** A104–A106.

Billard R. 1986. Spermatogenesis and spermatology of some teleost fish species. *Reprod Nutr Dev* **26:** 877–920.

Bisegna P, Caruso G, Andreucci D, Shen L, Gurevich VV, Hamm HE, DiBenedetto E. 2008. Diffusion of the second messengers in the cytoplasm acts as a variability suppressor of the single photon response in vertebrate phototransduction. *Biophys J* **94:** 3363–3383.

Biskup C, Kusch J, Schulz E, Nache V, Schwede F, Lehmann F, Hagen V, Benndorf K. 2007. Relating ligand binding to activation gating in CNGA2 channels. *Nature* **446:** 440–443.

Böhmer M, Van Q, Weyand I, Hagen V, Beyermann M, Matsumoto M, Hoshi M, Hildebrand E, Kaupp UB. 2005. Ca^{2+} spikes in the flagellum control chemotactic behavior of sperm. *EMBO J* **24:** 2741–2752.

Bönigk W, Loogen A, Seifert R, Kashikar N, Klemm C, Krause E, Hagen V, Kremmer E, Strünker T, Kaupp UB. 2009. An atypical CNG channel activated by a single cGMP molecule controls sperm chemotaxis. *Sci Signal* **2:** ra68.

Boryshpolets S, Perez-Cerezales S, Eisenbach M. 2015. Behavioral mechanism of human sperm in thermotaxis: A role for hyperactivation. *Hum Reprod* **30:** 884–892.

Bray D, Levin MD, Morton-Firth CJ. 1998. Receptor clustering as a cellular mechanism to control sensitivity. *Nature* **393:** 85–88.

Brelidze TI, Gianulis EC, DiMaio F, Trudeau MC, Zagotta WN. 2013. Structure of the C-terminal region of an ERG channel and functional implications. *Proc Natl Acad Sci* **110:** 11648–11653.

Brenker C, Goodwin N, Weyand I, Kashikar ND, Naruse M, Krähling M, Müller A, Kaupp UB, Strünker T. 2012. The CatSper channel: A polymodal chemosensor in human sperm. *EMBO J* **31:** 1654–1665.

Brenker C, Zhou Y, Müller A, Echeverry FA, Trötschel C, Poetsch A, Xia XM, Bönigk W, Lingle CJ, Kaupp UB, et al. 2014. The Ca^{2+}-activated K^+ current of human sperm is mediated by Slo3. *eLife* **3:** e01438.

Bretherton FP, Rothschild. 1961. Rheotaxis of spermatozoa. *Proc R Soc Lond B 2* **153:** 490–502.

Bukatin A, Kukhtevich I, Stoop N, Dunkel J, Kantsler V. 2015. Bimodal rheotactic behavior reflects flagellar beat asymmetry in human sperm cells. *Proc Natl Acad Sci* **112:** 15904–15909.

Burns ME, Pugh EN Jr. 2010. Lessons from photoreceptors: Turning off G-protein signaling in living cells. *Physiology (Bethesda)* **25:** 72–84.

Burton KA, McKnight GS. 2007. PKA, germ cells, and fertility. *Physiology (Bethesda)* **22:** 40–46.

Cai X, Clapham DE. 2008. Evolutionary genomics reveals lineage-specific gene loss and rapid evolution of a sperm-specific ion channel complex: CatSpers and CatSperβ. *PLoS ONE* **3:** e3569.

Carlson AE, Westenbroek RE, Quill T, Ren D, Clapham DE, Hille B, Garbers DL, Babcock DF. 2003. CatSper1 required for evoked Ca^{2+} entry and control of flagellar function in sperm. *Proc Natl Acad Sci* **100:** 14864–14868.

Carlson AE, Brelidze TI, Zagotta WN. 2013. Flavonoid regulation of EAG1 channels. *J Gen Physiol* **141:** 347–358.

Caruso G, Bisegna P, Andreucci D, Lenoci L, Gurevich VV, Hamm HE, DiBenedetto E. 2011. Identification of key factors that reduce the variability of the single photon response. *Proc Natl Acad* **108:** 7804–7807.

Chavez JC, Ferreira JJ, Butler A, De La Vega Beltran JL, Trevino CL, Darszon A, Salkoff L, Santi CM. 2014. SLO3 K^+ channels control calcium entry through CATSPER channels in sperm. *J Biol Chem* **289:** 32266–32275.

Chen Y, Cann MJ, Litvin TN, Iourgenko V, Sinclair ML, Levin LR, Buck J. 2000. Soluble adenylyl cyclase as an evolutionarily conserved bicarbonate sensor. *Science* **289:** 625–628.

Chen SR, Chen M, Deng SL, Hao XX, Wang XX, Liu YX. 2016. Sodium–hydrogen exchanger NHA1 and NHA2 control sperm motility and male fertility. *Cell Death Dis* **7:** e2152.

Cherr GN, Morisawa M, Vines CA, Yoshida K, Smith EH, Matsubara T, Pillai MC, Griffin FJ, Yanagimachi R. 2008. Two egg-derived molecules in sperm motility initiation and fertilization in the Pacific herring (*Clupea pallasi*). *Int J Dev Biol* **52:** 743–752.

Chung JJ, Navarro B, Krapivinsky G, Krapivinsky L, Clapham DE. 2011. A novel gene required for male fertility and functional CATSPER channel formation in spermatozoa. *Nat Commun* **2:** 153.

Chung JJ, Shim SH, Everley RA, Gygi SP, Zhuang X, Clapham DE. 2014. Structurally distinct Ca^{2+} signaling domains of sperm flagella orchestrate tyrosine phosphorylation and motility. *Cell* **157:** 808–822.

Cook SP, Babcock DF. 1993. Activation of Ca^{2+} permeability by cAMP is coordinated through the pH_i increase induced by speract. *J Biol Chem* **268:** 22408–22413.

Corkidi G, Taboada B, Wood CD, Guerrero A, Darszon A. 2008. Tracking sperm in three-dimensions. *Biochem Biophys Res Commun* **373:** 125–129.

Cosson J, Huitorel P, Gagnon C. 2003. How spermatozoa come to be confined to surfaces. *Cell Motil Cytoskel* **54:** 56–63.

Cosson J, Groison AL, Suquet M, Fauvel C, Dreanno C, Billard R. 2008. Marine fish spermatozoa: Racing ephemeral swimmers. *Reproduction* **136:** 277–294.

Crenshaw HC. 1990. Helical orientation—A novel mechanism for the orientation of microorganisms. In *Biological motion* (ed. Alt W, Hoffmann G), pp. 361–386. Springer, Berlin.

Crenshaw HC. 1993a. Orientation by helical motion. I: Kinematics of the helical motion of organisms with up to six degrees of freedom. *Bull Math Biol* **55:** 197–212.

Crenshaw HC. 1993b. Orientation by helical motion. III: Microorganisms can orient to stimuli by changing the direction of their rotational velocity. *Bull Math Biol* **55:** 231–255.

Crenshaw HC, Edelstein-Keshet L. 1993. Orientation by helical motion. II: Changing the direction of the axis of motion. *Bull Math Biol* **55:** 213–230.

Cukkemane A, Seifert R, Kaupp UB. 2011. Cooperative and uncooperative cyclic-nucleotide-gated ion channels. *Trends Biochem Sci* **36:** 55–64.

Darszon A, Nishigaki T, Beltran C, Trevino CL. 2011. Calcium channels in the development, maturation, and function of spermatozoa. *Physiol Rev* **91:** 1305–1355.

Denissenko P, Kantsler V, Smith DJ, Kirkman-Brown J. 2012. Human spermatozoa migration in microchannels reveals boundary-following navigation. *Proc Natl Acad Sci* **109:** 8007–8010.

Doan T, Mendez A, Detwiler PB, Chen J, Rieke F. 2006. Multiple phosphorylation sites confer reproducibility of the rod's single-photon responses. *Science* **313:** 530–533.

Duke TAJ, Bray D. 1999. Heightened sensitivity of a lattice of membrane receptors. *Proc Natl Acad Sci* **96:** 10104–10108.

Eisenbach M, Giojalas LC. 2006. Sperm guidance in mammals—An unpaved road to the egg. *Nat Rev Mol Cell Biol* **7:** 276–285.

Elgeti J, Gompper G. 2009. Self-propelled rods near surfaces *Eur Phys Lett* **85:** 38002.

Elgeti J, Kaupp UB, Gompper G. 2010. Hydrodynamics of sperm cells near surfaces. *Biophys J* **99:** 1018–1026.

Esposito G, Jaiswal BS, Xie F, Krajnc-Franken MAM, Robben TJAA, Strik AM, Kuil C, Philipsen RLA, van Duin M, Conti M, et al. 2004. Mice deficient for soluble adenylyl cyclase are infertile because of a severe sperm-motility defect. *Proc Natl Acad Sci* **101:** 2993–2998.

Fechner S, Alvarez L, Bönigk W, Müller A, Berger T, Pascal R, Trötschel C, Poetsch A, Stolting G, Siegfried KR, et al. 2015. A K^+-selective CNG channel orchestrates Ca^{2+} signalling in zebrafish sperm. *eLife* **4:** e07624.

Fotiadis D, Liang Y, Filipek S, Saperstein DA, Engel A, Palczewski K. 2003. Rhodopsin dimers in native disc membranes. *Nature* **421:** 127–128.

Galindo BE, Neill AT, Vacquier VD. 2005. A new hyperpolarization-activated, cyclic nucleotide-gated channel from sea urchin sperm flagella. *Biochem Biophys Res Commun* **334:** 96–101.

Gauss R, Seifert R, Kaupp UB. 1998. Molecular identification of a hyperpolarization-activated channel in sea urchin sperm. *Nature* **393:** 583–587.

Gestwicki JE, Kiessling LL. 2002. Inter-receptor communication through arrays of bacterial chemoreceptors. *Nature* **415:** 81–84.

Gross OP, Pugh EN Jr, Burns ME. 2012a. Calcium feedback to cGMP synthesis strongly attenuates single-photon responses driven by long rhodopsin lifetimes. *Neuron* **76:** 370–382.

Gross OP, Pugh EN Jr, Burns ME. 2012b. Spatiotemporal cGMP dynamics in living mouse rods. *Biophys J* **102:** 1775–1784.

Guerrero A, Nishigaki T, Carneiro J, Yoshiro T, Wood CD, Darszon A. 2010. Tuning sperm chemotaxis by calcium burst timing. *Dev Biol* **344:** 52–65.

Gunaratne HJ, Neill AT, Vacquier VD. 2006. Plasma membrane calcium ATPase is concentrated in the head of sea urchin spermatozoa. *J Cell Physiol* **207:** 413–419.

Gunkel M, Schöneberg J, Alkhaldi W, Irsen S, Noe F, Kaupp UB, Al-Amoudi A. 2015. Higher-order architecture of rhodopsin in intact photoreceptors and its implication for phototransduction kinetics. *Structure* **23:** 628–638.

Haitin Y, Carlson AE, Zagotta WN. 2013. The structural mechanism of KCNH-channel regulation by the *eag* domain. *Nature* **501:** 444–448.

Huber A, Sander P, Gobert A, Bahner M, Hermann R, Paulsen R. 1996. The transient receptor potential protein (Trp), a putative store-operated Ca^{2+} channel essential for phosphoinositide-mediated photoreception, forms a signaling complex with NorpA, InaC and InaD. *EMBO J* **15:** 7036–7045.

Hugentobler SA, Morris DG, Sreenan JM, Diskin MG. 2007. Ion concentrations in oviduct and uterine fluid and blood serum during the estrous cycle in the bovine. *Theriogenology* **68:** 538–548.

Jaiswal BS, Conti M. 2003. Calcium regulation of the soluble adenylyl cyclase expressed in mammalian spermatozoa. *Proc Natl Acad Sci* **100:** 10676–10681.

Jaiswal D, Singh V, Dwivedi US, Trivedi S, Singh K. 2014. Chromosome microarray analysis: A case report of infertile brothers with CATSPER gene deletion. *Gene* **542:** 263–265.

Jansen V, Alvarez L, Balbach M, Strünker T, Hegemann P, Kaupp UB, Wachten D. 2015. Controlling fertilization and cAMP signaling in sperm by optogenetics. *eLife* **4:** e05161.

Jikeli JF, Alvarez L, Friedrich BM, Wilson LG, Pascal R, Colin R, Pichlo M, Rennhack A, Brenker C, Kaupp UB. 2015. Sperm navigation along helical paths in 3D chemoattractant landscapes. *Nat Commun* **6:** 7985.

Kantsler V, Dunkel J, Blayney M, Goldstein RE. 2014. Correction: Rheotaxis facilitates upstream navigation of mammalian sperm cells. *eLife* **3:** e03521.

Kaupp UB. 2010. Olfactory signalling in vertebrates and insects: Differences and commonalities. *Nat Rev Neurosci* **11:** 188–200.

Kaupp UB, Seifert R. 2002. Cyclic nucleotide-gated ion channels. *Physiol Rev* **82:** 769–824.

Kaupp UB, Solzin J, Hildebrand E, Brown JE, Helbig A, Hagen V, Beyermann M, Pampaloni F, Weyand I. 2003. The signal flow and motor response controlling chemotaxis of sea urchin sperm. *Nat Cell Biol* **5:** 109–117.

Kirichok Y, Navarro B, Clapham DE. 2006. Whole-cell patch-clamp measurements of spermatozoa reveal an alkaline-activated Ca^{2+} channel. *Nature* **439:** 737–740.

Kleinboelting S, Diaz A, Moniot S, van den Heuvel J, Weyand M, Levin LR, Buck J, Steegborn C. 2014. Crystal structures of human soluble adenylyl cyclase reveal mechanisms of catalysis and of its activation through bicarbonate. *Proc Natl Acad Sci* **111:** 3727–3732.

Krähling AM, Alvarez L, Debowski K, Van Q, Gunkel M, Irsen S, Al-Amoudi A, Strünker T, Kremmer E, Krause E, et al. 2013. CRIS—A novel cAMP-binding protein controlling spermiogenesis and the development of flagellar bending. *PLoS Genet* **9:** e1003960.

Krasznai Z, Márián T, Izumi H, Damjanovich S, Balkay L, Trón L, Morisawa M. 2000. Membrane hyperpolarization removes inactivation of Ca^{2+} channels, leading to Ca^{2+} influx and subsequent initiation of sperm motility in the common carp. *Proc Natl Acad Sci* **97:** 2052–2057.

Lauga E, Powers TR. 2009. The hydrodynamics of swimming microorganisms. *Rep Prog Phys* **72:** 096601.

Lee HC. 1984a. A membrane potential-sensitive $Na^{+}-H^{+}$ exchange system in flagella isolated from sea urchin spermatozoa. *J Biol Chem* **259:** 15315–15319.

Lee HC. 1984b. Sodium and proton transport in flagella isolated from sea urchin spermatozoa. *J Biol Chem* **259:** 4957–4963.

Lee HC. 1985. The voltage-sensitive Na^{+}/H^{+} exchange in sea urchin spermatozoa flagellar membrane vesicles studied with an entrapped pH probe. *J Biol Chem* **260:** 10794–10799.

Lee HC, Garbers DL. 1986. Modulation of the voltage-sensitive Na^{+}/H^{+} exchange in sea urchin spermatozoa through membrane potential changes induced by the egg peptide speract. *J Biol Chem* **261:** 16026–16032.

Lishko PV, Kirichok Y. 2010. The role of Hv1 and CatSper channels in sperm activation. *J Physiol* **588:** 4667–4672.

Lishko PV, Botchkina IL, Fedorenko A, Kirichok Y. 2010. Acid extrusion from human spermatozoa is mediated by flagellar voltage-gated proton channel. *Cell* **140:** 327–337.

Lishko PV, Botchkina IL, Kirichok Y. 2011. Progesterone activates the principal Ca^{2+} channel of human sperm. *Nature* **471:** 387–391.

Lishko PV, Kirichok Y, Ren D, Navarro B, Chung JJ, Clapham DE. 2012. The control of male fertility by spermatozoan ion channels. *Annu Rev Physiol* **74:** 453–475.

Liu J, Xia J, Cho KH, Clapham DE, Ren D. 2007. CatSperβ, a novel transmembrane protein in the CatSper channel complex. *J Biol Chem* **282:** 18945–18952.

Liu T, Huang JC, Zuo WL, Lu CL, Chen M, Zhang XS, Li YC, Cai H, Zhou WL, Hu ZY, et al. 2010. A novel testis-specific Na^{+}/H^{+} exchanger is involved in sperm motility and fertility. *Front Biosci (Elite Ed)* **2:** 566–581.

Maddock JR, Shapiro L. 1993. Polar location of the chemoreceptor complex in the *Escherichia coli* cell. *Science* **259:** 1717–1723.

Mannowetz N, Naidoo NM, Choo SA, Smith JF, Lishko PV. 2013. Slo1 is the principal potassium channel of human spermatozoa. *eLife* **2:** e01009.

Matsumoto M, Solzin J, Helbig A, Hagen V, Ueno S-I, Kawase O, Maruyama Y, Ogiso M, Godde M, Minakata H, et al. 2003. A sperm-activating peptide controls a cGMP-signaling pathway in starfish sperm. *Dev Biol* **260:** 314–324.

Mendez A, Burns ME, Roca A, Lem J, Wu L-W, Simon MI, Baylor DA, Chen J. 2000. Rapid and reproducible deactivation of rhodopsin requires multiple phosphorylation sites. *Neuron* **28:** 153–164.

Miki K, Clapham DE. 2013. Rheotaxis guides mammalian sperm. *Curr Biol* **23:** 443–452.

Miller MR, Mansell SA, Meyers SA, Lishko PV. 2015. Flagellar ion channels of sperm: Similarities and differences between species. *Cell Calcium* **58:** 105–113.

Miller MR, Mannowetz N, Iavarone AT, Safavi R, Gracheva EO, Smith JF, Hill RZ, Bautista DM, Kirichok Y, Lishko PV. 2016. Unconventional endocannabinoid signaling governs sperm activation via sex hormone progesterone. *Science* **352:** 555–559.

Morisawa M. 2008. Adaptation and strategy for fertilization in the sperm of teleost fish. *J Appl Ichthyol* **24:** 362–370.

Mukherjee S, Jansen V, Jikeli JF, Hamzeh H, Alvarez L, Dombrowski M, Balbach M, Strünker T, Seifert R, Kaupp UB, et al. 2016. A novel biosensor to study cAMP dynamics in cilia and flagella. *eLife* **5:** e14052.

Navarro B, Kirichok Y, Clapham DE. 2007. KSper, a pH-sensitive K^+ current that controls sperm membrane potential. *Proc Natl Acad Sci* **104:** 7688–7692.

Nishigaki T, Chiba K, Miki W, Hoshi M. 1996. Structure and function of asterosaps, sperm-activating peptides from the jelly coat of starfish eggs. *Zygote* **4:** 237–245.

Nishigaki T, Jose O, Gonzalez-Cota AL, Romero F, Trevino CL, Darszon A. 2014. Intracellular pH in sperm physiology. *Biochem Biophys Res Commun* **450:** 1149–1158.

Nolan MA, Babcock DF, Wennemuth G, Brown W, Burton KA, McKnight GS. 2004. Sperm-specific protein kinase A catalytic subunit $C\alpha_2$ orchestrates cAMP signaling for male fertility. *Proc Natl Acad Sci* **101:** 13483–13488.

Nomura M, Beltrán C, Darszon A, Vacquier VD. 2005. A soluble adenylyl cyclase from sea urchin spermatozoa. *Gene* **353:** 231–238.

Ogawa H, Qui Y, Ogata CM, Misono KS. 2004. Crystal structure of hormone-bound atrial natriuretic peptide receptor extracellular domain: Rotation mechanism for transmembrane signal transduction. *J Biol Chem* **279:** 28625–28631.

Okunade GW, Miller ML, Pyne GJ, Sutliff RL, O'Connor KT, Neumann JC, Andringa A, Miller DA, Prasad V, Doetschman T, et al. 2004. Targeted ablation of plasma membrane Ca^{2+}-ATPase (PMCA) 1 and 4 indicates a major housekeeping function for PMCA1 and a critical role in hyperactivated sperm motility and male fertility for PMCA4. *J Biol Chem* **279:** 33742–33750.

Oren-Benaroya R, Orvieto R, Gakamsky A, Pinchasov M, Eisenbach M. 2008. The sperm chemoattractant secreted from human cumulus cells is progesterone. *Hum Reprod* **23:** 2339–2345.

Perez-Cerezales S, Boryshpolets S, Afanzar O, Brandis A, Nevo R, Kiss V, Eisenbach M. 2015. Involvement of opsins in mammalian sperm thermotaxis. *Sci Rep* **5:** 16146.

Pichlo M, Bungert-Plümke S, Weyand I, Seifert R, Bönigk W, Strünker T, Kashikar ND, Goodwin N, Müller A, Pelzer P, et al. 2014. High density and ligand affinity confer ultrasensitive signal detection by a guanylyl cyclase chemoreceptor. *J Cell Biol* **206:** 541–557.

Potter LR. 2011. Guanylyl cyclase structure, function and regulation. *Cell Signal* **23:** 1921–1926.

Pugh ENJ, Lamb TD. 2000. Phototransduction in vertebrate rods and cones: Molecular mechanisms of amplification, recovery and light adaptation. In *Handbook of biological physics* (ed. Stavenga DG, DeGrip WJ, Pugh ENJ), pp. 183–255. Elsevier, Amsterdam.

Qi H, Moran MM, Navarro B, Chong JA, Krapivinsky G, Krapivinsky L, Kirichok Y, Ramsey IS, Quill TA, Clapham DE. 2007. All four CatSper ion channel proteins are required for male fertility and sperm cell hyperactivated motility. *Proc Natl Acad Sci* **104:** 1219–1223.

Quill TA, Sugden SA, Rossi KL, Doolittle LK, Hammer RE, Garbers DL. 2003. Hyperactivated sperm motility driven by CatSper2 is required for fertilization. *Proc Natl Acad Sci* **100:** 14869–14874.

Ramsey IS, Moran MM, Chong JA, Clapham DE. 2006. A voltage-gated proton-selective channel lacking the pore domain. *Nature* **440:** 1213–1216.

Ren D, Navarro B, Perez G, Jackson AC, Hsu S, Shi Q, Tilly JL, Clapham DE. 2001. A sperm ion channel required for sperm motility and male fertility. *Nature* **413:** 603–609.

Rich TC, Fagan KA, Nakata H, Schaack J, Cooper DMF, Karpen JW. 2000. Cyclic nucleotide-gated channels colocalize with adenylyl cyclase in regions of restricted cAMP diffusion. *J Gen Physiol* **116:** 147–161.

Rich TC, Tse TE, Rohan JG, Schaack J, Karpen JW. 2001. In vivo assessment of local phosphodiesterase activity using tailored cyclic nucleotide-gated channels as cAMP sensors. *J Gen Physiol* **118:** 63–77.

Rothschild L. 1963. Non-random distribution of bull spermatozoa in a drop of sperm suspension. *Nature* **198:** 1221–1222.

Santi CM, Martinez-Lopez P, de la Vega-Beltran JL, Butler A, Alisio A, Darszon A, Salkoff L. 2010. The SLO3 sperm-specific potassium channel plays a vital role in male fertility. *FEBS Lett* **584:** 1041–1046.

Sasaki M, Takagi M, Okamura Y. 2006. A voltage sensor-domain protein is a voltage-gated proton channel. *Science* **312:** 589592.

Schiffer C, Müller A, Egeberg DL, Alvarez L, Brenker C, Rehfeld A, Frederiksen H, Wäschle B, Kaupp UB, Balbach M, et al. 2014. Direct action of endocrine disrupting chemicals on human sperm. *EMBO Rep* **15:** 758–765.

Schneeweis DM, Schnapf JL. 1995. Photovoltage of rods and cones in the macaque retina. *Science* **268:** 1053–1056.

Schreiber M, Wei A, Yuan A, Gaut J, Saito M, Salkoff L. 1998. Slo3, a novel pH-sensitive K^+ channel from mammalian spermatocytes. *J Biol Chem* **273:** 3509–3516.

Schuh K, Cartwright EJ, Jankevics E, Bundschu K, Liebermann J, Williams JC, Armesilla AL, Emerson M, Oceandy D, Knobeloch KP, et al. 2004. Plasma membrane Ca^{2+} ATPase 4 is required for sperm motility and male fertility. *J Biol Chem* **279:** 28220–28226.

Scott K, Zuker C. 1998. TRP, TRPL and trouble in photoreceptor cells. *Curr Opinion Neurobiol* **8:** 383–388.

Seifert R, Flick M, Bönigk W, Alvarez L, Trötschel C, Poetsch A, Müller A, Goodwin N, Pelzer P, Kashikar ND, et al. 2015. The CatSper channel controls chemosensation in sea urchin sperm. *EMBO J* **34:** 379–392.

Servin-Vences MR, Tatsu Y, Ando H, Guerrero A, Yumoto N, Darszon A, Nishigaki T. 2012. A caged progesterone analog alters intracellular Ca^{2+} and flagellar bending in human sperm. *Reproduction* **144:** 101–109.

Shiba K, Baba SA, Inoue T, Yoshida M. 2008. Ca^{2+} bursts occur around a local minimal concentration of attractant and trigger sperm chemotactic response. *Proc Natl Acad Sci* **105:** 19312–19317.

Smith JF, Syritsyna O, Fellous M, Serres C, Mannowetz N, Kirichok Y, Lishko PV. 2013. Disruption of the principal, progesterone-activated sperm Ca^{2+} channel in a CatSper2-deficient infertile patient. *Proc Natl Acad Sci* **110:** 6823–6828.

Sourjik V, Berg HC. 2004. Functional interactions between receptors in bacterial chemotaxis. *Nature* **428:** 437–441.

Steegborn C, Litvin TN, Levin LR, Buck J, Wu H. 2005. Bicarbonate activation of adenylyl cyclase via promotion of catalytic active site closure and metal recruitment. *Nat Struct Mol Biol* **12:** 32–37.

Strünker T, Weyand I, Bönigk W, Van Q, Loogen A, Brown JE, Kashikar N, Hagen V, Krause E, Kaupp UB. 2006. A K^+-selective cGMP-gated ion channel controls chemosensation of sperm. *Nat Cell Biol* **8:** 1149–1154.

Strünker T, Goodwin N, Brenker C, Kashikar ND, Weyand I, Seifert R, Kaupp UB. 2011. The CatSper channel mediates progesterone-induced Ca^{2+} influx in human sperm. *Nature* **471:** 382–386.

Su YH, Vacquier VD. 2002. A flagellar K^+-dependent Na^+/Ca^{2+} exchanger keeps Ca^{2+} low in sea urchin spermatozoa. *Proc Natl Acad Sci* **99:** 6743–6748.

Su YH, Vacquier VD. 2006. Cyclic GMP-specific phosphodiesterase-5 regulates motility of sea urchin spermatozoa. *Mol Biol Cell* **17:** 114–121.

Su TW, Xue L, Ozcan A. 2012. High-throughput lensfree 3D tracking of human sperms reveals rare statistics of helical trajectories. *Proc Natl Acad Sci* **109:** 16018–16022.

Suarez SS. 2008. Control of hyperactivation in sperm. *Hum Reprod Update* **14:** 647–657.

Suarez SS, Varosi SM, Dai X. 1993. Intracellular calcium increases with hyperactivation in intact, moving hamster sperm and oscillates with the flagellar beat cycle. *Proc Natl Acad Sci* **90:** 4660–4664.

Takai H, Morisawa M. 1995. Change in intracellular K^+ concentration caused by external osmolality change regulates sperm motility of marine and freshwater teleosts. *J Cell Sci* **108:** 1175–1181.

Tavares RS, Mansell S, Barratt CL, Wilson SM, Publicover SJ, Ramalho-Santos J. 2013. p,p′-DDE activates CatSper and compromises human sperm function at environmentally relevant concentrations. *Hum Reprod* **28:** 3167–3177.

Teves ME, Barbano F, Guidobaldi HA, Sanchez R, Miska W, Giojalas LC. 2006. Progesterone at the picomolar range is a chemoattractant for mammalian spermatozoa. *Fertil Steril* **86:** 745–749.

Teves ME, Guidobaldi HA, Unates DR, Sanchez R, Miska W, Publicover SJ, Morales Garcia AA, Giojalas LC. 2009. Molecular mechanism for human sperm chemotaxis mediated by progesterone. *PLoS ONE* **4:** e8211.

Tsunoda S, Sierralta J, Sun Y, Bodner R, Suzuki E, Becker A, Socolich M, Zuker CS. 1997. A multivalent PDZ-domain protein assembles signalling complexes in a G-protein-coupled cascade. *Nature* **388:** 243–249.

Vaandrager AB. 2002. Structure and function of the heat-stable enterotoxin receptor/guanylyl cyclase C. *Mol Cell Biochem* **230:** 73–83.

Vines CA, Yoshida K, Griffin FJ, Pillai MC, Morisawa M, Yanagimachi R, Cherr GN. 2002. Motility initiation in herring sperm is regulated by reverse sodium–calcium exchange. *Proc Natl Acad Sci* **99:** 2026–2031.

Visconti PE, Moore GD, Bailey JL, Leclerc P, Connors SA, Pan D, Olds-Clarke P, Kopf GS. 1995. Capacitation of mouse spermatozoa. II: Protein tyrosine phosphorylation and capacitation are regulated by a cAMP-dependent pathway. *Development* **121:** 1139–1150.

Wang D, King SM, Quill TA, Doolittle LK, Garbers DL. 2003. A new sperm-specific Na^+/H^+ exchanger required for sperm motility and fertility. *Nat Cell Biol* **5:** 1117–1122.

Wang D, Hu J, Bobulescu IA, Quill TA, McLeroy P, Moe OW, Garbers DL. 2007. A sperm-specific Na^+/H^+ exchanger (sNHE) is critical for expression and in vivo bicarbonate regulation of the soluble adenylyl cyclase (sAC). *Proc Natl Acad Sci* **104:** 9325–9330.

Wang H, Liu J, Cho KH, Ren D. 2009. A novel, single, transmembrane protein CATSPERG is associated with CATSPER1 channel protein. *Biol Reprod* **81:** 539–544.

Wennemuth G, Babcock DF, Hille B. 2003a. Calcium clearance mechanisms of mouse sperm. *J Gen Physiol* **122:** 115–128.

Wennemuth G, Carlson AE, Harper AJ, Babcock DF. 2003b. Bicarbonate actions on flagellar and Ca^{2+}-channel responses: initial events in sperm activation. *Development* **130:** 1317–1326.

Wertheimer E, Krapf D, de la Vega-Beltran JL, Sanchez-Cardenas C, Navarrete F, Haddad D, Escoffier J, Salicioni AM, Levin LR, Buck J, et al. 2013. Compartmentalization of distinct cAMP signaling pathways in mammalian sperm. *J Biol Chem* **288:** 35307–35320.

Whitlock GG, Lamb TD. 1999. Variability in the time course of single photon responses from toad rods: Termination of rhodopsin's activity. *Neuron* **23:** 337–351.

Wilson EM, Chinkers M. 1995. Identification of sequences mediating guanylyl cyclase dimerization. *Biochemistry* **34:** 4696–4701.

Winet H, Bernstein GS, Head J. 1984. Observations on the response of human spermatozoa to gravity, boundaries and fluid shear. *J Reprod Fertil* **70:** 511–523.

Wood CD, Nishigaki T, Furuta T, Baba SA, Darszon A. 2005. Real-time analysis of the role of Ca^{2+} in flagellar movement and motility in single sea urchin sperm. *J Cell Biol* **169:** 725–731.

Yanagimachi R, Cherr G, Matsubara T, Andoh T, Harumi T, Vines C, Pillai M, Griffin F, Matsubara H, Weatherby T, et al. 2013. Sperm attractant in the micropyle region of fish and insect eggs. *Biol Reprod* **88:** 47.

Yang C, Zeng XH, Zhou Y, Xia XM, Lingle CJ. 2011. LRRC52 (leucine-rich-repeat-containing protein 52), a testis-specific auxiliary subunit of the alkalization-activated Slo3 channel. *Proc Natl Acad Sci* **108:** 19419–19424.

Yau KW, Hardie RC. 2009. Phototransduction motifs and variations. *Cell* **139:** 246–264.

Yu H, Olshevskaya E, Duda T, Seno K, Hayashi F, Sharma RK, Dizhoor AM, Yamazaki A. 1999. Activation of retinal guanylyl cyclase-1 by Ca^{2+}-binding proteins involves its dimerization. *J Biol Chem* **274:** 15547–15555.

Zeng XH, Yang C, Kim ST, Lingle CJ, Xia XM. 2011. Deletion of the Slo3 gene abolishes alkalization-activated K^+ current in mouse spermatozoa. *Proc Natl Acad Sci* **108:** 5879–5884.

Zeng XH, Navarro B, Xia XM, Clapham DE, Lingle CJ. 2013. Simultaneous knockout of Slo3 and CatSper1 abolishes all alkalization- and voltage-activated current in mouse spermatozoa. *J Gen Physiol* **142:** 305–313.

Zeng XH, Yang C, Xia XM, Liu M, Lingle CJ. 2015. SLO3 auxiliary subunit LRRC52 controls gating of sperm KSPER currents and is critical for normal fertility. *Proc Natl Acad Sci* **112:** 2599–2604.

Zhang Y, Malekpour M, Al-Madani N, Kahrizi K, Zanganeh M, Lohr NJ, Mohseni M, Mojahedi F, Daneshi A, Najmabadi H, et al. 2007. Sensorineural deafness and male infertility: A contiguous gene deletion syndrome. *J Med Genet* **44:** 233–240.

Zhang Z, Liu J, Meriano J, Ru C, Xie S, Luo J, Sun Y. 2016. Human sperm rheotaxis: a passive physical process. *Sci Rep* **6:** 23553.

Cite this article as *Cold Spring Harb Perspect Biol* doi: 10.1101/cshperspect.a028225

Cilia and Mucociliary Clearance

Ximena M. Bustamante-Marin and Lawrence E. Ostrowski

Marsico Lung Institute, Cystic Fibrosis and Pulmonary Diseases Research and Treatment Center, University of North Carolina, Chapel Hill, North Carolina 27599

Correspondence: ostro@med.unc.edu

Mucociliary clearance (MCC) is the primary innate defense mechanism of the lung. The functional components are the protective mucous layer, the airway surface liquid layer, and the cilia on the surface of ciliated cells. The cilia are specialized organelles that beat in metachronal waves to propel pathogens and inhaled particles trapped in the mucous layer out of the airways. In health this clearance mechanism is effective, but in patients with primary cilia dyskinesia (PCD) the cilia are abnormal, resulting in deficient MCC and chronic lung disease. This demonstrates the critical importance of the cilia for human health. In this review, we summarize the current knowledge of the components of the MCC apparatus, focusing on the role of cilia in MCC.

The extensive epithelial surface of the respiratory tract between the nose and the alveoli is exposed daily to viral and bacterial pathogens, particulates, and gaseous material with potentially harmful effects. In response to these challenges, humans have developed a series of defense mechanisms to protect the airways from these insults, thereby maintaining the lungs in a nearly sterile condition (Dickson and Huffnagle 2015). Lung defense involves cough, anatomical barriers, aerodynamic changes, and immune mechanisms; however, the primary defense mechanism is mucociliary clearance (MCC). Healthy airway surfaces are lined by ciliated epithelial cells and covered with an airway surface layer (ASL), which has two components, a mucus layer that entraps inhaled particles and foreign pathogens, and a low viscosity periciliary layer (PCL) that lubricates airway surfaces and facilitates ciliary beating for efficient mucus clearance (Wanner et al. 1996; Knowles and Boucher 2002). The coordinated interaction of these components on the surface of the respiratory tract results in MCC.

Proper ciliary function is absolutely required for effective MCC. Cilia are specialized organelles that provide the force necessary to transport foreign materials in the respiratory tract toward the mouth where they can be swallowed or expectorated. To accomplish this crucial function, the cilia beat in coordinated metachronal waves at a beat frequency that has multiple physiological regulators. Much of what we know about cilia structure and function has been derived from studies performed using *Chlamydomonas* as a model, while the importance of cilia in maintaining airway clearance was revealed by clinical and pathological studies of genetic and acquired forms of chronic airway diseases, including primary ciliary dyskinesia

Cite this article as *Cold Spring Harb Perspect Biol* doi: 10.1101/cshperspect.a028241

(PCD), cystic fibrosis (CF), asthma, and chronic obstructive pulmonary disease (COPD). In this review, we summarize the current knowledge of the components of MCC, with an emphasis on the role of cilia in normal mucociliary transport. We then briefly describe the health effects of impaired MCC caused by genetic defects in cilia in patients with PCD.

ORIGIN AND ORGANIZATION OF THE RESPIRATORY TRACT

In humans, the formation of the respiratory system starts around the fourth week of gestation. In the head, the nasal placodes develop from the ectoderm at each side of the frontonasal prominence. They then become more concave forming the nasal pit. The mesenchyme proliferates around the placode establishing the medial and lateral nasal prominences, while the nasal pit becomes deeper, forming the nasal cavities (Kim et al. 2004). The paranasal sinuses develop during late fetal life and in infancy as diverticula of the lateral nasal walls. The sinuses extend into the maxilla and reach their mature size in the early 20s. In the neck, the laryngotracheal groove emerges from the primitive pharyngeal floor. The endoderm lining the laryngotracheal groove forms the epithelium and glands of the larynx, trachea, bronchi, and pulmonary lining epithelium (Panski 1982; Edgar et al. 2013). The primordial lung originates as a protrusion from the laryngotracheal groove of the ventral foregut endoderm, which then proliferates, forming two lung buds surrounded by the mesoderm in the primitive thoracic mesenchyme. Each lung bud develops into left and right lungs, respectively, by undergoing branching morphogenesis. This process is directed by signals between the epithelial endoderm and the surrounding mesoderm. Bronchial cartilage, smooth muscle, and other connective tissues are derived from the mesenchyme. As the lung progresses through its phases of development, it undergoes a complex series of epithelial–mesenchymal interactions regulated by homeobox genes, transcription factors, hormones, and growth factors (Chinoy 2003). During the postnatal phase, lung growth is geometric, and there

is no increase in airway number. The alveoli increase in number after birth to reach the adult range of 300 million by 2 years of age and the surface area of 75 to 100 m^2 by adulthood (Deutsch and Pinar 2002).

The result is a complex arrangement of organs and tissues that form the respiratory system, which is divided into two parts. The upper airway or upper respiratory tract includes the nose and nasal passages, paranasal sinuses, the oral cavity, the pharynx, and the portion of the larynx above the vocal cords. The nose and nasal cavity are the main external openings of the respiratory system. The oral cavity (mouth) acts as an alternative entry to the air, but it cannot filter the air of unwanted contaminants as the nose does. Although the upper respiratory tract represents the entryway to the conducting zone of the respiratory tract and plays an important role in trapping and removing particulate matter, in this review we will focus on the lower respiratory tract.

The lower airways, or lower respiratory tract, include the continuation of the conducting zone formed by the portion of the larynx below the vocal cords, trachea, and within the lungs, the bronchi, bronchioles, and the respiratory zone formed by respiratory bronchioles, alveolar ducts, and alveoli (Fig. 1A). On the surface of each alveolus, the oxygen from the atmosphere required for cellular respiration is exchanged for the carbon dioxide in the blood, which is released into the atmosphere.

CELL TYPES IN THE LOWER RESPIRATORY TRACT

Studies in mice during branching morphogenesis of the lung showed that the terminal buds contain a population of multipotent epithelial progenitors characterized by high levels of Nmyc, Sox9, and cyclin D1 proteins (Okubo et al. 2005). These multipotent cells will give rise to the major cell types of the conducting airways, including basal stem cells (Rock et al. 2009), neurosecretory cells (Kultschitsky or K-cells) (Becker and Silva 1981), ciliated cells, club cells, serous cells, goblet cells, intermediate cells, and brush cells (Reid and Jones 1980). The mu-

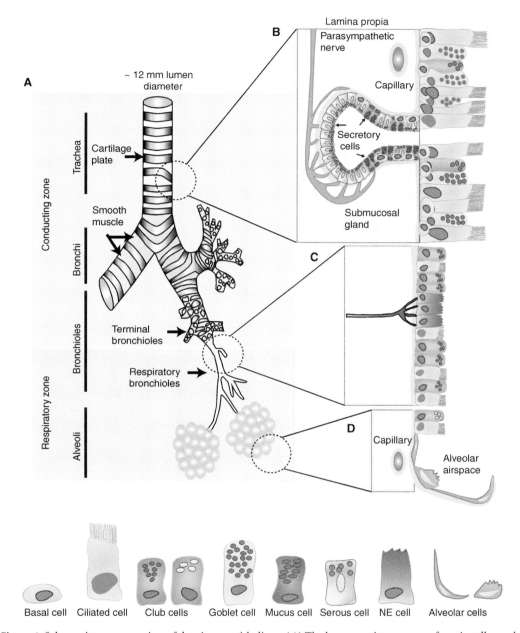

Figure 1. Schematic representation of the airway epithelium. (*A*) The lower respiratory tract, functionally can be divided into conducting and respiratory zones. The adult human trachea has an internal diameter of ~12 mm, cartilage plates, and smooth muscle. The trachea divides into right and left primary bronchi. A bronchus enters the lung at the hilum and then divides into bronchioles. After multiple bronchiolar branches (~23 generations in humans), at the end of each respiratory bronchiole, the alveoli are found. (*B*) The trachea and most proximal airways are lined by a pseudostratified epithelium formed by ciliated and secretory cells. Basal cells are located in this region and they can generate secretory and ciliated cell lineages. (*C*) The small airways are lined by a simple cuboidal epithelium with fewer goblet cells, but are rich in club cells. (*D*) The alveoli are made of type I and type II alveolar cells.

cosa of the lower respiratory tract is a pseudo-stratified columnar epithelium with numerous ciliated and goblet cells (Fig. 1B). During differentiation, Notch signaling plays a major role in controlling cell fate commitment by selectively suppressing ciliation to allow secretory cell differentiation (Tsao et al. 2008). In fact, the use of Notch pathway inhibitors (e.g., N-[(3,5-difluorophenyl)acetyl]-L-alanyl-2-phenyl]glycine-1,1-dimethylethyl ester [DAPT], a γ-secretase inhibitor that blocks the Notch pathway), has allowed researchers to direct the differentiation of airway cells into the ciliated epithelial cell lineage in culture (Konishi et al. 2016). In addition, the development of antibodies directed against members of this pathway can induce the transdifferentiation of goblet and club cells into ciliated cells. This observation may be useful for the development of therapeutic approaches to inhibit excess mucus production in airway diseases, including asthma (Lafkas et al. 2015).

The epithelium is supported by a collagenous lamina propia containing serous and mucus glands (Fig. 1B). The supporting lamina propia underneath the epithelium also contains elastin that plays a role in the elastic recoil of the trachea during inspiration and expiration. The smooth muscle of the airways plays a functional role in regulating airflow and is arranged in a complex spiral pattern that becomes progressively less prominent in the distal conducting airways.

The small airways (i.e., those <2 mm in internal diameter), lack cartilaginous support and mucous glands. The epithelium progressively transitions to a less tall, simple cuboidal, less ciliated epithelium with few goblet cells but with an increased number of club cells (Fig. 1C). The division of respiratory bronchioles originates the alveolar ducts that connect to the alveolar sacs, which contain the alveoli. The alveoli are formed by two types of alveolar cells; type I alveolar cells are thin cells involved in the process of gas exchange and type II alveolar cells that secrete pulmonary surfactant (Fig. 1D).

Although the airway epithelium contains different types of cells, the ciliated cells and secretory cells in the surface epithelium and sub-

mucosal glands contribute directly to mucociliary function.

Secretory Cells

The goblet cells are the principal secretory cells in the superficial epithelium of the tracheobronchial airway (Sleigh et al. 1988). They are intercalated among ciliated cells and connected to adjacent cells by tight junctions; together these cells form a selective barrier lining the respiratory tract. Of the cell population in the trachea, approximately 60% are ciliated cells and 20% are goblet cells. As the airways branch, the percentage of ciliated and goblet cells decreases, whereas the percentage of serous and club cells increases (Wanner et al. 1996; Davis and Randell 2001). The morphology of goblet cells is highly polarized, with the nucleus and other organelles localized to the base of the cell. Most of the apical cytoplasm contains membrane-bound secretory granules containing high molecular weight and gel-forming glycoproteins called mucins. In the small airways, serous and club cells contain small secretory granules and are thought to produce watery secretions. However, club cells can produce mucins and act as progenitors of goblet cells after bronchiolar injury (Sleigh et al. 1988; Zhu et al. 2008), suggesting there is significant functional overlap between these types of secretory cells.

The submucosa layer, present until the termination of the cartilaginous bronchioles, contains mixed seromucous glands that rapidly produce mucus in response to neural signals (Maggi et al. 1995). In the human trachea, the seromucous glands are found at a frequency of one gland per mm^2 and are even more abundant as the airway lumens decrease in diameter until the termination of the cartilaginous bronchioles (Tos 1966; Wine and Joo 2004). The watery secretions from the serous glands humidify inspired air and, together with mucus from the goblet cells, comprise the bulk of the airway surface layer.

Ciliated Cells

Ciliated airway epithelial cells are elongated columnar cells that make limited contact with the basement membrane (Mercer et al. 1994). Unlike other cells types described above, ciliated cells are terminally differentiated epithelial cells (Rawlins and Hogan 2008). In the apical region, they have abundant mitochondria to ensure the availability of ATP to sustain ciliary motion mediated by the motor activity of axonemal dynein (Kikkawa 2013).

The differentiation program of ciliated cells starts during lung morphogenesis. The inhibition of Notch signaling activates a transcription program that includes the nuclear protein geminin coiled-coil containing (GMNC), which turns on *MCIDAS* expression (Zhou et al. 2015). *MCIDAS* encodes the transcriptional cofactor MULTICILIN (Stubbs et al. 2012; Boon et al. 2014), which induces the expression of *FOXJ1*, the master regulator for basal body docking, cilia formation, and motility (You et al. 2004; Vladar and Mitchell 2016).

Airway ciliated epithelial cells form >100 centrioles that will dock to the apical membrane and become the basal bodies that will allow the growth of the cilia (Sorokin 1968). Using live imaging combined with superresolution and electron microscopy, Al Jord et al. (2014) showed that in multiciliated ependymal cells only the daughter centriole contributes to amplification of basal bodies. Once the basal body has correctly positioned on the apical surface of the plasma membrane, axonemal extension occurs exclusively at the plus ends of the microtubules in a process mediated by the intraflagellar transport (IFT) system (Kozminski et al. 1993; Rosenbaum and Witman 2002). Proteins are loaded onto the IFT system at the ciliary base within the cytoplasm and transferred across the ciliary compartment border in a process known as ciliogenesis (Avidor-Reiss et al. 2004). As a result, fully differentiated ciliated airway epithelial cells have >100 cilia on their surface (Wanner et al. 1996).

COMPONENTS OF THE MUCOCILIARY CLEARANCE APPARATUS

The cilia that reach the surface of epithelium interact with a thin layer of fluid covering the air-facing surface, the airway surface layer (ASL). The ASL includes a low viscosity periciliary layer (PCL) that lubricates airway surfaces and facilitates ciliary beating, and an overlaying mucous layer. The ASL may also include a surfactant layer to facilitate the spreading of mucus over the epithelial surface (Fig. 2A). Although the exact composition of the ASL is still unclear and a topic of debate, there is no doubt that the physical characteristics of the ASL are crucial to allow normal ciliary activity to maintain airway health.

The Mucous Layer

The major macromolecular constituents of the mucous layer are the mucin glycoproteins. The amino- and carboxy-terminal regions of these large proteins ($2-20 \times 10^5$ Da) are lightly glycosylated but rich in cysteines that establish disulfide links between and among mucin monomers. In the central region, mucins contain multiple tandem repeats of serine and threonine, which are the sites for extensive O-linked glycosylation. The mucin proteins are encoded by *MUC* genes, and many of them are expressed in the airways, including the gel-forming mucins MUC5AC, MUC5B, MUC2, MUC8, and MUC19. The membrane-associated mucins, MUC1, MUC4, MUC11, MUC13, MUC15, MUC16, and MUC20, are also expressed on the surface of the airway epithelium. Muc7, a small mucin that lacks domains and does not form a gel, is secreted by a subset of serous cells in submucosal glands (Rose and Voynow 2006; Fan and Bobek 2010). In healthy airways, the goblet cells typically express MUC5AC, whereas mucosal cells of the submucosal glands express primarily MUC5B (Groneberg et al. 2002). The secretion of mucin can be stimulated by many factors, including paracrine and autocrine mediators, especially ATP (Chen et al. 2003; Rose and Voynow 2005). Increased production and secretion of mucins is a feature of many chronic airway diseases, including asthma, COPD, and cystic fibrosis (Rose and Voynow 2006)

The rheological properties of mucus (i.e., the capacity to undergo flow and deformation in response to the forces applied to it), and, therefore, the transportability of the mucous

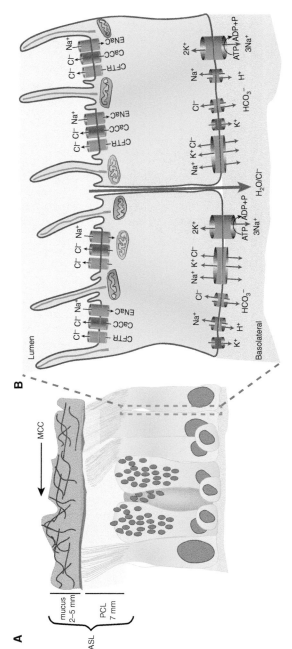

Figure 2. Representation of the components of the mucociliary clearance (MCC) apparatus. (*A*) An efficient MCC requires the cilia to interact with the periciliary layer (PCL) (~7 mm) and propel the overlaying mucous layer (~2 to 5 mm). A thin surfactant layer (in blue) could be present. The hydration of the airway surface layer (ASL) for optimal cilia performance is maintained by the active transmembrane ionic transport of the ciliated epithelia (dashed red box). (*B*) At the apical membrane, the Na^+ reabsorption is mediated by epithelial Na^+ channel (ENaC) with H_2O/Cl^- following passively the osmotic gradient. The Cl^- secretion is regulated by cystic fibrosis transmembrane conductance regulator (CFTR) and calcium-activated chloride channels (CaCC). The basolateral activity of the Na^+/K^+ ATPase, $Na^+/K^+/2Cl^-$ cotransporter, voltage-dependent K^+ channels, HCO3-/Cl^- exchanger, and other basolateral Cl^- channels, maintain the electrochemical gradient.

layer, are determined by the composition of the mucus and its hydration state. Normal mucus is composed of ~1% mucins, ~1% salt, ~1% other proteins, and ~97% water (Matthews et al. 1963; Hamed and Fiegel 2014). The hydration status is principally regulated by the export of Cl^- through the cystic fibrosis transmembrane conductance regulator (CFTR) and Ca^{+2}-activated chloride channels (CaCC), and by the influx of Na^+ through the epithelial Na^+ channel (ENaC) (Fig. 2B) (Tarran et al. 2005). By regulating these two processes, the epithelium controls the amount of water on the airway surface. Thus, in normal, healthy lungs, after secretion and hydration of mucins, a thin layer of mucus (2 to 5 μm thick in the trachea) is formed above the cilia (Fig. 2A) from the bronchioles to the upper airway to protect the epithelium.

Periciliary Layer and Surfactant

The periciliary layer (PCL) is a polyanionic gel that has a height approximately equal to the height of the extended cilium (7 μm, Fig. 2A) (Button et al. 2012). It contains the membrane-associated mucins (MUC1, MUC4, and MUC16) and other molecules including glycolipids (Randell and Boucher 2006; Button et al. 2012). It provides an efficient lubricating layer for ciliary beating, and serves as a barrier to restrict access of particles from the cell surface. As with the overlying mucus layer, the hydration state of the PCL reflects the balance of Na^+ and Cl^- ion transport activities (Tarran et al. 2005). Maintenance of the depth of PCL is important for effective mucociliary clearance. If the PCL is not sufficiently hydrated, the mucus layer collapses on the cilia and they become trapped. (Sleigh et al. 1988; Randell and Boucher 2006). In patients with pseudohypoaldosteronism, with no active absorption of sodium in the airways because of mutations in ENaC, the PCL is too deep. During the first year of life, children have frequent lower respiratory tract infections, possibly caused by inefficient MCC. Later in life, they have less frequent respiratory illness and, in a small sample, actually showed a faster than normal rate of MCC (Kerem et al. 1999).

Pulmonary surfactant is a surface-active material able to reduce surface tension at the alveolar and bronchiolar air–liquid interface. It originates mainly from the alveolar compartment, although specific components, including SP-A, SP-B, and SP-D, are also synthetized and secreted by nonciliated cells of the airway mucosa (Calkovska 2000). Some research groups have described a film of surfactants between the mucous and the PCL to permit the transfer of energy from the cilia to the mucus, while preventing ciliary entanglement in the mucus, thereby facilitating MCC (Girod de Bentzmann et al. 1993; Wanner et al. 1996). Surfactants could also be an integral component of the PCL, the mucus layer, or both.

The Structure of Airway Cilia

The adult lung is estimated to contain ~3 × 10^{12} total motile cilia. On average, a typical cilium is 6.5- to 7-μm long and has a diameter of 0.1 μm (Wanner et al. 1996; Brekman et al. 2014). The core structure is the highly conserved 9+2 axoneme that extends from the basal body in the apical region of ciliated cells into the PCL, with the cilia tips reaching the mucous layer of the airway lumen. The ciliary axoneme (Fig. 3A) consists of nine outer doublet microtubules surrounding two central singlet microtubules, the central pair (Satir and Christensen 2007). The outer doublets consist of a complete tubule, the A tubule, containing 13 tubulin subunits and a partial tubule, the B tubule, containing 11 subunits. These doublets are connected to each other by a large protein complex, the nexin–dynein regulatory complex (N-DRC) (Heuser et al. 2009). The radial spokes project from the outer doublets toward the central-pair complex and play a role in both the mechanical stability of the axoneme and the regulation of ciliary activity. Attached to the A tubule are the multisubunit protein complexes, the inner and outer dynein arms (IDA and ODA, respectively) (Fig. 3A). Based on studies of other model systems, especially the flagella of the unicellular, biflagellate green alga, *Chlamydomonas reinhardtii*, it is known that the basic structure of ODA and IDA consists of one or

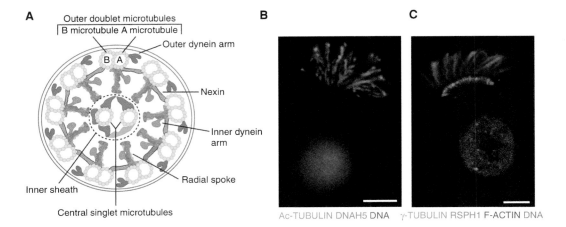

Figure 3. Axonemal structure of the cilia. (*A*) The core structure of the cilia is the axoneme, as shown in the cross-sectional schematic diagram of a motile cilium displaying the characteristic 9+2 pattern, nine peripheral doublets microtubules surrounding a central pair of single microtubules. The cilia contain multiple protein complexes, including dynein arms, radial spokes, nexin–dynein regulatory complex (N-DRC), and inner sheath that connect the microtubules to each other. (*B,C*) Examples of single-cell immunofluorescence of human ciliated cells showing the localization of dynein heavy chain 5 (DNAH5) and radial spoke head 1 (RSPH1). Scale bars, 5 μm.

more dynein heavy chain (DHC) proteins (470 to 540 kDa), dynein intermediate chain proteins (57 to 140 kDa), and dynein light chain proteins (6 to 22 kDa) (Holzbaur and Vallee 1994). Detailed structural analysis of the DHC identified four distinct domains, including (1) the "tail," that acts as the cargo-binding domain, (2) the stalk region that binds to the microtubule, (3) a linker region, and (4) the head domain, which is the site where ATP is hydrolyzed to provide the force for ciliary beating (Kikkawa 2013) (for a more detailed structure of the axoneme, see Ishikawa 2016). In the human genome, there are 14 putative genes that encode for axonemal DHC (Yagi 2009). Although the detailed structure and composition of the dynein arms from human cilia is not as well described as that of *Chlamydomonas*, the development of specific antibodies has allowed researchers to determine that the distribution of the DHC proteins varies between the proximal and distal regions of the cilium (Fliegauf et al. 2005), as well as to localize other proteins to the ciliary axoneme (e.g., dynein heavy chain 5 [DNAH5] and radial spoke head 1 [RSPH1]) (Fig. 3B,C).

The activation of the DHC by ATP (Wanner et al. 1996; Porter and Sale 2000; Roberts et al. 2012) causes the doublet microtubules to slide with respect to one another. The presence of the interdoublet N-DRC and radial-spoke–central-pair interactions produce a controlled bending (Yeates et al. 1981; Houtmeyers et al. 1999), yielding a ciliary beat with an effective stroke and a recovery stroke within the same plane (Chilvers and O'Callaghan 2000). Studies in *Chlamydomonas* have provided evidence that the ODA are primarily responsible for adjusting ciliary beat frequency (CBF), whereas the IDA are responsible for bend formation and waveform (Brokaw and Kamiya 1987).

Comparative genomic and proteomic studies have shown that the complete cilium contains ~600 proteins. Many of the ciliary proteins are highly conserved, specifically those that constitute the axonemal structure (Ostrowski et al. 2002; Avidor-Reiss et al. 2004; Lindskog et al. 2014). About one-third of these proteins have been correlated with a specific structure and function, and the majority of these studies were performed using *Chlamydomonas* flagella. Although there is tremendous conservation of both individual proteins and the overall axonemal structure between species, there are important structural differences between human cilia and *Chlamydomonas* flagella axonemes. For example, the ODA in *Chlamydomonas* is a three-headed structure containing three dynein heavy chains (α, β, γ), whereas human ODA contains two dynein heavy chains (Pazour et al. 2006). Human cilia have three radial spokes, whereas *Chlamydomonas* flagella have two and a short structure known as the radial spoke stand-in (Lin et al. 2014). This structure corresponds to the base of the third radial spoke in human cilia. The overall arrangement of inner dynein arms within the 96 nm axoneme repeat and the central pair is remarkably similar, but there are slight differences in size and density between the two species (O'Toole et al. 2012). The cilia and flagella are also functionally different. The 12 μm *Chlamydomonas* flagellum beats with two different waveforms at frequencies of up to 60 Hz, enabling propulsion in opposite directions. Human airway cilia are approximately 7 μm long and beat with a single waveform at lower frequencies (10–20 Hz) (Satir and Christensen 2007) to propel mucus out of the airways. The regulation of axonemal activity is also different between human airway cilia and other motile axonemes. For example, an increase in cAMP stimulates ciliary beat frequency in human cilia, while causing a decrease in *Chlamydomonas* (Wanner et al. 1996). Similarly, an increase in Ca^{2+} stimulates ciliary beat frequency in human cilia, while causing *Chlamydomonas* flagella to change from an asymmetric waveform to a symmetric waveform (Silflow and Lefebvre 2001). In spite of these differences, *Chlamydomonas* flagella have been an extremely useful model organism to study motile ciliopathies, and provided a wealth of biochemical, molecular, and structural information regarding the axoneme. The identification of a large number of mutations in *Chlamydomonas* that affect the assembly or function of specific axonemal structures (Porter and Sale 2000) have revealed that small defects in axoneme structure can severely impact cilia motility, and have helped to define the genetic defects responsible for PCD (e.g., Pennarun et al. 1999; see below).

ROLE OF AIRWAY CILIA IN MCC AND ITS REGULATION

In the mouse trachea, ciliary motion starts soon after birth. Analysis ex vivo of flow directionality in mouse tracheas measuring the net displacement of fluorescent microspheres by Francis et al. (2009), detected patches of ciliary motion as early as postnatal day 3. This becomes uniform with maximal mucociliary flow by postnatal day 9. This study also showed that tracheal CBF was elevated immediately after birth compared to other time points. This could be a physiological mechanism to compensate for the lack of ciliary coordination and ensure the clearance of amniotic fluid from the lungs (Francis et al. 2009). In humans, ciliogenesis of the airways starts at 7 weeks of embryonic development (Moscoso et al. 1988) and is complete before birth; however, whether this ciliated epithelium is active and functional is unknown (Gaillard et al. 1989; Ostrowski 2002). Although it is interesting to note that neonatal respiratory distress is a common feature of PCD (Shapiro et al. 2016), the mechanism responsible for this is still unclear.

The cilia of the airways beat in a coordinated fashion that results in metachronal waves. Each cilium beats at the same frequency but in a phase-shifted manner with its neighbors along the axis of the effective stroke, and in phase-synchronously with the cilia in the perpendicular axis (Sanderson and Sleigh 1981; Brooks and Wallingford 2014). This generates a wave that travels across the epithelium propelling the overlying mucus in a cephalad direction. The basal CBF ranges between 10 and 20 Hz (Satir and Christensen 2007), yielding a mucociliary clearance velocity of ~5.5 mm/min (Hofmann and Asgharian 2003). Studies in mouse airways and *Xenopus laevis* showed that, to establish a coordinated ciliary beat pattern, the cilia need to be physically oriented according to the tissue axis, which is specified by the planar cell polarity pathway (Vladar et al. 2015). This orientation allows the cilia to propagate the wave of ciliary beat using the underlying cytoskeletal network. Actin bridges, connecting neighboring basal bodies, are necessary for intracellular propagation of the metachronal wave (Werner et al. 2011), whereas cytoplasmic microtubules that emanate from the basal foot cap, anchor the basal foot. These interactions are required for the maintenance of local coordination of the cilia (Werner et al. 2011; Clare et al. 2014). In human airways, a similar mechanism is likely responsible for controlling the orientation of the cilia, which is crucial for coordinated ciliary beating and effective MCC.

Although the mechanisms that regulate basal CBF have been studied extensively (Sanderson and Sleigh 1981; Sleigh et al. 1988; Satir and Sleigh 1990), they are still unclear. However, the identification of factors that enhance CBF have been useful for the development of therapies to improve MCC in patients with chronic airway diseases. For example, CBF can be stimulated by β-adrenergic agonists (Bennett 2002), intracellular increase of nucleotides, including cAMP and cyclic guanosine monophosphate (cGMP) (Satir and Sleigh 1990; Wanner et al. 1996; Wyatt 2015), and an intracellular increase of Ca^{2+}. Ciliated cells express multiple bitter taste receptors (R2T) that localize at different places along the cilium, making it a chemosensory organelle. These receptors sense noxious stimuli and stimulate CBF in a Ca^{2+}-dependent manner (Shah et al. 2009). Studies performed in human nasal epithelial cultures showed that extracellular adenosine and uridine nucleotides, acting through the P2Y2 receptor, are one of the most potent CBF agonists. Interestingly, ATP produces a rapid increase in CBF, but the hydrolysis of ATP to ADP results in a more prolonged response through the activation of A2B receptor (Morse et al. 2001). ATP is constantly released during normal and induced mechanical stimulation of the airways (Button et al. 2007), and helps maintain proper hydration of the mucus layer in addition to stimulating CBF (Button et al. 2013).

CBF is temperature dependent (Clary-Meinesz et al. 1992; Sears et al. 2015); because there is an optimal temperature for the enzymatic hydrolysis of ATP by dynein, low temperatures decrease CBF. The manipulation of temperature is a useful tool to study the relationship between CBF and mucociliary transport in controlled systems of airway cultures (Sears et al. 2015). CBF can also be affected by pH changes. Although the human airway can tolerate variations in local pH (6.9 to 7.0) induced during the respiratory cycle (Clary-Meinesz et al. 1998), intracellular alkalization stimulates, whereas intracellular acidification decreases, CBF (Sutto et al. 2004). Cigarette smoke is known to be a major risk factor for the development of COPD. Studies in humans and mice have shown that this common air pollutant significantly decreases CBF (Elliott et al. 2007; Simet et al. 2010), increases mucus secretion (Gensch et al. 2004), and leads to the generation of reactive oxygen species that reduces the number of ciliated cells (Milara et al. 2012).

MEASUREMENT OF MCC

The standard measurement of MCC rates in humans requires the inhalation of a nonpermeating radiolabeled marker that deposits on the airway surface. This method assumes that the marker moves out of the lung at the same rate as the airway secretions in which it is immersed. The short half-life isotope ^{99m}technetium (^{99m}Tc) is used to label the inhaled marker, which is then monitored by γ camera over periods of 1 to 24 hr (Bennett et al. 2013). Although this method provides a direct measurement of MCC, it is not well suited for routine experimental studies. Various methods have been used to study MCC in many different animal models, including sheep, dogs, pigs, and mice (Sabater et al. 1999; Foster et al. 2001; Lay et al. 2003; Bhashyam et al. 2012; Hoegger et al. 2014). The use of genetically modified mouse models has provided a valuable tool to study the role of motile cilia; however, the small size of the mouse airways has made measurements of in vivo clearance rates using radioactivity challenging (Foster et al. 2001; Bhashyam et al. 2012). Other investigators studying mouse models have recorded the movement of fluorescent particles along the trachea or nasopharynx of the mouse (e.g., Ostrowski et al. 2010) or measured the recovery of fluorescent beads from the lungs after a specified time (Look et al. 2001; Grubb et al. 2016). Alternatively, investigators have also used ex vivo preparations of trachea or bronchi to investigate MCC (Cooper et al. 2013; Francis and Lo 2013). Although these studies provide valuable information, the techniques are invasive and the system is not easy to manipulate.

The development of well-differentiated human airway cultures (Gray et al. 1996; Matsui et al. 1998b), that showed coordinated ciliary activity, provided a powerful tool to study the integrated process of MCC, including cilia coordination and transport. Air–liquid interface cultures of human airway cells with areas of coordinated ciliary activity spontaneously transport mucus in a circular pattern, forming what is commonly referred to as mucus hurricanes, which have been useful for the study of CF (Matsui et al. 1998a; Tarran et al. 2005; Button and Boucher 2008; Zhang et al. 2009). However, in these cultures, the overall height of the mucus layer is variable and the areas of coordination are of various sizes. To avoid these disadvantages, a modified culture system was developed by gluing a 15-mm plastic cylinder in the center of a 30-mm millicell culture insert, generating a mucociliary transport device (MCTD), in which human airway cells were seeded and allowed to differentiate (Sears et al. 2015). In this arrangement, cilia coordinate their activity to transport mucus in a continuous circular path around the culture track. This culture system enables the study of mucociliary transport, the visualization of ciliary coordination over time, and the evaluation of different factors that affect the rate of mucociliary transport. For example, using this system, it was shown that increasing CBF by increasing temperature produced a linear increase in mucociliary transport speed (Sears et al. 2015).

Factors Involved in MCC

A number of studies measuring the basal rates of MCC in healthy nonsmoking subjects showed that physiological factors as age, sex, posture, sleep, and exercise affect MCC (Houtmeyers et al. 1999). Efficient MCC is optimal at core temperature and 100% relative humidity conditions (Williams et al. 1996). Under low temper-

ature and humidity, the depth of the PCL is reduced, ciliary cells decrease their activity, and MCC slows, potentially allowing for increased bacterial infection. Conversely, high humidity enhances MCC (Clary-Meinesz et al. 1992; Daviskas et al. 1995; Oozawa et al. 2012). The ciliary factors that affect MCC rate include the number and length of the cilia, the coordination of ciliary beating, and the ciliary beat frequency and amplitude, which together determine the maximal velocity at the tips of the cilia and, therefore, the forward velocity of the mucous layer (King 2006). However, a faster ciliary beat frequency does not necessarily imply improvement of MCC rates, as a lack of coordination can severely affect ciliary transport (see below). Serous factors include mucus rheology and hydration. If the mucous layer is too thick because of mucin hypersecretion, as in asthma (Kuyper et al. 2003), or the mucus is too viscous because of dehydration, as in CF (Knowles and Boucher 2002), then the rate of MCC decreases because the cilia are not able to beat properly. The transference of momentum between the cilia and the mucous layer during their forward stroke while minimally interacting with the mucus during the return is disrupted. As mentioned above, another important factor in maintaining normal MCC is the release of ATP onto the airway lumen to maintain or increase ASL hydration and accelerate CBF, thus promoting lung health (Button et al. 2007).

Most of the current knowledge about how MCC is endogenously regulated at the organ and cellular level came from studies designed to explore the pathogenesis of chronic lung diseases. For example, studies of CF have highlighted the role of ion transport and airway hydration in MCC. Studies of asthma and COPD have led to a better understanding of the effects of chronic inflammation and mucus hypersecretion (Kuyper et al. 2003; Ramos et al. 2014). Studies of the effect of cigarette smoke on cilia performance provided insight into its effect on cilia length and its consequences for COPD (Gensch et al. 2004; Elliott et al. 2007; Simet et al. 2010; Milara et al. 2012; Brekman et al. 2014). Finally, studies of PCD not only clearly showed the key role of ciliary motility in MCC, but also delineated the physiological role that motile cilia play elsewhere in the body (Fliegauf et al. 2007).

Cilia and MCC in PCD

Primary ciliary dyskinesia is a genetically heterogeneous autosomal-recessive disorder with an incidence of 1:15,000 live births. This disorder affects the proper biogenesis, assembly and activity of cilia, resulting in dysfunction of motile cilia and impairment of MCC. The clinical features include *situs inversus* in 50% of the patients, infertility, and chronic ear infections. The disease affects the entire respiratory tract, with symptoms appearing soon after birth. Over 80% of PCD patients experience neonatal respiratory distress during the first 24 h of life (Shapiro et al. 2016). Because of impaired MCC, mucus and pathogens accumulate in the upper and lower airways causing year-round daily cough and nasal congestion, chronic sinusitis, and recurrent lower respiratory infections, leading to bronchiectasis, and, in severe cases, lung transplantation.

Currently, mutations in 33 genes have been associated with PCD (Zariwala et al. 2007; Shapiro et al. 2016). The genetic variants of PCD can affect the pattern of ciliary beating, CBF, or both (Raidt et al. 2014). The most prevalent ultrastructural defect in the ciliary axoneme of patients with PCD (as evaluated by transmission electron microscopy), is the lack of ODAs (Fig. 4). The structural defect in the ODAs can be caused by mutations in genes encoding proteins that are structural components of the ODA (e.g., *DNAH5* and *DNAH1*) or mutation in genes encoding for ODA-docking complex components (e.g., *CCDC14* and *CCDC151*). Consistent with evidence that ODAs primarily control CBF, PCD patients with ODA defects have significantly lower CBF (Raidt et al. 2014).

Other PCD-causing mutations have been identified in genes that code for proteins involved in the cytoplasmic assembly of dynein arms (e.g., SPAG1, DNAF1-3, and DYXC1). In these patients, the ciliary axoneme lacks ODA and IDA; as a result, they typically have completely immotile cilia (Fig. 4) (Knowles et al. 2013; Tarkar et al. 2013). Mutations have also been identified in genes that encode for radial spoke components, including RSPH1, RSPH4A, and RSPH9 (Fig. 4). When examined by transmission electron microscopy, the majority of cilia from these patients appear normal;

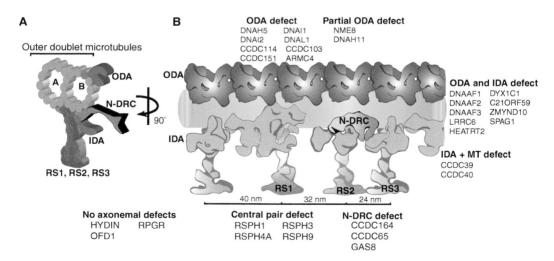

Figure 4. Mutations found in PCD patients and their association to ultrastructural axonemal defects. Schematic representation of the axoneme in (*A*) cross-sectional, and (*B*) the longitudinal views of the 96 nm repeat. Mutations in >30 genes have been identified in patients with PCD. The affected gene and the corresponding axoneme defect are listed.

a subset of ciliary axonemes display abnormalities of the central pair (Knowles et al. 2014). In these patients, the CBF is in the normal range but parameters related to the amplitude and velocity of cilia bend are affected; as a result, ciliary beat is abnormal with a circular beat pattern (Castleman et al. 2009; Knowles et al. 2014). Analysis of isolated cilia from a patient with RSPH1 mutations by cryoelectron tomography revealed that, although the RSPH1 mutation resulted in the loss of radial spoke heads 1 and 2, the third radial spoke head remained intact (Lin et al. 2014). Interestingly, patients with RSPH1 mutations have less severe clinical features of disease than typical PCD patients (Knowles et al. 2014). These data suggest that, although mutations in RSPH1 cause disease, the atypical circular beat pattern may be able to provide a low level of MCC. In addition, PCD may occur without discernible axonemal ultrastructural defects. For example, cilia from patients with mutations in genes that encode for HYDIN, a large central-pair apparatus protein, or DNAH11, an outer arm heavy chain, appeared normal by TEM (Fig. 4) (Knowles et al. 2012; Olbrich et al. 2012). However, immunofluorescence analysis have localized DNAH11 to the proximal region of the cilia and TEM tomography analysis of cilia from a patient carrying DNAH11 mutations detected a partial reduction of ODAs in the proximal but not distal regions (Dougherty et al. 2016). This new observation highlights the difficulties of diagnosing PCD by TEM, and shows the importance of using other techniques, including whole exome sequencing, to obtain a definitive diagnosis. Cilia from patients with DNAH11 mutations have a stiff but typically hyperkinetic beat frequency (Schwabe et al. 2008), while patients with HYDIN mutations have a stiff but typically reduced beat frequency. In these cases, although the axonemal structure appears normal and ciliary beat frequency ranges from immotile to hyperkinetic, the abnormal ciliary motion results in inefficient MCC. Future studies will no doubt uncover mutations in additional genes that cause PCD.

As expected, PCD patients show markedly reduced tracheobronchial clearance compared with healthy individuals (Walker et al. 2014; Munkholm et al. 2015). To compensate for the deficient mucociliary transport, patients with PCD rely almost entirely on cough clearance to remove secretions from their bronchial airways (Noone et al. 1999). Although the ability to clear mucus by coughing partially compensates for the lack of effective MCC, PCD patients suffer from chronic, life-long, respiratory disease, and additional research into new therapies is warranted.

CONCLUSIONS AND FUTURE PERSPECTIVES

In the last decade, important advances have been made in understanding the structure of the cilia as well as the regulation of ciliary activity. The standardization of techniques to measure MCC in humans in health and disease, the development of mouse models of PCD and other airway diseases, and the development of well-differentiated culture systems have yielded considerable knowledge about the role of cilia in MCC. The studies performed in patients with PCD highlight the crucial role that ciliary motility plays in MCC, and have identified many mutations that can affect ciliary function. However, there are still many important questions remaining, including (1) How are the cilia of the airway coordinated at both the cellular and organelle level? (2) What are the proteins localized at the tip of the cilia, and how do they interact with the mucous layer? (3) What is the contribution of the ciliated cells to the establishment and maintenance of the PCL? The collaborative efforts of many scientists from different fields will be required to uncover the answers to these questions.

REFERENCES

*Reference is also in this collection.

Al Jord A, Lemaitre AI, Delgehyr N, Faucourt M, Spassky N, Meunier A. 2014. Centriole amplification by mother and daughter centrioles differs in multiciliated cells. *Nature* **516**: 104–107.

Avidor-Reiss T, Maer AM, Koundakjian E, Polyanovsky A, Keil T, Subramaniam S, Zuker CS. 2004. Decoding cilia function: Defining specialized genes required for compartmentalized cilia biogenesis. *Cell* **117**: 527–539.

Becker KL, Silva OL. 1981. Hypothesis: The bronchial Kulchitsky (K) cell as a source of humoral biologic activity. *Med Hypotheses* 7: 943–949.

Bennett WD. 2002. Effect of β-adrenergic agonists on mucociliary clearance. *J Allergy Clin Immunol* 110: S291–S297.

Bennett WD, Laube BL, Corcoran T, Zeman K, Sharpless G, Thomas K, Wu J, Mogayzel PJ Jr, Pilewski J, Donaldson S. 2013. Multisite comparison of mucociliary and cough clearance measures using standardized methods. *J Aerosol Med Pulm Drug Deliv* 26: 157–164.

Bhashyam AR, Mogayzel PJ Jr, McGrath-Morrow S, Neptune E, Malinina A, Fox J, Laube BL. 2012. A pilot study to examine the effect of chronic treatment with immunosuppressive drugs on mucociliary clearance in a vagotomized murine model. *PloS ONE* 7: e45312.

Boon M, Wallmeier J, Ma L, Loges NT, Jaspers M, Olbrich H, Dougherty GW, Raidt J, Werner C, Amirav I, et al. 2014. MCIDAS mutations result in a mucociliary clearance disorder with reduced generation of multiple motile cilia. *Nat Commun* doi: 10.1038/ncomms5418.

Brekman A, Walters MS, Tilley AE, Crystal RG. 2014. FOXJ1 prevents cilia growth inhibition by cigarette smoke in human airway epithelium in vitro. *Am J Respir Cell Mol Biol* 51: 688–700.

Brokaw CJ, Kamiya R. 1987. Bending patterns of *Chlamydomonas* flagella. IV: Mutants with defects in inner and outer dynein arms indicate differences in dynein arm function. *Cell Motil Cytoskeleton* 8: 68–75.

Brooks ER, Wallingford JB. 2014. Multiciliated cells: A review. *Curr Biol* 24: R973–R982.

Button B, Boucher RC. 2008. Role of mechanical stress in regulating airway surface hydration and mucus clearance rates. *Respir Physiol Neurobiol* 163: 189–201.

Button B, Picher M, Boucher RC. 2007. Differential effects of cyclic and constant stress on ATP release and mucociliary transport by human airway epithelia. *J Physiol* 580: 577–592.

Button B, Cai LH, Ehre C, Kesimer M, Hill DB, Sheehan JK, Boucher RC, Rubinstein M. 2012. A periciliary brush promotes the lung health by separating the mucus layer from airway epithelia. *Science* 337: 937–941.

Button B, Okada SF, Frederick CB, Thelin WR, Boucher RC. 2013. Mechanosensitive ATP release maintains proper mucus hydration of airways. *Sci Signal* 6: ra46.

Calkovska A. 2000. Pulmonary surfactant in the respiratory tract. *Cesk Fysiol* 49: 145–151.

Castleman VH, Romio L, Chodhari R, Hirst RA, de Castro SC, Parker KA, Ybot-Gonzalez P, Emes RD, Wilson SW, Wallis C, et al. 2009. Mutations in radial spoke head protein genes RSPH9 and RSPH4A cause primary ciliary dyskinesia with central-microtubular-pair abnormalities. *Am J Hum Genet* 84: 197–209.

Chen Y, Thai P, Zhao YH, Ho YS, DeSouza MM, Wu R. 2003. Stimulation of airway mucin gene expression by interleukin (IL)-17 through IL-6 paracrine/autocrine loop. *J Biol Chem* 278: 17036–17043.

Chilvers MA, O'Callaghan C. 2000. Analysis of ciliary beat pattern and beat frequency using digital high speed imaging: Comparison with the photomultiplier and photodiode methods. *Thorax* 55: 314–317.

Chinoy MR. 2003. Lung growth and development. *Front Biosci* 8: d392–d415.

Clare DK, Magescas J, Piolot T, Dumoux M, Vesque C, Pichard E, Dang T, Duvauchelle B, Poirier F, Delacour D. 2014. Basal foot MTOC organizes pillar MTs required for coordination of beating cilia. *Nat Commun* 5: 4888.

Clary-Meinesz CF, Cosson J, Huitorel P, Blaive B. 1992. Temperature effect on the ciliary beat frequency of human nasal and tracheal ciliated cells. *Biol Cell* 76: 335–338.

Clary-Meinesz C, Mouroux J, Cosson J, Huitorel P, Blaive B. 1998. Influence of external pH on ciliary beat frequency in human bronchi and bronchioles. *Eur Respir J* 11: 330–333.

Cooper JL, Quinton PM, Ballard ST. 2013. Mucociliary transport in porcine trachea: Differential effects of inhibiting chloride and bicarbonate secretion. *Am J Physiol Lung Cell Mol Physiol* 304: L184–L190.

Davis CW, Randell SH. 2001. Airway goblet and mucous cells: Identical, similar, or different? In *Cilia and mucus: From development to respiratory defense* (ed. Salatheed M), pp. 195–210. Marcel Dekker, New York.

Daviskas E, Anderson SD, Gonda I, Chan HK, Cook P, Fulton R. 1995. Changes in mucociliary clearance during and after isocapnic hyperventilation in asthmatic and healthy subjects. *Eur Respir J* 8: 742–751.

Deutsch GH, Pinar H. 2002. Prenatal lung development. In *Chronic obstructive lung diseases* (ed. Voelkel NCF, MacNee W), pp. xi, 428. BC Decker, Hamilton, Ontario

Dickson RP, Huffnagle GB. 2015. The lung microbiome: New principles for respiratory bacteriology in health and disease. *PLoS Pathog* 11: e1004923.

Dougherty GW, Loges NT, Klinkenbusch JA, Olbrich H, Pennekamp P, Menchen T, Raidt J, Wallmeier J, Werner C, Westermann C, et al. 2016. DNAH11 localization in the proximal region of respiratory cilia defines distinct outer dynein arm complexes. *Am J Respir Cell Mol Biol* 55: 213–224.

Edgar R, Mazor Y, Rinon A, Blumenthal J, Golan Y, Buzhor E, Livnat I, Ben-Ari S, Lieder I, Shitrit A, et al. 2013. LifeMap discovery: The embryonic development, stem cells, and regenerative medicine research portal. *PloS ONE* 8: e66629.

Elliott MK, Sisson JH, Wyatt TA. 2007. Effects of cigarette smoke and alcohol on ciliated tracheal epithelium and inflammatory cell recruitment. *Am J Respir Cell Mol Biol* 36: 452–459.

Fan H, Bobek LA. 2010. Regulation of human MUC7 mucin gene expression by cigarette smoke extract or cigarette smoke and pseudomonas aeruginosa lipopolysaccharide in human airway epithelial cells and in MUC7 transgenic mice. *Open Respir Med J* 4: 63–70.

Fliegauf M, Olbrich H, Horvath J, Wildhaber JH, Zariwala MA, Kennedy M, Knowles MR, Omran H. 2005. Mislocalization of DNAH5 and DNAH9 in respiratory cells from patients with primary ciliary dyskinesia. *Am J Respir Crit Care Med* 171: 1343–1349.

Fliegauf M, Benzing T, Omran H. 2007. When cilia go bad: Cilia defects and ciliopathies. *Nat Rev Mol Cell Biol* 8: 880–893.

Foster WM, Walters DM, Longphre M, Macri K, Miller LM. 2001. Methodology for the measurement of mucociliary function in the mouse by scintigraphy. *J Appl Physiol* **90:** 1111–1117.

Francis R, Lo C. 2013. Ex vivo method for high resolution imaging of cilia motility in rodent airway epithelia. *J Vis Exp* doi: 10.3791/50343.

Francis RJ, Chatterjee B, Loges NT, Zentgraf H, Omran H, Lo CW. 2009. Initiation and maturation of cilia-generated flow in newborn and postnatal mouse airway. *Am J Physiol Lung Cell Mol Physiol* **296:** L1067–L1075.

Gaillard DA, Lallement AV, Petit AF, Puchelle ES. 1989. In vivo ciliogenesis in human fetal tracheal epithelium. *Am J Anat* **185:** 415–428.

Gensch E, Gallup M, Sucher A, Li D, Gebremichael A, Lemjabbar H, Mengistab A, Dasari V, Hotchkiss J, Harkema J, et al. 2004. Tobacco smoke control of mucin production in lung cells requires oxygen radicals AP-1 and JNK. *J Biol Chem* **279:** 39085–39093.

Girod de Bentzmann S, Pierrot D, Fuchey C, Zahm JM, Morancais JL, Puchelle E. 1993. Distearoyl phosphatidylglycerol liposomes improve surface and transport properties of CF mucus. *Eur Respir J* **6:** 1156–1161.

Gray TE, Guzman K, Davis CW, Abdullah LH, Nettesheim P. 1996. Mucociliary differentiation of serially passaged normal human tracheobronchial epithelial cells. *Am J Respi Cell Mol Biol* **14:** 104–112.

Groneberg DA, Eynott PR, Oates T, Lim S, Wu R, Carlstedt I, Nicholson AG, Chung KF. 2002. Expression of MUC5AC and MUC5B mucins in normal and cystic fibrosis lung. *Respir Med* **96:** 81–86.

Grubb BR, Livraghi-Butrico A, Rogers TD, Yin W, Ostrowski LE. 2016. Reduced mucociliary clearance in old mice is associated with a decrease in Muc5B mucin. *Am J Physiol Lung Cell Mol Physiol* doi: 10.1152/ajplung.00015.2016.

Hamed R, Fiegel J. 2014. Synthetic tracheal mucus with native rheological and surface tension properties. *J Biomed Mater Res A* **102:** 1788–1798.

Heuser T, Raytchev M, Krell J, Porter ME, Nicastro D. 2009. The dynein regulatory complex is the nexin link and a major regulatory node in cilia and flagella. *J Cell Biol* **187:** 921–933.

Hoegger MJ, Awadalla M, Namati E, Itani OA, Fischer AJ, Tucker AJ, Adam RJ, McLennan G, Hoffman EA, Stoltz DA, et al. 2014. Assessing mucociliary transport of single particles in vivo shows variable speed and preference for the ventral trachea in newborn pigs. *Proc Natl Acad Sci* **111:** 2355–2360.

Hofmann W, Asgharian B. 2003. The effect of lung structure on mucociliary clearance and particle retention in human and rat lungs. *Toxicol Sci* **73:** 448–456.

Holzbaur EL, Vallee RB. 1994. Dyneins: Molecular structure and cellular function. *Annu Rev Cell Biol* **10:** 339–372.

Houtmeyers E, Gosselink R, Gayan-Ramirez G, Decramer M. 1999. Regulation of mucociliary clearance in health and disease. *Eur Respir J* **13:** 1177–1188.

* Ishikawa T. 2016. Axoneme structure from motile cilia. *Cold Spring Harb Perspect Biol* doi: 10.1101/cshperspect. a028076.

Kerem E, Bistritzer T, Hanukoglu A, Hofmann T, Zhou Z, Bennett W, MacLaughlin E, Barker P, Nash M, Quittell L, et al. 1999. Pulmonary epithelial sodium-channel dys-

function and excess airway liquid in pseudohypoaldosteronism. *N Engl J Med* **341:** 156–162.

Kikkawa M. 2013. Big steps toward understanding dynein. *J Cell Biol* **202:** 15–23.

Kim CH, Park HW, Kim K, Yoon JH. 2004. Early development of the nose in human embryos: A stereomicroscopic and histologic analysis. *Laryngoscope* **114:** 1791–1800.

King M. 2006. Physiology of mucus clearance. *Paediatr Respir Rev* **7:** S212–S214.

Knowles MR, Boucher RC. 2002. Mucus clearance as a primary innate defense mechanism for mammalian airways. *J Clin Invest* **109:** 571–577.

Knowles MR, Leigh MW, Carson JL, Davis SD, Dell SD, Ferkol TW, Olivier KN, Sagel SD, Rosenfeld M, Burns KA, et al. 2012. Mutations of DNAH11 in patients with primary ciliary dyskinesia with normal ciliary ultrastructure. *Thorax* **67:** 433–441.

Knowles MR, Ostrowski LE, Loges NT, Hurd T, Leigh MW, Huang L, Wolf WE, Carson JL, Hazucha MJ, Yin W, et al. 2013. Mutations in SPAG1 cause primary ciliary dyskinesia associated with defective outer and inner dynein arms. *Am J Hum Genet* **93:** 711–720.

Knowles MR, Ostrowski LE, Leigh MW, Sears PR, Davis SD, Wolf WE, Hazucha MJ, Carson JL, Olivier KN, Sagel SD, et al. 2014. Mutations in RSPH1 cause primary ciliary dyskinesia with a unique clinical and ciliary phenotype. *Am J Respir Crit Care Med* **189:** 707–717.

Konishi S, Gotoh S, Tateishi K, Yamamoto Y, Korogi Y, Nagasaki T, Matsumoto H, Muro S, Hirai T, Ito I, et al. 2016. Directed induction of functional multi-ciliated cells in proximal airway epithelial spheroids from human pluripotent stem cells. *Stem Cell Rep* **6:** 18–25.

Kozminski KG, Johnson KA, Forscher P, Rosenbaum JL. 1993. A motility in the eukaryotic flagellum unrelated to flagellar beating. *Proc Natl Acad Sci* **90:** 5519–5523.

Kuyper LM, Paré PD, Hogg JC, Lambert RK, Ionescu D, Woods R, Bai TR. 2003. Characterization of airway plugging in fatal asthma. *Am J Med* **115:** 6–11.

Lafkas D, Shelton A, Chiu C, de Leon Boenig G, Chen Y, Stawicki SS, Siltanen C, Reichelt M, Zhou M, Wu X, et al. 2015. Therapeutic antibodies reveal Notch control of transdifferentiation in the adult lung. *Nature* **528:** 127–131.

Lay JC, Stang MR, Fisher PE, Yankaskas JR, Bennett WD. 2003. Airway retention of materials of different solubility following local intrabronchial deposition in dogs. *J Aerosol Med* **16:** 153–166.

Lin J, Yin W, Smith MC, Song K, Leigh MW, Zariwala MA, Knowles MR, Ostrowski LE, Nicastro D. 2014. Cryo-electron tomography reveals ciliary defects underlying human RSPH1 primary ciliary dyskinesia. *Nat Commun* doi: 10.1038/ncomms6727.

Lindskog C, Fagerberg L, Hallström B, Edlund K, Hellwig B, Rahnenführer J, Kampf C, Uhlén M, Pontén F, Micke P. 2014. The lung-specific proteome defined by integration of transcriptomics and antibody-based profiling. *FASEB J* **28:** 5184–5196.

Look DC, Walter MJ, Williamson MR, Pang L, You Y, Sreshta JN, Johnson JE, Zander DS, Brody SL. 2001. Effects of paramyxoviral infection on airway epithelial cell Foxj1 expression, ciliogenesis, and mucociliary function. *Am J Pathol* **159:** 2055–2069.

Maggi CA, Giachetti A, Dey RD, Said SI. 1995. Neuropeptides as regulators of airway function: Vasoactive intestinal peptide and the tachykinins. *Physiol Rev* **75:** 277–322.

Matsui H, Grubb BR, Tarran R, Randell SH, Gatzy JT, Davis CW, Boucher RC. 1998a. Evidence for periciliary liquid layer depletion, not abnormal ion composition, in the pathogenesis of cystic fibrosis airways disease. *Cell* **95:** 1005–1015.

Matsui H, Randell SH, Peretti SW, Davis CW, Boucher RC. 1998b. Coordinated clearance of periciliary liquid and mucus from airway surfaces. *J Clin Invest* **102:** 1125–1131.

Matthews LW, Spector S, Lemm J, Potter JL. 1963. Studies on pulmonary secretions. *Am Rev Respir Dis* **88:** 199–204.

Mercer RR, Russell ML, Roggli VL, Crapo JD. 1994. Cell number and distribution in human and rat airways. *Am J Respir Cell Mol Biol* **10:** 613–624.

Milara J, Armengot M, Bañuls P, Tenor H, Beume R, Artigues E, Cortijo J. 2012. Roflumilast *N*-oxide, a PDE4 inhibitor, improves cilia motility and ciliated human bronchial epithelial cells compromised by cigarette smoke in vitro. *Br J Pharmacol* **166:** 2243–2262.

Morse DM, Smullen JL, Davis CW. 2001. Differential effects of UTP, ATP, and adenosine on ciliary activity of human nasal epithelial cells. *Am J Physiol Cell Physiol* **280:** C1485–C1497.

Moscoso GJ, Driver M, Codd J, Whimster WF. 1988. The morphology of ciliogenesis in the developing fetal human respiratory epithelium. *Pathol Res Pract* **183:** 403–411.

Munkholm M, Nielsen KG, Mortensen J. 2015. Clinical value of measurement of pulmonary radioaerosol mucociliary clearance in the work up of primary ciliary dyskinesia. *EJNMMI Res* **5:** 118.

Noone PG, Bennett WD, Regnis JA, Zeman KL, Carson JL, King M, Boucher RC, Knowles MR. 1999. Effect of aerosolized uridine-5′-triphosphate on airway clearance with cough in patients with primary ciliary dyskinesia. *Am J Respir Crit Care Med* **160:** 144–149.

Okubo T, Knoepfler PS, Eisenman RN, Hogan BLM. 2005. Nmyc plays an essential role during lung development as a dosage-sensitive regulator of progenitor cell proliferation and differentiation. *Development* **132:** 1363–1374.

Olbrich H, Schmidts M, Werner C, Onoufriadis A, Loges NT, Raidt J, Banki NF, Shoemark A, Burgoyne T, Al Turki S, et al. 2012. Recessive HYDIN mutations cause primary ciliary dyskinesia without randomization of left–right body asymmetry. *Am J Hum Genet* **91:** 672–684.

Oozawa H, Kimura H, Noda T, Hamada K, Morimoto T, Majima Y. 2012. Effect of prehydration on nasal mucociliary clearance in low relative humidity. *Auris Nasus Larynx* **39:** 48–52.

Ostrowski LE. 2002. Ciliogenesis of airway epithelium. In *Basic mechanisms of pediatric respiratory disease* (ed. Haddad GG, Abman SH, Chernick V). BC Decker, Hamilton, ON, Canada.

Ostrowski LE, Blackburn K, Radde KM, Moyer MB, Schlatzer DM, Moseley A, Boucher RC. 2002. A proteomic analysis of human cilia: Identification of novel components. *Mol Cell Proteomics* **1:** 451–465.

Ostrowski LE, Yin W, Rogers TD, Busalacchi KB, Chua M, O'Neal WK, Grubb BR. 2010. Conditional deletion of dnaic1 in a murine model of primary ciliary dyskinesia causes chronic rhinosinusitis. *Am J Respir Cell Mol Biol* **43:** 55–63.

O'Toole ET, Giddings TH Jr, Porter ME, Ostrowski LE. 2012. Computer-assisted image analysis of human cilia and *Chlamydomonas* flagella reveals both similarities and differences in axoneme structure. *Cytoskeleton (Hoboken, NJ)* **69:** 577–590.

Panski B. 1982. *Review of medical embryology.* Embryome Sciences, Alameda, CA.

Pazour GJ, Agrin N, Walker BL, Witman GB. 2006. Identification of predicted human outer dynein arm genes: Candidates for primary ciliary dyskinesia genes. *J Med Genet* **43:** 62–73.

Pennarun G, Escudier E, Chapelin C, Bridoux AM, Cacheux V, Roger G, Clement A, Goossens M, Amselem S, Duriez B. 1999. Loss-of-function mutations in a human gene related to *Chlamydomonas reinhardtii* dynein IC78 result in primary ciliary dyskinesia. *Am J Hum Genet* **65:** 1508–1519.

Porter ME, Sale WS. 2000. The 9+2 axoneme anchors multiple inner arm dyneins and a network of kinases and phosphatases that control motility. *J Cell Biol* **151:** F37–F42.

Raidt J, Wallmeier J, Hjeij R, Onnebrink JG, Pennekamp P, Loges NT, Olbrich H, Häffner K, Dougherty GW, Omran H, et al. 2014. Ciliary beat pattern and frequency in genetic variants of primary ciliary dyskinesia. *Eur Respir J* **44:** 1579–1588.

Ramos FL, Krahnke JS, Kim V. 2014. Clinical issues of mucus accumulation in COPD. *Int J Chron Obstruct Pulmon Dis* **9:** 139–150.

Randell SH, Boucher RC. 2006. Effective mucus clearance is essential for respiratory health. *Am J Respir Cell Mol Biol* **35:** 20–28.

Rawlins EL, Hogan BLM. 2008. Ciliated epithelial cell lifespan in the mouse trachea and lung. *Am J Physiol Lung Cell Mol Physiol* **295:** L231–L234.

Reid LM, Jones R. 1980. Mucous membrane of respiratory epithelium. *Environ Health Perspect* **35:** 113–120.

Roberts Anthony J, Malkova B, Walker Matt L, Sakakibara H, Numata N, Kon T, Ohkura R, Edwards Thomas A, Knight Peter J, Sutoh K, et al. 2012. ATP-driven remodeling of the linker domain in the dynein motor. *Structure* **20:** 1670–1680.

Rock JR, Onaitis MW, Rawlins EL, Lu Y, Clark CP, Xue Y, Randell SH, Hogan BLM. 2009. Basal cells as stem cells of the mouse trachea and human airway epithelium. *Proc Natl Acad Sci* **106:** 12771–12775.

Rose MC, Voynow JA. 2005. Respiratory tract mucin genes and mucin glycoproteins in health and disease. *Physiol Rev* **86:** 245–278.

Rose MC, Voynow JA. 2006. Respiratory tract mucin genes and mucin glycoproteins in health and disease. *Physiol Rev* **86:** 245–278.

Rosenbaum JL, Witman GB. 2002. Intraflagellar transport. *Nat Rev Mol Cell Biol* **3:** 813–825.

Cite this article as *Cold Spring Harb Perspect Biol* doi: 10.1101/cshperspect.a028241

Sabater JR, Mao YM, Shaffer C, James MK, O'Riordan TG, Abraham WM. 1999. Aerosolization of P2Y$_2$-receptor agonists enhances mucociliary clearance in sheep. *J Appl Physiol* **87:** 2191–2196.

Sanderson MJ, Sleigh MA. 1981. Ciliary activity of cultured rabbit tracheal epithelium: Beat pattern and metachrony. *J Cell Sci* **47:** 331–347.

Satir P, Christensen ST. 2007. Overview of structure and function of mammalian cilia. *Annu Rev Physiol* **69:** 377–400.

Satir P, Sleigh MA. 1990. The physiology of cilia and mucociliary interactions. *Annu Rev Physiol* **52:** 137–155.

Schwabe GC, Hoffmann K, Loges NT, Birker D, Rossier C, de Santi MM, Olbrich H, Fliegauf M, Failly M, Liebers U, et al. 2008. Primary ciliary dyskinesia associated with normal axoneme ultrastructure is caused by DNAH11 mutations. *Hum Mutat* **29:** 289–298.

Sears PR, Yin WN, Ostrowski LE. 2015. Continuous mucociliary transport by primary human airway epithelial cells in vitro. *Am J Physiol Lung Cell Mol Physiol* **309:** L99–L108.

Shah AS, Ben-Shahar Y, Moninger TO, Kline JN, Welsh MJ. 2009. Motile cilia of human airway epithelia are chemosensory. *Science* **325:** 1131–1134.

Shapiro AJ, Zariwala MA, Ferkol T, Davis SD, Sagel SD, Dell SD, Rosenfeld M, Olivier KN, Milla C, Daniel SJ, et al. 2016. Diagnosis, monitoring, and treatment of primary ciliary dyskinesia: PCD foundation consensus recommendations based on state of the art review. *Pediatr Pulmonol* **51:** 115–132.

Silflow CD, Lefebvre PA. 2001. Assembly and motility of eukaryotic cilia and flagella. Lessons from *Chlamydomonas reinhardtii*. *Plant Physiol* **127:** 1500–1507.

Simet SM, Sisson JH, Pavlik JA, DeVasure JM, Boyer C, Liu X, Kawasaki S, Sharp JG, Rennard SI, Wyatt TA. 2010. Long-term cigarette smoke exposure in a mouse model of ciliated epithelial cell function. *Am J Respir Cell Mol Biol* **43:** 635–640.

Sleigh MA, Blake JR, Liron N. 1988. The propulsion of mucus by cilia. *Am Rev Respir Dis* **137:** 726–741.

Sorokin SP. 1968. Reconstructions of centriole formation and ciliogenesis in mammalian lungs. *J Cell Sci* **3:** 207–230.

Stubbs JL, Vladar EK, Axelrod JD, Kintner C. 2012. Multicilin promotes centriole assembly and ciliogenesis during multiciliate cell differentiation. *Nat Cell Biol* **14:** 140–147.

Sutto Z, Conner GE, Salathe M. 2004. Regulation of human airway ciliary beat frequency by intracellular pH. *J Physiol* **560:** 519–532.

Tarkar A, Loges NT, Slagle CE, Francis R, Dougherty GW, Tamayo JV, Shook B, Cantino M, Schwartz D, Jahnke C, et al. 2013. DYX1C1 is required for axonemal dynein assembly and ciliary motility. *Nat Genet* **45:** 995–1003.

Tarran R, Button B, Picher M, Paradiso AM, Ribeiro CM, Lazarowski ER, Zhang L, Collins PL, Pickles RJ, Fredberg JJ, et al. 2005. Normal and cystic fibrosis airway surface liquid homeostasis: The effects of phasic shear stress and viral infections. *J Biol Chem* **280:** 35751–35759.

Tos M. 1966. Development of the tracheal glands in man. Number, density, structure, shape, and distribution of mucous glands elucidated by quantitative studies of whole mounts. *Acta Pathol Microbiol Scand* **68.**

Tsao PN, Chen F, Izvolsky KI, Walker J, Kukuruzinska MA, Lu J, Cardoso WV. 2008. γ-Secretase activation of notch signaling regulates the balance of proximal and distal fates in progenitor cells of the developing lung. *J Biol Chem* **283:** 29532–29544.

Vladar EK, Mitchell BJ. 2016. It's a family act: The geminin triplets take center stage in motile ciliogenesis. *EMBO J* **35:** 904–906.

Vladar EK, Lee YL, Stearns T, Axelrod JD. 2015. Observing planar cell polarity in multiciliated mouse airway epithelial cells. *Method Cell Biol* **127:** 37–54.

Walker WT, Young A, Bennett M, Guy M, Carroll M, Fleming J, Conway J, Lucas JS. 2014. Pulmonary radioaerosol mucociliary clearance in primary ciliary dyskinesia. *Eur Respir J* **44:** 533–535.

Wanner A, Salathe M, O'Riordan TG. 1996. Mucociliary clearance in the airways. *Am J Respir Crit Care Med* **154:** 1868–1902.

Werner ME, Hwang P, Huisman F, Taborek P, Yu CC, Mitchell BJ. 2011. Actin and microtubules drive differential aspects of planar cell polarity in multiciliated cells. *J Cell Biol* **195:** 19–26.

Williams R, Rankin N, Smith T, Galler D, Seakins P. 1996. Relationship between the humidity and temperature of inspired gas and the function of the airway mucosa. *Crit Care Med* **24:** 1920–1929.

Wine JJ, Joo NS. 2004. Submucosal glands and airway defense. *Proc Am Thorac Soc* **1:** 47–53.

Wyatt TA. 2015. Cyclic GMP and cilia motility. *Cells* **4:** 315–330.

Yagi T. 2009. Bioinformatic approaches to dynein heavy chain classification. In *Methods in cell biology* (ed. Stephen MK, Gregory JP), pp. 1–9. Academic, New York.

Yeates DB, Pitt BR, Spektor DM, Karron GA, Albert RE. 1981. Coordination of mucociliary transport in human trachea and intrapulmonary airways. *J Appl Physiol* **51:** 1057–1064.

You Y, Huang T, Richer EJ, Schmidt JEH, Zabner J, Borok Z, Brody SL. 2004. Role of f-box factor foxj1 in differentiation of ciliated airway epithelial cells. *Am J Physiol Lung Cell Mol Physiol* **286:** L650–L657.

Zariwala MA, Knowles MR, Leigh MW. 2007. Primary ciliary dyskinesia. In *Gene reviews* (ed. Pagon RA, Adam MP, Ardinger HH, Wallace SE, Amemiya A, Bean LJH, Bird TD, Fong CT, Mefford HC, Smith RJH, et al.), pp. 1993–2016. University of Washington, Seattle, WA.

Zhang L, Button B, Gabriel SE, Burkett S, Yan Y, Skiadopoulos MH, Dang YL, Vogel LN, McKay T, Mengos A, et al. 2009. CFTR delivery to 25% of surface epithelial cells restores normal rates of mucus transport to human cystic fibrosis airway epithelium. *PLoS Biol* **7:** e1000155.

Zhou F, Narasimhan V, Shboul M, Chong Yan L, Reversade B, Roy S. 2015. GMNC is a master regulator of the multiciliated cell differentiation program. *Curr Biol* **25:** 3267–3273.

Zhu Y, Ehre C, Abdullah LH, Sheehan JK, Roy M, Evans CM, Dickey BF, Davis CW. 2008. Munc13-2$^{-/-}$ baseline secretion defect reveals source of oligomeric mucins in mouse airways. *J Physiol* **586:** 1977–1992.

Discovery, Diagnosis, and Etiology of Craniofacial Ciliopathies

Elizabeth N. Schock and Samantha A. Brugmann

Division of Plastic Surgery, Department of Surgery, and Division of Developmental Biology, Department of Pediatrics, Cincinnati Children's Hospital Medical Center, Cincinnati, Ohio 45229

Correspondence: samantha.brugmann@cchmc.org

Seventy-five percent of congenital disorders present with some form of craniofacial malformation. The frequency and severity of these malformations makes understanding the etiological basis crucial for diagnosis and treatment. A significant link between craniofacial malformations and primary cilia arose several years ago with the determination that ~30% of ciliopathies could be primarily defined by their craniofacial phenotype. The link between the cilium and the face has proven significant, as several new "craniofacial ciliopathies" have recently been diagnosed. Herein, we reevaluate public disease databases, report several new craniofacial ciliopathies, and propose several "predicted" craniofacial ciliopathies. Furthermore, we discuss why the craniofacial complex is so sensitive to ciliopathic dysfunction, addressing tissue-specific functions of the cilium as well as its role in signal transduction relevant to craniofacial development. As a whole, these analyses suggest a characteristic facial phenotype associated with craniofacial ciliopathies that can perhaps be used for rapid discovery and diagnosis of similar disorders in the future.

In 2009, Baker and Beales published a review characterizing human ciliopathies. This work detailed known, likely, and possible ciliopathies based on the presentation of nine core phenotypic features: retinitis pigmentosa, polydactyly, situs inversus, mental retardation, agenesis of the corpus callosum, Dandy–Walker malformation, posterior encephalocele, renal cystic disease, and hepatic disease. Using these core characteristics, they identified 15 known ciliopathies and 88 potential ciliopathies (Baker and Beales 2009). Two years later, we analyzed the ciliopathies put forth by Baker and Beales and determined that ~30% of known or proposed ciliopathies were primarily defined by their cra-

niofacial phenotype, which included cleft lip/ palate, increased/decreased midfacial width (hyper-/hypotelorism), small lower jaw (micrognathia), and prematurely fused cranial sutures (craniosynostosis) (Zaghloul and Brugmann 2011).

Much has been learned about ciliopathies in the last 5 years. In this review, we reexamine what is known regarding craniofacial ciliopathies, those ciliopathies defined by their craniofacial phenotype. To do so, we modified the original search methods used by Baker and Beales. We searched the Online Mendelian Inheritance in Man (OMIM) database for any disease that met two requirements: presentation of one of the ciliopathic craniofacial phenotypes (cleft lip/palate, hyper-/hypotelorism, micrognathia, or craniosynostosis) and presentation of two core ciliopathic phenotypes (excluding renal cystic disease and hepatic disease) (Table 1). We excluded searches for "cleft lip/palate AND mental retardation" and "hypertelorism AND mental retardation" because these searches resulted in exceedingly large lists that are likely not specific to ciliopathies. In total, we identified more than 400 possible ciliopathies using this unbiased method. We then manually inspected this list of diseases and excluded any syndrome caused by large-scale chromosomal duplication, deletion, or rearrangements, as a specific genetic etiology would be difficult to determine. The genetic cause of the remaining diseases, if known, was examined and lists of known and predicted ciliopathies were generated based on gene function (Tables 2 and 3). Unlike Baker and Beales, who included transcription factors thought to be downstream targets of ciliary signaling, we did not include these, as ciliary function is not dependent on these genes.

Our search identified 26 known ciliopathies defined by craniofacial phenotypes (Table 2). We confirmed 20 previously known craniofacial ciliopathies and reclassified five craniofacial ciliopathies formerly categorized by Baker and Beales as "likely" ciliopathies. Furthermore, our analysis identified one new craniofacial ciliopathy, Hydrolethalus syndrome 2, not previously reported in any search. Within our updated list of known ciliopathies, we found that cleft lip/palate was the most common craniofacial phenotype (Fig. 1A,B), followed by hypertelorism (Fig. 1C), micrognathia (Fig. 1D), craniosynostosis (Fig. 1E), and hypotelorism (Fig. 1C). Interestingly, the most prevalent craniofacial phenotypes were directly associated either widening or deficiencies of the midface (cleft lip/palate, hypertelorism, micrognathia). Finally, we cross-referenced our list of 26 known craniofacial ciliopathies against the seven core ciliopathic phenotypes. This analysis revealed that, within craniofacial ciliopathies, polydactyly most commonly presented in combination with craniofacial phenotypes (Fig. 1F), followed

by phenotypes associated with the brain: agenesis of the corpus callosum, Dandy–Walker malformation, posterior encephalocele, and mental retardation.

In addition to our list of "known" craniofacial ciliopathies, we also used our analyses to generate a list of "predicted" craniofacial ciliopathies, disorders that meet the phenotypic criteria, yet do not have a known genetic cause linked to the structure or function of the primary cilia. We identified 25 predicted craniofacial ciliopathies (Table 3). Of the 25 disorders, 13 were previously predicted (Baker and Beales 2009), whereas the remaining 12 were newly classified as predicted craniofacial ciliopathies. Many of these diseases share a striking phenotypic resemblance with known craniofacial ciliopathies, presenting with severe micrognathia, hypertelorism, and cleft lip/palate.

Why Is the Craniofacial Complex So Sensitive to Ciliary Dysfunction?

Primary cilia are ubiquitous organelles that are present on almost every cell type throughout development, yet the phenotypes of ciliopathies are not identical and can be confined to individual organ systems. The craniofacial complex is the primary organ system affected in almost 30% of ciliopathies. This metric begs the question—what makes the face so sensitive to ciliary dysfunction? There are perhaps two main reasons the face is so sensitive to defects in the primary cilium. First, the face is formed by intricate tissue–tissue interactions between cranial neural crest cells (CNCCs), neuroectoderm, facial ectoderm, and pharyngeal endoderm. Thus, if any one of these tissues requires ciliary function for proper development, the face, as a whole, will be negatively impacted. Second, several of the signaling pathways essential for proper craniofacial patterning require the cilium for signal transduction. In the following sections, we will explain the tissues and signaling pathways required for craniofacial development. Furthermore, we will comment on how ciliary defects can impact facial development.

Table 1. Pairwise combinations of phenotypes

	Cleft lip and/or palate	Hypertelorism	Hypotelorism	Micrognathia	Craniosynostosis
Retinitis pigmentosa + polydactyly	1	2	3	4	5
Retinitis pigmentosa + mental retardation	6[a]	7[a]	8	9	10
Retinitis pigmentosa + situs inversus	11	12	13	14	15
Retinitis pigmentosa + agenesis of corpus callosum	16	17	18	19	20
Retinitis pigmentosa + Dandy–Walker malformation	21	22	23	24	25
Retinitis pigmentosa + posterior encephalocele	26	27	28	29	30
Polydactyly + mental retardation	31[a]	32[a]	33	34	35
Polydactyly + situs inversus	36	37	38	39	40
Polydactyly + agenesis of corpus callosum	41	42	43	44	45
Polydactyly + Dandy–Walker malformation	46	47	48	49	50
Polydactyly + posterior encephalocele	51	52	53	54	55
Mental retardation + situs inversus	56[a]	57[a]	58	59	60
Mental retardation + agenesis of corpus callosum	61[a]	62[a]	63	64	64
Mental retardation + Dandy–Walker malformation	66[a]	67[a]	68	69	70
Mental retardation + posterior encephalocele	71[a]	72[a]	73	74	75
Situs inversus + agenesis of corpus callosum	76	77	78	79	80
Situs inversus + Dandy–Walker malformation	81	82	83	84	85
Situs inversus + posterior encephalocele	86	87	88	89	90
Agenesis of corpus callosum + Dandy–Walker malformation	91	92	93	94	95
Agenesis of corpus callosum + posterior encephalocele	96	97	98	99	100
Dandy–Walker malformation + posterior encephalocele	101	102	103	104	105

[a]These comparisons could not be evaluated because initial searches for "cleft lip and/or palate AND mental retardation" and "hypertelorism AND mental retardation" were not completed.

Table 2. Known craniofacial ciliopathies

Syndrome	Genes	Phenotypes	Classification in Baker and Beales
Bardet−Beidl syndrome 1	BBS1	1, 11, 36	Known ciliopathy
Cranioectodermal dysplasia	IFT43	4, 5	Known ciliopathy
Ellis−van Crevald syndrome	EVC, EVC2	36, 46, 81	Known ciliopathy
Joubert syndrome 1	INPP5E	34, 44, 49, 69, 94	Known ciliopathy
Joubert syndrome 2	TMEM216	35, 41, 42, 45, 46, 47, 51, 52, 53, 55, 65, 70, 75, 91, 92, 95, 96, 97, 100, 101, 102, 105	Known ciliopathy[a]
Joubert syndrome 14	TMEM237	47	Known ciliopathy[a]
Meckel syndrome, type I	MKS1	41, 42, 43, 44, 46, 47, 48, 49, 51, 52, 53, 54 91, 92, 93, 94 96, 97, 98, 99, 101, 102, 103, 104	Known ciliopathy
Meckel syndrome, type II	TMEM216	46	Known ciliopathy[a]
Meckel syndrome, type III	TMEM67	41, 46, 91	Known ciliopathy[a]
Meckel syndrome, type IV	CEP290	46	Known ciliopathy[a]
Oral−facial−digital syndrome I	OFD1	34, 41, 44, 64	Known ciliopathy
Oral−facial−digital syndrome IV	TCTN3	41, 42, 44	Known ciliopathy[a]
Oral−facial−digital syndrome V	DDX59	41, 42	Known ciliopathy[a]
Oral−facial−digital syndrome VI	C5ORF42	34, 41, 42, 44, 46, 47, 49, 64, 69, 91, 92, 94	Known ciliopathy[a]
Oral−facial−digital syndrome IX	Unknown	46, 47	Known ciliopathy[a]
Short-rib thoracic dysplasia 3 with or without polydactyly	DYNC2H1	41	Known ciliopathy
Short-rib thoracic dysplasia 6 with or without polydactyly	NEK11	40	Known ciliopathy[a]
Short-rib thoracic dysplasia 9 with or without polydactyly	IFT140	1	Known ciliopathy[a]
Short-rib thoracic dysplasia 13 with or without polydactyly	CEP120	46, 47	Known ciliopathy[a]
Short-rib thoracic dysplasia 14 with polydactyly	KIAA0586	41	Known ciliopathy[a]
Acrocallosal syndrome	KIF7	41, 42, 46, 47, 91, 92	Likely ciliopathy
Coach syndrome	CC2D2A, TMEM67, RPGRIP1L	42, 52, 98	Likely ciliopathy
Hydrolethalus syndrome 1	HYLS1	41, 44, 46, 49, 91, 94	Likely ciliopathy
Mohr syndrome (OFD2)	Unknown	42, 46, 47, 102	Likely ciliopathy
Renal−hepatic−pancreatic dysplasia	NPHP3	37, 47, 82	Likely ciliopathy
Hydrolethalus syndrome 2	KIF7	41, 44	N/A

N/A, Not applicable.

[a]These specific subforms were not distinguished among in the original Baker and Beales review.

CRANIOFACIAL TISSUES

The developing craniofacial complex is comprised of distinct prominences that fuse together to form a recognizable face (Fig. 2). At the midline is a single frontonasal prominence (FNP) that gives rise to the forehead, bridge, and tip of the nose, the philtrum, the medial

Cite this article as *Cold Spring Harb Perspect Biol* doi: 10.1101/cshperspect.a028258

Table 3. Predicted craniofacial ciliopathies

Syndromes	Gene	Phenotypes	Classification in Baker and Beales
Acromelic frontonasal dystosis	ZSWIM6	41, 42, 51, 52, 96, 97	Likely ciliopathy
Opitz GBBB syndrome, type II	SPECC1L	11, 16, 17, 19, 20, 22, 24, 25, 64, 69, 70, 86, 91, 92, 94, 95	Likely ciliopathy
Adams–Oliver syndrome 1	ARHGAP31	64, 65, 96	Potential ciliopathy
Carpenter syndrome 1	RAB23	36	Potential ciliopathy
Cerebrofaciothoracic dysplasia	TMCO1	64, 65	Potential ciliopathy
Dandy–Walker syndrome	Unknown	91	Potential ciliopathy
Fryns syndrome	Unknown	64, 65, 69, 70, 91, 92, 94, 95	Potential ciliopathy
Hemifacial microsomia	Unknown	36, 41, 51, 86, 96	Potential ciliopathy
Johnson neuroectodermal syndrome	Unknown	34, 35	Potential ciliopathy
Marden–Walker syndrome	PIEZO2	64, 65, 69, 70, 91, 92, 94, 95	Potential ciliopathy
Oculoauriculofrontonasal syndrome	Unknown	96, 97, 99, 100, 101	Potential ciliopathy
Oculocerebrocutaneous syndrome	Unknown	91, 96	Potential ciliopathy
Split-hand/foot malformations	Unknown	34	Potential ciliopathy
Agenesis of the corpus callosum with facial anomalies and Robin sequence	Unknown	64, 65	N/A
Aicardi syndrome	Unknown	91	N/A
Baraitser–Winter syndrome 1	ACTB	91	N/A
Craniosynostosis-mental retardation syndrome of Lin and Gettig	Unknown	35, 63, 64, 65	N/A
Dandy–Walker malformation with occipital cepalocele	Unknown	87	N/A
Faciocardiomelic syndrome	Unknown	34, 35	N.A
Galloway–Mowat syndrome	WDR73	69, 70	N/A
Gordon syndrome	PIEZO2	69, 70	N/A
Proliferative vasculopathy and hydranencephaly–hydrocephaly syndrome (PVHH)	FLVCR2	91, 94, 95	N/A
Pseudoaminopterin syndrome	Unknown	34, 35	N/A
Simpson–Golabi–Behmel syndrome, type I	GPC3	34, 35, 41, 42, 44, 45, 64, 89	N/A
Temtamy syndrome	C12ORF57	64, 65	N/A

N/A, Not applicable.

portion of the upper lip, and the primary palate. Lateral to the FNP are paired maxillary prominences (MXPs) that fuse at the midline with the FNP. The MXPs gives rise to the upper jaw and the sides of the face, the sides of the upper lip, and the secondary palate. Finally, the mandibular prominences (MNPs) fuse and form the base of the stomodium (presumptive mouth), below the FNP. The MNPs give rise to the lower jaw, lower lip, and the anterior portion of the tongue. The development of each of these prominences requires complex interac-

tions from different tissue types such as CNCCs, neuroectoderm, surface ectoderm, and pharyngeal endoderm. In this section, we review how each tissue shapes the developing face and comment on how loss of cilia effects the development of each tissue.

Cranial Neural Crest Cells

CNCCs are a transient population of cells that give rise to a variety of cells types, including the chondrocytes and osteocytes of the facial skel-

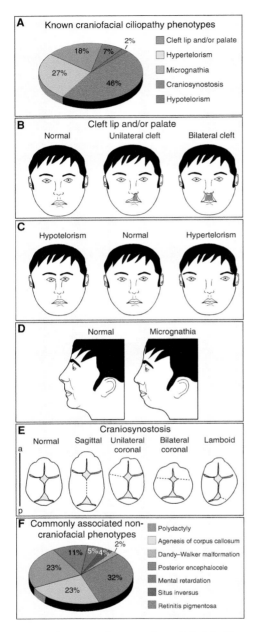

Figure 1. Common features of craniofacial ciliopathies. (*A*) Frequency of the five craniofacial phenotypes common to craniofacial ciliopathies. (*B*) Frontal diagram of cleft lip/palate phenotypes. (*C*) Frontal diagram of the midline defects hypotelorism and hypertelorism. (*D*) Sagittal diagram of micrognathia. (*E*) Dorsal view of skulls with multiple types of craniosynostosis. (*F*) Frequency of core ciliopathic phenotypes among known craniofacial ciliopathies. a, anterior; p, posterior.

eton (Le Douarin et al. 2004). This unique population of cells originates at the dorsal most portion of neural tube, undergoes an epithelial to mesenchymal transition, and migrates toward the developing facial prominences. Throughout their development and migration, CNCCs receive queues from adjacent tissues instructing them to proliferate and/or differentiate (Bhatt et al. 2013). CNCCs can be thought of as the "keystone" of the craniofacial complex, as they sit adjacent to, and participate in, tissue–tissue interactions with all other tissues making up the craniofacial complex.

Primary cilia extend from CNCCs throughout all developmental stages (Schock et al. 2015). Several pieces of data from zebrafish, chick, and mouse ciliopathic models strongly suggest that cilia are required on CNCCs for proper craniofacial development (Tobin et al. 2008; Brugmann et al. 2010; Schock et al. 2015). First, numerous ciliopathic mutants have aberrant CNCC migration characterized by a disorganized actin cytoskeleton, increased cell dispersion, ectopic CNCC migration, and/ or lack of directional persistence (Tobin et al. 2008; Tabler et al. 2013; Schock et al. 2015). Because disrupted CNCC migration is associated with facial clefting (He and Soriano 2013), it is plausible that aberrant CNCC migration contributes to the high frequency of facial clefts observed in ciliopathic patients. Second, loss of primary cilia on CNCCs is often accompanied by dysmorphologies of the CNCC-derived facial skeleton. Patients with oral–facial–digital syndrome frequently present with a cleft secondary palate, micrognathia and underdeveloped cheek bones (zygomatic hypoplasia) due to decreased bone development (Gorlin et al. 1990). Furthermore, cranioectodermal dysplasia patients have multiple skeletal defects with generalized osteopenia (Gorlin et al. 1990). Development of CNCC-derived cartilaginous elements is also aberrant in craniofacial ciliopathies. Both $Kif3a^{fl/fl};Wnt1\text{-}Cre$ and $Ofd^{-/-}$ murine mutants have a bifid nasal septum (Brugmann et al. 2010; Khonsari et al. 2013) and facial cartilages are enlarged in the $talpid^2$ ciliopathic avian mutant (Schock et al. 2015).

Cite this article as *Cold Spring Harb Perspect Biol* doi: 10.1101/cshperspect.a028258

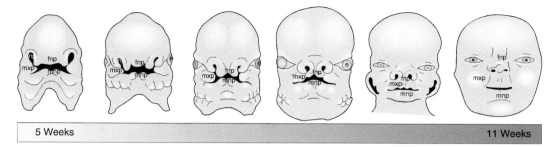

Figure 2. Human craniofacial development. Diagram showing progression of human craniofacial development beginning at 5 weeks of gestation and ending at 11 weeks of gestation. fnp, Frontonasal prominence; mxp, maxillary prominence; mnp, mandibular prominence; np, nasal pits; e, eye.

Together, these data suggests that primary cilia play an essential role in the regulation of both CNCC migration and differentiation into skeletal elements.

Neuroectoderm

Several studies have established a strong relationship between the development of the face and brain (Hu and Marcucio 2009a; Chong et al. 2012; Hu et al. 2015; Marcucio et al. 2015). The forebrain structurally supports the developing face and also serves as the source of many signals that are essential for patterning and outgrowth of the facial prominences, specifically the FNP. One of the most important signaling centers for craniofacial development is the ventral forebrain. This region robustly expresses *Sonic hedgehog* (*Shh*). When *Shh* is absent from this region of the brain, the telencephalon fails to develop into two distinct lobes, resulting in holoprosencephaly (Roessler et al. 1996). In many ciliopathic mutants, forebrain development is aberrant. There are several ciliary mouse mutants with forebrain ventralization and shifts in the diencephalon–telencephalon boundary (Tran et al. 2008; Willaredt et al. 2008; Stottmann et al. 2009; Besse et al. 2011). These data indicate that primary cilia play critical roles in early neuroectoderm development and patterning.

Forebrain abnormalities are frequently accompanied by defects in the facial midline. Blocking SHH in the neural tube causes a narrowing of the FNP (Chong et al. 2012). Addi-

tionally, the width of *Shh* domain in the ventral forebrain correlates with midfacial width. Ducks have a wider domain of *Shh* expression than chickens, and this correlates with a wider midline in ducks (bill vs. beak) (Hu et al. 2015). Although no study has been conducted specifically examining the role of primary cilia in the forebrain to face signaling cascade, one would expect that cilia play a critical role in this process, especially given the high prevalence of hypertelorism as a commonly observed craniofacial phenotype among ciliopathic patients.

Facial Ectoderm

The facial ectoderm is a thin layer of tissue that covers the surface of the developing facial prominences. Early in craniofacial development, this tissue abuts the neuroectoderm and endoderm; however, migration of CNCCs into the facial prominences causes a separation of tissues. The facial ectoderm contains many localized signaling centers, including the frontonasal ectodermal zone (FEZ) (Hu and Marcucio 2009b) and olfactory ectoderm (Szabo-Rogers et al. 2008). The FEZ is defined by robust *Shh* expression and is essential for driving facial morphogenesis. The FEZ signals to the underlying CNCC-derived mesenchyme to regulate *Bmp* expression. *Bmps* regulate both the shape and size of the upper jaw (Abzhanov et al. 2004; Hu and Marcucio 2009b). The olfactory ectoderm expresses *Fgf8, Wnts,* and synthesizes retinoic acid. These signals are important for both craniofacial morphogenesis and skeletal patterning (Song et al. 2004; Szabo-Rogers et al. 2008). Inhibition of fibroblast growth factor (FGF) signaling in the nasal pits results in facial clefting (Szabo-Rogers et al. 2008). The role of cilia on the facial ectoderm that makes up these signaling centers has not been evaluated; however, conditionally knocking out cilia in more specialized facial ectoderm did not affect development of ectodermal-derived structures, such as the teeth (Ohazama et al. 2009). It is possible that this lack of phenotype was due to late onset of recombination. Perhaps ablating cilia in the ectoderm at an earlier time point would prevent the establishment of these key ectodermal signaling centers and result in developmental defects.

Pharyngeal Endoderm

The pharyngeal endoderm lines the interior aspect of the pharyngeal arches and is the source of signals important for the patterning and proliferation of the lower face and neck. In the absence of pharyngeal endoderm, normal craniofacial development fails to occur indicating that it is an essential component for craniofacial development (Couly et al. 2002). The pharyngeal endoderm supplies the environment with many important signals, including BMPs, FGFs, and SHH (Graham and Smith 2001; Graham 2003). These signals are especially important for instructing CNCC development once they have migrated into the pharyngeal arches (Couly et al. 2002; Walshe and Mason 2003). Most studies on pharyngeal endoderm indicate that the endoderm provides key developmental signals, rather than being the recipient of such signals. The specific role for cilia in the pharyngeal endoderm has not been evaluated; however, endoderm in other areas of the body requires cilia for proper development. In *Tmem67* mutant mice, the developing lungs fail to undergo proper epithelial branching morphogenesis (Abdelhamed et al. 2015). Additionally, *orpk* mice (*Ift88* mutant) have an abnormal pancreas characterized by ductal hyperplasia and acinar cell apoptosis (Cano et al. 2004). Together, these data suggest that primary cilia are required for proper development of some endoderm-derived organs.

PRIMARY CILIA AND KEY SIGNALING PATHWAYS IN CRANIOFACIAL DEVELOPMENT

Proper craniofacial development depends on the integration of multiple signaling pathways in a spatially and temporally regulated manner. The primary cilium serves as a cellular hub for several signaling pathways (Goetz and Anderson 2010), specifically the Hedgehog (Hh) pathway. In the next section, we will review how two major signaling pathways drive craniofacial development and the known or potential roles of primary cilia in the transduction of these pathways.

Hedgehog Signaling and the Primary Cilium

The correlation between the cilium and Hh signaling is based on two pieces of data. First, receptors for the Hh pathway localize to the primary cilium (Rohatgi et al. 2007). In the absence of an Hh ligand, Patched is localized to the cilium, inhibiting Smoothened entry into the axoneme. Upon binding of an Hh ligand, Patched-mediated inhibition of Smoothened is alleviated, and Smoothened enters the ciliary axoneme. Second, the transcription factors used to transduce the Hh signal, the Gli family transcription factors, traffic through the ciliary axoneme. In mechanisms still not fully understood, it is believed this trafficking plays a role in converting the full-length Gli proteins into either full-length activators or a cleaved repressor (Haycraft et al. 2005; Huangfu and Anderson 2005). Loss of functional cilia prevents proper Gli processing, frequently resulting in increased levels of full-length Gli and reduced levels of cleaved Gli repressor (Haycraft et al. 2005; Huangfu and Anderson 2005; Liu et al. 2005; May et al. 2005; Hoover et al. 2008; Tran et al. 2008; Willaredt et al. 2008; Chang et al. 2014).

The craniofacial anomalies that result from loss of functional cilia often phenocopy those caused by mutation in the Gli proteins. Greig cephalopolysyndactyly syndrome is caused by mutations in GLI3. Individuals with this disease frequently present with polydactyly, syndactyly (digit fusion), macrocephaly, prominent forehead, and hypertelorism. Similarly, Pallister–Hall syndrome is also caused by mutations in GLI3 and features phenotypes such as polydactyly, syndactyly, hypothalamic hamertomas (benign brain tumor of the hypothalamus), microglossia micrognathia, and cleft palate. Furthermore, mutations in GLI2 are known to cause Culler–Jones syndrome. This disease is characterized by polydactyly, midline facial defects (including holoprosencephaly), and hypopituitarism (Roessler et al. 2003; Franca et al. 2010). Phenotypic presentation of individuals with GLI2 mutations strongly overlaps with those of craniofacial ciliopathies, including the frequent occurrence of dysmorphic and displaced pituitary glands (Khonsari et al. 2013). Although there is not a complete phenotypic overlap between diseases caused by Gli mutations and ciliopathies, there are numerous parallels. The discovery that primary cilia regulate Shh signaling via Gli protein processing explains the phenotypic overlap between ciliopathies and diseases caused by mutation in Gli proteins.

Common Craniofacial Ciliopathic Phenotypes Are Linked to Defects in Shh Signaling

Cleft lip/palate is one of the most common congenital defects, with an occurrence of one in every 500–750 live births (see cdc.gov/NCBDDD/birthdefects/facts.html). Clefting is also the most common defect in craniofacial ciliopathies, with 45% of craniofacial ciliopathies presenting with the condition (Fig. 1A). Clefting occurs when there is a failure of two or more facial prominences to fuse. Onset of clefting of either the lip or palate can result from aberrant cell migration, proliferation, or fusion. Spatial and temporal regulation of multiple signaling pathways is essential for normal palatogenesis. *Shh* is expressed in the epithelium of the developing palate in a series of stripes known as rugae and plays critical roles in induction of *Fgf10* and *Foxf1/2* in the palatal mesenchyme.

These factors promote mesenchymal proliferation and also work through intermediates to regulate *Shh* expression in the epithelium. When *Shh* or these downstream pathway targets are lost, a cleft in the secondary palate occurs (Cobourne and Green 2012; Xu et al. 2016). Clefting may also be secondary to other facial anomalies that alter the width or growth of the face. For example, a widened midline may prevent the MXPs from being able to fuse to form the secondary palate. Given the requirement for complex tissue–tissue interactions and the intricacies of signaling networks during palatogenesis, it is likely that ciliary-dependent Hh signaling is necessary for proper palatogenesis.

The Hh pathway is undisputedly the pathway most closely associated with aberrant midline patterning (Hu and Helms 1999) due to ligand expression in key signal centers, such as the ventral neuroectoderm and FEZ. In the simplest of terms, loss of Hh signaling results in a reduced or collapsed midline (hypotelorism, cyclopia), whereas gain of Hh activity causes an expanded midline (hypertelorism) (Chiang et al. 1996) or, in severe instances, facial duplication (Hu and Helms 1999). These data strongly indicate that levels of Shh are a key determinant of midfacial width. Several ciliary mutants with aberrant Shh signaling present with midfacial defects, strongly suggesting midfacial development, requires ciliary-dependent Hh signaling (Buxton et al. 2004; May et al. 2005; Beales et al. 2007; Gorivodsky et al. 2009; Brugmann et al. 2010). Currently, it is not understood why some ciliary mutations result in a loss of Hh function, whereas others result in a gain of Hh function.

Micrognathia occurs in almost 20% of craniofacial ciliopathies. Jaw deficiencies can be caused by a number of things, including failure of CNCCs to migrate and populate the MNP, aberrant CNCC differentiation into skeletal elements, or disruptions in molecular signaling. Defects in the Hh signaling pathway have been associated with the onset of micrognathia. Early in craniofacial development, the oral ectoderm and pharyngeal endoderm express *Shh*. Expression of *Shh* in these domains is required for CNCC survival within the MNP (Billmyre and

Klingensmith 2015). Increased death of CNCC in the MNP reduces the pool of skeletal progenitors, thus resulting in micrognathia. Additionally, at slightly later stages, Shh from the mandibular epithelium directs early differentiation of mandibular chondrogenic elements, such as Meckel's cartilage (Billmyre and Klingensmith 2015). Once again, aberrant mandibular chondrogenesis could ultimately manifest as micrognathia. Disruptions in ciliary-dependent Hh signaling at either of these stages of mandibular development could lead to severe micrognathia.

Finally, craniosysnostosis presents in a modest 7% of known craniofacial ciliopathies. Craniosynostosis is most commonly associated with mutations in Twist transcription factors (Howard et al. 1997) and members of the FGF signaling pathway (Muenke and Schell 1995; Rice et al. 2000), which may explain why this is not a prevalent phenotype in ciliopathy patients. Recent studies have investigated potential roles for Hh signaling in suture maintenance and calvaria formation. Gli1-positive cells reside at the sutures in postnatal mice and Indian Hedgehog (Ihh) promotes the differentiation of these cells into craniofacial bones, dura, periosteum and the osteogenic front (Zhao et al. 2015). Given these newly identified roles for Hh signaling in suture formation and maintenance, it is quite possible that cilia may be playing a role in this process as well.

Craniofacial ciliopathies are often paired with defects in the brain (Fig. 1F). A long-asked question in developmental biology is does the face predict the brain? Or the converse question of does the brain predict the face? Given the high correlation between brain and face phenotypes observed in ciliopathies, it is likely that the relationship between brain and face development is not unidirectional, but rather the development of one affects the development of the other. Similar to what is observed in the face, most of the brain phenotypes found in ciliopathic models are specific to the midline of the brain. Agenesis of the corpus callosum, defects in dopaminergic neurogenesis at the midbrain midline, aberrant hippocampal neurogenesis, and hypoplasia of the cerebellum are all midline brain defects that have been identified in ciliopathic models (Breunig et al. 2008; Han et al. 2008; Laclef et al. 2015; Schock et al. 2015; Gazea et al. 2016). Given that Shh is expressed throughout the midline of the brain, it is not surprising that many of the above phenotypes have been attributed to defective cilia-dependent Hh signaling.

An Alternative to Hh-Mediated Ciliopathic Facial Phenotypes: Wnt Signaling during Craniofacial Development

The Wnt pathway represents one of the major developmental signaling pathways regulating several cellular processes, including cell proliferation, determination, differentiation, and survival (reviewed in Cadigan and Nusse 1997). In terms of craniofacial development, the Wnt pathway is required at several crucial time points and for the specialization of numerous craniofacial features (midface, oral cavity, facial skeleton, etc.). Most notably, Wnt signaling plays an important role in the generation and migration of CNCCs (Schmidt and Patel 2005; Basch and Bronner-Fraser 2006), and the development, regionalization, and species-specification of the face (Brugmann et al. 2007). As shown by various Wnt reporter mice (BAT-gal and TOP-gal) (DasGupta and Fuchs 1999; Maretto et al. 2003), Wnt pathway activity is localized to facial ectoderm and the underlying mesenchyme in the lateral nasal prominence, MXP and MNP.

Relative to the Hh pathway, much less is known regarding cilium-dependent transduction of a Wnt signal. There are several contradictory reports regarding the nature of the relationship between Wnt and primary cilia. Some reports suggest a requirement for cilia in Wnt signal transduction (Otto et al. 2003; Watanabe et al. 2003; Simons et al. 2005; Corbit et al. 2008; Tran et al. 2008; McDermott et al. 2010; Lienkamp et al. 2012). Many of these studies support the idea that the cilia or cilia-related proteins play a role in restraining a Wnt signal, albeit in a mechanism that is not fully understood. Other studies, however, suggest that loss of cilia does not have an effect on Wnt signaling (Huang and Schier 2009; Ocbina et al. 2009;

Brugmann et al. 2010). Several explanations exist for these seemingly contradictory results, including spatial and/or temporal gene expression differences and species specificity. Finally, and not mutually exclusive from the other possibilities, these contradictory findings could be the result of examining mutations in different classes of ciliary proteins. Thus, different results could be acquired depending on whether an axonemal, transition zone, or basal body protein was mutated.

More recently, numerous studies have addressed how the primary cilium may coordinate the balance between canonical and noncanonical Wnt signaling during development and in tissue homeostasis (Wallingford and Mitchell 2011; Oh and Katsanis 2013). Localization of the Wnt signaling component Inversin at the base of the cilium is hypothesized to act as a molecular switch between canonical and non-canonical/PCP signaling (Otto et al. 2003; Watanabe et al. 2003; Simons et al. 2005). The link between the cilium and noncanonical/PCP continues to grow. Loss of several ciliary proteins causes phenotypes indicative of a PCP mutation (Ross et al. 2005; Gerdes et al. 2007; May-Simera et al. 2010). Yet, despite the fact that cilia and/or some cilia-related proteins can influence planar cell polarization, the mechanism of how PCP signaling and ciliary function are related remains a "black box."

Several facial phenotypes associated with craniofacial ciliopathies also result from defects in Wnt signaling. Susceptibility to clefting is linked to disruptions in various Wnt genes (Juriloff et al. 1996, 2001, 2005) and disruptions to the Wnt pathway produce mild to severe facial clefting in both animal models and human patients (Miller et al. 2006; Brugmann et al. 2007; Chiquet et al. 2008; Song et al. 2009; Ferretti et al. 2011). Most Wnt-related clefting defects arise in one of two ways: CNCC insufficiency or a lack of fusion between facial prominences. First, CNCC insufficiency is common in Wnt mutants because Wnts promote proliferation of CNCC-derived mesenchyme, and subsequent growth of the facial prominences (Brugmann et al. 2007). Loss of Wnt activity in this tissue results in hypoplastic prominences which fail to abut one another, causing a wide cleft. Second, failure of fusion is common in Wnt mutants, as multiple Wnt ligands are expressed in the facial ectoderm and essential for the fusion of facial prominences (Lan et al. 2006; Geetha-Loganathan et al. 2009; Song et al. 2009). Based on the roles for both Wnt and cilia in these cellular processes, it is not difficult to hypothesize that onset of clefting in craniofacial ciliopathies is caused by a defect in cilia-dependent Wnt signaling.

In addition to clefting, both micrognathia and hypertelorism are common phenotypes in Wnt signaling defects. Based on the suggested role for Wnt in promoting mesenchymal proliferation in both the MXP and MNP, it is not difficult to imagine how a disruption in the pathway could lead to reduced growth of the upper and lower jaw. Disruptions to Wnt signaling in the ectoderm also result in ciliopathic craniofacial phenotypes. Loss of β-catenin causes loss of CNCC-derived bone and ectopic facial cartilages (Reid et al. 2011). These data indicate that Wnt/β-catenin signaling within the ectoderm is critical for facial development. Determining whether the onset of these ciliopathic craniofacial phenotypes is due to ciliary-dependent Wnt signaling will be an important future step in formulating potential therapeutic treatments.

CONCLUSIONS

Our analyses revealed the most common phenotypes present in craniofacial ciliopathies are cleft lip/palate, hyper-/hypotelorism, micrognathia, and craniosynostosis. With this information in hand, it is worthwhile to reexamine "orphan" craniofacial anomalies with no known genetic cause for defects in ciliary proteins. Performing array analysis for known ciliary genes or examining ciliary extension on cells from patients that present with phenotypes typical of a craniofacial ciliopathy could greatly inform diagnosis and provide a novel manner to examine craniofacial anomalies with no known genetic cause.

Ciliopathies are not simply a class of diseases in which only one organ or organ system is affected; they are multifaceted and ciliopathic individuals often present with multiple phenotypes. The face is a prime example of the myriad of phenotypes that can be present within a single ciliopathy. Perhaps this is due to a high dependence on tissue–tissue interactions during craniofacial development. Whether signaling between tissues is unidirectional or bidirectional, it stands to reason that primary cilia may be key mediators of developmental processes that require tissue–tissue interactions. Understanding the mechanism by which signaling pathways are propagated through the primary cilium will help us to understand how specific phenotypes arise.

ACKNOWLEDGMENTS

Data reviewed in this work is supported by grants from the National Institutes of Health (NIH)/National Institute of Dental and Craniofacial Research (NIDCR) R01DE023804 (S.A.B.) and F31DE025537 (E.N.S).

REFERENCES

Abdelhamed ZA, Natarajan S, Wheway G, Inglehearn CF, Toomes C, Johnson CA, Jagger DJ. 2015. The Meckel–Gruber syndrome protein TMEM67 controls basal body positioning and epithelial branching morphogenesis in mice via the non-canonical Wnt pathway. *Dis Model Mech* **8**: 527–541.

Abzhanov A, Protas M, Grant RB, Grant PR, Tabin CJ. 2004. Bmp4 and morphological variation of beaks in Darwin's finches. *Science* **305**: 1462–1465.

Baker K, Beales P. 2009. Making sense of cilia in disease: The human ciliopathies. *Am J Med Genet* **151C**: 281–295.

Basch ML, Bronner-Fraser M. 2006. Neural crest inducing signals. *Adv Exp Med Biol* **589**: 24–31.

Beales PL, Bland E, Tobin JL, Bacchelli C, Tuysuz B, Hill J, Rix S, Pearson CG, Kai M, Hartley J, et al. 2007. IFT80, which encodes a conserved intraflagellar transport protein, is mutated in Jeune asphyxiating thoracic dystrophy. *Nat Genet* **39**: 727–729.

Besse L, Neti M, Anselme I, Gerhardt C, Ruther U, Laclef C, Schneider-Maunoury S. 2011. Primary cilia control telencephalic patterning and morphogenesis via Gli3 proteolytic processing. *Development* **138**: 2079–2088.

Bhatt S, Diaz R, Trainor PA. 2013. Signals and switches in mammalian neural crest cell differentiation. *Cold Spring Harb Perspect Biol* **5**: a008326.

Billmyre KK, Klingensmith J. 2015. Sonic Hedgehog from pharyngeal arch 1 epithelium is necessary for early mandibular arch cell survival and later cartilage condensation differentiation. *Dev Dyn* **244**: 564–576.

Breunig JJ, Sarkisian MR, Arellano JI, Morozov YM, Ayoub AE, Sojitra S, Wang B, Flavell RA, Rakic P, Town T. 2008. Primary cilia regulate hippocampal neurogenesis by mediating sonic hedgehog signaling. *Proc Natl Acad Sci* **105**: 13127–13132.

Brugmann SA, Goodnough LH, Gregorieff A, Leucht P, ten Berge D, Fuerer C, Clevers H, Nusse R, Helms JA. 2007. Wnt signaling mediates regional specification in the vertebrate face. *Development* **134**: 3283–3295.

Brugmann SA, Allen NC, James AW, Mekonnen Z, Madan E, Helms JA. 2010. A primary cilia-dependent etiology for midline facial disorders. *Hum Mol Genet* **19**: 1577–1592.

Buxton P, Davey MG, Paton IR, Morrice DR, Francis-West PH, Burt DW, Tickle C. 2004. Craniofacial development in the *talpid3* chicken mutant. *Differentiation* **72**: 348–362.

Cadigan KM, Nusse R. 1997. Wnt signaling: A common theme in animal development. *Genes Dev* **11**: 3286–3305.

Cano DA, Murcia NS, Pazour GJ, Hebrok M. 2004. Orpk mouse model of polycystic kidney disease reveals essential role of primary cilia in pancreatic tissue organization. *Development* **131**: 3457–3467.

Chang CF, Schock EN, O'Hare EA, Dodgson J, Cheng HH, Muir WM, Edelmann RE, Delany ME, Brugmann SA. 2014. The cellular and molecular etiology of the craniofacial defects in the avian ciliopathic mutant *talpid2*. *Development* **141**: 3003–3012.

Chiang C, Litingtung Y, Lee E, Young KE, Corden JL, Westphal H, Beachy PA. 1996. Cyclopia and defective axial patterning in mice lacking Sonic Hedgehog gene function. *Nature* **383**: 407–413.

Chiquet BT, Blanton SH, Burt A, Ma D, Stal S, Mulliken JB, Hecht JT. 2008. Variation in WNT genes is associated with non-syndromic cleft lip with or without cleft palate. *Hum Mol Genet* **17**: 2212–2218.

Chong HJ, Young NM, Hu D, Jeong J, McMahon AP, Hallgrimsson B, Marcucio RS. 2012. Signaling by SHH rescues facial defects following blockade in the brain. *Dev Dyn* **241**: 247–256.

Cobourne MT, Green JB. 2012. Hedgehog signalling in development of the secondary palate. *Front Oral Biol* **16**: 52–59.

Corbit KC, Shyer AE, Dowdle WE, Gaulden J, Singla V, Chen MH, Chuang PT, Reiter JF. 2008. Kif3a constrains β-catenin-dependent Wnt signalling through dual ciliary and non-ciliary mechanisms. *Nat Cell Biol* **10**: 70–76.

Couly G, Creuzet S, Bennaceur S, Vincent C, Le Douarin NM. 2002. Interactions between Hox-negative cephalic neural crest cells and the foregut endoderm in patterning the facial skeleton in the vertebrate head. *Development* **129**: 1061–1073.

DasGupta R, Fuchs E. 1999. Multiple roles for activated LEF/TCF transcription complexes during hair follicle development and differentiation. *Development* **126**: 4557–4568.

Ferretti E, Li B, Zewdu R, Wells V, Hebert JM, Karner C, Anderson MJ, Williams T, Dixon J, Dixon MJ, et al. 2011. A conserved Pbx-Wnt-p63-Irf6 regulatory module controls face morphogenesis by promoting epithelial apoptosis. *Dev Cell* **21**: 627–641.

Franca MM, Jorge AA, Carvalho LR, Costalonga EF, Vasques GA, Leite CC, Mendonca BB, Arnhold IJ. 2010. Novel heterozygous nonsense *GLI2* mutations in patients with hypopituitarism and ectopic posterior pituitary lobe without holoprosencephaly. *J Clin Endocrinol Metab* **95:** E384–E391.

Gazea M, Tasouri E, Tolve M, Bosch V, Kabanova A, Gojak C, Kurtulmus B, Novikov O, Spatz J, Pereira G, et al. 2016. Primary cilia are critical for Sonic hedgehog–mediated dopaminergic neurogenesis in the embryonic midbrain. *Devel Biol* **409:** 55–71.

Geetha-Loganathan P, Nimmagadda S, Antoni L, Fu K, Whiting CJ, Francis-West P, Richman JM. 2009. Expression of WNT signalling pathway genes during chicken craniofacial development. *Dev Dyn* **238:** 1150–1165.

Gerdes JM, Liu Y, Zaghloul NA, Leitch CC, Lawson SS, Kato M, Beachy PA, Beales PL, DeMartino GN, Fisher S, et al. 2007. Disruption of the basal body compromises proteasomal function and perturbs intracellular Wnt response. *Nat Genet* **39:** 1350–1360.

Goetz SC, Anderson KV. 2010. The primary cilium: A signalling centre during vertebrate development. *Nat Rev Genet* **11:** 331–344.

Gorivodsky M, Mukhopadhyay M, Wilsch-Braeuninger M, Phillips M, Teufel A, Kim C, Malik N, Huttner W, Westphal H. 2009. Intraflagellar transport protein 172 is essential for primary cilia formation and plays a vital role in patterning the mammalian brain. *Devel Biol* **325:** 24–32.

Gorlin RJ, Cohen MM, Levin LS. 1990. *Syndromes of the head and neck.* Oxford University Press, New York.

Graham A. 2003. Development of the pharyngeal arches. *Am J Med Genet* **119A:** 251–256.

Graham A, Smith A. 2001. Patterning the pharyngeal arches. *Bioessays* **23:** 54–61.

Han YG, Spassky N, Romaguera-Ros M, Garcia-Verdugo JM, Aguilar A, Schneider-Maunoury S, Alvarez-Buylla A. 2008. Hedgehog signaling and primary cilia are required for the formation of adult neural stem cells. *Nat Neurosci* **11:** 277–284.

Haycraft CJ, Banizs B, Aydin-Son Y, Zhang Q, Michaud EJ, Yoder BK. 2005. Gli2 and Gli3 localize to cilia and require the intraflagellar transport protein polaris for processing and function. *PLoS Genet* **1:** e53.

He F, Soriano P. 2013. A critical role for PDGFRα signaling in medial nasal process development. *PLoS Genet* **9:** e1003851.

Hoover AN, Wynkoop A, Zeng H, Jia J, Niswander LA, Liu A. 2008. C2cd3 is required for cilia formation and Hedgehog signaling in mouse. *Development* **135:** 4049–4058.

Howard TD, Paznekas WA, Green ED, Chiang LC, Ma N, Ortiz de Luna RI, Garcia Delgado C, Gonzalez-Ramos M, Kline AD, Jabs EW. 1997. Mutations in *TWIST*, a basic helix–loop–helix transcription factor, in Saethre–Chotzen syndrome. *Nat Genet* **15:** 36–41.

Hu D, Helms JA. 1999. The role of sonic hedgehog in normal and abnormal craniofacial morphogenesis. *Development* **126:** 4873–4884.

Hu D, Marcucio RS. 2009a. A SHH-responsive signaling center in the forebrain regulates craniofacial morphogenesis via the facial ectoderm. *Development* **136:** 107–116.

Hu D, Marcucio RS. 2009b. Unique organization of the frontonasal ectodermal zone in birds and mammals. *Dev Biol* **325:** 200–210.

Hu D, Young NM, Xu Q, Jamniczky H, Green RM, Mio W, Marcucio RS, Hallgrimsson B. 2015. Signals from the brain induce variation in avian facial shape. *Dev Dyn* doi: 10.1002/dvdy.24284.

Huang P, Schier AF. 2009. Dampened Hedgehog signaling but normal Wnt signaling in zebrafish without cilia. *Development* **136:** 3089–3098.

Huangfu D, Anderson KV. 2005. Cilia and Hedgehog responsiveness in the mouse. *Proc Natl Acad Sci* **102:** 11325–11330.

Juriloff DM, Harris MJ, Mah DG. 1996. The *clf1* gene maps to a 2- to 3-cM region of distal mouse chromosome 11. *Mamm Genome* **7:** 789.

Juriloff DM, Harris MJ, Brown CJ. 2001. Unravelling the complex genetics of cleft lip in the mouse model. *Mamm Genome* **12:** 426–435.

Juriloff DM, Harris MJ, Dewell SL, Brown CJ, Mager DL, Gagnier L, Mah DG. 2005. Investigations of the genomic region that contains the *clf1* mutation, a causal gene in multifactorial cleft lip and palate in mice. *Birth Defects Res A Clin Mol Teratol* **73:** 103–113.

Khonsari RH, Seppala M, Pradel A, Dutel H, Clement G, Lebedev O, Ghafoor S, Rothova M, Tucker A, Maisey JG, et al. 2013. The buccohypophyseal canal is an ancestral vertebrate trait maintained by modulation in sonic hedgehog signaling. *BMC Biol* **11:** 27.

Laclef C, Anselme I, Besse L, Catala M, Palmyre A, Baas D, Paschaki M, Pedraza M, Metin C, Durand B, Schneider-Maunoury S. 2015. The role of primary cilia in corpus callosum formation is mediated by production of the Gli3 repressor. *Hum Mol Genet* **24:** 4997–5014.

Lan Y, Ryan RC, Zhang Z, Bullard SA, Bush JO, Maltby KM, Lidral AC, Jiang R. 2006. Expression of *Wnt9b* and activation of canonical Wnt signaling during midfacial morphogenesis in mice. *Dev Dyn* **235:** 1448–1454.

Le Douarin NM, Creuzet S, Couly G, Dupin E. 2004. Neural crest cell plasticity and its limits. *Development* **131:** 4637–4650.

Lienkamp S, Ganner A, Walz G. 2012. Inversin, Wnt signaling and primary cilia. *Differentiation* **83:** S49–S55.

Liu A, Wang B, Niswander LA. 2005. Mouse intraflagellar transport proteins regulate both the activator and repressor functions of Gli transcription factors. *Development* **132:** 3103–3111.

Marcucio R, Hallgrimsson B, Young NM. 2015. Facial morphogenesis: Physical and molecular interactions between the brain and the face. *Curr Topics Dev Biol* **115:** 299–320.

Maretto S, Cordenonsi M, Dupont S, Braghetta P, Broccoli V, Hassan AB, Volpin D, Bressan GM, Piccolo S. 2003. Mapping Wnt/β-catenin signaling during mouse development and in colorectal tumors. *Proc Natl Acad Sci* **100:** 3299–3304.

May SR, Ashique AM, Karlen M, Wang B, Shen Y, Zarbalis K, Reiter J, Ericson J, Peterson AS. 2005. Loss of the retrograde motor for IFT disrupts localization of Smo to cilia and prevents the expression of both activator and repressor functions of Gli. *Dev Biol* **287:** 378–389.

May-Simera HL, Kai M, Hernandez V, Osborn DP, Tada M, Beales PL. 2010. Bbs8, together with the planar cell polarity protein Vangl2, is required to establish left–right asymmetry in zebrafish. *Dev Biol* **345:** 215–225.

McDermott KM, Liu BY, Tlsty TD, Pazour GJ. 2010. Primary cilia regulate branching morphogenesis during mammary gland development. *Curr Biol* **20:** 731–737.

Miller LA, Smith AN, Taketo MM, Lang RA. 2006. Optic cup and facial patterning defects in ocular ectoderm β-catenin gain-of-function mice. *BMC Dev Biol* **6:** 14.

Muenke M, Schell U. 1995. Fibroblast-growth-factor receptor mutations in human skeletal disorders. *Trends Genet* **11:** 308–313.

Ocbina PJ, Tuson M, Anderson KV. 2009. Primary cilia are not required for normal canonical Wnt signaling in the mouse embryo. *PLoS ONE* **4:** e6839.

Oh EC, Katsanis N. 2013. Context-dependent regulation of Wnt signaling through the primary cilium. *J Am Soc Nephrol* **24:** 10–18.

Ohazama A, Haycraft CJ, Seppala M, Blackburn J, Ghafoor S, Cobourne M, Martinelli DC, Fan CM, Peterkova R, Lesot H, et al. 2009. Primary cilia regulate Shh activity in the control of molar tooth number. *Development* **136:** 897–903.

Otto EA, Schermer B, Obara T, O'Toole JF, Hiller KS, Mueller AM, Ruf RG, Hoefele J, Beekmann F, Landau D, et al. 2003. Mutations in *INVS* encoding inversin cause nephronophthisis type 2, linking renal cystic disease to the function of primary cilia and left–right axis determination. *Nat Genet* **34:** 413–420.

Reid BS, Yang H, Melvin VS, Taketo MM, Williams T. 2011. Ectodermal Wnt/β-catenin signaling shapes the mouse face. *Dev Biol* **349:** 261–269.

Rice DP, Aberg T, Chan Y, Tang Z, Kettunen PJ, Pakarinen L, Maxson RE, Thesleff I. 2000. Integration of FGF and TWIST in calvarial bone and suture development. *Development* **127:** 1845–1855.

Roessler E, Belloni E, Gaudenz K, Jay P, Berta P, Scherer SW, Tsui LC, Muenke M. 1996. Mutations in the human Sonic Hedgehog gene cause holoprosencephaly. *Nat Genet* **14:** 357–360.

Roessler E, Du YZ, Mullor JL, Casas E, Allen WP, Gillessen-Kaesbach G, Roeder ER, Ming JE, Ruiz i Altaba A, Muenke M. 2003. Loss-of-function mutations in the human *GLI2* gene are associated with pituitary anomalies and holoprosencephaly-like features. *Proc Natl Acad Sc* **100:** 13424–13429.

Rohatgi R, Milenkovic L, Scott MP. 2007. Patched1 regulates hedgehog signaling at the primary cilium. *Science* **317:** 372–376.

Ross AJ, May-Simera H, Eichers ER, Kai M, Hill J, Jagger DJ, Leitch CC, Chapple JP, Munro PM, Fisher S, et al. 2005. Disruption of Bardet-Biedl syndrome ciliary proteins perturbs planar cell polarity in vertebrates. *Nat Genet* **37:** 1135–1140.

Schmidt C, Patel K. 2005. Wnts and the neural crest. *Anat Embryol* **209:** 349–355.

Schock EN, Chang CF, Struve JN, Chang YT, Chang J, Delany ME, Brugmann SA. 2015. Using the avian mutant *talpid2* as a disease model for understanding the oral–

facial phenotypes of oral–facial–digital syndrome. *Dis Models Mech* **8:** 855–866.

Simons M, Gloy J, Ganner A, Bullerkotte A, Bashkurov M, Kronig C, Schermer B, Benzing T, Cabello OA, Jenny A, et al. 2005. Inversin, the gene product mutated in nephronophthisis type II, functions as a molecular switch between Wnt signaling pathways. *Nat Genet* **37:** 537–543.

Song Y, Hui JN, Fu KK, Richman JM. 2004. Control of retinoic acid synthesis and FGF expression in the nasal pit is required to pattern the craniofacial skeleton. *Dev Biol* **276:** 313–329.

Song L, Li Y, Wang K, Wang YZ, Molotkov A, Gao L, Zhao T, Yamagami T, Wang Y, Gan Q, et al. 2009. Lrp6-mediated canonical Wnt signaling is required for lip formation and fusion. *Development* **136:** 3161–3171.

Stottmann RW, Tran PV, Turbe-Doan A, Beier DR. 2009. *Ttc21b* is required to restrict sonic hedgehog activity in the developing mouse forebrain. *Dev Biol* **335:** 166–178.

Szabo-Rogers HL, Geetha-Loganathan P, Nimmagadda S, Fu KK, Richman JM. 2008. FGF signals from the nasal pit are necessary for normal facial morphogenesis. *Dev Biol* **318:** 289–302.

Tabler JM, Barrell WB, Szabo-Rogers HL, Healy C, Yeung Y, Perdiguero EG, Schulz C, Yannakoudakis BZ, Mesbahi A, Wlodarczyk B, et al. 2013. *Fuz* mutant mice reveal shared mechanisms between ciliopathies and FGF-related syndromes. *Dev Cell* **25:** 623–635.

Tobin JL, Di Franco M, Eichers E, May-Simera H, Garcia M, Yan J, Quinlan R, Justice MJ, Hennekam RC, Briscoe J, et al. 2008. Inhibition of neural crest migration underlies craniofacial dysmorphology and Hirschsprung's disease in Bardet–Biedl syndrome. *Proc Natl Acad Sci* **105:** 6714–6719.

Tran PV, Haycraft CJ, Besschetnova TY, Turbe-Doan A, Stottmann RW, Herron BJ, Chesebro AL, Qiu H, Scherz PJ, Shah JV, et al. 2008. THM1 negatively modulates mouse sonic hedgehog signal transduction and affects retrograde intraflagellar transport in cilia. *Nat Genet* **40:** 403–410.

Wallingford JB, Mitchell B. 2011. Strange as it may seem: The many links between Wnt signaling, planar cell polarity, and cilia. *Genes Dev* **25:** 201–213.

Walshe J, Mason I. 2003. Fgf signalling is required for formation of cartilage in the head. *Dev Biol* **264:** 522–536.

Watanabe D, Saijoh Y, Nonaka S, Sasaki G, Ikawa Y, Yokoyama T, Hamada H. 2003. The left–right determinant Inversin is a component of node monocilia and other 9+0 cilia. *Development* **130:** 1725–1734.

Willaredt MA, Hasenpusch-Theil K, Gardner HA, Kitanovic I, Hirschfeld-Warneken VC, Gojak CP, Gorgas K, Bradford CL, Spatz J, Wolfl S, et al. 2008. A crucial role for primary cilia in cortical morphogenesis. *J Neurosci* **28:** 12887–12900.

Xu J, Liu H, Lan Y, Aronow BJ, Kalinichenko VV, Jiang R. 2016. A Shh-Foxf-Fgf18-Shh molecular circuit regulating palate development. *PLoS Genet* **12:** e1005769.

Zaghloul NA, Brugmann SA. 2011. The emerging face of primary cilia. *Genesis* **49:** 231–246.

Zhao H, Feng J, Ho TV, Grimes W, Urata M, Chai Y. 2015. The suture provides a niche for mesenchymal stem cells of craniofacial bones. *Nat Cell Biol* **17:** 386–396.

Cite this article as *Cold Spring Harb Perspect Biol* doi: 10.1101/cshperspect.a028258

Cilia and Ciliopathies in Congenital Heart Disease

Nikolai T. Klena, Brian C. Gibbs, and Cecilia W. Lo

Department of Developmental Biology, University of Pittsburgh School of Medicine, Pittsburgh, Pennsylvania 15201

Correspondence: cel36@pitt.edu

A central role for cilia in congenital heart disease (CHD) was recently identified in a large-scale mouse mutagenesis screen. Although the screen was phenotype-driven, the majority of genes recovered were cilia-related, suggesting that cilia play a central role in CHD pathogenesis. This partly reflects the role of cilia as a hub for cell signaling pathways regulating cardiovascular development. Consistent with this, many cilia-transduced cell signaling genes were also recovered, and genes regulating vesicular trafficking, a pathway essential for ciliogenesis and cell signaling. Interestingly, among CHD-cilia genes recovered, some regulate left–right patterning, indicating cardiac left–right asymmetry disturbance may play significant roles in CHD pathogenesis. Clinically, CHD patients show a high prevalence of ciliary dysfunction and show enrichment for de novo mutations in cilia-related pathways. Combined with the mouse findings, this would suggest CHD may be a new class of ciliopathy.

Congenital heart disease (CHD) is one of the most common birth defects, found in an estimated 1% of live births (Hoffman and Kaplan 2002). With advances in surgical palliation, most patients with CHD now survive their critical heart disease such that currently there are more adults with CHD than infants born with CHD each year (van der Bom et al. 2012). However, CHD patient prognosis is variable, with long-term outcome shown to be dependent on patient intrinsic factors rather than surgical parameters (Newburger et al. 2012; Marelli et al. 2016). This is likely driven by genetic factors, given CHD is highly associated with chromosomal anomalies (Fahed et al. 2013), and with copy number variants (Glessner et al. 2014). In addition, CHD has been shown to have a high recurrence risk, with familial clustering indicating a genetic contribution (Gill et al. 2003; Oyen et al. 2009). The identification of the genetic causes of CHD may provide mechanistic insights that can help stratify patients for guiding the therapeutic management of their clinical care.

Investigations into the genetic causes of CHD in human clinical studies have been challenging given the high degree of genetic diversity in the human population. This has made a compelling case for pursuing the use of a systems genetic approach with large-scale forward genetic screens in animal models to investigate the genetic etiology of CHD. Although many animal models have provided invaluable insights into the developmental regulation of cardiovascular development, investigations into the genetic etiology of CHD must be conducted in a model system with the same four-chamber cardiac anatomy that is the substrate of human CHD. The mouse is one such model system, advantageous not only given its similar four-chamber cardiac anatomy, but also inbred mouse strains are readily available with genomes that are fully sequenced and annotated that would facilitate mutation recovery. Moreover, cardiovascular development in the mouse embryo is well studied, providing a strong foundation to interrogate the developmental and genetic etiology of CHD.

DEVELOPMENT OF THE CARDIOVASCULAR SYSTEM

Congenital heart defect is a structural birth defect arising from disruption of cardiovascular development. Formation of the four-chamber heart in mammals is orchestrated by the highly coordinated specification and migration of different cell populations in the embryo that together form the complex left–right asymmetric anatomy of the cardiovascular system. In the mouse embryo, ingression of cells through the primitive streak at E7.5 generates the anterior mesoderm forming the cardiac crescent–containing cells of the first heart field (FHF) and adjacent to it, the second heart field (SHF) (Fig. 1) (Buckingham 2016). Cells of the FHF migrate toward the midline, fusing to form the linear heart tube at E8.0 (Fig. 1). Pharyngeal mesoderm located anterior and medially continues to be added to the expanding heart tube, as the heart tube undergoes rightward looping at E8.5, delineating the primitive anlage of the left ventricle (LV) (Fig. 1). This is followed by addition of SHF cells to the anterior and posterior poles of the heart tube, giving rise to the outflow tract (OFT), right ventricle (RV), and most of the left and right atria (LA, RA) (Fig. 1).

Normal development of the heart also requires the contribution and activity of several other extracardiac cell lineages, including the cardiac neural crest cells derived from the dorsal hindbrain neural fold. The cardiac neural crest cells migrate into the cardiac OFT in two spiral streams, helping to remodel the pharyngeal arch arteries and orchestrating OFT septation to form the two great arteries—the aorta and pulmonary artery (Kirby and Waldo 1990). The pharyngeal endoderm and ectoderm also play an important regulatory function in developmental patterning of the aortic arch arteries and the OFT. Dynamic processes mediating endocardial epithelial–mesenchyme transformation (EMT) lead to formation of the cushion mesenchyme that provides early valve function in the embryonic heart. These endocardial cushion tissues later remodel to form the mature leaflets of the outflow semilunar and atrioventricular valves (Fig. 1). Another extracardiac cell population required for heart development are the pro-epicardial cells that originate near the septum transversum. These cells migrate to the heart via the sinus venosus, delaminating onto the surface of the heart, and forming the epicardium that plays an essential role in development of the coronary arteries. Together, these diverse cell populations are recruited to orchestrate formation of the mammalian heart, an organ that is an unexpected mosaic of distinct cell lineages.

FOUR-CHAMBER HEART—THE ANATOMICAL SUBSTRATE FOR CONGENITAL HEART DISEASE

The cardiovascular system in mouse and human is adapted for breathing air, being comprised of four chambers organized into functionally distinct left versus right sides. This allows the formation of a separate pulmonary circuit that pumps deoxygenated blood from the body to the lungs via the RV and a systemic circuit pumping oxygenated blood from the lung to the body via the LV. This left–right asymmetric organization is critically dependent on appropriate patterning of the left–right body axis and entails formation of an atrial and ventricular septum separating the right versus left sides of the heart. This allows for compartmentalization of the heart into four chambers, LA versus RA and LV versus RV. This is coupled with septation of the OFT into two great arteries, the aorta, which is inserted into the LV and pulmonary artery into the RV, and formation of the atrio-

Cite this article as *Cold Spring Harb Perspect Biol* doi: 10.1101/cshperspect.a028266

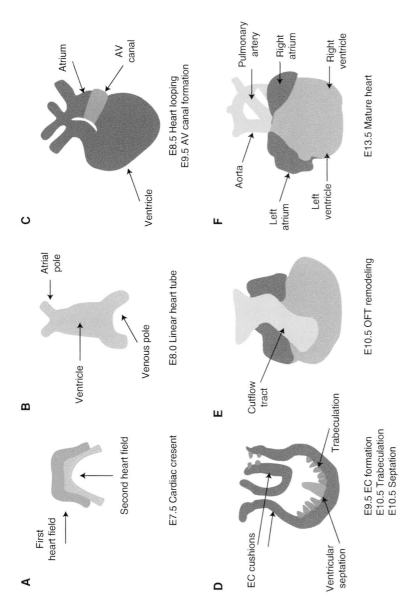

Figure 1. Diagram of mouse cardiovascular development. (*A*) Cardiac crescent formation containing first heart field (FHF) and second heart field (SHF) cells. (*B*) Cardiac crescent cells migrate toward the midline creating the linear heart tube with its arterial and venous poles and a primitive ventricle. (*C*) At E8.5, dextral looping of the heart tube leads to formation of the primitive atrial and ventricular chambers in the morphologically correct position. (*D*) At E9.5, the endocardial cushion cells pinch inward, creating the atrioventricular canal. At E9.5, the endocardial cushions form at the dorsal and ventral lumen of the atrial canal as the endocardial cells undergo epithelial to mesenchymal transition. Cardiac trabeculation initiates at E10.5, creating bundles of cardiomyocytes that extend into the primitive cardiac chambers. Septation initiates at E10.5, starting division of the chambers into the four-chamber anatomy. (*E*) At E10.5, the outflow tract (OFT) is remodeled leading to the primitive connection of the aorta and pulmonary artery from the primitive ventricle. (*F*) By E13.5, the heart is fully developed into four distinct chambers with appropriate aorta and pulmonary artery connections to the morphological left and right ventricles (RVs), respectively. Dark pink, FHF; light pink, SHF; light green, atrioventricular (AV) canal; dark blue, endocardial (EC) cushions; yellow green, septation; purple, trabeculations; yellow, OFT.

ventricular and outflow valves that allow unidirectional blood flow. It is this complex left–right asymmetric developmental patterning of the cardiovascular anatomy that ensures efficient oxygenation of blood with air exchange via the lungs. The perturbation of this distinct four-chamber cardiac anatomy in CHD invariably results in neonatal mortality unless surgical intervention is provided to palliate the structural heart defects. Identifying the genetic causes of CHD may help elucidate the developmental processes contributing to CHD and suggests new avenues for prevention or intervention.

CENTRAL ROLE OF CILIA IN CARDIOVASCULAR DEVELOPMENT AND CONGENITAL HEART DISEASE

To elucidate the genetic etiology of CHD, a large-scale, near-saturation level forward genetic screen with ethylnitrosourea (ENU) chemical mutagenesis was conducted (Li et al. 2015b). This phenotype-driven cardiovascular screen used fetal echocardiography, a noninvasive imaging modality routinely used clinically for CHD diagnosis (Fig. 2). This allowed high detection sensitivity and specificity for CHD diagnosis and allowed the recovery of a wide spectrum of CHD in the mouse screen similar to those observed clinically (Figs. 2 and 3) (Liu et al. 2014). From ultrasound screening of ~100,000 mouse fetuses, we recovered >200 mutant mouse lines with a wide variety of CHD.

Using exome-sequencing analysis, ~100 CHD-causing mutations were recovered in 61 genes, with more than half being cilia-related (Fig. 2) (Li et al. 2015). The cilia genes recovered included proteins localized in the cilia transition zone, basal body/centrosome, ciliary axoneme, and also multiprotein complexes in the cytoplasm required for cilia assembly (Fig. 2). Most of the proteins recovered are expressed in both motile (9 + 2) and primary cilia (9 + 0), such as components of the cilia transition zone. However, some genes encode proteins unique to motile cilia, such as the motor dyneins *Dnah5* and *Dnah11* localized in the outer dynein arm required for motile cilia function (Fig. 2). Many of these cilia protein components are known to cause various human ciliopathies, mostly involving nonmotile, primary cilia defects, such as in Joubert syndrome (JBTS), Jeune syndrome, nephronophthisis, Meckel–Gruber syndrome, and others. The motile cilia mutations recovered in the screen are linked to the sinopulmonary disease primary ciliary dyskinesia (PCD). Although CHD is not an essential feature of ciliopathies, it is notable that the mutants we recovered were all based on having CHD phenotypes.

Further indicating the important role of cilia in CHD pathogenesis, we also recovered mutations in 12 CHD genes that are in cilia-transduced cell signaling pathways, including genes mediating sonic hedgehog (Shh), transforming growth factor β and bone morphogenetic proteins (TGF-β/BMPs), and Wnt signaling (Fig. 2). This enrichment of genes mediating cell signaling reflects the central role of cilia as a hub for signal transduction pathways essential to the regulation of key cardiovascular developmental processes. Also unexpected was the recovery of 10 CHD genes involved in vesicular trafficking. This included *Dynamin 2* and *Ap2b1* required for clathrin-mediated endocytosis, adaptin proteins *Ap1b1* and *Ap2b1*, and *Lrp1*, *Lrp2*, and *Snx17* mediating endocytic receptor recycling (Li et al. 2015b). Significantly, vesicular trafficking plays an essential role in cilia biology, with ciliogenesis initiated with capture of a ciliary vesicle by the mother centriole followed by docking of the basal body to the cell membrane and fusion of additional secondary vesicles that allow lengthening of the ciliary axoneme (Sorokin 1962; Kobayashi and Dynlacht 2011; Reiter et al. 2012). Vesicular trafficking and receptor recycling also play important roles in the regulation of cell signaling. Although the endocytic pathway was not previously known to play a role in CHD, its importance can be easily appreciated in the context of its role in regulating ciliogenesis and cilia-transduced cell signaling.

CILIA AND CILIA-TRANSDUCED CELL SIGNALING IN HEART DEVELOPMENT

The overall finding that the large majority of the CHD genes recovered were cilia or cilia-related was unexpected, given the screen was entirely

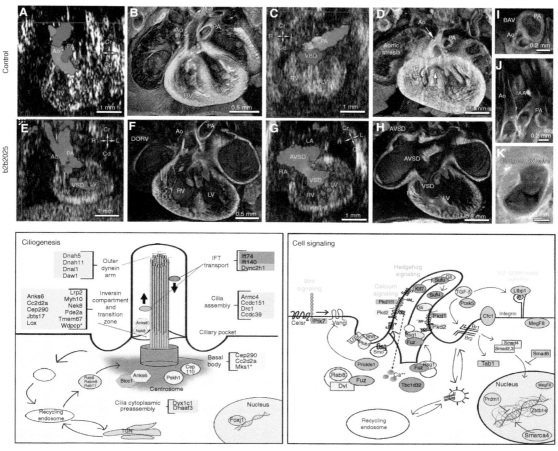

Figure 2. Congenital heart disease (CHD) mutants recovered from mouse mutagenesis screen by fetal echocardiography show preponderance of cilia-related mutations. Vevo 2100 color flow Doppler imaging showed criss-cross pattern of blood flow indicating normal aorta (Ao) and pulmonary artery (PA) alignment (*A*) confirmed by histopathology (*B*). E16.5 mutant (line b2b327) showed blood flow pattern indicating single great artery (PA) and ventricular septal defect (VSD) (*C*), suggesting aortic atresia with VSD, confirmed by histopathology (*D*). Color flow imaging of E15.5 mutant (line b2b2025) with heterotaxy (stomach on *right*) showed Ao/PA side-by-side with Ao emerging from right ventricle (RV) (*E*), indicating double outlet right ventricle (DORV)/VSD (*F*) and presence of atrioventricular septal defect (AVSD) (*G,H*). Histopathology also showed bicuspid aortic valve (BAV) (*I*), interrupted aortic arch (IAA) (*J*), and common atrioventricular (AV) valve (*K*). (*Bottom*) Diagrams summarize genes recovered causing CHD that are related to cilia or cell signaling, providing biological context of CHD gene function. Color highlighting indicates CHD genes recovered; asterisks denote CHD genes recovered from previous screen (Shen et al. 2005). R, Receptor; TGN, *trans*-Golgi network (adapted from data in Li et al. 2015b).

phenotype-driven. Hence, they point to a central role for cilia biology in regulating cardiovascular development and the pathogenesis of CHD. Primary cilia in the developing heart were first identified via electron microscopy in the chick, rabbit, mouse, and lizard embryos (Rash et al. 1969). These were observed only in nonmitotic cardiomyocytes or myoblasts, whereas in the adult heart tissue, cilia were only observed in fibroblasts. A more recent study of the mouse embryo showed that cilia can be found throughout the early E9.5 heart tube (Slough et al. 2008). As development progresses to E12.5, cilia continue to be expressed in the atria and in the trabeculated myocardium (Fig. 3J). Cilia are also found in the

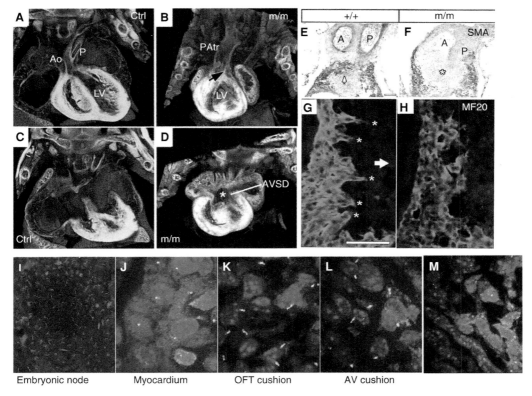

Figure 3. Congenital heart defects in a *Wdpcp* mutant and cilia localization. (*A–H*) Episcopic confocal histopathology showed a *Wdpcp*[Cys40] mutant with an incomplete septum unevenly dividing the outflow tract (OFT) into one large and one small chamber, indicating pulmonary atresia (PAtr; black arrow in *B*). Also observed was an atrioventricular septal defect (AVSD; asterisk in *D*). Shown in *A* and *C* are comparable views of a control heart. In wild-type hearts (*E*), cardiomyocytes were observed in the OFT cushion (arrow), but in*Wdpcp*[Cys40]mutants, cardiomyocytes were mostly absent in the cushion tissue (asterisk in *F*). Cardiomyocytes in outflow cushion of wild-type embryos (*G*) visualized with MF20 immunostaining showed polarized cell morphology with distinct elongated finger-like projections (asterisks) aligned with direction of cell migration and projecting into forming outflow septum (arrow in *G*). In contrast, in *Wdpcp*[Cys40]mutant embryos (*H*), the cardiomyocytes showed rounded morphology without obvious cell polarity, nor the elongated cell projections seen in wild-type embryos. (From Cui et al. 2013; reprinted under the Creative Commons CCO public domain dedication.) (*I–M*) Immunofluorescence staining of cilia with acetylated tubulin (green) and γ-tubulin (red) antibodies (from data in Cui et al. 2013). Shown are the detection of cilia in the mouse embryonic node (*I*), and in the myocardium (*J*), outflow (OFT) cushions (*K*), and atrioventricular cushions (*L*) of wild-type E12.5 embryonic mouse heart. In contrast, cilia are missing in the atrioventricular cushion tissue of *Cc2d2a* mutant known to develop atrioventricular septal defect (adapted from data in Li et al. 2015b). Scale bars, 100 μm (*E*, *G*).

atrial endocardial layer and more prominently in the endocardial cushion mesenchyme (Fig. 3K,L) and in the epicardium (Slough et al. 2008; Willaredt et al. 2012; Li et al. 2015b). A number of cilia-transduced cell signaling pathways have been shown to play essential roles in regulating cardiovascular development and may contribute to the pathogenesis of CHD. These include Shh, TGF-β, BMP, and Wnt signaling. Four genes involved in Shh signaling were recovered from the mouse CHD screen, including *Sufu*, *Fuz*, *Tbc1d32*, and *Kif7* (Fig. 2). Also recovered were six genes involved in TGF-β /BMP signaling, including *Cfc1*, *Megf8*, *Tab1*, *Ltbp1*, *Smad6*, and *Pcsk5* and three mediating Wnt signaling— *Ptk7*, *Prickle1*, and *Fuz* (Fig. 2).

Role of Shh Signaling in Cardiac Development and CHD

Shh signaling is the best-described cilia-transduced cell signaling pathway. Numerous studies have shown that ablation of cilia can result in a drastic reduction of Shh signaling (Huangfu et al. 2003; Han et al. 2008; Goetz and Anderson 2010). During heart development, Shh is expressed in the pharyngeal endoderm and in the foregut endoderm adjacent to incoming SHF derivatives in the dorsal mesenchyme protrusion (Dyer and Kirby 2009). Shh knockout mice show atrial and atrioventricular septation defects, defects in OFT septation, and abnormal pharyngeal arch artery patterning (Washington Smoak et al. 2005). The outflow septation defects are characterized by the aorta shifted rightward overriding the septum, and with either pulmonary atresia or a hypoplastic pulmonary artery observed in conjunction with a variable degree of ventricular hypertrophy. This constellation of defects is reminiscent of tetralogy of Fallot (TOF), one of the most common complex CHD observed clinically (Washington Smoak et al. 2005). Using Cre targeted deletion analysis, it was shown that these outflow defects reflect a dual requirement for pharyngeal endodermal-derived Shh in the cardiac neural crest cells and the SHF derivatives (Goddeeris et al. 2007). These studies showed Shh signaling to the SHF and cardiac neural crest cells are required for OFT septation, but not for either OFT lengthening, or cushion formation, respectively. As the Shh knockout embryos showed a reduction in the number of SHF derivatives, this suggested a requirement for Shh in the specification of the SHF (Hildreth et al. 2009).

Role of Wnt Signaling in Cardiac Development and CHD

Primary cilia also play a role in the transduction of canonical and noncanonical Wnt signaling (Clevers 2006; MacDonald et al. 2009; Wallingford and Mitchell 2011; May-Simera and Kelley 2012) pathways that are also essential for normal heart development. One early evidence linking cilia with β-catenin-dependent canonical Wnt signaling was the observation that knockdown of basal body components bbs1, bbs4, and bss6 resulted in several-fold increase in Wnt activity in zebrafish (Gerdes et al. 2007). The functional link between Wnt signaling and cilia was also shown by the observed localization of noncanonical Wnt/planar cell polarity (PCP) components, such as *Inversin, Dishevelled, Vangl2,* and *Wdpcp* in the basal body and/or ciliary axoneme (Fig. 4A,B) (Montcouquiol et al. 2003; May-Simera and Kelley 2012; Cui et al. 2013). Other studies also showed a role for cilia as a switch that can constrain canonical versus noncanonical Wnt signaling (Ross et al. 2005; Simons et al. 2005; Barrow et al. 2007; Gerdes et al. 2007; Corbit et al. 2008; Huang and Schier 2009; Stottmann et al. 2009; Lienkamp et al. 2012; Oh and Katsanis 2013). However, the precise mechanism by which cilia regulate Wnt signaling is not well understood.

In mice, the noncanonical Wnt/PCP genes such as *Celsr, Frizzled3 (Fzd3), Fzd6, Vangl1-2,* and *Dvl1-3* are highly expressed in the OFT (Etheridge et al. 2008; Paudyal et al. 2010). Mice with mutations in the PCP genes *Vangl2, Scrib* (Phillips et al. 2007), *Dvl 1, 2,* and *3* (Hamblet et al. 2002; Etheridge et al. 2008; Sinha et al. 2012), *Wdpcp* (Cui et al. 2013), and *Pk1* (Gibbs et al. 2016) show a spectrum of CHD phenotypes involving OFT malalignment and septation defects, such as double outlet RV (Fig. 2E,F), overriding aorta, pulmonary atresia (Fig. 3B), and persistent truncus arteriosus (Henderson et al. 2006; Cui et al. 2013; Boczonadi et al. 2014; Gibbs et al. 2016). These cardiac defects likely reflect a role for noncanonical Wnt/PCP pathway in regulating the polarized migration of cardiac neural crest and SHF derivatives (Tada and Smith 2000; Montcouquiol et al. 2003; Simons et al. 2005; Verzi et al. 2005; Cohen et al. 2007; Simons and Mlodzik 2008; Schlessinger et al. 2009; Gibbs et al. 2016). Consistent with this, examination of mouse embryonic fibroblasts derived from the *Wdpcp* or *Pk1* mutant embryos showed inability of the cells to polarize and engage in directional cell migration (Figs. 4C–G, 5K–N). In contrast to Shh deficiency, Wnt/PCP disruption caused failure of the OFT to appropriately lengthen

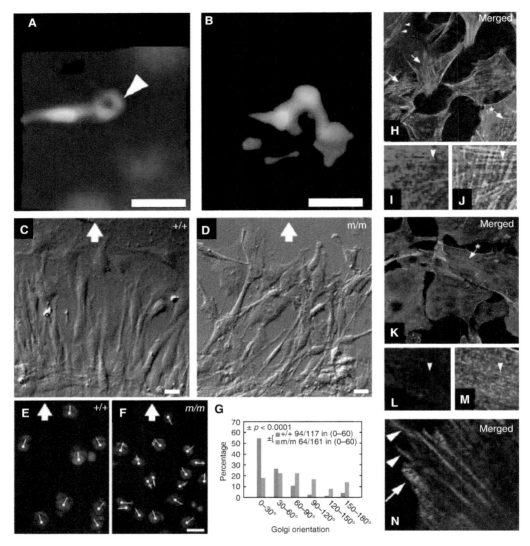

Figure 4. Wdpcp is a cilia protein regulating cell polarity, directional cell migration, and the actin cytoskeleton. (*A,B*) IMCD3 cells immunostained with Wdpcp (green) and acetylated α-tubulin (red) antibodies show Wdpcp localization in the axoneme and ring-like structure (arrowhead) at the ciliary base. Localization of Wdpcp (red) in this ring-like structure is better seen with a 3D isosurface reconstruction, which also shows some colocalization of Septin-2 (green) with the Wdpcp ring. (*C–G*) In a wound-healing assay, control mouse embryonic fibroblasts (MEFs) (*A*) show good alignment with the direction of wound closure (indicated by white arrow). In contrast, *Wdpcp*^{Cys40} mutant MEFs (*B*) showed a disorganized distribution. These differences in cell polarity were also reflected in the Golgi orientation (white line drawn through the center of the Golgi stained green) (*E,F*). In wild-type MEFs, the Golgi (green) was mostly situated at the cell's leading edge (*E,G*), aligned with the direction of wound closure (white arrow), whereas the *Wdpcp*^{Cys40} mutant MEFs show randomized Golgi orientation (*F,G*). Scale bars, 20 μm (*A, B, C, D, F*). (*H–N*) Confocal imaging of Sept2 (red) and Wdpcp (green) showed they are colocalized in actin stress fiber (phalloidin stained, blue) in wild-type MEFs (*H–J*), but in the *Wdpcp*^{Cys40} mutant MEFs, Wdpcp expression was lost (blue, *L*), whereas Sept2 immunostaining (red, *K,M*) showed the loss of colocalization with actin (blue) (*K,L,M*). (*L*) Wdpcp (green) is enriched at the cell cortex where actin filaments (phalloidin) insert into vinculin (red)-containing focal adhesions (*N*) in wild-type MEFs. (Adapted from Cui et al. 2013 under the Creative Commons CC0 public domain dedication.)

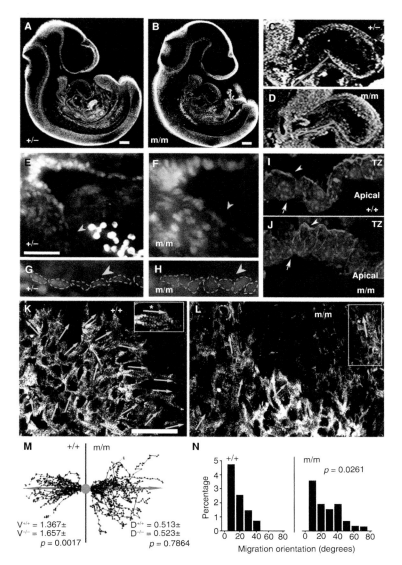

Figure 5. Shortened outflow tract (OFT) and defects in cell polarity and directional cell migration in the $Pk1^{Bj}$ mutant. $(A–D)$ E10.5 $Pk1^{Bj}$ mutant embryo. B and D show shortened OFT as compared with that of heterozygous embryo (C,D). $(E–H)$ Islet1 immunostaining show distribution of SHF cells in the dorsal pericardial wall of the OFT of control (E,G) and $Pk1^{Bj}$ mutant embryos (F,H). Magnified views of region denoted by arrowheads in E and F revealed a cuboidal (H) rather than flat squamous (G) epithelial morphology in the homozygous mutant versus heterozygous embryo. (I,J) β-Catenin (green) and laminin (red) antibody staining of wild-type (I) and Bj mutant embryos (J) shown in the E10.5 Bj mutant embryo, marked disorganization of the epithelium in the transition zone (TZ) of the pericardial wall where SHF derivatives are found. Confocal imaging showed laminin (red) is localized basally (arrowhead I) in the TZ of the control embryo, but in the mutant embryo, it is localized apically (arrow) and basally (arrowhead J), indicating a loss of normal epithelial polarity. The distribution of β-catenin (green) remains at the cell surface in both the control and Bj mutant embryos. (K,L) Myocardiolization defect in the OFT of $Pk1^{Bj}$ mutants. Examination of the striated banding pattern from MF20 immunostain showed the developing myofilaments in the heart are closely aligned and oriented toward the direction of myocardialization in the wild-type E14.5 embryo (K), but in the Bj mutant, the myofilaments are sparse and are largely oriented perpendicular to the direction of myocardialization and septum formation (L). (M,N) Wound closure assay shows a defect in directional cell migration in $Pk1^{Bj}$ mutant mouse embryonic fibroblasts (MEFs). The migration path of MEFs 8 h after wound scratch were well aligned with the direction of wound closure, but tortuous paths were observed with increased velocity for the $Pk1^{Bj}$ mutant MEFs (M,N) (adapted from data in Gibbs et al. 2016). Scale bars, 0.5 μm (A, B); 50 μm (E); 20 μm (K).

(Fig. 5A–D). In the *Pk1* mutant, the epithelial organization and apical-basal polarity of the SHF derivatives in the OFT are disrupted. This would suggest a defect in convergent-extension cell movement required for delamination of a cohesive epithelial sheet mediating OFT lengthening (Fig. 5E–J). This is followed later by a myocardialization defect of the OFT (Figs. 3E–H, 5K–L), that together with the shortened OFT likely account for the great artery malalignment defect in the *Pk1* mutant.

Role of TGF-β Signaling in Cardiac Development and CHD

A role for cilia in mediating TGF-β signaling was recently shown with the finding that ligand binding causes accumulation of TGF-β receptors at the base of the cilium, in a region known as the ciliary pocket (Clement et al. 2013). This triggers receptor-mediated endocytosis involving clathrin-coated vesicles, leading to downstream activation of SMAD phosphorylation (Clement et al. 2013). The essential role of TGF-β/BMP signaling in CHD is well described via in vitro and in vivo analyses of chick and mouse embryos, and also with the examination of knockout mouse models (Combs and Yutzey 2009; de Vlaming et al. 2012; Kruithof et al. 2012; von Gise and Pu 2012). These studies show TGF-β/BMP signaling has multiple roles in cardiovascular development that include the regulation of both endocardial EMT and endocardial cushion development (Potts and Runyan 1989; Camenisch et al. 2002). For example, early endocardial cushion development to acquire critical valve-like function requires BMP signaling in cardiac neural crest cells via the BMPRIA receptors (Nomura-Kitabayashi et al. 2009). A role for *Tgfb2* in OFT and aortic arch remodeling is indicated by the finding that *Tgfb2* knockout mice die perinatally with double outlet RV and interrupted aortic arch (Sanford et al. 1997).

The disturbance of TGF-β/BMP signaling is likely to play a major role in the valvular defects seen in mice harboring mutations disrupting clathrin-mediated endocytosis and endocytic receptor recycling (*Ap2b1, Dnm2, Ap1b1,* *Snx17, LRP1, LRP2*). These endocytic mutants all show OFT malalignment and endocardial cushion defects, phenotypes reminiscent of those observed in mutants with disruption of TGF-β/BMP signaling (Li et al. 2015b). Similarly, mutations affecting cilia integrity in the endocardial cushions may cause disruption of cilia-transduced TGF-β/BMP signaling required for normal valve development. Thus, mutation in *Cc2d2a*, a cilia transition zone component, causes selective loss of cilia in the atrioventricular (AV) but not outflow cushions, and as might be expected, such mutants showed AV valve defects, while the outflow valves were spared (Fig. 3K–M).

ROLE OF CILIA IN SPECIFICATION OF CELL POLARITY AND POLARIZED CELL MIGRATION

Some cilia proteins may help regulate cardiovascular development through cross talk, directly or indirectly, with the cytoskeleton to specify cell polarity and directional cell migration, morphogenetic cell movements, and epithelial–mesenchyme cell transformation. Given the basal body is a microtubule organizing center that can regulate nucleation and organization of microtubule outgrowth, one concept that has emerged is that cilia may regulate the cytoskeleton through dynamic interactions with PCP components and, in this manner, specify cell polarity and polarized cell migration (Figs. 4 and 5) (Wallingford and Mitchell 2011; May-Simera and Kelley 2012). These dynamic cell processes may help to direct the long-distance migration of multiple extracardiac cell populations to the embryonic heart that are required for normal heart development. This includes cells from the SHF, neural crest cells, and the pro-epicardial cells. In addition, cilia directed reorganization of the actin cytoskeleton may also contribute to the regulation of EMT, such as required for the emergence of cardiac neural crest cells from the dorsal neural fold, endocardial EMT mediating formation of the cardiac cushions and valves, or epicardial EMT that generate the epicardially derived cells forming the coronary vessels. These developmental processes involving dynamic reorganization of the cytoskeleton is impacted by cilia and, in conjunction with cilia-transduced cell signaling, may help orchestrate development of the cardiovascular system.

Although the role of cilia in the regulation of cell polarity and directional cell migration in the cardiovascular development is well described in the context of OFT morphogenesis (see above), the precise mechanism and role of the cilia in modulating cell polarity is less understood. In this regard, it is worth pointing out that Wdpcp, a PCP component also known as Fritz, is localized not only in the cilia, but it is also colocalized with septins in the cilia (Kim et al. 2010; Cui et al. 2013) and in the actin cytoskeleton (Kim et al. 2010; Cui et al. 2013). In mouse embryonic fibroblast (MEF) cells deficient in *Wdpcp*, a marked reorganization of the actin cytoskeleton is observed (Fig. 4H–L), and this is associated with altered focal contacts (Fig. 4N) inability to establish cell polarity and engage in directional cell migration (Fig. 4C–G). Similar studies of MEFs harboring a mutation in the PCP component *Pk1* also showed a similar loss of cell polarity and defect in directional cell migration (Fig. 5M,N) (Gibbs et al. 2016). Together, these findings suggest that cilia mutations may cause CHD not only via the disruption of cilia-transduced cell signaling, but cilia mutations also may disrupt the cytoarchitecture and perturb the establishment of cell polarity, polarized cell migration, and/or EMT.

CILIA IN LEFT–RIGHT PATTERNING AND CONGENITAL HEART DISEASE

The enrichment of cilia genes was also notable in that it included a subset of genes that caused CHD in conjunction with left–right patterning defects. This likely reflects the known requirement for cilia in left–right patterning, with previous studies indicating that motile cilia at the embryonic node is required to break symmetry (Fig. 3I) (Hirokawa et al. 2009; Nakamura and Hamada 2012). Analysis of motile cilia mutant mice revealed CHD is typically observed in conjunction with heterotaxy, the randomization of left–right patterning (Tan et al. 2007). This is consistent with the well-described clinical association of complex CHD with heterotaxy (Lin et al. 2014). As the heart is the most left–right asymmetric organ, and this asymmetry is essential for efficient oxygenation of blood, it is perhaps not surprising that left–right patterning defects may play a major role in CHD pathogenesis.

Among 34 cilia mutations recovered causing laterality defects, 22 genes perturbed the primary cilia (*Cc2d2a, Anks6, Nek8, Mks1, Cep290, Bicc1*) versus 12 genes that disrupted motile cilia (*Dnah5, Dnah11, Dnai1, Daw1, Armc4, Ccdc151, Drc1, Ccdc39, Dyxc1x1, Dnaaf3*) (Li et al. 2015b). The latter genes are known to cause PCD, a ciliopathy that is autosomal recessive (Collins et al. 2014; Horani et al. 2014; Lobo et al. 2015). In PCD, immotile/dyskinetic cilia in the airway cause mucociliary clearance defects that can lead to severe sinopulmonary disease. Approximately half of PCD patients show situs solitus, half situs inversus totalis, and varying numbers up to 8% may show CHD with heterotaxy (Kennedy et al. 2007; Shapiro et al. 2014). The disturbance of laterality with PCD reflects the essential role of motile cilia in left–right patterning. Studies in the PCD mutant mouse models showed each PCD mutation can give rise to three phenotypes—approximately half with situs solitus or situs inversus and half with heterotaxy, with complex CHD observed only with heterotaxy (Tan et al. 2007). Although the heterotaxy mutants mostly die prenatally or neonatally from the CHD, mutants with situs solitus or inversus are largely viable postnatally without CHD. Videomicroscopy showed most of these PCD mutants have immotile cilia in the embryonic node, even as half of the mutants show normal or inverted concordant situs that indicate the breaking of symmetry.

These striking observations suggest that motile cilia are not absolutely required for breaking symmetry, nor for left–right axis specification, although motile cilia are clearly required for high-fidelity situs solitus specification. As CHD is only seen with heterotaxy, this provides a clue that patterning of the cardiovascular system may occur very early in development, at the time the left–right body axis is specified. Even as these findings show that motile cilia play an important role in left–right patterning, the recovery of 24 mutations affecting primary cilia suggests nonmotile cilia also play an essential role in laterality specification (Li et al. 2015b). Previous studies suggested a two-cilia hypothesis in which motile cilia at the node generated right to left flow (for additional information, see Shinohara and Hamada 2016). This is proposed to trigger mechanosensory transduction of primary cilia in the perinodal crown cells, causing left-sided calcium release that is propagated into the surrounding lateral plate mesoderm, causing the breaking of symmetry (Nonaka et al. 2002; McGrath et al. 2003; Bruekner 2007; Yoshiba et al. 2012). However, this model has been called into question recently given the failure to detect cilia-mediated mechanosensation and calcium release (Delling et al. 2016).

A role for primary cilia in left–right patterning could be easily understood nevertheless without invoking mechanosensation, because Shh and TGF-β signaling, both cilia-transduced pathways, play important roles in left–right patterning. Although *Shh* knockout mice do not show overt laterality defects, they show LA isomerism (Hildreth et al. 2009). Furthermore, the single outflow vessel seen in the *Shh* knockout mouse is said to represent pulmonary atresia, as the single great artery shows *Pitx2c*, indicating a left-sided identity (Washington Smoak et al. 2005). It is interesting to note in chick embryos where Shh plays a much more primary role in left–right patterning, the experimental manipulation of left–right expression of Shh can cause CHD, confirming its importance of left–right patterning in the pathogenesis of CHD (Levin et al. 1995). Signaling mediated by the TGF-β family of growth factors, including nodal, lefty1, and lefty2, are well described to specify the left–right axis. This nodal signaling cascade is believed to propagate left–right specification initiated at the node. How mutations affecting primary cilia may contribute to the disruption of left–right patterning is not known, but it is thought to cause disturbance in the propagation of this nodal signaling cascade.

CILIARY DYSFUNCTION AND CILIOME MUTATIONS IN CHD PATIENTS

The unexpected enrichment for mutations in cilia-related (ciliome) genes and genes involved in endocytic trafficking and in cilia-transduced cell signaling (Shh, WNT/Pcp, TGF-β) in the mouse mutagenesis screen point to a central role for cilia in CHD pathogenesis. To assess the relevance of these findings to human CHD, we investigated the findings from exome-sequencing analysis of CHD patients by the Pediatric Cardiac Genomics Consortium (PCGC) (Zaidi et al. 2013). In this analysis, the focus was on examining de novo predicted pathogenic coding variants. Although the PCGC publication focused on the recovery of de novo variants in a number of chromatin-modifying genes, interestingly, we noted among the 28 de novo damaging mutations identified in the PCGC CHD patient cohort, 13 or nearly half were in genes associated with pathways identified in the mouse forward genetic screen—that is, ciliogenesis, endocytic trafficking, and cilia-transduced cell signaling (SHH, WNT, TGF-β) (Table 1), with *LRP2* being a gene recovered in both the PCGC CHD patients and the mouse CHD mutants recovered in our screen. We also noted the recovery in the PCGC cohort of a de novo variant in *Pitx2*, a gene known to play an essential role in left–right patterning, supporting an important role for left–right patterning disturbance in CHD pathogenesis.

Further supporting a central role for cilia in the pathogenesis of CHD are clinical studies showing a high prevalence of ciliary dysfunction in CHD patients (Nakhleh et al. 2012; Garrod et al. 2014). Given that respiratory complications are among the biggest postsurgical complications for CHD patients, we previously hypothesized that some CHD patients with respiratory complications may have undiagnosed PCD. These studies were initiated with an examination of CHD patients with heterotaxy. Nasal scrapes were conducted and video microscopy was used to examine cilia motility in the nasal epithelium. This analysis showed a high prevalence of ciliary dysfunction in CHD patients with heterotaxy. The ciliary motion defects

Table 1. Functional annotation for 13 PCGC patients with de novo mutations

Patient ID	CHD type[a]	Gene	Mutation	Gene function annotation
1-00638	CTD	*FBN2*	p.D2191N	TGF-β signaling
1-02020	HTX	*SMAD2*	p.IVS12 + 1G > A	TGF-β signaling
1-02621	HTX	*SMAD2*	p.W244C	TGF-β signaling
1-00197	LVO	*BCL9*	p.M1395K	Wnt signaling
1-01828	CTD	*DAPK3*	p.P193L	Wnt signaling
1-01138	LVO	*USP34*	p.L432P	Wnt signaling
1-00802	LVO	*PTCH1*	p.R831Q	Shh signaling/ciliome
1-02598	HTX	*LRP2*[b]	p.E4372K	Shh signaling/endocytic trafficking
1-01913	Other	*RAB10*	p.N112S	Endocytic trafficking
1-00750	LVO	*HUWE1*	p.R3219C	Ciliome
1-01151	CTD	*SUV420H1*	p.R143C	Ciliome
1-00853	CTD	*WDR5*	p.K7Q	Ciliome
1-02952	LVO	*PITX2*	p.A47V	Laterality-related

Based on exome-sequencing analysis of congenital heart disease (CHD) patients by Pediatric Cardiac Genomics Consortium (Data from Zaidi et al. 2013).

[a]CTD, Conotruncal defect; HTX, heterotaxy; LVO, left ventricular obstruction.

[b]LRP2 is an endocytic gene also recovered from our mouse screen.

observed span a spectrum that included some showing dyskinetic ciliary motion to slow or even immotile cilia. Overall, >40% of the patients showed ciliary dysfunction (Nakhleh et al. 2012). Although this was associated with an enrichment for coding variants in PCD genes, no patient was either homozygous or compound heterozygous for any PCD gene mutations. Thus, although CHD patients with heterotaxy are at high risk for ciliary dysfunction, these patients largely do not have PCD. Since this initial study, a large study has been conducted comprising >200 patients with CHD of a broad spectrum, mostly without heterotaxy. This analysis showed a similar high prevalence of ciliary dysfunction and this was correlated with increased risk of having PCD-related respiratory symptoms (Garrod et al. 2014). Together, these findings suggest ciliary dysfunction is commonly associated with CHD in the human population.

Although these studies focused on assessing motile cilia function, we note many cilia genes are expressed in both motile and primary cilia. Hence, the high prevalence of ciliary dysfunction in CHD patients may reflect not only the perturbation of motile cilia genes, but also genes required for primary cilia function. Indeed, we recently showed a patient harboring compound heterozygous mutations in *WDR35* causing

Sensenbrenner syndrome, a ciliopathy thought to affect only the primary cilia, showed motile cilia dysfunction. Pulmonary function assessments indicated obstructive airway disease that suggested possible mucociliary clearance defects in the airway (Li et al. 2015a). Indeed, several clinical studies have shown an increase in respiratory symptoms and disease in patients with other ciliopathies thought to affect only the primary cilia, indicating the distinction between ciliopathies involving motile versus primary cilia may not be so clear cut (Tobin and Beales 2009). These findings suggest further studies are warranted to assess ciliopathy patients of a wide spectrum for potential pulmonary complications, especially for those who will undergo high-risk surgeries, such as those involving cardiopulmonary bypass.

CONGENITAL HEART DISEASE AND CILIOPATHIES

It is notable that many cilia genes recovered in the mouse forward genetic screen for CHD-causing mutations are genes clinically known to cause various human ciliopathies. This includes not only motile cilia genes associated with PCD, but also cilia genes linked to various ciliopathies thought to affect the primary cilia, such as in JBTS, polycystic kidney disease, acrocallosal syndrome, hydroelethalus, Leber congential amaurosis, Meckel–Gruber syndrome, Bardet–Biedl syndrome, etc. (Li et al. 2015b). While in our mouse screen, ciliopathy genes were recovered based on mutations causing CHD phenotypes, clinically these ciliopathies are not commonly associated with CHD. This may reflect ascertainment bias given that the patient population represent only human fetuses that can survive to term and, hence, are less likely to have severe cardiac anomalies. Indeed, clinical studies of aborted or stillborn fetuses have shown that the human fetal population has more than ten times higher incidence of CHD as compared with those in the clinical patient population (Hoffman and Kaplan 2002). Consistent with this, most of the CHD ciliopathy mutants recovered from our screen were inviable to term and were harvested preterm after in utero phenotyping by fetal echocardiography. On the flip side, there is undoubtedly ascertainment bias in our screen in the recovery of mutations in ciliopathy genes that specifically can cause CHD. That different ciliopathy mutations may have varying levels of penetrance for CHD phenotypes is suggested by observations of our mutant *Hug* (Damerla et al. 2015). This mutant has a mutation in *Jbts17*, a gene encoding a cilia transition zone protein known to cause JBTS (Srour et al. 2012). *Hug* mutants show cerebellar defects expected for JBTS and they also can show CHD comprising of pulmonary atresia. However, the CHD phenotype is incomplete in penetrance, as some *Hug* mutants show no heart defects (Damerla et al. 2015). These observations suggest that different mutations in the same ciliopathy gene may generate different phenotypic outcome and this perhaps can be further modified by the genetic background of the individual.

In light of these observations, we suggest that, clinically, CHD may be considered a structural birth defect related to ciliopathies. However, unlike other ciliopathies, which are relatively rare (<1 in 10,000) and with a Mendelian recessive inheritance, the much higher prevalence of CHD (up to 1%) and its sporadic occurrence would suggest the contribution of cilia-related

or ciliome genes in CHD will be multigenic and highly genetically heterogeneous. Such complex genetics is expected to reflect the complexity of cilia biology in which sequence variants found among different "ciliome" genes may affect the function of large multiprotein complexes that regulate ciliogenesis and cilia structure and function. Given that there are hundreds of ciliome genes that contribute to cilia assembly and cilia structure and function, it is perhaps not surprising that CHD patients are observed to have a high prevalence of ciliary dysfunction. While the CHD genes recovered from the mouse screen were by design recessive mutations, we expect mutations in these same genes can contribute to more complex genetic models of disease. Such complex genetics may also contribute to classic ciliopathies, as there are clinical reports of PCD patients and patients with other ciliopathies that have no homozygous or compound heterozygous ciliopathy mutations, but instead show multiple heterozygous mutations in known PCD or other ciliopathy genes (de Pontual et al. 2009; Li et al. 2016). A future challenge is to develop an effective bioinformatics pipeline for modeling and interrogating such complex genetics and assess the contribution of ciliome mutations in the pathogenesis of CHD and other structural birth defects.

CONCLUSIONS

CHDs are the most common structural birth defects, and despite its prevalence, the genetic etiology of CHD remains poorly understood. Interrogations into the genetic landscape for CHD using a large-scale forward genetic screen in mice unveiled a central role for ciliome genes in the pathogenesis of CHD. These studies suggest the perturbation of cilia and cilia-transduced cell signaling pathways may play a central role in the pathogenesis of CHD. The future challenge is to clinically translate these findings in mice to patients with CHD. The finding of a high prevalence of ciliary dysfunction in CHD patients and the enrichment of de novo pathogenic variants in cilia and cilia-related pathways in CHD patients would suggests such studies will be fruitful and may provide the basis for stratifying patients to optimize the clinical management of patient care. The recent finding of primary cilia in the endothelial cells of the aorta regulating anti-atherosclerotic responses also point to a potential role for cilia in adult cardiac disease (Dinsmore and Reiter 2016). Further work in the future will be needed to clarify the role of cilia biology in human CHD and perhaps other cardiovascular diseases, and with such insights may come new avenues of therapeutic intervention to improve the outcome for patients with critical heart disease.

REFERENCES

*Reference is also in this collection.

Barrow JR, Howell WD, Rule M, Hayashi S, Thomas KR, Capecchi MR, McMahon AP. 2007. Wnt3 signaling in the epiblast is required for proper orientation of the anteroposterior axis. *Dev Biol* **312:** 312–320.

Boczonadi V, Gillespie R, Keenan I, Ramsbottom SA, Donald-Wilson C, Al Nazer M, Humbert P, Schwarz RJ, Chaudhry B, Henderson DJ. 2014. Scrib:Rac1 interactions are required for the morphogenesis of the ventricular myocardium. *Cardiovasc Res* **104:** 103–115.

Brueckner M. 2007. Heterotaxia, congenital heart disease, and primary ciliary dyskinesia. *Circulation* **115:** 2793–2795.

Buckingham M. 2016. First and second heart field. In *Congenital heart diseases: The broken heart: Clinical features, human genetics and molecular pathways* (ed. Rickert-Sperling S, Kelly GR, Driscoll JD), pp. 25–40. Springer, Vienna.

Camenisch TD, Molin DG, Person A, Runyan RB, Gittenberger-de Groot AC, McDonald JA, Klewer SE. 2002. Temporal and distinct TGF-β ligand requirements during mouse and avian endocardial cushion morphogenesis. *Dev Biol* **248:** 170–181.

Clement CA, Ajbro KD, Koefoed K, Vestergaard ML, Veland IR, Henriques de Jesus MP, Pedersen LB, Benmerah A, Andersen CY, Larsen LA, et al. 2013. TGF-β signaling is associated with endocytosis at the pocket region of the primary cilium. *Cell Rep* **3:** 1806–1814.

Clevers H. 2006. Wnt/β-catenin signaling in development and disease. *Cell* **127:** 469–480.

Cohen ED, Wang Z, Lepore JJ, Lu MM, Taketo MM, Epstein DJ, Morrisey EE. 2007. Wnt/β-catenin signaling promotes expansion of Isl-1-positive cardiac progenitor cells through regulation of FGF signaling. *J Clin Invest* **117:** 1794–1804.

Collins SA, Walker WT, Lucas JS. 2014. Genetic testing in the diagnosis of primary ciliary dyskinesia: State-of-the-art and future perspectives. *J Clin Med* **3:** 491–503.

Combs MD, Yutzey KE. 2009. Heart valve development: Regulatory networks in development and disease. *Circ Res* **105:** 408–421.

Corbit KC, Shyer AE, Dowdle WE, Gaulden J, Singla V, Chen MH, Chuang PT, Reiter JF. 2008. Kif3a constrains β-catenin-dependent Wnt signalling through dual ciliary and non-ciliary mechanisms. *Nat Cell Biol* **10:** 70–76.

Cui C, Chatterjee B, Lozito TP, Zhang Z, Francis RJ, Yagi H, Swanhart LM, Sanker S, Francis D, Yu Q, et al. 2013. Wdpcp, a PCP protein required for ciliogenesis, regulates directional cell migration and cell polarity by direct modulation of the actin cytoskeleton. *PLoS Biol* **11:** e1001720.

Damerla RR, Cui C, Gabriel GC, Liu X, Craige B, Gibbs BC, Francis R, Li Y, Chatterjee B, San Agustin JT, et al. 2015. Novel *Jbts17* mutant mouse model of Joubert syndrome with cilia transition zone defects and cerebellar and other ciliopathy related anomalies. *Hum Mol Genet* **24:** 3994–4005.

Delling M, Indzhykulian AA, Liu X, Li Y, Xie T, Corey DP, Clapham DE. 2016. Primary cilia are not calcium-responsive mechanosensors. *Nature* **531:** 656–660.

de Pontual L, Zaghloul NA, Thomas S, Davis EE, Mcgaughey DM, Dollfus H, Baumann C, Bessling SL, Babarit C, Pelet A, et al. 2009. Epistasis between *RET* and *BBS* mutations modulates enteric innervation and causes syndromic Hirschsprung disease. *Proc Natl Acad Sci* **106:** 13921–13926.

de Vlaming A, Sauls K, Hajdu Z, Visconti RP, Mehesz AN, Levine RA, Slaugenhaupt SA, Hagege A, Chester AH, Markwald RR, et al. 2012. Atrioventricular valve development: New perspectives on an old theme. *Differentiation* **84:** 103–116.

Dinsmore C, Reiter JF. 2016. Endothelial primary cilia inhibit atherosclerosis. *EMBO Rep* **17:** 156–166.

Dyer LA, Kirby ML. 2009. Sonic Hedgehog maintains proliferation in secondary heart field progenitors and is required for normal arterial pole formation. *Dev Biol* **330:** 305–317.

Etheridge SL, Ray S, Li S, Hamblet NS, Lijam N, Tsang M, Greer J, Kardos N, Wang J, Sussman DJ, et al. 2008. Murine dishevelled 3 functions in redundant pathways with dishevelled 1 and 2 in normal cardiac outflow tract, cochlea, and neural tube development. *PLoS Genet* **4:** e1000259.

Fahed AC, Gelb BD, Seidman JG, Seidman CE. 2013. Genetics of congenital heart disease: The glass half empty. *Circ Res* **112:** 707–720.

Garrod AS, Zahid M, Tian X, Francis RJ, Khalifa O, Devine W, Gabriel GC, Leatherbury L, Lo CW. 2014. Airway ciliary dysfunction and sinopulmonary symptoms in patients with congenital heart disease. *Ann Am Thorac Soc* **11:** 1426–1432.

Gerdes JM, Liu Y, Zaghloul NA, Leitch CC, Lawson SS, Kato M, Beachy PA, Beales PL, DeMartino GN, Fisher S, et al. 2007. Disruption of the basal body compromises proteasomal function and perturbs intracellular Wnt response. *Nat Genet* **39:** 1350–1360.

Gibbs BC, Damerla RR, Vladar EK, Chatterjee B, Wan Y, Liu X, Cui C, Gabriel GC, Zahid M, Yagi H, et al. 2016. Prickle1 mutation causes planar cell polarity and directional cell migration defects associated with cardiac outflow tract anomalies and other structural birth defects. *Biol Open* **5:** 323–335.

Gill HK, Splitt M, Sharland GK, Simpson JM. 2003. Patterns of recurrence of congenital heart disease: An analysis of 6,640 consecutive pregnancies evaluated by detailed fetal echocardiography. *J Am Coll Cardiol* **42:** 923–929.

Glessner JT, Bick AG, Ito K, Homsy JG, Rodriguez-Murillo L, Fromer M, Mazaika E, Vardarajan B, Italia M, Leipzig J, et al. 2014. Increased frequency of de novo copy number variants in congenital heart disease by integrative analysis of single nucleotide polymorphism array and exome sequence data. *Circ Res* **115:** 884–896.

Goddeeris MM, Schwartz R, Klingensmith J, Meyers EN. 2007. Independent requirements for Hedgehog signaling by both the anterior heart field and neural crest cells for outflow tract development. *Development* **134:** 1593–1604.

Goetz SC, Anderson KV. 2010. The primary cilium: A signalling centre during vertebrate development. *Nat Rev Genet* **11:** 331–344.

Hamblet NS, Lijam N, Ruiz-Lozano P, Wang J, Yang Y, Luo Z, Mei L, Chien KR, Sussman DJ, Wynshaw-Boris A. 2002. Dishevelled 2 is essential for cardiac outflow tract development, somite segmentation and neural tube closure. *Development* **129:** 5827–5838.

Han YG, Spassky N, Romaguera-Ros M, Garcia-Verdugo JM, Aguilar A, Schneider-Maunoury S, Alvarez-Buylla A. 2008. Hedgehog signaling and primary cilia are required for the formation of adult neural stem cells. *Nat Neurosci* **11:** 277–284.

Henderson DJ, Phillips HM, Chaudhry B. 2006. Vang-like 2 and noncanonical Wnt signaling in outflow tract development. *Trends Cardiovasc Med* **16:** 38–45.

Hildreth V, Webb S, Chaudhry B, Peat JD, Phillips HM, Brown N, Anderson RH, Henderson DJ. 2009. Left cardiac isomerism in the Sonic Hedgehog null mouse. *J Anat* **214:** 894–904.

Hirokawa N, Tanaka Y, Okada Y. 2009. Left–right determination: Involvement of molecular motor KIF3, cilia, and nodal flow. *Cold Spring Harb Perspect Biol* **1:** a000802.

Hoffman JI, Kaplan S. 2002. The incidence of congenital heart disease. *J Am Coll Cardiol* **39:** 1890–1900.

Horani A, Brody SL, Ferkol TW. 2014. Picking up speed: Advances in the genetics of primary ciliary dyskinesia. *Pediatr Res* **75:** 158–164.

Huang P, Schier AF. 2009. Dampened Hedgehog signaling but normal Wnt signaling in zebrafish without cilia. *Development* **136:** 3089–3098.

Huangfu D, Liu A, Rakeman AS, Murcia NS, Niswander L, Anderson KV. 2003. Hedgehog signalling in the mouse requires intraflagellar transport proteins. *Nature* **426:** 83–87.

Kennedy MP, Omran H, Leigh MW, Dell S, Morgan L, Molina PL, Robinson BV, Minnix SL, Olbrich H, Severin T, et al. 2007. Congenital heart disease and other heterotaxic defects in a large cohort of patients with primary ciliary dyskinesia. *Circulation* **115:** 2814–2821.

Kim SK, Shindo A, Park TJ, Oh EC, Ghosh S, Gray RS, Lewis RA, Johnson CA, Attie-Bittach T, Katsanis N, et al. 2010. Planar cell polarity acts through septins to control collective cell movement and ciliogenesis. *Science* **329:** 1337–1340.

Kirby ML, Waldo KL. 1990. Role of neural crest in congenital heart disease. *Circulation* **82:** 332–340.

Kobayashi T, Dynlacht BD. 2011. Regulating the transition from centriole to basal body. *J Cell Biol* **193:** 435–444.

Kruithof BP, Duim SN, Moerkamp AT, Goumans MJ. 2012. TGF-β and BMP signaling in cardiac cushion formation: Lessons from mice and chicken. *Differentiation* **84:** 89–102.

Levin M, Johnson RL, Stern CD, Kuehn M, Tabin C. 1995. A molecular pathway determining left–right asymmetry in chick embryogenesis. *Cell* **82:** 803–814.

Li Y, Garrod AS, Madan-Khetarpal S, Sreedher G, McGuire M, Yagi H, Klena NT, Gabriel GC, Khalifa O, Zahid M, et al. 2015a. Respiratory motile cilia dysfunction in a patient with cranioectodermal dysplasia. *Am J Med Genet A* **167A:** 2188–2196.

Li Y, Klena NT, Gabriel GC, Liu X, Kim AJ, Lemke K, Chen Y, Chatterjee B, Devine W, Damerla RR, et al. 2015b. Global genetic analysis in mice unveils central role for cilia in congenital heart disease. *Nature* **521:** 520–524.

Li Y, Yagi H, Onuoha EO, Damerla RR, Francis R, Furutani Y, Tariq M, King SM, Hendricks G, Cui C, et al. 2016. *DNAH6* and its interactions with PCD genes in heterotaxy and primary ciliary dyskinesia. *PLoS Genet* **12:** e1005821.

Lienkamp S, Ganner A, Walz G. 2012. Inversin, Wnt signaling and primary cilia. *Differentiation* **83:** S49–S55.

Lin AE, Krikov S, Riehle-Colarusso T, Frias JL, Belmont J, Anderka M, Geva T, Getz KD, Botto LD; National Birth Defects Prevention S. 2014. Laterality defects in the national birth defects prevention study (1998–2007): Birth prevalence and descriptive epidemiology. *Am J Med Genet A* **164:** 2581–2591.

Liu X, Francis R, Kim AJ, Ramirez R, Chen G, Subramanian R, Anderton S, Kim Y, Wong L, Morgan J, et al. 2014. Interrogating congenital heart defects with noninvasive fetal echocardiography in a mouse forward genetic screen. *Circ Cardiovasc Imaging* **7:** 31–42.

Lobo J, Zariwala MA, Noone PG. 2015. Primary ciliary dyskinesia. *Semin Respir Crit Care Med* **36:** 169–179.

MacDonald BT, Tamai K, He X. 2009. Wnt/β-catenin signaling: Components, mechanisms, and diseases. *Dev Cell* **17:** 9–26.

Marelli A, Miller SP, Marino BS, Jefferson AL, Newburger JW. 2016. Brain in congenital heart disease across the lifespan: The cumulative burden of injury. *Circulation* **133:** 1951–1962.

May-Simera HL, Kelley MW. 2012. Cilia, Wnt signaling, and the cytoskeleton. *Cilia* **1:** 7.

McGrath J, Somlo S, Makova S, Tian X, Brueckner M. 2003. Two populations of node monocilia initiate left–right asymmetry in the mouse. *Cell* **114:** 61–73.

Montcouquiol M, Rachel RA, Lanford PJ, Copeland NG, Jenkins NA, Kelley MW. 2003. Identification of *Vangl2* and *Scrb1* as planar polarity genes in mammals. *Nature* **423:** 173–177.

Nakamura T, Hamada H. 2012. Left–right patterning: Conserved and divergent mechanisms. *Development* **139:** 3257–3262.

Nakhleh N, Francis R, Giese RA, Tian X, Li Y, Zariwala MA, Yagi H, Khalifa O, Kureshi S, Chatterjee B, et al. 2012.

High prevalence of respiratory ciliary dysfunction in congenital heart disease patients with heterotaxy. *Circulation* **125:** 2232–2242.

Newburger JW, Sleeper LA, Bellinger DC, Goldberg CS, Tabbutt S, Lu M, Mussatto KA, Williams IA, Gustafson KE, Mital S, et al. 2012. Early developmental outcome in children with hypoplastic left heart syndrome and related anomalies: The single ventricle reconstruction trial. *Circulation* **125:** 2081–2091.

Nomura-Kitabayashi A, Phoon CK, Kishigami S, Rosenthal J, Yamauchi Y, Abe K, Yamamura K, Samtani R, Lo CW, Mishina Y. 2009. Outflow tract cushions perform a critical valve-like function in the early embryonic heart requiring BMPRIA-mediated signaling in cardiac neural crest. *Am J Physiol Heart Circ Physiol* **297:** H1617–H1628.

Nonaka S, Shiratori H, Saijoh Y, Hamada H. 2002. Determination of left–right patterning of the mouse embryo by artificial nodal flow. *Nature* **418:** 96–99.

Oh EC, Katsanis N. 2013. Context-dependent regulation of Wnt signaling through the primary cilium. *J Am Soc Nephrol* **24:** 10–18.

Oyen N, Poulsen G, Boyd HA, Wohlfahrt J, Jensen PK, Melbye M. 2009. Recurrence of congenital heart defects in families. *Circulation* **120:** 295–301.

Paudyal A, Damrau C, Patterson VL, Ermakov A, Formstone C, Lalanne Z, Wells S, Lu X, Norris DP, Dean CH, et al. 2010. The novel mouse mutant, *chuzhoi*, has disruption of Ptk7 protein and exhibits defects in neural tube, heart and lung development and abnormal planar cell polarity in the ear. *BMC Dev Biol* **10:** 87.

Phillips HM, Rhee HJ, Murdoch JN, Hildreth V, Peat JD, Anderson RH, Copp AJ, Chaudhry B, Henderson DJ. 2007. Disruption of planar cell polarity signaling results in congenital heart defects and cardiomyopathy attributable to early cardiomyocyte disorganization. *Circ Res* **101:** 137–145.

Potts JD, Runyan RB. 1989. Epithelial-mesenchymal cell transformation in the embryonic heart can be mediated, in part, by transforming growth factor β. *Dev Biol* **134:** 392–401.

Rash JE, Shay JW, Biesele JJ. 1969. Cilia in cardiac differentiation. *J Ultrastruct Res* **29:** 470–484.

Reiter JF, Blacque OE, Leroux MR. 2012. The base of the cilium: Roles for transition fibres and the transition zone in ciliary formation, maintenance and compartmentalization. *EMBO Rep* **13:** 608–618.

Ross AJ, May-Simera H, Eichers ER, Kai M, Hill J, Jagger DJ, Leitch CC, Chapple JP, Munro PM, Fisher S, et al. 2005. Disruption of Bardet–Biedl syndrome ciliary proteins perturbs planar cell polarity in vertebrates. *Nat Genet* **37:** 1135–1140.

Sanford LP, Ormsby I, Gittenberger-de Groot AC, Sariola H, Friedman R, Boivin GP, Cardell EL, Doetschman T. 1997. TGF-β2 knockout mice have multiple developmental defects that are non-overlapping with other TGF-β knockout phenotypes. *Development* **124:** 2659–2670.

Schlessinger K, Hall A, Tolwinski N. 2009. Wnt signaling pathways meet Rho GTPases. *Genes Dev* **23:** 265–277.

Shapiro AJ, Davis SD, Ferkol T, Dell SD, Rosenfeld M, Olivier KN, Sagel SD, Milla C, Zariwala MA, Wolf W, et al. 2014. Laterality defects other than situs inversus totalis in primary ciliary dyskinesia: Insights into situs ambiguus and heterotaxy. *Chest* **146:** 1176–1186.

Shen Y, Leatherbury L, Rosenthal J, Yu Q, Pappas MA, Wessels A, Lucas J, Siegfried B, Chatterjee B, Svenson K, et al. 2005. Cardiovascular phenotyping of fetal mice by noninvasive high-frequency ultrasound facilitates recovery of ENU-induced mutations causing congenital cardiac and extracardiac defects. *Physiol Genomics* **24:** 23–36.

* Shinohara K, Hamada H. 2016. Cilia in left–right symmetry breaking. *Cold Spring Harb Perspect Biol* doi: 10.1101/cshperspect.a028282.

Simons M, Mlodzik M. 2008. Planar cell polarity signaling: From fly development to human disease. *Annu Rev Genet* **42:** 517–540.

Simons M, Gloy J, Ganner A, Bullerkotte A, Bashkurov M, Kronig C, Schermer B, Benzing T, Cabello OA, Jenny A, et al. 2005. Inversin, the gene product mutated in nephronophthisis type II, functions as a molecular switch between Wnt signaling pathways. *Nat Genet* **37:** 537–543.

Sinha T, Wang B, Evans S, Wynshaw-Boris A, Wang J. 2012. Disheveled mediated planar cell polarity signaling is required in the second heart field lineage for outflow tract morphogenesis. *Dev Biol* **370:** 135–144.

Slough J, Cooney L, Brueckner M. 2008. Monocilia in the embryonic mouse heart suggest a direct role for cilia in cardiac morphogenesis. *Dev Dyn* **237:** 2304–2314.

Sorokin S. 1962. Centrioles and the formation of rudimentary cilia by fibroblasts and smooth muscle cells. *J Cell Biol* **15:** 363–377.

Srour M, Schwartzentruber J, Hamdan FF, Ospina LH, Patry L, Labuda D, Massicotte C, Dobrzeniecka S, Capo-Chichi JM, Papillon-Cavanagh S, et al. 2012. Mutations in *C5ORF42* cause Joubert syndrome in the French Canadian population. *Am J Hum Genet* **90:** 693–700.

Stottmann RW, Tran PV, Turbe-Doan A, Beier DR. 2009. *Ttc21b* is required to restrict Sonic Hedgehog activity in the developing mouse forebrain. *Dev Biol* **335:** 166–178.

Tada M, Smith JC. 2000. Xwnt11 is a target of *Xenopus* Brachyury: Regulation of gastrulation movements via Dishevelled, but not through the canonical Wnt pathway. *Development* **127:** 2227–2238.

Tan SY, Rosenthal J, Zhao XQ, Francis RJ, Chatterjee B, Sabol SL, Linask KL, Bracero L, Connelly PS, Daniels MP, et al. 2007. Heterotaxy and complex structural heart defects in a mutant mouse model of primary ciliary dyskinesia. *J Clin Invest* **117:** 3742–3752.

Tobin JL, Beales PL. 2009. The nonmotile ciliopathies. *Genet Med* **11:** 386–402.

van der Bom T, Bouma BJ, Meijboom FJ, Zwinderman AH, Mulder BJ. 2012. The prevalence of adult congenital heart disease, results from a systematic review and evidence based calculation. *Am Heart J* **164:** 568–575.

Verzi MP, McCulley DJ, De Val S, Dodou E, Black BL. 2005. The right ventricle, outflow tract, and ventricular septum comprise a restricted expression domain within the secondary/anterior heart field. *Dev Biol* **287:** 134–145.

von Gise A, Pu WT. 2012. Endocardial and epicardial epithelial to mesenchymal transitions in heart development and disease. *Circ Res* **110:** 1628–1645.

Wallingford JB, Mitchell B. 2011. Strange as it may seem: The many links between Wnt signaling, planar cell polarity, and cilia. *Genes Dev* **25:** 201–213.

Washington Smoak I, Byrd NA, Abu-Issa R, Goddeeris MM, Anderson R, Morris J, Yamamura K, Klingensmith J, Meyers EN. 2005. Sonic hedgehog is required for cardiac outflow tract and neural crest cell development. *Dev Biol* **283:** 357–372.

Willaredt MA, Gorgas K, Gardner HA, Tucker KL. 2012. Multiple essential roles for primary cilia in heart development. *Cilia* **1:** 23.

Yoshiba S, Shiratori H, Kuo IY, Kawasumi A, Shinohara K, Nonaka S, Asai Y, Sasaki G, Belo JA, Sasaki H, et al. 2012. Cilia at the node of mouse embryos sense fluid flow for left–right determination via Pkd2. *Science* **338:** 226–231.

Zaidi S, Choi M, Wakimoto H, Ma L, Jiang J, Overton JD, Romano-Adesman A, Bjornson RD, Breitbart RE, Brown KK, et al. 2013. De novo mutations in histone-modifying genes in congenital heart disease. *Nature* **498:** 220–223.

Cite this article as *Cold Spring Harb Perspect Biol* doi: 10.1101/cshperspect.a028266

Photoreceptor Cilia and Retinal Ciliopathies

Kinga M. Bujakowska, Qin Liu, and Eric A. Pierce

Ocular Genomics Institute, Massachusetts Eye and Ear Infirmary, Department of Ophthalmology, Harvard Medical School, Boston, Massachusetts 02114

Correspondence: eric_pierce@meei.harvard.edu

Photoreceptors are sensory neurons designed to convert light stimuli into neurological responses. This process, called phototransduction, takes place in the outer segments (OS) of rod and cone photoreceptors. OS are specialized sensory cilia, with analogous structures to those present in other nonmotile cilia. Deficient morphogenesis and/or dysfunction of photoreceptor sensory cilia (PSC) caused by mutations in a variety of photoreceptor-specific and common cilia genes can lead to inherited retinal degenerations (IRDs). IRDs can manifest as isolated retinal diseases or syndromic diseases. In this review, we describe the structure and composition of PSC and different forms of ciliopathies with retinal involvement. We review the genetics of the IRDs, which are monogenic disorders but genetically diverse with regard to causality.

Photoreceptors are sensory neurons designed to convert light stimuli into electrical responses, a process called phototransduction. Phototransduction takes place in the highly specialized compartment of photoreceptors, the outer segment (OS) (Pearring et al. 2013; Molday and Moritz 2015). The OS of the rod and cone photoreceptors differ in structure and protein composition, related to their functional adaptation, in which rods have high sensitivity necessary in dim light and cones are responsible for the high-resolution color vision working in bright light (Lamb and Pugh 2006; Lamb et al. 2007). Research over the past decade on the genetic and molecular components of photoreceptors in vertebrate retinae has led to the clear recognition that photoreceptor OS are specialized sensory cilia (Liu et al. 2007a; Ramamurthy and Cayouette 2009; Khanna 2015). Deficient

morphogenesis and/or dysfunction of photoreceptor sensory cilia (PSC) caused by mutations in a variety of photoreceptor-specific and common cilia genes can lead to a group of clinical manifestations, called inherited retinal degenerations (IRDs). In this review, we will discuss the structure and composition of PSC and different forms of ciliopathies with retinal involvement.

SPECIALIZED PHOTORECEPTOR SENSORY CILIA

In vertebrate retina, the visual function depends on the formation of complex sensory cilia of rod and cone photoreceptors. Photoreceptors are highly polarized neurons, composed of four distinct compartments: the OS, the inner segment (IS), the nucleus and a short axon extending to second order neurons (bipolar and horizontal cells) (Fig. 1) (Kennedy and Malicki 2009; Pearring et al. 2013; Molday and Moritz 2015). OS are ciliary organelles with analogous structure to the primary sensory cilia in other cell types (Rosenbaum and Witman 2002; Liu

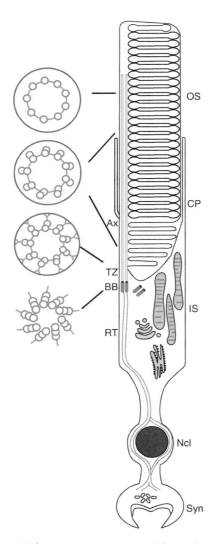

Figure 1. Photoreceptor structure. Schematic representation of the rod photoreceptor with sensory cilia components indicated. The drawings to the *left* of the photoreceptor represent the cross-sectional view of the microtubule structure of the distal axoneme (Ax), proximal Ax, transition zone (TZ), and basal body (BB). RT, rootlet; OS, outer segment; CP, calyceal process; IS, inner segment; Ncl, nucleus; Syn, synapse.

et al. 2007a; Ramamurthy and Cayouette 2009; Khanna 2015). The recognition of PSCs as distinct morphological structures is valuable for the study of photoreceptor cell biology and disease pathogenesis. Studies of genes involved in IRDs and other ciliopathies have identified dozens of novel components of the PSCs. A comprehensive proteomic study of mouse PSCs has identified ~2000 proteins in this organelle, out of which hundreds are present in other cilia (Liu et al. 2007a). These findings have greatly improved our understanding of how photoreceptor cilia are built and maintained, and how these processes are disrupted in disease.

Ciliary Backbone of PSC

The structure of the PSC is analogous to other cilia, where the axoneme arises from the basal body through the transition zone (also called the "connecting cilium") and extends up to two-thirds of the OS (Fig. 1) (De Robertis 1956; Kaplan et al. 1987). The basal body also nucleates the ciliary rootlet, which extends into the IS, and which is covalently linked to the PSC structure (Yang et al. 2005; Liu et al. 2007a). The basal body contains nine triplet microtubules, two of which extend further to form the axoneme and the third anchors transition fibers linking the basal body to the plasma membrane. The nine doublet microtubules in the transition zone are cross-linked to the surrounding plasma membrane by Y-link structures (Besharse et al. 1985; Horst et al. 1990). These Y-link structures are absent in the rest of the axoneme. The transition zone of rods and cones measures ~0.3 μm in diameter and 1–1.5 μm in length, which is fairly consistent throughout the species (Besharse et al. 1985). This structure was originally called the "connecting cilium" by De Robertis in 1956, when he was studying some of the first electron micrographs of photoreceptor cells (De Robertis 1956). However, as an analogy with other primary sensory cilia, we refer to this region as a transition zone. Above the transition zone, disc morphogenesis takes place where the surrounding plasma membrane transforms into the disc precursors through membrane evagination (Steinberg et al. 1980; Ding et al. 2015; Pugh 2015). At the distal part of the PSC axoneme, the double microtubules are reduced to singlets (Fig. 1) (Rosenbaum and Witman 2002; Pearring et al. 2013).

Other Structural Components of Outer Segments

PSCs are highly specialized sensory cilia, adapted for light detection by the presence of tightly packed membranous discs containing visual pigments and other phototransduction proteins (Sjostrand 1953; Nickell et al. 2007; Gilliam et al. 2012). PSCs are among the largest of mammalian cilia (Pan et al. 2005) and, like other cilia, they are comprised of a cytoskeleton backbone and a membrane domain, which is distinct from the surrounding plasma membrane (Steinberg et al. 1980; Molday and Molday 1987). In murine rods, the numerous membranous discs in the PSC compartment are stacked at a density of ∼30 discs per micrometer, which is thought to be constant throughout species (Nickell et al. 2007; Gilliam et al. 2012). Such OS organization provides a large surface area for optimized photon capture and rapid signal transduction reactions to occur. Rhodopsin is the most abundant disc membrane protein, organized as rows of dimers with a density of ∼48,000 monomers per μm^2 (Fotiadis et al. 2003). With this high density in the disc membranes, rhodopsin plays an important structural role apart from being the main visual pigment in the retina (Wang and Deretic 2014). The rim of the photoreceptor discs contains two tetraspanins: Rds/peripherin-2 (PRPH2) and retinal OS membrane protein 1 (ROM1), which facilitate the folding of the OS discs and are crucial for rim formation and sorting of the OS proteins during the OS biogenesis (Molday et al. 1987; Goldberg and Molday 1996; Arikawa et al. 2011). PRPH2 and its homolog ROM1 both form homodimers and then associate together to form tetrameric complexes, exclusively present at the disc rims (Molday et al. 1987; Goldberg and Molday 1996; Arikawa et al. 2011). Two other membrane proteins prominin 1 (PROM1) and cadherin-related family member 1 (CDHR1) were associated with the open lamellar evaginations in rod and cone discs in *Xenopus laevis* and mice, respectively (Rattner et al. 2001; Han et al. 2012). PSC are responsible for mediating the sensory transduction of the visual system with a number of proteins involved in this process, including the abovementioned rhodopsin. Most of these proteins are expressed specifically in PSC and, when mutated, cause nonsyndromic IRDs (Table 1) (Dryja et al. 1990; Farrar et al. 1990; Kajiwara et al. 1991, 1994; Travis et al. 1991; Bascom et al. 1992; Rosenfeld et al. 1992; Dryja et al. 1993; Maw et al. 2000; Yang et al. 2008). Studies of mutant animals have shown that the abovementioned proteins are essential for OS disc morphogenesis and maintenance (Sanyal et al. 1980; Clarke et al. 2000; Rattner et al. 2001; Dellett et al. 2015).

PROTEIN TRANSPORT TO PSC

A unique feature of the photoreceptor OS is the high level of its renewal. Each day ∼10% of the OS is shed from the distal tip, which is replaced by new disc formation at the base of the PSC (Young 1967). This necessitates a robust system of protein synthesis in the IS and efficient trafficking of selected proteins to the photoreceptor OS.

Intraflagellar Transport in PSC

The axoneme, initiated at the mother centriole, is built and maintained by extending its distal (+) end (Pedersen and Rosenbaum 2008). Because protein synthesis occurs in the IS, the axoneme building blocks need to be transported to the distal end via intraflagellar transport (IFT) (Rosenbaum and Witman 2002; Pedersen and Rosenbaum 2008; Taschner et al. 2012). The anterograde transport from the base to the tip of the axoneme is mediated by IFT complex B (IFT-B), where kinesin-2 is the motor protein (Rosenbaum and Witman 2002). Kinesin-2 is a heterotrimeric protein composed of Kif3A, Kif3B, and KAP, which is further associated with 14 other IFT proteins that bind cargo molecules (Taschner et al. 2012). Once the axoneme

Table 1. Genes associated with nonsyndromic retinal degeneration

Gene	Inheritance pattern	Nonsyndromic form	Syndromic form	Notes and references
Genes coding axoneme-associated proteins				
ARL6	AR	RP	BBS	New retina-specific exon present (Chiang et al. 2004; Fan et al. 2004; Aldahmesh et al. 2009; Pretorius et al. 2010)
BBS1	AR	RP	BBS	Nishimura et al. 2001; Mykytyn et al. 2002; Estrada-Cuzcano et al. 2012a
BBS2	AR	RP	BBS	Consugar et al. 2014; Shevach et al. 2015
BBS9	AR	RP	BBS	Nishimura et al. 2005; Abu-Safieh et al. 2012
C2orf71	AR	RP	-	Putative cilia function, exclusive eye expression (Collin et al. 2010; Nishimura et al. 2010; Kevany et al. 2015)
C8orf37	AR	RP	BBS	Estrada-Cuzcano et al. 2012b; Heon et al. 2016; Khan et al. 2016
CEP164	AR	LCA	SLS	Chaki et al. 2012
CEP290	AR	LCA	BBS, JBS, MKS, SLS	den Hollander et al. 2006; Sayer et al. 2006; Valente et al. 2006; Baala et al. 2007a; Helou et al. 2007; Frank et al. 2008; Leitch et al. 2008
CLRN1	AR	RP	USH	Joensuu et al. 2001; Khan et al. 2011
FAM161A	AR	RP	-	Bandah-Rozenfeld et al. 2010; Langmann et al. 2010
IFT140	AR	RP, LCA	JATD, MZSDS	Perrault et al. 2012; Schmidts et al. 2013a; Bifari et al. 2015; Xu et al. 2015
IFT172	AR	RP	BBS, JATD, MZSDS	Halbritter et al. 2013; Bujakowska et al. 2014
IQCB1	AR	LCA	SLS	Otto et al. 2005; Estrada-Cuzcano et al. 2011; Stone et al. 2011
KIZ	AR	RP	-	El Shamieh et al. 2014
LAC5	AR	LCA	-	den Hollander et al. 2007
MAK	AR	RP	-	Ozgül et al. 2011; Tucker et al. 2011
NEK2	AR	RP		Nishiguchi et al. 2013
OFD1	XL	RP	OFD, JBS	Ferrante et al. 2001; Coene et al. 2009; Webb et al. 2012
RAB28	AR	CRD	-	Roosing et al. 2013
RP1	AR, AD	RP	-	Guillonneau et al. 1999; Pierce et al. 1999
RP1L1	AR, AD	RP, OMD	-	There are some doubts about this gene being truly associated with IRD, because this gene is highly polymorphic and some mutations were seen in the

Continued

Table 1. *Continued*

Gene	Inheritance pattern	Nonsyndromic form	Syndromic form	Notes and references
				controls (Bowne et al. 2003; Yamashita et al. 2009; Akahori et al. 2010; Davidson et al. 2013)
RP2	XL	RP	-	Hardcastle et al. 1999; Mears et al. 1999
RPGR	XL	RP, CD, MD	RP with hearing loss and sinorespiratory infections	Extraocular phenotypes may not be related to *RPGR*, because it mapped to a 43.6-Mb interval with 215 genes including *OFD1* (Meindl et al. 1996; Roepman et al. 1996; Ayyagari et al. 2002; Yang et al. 2002; Zito et al. 2003)
RPGRIP1	AR	LCA, CRD	-	Dryja et al. 2001; Hameed et al. 2003
SPATA7	AR	LCA, RP	-	Wang et al. 2009
TOPORS	AD	RP	-	Chakarova et al. 2007, 2011
TTC8	AR	RP	BBS	Ansley et al. 2003; Riazuddin et al. 2010
USH2A	AR	RP	USH	Eudy et al. 1998; Rivolta et al. 2000
WDR19	AR	RP	SLS, CED, JATD	Bredrup et al. 2011; Coussa et al. 2013

Other structural OS proteins

Gene	Inheritance pattern	Nonsyndromic form	Syndromic form	Notes and references
CDHR1	AR	CRD		Henderson et al. 2010; Ostergaard et al. 2010
EYS	AR	RP		Abd El-Aziz et al. 2008; Collin et al. 2008
FSCN2	AD	RP, MD		There are some doubts about this gene being truly associated with IRD, because a frameshift c.208delG is a common polymorphism in the Asian population (Wada et al. 2001, 2003; Zhang et al. 2007; Shin et al. 2010)
PROM1	AR, AD	RP, MD		Maw et al. 2000; Yang et al. 2008
PRPH2	AD, digenic with ROM1	RP, MD		Kajiwara et al. 1991, 1994; Travis et al. 1991
ROM1	AD, digenic with ROM1	RP		Bascom et al. 1992; Kajiwara et al. 1994
TULP1	AR	RP		Banerjee et al. 1998; Hagstrom et al. 1998; Larsson et al. 1998

Genes involved with the POS sensory function (phototransduction cascade and retinoid cycle in the photoreceptors)

Gene	Inheritance pattern	Nonsyndromic form	Syndromic form	Notes and references
ABCA4	AR	STGD, RP, CRD		Allikmets et al. 1997; Sun and Nathans 1997; Cremers et al. 1998; Martínez-Mir et al. 1998
CNGA1	AR	RP		Dryja et al. 1995
CNGA3	AR	ACHR		Kohl et al. 1998

Continued

Table 1. *Continued*

Gene	Inheritance pattern	Nonsyndromic form	Syndromic form	Notes and references
CNGB1	AR	RP		Bareil et al. 2001
CNGB3	AR	ACHM, CD		Kohl et al. 2000
GNAT1	AD, AR	CSNB		Dryja et al. 1996
GNAT2	AR	ACHM		Aligianis et al. 2002; Kohl et al. 2002
GRK	AR	CSNB		Yamamoto et al. 1997
GUCA1A	AD	CD, CRD		Payne et al. 1998; Sokal et al. 1998
GUCA1B	AD	RP, MD		Sato et al. 2005
GUCY2D	AR, AD	LCA, CRD		Perrault et al. 1996; Kelsell et al. 1998
OPN1LW	XL	Deuteranopia, blue cone monochromacy		Nathans et al. 1986; Winderickx et al. 1992; Ayyagari et al. 1999
OPN1MW	XL	Protanopia, blue cone monochromacy		Nathans et al. 1986; Ayyagari et al. 1999
OPN1SW	AD	Tritanopia		Nathans et al. 1992; Weitz et al. 1992a,b
PDE6A	AR	RP		Huang et al. 1995
PDE6B	AR, AD	RP, CSNB		McLaughlin et al. 1993; Gal et al. 1994
PDE6C	AR	CD, ACHM		Thiadens et al. 2009
PDE6G	AR	RP		Dvir et al. 2010
RDH12	AR, AD	LCA, RP		Janecke et al. 2004; Perrault et al. 2004; Fingert et al. 2008
RGS9	AR	Delayed cone adaptation		Nishiguchi et al. 2004
RGS9BP	AR	Delayed cone adaptation		Nishiguchi et al. 2004
RHO	AD, AR	RP, CSNB		Dryja et al. 1990; Farrar et al. 1990; Rosenfeld et al. 1992; Dryja et al. 1993
SAG	AR	RP, CSNB		Fuchs et al. 1995; Nakazawa et al. 1998

ACHM, Achromatopsia; BBS, Bardet–Biedl syndrome; CD, cone dystrophy; CED, cranioectodermal dysplasia, also known as Sensenbrenner syndrome; CRD, cone–rod dystrophy; CSNB, congenital stationary night blindness; JBS, Joubert syndrome; JATD, Jeune asphyxiating thoracic dystrophy; LCA, Leber congenital amaurosis; MKS, Meckel–Gruber syndrome; MZSDS, Mainzer–Saldino syndrome; OFD, oral-facial-digital syndrome; OMD, occult macular dystrophy; RP, retinitis pigmentosa; SLS, Senior–Løken syndrome; STGD, Stargardt disease; USH, Usher syndrome.

and other PSC components have been delivered to the tip of the cilium, the IFT-B components are recycled back to the base of the cilium by retrograde transport mediated by IFT complex A (IFT-A) (Rosenbaum and Witman 2002). Dynein-2 is the motor protein of IFT-A and it is associated with six other IFT proteins (Taschner et al. 2012). Apart from the IFT complexes, Bardet–Biedl syndrome proteins (BBSome) are also involved in the transport of membrane proteins to the cilium (Taschner et al. 2012; Williams et al. 2014).

Because 10% of the PSC is shed and renewed every day, the necessity for the retrograde transport in this cell type was not clear. However, identification of IRD patients with mutations in genes coding for retrograde transport proteins (e.g., *TTC21B*) and proteins involved in switching from anterograde to retrograde IFT direction (e.g., *IFT172*) underlines the importance of transport in both directions for the development and maintenance of PSC (Liu et al. 2010; Davis et al. 2011; Halbritter et al. 2013; Bujakowska et al. 2014).

Transport of Membrane Proteins to PSC

Even though the plasma membrane surrounding the photoreceptor cilium is continuous with the plasma membrane of the cell body, its protein composition is different. In addition, the membranous discs in the rod photoreceptors are distinct from the surrounding plasma membrane (Steinberg et al. 1980; Molday and Molday 1987). This selective protein content in PSC membranes is established by diffusional barriers present at the base of the cilium and within the transition zone (Pearring et al. 2013; Wang and Deretic 2014; Khanna 2015). The molecular composition of the diffusional barrier is not fully understood, although certain proteins like Septin 2 and CEP290 are thought to play an important role (Pearring et al. 2013; Wang and Deretic 2014).

Because rhodopsin is the most abundant protein in the PSC, its photoreceptor OS transport has been studied in detail. After synthesis in the IS endoplasmic reticulum and transport through the Golgi and *trans*-Golgi network, rhodopsin is sorted into vesicles destined for the OS. This is achieved thanks to the presence of specific sequence signatures (e.g., VXPX and FR motifs), which facilitate interaction with a ciliary targeting molecules Arf4 and ASAP1 (Deretic et al. 2005; Wang and Deretic 2014). Further interaction with FIP3 and small GTPases Rab6, Rab8, and Rab11 directs the vesicle to the base of the OS for fusion with the membrane (Deretic et al. 2005; Pearring et al. 2013; Wang and Deretic 2014). Further, rhodopsin molecules are transported through the transition zone to the site of the disc morphogenesis by two motor proteins, kinesin II and myosin VIIa, as shown in *Kif3a* and *Myo7a* knockout mice (Liu et al. 1999; Williams 2002). As mentioned before, kinesin II mediates

a microtubule-dependent anterograde IFT (Rosenbaum and Witman 2002). Myosin VIIa is an actin-dependent motor molecule and, in mouse photoreceptors, it locates to the periciliary membrane complex; however, in primates, it locates to the calyceal processes (Sahly et al. 2012). It is, therefore, unclear whether mutations in *Myosin VIIa* in humans also lead to the aberrant opsin trafficking as shown in mice (Liu et al. 1999). Immunoelectron microscope studies of *Rana pipiens* frog photoreceptors, revealed that actin is present not only in calyceal processes but also at the sites of disc morphogenesis, suggesting involvement of actin-mediated transport in protein delivery to the forming discs (Chaitin et al. 1984). Little is known about OS targeting of other OS-specific transmembrane proteins, apart from retinol dehydrogenase (RDH8), which also contains the VXPX motif and PRPH2, which has its own OS-targeting sequence (Pearring et al. 2013).

Photoactivated Protein Diffusion

The base of the OS does not contain a selective barrier for the soluble proteins as shown in mice by the light-activated translocation of phototransduction proteins: transducin, arrestin, and recoverin (Sokolov et al. 2002; Calvert et al. 2006). Furthermore, this translocation is thought to be energy-independent, implying that the protein movement occurs by simple diffusion (Nair et al. 2005; Calvert et al. 2010). The diffusion of the proteins is dependent, however, on the steric interactions between the molecules and cell structures, termed steric volume exclusion, which reduces the entry of larger molecular weight proteins to the OS (Najafi and Calvert 2012). Light-mediated translocation of these proteins is thought to play a role in adaptation to different light conditions, in which for instance concentrating transducin in the rod OS in the darkness amplifies the phototransduction signal, and translocation of arrestin to OS in light conditions terminates transducin activation and accelerates photopigment recovery (Pearring et al. 2013). A neuroprotective role for the light-induced protein translocation has also been suggested (Pearring et al. 2013).

RETINAL CILIOPATHIES

Mutations in genes coding for ciliary proteins lead to ciliopathies, rare genetic disorders that may affect one or more organs, including the retina, central nervous system, olfactory epithelium, cardiovascular system, liver, kidney, skeletal system, gonads, and adipose tissue (Goetz and Anderson 2010; Patel and Honoré 2010; Mockel et al. 2011; Waters and Beales 2011). In this review, we will focus on ciliopathies that involve the retina, manifesting most commonly as retinitis pigmentosa (RP) (Hamel 2006; Hartong et al. 2006; Berger et al. 2010) or Leber congenital amaurosis (LCA) (Weleber 2002; Chung and Traboulsi 2009). RP is a condition that primarily affects rod photoreceptors and retinal pigment epithelium. It is the most frequent cause of the IRDs, with a prevalence of ~1/3500 and accounting for roughly 25% of vision loss in adults (Hamel 2006; Hartong et al. 2006; Berger et al. 2010). It may start in the first or second decade of life, often with nyctalopia and peripheral vision loss as early symptoms, because of the dysfunction of PSCs and photoreceptor cell death in the peripheral retina. In many cases, the disease progresses to include central vision loss as well, because of eventual dysfunction of PSCs and death of photoreceptor cells in the macula (central retina) (Fig. 2) (Hamel 2006; Hartong et al. 2006; Berger et al. 2010). LCA affects rods and cones and leads to vision loss in infancy or early childhood (Weleber 2002; den Hollander et al. 2008; Chung and Traboulsi 2009). LCA is rare, with a population frequency of ~1/50,000, yet affecting ~20% of children attending schools for the blind (Weleber 2002; Koenekoop 2004; Berger et al. 2010). Other subtypes of IRD are present in ciliopathy patients and often involve cone photoreceptors and the macula (Michaelides et al. 2006; Estrada-Cuzcano et al. 2012c).

Nonsyndromic Retinal Ciliopathies

As mentioned above, photoreceptor OS can be regarded as specialized cilia designed to detect light and to convert this information into a biochemical signal. Therefore, we consider that all proteins that participate in this sensory function, as well as proteins that build the PSC structure, are in effect cilia proteins. Consequently, we can distinguish two groups of retinal ciliopathies: (1) affecting the structure, and (2) the sensory function of photoreceptor OS.

Mutations in Genes Disrupting POS Structure

There are currently 36 known genes that have been identified to harbor mutations that disrupt PSC structure and can lead to an isolated or syndromic retinal degeneration (Table 1). Thirteen genes that encode axoneme or basal body-associated proteins (*C2ORF71, FAM161A, KIZ, LCA5, MAK, NEK2, RAB28, RPGRIP1, RP1, RP1L1, RP2*, SPATA7, *TOPORS*) and seven genes coding for other structural PSC components (*CDHR1, EYS, FSCN2, PROM1, PRPH2, ROM1, TULP1*) have been exclusively associated with nonsyndromic retinal degeneration (Kajiwara et al. 1991, 1994; Bascom et al. 1992; Banerjee et al. 1998; Hagstrom et al. 1998; Guillonneau et al. 1999; Hardcastle et al. 1999; Mears et al. 1999; Pierce et al. 1999; Maw et al. 2000; Dryja et al. 2001; Wada et al. 2001, 2003; Chakarova et al. 2007; den Hollander et al. 2007; Abd El-Aziz et al. 2008; Collin et al. 2008; Wang et al. 2009; Akahori et al. 2010; Bandah-Rozenfeld et al. 2010; Collin et al. 2010; Henderson et al. 2010; Langmann et al. 2010; Nishimura et al. 2010; Ostergaard et al. 2010; Ozgül et al. 2011; Tucker et al. 2011; Estrada-Cuzcano et al. 2012b; Davidson et al. 2013; Nishiguchi et al. 2013; Roosing et al. 2013; El Shamieh et al. 2014). A query of the human proteome map (Kim et al. 2014) shows that nine of these genes (*CDHR1, FSCN2, MAK, PROM1, PRPH2, ROM1, RP1, RP1L1, TULP1*) are predominantly expressed in the human retina, which corroborates with the retina-specific phenotype. With the exception of *LCA5*, which shows no significant expression in any of the assayed tissues, the remaining genes are also significantly expressed in other human tissues, and it remains unclear why mutations in these genes affect specifically the retina.

Two genes stand apart in IRD ciliopathies, *USH2A* and *CLRN1*. Mutations in these genes

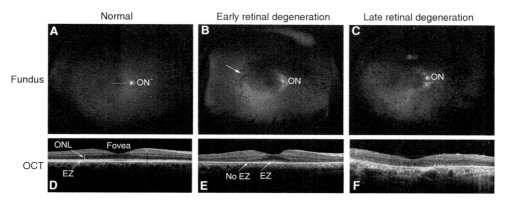

Figure 2. Clinical features of inherited retinal degeneration (IRD). Fundus photos and optical coherence tomography (OCT) images of normal and diseased retinas are shown. (*A*) Wide-field fundus photo shows the appearance of a normal retina; the optic nerve (ON) is visible. (*B*) Fundus image from a patient with early retinal degeneration because of retinitis pigmentosa (RP). The arrow shows the pigment changes, which are characteristic of this disorder. (*C*) Fundus image from a patient with advanced retinal degeneration because of RP. (*D*) A cross-sectional image of the center of the normal retina (macula; white line in *A*) obtained with OCT shows normal retinal layers, including the outer nuclear layer (ONL), where photoreceptor cell nuclei are located. The central indentation is normal, and indicates the fovea. The white band showing the elipsoid zone (EZ, arrow) is generated from the junction of the inner and outer segments of photoreceptor cells, and its presence indicates normal photoreceptor cell and thus PSC structure. (*E*) The OCT image shows loss of photoreceptor cells peripherally, with the ONL visible only near the fovea. The EZ is evident centrally, but is lost more peripherally, indicating loss of PSCs by the more peripheral photoreceptor cells present. (*F*) The OCT image shows loss of the ONL and thus all of the photoreceptor cells in the macula of this patient consistent with greatly reduced central vision.

lead to isolated retinal degeneration (Rivolta et al. 2000; Khan et al. 2011) or to deaf–blindness, called Usher syndrome (Eudy et al. 1998; Joensuu et al. 2001). There is only one report of nonsyndromic IRD because of mutations in *CLRN1* (Khan et al. 2011); however, mutations in *USH2A* are the leading cause of the nonsyndromic autosomal recessive IRD, accounting for ~9% of RP patients (Hartong et al. 2006). These two genes will be further discussed in the section on Usher syndrome.

The remaining 12 genes code for axoneme proteins and are associated with isolated retinopathy or syndromic disease (Table 1). In certain cases, the broad spectrum of phenotypes associated with mutations in a given gene can be explained by the primary mutations in the gene, where mutations in the retina-specific transcripts or hypomorphic alleles may lead to the isolated retinal phenotype (Ansley et al. 2003; den Hollander et al. 2006; Coene et al. 2009; Riazuddin et al. 2010; Webb et al. 2012;

Xu et al. 2015). In other cases, the relationship between the primary disease-causing mutation and the phenotype is not clear and epistatic effects of other alleles have been suggested (Badano et al. 2003b, 2006; Estrada-Cuzcano et al. 2012a; Bujakowska et al. 2014). (Both cases will be discussed in detail in the section Broad Phenotypic Spectrum of Ciliopathies and Genetic Modifiers.)

Mutations in Genes Impeding POS Sensory Function

A less obvious subgroup of nonsyndromic retinal ciliopathies are IRDs in which PSC sensory function is affected. This classification is analogous to another ciliopathy, the dominant form of polycystic kidney disease, because of mutations in *PKD1* or *PKD2*, which code for membrane proteins that act as a cilia sensor and a calcium channel, respectively (Ong and Harris 2015). Similarly, genetic mutations that affect phototransduction cascade and retinoid cycle in the photoreceptors are also considered as ciliopathies (Table 1). However, we do not consider as ciliopathies IRDs that are caused by mutations in genes that are expressed in the RPE or code for proteins that function in other compartments of the photoreceptor cell (e.g., splicing factor genes). A comprehensive list of all IRD genes can be found at the Retinal Information Network (RetNet) portal (sph.uth.edu/retnet/home.htm).

Usher Syndrome

Usher syndrome is an autosomal recessive dual impairment of vision and sensorineural hearing with a prevalence of ∼1/25,000 people (Kremer et al. 2006; Millán et al. 2011; Bonnet and El-Amraoui 2012). It is phenotypically and genetically heterogeneous and most of the patients fall into one of the three clinical subtypes of decreasing severity: Usher syndrome type I (USH1), type II (USH2), and type III (USH3).

The most severe form, USH1, is not strictly a ciliopathy because the proteins coded by the six associated genes: *MYO7A* (Weil et al. 1995), *USH1C* (Verpy et al. 2000), *CDH23* (Bolz et al. 2001), *PCDH15* (Ahmed et al. 2001), *USH1G* (Weil et al. 2003), and *CIB2* (Riazuddin et al. 2012), locate to actin-based structures in the periciliary region, called the calyceal processes, and not to cilia themselves (Sahly et al. 2012). The function of these structures is unclear and they might have a structural role supporting the photoreceptor outer segments or be involved in fine tuning of the photoreceptor signaling. Calyceal processes are not present in all of the vertebrates and their absence in rodents may account for the subtle retinal phenotypes in most of the USH1 murine models, in which hearing and vestibular phenotypes are profound (Liu et al. 1999; Di Palma et al. 2001; Libby and Steel 2001; Johnson et al. 2003; Ahmed et al. 2008; Williams 2008; Miyasaka et al. 2013). In the ear, USH1 proteins are involved in the maturation of the stereocilia in the auditory hair cells of the inner-ear cochlea, where they form transient links between the kinocilium and stereocilia during development and the tip links between the mature stereocilia (Kremer et al. 2006; Bonnet and El-Amraoui 2012; Riazuddin et al. 2012; Sahly et al. 2012).

USH2 is caused by mutations in one of three genes: *USH2A* (Eudy et al. 1998), *GPR98* (Weston et al. 2004), and *DFNB31* (Mburu et al. 2003). The products of these genes locate to the periciliary membrane complex adjacent to the transition zone of the photoreceptor cilia (Liu et al. 2007b; Sahly et al. 2012). In the cochlea, they form the transient ankle links between the developing sterocilia (reviewed in Bonnet and El-Amraoui 2012).

USH3, the least severe form of Usher, is caused by mutations in *CLRN1* or *HARS* (Joensuu et al. 2001; Puffenberger et al. 2012). In mice, Clarin-1 locates to the base of the photoreceptors cilia and to synaptic ribbons (Zallocchi et al. 2009); however, it does not seem to be essential for the photoreceptor function in rodents, because homozygous *Clrn1* knockout mice show no retinal phenotype (Geller et al. 2009). In the cochlea, Clarin-1 locates to the apical and basal aspects of stereocilia depending on the developmental stage (Zallocchi et al. 2009). *HARS* codes for a histidyl-tRNA synthetase and, currently, the mechanism of the disease is unknown (Puffenberger et al. 2012).

Of the three Usher types, the most frequent is USH2 (56%−57% cases), in which mutations in the *USH2A* gene are the most common (Millán et al. 2011). USH1 represents 33%−44% of all Usher cases, followed by USH3, with a prevalence of 2% (Yan and Liu 2010; Millán et al. 2011). However, in certain populations (Finland and Ashkenazi Jewish), USH3 reaches 40% of all Usher patients because of founder mutations (Yan and Liu 2010; Millán et al. 2011). Atypical Usher syndromes have also been reported in which the deaf−blindness phenotype is explained by mutations in two different genes: *USH2A* with *PDZD7* and *C2orf71* with *CEP250* (Table 2) (Ebermann et al. 2010; Khateb et al. 2014). Recently, two groups have reported another form of Usher syndrome because of mutations in *CEP78*, which manifest as a cone−rod dystrophy accompanied by sensorineural hearing loss involving mainly high frequencies (Nikopoulos et al. 2016; Sharon et al.

Table 2. Genetic modifiers of ciliopathies

Primary gene	Modifier	Phenotype	Evidence	References
PRPH2/RDS p.(Leu185Pro) ROM1 p.(Gly80fs) and p.(Leu114fs)	Digenic inheritance	RP	Three families (11 affected in total)	Kajiwara et al. 1994
BBS2 p.[Tyr24*] and p.[Gln59*] BBS6 p.(Gln147*)	Digenic inheritance	BBS	In one family, the affected sib carried three alleles, whereas the unaffected sib is the only compound heterozygous for the BBS2 mutations; three additional families with a probable triallelic BBS2/BBS6 inheritance	Katsanis et al. 2001
RPE65 p.(Glu102*)	GUCY2D p.(Ile539Val)	More severe RD	One family with two sibs, targeted analysis of: GUCY2D, RPE65, CRX, AIPL1, and RPGRIP1	Silva et al. 2004
BBS1 p.(Met390Arg)	ARL6 p.(Gly169Ala)	More severe ciliopathy	In one family of two affected sibs, the sister carrying the additional ALR6 allele showed a more severe phenotype	Fan et al. 2004
BBS1 and unknown BBS genotypes	CCDC28B (c.330C>T, p.(=)) (originally noted as C430T)	Increased severity of disease, general mutational load	In three families, variant associated with more severe BBS; variant also enriched in BBS patients (14/226) over controls (4/274)	Badano et al. 2006
Mixed ciliopathy cohort	RPGRIP1L p.(Ala229Thr)	Presence of RD in ciliopathies	Targeted screening of RPGRIP1L; enrichment of 226Thr variant in ciliopathy patients with RD (43/487) over ciliopathy patients without RD (0/115); functional data, showing a decreased biding activity to RPGR	Khanna et al. 2009
BBS mixed cohort	MKS1 p.(Arg123Gln) p.(Asp286Gly) p.(Ile450Thr) p.(Val339Met) (specific to short isoform)	More severe ciliopathy phenotype, seizures in five of six patients	Targeted screening of MKS1 in BBS cohort, heterozygous potentially modifying changes found in six patients from five families (5/155); functionally, variants ranged from mild hypomorphs to null	Leitch et al. 2008
BBS9 and CEP290	TMEM67 (c.2241G>A) and p.(Ser320Cys)	General mutational load	Potentially pathogenic alleles found in two families: c.2241G>A affects the canonical splice site and is predicted to lead to exon skipping; p.(Ser320Cys) was functionally null	Leitch et al. 2008
PRPH2	ROM1 p.(Arg229His) and/or ABCA4 p.(Val2050Leu)	Increased macular involvement in RD	One family, eight affected; microarray mutation detection in 16 genes, full sequencing of PRPH2, ROM1, and ABCA4	Poloschek et al. 2010

Continued

Table 2. *Continued*

Primary gene	Modifier	Phenotype	Evidence	References
USH2A	PDZD7 p.(Arg56fs)	More severe RD	One family with two sibs; targeted analysis of *PDZD7* and other Usher genes; digenic inheritance stated but insufficient genetic data to prove it	Ebermann et al. 2010
CEP290	AHI1 p.(Asn811Lys) and p.(His758Pro)	More severe neurological phenotype	Targeted screening of *AHI1* in eight patients; variants detected in two patients	Coppieters et al. 2010
NPHP1 and unknown NPHP genotypes	AHI1 p.(Arg830Trp)	Presence of RD	Targeted screening of *AHI1* in 153 NPHP $\pm$ RD patients	Louie et al. 2010
Mixed MKS and BBS cohort	C2ORF86 (various changes)	General mutational load	Targeted screening of *C2ORF86*; enrichment of nonsynonymous coding changes in patients (6/192) versus controls (0/384)	Kim et al. 2010
RPGR	RPGRIP1L p.(Arg744Gln) and IQCB1 p.(Ile393Asn)	Severity of RD	Targeted screening of *RPGRIP1L*, *RPGRIP1L*, *CEP290*, and *IQCB1* in 98 male patients; the results were marginally significant	Fahim et al. 2011
Mixed ciliopathy cohort	TTC21B (various changes)	General mutational load	Targeted screening of *TTC21B*; enrichment of pathogenic changes in ciliopathy patients (28/555) over controls (4/305)	Davis et al. 2011
C2orf71 p.(Gln1097*)	CEP250 p.(Arg1155*)	More severe retinal degeneration + hearing loss	Homozygosity mapping in family with seven affected, subsequent WES in two affected; homozygous *CEP250* mutation leads to an early-onset severe hearing loss with a mild retinal degeneration and an additional homozygous stop mutation in *C2orf71* exacerbated the retinal phenotype in three individuals	Khateb et al. 2014
Mixed BBS cohort	NPHP1 whole gene deletion and p.(Arg5Leu)	General mutational load in ciliopathies	Targeted analysis of *NPHP1* in a BBS cohort of 200 families, mutations enriched in the patient population compared with control (incidence of 1.5% (deletion) and 2.5% (missense); functional data for *NPHP1*-BBS genes interaction	Lindstrand et al. 2014
PRPH2 (c.828+3A>T)	PRPH2 p.[(Glu304;Lys310;Gly338)] haplotype in *trans* with the causal mutation	More severe retinal phenotype	p.[(Glu304;Lys310;Gly338)] haplotype in *trans* with the splice site mutation was associated with a more severe phenotype as investigated in 62 patients	Shankar et al. 2016

BBS, Bardet–Biedl syndrome; RD, retinal degeneration; RP, retinitis pigmentosa.

Cite this article as *Cold Spring Harb Perspect Biol* doi: 10.1101/cshperspect.a028274

2016). Hearing loss can also be part of other syndromes, such as Altrom, as will be discussed later (Mockel et al. 2011).

Other Syndromic IRDs

Based on the presence of particular symptoms, ciliopathies are subdivided into different subtypes, traditionally named after clinicians who first described them. Here, we will review the major syndromes, which involve the retina. Even though these conditions are considered as distinct clinical entities, it is being increasingly recognized that there is a large phenotypic overlap between these diseases, in which characteristic features of two different syndromes are present in the same patient or distinct ciliopathies co-occur in single families (Lehman et al. 2010; Zaki et al. 2011; Valente et al. 2014). Traditional naming of these conditions is therefore often inaccurate and depends on the clinicians' training and their specialty. To overcome this bias, we believe that it is crucial to include in the syndrome naming the underlying molecular cause of the disease (e.g., *AHI1*-associated ciliopathy). However, for the purpose of this review, we describe each syndrome as traditionally called and present the genes associated with them.

Senior–Løken syndrome (SLS) is an autosomal recessive disease characterized by juvenile nephronophtitis (NPHP) and early-onset retinal degeneration (Løken et al. 1961; Senior et al. 1961). NPHP is a medullary cystic kidney disease leading to the end-stage renal failure later in childhood or in adolescence (Ronquillo et al. 2012). About 10% of NPHP patients also have retinal degeneration (Otto et al. 2005; Mockel et al. 2011). So far, eight genes are associated with SLS (*CEP164, CEP290, IQCB1, NPHP1, NPHP4, SDCCAG8, TRAF3IP1, WDR19*), although other NPHP-associated genes are mutated in different syndromes involving the retina (Caridi et al. 1998; Otto et al. 2002, 2005, 2010; Sayer et al. 2006; Chaki et al. 2012; Coussa et al. 2013; Bizet et al. 2015). There is a considerable overlap between SLS and other ciliopathies, where almost all SLS genes are associated with other diseases (Fig. 3).

Joubert syndrome (JBS) is a neurological condition characterized by a distinctive abnormality of the midbrain–hindbrain junction and cerebellar vermis hypoplasia, presenting as the molar tooth sign (MTS) on brain imaging (Maria et al. 1997; Valente et al. 2013). These neurological defects correlate with the clinical presentation of hypotonia, ataxia, abnormal breathing, developmental delay, and abnormal ocular movements. JBS patients may also present with retinal degeneration, renal or hepatic defects, polydactyly, and orofacial dysmorphism (Mockel et al. 2011; Valente et al. 2013). The prevalence of JBS is estimated to be between 1/80,000 and 1/100,000 of live births (Valente et al. 2014). Mutations in 26 genes have been reported to cause JBS (*AHI1, ARL13B, B9D1, C5orf42, CC2D2A, CEP104, CEP290, CEP41, CSPP1, INPP5E, KIAA0556, KIAA0586, KIF7, NPHP1, OFD1, RPGRIP1L, SRTD1, TCTN2, TCTN3, TECT1, TMEM67, TMEM138, TMEM216, TMEM231, TMEM237, ZNF423*) (Dixon-Salazar et al. 2004; Ferland et al. 2004; Parisi et al. 2004; Sayer et al. 2006; Baala et al. 2007b; Delous et al. 2007; Cantagrel et al. 2008; Gorden et al. 2008; Noor et al. 2008; Bielas et al. 2009; Coene et al. 2009; Edvardson et al. 2010; Dafinger et al. 2011; Garcia-Gonzalo et al. 2011; Huang et al. 2011; Chaki et al. 2012; Lee et al. 2012a,b; Srour et al. 2012a,b; Thomas et al. 2012; Romani et al. 2014; Shaheen et al. 2014; Thomas et al. 2014; Tuz et al. 2014; Sanders et al. 2015). All except one of the above genes cause autosomal-recessive or X-linked disease; *ZNF423* has been associated with a dominant JBS form, although this association showed limited genetic evidence (Chaki et al. 2012). *ZNF423* is also the only JBS gene, which is not associated with the cilium but with the DNA damage response pathway (Chaki et al. 2012).

Meckel–Gruber syndrome, also known as Meckel syndrome (MKS) is a neonatal lethal autosomal recessive disorder defined by the malformation of the central nervous system (occipital encephalocele), cystic kidneys, and liver fibrosis (Wright et al. 1994; Logan et al. 2011). Other features that may be present are postaxial or preaxial polydactyly, skeletal dys-

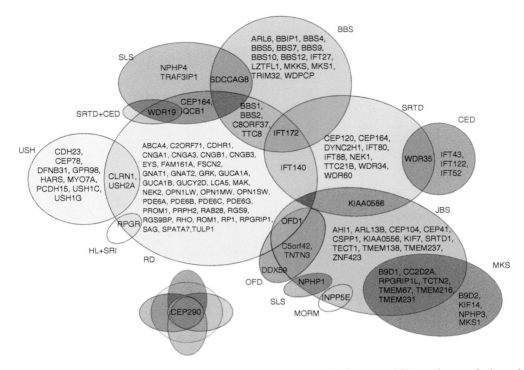

Figure 3. A Venn diagram showing a phenotypic and genetic overlap between different forms of ciliopathy. BBS, Bardet−Biedl syndrome; CED, cranioectodermal dysplasia (also known as Sensenbrenner syndrome); HL + SRI, hearing loss and sinorespiratory infections; JBS, Joubert syndrome; MKS, Meckel−Gruber syndrome; MORM, mental retardation, truncal obesity, retinal degeneration, and micropenis; OFD, oral-facial-digital syndrome; RD, retinal degeneration—nonsyndromic; SLS, Senior−Løken syndrome; SRTD, short-rib thoracic dysplasia; USH, Usher syndrome.

plasia, cleft lip/palate, microphthalmia, optic nerve coloboma, heart defects, genital anomalies, and complete or partial situs inversus (Logan et al. 2011). Mutations in 12 genes have been associated with MKS (*B9D1, B9D2, CC2D2A, CEP290, KIF14, MKS1, NPHP3, RPGRIP1L, TCTN2, TMEM67, TMEM216, TMEM231*) (Kyttälä et al. 2006; Smith et al. 2006; Baala et al. 2007a; Delous et al. 2007; Bergmann et al. 2008; Tallila et al. 2008; Valente et al. 2010; Dowdle et al. 2011; Hopp et al. 2011; Shaheen et al. 2011, 2013; Filges et al. 2014). There is a considerable genetic overlap between JBS and MKS, in which eight of the genes are shared between the two syndromes and co-occurrence of the two diseases was reported in the same families (Valente et al. 2014). The incidence of MKS varies among populations and it has been estimated as 1/13,250 in the United

States, 1/140,000 in the United Kingdom, and 1/9000 in Finland (Logan et al. 2011).

Bardet−Biedl syndrome (BBS) is an autosomal recessive condition defined by rod−cone degeneration, postaxial polydactyly, central obesity, mental retardation, hypogonadism, and renal dysfunction (Beales et al. 1999; Mockel et al. 2011). Other features such as hepatic fibrosis, diabetes mellitus, endocrinological disturbances, heart disease, and short stature may also be present (Beales et al. 1999). Twenty genes have been associated with BBS (*ARL6, BBIP1, BBS1, BBS2, BBS4, BBS5, BBS7, BBS9, BBS10, BBS12, CEP290, IFT27, IFT172, LZTFL1, MKKS, MKS1, SDCCAG8, TRIM32, TTC8, WDPCP*) (Katsanis et al. 2000; Slavotinek et al. 2000; Mykytyn et al. 2001, 2002; Nishimura et al. 2001, 2005; Ansley et al. 2003; Badano et al. 2003a; Chiang et al. 2004, 2006;

  Cite this article as *Cold Spring Harb Perspect Biol* doi: 10.1101/cshperspect.a028274

Fan et al. 2004; Li et al. 2004; Stoetzel et al. 2006, 2007; Leitch et al. 2008; Kim et al. 2010; Otto et al. 2010; Marion et al. 2012; Aldahmesh et al. 2014; Bujakowska et al. 2014; Scheidecker et al. 2014). Apart from the BBS genes associated with the nonsyndromic IRD (Table 1), a homozygous nonsense mutation (p.S701X) in *BBS12* was shown to lead to a late-onset retinal degeneration and postaxial polydactyly but no other BBS-associated clinical features (Pawlik et al. 2010).

There are two phenotypically similar diseases to BBS: Alstrom syndrome (ALMS) and MORM syndrome (mental retardation, truncal obesity, retinal degeneration, and micropenis). ALMS is characterized by cone–rod degeneration, sensorineural hearing loss, childhood obesity, and type 2 diabetes mellitus. ALMS patients often present with cardiomyopathy and other features such as renal, pulmonary, or hepatic disease may also be present. In contrast to BBS, ALMS is not associated with mental retardation, polydactyly, or hypogonadism (Mockel et al. 2011). Mutations in only one gene, *ALMS1*, have been associated with Alstrom syndrome (Collin et al. 2002). MORM has been described in only one Pakistani family of 14 individuals, in whom a homozygous truncating mutation in *INPP5E*, was found to cause the disease (Hampshire et al. 2006; Jacoby et al. 2009).

Short-rib thoracic dysplasia (SRTD) with or without polydactyly regroups syndromes formerly known as Mainzer–Saldino (MZSDS), Jeune asphyxiating thoracic dystrophy (JATD), and Ellis–van Creveld (EVC) syndromes. SRTDs are autosomal recessive skeletal ciliopathies, characterized by short ribs, constricted thoracic cage, shortened tubular bones, and a "trident" appearance of the acetabular roof (Huber and Cormier-Daire 2012). The severely constricted thoracic cage leads to respiratory insufficiency, often resulting in death in infancy. Other features that may be present are polydactyly, cleft lip/palate, retinal degeneration, and anomalies of the brain, heart, kidneys, liver, pancreas, intestines, and genitalia (Waters and Beales 2011; Huber and Cormier-Daire 2012). There is a phenotypic and genetic overlap between SRTDs and cranioectodermal dysplasia (CED), also known as Sensenbrenner syndrome. CED is characterized by sagittal craniosynostosis, narrow thorax, short limbs, brachydactyly, protuberant abdomen, and facial and ectodermal anomalies (Huber and Cormier-Daire 2012). Seventeen genes have been associated with these diseases (*CEP120, DYNC2H1, EVC, EVC2, IFT52, IFT122, IFT140, IFT172, IFT43, IFT80, KIAA0586, NEK1, TTC21B, WDR19, WDR34, WDR35, WDR60*) (Ruiz-Perez et al. 2000; Galdzicka et al. 2002; Beales et al. 2007; Dagoneau et al. 2009; Gilissen et al. 2010; Walczak-Sztulpa et al. 2010; Arts et al. 2011; Bredrup et al. 2011; Davis et al. 2011; Mill et al. 2011; Thiel et al. 2011; Perrault et al. 2012; Halbritter et al. 2013; McInerney-Leo et al. 2013; Schmidts et al. 2013b; Alby et al. 2015; Shaheen et al. 2015; Girisha et al. 2016). Interestingly, some of these genes have also been implicated with a nonsyndromic disease, for example, *TTC21B* and *WDR19* in NPHP (Bredrup et al. 2011; Davis et al. 2011) or *IFT172* and *IFT140* in RP (Fig. 3) (Bujakowska et al. 2014; Bifari et al. 2015; Xu et al. 2015).

BROAD PHENOTYPIC SPECTRUM OF CILIOPATHIES AND GENETIC MODIFIERS

One of the important aspects of the IRDs is that mutations in the same gene can lead to variable phenotypes (Ferrante et al. 2001; Ansley et al. 2003; Sayer et al. 2006; Perrault et al. 2007; Frank et al. 2008; Leitch et al. 2008; Coene et al. 2009; Riazuddin et al. 2010; Bujakowska et al. 2012; Estrada-Cuzcano et al. 2012a; Webb et al. 2012). In some cases, the severity of disease can be explained by the primary disease-causing mutation. For example, a splice site mutation of a retina-specific exon in *TTC8* leads to a nonsyndromic RP (Riazuddin et al. 2010), whereas the gene is most commonly associated with BBS (Ansley et al. 2003). The position of a mutation may also determine the phenotype as in the case of truncating mutations in *OFD1*. Nonsense mutations downstream from exon 17 lead to an X-linked dominant oral-facial-digital type 1 (OFD1) syndrome, manifesting with malformations of face, oral cavity, and digits in affected females and lethal in males (Ferrante et al. 2001). However, truncating mutations upstream of exon 17 lead to an X-linked recessive JBS (Coene et al. 2009). In addition, hypomorphic alleles can arise by mutations activating cryptic splice sites, which leads to severely reduced levels of wild-type transcripts as in the case of *CEP290* (den Hollander et al. 2006) and *OFD1* (Webb et al. 2012).

In many cases, however, even a precise genetic diagnosis does not yield a clear genotype–phenotype correlation and the severity of disease can vary greatly even between patients with the same genetic cause of disease. Examples of this include family members that share the same 3bp deletion (c.461_463del) in the *PRPH2* gene but show phenotypes varying from RP involving the peripheral retina to macular disease involving only the central retina (Weleber et al. 1993). Similarly, individuals with mutations in the *RP1* gene show variable phenotypes, ranging from near normal to profoundly affected by retinal degeneration (Jacobson et al. 2000; Berson et al. 2001). Several genetic modifiers have already been identified in IRD disease (Table 2), in which extreme examples are cases of digenic inheritance of nonsyndromic IRD (Kajiwara et al. 1994) and BBS (Katsanis et al. 2001) or the rescuing effect of the wild-type *PRPF31* allele in the dominant *PRPF31*-associated disease (McGee et al. 1997; Vithana et al. 2003; Rose et al. 2016). Even though more than a dozen of genetic modifiers of IRD disease severity have been reported, our knowledge about these variants is still limited because the studies were conducted on a limited number of patients (sometimes single families) targeting a small number of genes and functional validation was not always performed (Table 2). In addition, no study has yet shown the validity of the previously reported modifiers and therefore they remain to be scrutinized by future research.

CONCLUSIONS

In summary, mutations in many different genes can cause retinal ciliopathies, reflecting the diversity of protein functions required for normal PSC function. As indicated, it is increasingly clear that the phenotypes ascribed to specific genetic forms of disease overlap, and thus a revised system of disease definitions that includes the genetic etiology in the disease name would improve our understanding of these disorders, and their description for patients and clinicians. Further, as we have attempted to illustrate, studies of retinal ciliopathies have provided insights into syndromic disorders, and cilia function in general. Given the ubiquitous presence of cilia on mammalian cells, we anticipate that further study of these disorders and their pathogenesis will continue to inform us about cilia function broadly, and to be informed by the results of cilia in other contexts.

REFERENCES

Abd El-Aziz MM, Barragan I, O'Driscoll CA, Goodstadt L, Prigmore E, Borrego S, Mena M, Pieras JI, El-Ashry MF, Safieh LA, et al. 2008. *EYS*, encoding an ortholog of *Drosophila* spacemaker, is mutated in autosomal recessive retinitis pigmentosa. *Nat Genet* **40:** 1285–1287.

Abu-Safieh L, Al-Anazi S, Al-Abdi L, Hashem M, Alkuraya H, Alamr M, Sirelkhatim MO, Al-Hassnan Z, Alkuraya B, Mohamed JY, et al. 2012. In search of triallelism in Bardet–Biedl syndrome. *Eur J Hum Genet* **20:** 420–427.

Ahmed ZM, Riazuddin S, Bernstein SL, Ahmed Z, Khan S, Griffith AJ, Morell RJ, Friedman TB, Wilcox ER. 2001. Mutations of the protocadherin gene *PCDH15* cause Usher syndrome type 1F. *Am J Hum Genet* **69:** 25–34.

Ahmed ZM, Kjellstrom S, Haywood-Watson RJL, Bush RA, Hampton LL, Battey JF, Riazuddin S, Frolenkov G, Sieving PA, Friedman TB. 2008. Double homozygous waltzer and Ames waltzer mice provide no evidence of retinal degeneration. *Mol Vis* **14:** 2227–2236.

Akahori M, Tsunoda K, Miyake Y, Fukuda Y, Ishiura H, Tsuji S, Usui T, Hatase T, Nakamura M, Ohde H, et al. 2010. Dominant mutations in *RP1L1* are responsible for occult macular dystrophy. *Am J Hum Genet* **87:** 424–429.

Alby C, Piquand K, Huber C, Megarbané A, Ichkou A, Legendre M, Pelluard F, Encha-Razavi F, Abi-Tayeh G, Bessières B, et al. 2015. Mutations in *KIAA0586* cause lethal ciliopathies ranging from a hydrolethalus phenotype to short-rib polydactyly syndrome. *Am J Hum Genet* **97:** 311–318.

Aldahmesh MA, Safieh LA, Alkuraya H, Al-Rajhi A, Shamseldin H, Hashem M, Alzahrani F, Khan AO, Alqahtani F, Rahbeeni Z, et al. 2009. Molecular characterization of retinitis pigmentosa in Saudi Arabia. *Mol Vis* **15:** 2464–2469.

Aldahmesh MA, Li Y, Alhashem A, Anazi S, Alkuraya H, Hashem M, Awaji AA, Sogaty S, Alkharashi A, Alzahrani S, et al. 2014. *IFT27*, encoding a small GTPase component of IFT particles, is mutated in a consanguineous family with Bardet–Biedl syndrome. *Hum Mol Genet* **23:** 3307–3315.

Aligianis IA, Forshew T, Johnson S, Michaelides M, Johnson CA, Trembath RC, Hunt DM, Moore AT, Maher ER. 2002. Mapping of a novel locus for achromatopsia (*ACHM4*) to 1p and identification of a germline mutation in the alpha subunit of cone transducin (*GNAT2*). *J Med Genet* **39:** 656–660.

Allikmets R, Singh N, Sun H, Shroyer NF, Hutchinson A, Chidambaram A, Gerrard B, Baird L, Stauffer D, Peiffer A, et al. 1997. A photoreceptor cell-specific ATP-binding transporter gene (*ABCR*) is mutated in recessive Stargardt macular dystrophy. *Nat Genet* **15:** 236–246.

Ansley SJ, Badano JL, Blacque OE, Hill J, Hoskins BE, Leitch CC, Kim JC, Ross AJ, Eichers ER, Teslovich TM, et al. 2003. Basal body dysfunction is a likely cause of pleiotropic Bardet–Biedl syndrome. *Nature* **425:** 628–633.

Arikawa K, Molday LL, Molday RS, Williams DS. 2011. Localization of peripherin/RDS in the disk membranes of cone and rod photoreceptors: Relationship to disk membrane morphogenesis and retinal degeneration. *J Cell Biol* **116:** 659–667.

Arts HH, Bongers EMHF, Mans DA, van Beersum SEC, Oud MM, Bolat E, Spruijt L, Cornelissen EAM, Schuurs-Hoeijmakers JHM, de Leeuw N, et al. 2011. *C14ORF179* encoding IFT43 is mutated in Sensenbrenner syndrome. *J Med Genet* **48:** 390–395.

Ayyagari R, Kakuk LE, Coats CL, Bingham EL, Toda Y, Felius J, Sieving P. 1999. Bilateral macular atrophy in blue cone monochromacy (BCM) with loss of the locus control region (LCR) and part of the red pigment gene. *Mol Vis* **5:** 13.

Ayyagari R, Demirci FY, Liu J, Bingham EL, Stringham H, Kakuk LE, Boehnke M, Gorin MB, Richards JE, Sieving PA. 2002. X-linked recessive atrophic macular degeneration from *RPGR* mutation. *Genomics* **80:** 166–171.

Baala L, Audollent S, Martinovic J, Ozilou C, Babron M-C, Sivanandamoorthy S, Saunier S, Salomon R, Gonzales M, Rattenberry E, et al. 2007a. Pleiotropic effects of *CEP290* (*NPHP6*) mutations extend to Meckel syndrome. *Am J Hum Genet* **81:** 170–179.

Baala L, Romano S, Khaddour R, Saunier S, Smith UM, Audollent S, Ozilou C, Faivre L, Laurent N, Foliguet B, et al. 2007b. The Meckel–Gruber syndrome gene, *MKS3*, is mutated in Joubert syndrome. *Am J Hum Genet* **80:** 186–194.

Badano JL, Ansley SJ, Leitch CC, Lewis RA, Lupski JR, Katsanis N. 2003a. Identification of a novel Bardet–Biedl syndrome protein, BBS7, that shares structural features with BBS1 and BBS2. *Am J Hum Genet* **72:** 650–658.

Badano JL, Kim JC, Hoskins BE, Lewis RA, Ansley SJ, Cutler DJ, Castellan C, Beales PL, Leroux MR, Katsanis N. 2003b. Heterozygous mutations in *BBS1*, *BBS2* and *BBS6* have a potential epistatic effect on Bardet–Biedl patients with two mutations at a second BBS locus. *Hum Mol Genet* **12:** 1651–1659.

Badano JL, Leitch CC, Ansley SJ, May-Simera H, Lawson S, Lewis RA, Beales PL, Dietz HC, Fisher S, Katsanis N. 2006. Dissection of epistasis in oligogenic Bardet–Biedl syndrome. *Nature* **439:** 326–330.

Bandah-Rozenfeld D, Mizrahi-Meissonnier L, Farhy C, Obolensky A, Chowers I, Pe'er J, Merin S, Ben-Yosef T,

Ashery-Padan R, Banin E, et al. 2010. Homozygosity mapping reveals null mutations in *FAM161A* as a cause of autosomal-recessive retinitis pigmentosa. *Am J Hum Genet* **87:** 382–391.

Banerjee P, Kleyn PW, Knowles JA, Lewis CA, Ross BM, Parano E, Kovats SG, Lee JJ, Penchaszadeh GK, Ott J, et al. 1998. *TULP1* mutation in two extended Dominican kindreds with autosomal recessive retinitis pigmentosa. *Nat Genet* **18:** 177–179.

Bareil C, Hamel CP, Delague V, Arnaud B, Demaille J, Claustres M. 2001. Segregation of a mutation in CNGB1 encoding the β-subunit of the rod cGMP-gated channel in a family with autosomal recessive retinitis pigmentosa. *Hum Genet* **108:** 328–334.

Bascom RA, Manara S, Collins L, Molday RS, Kalnins VI, McInnes RR. 1992. Cloning of the cDNA for a novel photoreceptor membrane protein (rom-1) identifies a disk rim protein family implicated in human retinopathies. *Neuron* **8:** 1171–1184.

Beales PL, Elcioglu N, Woolf AS, Parker D, Flinter FA. 1999. New criteria for improved diagnosis of Bardet–Biedl syndrome: Results of a population survey. *J Med Genet* **36:** 437–446.

Beales PL, Bland E, Tobin JL, Bacchelli C, Tuysuz B, Hill J, Rix S, Pearson CG, Kai M, Hartley J, et al. 2007. *IFT80*, which encodes a conserved intraflagellar transport protein, is mutated in Jeune asphyxiating thoracic dystrophy. *Nat Genet* **39:** 727–729.

Berger W, Kloeckener-Gruissem B, Neidhardt J. 2010. The molecular basis of human retinal and vitreoretinal diseases. *Prog Retin Eye Res* **29:** 335–375.

Bergmann C, Fliegauf M, Brüchle NO, Frank V, Olbrich H, Kirschner J, Schermer B, Schmedding I, Kispert A, Kränzlin B, et al. 2008. Loss of nephrocystin-3 function can cause embryonic lethality, Meckel–Gruber-like syndrome, situs inversus, and renal-hepatic-pancreatic dysplasia. *Am J Hum Genet* **82:** 959–970.

Berson EL, Grimsby JL, Adams SM, McGee TL, Sweklo E, Pierce EA, Sandberg MA, Dryja TP. 2001. Clinical features and mutations in patients with dominant retinitis pigmentosa-1 (*RP1*). *Invest Ophthalmol Vis Sci* **42:** 2217–2224.

Besharse JC, Forestner DM, Defoe DM. 1985. Membrane assembly in retinal photoreceptors. III: Distinct membrane domains of the connecting cilium of developing rods. *J Neurosci* **5:** 1035–1048.

Bielas SL, Silhavy JL, Brancati F, Kisseleva MV, Al-Gazali L, Sztriha L, Bayoumi RA, Zaki MS, Abdel-Aleem A, Rosti RO, et al. 2009. Mutations in *INPP5E*, encoding inositol polyphosphate-5-phosphatase E, link phosphatidyl inositol signaling to the ciliopathies. *Nat Genet* **41:** 1032–1036.

Bifari IN, Elkhamary SM, Bolz HJ, Khan AO. 2015. The ophthalmic phenotype of *IFT140*-related ciliopathy ranges from isolated to syndromic congenital retinal dystrophy. *Br J Ophthalmol* **6:** 829–833.

Bizet AA, Becker-Heck A, Ryan R, Weber K, Filhol E, Krug P, Halbritter J, Delous M, Lasbennes M-C, Linghu B, et al. 2015. Mutations in TRAF3IP1/IFT54 reveal a new role for IFT proteins in microtubule stabilization. *Nat Commun* **6:** 8666.

Bolz H, von Brederlow B, Ramírez A, Bryda EC, Kutsche K, Nothwang HG, Seeliger M, del C-Salcedó Cabrera M, Vila MC, Molina OP, et al. 2001. Mutation of *CDH23*, encoding a new member of the cadherin gene family, causes Usher syndrome type 1D. *Nat Genet* 27: 108–112.

Bonnet C, El-Amraoui A. 2012. Usher syndrome (sensorineural deafness and retinitis pigmentosa): Pathogenesis, molecular diagnosis and therapeutic approaches. *Curr Opin Neurol* 25: 42–49.

Bowne SJ, Daiger SP, Malone KA, Heckenlively JR, Kennan A, Humphries P, Hughbanks-Wheaton D, Birch DG, Liu Q, Pierce EA, et al. 2003. Characterization of *RP1L1*, a highly polymorphic paralog of the retinitis pigmentosa 1 (*RP1*) gene. *Mol Vis* 9: 129–137.

Bredrup C, Saunier S, Oud MM, Fiskerstrand T, Hoischen A, Brackman D, Leh SM, Midtbø M, Filhol E, Bole-Feysot C, et al. 2011. Ciliopathies with skeletal anomalies and renal insufficiency due to mutations in the IFT-A gene *WDR19*. *Am J Hum Genet* 89: 634–643.

Bujakowska K, Audo I, Mohand-Saïd S, Lancelot ME, Antonio A, Germain A, Leveillard T, Letexier M, Saraiva JP, Lonjou C, et al. 2012. *CRB1* mutations in inherited retinal dystrophies. *Hum Mutat* 33: 306–315.

Bujakowska KM, Zhang Q, Siemiatkowska AM, Liu Q, Place E, Falk MJ, Consugar M, Lancelot ME, Antonio A, Lonjou C, et al. 2014. Mutations in *IFT172* cause isolated retinal degeneration and Bardet–Biedl syndrome. *Hum Mol Genet* 24: 230–242.

Calvert PD, Strissel KJ, Schiesser WE, Pugh EN, Arshavsky VY. 2006. Light-driven translocation of signaling proteins in vertebrate photoreceptors. *Trends Cell Biol* 16: 560–568.

Calvert PD, Schiesser WE, Pugh EN. 2010. Diffusion of a soluble protein, photoactivatable GFP, through a sensory cilium. *J Gen Physiol* 135: 173–196.

Cantagrel V, Silhavy JL, Bielas SL, Swistun D, Marsh SE, Bertrand JY, Audollent S, Attié-Bitach T, Holden KR, Dobyns WB, et al. 2008. Mutations in the cilia gene *ARL13B* lead to the classical form of Joubert syndrome. *Am J Hum Genet* 83: 170–179.

Caridi G, Murer L, Bellantuono R, Sorino P, Caringella DA, Gusmano R, Ghiggeri GM. 1998. Renal–retinal syndromes: Association of retinal anomalies and recessive nephronophthisis in patients with homozygous deletion of the NPH1 locus. *Am J Kidney Dis* 32: 1059–1062.

Chaitin MH, Schneider BG, Hall MO, Papermaster DS. 1984. Actin in the photoreceptor connecting cilium: Immunocytochemical localization to the site of outer segment disk formation. *J Cell Biol* 99: 239–247.

Chakarova CF, Papaioannou MG, Khanna H, Lopez I, Waseem N, Shah A, Theis T, Friedman J, Maubaret C, Bujakowska K, et al. 2007. Mutations in *TOPORS* cause autosomal dominant retinitis pigmentosa with perivascular retinal pigment epithelium atrophy. *Am J Hum Genet* 81: 1098–1103.

Chakarova CF, Khanna H, Shah AZ, Patil SB, Sedmak T, Murga-Zamalloa CA, Papaioannou MG, Nagel-Wolfrum K, Lopez I, Munro P, et al. 2011. TOPORS, implicated in retinal degeneration, is a cilia-centrosomal protein. *Hum Mol Genet* 20: 975–987.

Chaki M, Airik R, Ghosh AK, Giles RH, Chen R, Slaats GG, Wang H, Hurd TW, Zhou W, Cluckey A, et al. 2012.

Exome capture reveals *ZNF423* and *CEP164* mutations, linking renal ciliopathies to DNA damage response signaling. *Cell* 150: 533–548.

Chiang AP, Nishimura D, Searby C, Elbedour K, Carmi R, Ferguson AL, Secrist J, Braun T, Casavant T, Stone EM, et al. 2004. Comparative genomic analysis identifies an ADP-ribosylation factor-like gene as the cause of Bardet–Biedl syndrome (BBS3). *Am J Hum Genet* 75: 475–484.

Chiang AP, Beck JS, Yen HJ, Tayeh MK, Scheetz TE, Swiderski RE, Nishimura DY, Braun T, Kim KY, Huang J, et al. 2006. Homozygosity mapping with SNP arrays identifies *TRIM32*, an E3 ubiquitin ligase, as a Bardet–Biedl syndrome gene (BBS11). *Proc Natl Acad Sci* 103: 6287–6292.

Chung DC, Traboulsi EI. 2009. Leber congenital amaurosis: Clinical correlations with genotypes, gene therapy trials update, and future directions. *J AAPOS* 13: 587–592.

Clarke G, Goldberg AF, Vidgen D, Collins L, Ploder L, Schwarz L, Molday LL, Rossant J, Szél A, Molday RS, et al. 2000. Rom-1 is required for rod photoreceptor viability and the regulation of disk morphogenesis. *Nat Genet* 25: 67–73.

Coene KLM, Roepman R, Doherty D, Afroze B, Kroes HY, Letteboer SJF, Ngu LH, Budny B, van Wijk E, Gorden NT, et al. 2009. OFD1 is mutated in X-linked Joubert syndrome and interacts with *LCA5*-encoded lebercilin. *Am J Hum Genet* 85: 465–481.

Collin GB, Marshall JD, Ikeda A, So WV, Russell-Eggitt I, Maffei P, Beck S, Boerkoel CF, Sicolo N, Martin M, et al. 2002. Mutations in *ALMS1* cause obesity, type 2 diabetes and neurosensory degeneration in Alström syndrome. *Nat Genet* 31: 74–78.

Collin RWJ, Littink KW, Klevering BJ, van den Born LI, Koenekoop RK, Zonneveld MN, Blokland EAW, Strom TM, Hoyng CB, den Hollander AI, et al. 2008. Identification of a 2 Mb human ortholog of *Drosophila* eyes shut/spacemaker that is mutated in patients with retinitis pigmentosa. *Am J Hum Genet* 83: 594–603.

Collin RWJ, Safieh C, Littink KW, Shalev SA, Garzozi HJ, Rizel L, Abbasi AH, Cremers FPM, den Hollander AI, Klevering BJ, et al. 2010. Mutations in *C2ORF71* cause autosomal-recessive retinitis pigmentosa. *Am J Hum Genet* 86: 783–788.

Consugar MB, Navarro-Gomez D, Place EM, Bujakowska KM, Sousa ME, Fonseca-Kelly ZD, Taub DG, Janessian M, Wang DY, Au ED, et al. 2014. Panel-based genetic diagnostic testing for inherited eye diseases is highly accurate and reproducible, and more sensitive for variant detection, than exome sequencing. *Genet Med* 17: 253–261.

Coppieters F, Casteels I, Meire F, De Jaegere S, Hooghe S, van Regemorter N, Van Esch H, Matuleviciene A, Nunes L, Meersschaut V, et al. 2010. Genetic screening of LCA in Belgium: Predominance of *CEP290* and identification of potential modifier alleles in *AHI1* of *CEP290*-related phenotypes. *Hum Mutat* 31: E1709–E1766.

Coussa RG, Otto EA, Gee HY, Arthurs P, Ren H, Lopez I, Keser V, Fu Q, Faingold R, Khan A, et al. 2013. WDR19: An ancient, retrograde, intraflagellar ciliary protein is mutated in autosomal recessive retinitis pigmentosa and in Senior–Løken syndrome. *Clin Genet* 84: 150–159.

Cremers FP, van de Pol DJ, van Driel M, den Hollander AI, van Haren FJ, Knoers NV, Tijmes N, Bergen AA, Rohrschneider K, Blankenagel A, et al. 1998. Autosomal recessive retinitis pigmentosa and cone–rod dystrophy caused by splice site mutations in the Stargardt's disease gene *ABCR*. *Hum Mol Genet* **7:** 355–362.

Dafinger C, Liebau MC, Elsayed SM, Hellenbroich Y, Boltshauser E, Korenke GC, Fabretti F, Janecke AR, Ebermann I, Nürnberg G, et al. 2011. Mutations in *KIF7* link Joubert syndrome with Sonic Hedgehog signaling and microtubule dynamics. *J Clin Invest* **121:** 2662–2667.

Dagoneau N, Goulet M, Geneviève D, Sznajer Y, Martinovic J, Smithson S, Huber C, Baujat G, Flori E, Tecco L, et al. 2009. *DYNC2H1* mutations cause asphyxiating thoracic dystrophy and short rib-polydactyly syndrome, type III. *Am J Hum Genet* **84:** 706–711.

Davidson AE, Sergouniotis PI, Mackay DS, Wright GA, Waseem NH, Michaelides M, Holder GE, Robson AG, Moore AT, Plagnol V, et al. 2013. *RP1L1* variants are associated with a spectrum of inherited retinal diseases including retinitis pigmentosa and occult macular dystrophy. *Hum Mutat* **34:** 506–514.

Davis EE, Zhang Q, Liu Q, Diplas BH, Davey LM, Hartley J, Stoetzel C, Szymanska K, Ramaswami G, Logan CV, et al. 2011. *TTC21B* contributes both causal and modifying alleles across the ciliopathy spectrum. *Nat Genet* **43:** 189–196.

Dellett M, Sasai N, Nishide K, Becker S, Papadaki V, Astrid Limb G, Moore AT, Kondo T, Ohnuma SI. 2015. Genetic background and light-dependent progression of photoreceptor cell degeneration in *Prominin-1* knockout mice. *Investig Ophthalmol Vis Sci* **56:** 164–176.

Delous M, Baala L, Salomon R, Laclef C, Vierkotten J, Tory K, Golzio C, Lacoste T, Besse L, Ozilou C, et al. 2007. The ciliary gene *RPGRIP1L* is mutated in cerebello-oculo-renal syndrome (Joubert syndrome type B) and Meckel syndrome. *Nat Genet* **39:** 875–881.

den Hollander AI, Koenekoop RK, Yzer S, Lopez I, Arends ML, Voesenek KEJ, Zonneveld MN, Strom TM, Meitinger T, Brunner HG, et al. 2006. Mutations in the *CEP290* (*NPHP6*) gene are a frequent cause of Leber congenital amaurosis. *Am J Hum Genet* **79:** 556–561.

den Hollander AI, Koenekoop RK, Mohamed MD, Arts HH, Boldt K, Towns KV, Sedmak T, Beer M, Nagel-Wolfrum K, McKibbin M, et al. 2007. Mutations in *LCA5*, encoding the ciliary protein lebercilin, cause Leber congenital amaurosis. *Nat Genet* **39:** 889–895.

den Hollander AI, Roepman R, Koenekoop RK, Cremers FPM. 2008. Leber congenital amaurosis: Genes, proteins and disease mechanisms. *Prog Retin Eye Res* **27:** 391–419.

Deretic D, Williams AH, Ransom N, Morel V, Hargrave PA, Arendt A. 2005. Rhodopsin C terminus, the site of mutations causing retinal disease, regulates trafficking by binding to ADP-ribosylation factor 4 (ARF4). *Proc Natl Acad Sci* **102:** 3301–3306.

De Robertis E. 1956. Electron microscope observations on the submicroscopic organization of the retinal rods. *J Biophys Biochem Cytol* **2:** 319–330.

Ding JD, Salinas RY, Arshavsky VY. 2015. Discs of mammalian rod photoreceptors form through the membrane evagination mechanism. *J Cell Biol* **211:** 495–502.

Di Palma F, Holme RH, Bryda EC, Belyantseva IA, Pellegrino R, Kachar B, Steel KP, Noben-Trauth K. 2001. Mutations in *Cdh23*, encoding a new type of cadherin, cause stereocilia disorganization in waltzer, the mouse model for Usher syndrome type 1D. *Nat Genet* **27:** 103–107.

Dixon-Salazar T, Silhavy JL, Marsh SE, Louie CM, Scott LC, Gururaj A, Al-Gazali L, Al-Tawari AA, Kayserili H, Sztriha L, et al. 2004. Mutations in the *AHI1* gene, encoding jouberin, cause Joubert syndrome with cortical polymicrogyria. *Am J Hum Genet* **75:** 979–987.

Dowdle WE, Robinson JF, Kneist A, Sirerol-Piquer MS, Frints SGM, Corbit KC, Zaghloul NA, Zaghloul NA, van Lijnschoten G, Mulders L, et al. 2011. Disruption of a ciliary B9 protein complex causes Meckel syndrome. *Am J Hum Genet* **89:** 94–110.

Dryja TP, McGee TL, Reichel E, Hahn LB, Cowley GS, Yandell DW, Sandberg MA, Berson EL. 1990. A point mutation of the rhodopsin gene in one form of retinitis pigmentosa. *Nature* **343:** 364–366.

Dryja TP, Berson EL, Rao VR, Oprian DD. 1993. Heterozygous missense mutation in the rhodopsin gene as a cause of congenital stationary night blindness. *Nat Genet* **4:** 280–283.

Dryja TP, Finn JT, Peng YW, McGee TL, Berson EL, Yau KW. 1995. Mutations in the gene encoding the α subunit of the rod cGMP-gated channel in autosomal recessive retinitis pigmentosa. *Proc Natl Acad Sci* **92:** 10177–10181.

Dryja TP, Hahn LB, Reboul T, Arnaud B. 1996. Missense mutation in the gene encoding the α subunit of rod transducin in the Nougaret form of congenital stationary night blindness. *Nat Genet* **13:** 358–360.

Dryja TP, Adams SM, Grimsby JL, McGee TL, Hong DH, Li T, Andréasson S, Berson EL. 2001. Null *RPGRIP1* alleles in patients with Leber congenital amaurosis. *Am J Hum Genet* **68:** 1295–1298.

Dvir L, Srour G, Abu-Ras R, Miller B, Shalev S, Ben-Yosef T. 2010. Autosomal-recessive early-onset retinitis pigmentosa caused by a mutation in *PDE6G*, the gene encoding the gamma subunit of rod cGMP phosphodiesterase. *Am J Hum Genet* **87:** 258–264.

Ebermann I, Phillips JB, Liebau MC, Koenekoop RK, Schermer B, Lopez I, Schäfer E, Roux AF, Dafinger C, Bernd A, et al. 2010. PDZD7 is a modifier of retinal disease and a contributor to digenic Usher syndrome. *J Clin Invest* **120:** 1812–1823.

Edvardson S, Shaag A, Zenvirt S, Erlich Y, Hannon GJ, Shanske AL, Gomori JM, Ekstein J, Elpeleg O. 2010. Joubert syndrome 2 (JBTS2) in Ashkenazi Jews is associated with a *TMEM216* mutation. *Am J Hum Genet* **86:** 93–97.

El Shamieh S, Neuillé M, Terray A, Orhan E, Condroyer C, Démontant V, Michiels C, Antonio A, Boyard F, Lancelot ME, et al. 2014. Whole-exome sequencing identifies *KIZ* as a ciliary gene associated with autosomal-recessive rod–cone dystrophy. *Am J Hum Genet* **94:** 625–633.

Estrada-Cuzcano A, Koenekoop RK, Coppieters F, Kohl S, Lopez I, Collin RW, De Baere E, Roeleveld D, Marek J, Bernd A, et al. 2011. *IQCB1* mutations in patients with Leber congenital amaurosis. *Invest Ophthalmol Vis Sci* **52:** 834–839.

Estrada-Cuzcano A, Koenekoop RK, Senechal A, De Baere EBW, de Ravel T, Banfi S, Kohl S, Ayuso C, Sharon D, Hoyng CB, et al. 2012a. *BBS1* mutations in a wide spectrum of phenotypes ranging from nonsyndromic retinitis pigmentosa to Bardet−Biedl syndrome. *Arch Ophthalmol* **130:** 1425−1432.

Estrada-Cuzcano A, Neveling K, Kohl S, Banin E, Rotenstreich Y, Sharon D, Falik-Zaccai TC, Hipp S, Roepman R, Wissinger B, et al. 2012b. Mutations in *C8orf37*, encoding a ciliary protein, are associated with autosomal-recessive retinal dystrophies with early macular involvement. *Am J Hum Genet* **90:** 102−109.

Estrada-Cuzcano A, Roepman R, Cremers FPM, den Hollander AI, Mans D. 2012c. Non-syndromic retinal ciliopathies: Translating gene discovery into therapy. *Hum Mol Genet* **21:** R111−R124.

Eudy JD, Weston MD, Yao S, Hoover DM, Rehm HL, Ma-Edmonds M, Yan D, Ahmad I, Cheng JJ, Ayuso C, et al. 1998. Mutation of a gene encoding a protein with extracellular matrix motifs in Usher syndrome type IIa. *Science* **280:** 1753−1757.

Fahim AT, Bowne SJ, Sullivan LS, Webb KD, Williams JT, Wheaton DK, Birch DG, Daiger SP. 2011. Allelic heterogeneity and genetic modifier loci contribute to clinical variation in males with X-linked retinitis pigmentosa due to *RPGR* mutations. *PLoS ONE* **6:** e23021.

Fan Y, Esmail MA, Ansley SJ, Blacque OE, Boroevich K, Ross AJ, Moore SJ, Badano JL, May-Simera H, Compton DS, et al. 2004. Mutations in a member of the Ras superfamily of small GTP-binding proteins causes Bardet−Biedl syndrome. *Nat Genet* **36:** 989−993.

Farrar GJ, McWilliam P, Bradley DG, Kenna P, Lawler M, Sharp EM, Humphries MM, Eiberg H, Conneally PM, Trofatter JA. 1990. Autosomal dominant retinitis pigmentosa: Linkage to rhodopsin and evidence for genetic heterogeneity. *Genomics* **8:** 35−40.

Ferland RJ, Eyaid W, Collura RV, Tully LD, Hill RS, Al-Nouri D, Al-Rumayyan A, Topcu M, Gascon G, Bodell A, et al. 2004. Abnormal cerebellar development and axonal decussation due to mutations in *AHI1* in Joubert syndrome. *Nat Genet* **36:** 1008−1013.

Ferrante MI, Giorgio G, Feather SA, Bulfone A, Wright V, Ghiani M, Selicorni A, Gammaro L, Scolari F, Woolf AS, et al. 2001. Identification of the gene for oral-facial-digital type I syndrome. *Am J Hum Genet* **68:** 569−576.

Filges I, Nosova E, Bruder E, Tercanli S, Townsend K, Gibson WT, Röthlisberger B, Heinimann K, Hall JG, Gregory-Evans CY, et al. 2014. Exome sequencing identifies mutations in *KIF14* as a novel cause of an autosomal recessive lethal fetal ciliopathy phenotype. *Clin Genet* **86:** 220−228.

Fingert JH, Oh K, Chung M, Scheetz TE, Andorf JL, Johnson RM, Sheffield VC, Stone EM. 2008. Association of a novel mutation in the retinol dehydrogenase 12 (RDH12) gene with autosomal dominant retinitis pigmentosa. *Arch Ophthalmol* **126:** 1301−1307.

Fotiadis D, Liang Y, Filipek S, Saperstein DA, Engel A, Palczewski K. 2003. Rhodopsin dimers in native disc membranes. *Nature* **421:** 127−128.

Frank V, den Hollander AI, Brüchle NO, Zonneveld MN, Nürnberg G, Becker C, Du Bois G, Kendziorra H, Roosing S, Senderek J, et al. 2008. Mutations of the *CEP290* gene encoding a centrosomal protein cause Meckel−Gruber syndrome. *Hum Mutat* **29:** 45−52.

Fuchs S, Nakazawa M, Maw M, Tamai M, Oguchi Y, Gal A. 1995. A homozygous 1-base pair deletion in the arrestin gene is a frequent cause of Oguchi disease in Japanese. *Nat Genet* **10:** 360−362.

Gal A, Orth U, Baehr W, Schwinger E, Rosenberg T. 1994. Heterozygous missense mutation in the rod cGMP phosphodiesterase beta-subunit gene in autosomal dominant stationary night blindness. *Nat Genet* **7:** 64−68.

Galdzicka M, Patnala S, Hirshman MG, Cai JF, Nitowsky H, Egeland JA, Ginns EI. 2002. A new gene, *EVC2*, is mutated in Ellis−van Creveld syndrome. *Mol Genet Metab* **77:** 291−295.

Garcia-Gonzalo FR, Corbit KC, Sirerol-Piquer MS, Ramaswami G, Otto EA, Noriega TR, Seol AD, Robinson JF, Bennett CL, Josifova DJ, et al. 2011. A transition zone complex regulates mammalian ciliogenesis and ciliary membrane composition. *Nat Genet* **43:** 776−784.

Geller SF, Guerin KI, Visel M, Pham A, Lee ES, Dror AA, Avraham KB, Hayashi T, Ray CA, Reh TA, et al. 2009. CLRN1 is nonessential in the mouse retina but is required for cochlear hair cell development. *PLoS Genet* **5:** 17−19.

Gilissen C, Arts HH, Hoischen A, Spruijt L, Mans DA, Arts P, van Lier B, Steehouwer M, van Reeuwijk J, Kant SG, et al. 2010. Exome sequencing identifies *WDR35* variants involved in Sensenbrenner syndrome. *Am J Hum Genet* **87:** 418−423.

Gilliam JC, Chang JT, Sandoval IM, Zhang Y, Li T, Pittler SJ, Chiu W, Wensel TG. 2012. Three-dimensional architecture of the rod sensory cilium and its disruption in retinal neurodegeneration. *Cell* **151:** 1029−1041.

Girisha KM, Shukla A, Trujillano D, Bhavani GS, Hebbar M, Kadavigere R, Rolfs A. 2016. A homozygous nonsense variant in *IFT52* is associated with a human skeletal ciliopathy. *Clin Genet* doi: 10.1111/cge.12762.

Goetz SC, Anderson KV. 2010. The primary cilium: A signalling centre during vertebrate development. *Nat Rev Genet* **11:** 331−344.

Goldberg AFX, Molday RS. 1996. Subunit composition of the peripherin/RDS−rom-1 disk rim complex from rod photoreceptors: Hydrodynamic evidence for a tetrameric quaternary structure. *Biochemistry* **35:** 6144−6149.

Gorden NT, Arts HH, Parisi MA, Coene KLM, Letteboer SJF, van Beersum SEC, Mans DA, Hikida A, Eckert M, Knutzen D, et al. 2008. *CC2D2A* is mutated in Joubert syndrome and interacts with the ciliopathy-associated basal body protein CEP290. *Am J Hum Genet* **83:** 559−571.

Guillonneau X, Piriev NI, Danciger M, Kozak CA, Cideciyan AV, Jacobson SG, Farber DB. 1999. A nonsense mutation in a novel gene is associated with retinitis pigmentosa in a family linked to the RP1 locus. *Hum Mol Genet* **8:** 1541−1546.

Hagstrom SA, North MA, Nishina PL, Berson EL, Dryja TP. 1998. Recessive mutations in the gene encoding the tubby-like protein TULP1 in patients with retinitis pigmentosa. *Nat Genet* **18:** 174−176.

Halbritter J, Bizet AA, Schmidts M, Porath JD, Braun DA, Gee HY, McInerney-Leo AM, Krug P, Filhol E, Davis EE, et al. 2013. Defects in the IFT-B component IFT172 cause Jeune and Mainzer−Saldino syndromes in humans. *Am J Hum Genet* **93:** 1−11.

Hameed A, Abid A, Aziz A, Ismail M, Mehdi SQ, Khaliq S. 2003. Evidence of *RPGRIP1* gene mutations associated with recessive cone-rod dystrophy. *J Med Genet* **40**: 616–619.

Hamel C. 2006. Retinitis pigmentosa. *Orphanet J Rare Dis* **1**: 40.

Hampshire DJ, Ayub M, Springell K, Roberts E, Jafri H, Rashid Y, Bond J, Riley JH, Woods CG. 2006. MORM syndrome (mental retardation, truncal obesity, retinal dystrophy and micropenis), a new autosomal recessive disorder, links to 9q34. *Eur J Hum Genet* **14**: 543–548.

Han Z, Anderson DW, Papermaster DS. 2012. Prominin-1 localizes to the open rims of outer segment lamellae in *Xenopus laevis* rod and cone photoreceptors. *Investig Ophthalmol Vis Sci* **53**: 361–373.

Hardcastle AJ, Thiselton DL, Van Maldergem L, Saha BK, Jay M, Plant C, Taylor R, Bird AC, Bhattacharya S. 1999. Mutations in the *RP2* gene cause disease in 10% of families with familial X-linked retinitis pigmentosa assessed in this study. *Am J Hum Genet* **64**: 1210–1215.

Hartong DT, Berson EL, Dryja TP. 2006. Retinitis pigmentosa. *Lancet* **368**: 1795–1809.

Helou J, Otto EA, Attanasio M, Allen SJ, Parisi MA, Glass I, Utsch B, Hashmi S, Fazzi E, Omran H, et al. 2007. Mutation analysis of *NPHP6/CEP290* in patients with Joubert syndrome and Senior–Løken syndrome. *J Med Genet* **44**: 657–663.

Henderson RH, Li Z, Abd El Aziz MM, Mackay DS, Eljinini MA, Zeidan M, Moore AT, Bhattacharya SS, Webster AR. 2010. Biallelic mutation of protocadherin-21 (*PCDH21*) causes retinal degeneration in humans. *Mol Vis* **16**: 46–52.

Heon E, Kim G, Qin S, Garrison JE, Tavares E, Vincent A, Nuangchamnong N, Scott CA, Slusarski DC, Sheffield VC. 2016. Mutations in *C8ORF37* cause Bardet–Biedl syndrome (BBS21). *Hum Mol Genet* doi: 10.1093/hmg/ddw096.

Hopp K, Heyer CM, Hommerding CJ, Henke SA, Sundsbak JL, Patel S, Patel P, Consugar MB, Czarnecki PG, Gliem TJ, et al. 2011. *B9D1* is revealed as a novel Meckel syndrome (MKS) gene by targeted exon-enriched next-generation sequencing and deletion analysis. *Hum Mol Genet* **20**: 2524–2534.

Horst CJ, Johnson LV, Besharse JC. 1990. Transmembrane assemblage of the photoreceptor connecting cilium and motile cilium transition zone contain a common immunologic epitope. *Cell Motil Cytoskeleton* **17**: 329–344.

Huang SHH, Pittler SJJ, Huang X, Oliveira L, Berson ELL, Dryja TPP. 1995. Autosomal recessive retinitis pigmentosa caused by mutations in the α subunit of rod cGMP phosphodiesterase. *Nat Genet* **11**: 468–471.

Huang L, Szymanska K, Jensen VL, Janecke AR, Innes AM, Davis EE, Frosk P, Li C, Willer JR, Chodirker BN, et al. 2011. *TMEM237* is mutated in individuals with a Joubert syndrome related disorder and expands the role of the TMEM family at the ciliary transition zone. *Am J Hum Genet* **89**: 713–730.

Huber C, Cormier-Daire V. 2012. Ciliary disorder of the skeleton. *Am J Med Genet Semin Med Genet* **160C**: 165–174.

Jacobson SG, Cideciyan AV, Iannaccone A, Weleber RG, Fishman GA, Maguire AM, Affatigato LM, Bennett J, Pierce EA, Danciger M, et al. 2000. Disease expression of *RP1* mutations causing autosomal dominant retinitis pigmentosa. *Invest Ophthalmol Vis Sci* **41**: 1898–1908.

Jacoby M, Cox JJ, Gayral S, Hampshire DJ, Ayub M, Blockmans M, Pernot E, Kisseleva MV, Compère P, Schiffmann SN, et al. 2009. *INPP5E* mutations cause primary cilium signaling defects, ciliary instability and ciliopathies in human and mouse. *Nat Genet* **41**: 1027–1031.

Janecke AR, Thompson DA, Utermann G, Becker C, Hubner CA, Schmid E, McHenry CL, Nair AR, Ruschendorf F, Heckenlively J, et al. 2004. Mutations in *RDH12* encoding a photoreceptor cell retinol dehydrogenase cause childhood-onset severe retinal dystrophy. *Nat Genet* **36**: 850–854.

Joensuu T, Hämäläinen R, Yuan B, Johnson C, Tegelberg S, Gasparini P, Zelante L, Pirvola U, Pakarinen L, Lehesjoki AE, et al. 2001. Mutations in a novel gene with transmembrane domains underlie Usher syndrome type 3. *Am J Hum Genet* **69**: 673–684.

Johnson KR, Gagnon LH, Webb LS, Peters LL, Hawes NL, Zheng QY, Chang B, Zheng QY. 2003. Mouse models of USH1C and DFNB18: Phenotypic and molecular analyses of two new spontaneous mutations of the *Ush1c* gene. *Hum Mol Genet* **12**: 3075–3086.

Kajiwara K, Hahn LB, Mukai S, Travis GH, Berson EL, Dryja TP. 1991. Mutations in the human retinal degeneration slow gene in autosomal dominant retinitis pigmentosa. *Nature* **354**: 480–483.

Kajiwara K, Berson EL, Dryja TP. 1994. Digenic retinitis pigmentosa due to mutations at the unlinked peripherin/RDS and ROM1 loci. *Science* **264**: 1604–1608.

Kaplan MW, Iwata RT, Sears RC. 1987. Lengths of immunolabeled ciliary microtubules in frog photoreceptor outer segments. *Exp Eye Res* **44**: 623–632.

Katsanis N, Beales PL, Woods MO, Lewis RA, Green JS, Parfrey PS, Ansley SJ, Davidson WS, Lupski JR. 2000. Mutations in *MKKS* cause obesity, retinal dystrophy and renal malformations associated with Bardet–Biedl syndrome. *Nat Genet* **26**: 67–70.

Katsanis N, Ansley SJ, Badano JL, Eichers ER, Lewis RA, Hoskins BE, Scambler PJ, Davidson WS, Beales PL, Lupski JR. 2001. Triallelic inheritance in Bardet–Biedl syndrome, a Mendelian recessive disorder. *Science* **293**: 2256–2259.

Kelsell RE, Gregory-Evans K, Payne AM, Perrault I, Kaplan J, Yang RB, Garbers DL, Bird AC, Moore AT, Hunt DM. 1998. Mutations in the retinal guanylate cyclase (*RETGC-1*) gene in dominant cone–rod dystrophy. *Hum Mol Genet* **7**: 1179–1184.

Kennedy B, Malicki J. 2009. What drives cell morphogenesis—A look inside the vertebrate photoreceptor. *Dev Dyn* **238**: 2115–2138.

Kevany BM, Zhang N, Jastrzebska B, Palczewski K. 2015. Animals deficient in C2Orf71, an autosomal recessive retinitis pigmentosa-associated locus, develop severe early-onset retinal degeneration. *Hum Mol Genet* **24**: 2627–2640.

Khan MI, Kersten FFJ, Azam M, Collin RWJ, Hussain A, Shah STA, Keunen JEE, Kremer H, Cremers FPM, Qamar R, et al. 2011. *CLRN1* mutations cause nonsyndromic retinitis pigmentosa. *Ophthalmology* **118**: 1444–1448.

Khan AO, Decker E, Bachmann N, Bolz HJ, Bergmann C. 2016. *C8orf37* is mutated in Bardet–Biedl syndrome and constitutes a locus allelic to non-syndromic retinal dystrophies. *Ophthalmic Genet* **8:** 1–4.

Khanna H. 2015. Photoreceptor sensory cilium: Traversing the ciliary gate. *Cells* **4:** 674–686.

Khanna H, Davis EE, Murga-Zamalloa CA, Estrada-Cuzcano A, Lopez I, den Hollander AI, Zonneveld MN, Othman MI, Waseem N, Chakarova CF, et al. 2009. A common allele in RPGRIP1L is a modifier of retinal degeneration in ciliopathies. *Nat Genet* **41:** 739–745.

Khateb S, Zelinger L, Mizrahi-Meissonnier L, Ayuso C, Koenekoop RK, Laxer U, Gross M, Banin E, Sharon D. 2014. A homozygous nonsense CEP250 mutation combined with a heterozygous nonsense C2orf71 mutation is associated with atypical Usher syndrome. *J Med Genet* **51:** 460–469.

Kim SK, Shindo A, Park TJ, Oh EC, Ghosh S, Gray RS, Lewis RA, Johnson CA, Attie-Bittach T, Katsanis N, et al. 2010. Planar cell polarity acts through septins to control collective cell movement and ciliogenesis. *Science* **329:** 1337–40.

Kim MS, Pinto SM, Getnet D, Nirujogi RS, Manda SS, Chaerkady R, Madugundu AK, Kelkar DS, Isserlin R, Jain S, et al. 2014. A draft map of the human proteome. *Nature* **509:** 575–581.

Koenekoop RK. 2004. An overview of leber congenital amaurosis: A model to understand human retinal development. *Surv Ophthalmol* **49:** 379–398.

Kohl S, Marx T, Giddings I, Jägle H, Jacobson SG, Apfelstedt-Sylla E, Zrenner E, Sharpe LT, Wissinger B. 1998. Total colourblindness is caused by mutations in the gene encoding the α-subunit of the cone photoreceptor cGMP-gated cation channel. *Nat Genet* **19:** 257–259.

Kohl S, Baumann B, Broghammer M, Jägle H, Sieving P, Kellner U, Spegal R, Anastasi M, Zrenner E, Sharpe LT, et al. 2000. Mutations in the CNGB3 gene encoding the β-subunit of the cone photoreceptor cGMP-gated channel are responsible for achromatopsia (ACHM3) linked to chromosome 8q21. *Hum Mol Genet* **9:** 2107–2116.

Kohl S, Baumann B, Rosenberg T, Kellner U, Lorenz B, Vadalà M, Jacobson SG, Wissinger B. 2002. Mutations in the cone photoreceptor G-protein α-subunit gene GNAT2 in patients with achromatopsia. *Am J Hum Genet* **71:** 422–425.

Kremer H, van Wijk E, Märker T, Wolfrum U, Roepman R. 2006. Usher syndrome: Molecular links of pathogenesis, proteins and pathways. *Hum Mol Genet* **15:** R262–R270.

Kyttälä M, Tallila J, Salonen R, Kopra O, Kohlschmidt N, Paavola-Sakki P, Peltonen L, Kestilä M. 2006. MKS1, encoding a component of the flagellar apparatus basal body proteome, is mutated in Meckel syndrome. *Nat Genet* **38:** 155–157.

Lamb TD, Pugh EN. 2006. Phototransduction, dark adaptation, and rhodopsin regeneration the proctor lecture. *Invest Ophthalmol Vis Sci* **47:** 5137–5152.

Lamb TD, Collin SP, Pugh EN. 2007. Evolution of the vertebrate eye: Opsins, photoreceptors, retina and eye cup. *Nat Rev Neurosci* **8:** 960–976.

Langmann T, Di Gioia SA, Rau I, Stöhr H, Maksimovic NS, Corbo JC, Renner AB, Zrenner E, Kumaramanickavel G, Karlstetter M, et al. 2010. Nonsense mutations in FAM161A cause RP28-associated recessive retinitis pigmentosa. *Am J Hum Genet* **87:** 376–381.

Larsson NG, Wang J, Wilhelmsson H, Oldfors A, Rustin P, Lewandoski M, Barsh GS, Clayton DA. 1998. Recessive mutations in TULP1 in RP. *Nat Genet* **18:** 231–236.

Lee JE, Silhavy JL, Zaki MS, Schroth J, Bielas SL, Marsh SE, Olvera J, Brancati F, Iannicelli M, Ikegami K, et al. 2012a. CEP41 is mutated in Joubert syndrome and is required for tubulin glutamylation at the cilium. *Nat Genet* **44:** 193–199.

Lee JH, Silhavy JL, Lee JE, Al-Gazali L, Thomas S, Davis EE, Bielas SL, Hill KJ, Iannicelli M, Brancati F, et al. 2012b. Evolutionarily assembled cis-regulatory module at a human ciliopathy locus. *Science* **335:** 966–969.

Lehman AM, Eydoux P, Doherty D, Glass IA, Chitayat D, Chung BYH, Langlois S, Yong SL, Lowry RB, Hildebrandt F, et al. 2010. Co-occurrence of Joubert syndrome and Jeune asphyxiating thoracic dystrophy. *Am J Med Genet Part A* **152:** 1411–1419.

Leitch CC, Zaghloul NA, Davis EE, Stoetzel C, Diaz-Font A, Rix S, Alfadhel M, Al-Fadhel M, Lewis RA, Eyaid W, et al. 2008. Hypomorphic mutations in syndromic encephalocele genes are associated with Bardet–Biedl syndrome. *Nat Genet* **40:** 443–448.

Li JB, Gerdes JM, Haycraft CJ, Fan Y, Teslovich TM, May-Simera H, Li H, Blacque OE, Li L, Leitch CC, et al. 2004. Comparative genomics identifies a flagellar and basal body proteome that includes the BBS5 human disease gene. *Cell* **117:** 541–552.

Libby RT, Steel KP. 2001. Electroretinographic anomalies in mice with mutations in Myo7a, the gene involved in human Usher syndrome type 1B. *Invest Ophthalmol Vis Sci* **42:** 770–778.

Lindstrand A, Davis EE, Carvalho CMB, Pehlivan D, Willer JR, Tsai IC, Ramanathan S, Zuppan C, Sabo A, Muzny D, et al. 2014. Recurrent CNVs and SNVs at the NPHP1 locus contribute pathogenic alleles to Bardet–Biedl syndrome. *Am J Hum Genet* **94:** 745–754.

Liu X, Udovichenko IP, Brown SD, Steel KP, Williams DS. 1999. Myosin VIIa participates in opsin transport through the photoreceptor cilium. *J Neurosci* **19:** 6267–6274.

Liu Q, Tan G, Levenkova N, Li T, Pugh EN, Rux JJ, Speicher DW, Pierce EA. 2007a. The proteome of the mouse photoreceptor sensory cilium complex. *Mol Cell Proteomics* **6:** 1299–1317.

Liu X, Bulgakov OV, Darrow KN, Pawlyk B, Adamian M, Liberman MC, Li T. 2007b. Usherin is required for maintenance of retinal photoreceptors and normal development of cochlear hair cells. *Proc Natl Acad Sci* **104:** 4413–4418.

Liu Q, Zhang Q, Pierce EA. 2010. Photoreceptor sensory cilia and inherited retinal degeneration. *Adv Exp Med Biol* **664:** 223–232.

Logan CV, Abdel-Hamed Z, Johnson CA. 2011. Molecular genetics and pathogenic mechanisms for the severe ciliopathies: Insights into neurodevelopment and pathogenesis of neural tube defects. *Mol Neurobiol* **43:** 12–26.

Løken AC, Hanssen O, Halvorsen S, Jolster NJ. 1961. Hereditary renal dysplasia and blindness. *Acta Paediatr* **50:** 177–184.

Louie CM, Caridi G, Lopes VS, Brancati F, Kispert A, Lancaster MA, Schlossman AM, Otto EA, Leitges M, Gröne HJ, et al. 2010. AHI1 is required for photoreceptor outer

Cite this article as *Cold Spring Harb Perspect Biol* doi: 10.1101/cshperspect.a028274

segment development and is a modifier for retinal degeneration in nephronophthisis. *Nat Genet* **42:** 175–180.

Maria BL, Hoang KB, Tusa RJ, Mancuso AA, Hamed LM, Quisling RG, Hove MT, Fennell EB, Booth-Jones M, Ringdahl DM, et al. 1997. "Joubert syndrome" revisited: Key ocular motor signs with magnetic resonance imaging correlation. *J Child Neurol* **12:** 423–430.

Marion V, Stutzmann F, Gérard M, De Melo C, Schaefer E, Claussmann A, Hellé S, Delague V, Souied E, Barrey C, et al. 2012. Exome sequencing identifies mutations in *LZTFL1*, a BBSome and smoothened trafficking regulator, in a family with Bardet–Biedl syndrome with situs inversus and insertional polydactyly. *J Med Genet* **49:** 317–321.

Martínez-Mir A, Paloma E, Allikmets R, Ayuso C, del Rio T, Dean M, Vilageliu L, Gonzàlez-Duarte R, Balcells S. 1998. Retinitis pigmentosa caused by a homozygous mutation in the Stargardt disease gene *ABCR*. *Nat Genet* **18:** 11–12.

Maw MA, Corbeil D, Koch J, Hellwig A, Wilson-Wheeler JC, Bridges RJ, Kumaramanickavel G, John S, Nancarrow D, Röper K, et al. 2000. A frameshift mutation in prominin (mouse)-like 1 causes human retinal degeneration. *Hum Mol Genet* **9:** 27–34.

Mburu P, Mustapha M, Varela A, Weil D, El-Amraoui A, Holme RH, Rump A, Hardisty RE, Blanchard S, Coimbra RS, et al. 2003. Defects in whirlin, a PDZ domain molecule involved in stereocilia elongation, cause deafness in the whirler mouse and families with DFNB31. *Nat Genet* **34:** 421–428.

McGee TL, Devoto M, Ott J, Berson EL, Dryja TP. 1997. Evidence that the penetrance of mutations at the RP11 locus causing dominant retinitis pigmentosa is influenced by a gene linked to the homologous RP11 allele. *Am J Hum Genet* **61:** 1059–1066.

McInerney-Leo AM, Schmidts M, Cortés CR, Leo PJ, Gener B, Courtney AD, Gardiner B, Harris JA, Lu Y, Marshall M, et al. 2013. Short-rib polydactyly and Jeune syndromes are caused by mutations in *WDR60*. *Am J Hum Genet* **93:** 515–523.

McLaughlin MEE, Sandberg MAA, Berson ELL, Dryja TPP. 1993. Recessive mutations in the gene encoding the β-subunit of rod phosphodiesterase in patients with retinitis pigmentosa. *Nat Genet* **4:** 130–134.

Mears AJ, Gieser L, Yan D, Chen C, Fahrner S, Hiriyanna S, Fujita R, Jacobson SG, Sieving PA, Swaroop A. 1999. Protein-truncation mutations in the *RP2* gene in a North American cohort of families with X-linked retinitis pigmentosa. *Am J Hum Genet* **64:** 897–900.

Meindl A, Dry K, Herrmann K, Manson F, Ciccodicola A, Edgar A, Carvalho MR, Achatz H, Hellebrand H, Lennon A, et al. 1996. A gene (*RPGR*) with homology to the *RCC1* guanine nucleotide exchange factor is mutated in X-linked retinitis pigmentosa (RP3). *Nat Genet* **13:** 35–42.

Michaelides M, Hardcastle AJ, Hunt DM, Moore AT. 2006. Progressive cone and cone–rod dystrophies: Phenotypes and underlying molecular genetic basis. *Surv Ophthalmol* **51:** 232–258.

Mill P, Lockhart PJ, Fitzpatrick E, Mountford HS, Hall EA, Reijns MAM, Keighren M, Bahlo M, Bromhead CJ, Budd P, et al. 2011. Human and mouse mutations in *WDR35*

cause short-rib polydactyly syndromes due to abnormal ciliogenesis. *Am J Hum Genet* **88:** 508–515.

Millán JM, Aller E, Jaijo T, Blanco-Kelly F, Gimenez-Pardo A, Ayuso C. 2011. An update on the genetics of Usher syndrome. *J Ophthalmol* **2011:** 417217.

Miyasaka Y, Suzuki S, Ohshiba Y, Watanabe K, Sagara Y, Yasuda SP, Matsuoka K, Shitara H, Yonekawa H, Kominami R, et al. 2013. Compound heterozygosity of the functionally null $Cdh23^{v-ngt}$ and hypomorphic $Cdh23^{ahl}$ alleles leads to early-onset progressive hearing loss in mice. *Exp Anim* **62:** 333–346.

Mockel A, Perdomo Y, Stutzmann F, Letsch J, Marion V, Dollfus H. 2011. Retinal dystrophy in Bardet–Biedl syndrome and related syndromic ciliopathies. *Prog Retin Eye Res* **30:** 258–274.

Molday RS, Molday LL. 1987. Differences in the protein composition of bovine retinal rod outer segment disk and plasma membranes isolated by a ricin–gold–dextran density perturbation method. *J Cell Biol* **105:** 2589–2601.

Molday RS, Moritz OL. 2015. Photoreceptors at a glance. *J Cell Sci* **128:** 4039–4045.

Molday RS, Hicks D, Molday L. 1987. Peripherin: A rim specific membrane protein of rod outer segment discs. *Investig Opthalmology Vis Sci* **28:** 50–61.

Mykytyn K, Braun T, Carmi R, Haider NB, Searby CC, Shastri M, Beck G, Wright AF, Iannaccone A, Elbedour K, et al. 2001. Identification of the gene that, when mutated, causes the human obesity syndrome BBS4. *Nat Genet* **28:** 188–191.

Mykytyn K, Nishimura DY, Searby CC, Shastri M, Yen H, Beck JS, Braun T, Streb LM, Cornier AS, Cox GF, et al. 2002. Identification of the gene (*BBS1*) most commonly involved in Bardet–Biedl syndrome, a complex human obesity syndrome. *Nat Genet* **31:** 435–438.

Nair KS, Hanson SM, Mendez A, Gurevich EV, Kennedy MJ, Shestopalov VI, Vishnivetskiy SA, Chen J, Hurley JB, Gurevich VV, et al. 2005. Light-dependent redistribution of arrestin in vertebrate rods is an energy-independent process governed by protein–protein interactions. *Neuron* **46:** 555–567.

Najafi M, Calvert PD. 2012. Transport and localization of signaling proteins in ciliated cells. *Vision Res* **75:** 11–18.

Nakazawa M, Wada Y, Tamai M. 1998. Arrestin gene mutations in autosomal recessive retinitis pigmentosa. *Arch Ophthalmol* **116:** 498–501.

Nathans J, Thomas D, Hogness DS. 1986. Molecular genetics of human color vision: The genes encoding blue, green, and red pigments. *Science* **232:** 193–202.

Nathans J, Merbs S, Sung CH, Weitz C, Wang Y. 1992. Molecular genetics of human visual pigments. *Annu Rev Genet* **26:** 403–424.

Nickell S, Park PSH, Baumeister W, Palczewski K. 2007. Three-dimensional architecture of murine rod outer segments determined by cryoelectron tomography. *J Cell Biol* **177:** 917–925.

Nikopoulos K, Farinelli P, Royer-Bertrand B, Bedoni N, Kjellström U, Andreasson S, Tsilimbaris MK, Tsika C, Blazaki S, Rivolta C. 2016. Whole exome sequencing reveals CEP78 as a novel disease gene for cone-rod dystrophy. *ARVO Annual Meeting*, Abstract 1423. Seattle, WA, May 1–6.

Nishiguchi KM, Sandberg MA, Kooijman AC, Martemyanov KA, Pott JWR, Hagstrom SA, Arshavsky VY, Berson EL, Dryja TP. 2004. Defects in RGS9 or its anchor protein R9AP in patients with slow photoreceptor deactivation. *Nature* **427:** 75–78.

Nishiguchi KM, Tearle RG, Liu YP, Oh EC, Miyake N, Benaglio P, Harper S, Koskiniemi-Kuendig H, Venturini G, Sharon D, et al. 2013. Whole genome sequencing in patients with retinitis pigmentosa reveals pathogenic DNA structural changes and *NEK2* as a new disease gene. *Proc Natl Acad Sci* **110:** 16139–16144.

Nishimura DY, Searby CC, Carmi R, Elbedour K, Van Maldergem L, Fulton AB, Lam BL, Powell BR, Swiderski RE, Bugge KE, et al. 2001. Positional cloning of a novel gene on chromosome 16q causing Bardet–Biedl syndrome (BBS2). *Hum Mol Genet* **10:** 865–874.

Nishimura DY, Swiderski RE, Searby CC, Berg EM, Ferguson AL, Hennekam R, Merin S, Weleber RG, Biesecker LG, Stone EM, et al. 2005. Comparative genomics and gene expression analysis identifies *BBS9*, a new Bardet–Biedl syndrome gene. *Am J Hum Genet* **77:** 1021–1033.

Nishimura DY, Baye LM, Perveen R, Searby CC, Avila-Fernandez A, Pereiro I, Ayuso C, Valverde D, Bishop PN, Manson FDC, et al. 2010. Discovery and functional analysis of a retinitis pigmentosa gene, *C2ORF71*. *Am J Hum Genet* **86:** 686–695.

Noor A, Windpassinger C, Patel M, Stachowiak B, Mikhailov A, Azam M, Irfan M, Siddiqui ZK, Naeem F, Paterson AD, et al. 2008. CC2D2A, encoding a coiled-coil and C2 domain protein, causes autosomal-recessive mental retardation with retinitis pigmentosa. *Am J Hum Genet* **82:** 1011–1018.

Ong ACM, Harris PC. 2015. A polycystin-centric view of cyst formation and disease: The polycystins revisited. *Kidney Int* **88:** 699–710.

Ostergaard E, Batbayli M, Duno M, Vilhelmsen K, Rosenberg T. 2010. Mutations in *PCDH21* cause autosomal recessive cone–rod dystrophy. *J Med Genet* **47:** 665–669.

Otto E, Hoefele J, Ruf R, Mueller AM, Hiller KS, Wolf MTF, Schuermann MJ, Becker A, Birkenhäger R, Sudbrak R, et al. 2002. A gene mutated in nephronophthisis and retinitis pigmentosa encodes a novel protein, nephroretinin, conserved in evolution. *Am J Hum Genet* **71:** 1161–1117.

Otto EA, Loeys B, Khanna H, Hellemans J, Sudbrak R, Fan S, Muerb U, O'Toole JF, Helou J, Attanasio M, et al. 2005. Nephrocystin-5, a ciliary IQ domain protein, is mutated in Senior–Løken syndrome and interacts with RPGR and calmodulin. *Nat Genet* **37:** 282–288.

Otto EA, Hurd TW, Airik R, Chaki M, Zhou W, Stoetzel C, Patil SB, Levy S, Ghosh AK, Murga-Zamalloa CA, et al. 2010. Candidate exome capture identifies mutation of *SDCCAG8* as the cause of a retinal–renal ciliopathy. *Nat Genet* **42:** 840–850.

Ozgül RK, Siemiatkowska AM, Yücel D, Myers CA, Collin RWJ, Zonneveld MN, Beryozkin A, Banin E, Hoyng CB, van den Born LI, et al. 2011. Exome sequencing and *cis*-regulatory mapping identify mutations in *MAK*, a gene encoding a regulator of ciliary length, as a cause of retinitis pigmentosa. *Am J Hum Genet* **89:** 253–264.

Pan J, Wang Q, Snell WJ. 2005. Cilium-generated signaling and cilia-related disorders. *Lab Invest* **85:** 452–463.

Parisi MA, Bennett CL, Eckert ML, Dobyns WB, Gleeson JG, Shaw DWW, McDonald R, Eddy A, Chance PF, Glass IA. 2004. The *NPHP1* gene deletion associated with juvenile nephronophthisis is present in a subset of individuals with Joubert syndrome. *Am J Hum Genet* **75:** 82–91.

Patel A, Honoré E. 2010. Polycystins and renovascular mechanosensory transduction. *Nat Rev Nephrol* **6:** 530–538.

Pawlik B, Mir A, Iqbal H, Li Y, Nürnberg G, Becker C, Qamar R, Nürnberg P, Wollnik B. 2010. A novel familial *BBS12* mutation associated with a mild phenotype: Implications for clinical and molecular diagnostic strategies. *Mol Syndromol* **1:** 27–34.

Payne AM, Downes SM, Bessant DA, Taylor R, Holder GE, Warren MJ, Bird AC, Bhattacharya SS. 1998. A mutation in guanylate cyclase activator 1A (GUCA1A) in an autosomal dominant cone dystrophy pedigree mapping to a new locus on chromosome 6p21.1. *Hum Mol Genet* **7:** 273–277.

Pearring JN, Salinas RY, Baker SA, Arshavsky VY. 2013. Protein sorting, targeting and trafficking in photoreceptor cells. *Prog Retin Eye Res* **36:** 24–51.

Pedersen LB, Rosenbaum JL. 2008. Intraflagellar transport (IFT) role in ciliary assembly, resorption and signalling. *Curr Top Dev Biol* **85:** 23–61.

Perrault I, Rozet JM, Calvas P, Gerber S, Camuzat A, Dollfus H, Châtelin S, Souied E, Ghazi I, Leowski C, et al. 1996. Retinal-specific guanylate cyclase gene mutations in Leber's congenital amaurosis. *Nat Genet* **14:** 461–464.

Perrault I, Hanein S, Gerber S, Barbet F, Ducroq D, Dollfus H, Hamel C, Dufier JL, Munnich A, Kaplan J, et al. 2004. Retinal dehydrogenase 12 (RDH12) mutations in leber congenital amaurosis. *Am J Hum Genet* **75:** 639–646.

Perrault I, Delphin N, Hanein S, Gerber S, Dufier JL, Roche O, Defoort-Dhellemmes S, Dollfus H, Fazzi E, Munnich A, et al. 2007. Spectrum of *NPHP6/CEP290* mutations in Leber congenital amaurosis and delineation of the associated phenotype. *Hum Mutat* **28:** 416.

Perrault I, Saunier S, Hanein S, Filhol E, Bizet AA, Collins F, Salih MAM, Gerber S, Delphin N, Bigot K, et al. 2012. Mainzer–Saldino syndrome is a ciliopathy caused by *IFT140* mutations. *Am J Hum Genet* **90:** 864–870.

Pierce EA, Quinn T, Meehan T, McGee TL, Berson EL, Dryja TP. 1999. Mutations in a gene encoding a new oxygen-regulated photoreceptor protein cause dominant retinitis pigmentosa. *Nat Genet* **22:** 248–254.

Poloschek CM, Bach M, Lagrèze WA, Glaus E, Lemke JR, Berger W, Neidhardt J, Lagreze WA, Glaus E, Lemke JR, et al. 2010. *ABCA4* and *ROM1*: Implications for modification of the *PRPH2*-associated macular dystrophy phenotype. *Invest Ophthalmol Vis Sci* **51:** 4253–4265.

Pretorius PR, Baye LM, Nishimura DY, Searby CC, Bugge K, Yang B, Mullins RF, Stone EM, Sheffield VC, Slusarski DC. 2010. Identification and functional analysis of the vision-specific BBS3 (ARL6) long isoform. *PLoS Genet* **6:** e1000884.

Puffenberger EG, Jinks RN, Sougnez C, Cibulskis K, Willert RA, Achilly NP, Cassidy RP, Fiorentini CJ, Heiken KF, Lawrence JJ, et al. 2012. Genetic mapping and exome sequencing identify variants associated with five novel diseases. *PLoS ONE* **7:** e28936.

Pugh EN. 2015. Photoreceptor disc morphogenesis: The classical evagination model prevails. *J Cell Biol* **211:** 2–4.

Ramamurthy V, Cayouette M. 2009. Development and disease of the photoreceptor cilium. *Clin Genet* **76:** 137–145.

Rattner A, Smallwood PM, Williams J, Cooke C, Savchenko A, Lyubarsky A, Pugh EN, Nathans J. 2001. A photoreceptor-specific cadherin is essential for the structural integrity of the outer segment and for photoreceptor survival. *Neuron* **32:** 775–786.

Riazuddin SA, Iqbal M, Wang Y, Masuda T, Chen Y, Bowne S, Sullivan LS, Waseem NH, Bhattacharya S, Daiger SP, et al. 2010. A splice-site mutation in a retina-specific exon of *BBS8* causes nonsyndromic retinitis pigmentosa. *Am J Hum Genet* **86:** 805–812.

Riazuddin S, Belyantseva IA, Giese APJ, Lee K, Indzhykulian AA, Nandamuri SP, Yousaf R, Sinha GP, Lee S, Terrell D, et al. 2012. Alterations of the CIB2 calcium- and integrin-binding protein cause Usher syndrome type 1J and nonsyndromic deafness DFNB48. *Nat Genet* **44:** 1265–1271.

Rivolta C, Sweklo EA, Berson EL, Dryja TP. 2000. Missense mutation in the *USH2A* gene: Association with recessive retinitis pigmentosa without hearing loss. *Am J Hum Genet* **66:** 1975–1978.

Roepman R, van Duijnhoven G, Rosenberg T, Pinckers AJ, Bleeker-Wagemakers LM, Bergen AA, Post J, Beck A, Reinhardt R, Ropers HH, et al. 1996. Positional cloning of the gene for X-linked retinitis pigmentosa 3: Homology with the guanine-nucleotide-exchange factor RCC1. *Hum Mol Genet* **5:** 1035–1041.

Romani M, Micalizzi A, Kraoua I, Dotti MT, Cavallin M, Sztriha L, Ruta R, Mancini F, Mazza T, Castellana S, et al. 2014. Mutations in *B9D1* and *MKS1* cause mild Joubert syndrome: Expanding the genetic overlap with the lethal ciliopathy Meckel syndrome. *Orphanet J Rare Dis* **9:** 72.

Ronquillo CC, Bernstein PS, Baehr W. 2012. Senior-Løken syndrome: A syndromic form of retinal dystrophy associated with nephronophthisis. *Vision Res* **75:** 88–97.

Roosing S, Rohrschneider K, Beryozkin A, Sharon D, Weisschuh N, Staller J, Kohl S, Zelinger L, Peters TA, Neveling K, et al. 2013. Mutations in *RAB28*, encoding a farnesylated small GTPase, are associated with autosomal-recessive cone–rod dystrophy. *Am J Hum Genet* **93:** 110–117.

Rose AM, Shah AZ, Venturini G, Krishna A, Chakravarti A, Rivolta C, Bhattacharya SS. 2016. Transcriptional regulation of *PRPF31* gene expression by MSR1 repeat elements causes incomplete penetrance in retinitis pigmentosa. *Sci Rep* **6:** 19450.

Rosenbaum JL, Witman GB. 2002. Intraflagellar transport. *Nat Rev Mol Cell Biol* **3:** 813–825.

Rosenfeld PJ, Cowley GS, McGee TL, Sandberg MA, Berson EL, Dryja TP. 1992. A *null* mutation in the rhodopsin gene causes rod photoreceptor dysfunction and autosomal recessive retinitis pigmentosa. *Nat Genet* **1:** 209–213.

Ruiz-Perez VL, Ide SE, Strom TM, Lorenz B, Wilson D, Woods K, King L, Francomano C, Freisinger P, Spranger S, et al. 2000. Mutations in a new gene in Ellis–van Creveld syndrome and Weyers acrodental dysostosis. *Nat Genet* **24:** 283–286.

Sahly I, Dufour E, Schietroma C, Michel V, Bahloul A, Perfettini I, Pepermans E, Estivalet A, Carette D, Aghaie A, et al. 2012. Localization of Usher 1 proteins to the photoreceptor calyceal processes, which are absent from mice. *J Cell Biol* **199:** 381–399.

Sanders AAWM, de Vrieze E, Alazami AM, Alzahrani F, Malarkey EB, Sorusch N, Tebbe L, Kuhns S, van Dam TJP, Alhashem A, et al. 2015. KIAA0556 is a novel ciliary basal body component mutated in Joubert syndrome. *Genome Biol* **16:** 293.

Sanyal S, De Ruiter A, Hawkins R. 1980. Development and degeneration of retina in *rds* mutant mice: Light microscopy. *J Comp Neurol* **194:** 193–207.

Sato M, Nakazawa M, Usui T, Tanimoto N, Abe H, Ohguro H. 2005. Mutations in the gene coding for guanylate cyclase-activating protein 2 (*GUCA1B* gene) in patients with autosomal dominant retinal dystrophies. *Graefes Arch Clin Exp Ophthalmol* **243:** 235–242.

Sayer JA, Otto EA, O'Toole JF, Nurnberg G, Kennedy MA, Becker C, Hennies HC, Helou J, Attanasio M, Fausett BV, et al. 2006. The centrosomal protein nephrocystin-6 is mutated in Joubert syndrome and activates transcription factor ATF4. *Nat Genet* **38:** 674–681.

Scheidecker S, Etard C, Pierce NW, Geoffroy V, Schaefer E, Muller J, Chennen K, Flori E, Pelletier V, Poch O, et al. 2014. Exome sequencing of Bardet–Biedl syndrome patient identifies a null mutation in the BBSome subunit *BBIP1* (*BBS18*). *J Med Genet* **51:** 132–136.

Schmidts M, Frank V, Eisenberger T, Al Turki S, Bizet AA, Antony D, Rix S, Decker C, Bachmann N, Bald M, et al. 2013a. Combined NGS approaches identify mutations in the intraflagellar transport gene *IFT140* in skeletal ciliopathies with early progressive kidney disease. *Hum Mutat* **34:** 714–724.

Schmidts M, Vodopiutz J, Christou-Savina S, Cortés CR, McInerney-Leo AM, Emes RD, Arts HH, Tüysüz B, D'Silva J, Leo PJ, et al. 2013b. Mutations in the gene encoding IFT dynein complex component WDR34 cause Jeune asphyxiating thoracic dystrophy. *Am J Hum Genet* **93:** 932–944.

Senior B, Friedmann AI, Braudo JL. 1961. Juvenile familial nephropathy with tapetoretinal degeneration. A new oculorenal dystrophy. *Am J Ophthalmol* **52:** 625–633.

Shaheen R, Faqeih E, Seidahmed MZ, Sunker A, Alali FE, AlQahtani K, Alkuraya FS. 2011. A *TCTN2* mutation defines a novel Meckel–Gruber syndrome locus. *Hum Mutat* **32:** 573–578.

Shaheen R, Ansari S, Mardawi EAl, Alshammari MJ, Alkuraya FS. 2013. Mutations in *TMEM231* cause Meckel–Gruber syndrome. *J Med Genet* **50:** 160–162.

Shaheen R, Shamseldin HE, Loucks CM, Seidahmed MZ, Ansari S, Ibrahim Khalil M, Al-Yacoub N, Davis EE, Mola NA, Szymanska K, et al. 2014. Mutations in *CSPP1*, encoding a core centrosomal protein, cause a range of ciliopathy phenotypes in humans. *Am J Hum Genet* **94:** 73–79.

Shaheen R, Schmidts M, Faqeih E, Hashem A, Lausch E, Holder I, Superti-Furga A, Mitchison HM, Almoisheer A, Alamro R, et al. 2015. A founder *CEP120* mutation in Jeune asphyxiating thoracic dystrophy expands the role of centriolar proteins in skeletal ciliopathies. *Hum Mol Genet* **24:** 1410–1419.

Shankar SP, Hughbanks-Wheaton DK, Birch DG, Sullivan LS, Conneely KN, Bowne SJ, Stone EM, Daiger SP. 2016. Autosomal dominant retinal dystrophies caused by a founder splice site mutation, c.828+3A>T, in *PRPH2* and protein haplotypes in *trans* as modifiers. *Investig Opthalmology Vis Sci* **57**: 349.

Sharon D, Namburi P, Ratnapriya R, Lazar C, Obolensky A, Ben-Yosef T, Pras E, Gross M, Banin E, Swaroop A. 2016. Whole exome sequencing reveals a homozygous splicing mutation in CEP78 as the cause of atypical Usher syndrome in Eastern Jewish patients. *ARVO Annual Meeting*, Abstract B0253. Seattle, WA, May 1–6.

Shevach E, Ali M, Mizrahi-Meissonnier L, McKibbin M, El-Asrag M, Watson CM, Inglehearn CF, Ben-Yosef T, Blumenfeld A, Jalas C, et al. 2015. Association between missense mutations in the *BBS2* gene and nonsyndromic retinitis pigmentosa. *JAMA Ophthalmol* **133**: 312.

Shin JB, Longo-Guess CM, Gagnon LH, Saylor KW, Dumont RA, Spinelli KJ, Pagana JM, Wilmarth PA, David LL, Gillespie PG, et al. 2010. The R109H variant of fascin-2, a developmentally regulated actin crosslinker in hair-cell stereocilia, underlies early-onset hearing loss of DBA/2J mice. *J Neurosci* **30**: 9683–9694.

Silva E, Dharmaraj S, Li YY, Pina AL, Carter RC, Loyer M, Traboulsi E, Theodossiadis G, Koenekoop R, Sundin O, et al. 2004. A missense mutation in *GUCY2D* acts as a genetic modifier in *RPE65*-related Leber congenital amaurosis. *Ophthalmic Genet* **25**: 205–217.

Sjostrand FS. 1953. The ultrastructure of the outer segments of rods and cones of the eye as revealed by the electron microscope. *J Cell Physiol* **42**: 15–44.

Slavotinek AM, Stone EM, Mykytyn K, Heckenlively JR, Green JS, Heon E, Musarella MA, Parfrey PS, Sheffield VC, Biesecker LG. 2000. Mutations in *MKKS* cause Bardet–Biedl syndrome. *Nat Genet* **26**: 15–16.

Smith UM, Consugar M, Tee LJ, McKee BM, Maina EN, Whelan S, Morgan NV, Goranson E, Gissen P, Lilliquist S, et al. 2006. The transmembrane protein meckelin (MKS3) is mutated in Meckel–Gruber syndrome and the wpk rat. *Nat Genet* **38**: 191–196.

Sokal I, Li N, Surgucheva I, Warren MJ, Payne AM, Bhattacharya SS, Baehr W, Palczewski K. 1998. GCAP1 (Y99C) mutant is constitutively active in autosomal dominant cone dystrophy. *Mol Cell* **2**: 129–133.

Sokolov M, Lyubarsky AL, Strissel KJ, Savchenko AB, Govardovskii VI, Pugh EN, Arshavsky VY. 2002. Massive light-driven translocation of transducin between the two major compartments of rod cells: A novel mechanism of light adaptation. *Neuron* **34**: 95–106.

Srour M, Hamdan FF, Schwartzentruber JA, Patry L, Ospina LH, Shevell MI, Désilets V, Dobrzeniecka S, Mathonnet G, Lemyre E, et al. 2012a. Mutations in *TMEM231* cause Joubert syndrome in French Canadians. *J Med Genet* **49**: 636–641.

Srour M, Schwartzentruber J, Hamdan FF, Ospina LH, Patry L, Labuda D, Massicotte C, Dobrzeniecka S, Capo-Chichi J-M, Papillon-Cavanagh S, et al. 2012b. Mutations in *C5ORF42* cause Joubert syndrome in the French Canadian population. *Am J Hum Genet* **90**: 693–700.

Steinberg RH, Fisher SK, Anderson DH. 1980. Disc morphogenesis in vertebrate photoreceptors. *J Comp Neurol* **190**: 501–518.

Stoetzel C, Laurier V, Davis EE, Muller J, Rix S, Badano JL, Leitch CC, Salem N, Chouery E, Corbani S, et al. 2006.

BBS10 encodes a vertebrate-specific chaperonin-like protein and is a major BBS locus. *Nat Genet* **38**: 521–524.

Stoetzel C, Muller J, Laurier V, Davis EE, Zaghloul NA, Vicaire S, Jacquelin C, Plewniak F, Leitch CC, Sarda P, et al. 2007. Identification of a novel BBS gene (*BBS12*) highlights the major role of a vertebrate-specific branch of chaperonin-related proteins in Bardet–Biedl syndrome. *Am J Hum Genet* **80**: 1–11.

Stone EM, Cideciyan AV, Aleman TS, Scheetz TE, Sumaroka A, Ehlinger MA, Schwartz SB, Fishman GA, Traboulsi EI, Lam BL, et al. 2011. Variations in NPHP5 in patients with nonsyndromic leber congenital amaurosis and Senior–Løken syndrome. *Arch Ophthalmol* **129**: 81–87.

Sun H, Nathans J. 1997. Stargardt's ABCR is localized to the disc membrane of retinal rod outer segments. *Nat Genet* **17**: 15–16.

Tallila J, Jakkula E, Peltonen L, Salonen R, Kestilä M. 2008. Identification of *CC2D2A* as a Meckel syndrome gene adds an important piece to the ciliopathy puzzle. *Am J Hum Genet* **82**: 1361–1367.

Taschner M, Bhogaraju S, Lorentzen E. 2012. Architecture and function of IFT complex proteins in ciliogenesis. *Differentiation* **83**: S12–S22.

Thiadens AAHJ, den Hollander AI, Roosing S, Nabuurs SB, Zekveld-Vroon RC, Collin RWJ, De Baere E, Koenekoop RK, van Schooneveld MJ, Strom TM, et al. 2009. Homozygosity mapping reveals *PDE6C* mutations in patients with early-onset cone photoreceptor disorders. *Am J Hum Genet* **85**: 240–247.

Thiel C, Kessler K, Giessl A, Dimmler A, Shalev SA, von der Haar S, Zenker M, Zahnleiter D, Stöss H, Beinder E, et al. 2011. *NEK1* mutations cause short-rib polydactyly syndrome type majewski. *Am J Hum Genet* **88**: 106–114.

Thomas S, Legendre M, Saunier S, Bessières B, Alby C, Bonnière M, Toutain A, Loeuillet L, Szymanska K, Jossic F, et al. 2012. *TCTN3* mutations cause Mohr–Majewski syndrome. *Am J Hum Genet* **91**: 372–378.

Thomas S, Wright KJ, Le Corre S, Micalizzi A, Romani M, Abhyankar A, Saada J, Perrault I, Amiel J, Litzler J, et al. 2014. A homozygous *PDE6D* mutation in Joubert syndrome impairs targeting of farnesylated INPP5E protein to the primary cilium. *Hum Mutat* **35**: 137–146.

Travis GH, Sutcliffe JG, Bok D. 1991. The *retinal degeneration slow* (*rds*) gene product is a photoreceptor disc membrane-associated glycoprotein. *Neuron* **6**: 61–70.

Tucker BA, Scheetz TE, Mullins RF, DeLuca AP, Hoffmann JM, Johnston RM, Jacobson SG, Sheffield VC, Stone EM. 2011. Exome sequencing and analysis of induced pluripotent stem cells identify the cilia-related gene *male germ cell-associated kinase* (*MAK*) as a cause of retinitis pigmentosa. *Proc Natl Acad Sci* **108**: E569–E576.

Tuz K, Bachmann-Gagescu R, O'Day DR, Hua K, Isabella CR, Phelps IG, Stolarski AE, O'Roak BJ, Dempsey JC, Lourenco C, et al. 2014. Mutations in *CSPP1* cause primary cilia abnormalities and Joubert syndrome with or without Jeune asphyxiating thoracic dystrophy. *Am J Hum Genet* **94**: 62–72.

Valente EM, Silhavy JL, Brancati F, Barrano G, Krishnaswami SR, Castori M, Lancaster MA, Boltshauser E, Boccone L, Al-Gazali L, et al. 2006. Mutations in *CEP290*, which encodes a centrosomal protein, cause pleiotropic forms of Joubert syndrome. *Nat Genet* **38:** 623–625.

Valente EM, Logan CV, Mougou-Zerelli S, Lee JH, Silhavy JL, Brancati F, Iannicelli M, Travaglini L, Romani S, Illi B, et al. 2010. Mutations in *TMEM216* perturb ciliogenesis and cause Joubert, Meckel and related syndromes. *Nat Genet* **42:** 619–625.

Valente EM, Dallapiccola B, Bertini E. 2013. Joubert syndrome and related disorders. *Handb Clin Neurol* **113:** 1879–1888.

Valente EM, Rosti RO, Gibbs E, Gleeson JG. 2014. Primary cilia in neurodevelopmental disorders. *Nat Rev Neurol* **10:** 27–36.

Verpy E, Leibovici M, Zwaenepoel I, Liu XZ, Gal A, Salem N, Mansour A, Blanchard S, Kobayashi I, Keats BJ, et al. 2000. A defect in harmonin, a PDZ domain-containing protein expressed in the inner ear sensory hair cells, underlies Usher syndrome type 1C. *Nat Genet* **26:** 51–55.

Vithana EN, Safieh LA, Pelosini L, Winchester E, Hornan D, Bird AC, Hunt DM, Bustin SA, Bhattacharya SS. 2003. Expression of *PRPF31* mRNA in patients with autosomal dominant retinitis pigmentosa: A molecular clue for incomplete penetrance? *Invest Ophthalmol Vis Sci* **44:** 4204–4209.

Wada Y, Abe T, Takeshita T, Sato H, Yanashima K, Tamai M. 2001. Mutation of human retinal fascin gene (*FSCN2*) causes autosomal dominant retinitis pigmentosa. *Invest Ophthalmol Vis Sci* **42:** 2395–2400.

Wada Y, Abe T, Itabashi T, Sato H, Kawamura M, Tamai M. 2003. Autosomal dominant macular degeneration associated with 208delG mutation in the FSCN2 gene. *Arch Ophthalmol* **121:** 1613–1620.

Walczak-Sztulpa J, Eggenschwiler J, Osborn D, Brown DA, Emma F, Klingenberg C, Hennekam RC, Torre G, Garshasbi M, Tzschach A, et al. 2010. Cranioectodermal Dysplasia, Sensenbrenner syndrome, is a ciliopathy caused by mutations in the *IFT122* gene. *Am J Hum Genet* **86:** 949–956.

Wang J, Deretic D. 2014. Molecular complexes that direct rhodopsin transport to primary cilia. *Prog Retin Eye Res* **38:** 1–19.

Wang H, den Hollander AI, Moayedi Y, Abulimiti A, Li Y, Collin RWJ, Hoyng CB, Lopez I, Abboud EB, Al-Rajhi AA, et al. 2009. Mutations in *SPATA7* cause Leber congenital amaurosis and juvenile retinitis pigmentosa. *Am J Hum Genet* **84:** 380–387.

Waters AM, Beales PL. 2011. Ciliopathies: An expanding disease spectrum. *Pediatr Nephrol* **26:** 1039–1056.

Webb TR, Parfitt DA, Gardner JC, Martinez A, Bevilacqua D, Davidson AE, Zito I, Thiselton DL, Ressa JHC, Apergi M, et al. 2012. Deep intronic mutation in *OFD1*, identified by targeted genomic next-generation sequencing, causes a severe form of X-linked retinitis pigmentosa (RP23). *Hum Mol Genet* **21:** 3647–3654.

Weil D, Blanchard S, Kaplan J, Guilford P, Gibson G, Walsh J, Mburu P, Varela A, Levilliers J, Weston M, et al. 1995. Defective myosin VIIA gene responsible for Usher syndrome type 1B. *Nature* **374:** 60–61.

Weil D, El-Amraoui A, Masmoudi S, Mustapha M, Kikkawa Y, Laine S, Delmaghani S, Adato A, Nadifi S, Ben Zina Z, et al. 2003. Usher syndrome type I G (USH1G) is caused by mutations in the gene encoding SANS, a protein that associates with the USH1C protein, harmonin. *Hum Mol Genet* **12:** 463–471.

Weitz CJ, Miyake Y, Shinzato K, Montag E, Zrenner E, Went LN, Nathans J. 1992a. Human tritanopia associated with two amino acid substitutions in the blue-sensitive opsin. *Am J Hum Genet* **50:** 498–507.

Weitz CJ, Went LN, Nathans J. 1992b. Human tritanopia associated with a third amino acid substitution in the blue-sensitive visual pigment. *Am J Hum Genet* **51:** 444–446.

Weleber RG. 2002. Infantile and childhood retinal blindness: A molecular perspective (The Franceschetti Lecture). *Ophthalmic Genet* **23:** 71–97.

Weleber RG, Carr RE, Murphey WH, Sheffield VC, Stone EM. 1993. Phenotypic variation including retinitis pigmentosa, pattern dystrophy, and fundus flavimaculatus in a single family with a deletion of codon 153 or 154 of the peripherin/RDS gene. *Arch Ophthalmol* **111:** 1531–1542.

Weston MD, Luijendijk MWJ, Humphrey KD, Möller C, Kimberling WJ. 2004. Mutations in the *VLGR1* gene implicate G-protein signaling in the pathogenesis of Usher syndrome type II. *Am J Hum Genet* **74:** 357–366.

Williams DS. 2002. Transport to the photoreceptor outer segment by myosin VIIa and kinesin II. *Vision Res* **42:** 455–462.

Williams D. 2008. Usher syndrome: Animal models, retinal function of Usher proteins, and prospects for gene therapy. *Vision Res* **48:** 433–441.

Williams CL, McIntyre JC, Norris SR, Jenkins PM, Zhang L, Pei Q, Verhey K, Martens JR. 2014. Direct evidence for BBSome-associated intraflagellar transport reveals distinct properties of native mammalian cilia. *Nat Commun* **5:** 5813.

Winderickx J, Sanocki E, Lindsey DT, Teller DY, Motulsky AG, Deeb SS. 1992. Defective colour vision associated with a missense mutation in the human green visual pigment gene. *Nat Genet* **1:** 251–256.

Wright C, Healicon R, English C, Burn J. 1994. Meckel syndrome: What are the minimum diagnostic criteria? *J Med Genet* **31:** 482–485.

Xu M, Yang L, Wang F, Li HH, Wang X, Wang W, Ge Z, Wang K, Zhao L, Li HH, et al. 2015. Mutations in human *IFT140* cause non-syndromic retinal degeneration. *Hum Genet* **134:** 1069–1078.

Yamamoto S, Sippel KC, Berson EL, Dryja TP. 1997. Defects in the rhodopsin kinase gene in the Oguchi form of stationary night blindness. *Nat Genet* **15:** 175–178.

Yamashita T, Liu J, Gao J, LeNoue S, Wang C, Kaminoh J, Bowne SJ, Sullivan LS, Daiger SP, Zhang K, et al. 2009. Essential and synergistic roles of RP1 and RP1L1 in rod photoreceptor axoneme and retinitis pigmentosa. *J Neurosci* **29:** 9748–9760.

Yan D, Liu XZ. 2010. Genetics and pathological mechanisms of Usher syndrome. *J Hum Genet* **55:** 327–335.

Yang Z, Peachey NS, Moshfeghi DM, Thirumalaichary S, Chorich L, Shugart YY, Fan K, Zhang K. 2002. Mutations in the *RPGR* gene cause X-linked cone dystrophy. *Hum Mol Genet* **11:** 605–611.

Yang J, Gao J, Adamian M, Wen XH, Pawlyk B, Zhang L, Sanderson MJ, Zuo J, Makino CL, Li T. 2005. The ciliary rootlet maintains long-term stability of sensory cilia. *Mol Cell Biol* **25**: 4129–4137.

Yang Z, Chen Y, Lillo C, Chien J, Yu Z, Michaelides M, Klein M, Howes KA, Li Y, Kaminoh Y, et al. 2008. Mutant prominin 1 found in patients with macular degeneration disrupts photoreceptor disk morphogenesis in mice. *J Clin Invest* **118**: 2908–2916.

Young RW. 1967. The renewal of photoreceptor cell outer segments. *J Cell Biol* **33**: 61–72.

Zaki MS, Sattar S, Massoudi RA, Gleeson JG. 2011. Co-occurrence of distinct ciliopathy diseases in single families suggests genetic modifiers. *Am J Med Genet Part A* **155**: 3042–3049.

Zallocchi M, Meehan DT, Delimont D, Askew C, Garige S, Gratton MA, Rothermund-Franklin CA, Cosgrove D. 2009. Localization and expression of clarin-1, the *Clrn1* gene product, in auditory hair cells and photoreceptors. *Hear Res* **255**: 109–120.

Zhang Q, Li S, Xiao X, Jia X, Guo X. 2007. The 208delG mutation in *FSCN2* does not associate with retinal degeneration in Chinese individuals. *Invest Ophthalmol Vis Sci* **48**: 530–533.

Zito I, Downes SM, Patel RJ, Cheetham ME, Ebenezer ND, Jenkins SA, Bhattacharya SS, Webster AR, Holder GE, Bird AC, et al. 2003. *RPGR* mutation associated with retinitis pigmentosa, impaired hearing, and sinorespiratory infections. *J Med Genet* **40**: 609–615.

Evolution of Cilia

David R. Mitchell

Department of Cell and Developmental Biology, SUNY Upstate Medical University, Syracuse, New York 13210
Correspondence: mitcheld@upstate.edu

Anton van Leeuwenhoek's startling microscopic observations in the 1600s first stimulated fascination with the way that cells use cilia to generate currents and to swim in a fluid environment. Research in recent decades has yielded deep knowledge about the mechanical and biochemical nature of these organelles but only opened a greater fascination about how such beautifully intricate and multifunctional structures arose during evolution. Answers to this evolutionary puzzle are not only sought to satisfy basic curiosity, but also, as stated so eloquently by Dobzhansky (*Am Zool* 4: 443 [1964]), because "nothing in biology makes sense except in the light of evolution." Here I attempt to summarize current knowledge of what ciliary organelles of the last eukaryotic common ancestor (LECA) were like, explore the ways in which cilia have evolved since that time, and speculate on the selective processes that might have generated these organelles during early eukaryotic evolution.

Several excellent papers have appeared in the recent past that follow evolution of structures or complexes essential to ciliary function, including reviews on basal bodies and centrioles (Hodges et al. 2010; Ross and Normark 2015) and centrosomes (Azimzadeh 2014), trafficking and signaling modules (Johnson and Leroux 2010; Lim et al. 2011; Sung and Leroux 2013; Malicki and Avidor-Reiss 2014), tubulins (Findeisen et al. 2014), transition-zone complexes (Barker et al. 2014), and cilia themselves (Satir et al. 2008; Carvalho-Santos et al. 2011). Rather than attempting to cover all of the same ground in a single review, I will refer the reader to other sources where appropriate and instead try to look at the bigger pictures that are emerging from these details. Importantly, many model organisms have contributed to studies of cilia-related processes such as intraflagellar transport (IFT) trafficking, centriole assembly,

and motility regulation, and these organisms include representatives of phylogenetically distant groups (green algae, ciliates, excavates, metazoa). Together with genomic data, these functional and structural studies provide a basis for understanding common features of cilia that would have been present in the last common ancestor as well as changes that may have occurred since that time. These aspects of ciliary evolution can, therefore, be presented with only a modest level of speculation, and speculative aspects are likely to be short-lived as additional data is collected.

Knowing how cilia evolved in the first place, during those dark ages of eukaryotic evolution after formation of true eukaryotes but before diversification into currently recognized supergroups, provides greater challenges. Evolutionary change includes continuous (as well as sudden and catastrophic) extinctions, which for the most part leave no trace in the genomic record of extant organisms. For larger, multicellular species and for those with hard shells or tests, fossils may provide some clues to morphologic features of extinct clades. For single-celled eukaryotes that existed before the last eukaryotic common ancestor (LECA), and especially for evidence of their motile mechanisms, fossils provide little information. We are, therefore, reduced to analysis of genomic changes between the closest living prokaryotic relatives of eukaryotes, as a proxy for the first eukaryotic ancestor, and the inferred common set of genes of the LECA, to learn how cilia evolved. Missing from this equation are data on other evolutionary paths leading to development of motile organelles (perhaps resembling cilia or perhaps not) that might have flourished for some time but that became extinct and left no trace in the current record.

THE CILIATED ORGANELLE OF THE LECA

When it comes to single-celled organisms, there is only so much to be learned from morphological features. Many species that were once thought to be closely related, based on outward appearance, have turned out to belong to quite

distant branches of the tree of life. Such is certainly the case with prokaryotes, whose division into eubacteria and archaea was largely dependent on sequence comparisons, not shape, but similar discoveries have been made among eukaryotes such as oomycetes, which look like fungi but clearly belong in stramenopiles, and more recently with apusozoa, which resemble amoebae but are more closely related to us (Paps et al. 2013; Cavalier-Smith et al. 2014). The tree in Figure 1 summarizes many recent studies on the relationships among extant eukaryotes. Disagreement remains on some of the deep connections in this tree, in part because we still lack genomic sequences of some well-known eukaryotes and in part because some genomes appear to be evolving so fast that relationships are difficult to establish. Of course many other species remain to be discovered and some of these could reveal previously unappreciated relationships, but such trees remain useful because structures, mechanisms, and individual proteins that are found in multiple branches must have existed in the last common ancestor of those divergent species, regardless of their true phylogenetic relationships.

Comparison of many recently generated trees of eukaryotic phylogeny leads to two strik-

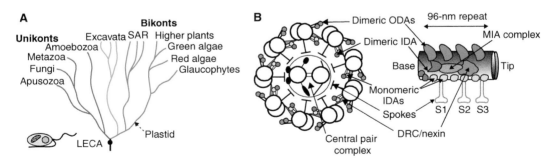

Figure 1. (A) Diagram of major eukaryotic clades diverging from a last eukaryotic common ancestor (LECA), with the root placed between unikont and bikont superclades. Branches are color-coded according to differences in orientation of ciliary central-pair microtubules, which appear perpendicular to the bend plane and fixed in unikonts (blue), perpendicular to the bend plane but floating in excavates (orange), and twisted (helical) and rotating in other bikonts (green). The LECA is cartooned as a single cell with a nucleus, a mitochondrion, and two flagella: an anterior motile flagellum and a posterior gliding flagellum. The dashed arrow indicates the first acquisition of a plastid through endosymbiosis of a cyanobacterium. (B) Diagrams of ciliary axonemal structures that were present in the LECA. (Left) Cross-sectional view from inside the cell. (Right) Longitudinal view of one 96-nm repeat along an outer doublet. MIA, Modifier of inner arms; ODA, outer dynein arm; IDA, inner dynein arm; DRC, dynein regulatory complex.

 Cite this article as *Cold Spring Harb Perspect Biol* doi: 10.1101/cshperspect.a028290

ing conclusions. First, it is very difficult to sort out the branch order at the base of the tree, because all the major branches of eukaryotes seem to have diverged from a common ancestor within a short time (Fig. 1A). As with evolution anywhere, anytime, branching is a continuous process matched only by the frequency of extinction, so trees need to be interpreted by imagining what has been lost as well as what remains (Jablonski 1986). The effects of horizontal gene transfer can sometimes make trees look more like webs (Soucy et al. 2015), but most existing eukaryotes fit into one of several major clades related primarily by their decent from a common ancestor. These include the Stramenopiles, Alveolates, Rhizaria (SAR); the Plantae (probably a sister group to the SAR); the excavates (Heterolobosea, Euglenozoa, Diplomonads), which may be more closely related to the SAR than to the Plantae; the Amoebozoa; and the opisthokonts (Fungi, Choanozoa, Metazoa), which are more closely related to Amoebozoa than to the other clades. These branches appear to have originated in an evolutionary big bang, probably about one billion years ago (Chernikova et al. 2011; Parfrey et al. 2011). Second, the last common ancestor of all of these branches was highly complex (Koumandou et al. 2013). It had to have everything that is now found, in at least some representatives, in every branch. Based on features common to eukaryotes in all of these clades, the LECA must have divided by mitosis but also reproduced through a meiotic sexual cycle. It had a nucleus, endomembrane system, cytoskeleton, vesicular transport, mitochondria, and quite surely a cilium.

Fortunately model organisms from distantly related branches have been used extensively in studies of ciliary structure, function, and biochemistry, so that features common to the LECA and the extent of evolutionary change since that time can be reconstructed. Particularly useful has been the intense focus on *Trypanosoma* (excavata), *Tetrahymena* (SAR), *Chlamydomonas* (plantae), and several metazoa (opisthokonts). Based on features common to cilia in these distantly related organisms, the LECA had (most likely two) cilia with a typical 9+2 architecture of outer doublet and central

singlet microtubules, assembled on basal bodies built from nine triplet microtubules. Motility generated by both outer and inner rows of dynein arms was regulated by interactions between radial spokes and a central-pair complex, and by nexin/dynein regulatory complex (DRC) interdoublet links (Fig. 1B). Ciliary assembly required trafficking on IFT complexes, the resulting organelles possessed sensory capabilities maintained in part by Bardet–Beidl syndrome (BBS) complex-dependent receptor trafficking, and movement of transmembrane proteins could be used for transport of material along the ciliary surface or gliding motility of the cell. In short, these cilia would have been structurally and functionally indistinguishable from typical cilia seen today.

TUBULINS, DOUBLETS, AND TRIPLETS

Phylogenomics shows that this ciliated LECA expressed α-, β-, γ-, δ-, ε-, and ζ-tubulin (Dutcher 2003; Findeisen et al. 2014). α-, β-, and γ-tubulin are essentially universal in eukaryotes, whereas δ-, ε-, and ζ-tubulins are only found in species with cilia. Although not all organisms with cilia retain all three of these more divergent tubulin isoforms, they occur in diverse taxa, supporting their presence in the LECA (see Azimzadeh 2014 and Turk et al. 2015 for other recent overviews of centrioles and tubulin diversity). δ- and ε-tubulin have been implicated in formation of basal bodies (Dutcher et al. 2002), and recent evidence suggests that ζ-tubulin is important for formation of basal feet, structures equivalent to subdistal appendages, which organize interactions of basal bodies with other cell structures and maintain basal body orientation in multiciliated cells (Turk et al. 2015).

Perhaps one of the most elusive features of basal bodies is the way in which triplet microtubules form. High-resolution electron microscopy of ciliary doublet (Nicastro et al. 2011; Maheshwari et al. 2015) and basal body triplet microtubules (Li et al. 2012; Guichard and Gonczy 2016) has revealed not only the underlying tubulin lattice structure but also many nontubulin components, including structures important for closing the walls of these doublet and triplet microtubules. The protein that forms one of these nontubulin "protofilaments," the elusive protofilament 11 that closes the wall of the B-tubule, has recently been identified through a combined genetic and structural study in *Chlamydomonas* (Yanagisawa et al. 2014). Equally elusive until relatively recently was the basis for ninefold symmetry of basal bodies, which depends on self-association of SAS-6 into a cartwheel (Kitagawa et al. 2011; van Breugel et al. 2011) and on its attachment to intertriplet linkers (Hiraki et al. 2007; for more on centriole assembly, see reviews by Azimzadeh and Marshall 2010; Hirono 2014; Winey and O'Toole 2014).

More than a dozen phylogenetically widespread nontubulin proteins have also been identified as common components of centrioles and basal bodies, and many of these play specific roles in centriolar assembly or function (Hodges et al. 2010; Carvalho-Santos et al. 2011). Others have a more limited phylogenetic distribution and likely evolved to fill organism-specific functions. Despite a number of proteomic and transcriptomic studies (Keller et al. 2005; Kilburn et al. 2007; Fritz-Laylin and Cande 2010; Firat-Karalar et al. 2014) and genetic screens (Dobbelaere et al. 2008; Lin and Dutcher 2015), additional proteins involved in evolutionarily conserved basal body functions likely remain to be discovered.

MOTILITY

Motility in cilia of the LECA would have been generated by a two-headed outer row of dynein arms (ODAs) and an inner row that had both a two-headed inner dynein arm (similar to the I1 dynein of *Chlamydomonas*) and single-headed inner dynein arms (IDAs) generated from three different heavy chain isoforms (Fig. 1B) (Wickstead and Gull 2007; Wilkes et al. 2008). A pair of WD-repeat-containing intermediate chains was likely associated with each of the two-headed axonemal dyneins (Wickstead and Gull 2007; Patel-King et al. 2013). These intermediate chains (ICs) were similar to, and likely evolved from, the intermediate chain homodimers that are associated with all cytoplasmic dyneins.

Both ODA and I1 axonemal dyneins have heterodimeric intermediate chains. Sequence comparisons suggest the formation first of an axonemal IC heterodimer, followed by a round of duplication to create separate heterodimers for outer dynein arms and I1 inner dynein arms. Interestingly, IFT dynein (unlike cytoplasmic dynein) appears to have heterodimeric intermediate chains that most closely resemble those of axonemal dyneins in the ODA-IC2 family (Patel-King et al. 2013). A potential evolutionary pathway would start with a two-headed IFT dynein that had an IC2-like homodimer, followed by evolution of a two-headed axonemal dynein from association of IFT dynein heads with a new heterodimeric IC base. This new base with its new IC1/IC2 heterodimer would allow IFT dynein to associate with A-tubules as cargoes and thus become the first axonemal dynein. Through a later duplication of the single IC2-like IFT dynein intermediate chain gene, IFT dynein itself evolved to its present form with a base made from heterodimeric intermediate chains.

Outer dynein arms of the LECA were probably anchored onto doublets with the aid of a docking complex made from a heterodimer of coiled-coil proteins, based on localization studies and similarities between the phenotypes of vertebrate (Jerber et al. 2013; Knowles et al. 2013; Onoufriadis et al. 2013; Hjeij et al. 2014) and algal (Koutoulis et al. 1997; Takada et al. 2002) docking complex mutants. However, homologs of one of the two proteins thought to function as a docking complex in vertebrates (CCDC151) is more closely related at the primary sequence level to a protein (ODA10) that functions in outer dynein assembly rather than docking in the alga (Dean and Mitchell 2013, 2015). Thus some caution is needed before all proteins identified on the basis of sequence similarity are assumed to be functionally equivalent in different organisms, including the LECA. Many additional proteins associated with axonemal dyneins in *Chlamydomonas* that presumably act as axonemal dynein cargo adaptors (Yamamoto et al. 2006, 2008) or regulatory complexes (Yamamoto et al. 2013) have easily identified homologs in the genomes of other ciliated organisms (Hom et al. 2011), but direct tests of their functional equivalency are lacking. As a further example of more recently acquired diversity underlying apparent homology, regulation of ciliary motility by calcium-dependent modulation of outer dynein arm activity is widespread, but the actual calcium sensors likely evolved independently in different clades (Inaba 2015).

An interesting point of speculation is how the three single-headed IDAs expressed in the LECA (Wickstead and Gull 2007) were arranged along the axoneme. Conservation of 96-nm periodicity (Oda et al. 2014) and of multiple structures associated with IDAs in such diverse organisms as *Chlamydomonas* and sea urchins (Nicastro et al. 2006; O'Toole et al. 2012; Pigino et al. 2012) argues for the presence in the LECA of not three, but six single-headed dyneins in each repeating unit (Fig. 1B). Perhaps future studies of organisms with simplified genomes that have reduced numbers of single-headed dynein genes will provide clues to how three single-headed inner arm dyneins may have been anchored in these six sites in cilia of the LECA.

Ancestral axonemal dyneins would have been regulated by at least three complexes (Fig. 1B): the modifier of inner arms (MIA) complex associated with I1 dynein (Yamamoto et al. 2013), the DRC/nexin link that joins adjacent doublets and likely transmits information about the extent or rate of interdoublet sliding (Heuser et al. 2009; Bower et al. 2013), and the radial spoke–central-pair complex that may transmit information across the axoneme and provide a more global level of control (Oda et al. 2014). The MIA complex and interdoublet DRC/nexin links occur once every 96 nm along each doublet. They form interactions with multiple dyneins as well as with radial spokes (Heuser et al. 2009; Oda et al. 2015) and play important, if poorly understood, roles in dynein regulation. Similar links have been seen by electron microscopy in motile cilia from many diverse organisms, and mutations in DRC subunits have been shown to alter the activity of axonemal dyneins in *Chlamydomonas* (Brokaw et al. 1982; Awata et al. 2015) and vertebrates (Wirschell et al. 2013; Olbrich et al.

2015). Most of the known *Chlamydomonas* DRC subunits have clear homologs in vertebrate genomes, confirming the overall high level of conservation of this complex (Bower et al. 2013).

The regulatory network supplied by interactions between the central-pair complex and radial spokes must also have evolved before the LECA, based on overall similarities between central-pair complexes (Pigino et al. 2012; Carbajal-Gonzalez et al. 2013; Oda et al. 2014) and radial spokes (Lin et al. 2012; Pigino et al. 2012) of sea urchin, *Tetrahymena* and *Chlamydomonas*, as well as conservation of many central-pair and radial spoke proteins at the primary sequence level. We lack biochemical and genetic data on the functions of many of these central-pair proteins, but continued exploration of mutations that generate ciliary motility defects in humans and in model organisms such as mice, zebrafish, *Trypanosoma*, and *Chlamydomonas* promises further tests of functional as well as sequence-level evolutionary conservation.

TRAFFICKING, ASSEMBLY, AND SIGNALING

The way in which cilia are assembled and maintained as segregated cytoplasmic and membrane compartments requires specialized docking of basal bodies to the plasma membrane through interactions that involve the centriolar proximal and distal appendages, and functions related to endosomal trafficking. Once an axoneme begins to assemble, transition zone components are added that provide a selective barrier, and specialized transport systems are needed to supply and retain material targeted to cilia. Many known components of each of these systems have been conserved in multiple eukaryotic clades and were therefore already present in the LECA. These include proteins associated with distal appendages needed for membrane anchoring and structures such as basal feet and rootlets involved in anchoring motile cilia in the cytoplasm (Carvalho-Santos et al. 2011).

IFT complexes and their associated dynein and kinesin motors are needed both for assembling the organelle and for maintaining its normal length and composition in many diverse species (Briggs et al. 2004; van Dam et al. 2013). The gliding motility associated with flagella in many single-celled eukaryotes likely depends on IFT motors as well (Collingridge et al. 2013; Shih et al. 2013), although as yet direct analysis has unfortunately been limited to one organism. The BBS complex, which plays a conserved role in maintaining normal concentrations of ciliary membrane-associated proteins, likely evolved from gene duplication and repurposing of IFT components, which in turn were derived from vesicle coat proteins (Satir et al. 2008; van Dam et al. 2013), before evolution of the LECA.

Signaling pathways associated with ciliary membranes have been modified extensively to fit the needs of each organism or cell type, with relatively few signaling modules retained since the LECA. One that appears to have associated with cilia comparatively early is the transient receptor potential (TRP) channel-linked mechanical gating, including PKD1/2-like channels. In vertebrates, polycystic kidney disease (PKD) channels have been linked to signaling in embryonic nodal cilia (Field et al. 2011; Kamura et al. 2011; Yoshiba et al. 2012) as well as in purely sensory cilia such as those in kidney tubules that give these proteins the PKD label (Nauli et al. 2003). In *Chlamydomonas*, a PKD2-related TRP channel has been shown to play a role in the mating reaction, which involves adhesion-dependent ciliary signaling (Huang et al. 2007), and similar channels are localized to ciliary membranes in *Paramecium* (Valentine et al. 2012). *Chlamydomonas* CAV2 is a voltage-dependent calcium channel important for control of waveform (Fujiu et al. 2009), and another *Chlamydomonas* TRPV-related protein, which shows homology with the mechanosensitive *Drosophila* TRPV ciliary channels, is involved in ciliary mechanosensation (Fujiu et al. 2011).

The regulation of trafficking into this compartment by small G proteins of the Arl/Rab/Arf family has been well documented in diverse organisms, and must therefore have also been present in the LECA (Sung and Leroux 2013). Some other modules associated with trafficking and signaling likely evolved in association with cilia before or soon after the LECA, such as the

RJL family of small G proteins (Elias and Archibald 2009) and some cyclic nucleotide-based sensory systems (Johnson and Leroux 2010; Sung and Leroux 2013), whereas prominent cilia-associated developmental signaling pathways such as Hedgehog are widespread among metazoa, but not universally associated with cilia (Roy 2012).

EVOLUTIONARY TRENDS SINCE THE DIVERGENCE OF EUKARYOTES

Evolutionary change at the level of individual proteins or protein complexes, such as those providing the functional elements of cilia, can be conceptually divided into simple categories. Gene or whole genome duplication can result in multiple copies of slightly divergent proteins that provide greater functional diversity, either within one organelle or as a way to differentiate organelles that provide unique abilities to a cell. Examples within an organelle include the frequent expansion of the number of genes encoding one or more of the three types of single-headed IDAs. Different eukaryotic clades have independently expanded the number of these single-headed axonemal dyneins, and at least in *Chlamydomonas* we know that this expansion results in a single cilium that uses multiple related dyneins within a single organelle, presumably to generate subtle levels of motility control (Kamiya and Yagi 2014). Even greater expansion of the axonemal dynein family in some organisms may allow assembly of cilia with unique traits, such as the different cilia types present in the oral zone, body wall, and caudal region of ciliates (Rajagopalan and Wilkes 2016), or the different cilia found on epithelia of mammalian airways, brain ventricles, fallopian tubes, and the embryonic node.

Alternatively, gene duplication can lead to diversification into proteins used in very different ways. Many proteins in cilia appear to be more or less closely related to cytoplasmic proteins, but in some cases these proteins have clearly acquired new functions quite independently in different eukaryotic clades, and were not likely present in cilia of the LECA. An example is the presence of metabolic enzymes involved in adenosine triphosphate (ATP) generation. Mammalian sperm have glycolytic enzymes and can use directly imported glucose for ATP generation (Mukai and Okuno 2004), whereas ATP concentrations in other cilia are maintained through phosphate shuttles (phosphocreatine in sea urchin sperm [Tombes et al. 1987], phosphoarginine in *Paramecium* cilia [Noguchi et al. 2001] and *Trypanosoma* flagella [Ooi et al. 2015]), or through independently acquired steps of the glycolytic pathway (*Chlamydomonas* flagella [Mitchell et al. 2005]).

Equally commonly seen is the loss of proteins or structures, including cilia themselves, when the energy inherent in maintaining such complexity is no longer needed. Examples include the loss of IFT complexes in organisms that only assemble cilia in the cytoplasm and then extrude them preassembled (e.g., apicomplexans such as *Plasmodium*); loss of outer row (e.g., mosses and ferns) or inner row (diatoms) dyneins; and loss of the central-pair/radial spoke regulatory network (e.g., eel sperm, mammalian nodal cilia). Comparative genomics backed by electron microscopy has also shown that transition zone structures, essential for maintaining correct abundance of signaling modules on ciliary membranes (Li et al. 2016), have been lost in some organisms (Barker et al. 2014).

Evolutionary changes in motility regulation are more difficult to follow phylogenetically, but one interesting case may add clarity to weakly supported deep relationships among major eukaryotic clades. Orientation of an asymmetric central-pair complex has been linked to planar beating of cilia. In many unrelated instances, loss of the central-pair has been correlated with a switch to a circular or helical bending pattern, as seen in mammalian nodal cilia, eel sperm, and the simplified cilia of some parasitic protozoa (Satir et al. 2008). Although detailed structural comparisons support a remarkable conservation of the central-pair complex, other evidence suggests that central-pair function changed in a dramatic way during divergence from the LECA (Fig. 1A). In all unikonts examined thus far, orientation of the central pair is fixed along the length of the cilium, with a plane through the two central singlets perpendicular to the bend plane. In all members of the SAR or Plantae superclades, including ciliates, stramenopiles, and green algae, the central pair rotates during bend propagation so that, within each bend, a plane through the two central singlets is parallel to the bend plane (Omoto et al. 1999; Mitchell 2003; Mitchell and Nakatsugawa 2004). In organisms with a rotating central pair this complex is inherently helical, twisting in interbend regions. Finally, in members of the Excavata superclade such as *Euglena* and *Trypanosoma*, central-pair orientation resembles that in unikonts, but is not as tightly fixed (Melkonian et al. 1982) and can "float" through as much as 180° when released by mutations that affect flagellar motility (Branche et al. 2006; Gadelha et al. 2006). Thus, rotation of a helical central pair is a synapomorphy of the SAR and Plantae that supports the early branching of excavates before separation of these two superclades, whereas endosymbiosis of a plastid precursor occurred later, after their separation.

EVOLUTION OF CILIA BEFORE EUKARYOTIC DIVERGENCE

Evolution of ciliary proteins in the pre-LECA era must have involved transformation of microtubule-based trafficking systems (single microtubules, cytoplasmic dynein and kinesin, and vesicle coat proteins) into cilia-specific forms (doublet microtubules, axonemal dyneins, IFT proteins). One view of this process starts with formation of doublet microtubules that act as primitive centrioles, followed by evolution of mechanisms to dock a protocentriole to the cell membrane and to extend the centriolar doublet into a cilium-like structure (Marshall 2009). Alternatively, projections of the cell membrane generated by assembly of multiple singlet microtubules may have come first, and doublet formation may have evolved afterward (Satir et al. 2008). In this latter scenario, microtubule-based membrane extensions function as both sensory antennae and as gliding organelles, based on properties common to cytoplasmic trafficking by dyneins and kinesins. Doublets would evolve later, with the selective advantage deriving from their differentiation into two sides that can uniquely act as dynein cargos (A-tubules) and tracks (B-tubules) so that unidirectional bends could form between doublet pairs.

Many steps must have been needed to generate the 600 or so proteins and 9+2 architecture traceable to cilia of the LECA. Presumed intermediate steps included trials of many alternative mechanisms of microtubule-based cell motility, as well as more primitive versions of cilia, before a 9+2 organelle appeared. The puzzle therefore is why no eukaryotes exist today that descended from cells with any of these alternative proto-cilia. Why have no clades survived from earlier branches during the long evolutionary period needed to get from the first eukaryote common ancestor (FECA) to the last common ancestor of all eukaryotes? Was there an extreme bottleneck in eukaryotic evolution at the time of the LECA, which eliminat-

ed any traces of the previous evolutionary divergence of eukaryotes? Such an event could have opened multiple ecological niches for rapid expansion of the LECA into the major branches represented by existing species. If so, did this bottleneck result from a normal level of fluctuation in a relatively small community of eukaryotic organisms (Fig. 2A), a cataclysmic extinction event (Fig. 2B), or a sudden advantage that appeared in only one lineage (Fig. 2C)?

The spectacular finding of Lokiarchaea and their kin in deep sea vents has opened a new vista on the origins of eukaryotes (Spang et al. 2015). These archaeal genomes reveal present-day organisms that have all the components needed for vesicular trafficking, including coat proteins and small G proteins. However, these trafficking components have not differentiated into the specific families of coat proteins (IFT and BBS complex subunits) and G proteins

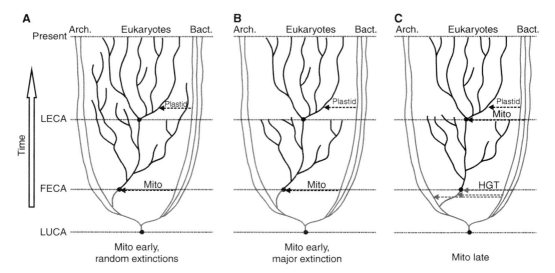

Figure 2. Three alternative branching and extinction pathways leading from the last universal common ancestor (LUCA) to the first eukaryotic common ancestor (FECA), the last eukaryotic common ancestor (LECA), and present eukaryotic clades are diagrammed to explore mechanisms that selected for a LECA with 9+2 cilia. (*A*) Mitochondrial (Mito) endosymbiosis early, together with formation of a nucleus, transforms an archaeum into the FECA. Present diversity results from random branching and extinction, and alternate ciliary architectures are lost by chance. (*B*) Mitochondria are acquired early. Present clades result from rapid divergence from the LECA following a major extinction event, which chanced to preserve an organism with 9+2 cilia. (*C*) Horizontal gene transfer (HGT) between eubacteria and archaea generates an amitochondriate FECA. Much later, endosymbiosis of an α-proteobacterium creates the LECA, which rapidly diverges into present clades, whereas clades that lack mitochondria become extinct. The host for endosymbiosis, by chance, has 9+2 cilia. Arch., Archaea; Bact., bacteria.

 Cite this article as *Cold Spring Harb Perspect Biol* doi: 10.1101/cshperspect.a028290

(Arls, Rabs, IFT22, IFT27) that are known to function in ciliary trafficking. Likewise, Lokiarchaeotes have actin-related sequences, but lack canonical eukaryotic actin and Arps. The available genomic sequences also lack homologs of the GTPase-activating proteins (GAPs) and guanine nucleotide exchange factors (GEFs) that regulate trafficking in modern eukaryotes, and, most important, have no obvious homologs of tubulin. Based on recent analysis of orthologous groups, proto-eukaryotes likely formed through acquisition of a large set of bacterial genes by an archaeote closely related to existing Lokiarchaeota, perhaps through multiple HGTs (Spang et al. 2015; Pittis and Gabaldon 2016) as indicated in Figure 2C. Then, at a much later time, endosymbiosis of an α proteobacterium in one descendant branch led to mitochondrial formation (Pittis and Gabaldon 2016). The Big Bang of eukaryotic diversification, which probably occurred about one billion years ago, may thus have been stimulated by the energetic and metabolic benefits conferred by mitochondria (Lane 2014).

Because eukaryotic tubulins are not specifically more closely related to α-proteobacterial tubulin homologs than to tubulin homologs from other prokaryotes (Findeisen et al. 2014), an early HGT (perhaps one leading specifically to eukaryote formation, as in Fig. 2C) likely included a bacterial FtsZ homolog that became the ancestor of all tubulins. Duplication and diversification resulted in the full repertoire of tubulins of the LECA. Multiple tubulin-based motility mechanisms may have coexisted in divergent pre-LECA branches, with dominance of 9+2 cilia only occurring post-LECA. One apparent difficulty with this view is that if an ATP-rich intracellular environment is needed to drive ciliary motility (Chen et al. 2015), then mitochondria would need to already be present to support evolutionary selection of an energy-demanding motility mechanism. However, although abundant ATP might have been necessary to support initial selection of the many steps needed during cilia evolution, flagellated anaerobic (and virtually amitochondriate) organisms are known among both parasitic (e.g., *Giardia*) and free-living (e.g., *Hexamita* and

Mastigamoeba) genera, so mitochondrial-based ATP formation is not essential for a cell to maintain flagellar function. Thus, an organism that had already evolved a primitive cilium might have had a selective advantage over competing organisms when its motility was further enhanced by acquisition of an endosymbiotic α-proteobacterial power source.

CONCLUDING REMARKS

The last eukaryotic common ancestor was a remarkably complex cell with motile cilia built on a typical 9+2 microtubule array, capable of gliding motility, beating motility, and display of sensory receptors. Genomic, metagenomic, and transcriptomic sequence analyses have revealed that most of the proteins present in cilia today were also present in this common ancestor, pushing the initial evolution of cilia into the interval between divergence of eukaryotes from archaeal prokaryotes and divergence of extant eukaryotic clades beginning about one billion years ago. Diversification into modern organisms may be traced by relatively subtle changes in ciliary architecture but by more radical changes in regulation of bending patterns and by many instances of reduction or loss of cilia. Ciliary based signaling through sensory receptors, although present ancestrally, has shown the greatest clade- and cell-type-specific expansion in the repertoire of these versatile organelles.

REFERENCES

Awata J, Song K, Lin J, King SM, Sanderson MJ, Nicastro D, Witman GB. 2015. DRC3 connects the N-DRC to dynein g to regulate flagellar waveform. *Mol Biol Cell* **26:** 2788–2800.

Azimzadeh J. 2014. Exploring the evolutionary history of centrosomes. *Philos Trans R Soc Lond B Biol Sci* **369:** 20130453.

Azimzadeh J, Marshall WF. 2010. Building the centriole. *Curr Biol* **20:** R816–R825.

Barker AR, Renzaglia KS, Fry K, Dawe HR. 2014. Bioinformatic analysis of ciliary transition zone proteins reveals insights into the evolution of ciliopathy networks. *BMC Genomics* **15:** 531.

Bower R, Tritschler D, Vanderwaal K, Perrone CA, Mueller J, Fox L, Sale WS, Porter ME. 2013. The N-DRC forms a conserved biochemical complex that maintains outer doublet alignment and limits microtubule sliding in motile axonemes. *Mol Biol Cell* **24:** 1134–1152.

Branche C, Kohl L, Toutirais G, Buisson J, Cosson J, Bastin P. 2006. Conserved and specific functions of axoneme components in trypanosome motility. *J Cell Sci* **119:** 3443–3455.

Briggs LJ, Davidge JA, Wickstead B, Ginger ML, Gull K. 2004. More than one way to build a flagellum: Comparative genomics of parasitic protozoa. *Curr. Biol.* **14:** R611–R612.

Brokaw CJ, Luck DJL, Huang B. 1982. Analysis of the movement of *Chlamydomonas* flagella: The function of the radial-spoke system is revealed by comparison of wild-type and mutant flagella. *J Cell Biol* **92:** 722–732.

Carbajal-Gonzalez BI, Heuser T, Fu X, Lin J, Smith BW, Mitchell DR, Nicastro D. 2013. Conserved structural motifs in the central pair complex of eukaryotic flagella. *Cytoskeleton (Hoboken)* **70:** 101–120.

Carvalho-Santos Z, Azimzadeh J, Pereira-Leal JB, Bettencourt-Dias M. 2011. Evolution: Tracing the origins of centrioles, cilia, and flagella. *J Cell Biol* **194:** 165–175.

Cavalier-Smith T, Chao EE, Snell EA, Berney C, Fiore-Donno AM, Lewis R. 2014. Multigene eukaryote phylogeny reveals the likely protozoan ancestors of opisthokonts (animals, fungi, choanozoans) and Amoebozoa. *Mol Phylogenet Evol* **81:** 71–85.

Chen DT, Heymann M, Fraden S, Nicastro D, Dogic Z. 2015. ATP consumption of eukaryotic flagella measured at a single-cell level. *Biophys J* **109:** 2562–2573.

Chernikova D, Motamedi S, Csuros M, Koonin EV, Rogozin IB. 2011. A late origin of the extant eukaryotic diversity: Divergence time estimates using rare genomic changes. *Biol Direct* **6:** 26.

Collingridge P, Brownlee C, Wheeler GL. 2013. Compartmentalized calcium signaling in cilia regulates intraflagellar transport. *Curr Biol* **23:** 2311–2318.

Dean AB, Mitchell DR. 2013. Chlamydomonas ODA10 is a conserved axonemal protein that plays a unique role in outer dynein arm assembly. *Mol Biol Cell* **24:** 3689–3696.

Dean AB, Mitchell DR. 2015. Late steps in cytoplasmic maturation of assembly-competent axonemal outer arm dynein in *Chlamydomonas* require interaction of ODA5 and ODA10 in a complex. *Mol Biol Cell* **26:** 3596–3605.

Dobbelaere J, Josue F, Suijkerbuijk S, Baum B, Tapon N, Raff J. 2008. A genome-wide RNAi screen to dissect centriole duplication and centrosome maturation in *Drosophila*. *PLoS Biol* **6:** e224.

Dutcher SK. 2003. Long-lost relatives reappear: Identification of new members of the tubulin superfamily. *Curr Opin Microbiol* **6:** 634–640.

Dutcher SK, Morrissette NS, Preble AM, Rackley C, Stanga J. 2002. ε-Tubulin is an essential component of the centriole. *Mol Biol Cell* **13:** 3859–3869.

Elias M, Archibald JM. 2009. The RJL family of small GTPases is an ancient eukaryotic invention probably functionally associated with the flagellar apparatus. *Gene* **442:** 63–72.

Field S, Riley KL, Grimes DT, Hilton H, Simon M, Powles-Glover N, Siggers P, Bogani D, Greenfield A, Norris DP. 2011. Pkd1l1 establishes left–right asymmetry and physically interacts with Pkd2. *Development* **138:** 1131–1142.

Findeisen P, Muhlhausen S, Dempewolf S, Hertzog J, Zietlow A, Carlomagno T, Kollmar M. 2014. Six subgroups and extensive recent duplications characterize the evolution of the eukaryotic tubulin protein family. *Genome Biol Evol* **6:** 2274–2288.

Firat-Karalar EN, Rauniyar N, Yates JR III, Stearns T. 2014. Proximity interactions among centrosome components identify regulators of centriole duplication. *Curr Biol* **24:** 664–670.

Fritz-Laylin LK, Cande WZ. 2010. Ancestral centriole and flagella proteins identified by analysis of *Naegleria* differentiation. *J Cell Sci* **123:** 4024–4031.

Fujiu K, Nakayama Y, Yanagisawa A, Sokabe M, Yoshimura K. 2009. *Chlamydomonas* CAV2 encodes a voltage-dependent calcium channel required for the flagellar waveform conversion. *Curr Biol* **19:** 133–139.

Fujiu K, Nakayama Y, Iida H, Sokabe M, Yoshimura K. 2011. Mechanoreception in motile flagella of *Chlamydomonas*. *Nat Cell Biol* **13:** 630–632.

Gadelha C, Wickstead B, McKean PG, Gull K. 2006. Basal body and flagellum mutants reveal a rotational constraint of the central pair microtubules in the axonemes of trypanosomes. *J Cell Sci* **119:** 2405–2413.

Guichard P, Gonczy P. 2016. Basal body structure in *Trichonympha*. *Cilia* **5:** 9.

Heuser T, Raytchev M, Krell J, Porter ME, Nicastro D. 2009. The dynein regulatory complex is the nexin link and a major regulatory node in cilia and flagella. *J Cell Biol* **187:** 921–933.

Hiraki M, Nakazawa Y, Kamiya R, Hirono M. 2007. Bld10p constitutes the cartwheel-spoke tip and stabilizes the 9-fold symmetry of the centriole. *Curr Biol* **17:** 1778–1783.

Hirono M. 2014. Cartwheel assembly. *Philos Trans R Soc Lond B Biol Sci* **369:** 20130458.

Hjeij R, Onoufriadis A, Watson CM, Slagle CE, Klena NT, Dougherty GW, Kurkowiak M, Loges NT, Diggle CP, Morante NF, Gabriel GC, et al. 2014. CCDC151 mutations cause primary ciliary dyskinesia by disruption of the outer dynein arm docking complex formation. *Am J Hum Genet* **95:** 257–274.

Hodges ME, Scheumann N, Wickstead B, Langdale JA, Gull K. 2010. Reconstructing the evolutionary history of the centriole from protein components. *J Cell Sci* **123:** 1407–1413.

Hom EF, Witman GB, Harris EH, Dutcher SK, Kamiya R, Mitchell DR, Pazour GJ, Porter ME, Sale WS, Wirschell M, et al. 2011. A unified taxonomy for ciliary dyneins. *Cytoskeleton* **68:** 555–565.

Huang K, Diener DR, Mitchell A, Pazour GJ, Witman GB, Rosenbaum JL. 2007. Function and dynamics of PKD2 in *Chlamydomonas reinhardtii* flagella. *J Cell Biol* **179:** 501–514.

Inaba K. 2015. Calcium sensors of ciliary outer arm dynein: Functions and phylogenetic considerations for eukaryotic evolution. *Cilia* **4:** 6.

Jablonski D. 1986. Background and mass extinctions: The alternation of macroevolutionary regimes. *Science.* **231:** 129–133.

Jerber J, Baas D, Soulavie F, Chhin B, Cortier E, Vesque C, Thomas J, Durand B. 2013. The coiled-coil domain containing protein CCDC151 is required for the function of

Cite this article as *Cold Spring Harb Perspect Biol* doi: 10.1101/cshperspect.a028290

IFT-dependent motile cilia in animals. *Hum Mol Genet* **23:** 563–577.

Johnson JL, Leroux MR. 2010. cAMP and cGMP signaling: Sensory systems with prokaryotic roots adopted by eukaryotic cilia. *Trends Cell Biol* **20:** 435–444.

Kamiya R, Yagi T. 2014. Functional diversity of axonemal dyneins as assessed by in vitro and in vivo motility assays of *Chlamydomonas* mutants. *Zoolog Sci* **31:** 633–644.

Kamura K, Kobayashi D, Uehara Y, Koshida S, Iijima N, Kudo A, Yokoyama T, Takeda H. 2011. Pkd1l1 complexes with Pkd2 on motile cilia and functions to establish the left–right axis. *Development* **138:** 1121–1129.

Keller LC, Romijn EP, Zamora I, Yates JR III, Marshall WF. 2005. Proteomic analysis of isolated *Chlamydomonas* centrioles reveals orthologs of ciliary-disease genes. *Curr Biol* **15:** 1090–1098.

Kilburn CL, Pearson CG, Romijn EP, Meehl JB, Giddings TH Jr, Culver BP, Yates JR III, Winey M. 2007. New *Tetrahymena* basal body protein components identify basal body domain structure. *J Cell Biol* **178:** 905–912.

Kitagawa D, Vakonakis I, Olieric N, Hilbert M, Keller D, Olieric V, Bortfeld M, Erat MC, Fluckiger I, Gonczy P, et al. 2011. Structural basis of the 9-fold symmetry of centrioles. *Cell* **144:** 364–375.

Knowles MR, Leigh MW, Ostrowski LE, Huang L, Carson JL, Hazucha MJ, Yin W, Berg JS, Davis SD, Dell SD, et al. 2013. Exome sequencing identifies mutations in CCDC114 as a cause of primary ciliary dyskinesia. *Am J Hum Genet* **92:** 99–106.

Koumandou VL, Wickstead B, Ginger ML, van der GM, Dacks JB, Field MC. 2013. Molecular paleontology and complexity in the last eukaryotic common ancestor. *Crit Rev Biochem Mol Biol* **48:** 373–396.

Koutoulis A, Pazour GJ, Wilkerson CG, Inaba K, Sheng H, Takada S, Witman GB. 1997. The *Chlamydomonas reinhardtii ODA3* gene encodes a protein of the outer dynein arm docking complex. *J Cell Biol* **137:** 1069–1080.

Lane N. 2014. Bioenergetic constraints on the evolution of complex life. *Cold Spring Harb Perspect Biol* **6:** a015982.

Li S, Fernandez JJ, Marshall WF, Agard DA. 2012. Three-dimensional structure of basal body triplet revealed by electron cryo-tomography. *EMBO J* **31:** 552–562.

Li C, Jensen VL, Park K, Kennedy J, Garcia-Gonzalo FR, Romani M, De MR, Bruel AL, Gaillard D, Doray B, et al. 2016. MKS5 and CEP290 dependent assembly pathway of the ciliary transition zone. *PLoS Biol* **14:** e1002416.

Lim YS, Chua CE, Tang BL. 2011. Rabs and other small GTPases in ciliary transport. *Biol Cell* **103:** 209–221.

Lin J, Heuser T, Carbajal-Gonzalez BI, Song K, Nicastro D. 2012. The structural heterogeneity of radial spokes in cilia and flagella is conserved. *Cytoskeleton (Hoboken)* **69:** 88–100.

Lin H, Dutcher SK. 2015. Genetic and genomic approaches to identify genes involved in flagellar assembly in *Chlamydomonas reinhardtii*. *Methods Cell Biol* **127:** 349–386.

Maheshwari A, Obbineni JM, Bui KH, Shibata K, Toyoshima YY, Ishikawa T. 2015. α- and β-tubulin lattice of the axonemal microtubule doublet and binding proteins revealed by single particle cryo-electron microscopy and tomography. *Structure* **23:** 1584–1595.

Malicki J, Avidor-Reiss T. 2014. From the cytoplasm into the cilium: Bon voyage. *Organogenesis* **10:** 138–157.

Marshall WF. 2009. Centriole evolution. *Curr Opin Cell Biol* **21:** 14–19.

Melkonian M, Robenek H, Rassat J. 1982. Flagellar membrane specializations and their relationship to mastigonemes and microtubules in *Euglena gracilis*. *J Cell Sci* **55:** 115–135.

Mitchell DR. 2003. Orientation of the central pair complex during flagellar bend formation in *Chlamydomonas*. *Cell Motil Cytoskeleton* **56:** 120–129.

Mitchell DR, Nakatsugawa M. 2004. Bend propagation drives central pair rotation in *Chlamydomonas reinhardtii* flagella. *J Cell Biol* **166:** 709–715.

Mitchell BF, Pedersen LB, Feely M, Rosenbaum JL, Mitchell DR. 2005. ATP production in *Chlamydomonas reinhardtii* flagella by glycotytic enzymes. *Mol Biol Cell* **16:** 4509–4518.

Mukai C, Okuno M. 2004. Glycolysis plays a major role for adenosine triphosphate supplementation in mouse sperm flagellar movement. *Biol Reprod* **71:** 540–547.

Nauli SM, Alenghat FJ, Luo Y, Williams E, Vassilev P, Li X, Elia AE, Lu W, Brown EM, Quinn SJ, et al. 2003. Polycystins 1 and 2 mediate mechanosensation in the primary cilium of kidney cells. *Nat Genet* **33:** 129–137.

Nicastro D, Schwartz C, Pierson J, Gaudette R, Porter ME, McIntosh JR. 2006. The molecular architecture of axonemes revealed by cryoelectron tomography. *Science* **313:** 944–948.

Nicastro D, Fu X, Heuser T, Tso A, Porter ME, Linck RW. 2011. Cryo-electron tomography reveals conserved features of doublet microtubules in flagella. *Proc Natl Acad Sci* **108:** E845–E853.

Noguchi M, Sawadas T, Akazawa T. 2001. ATP-regenerating system in the cilia of *Paramecium caudatum*. *J Exp Biol* **204:** 1063–1071.

Oda T, Yanagisawa H, Yagi T, Kikkawa M. 2014. Mechano-signaling between central apparatus and radial spokes controls axonemal dynein activity. *J Cell Bio.* **204:** 807–819.

Oda T, Yanagisawa H, Kikkawa M. 2015. Detailed structural and biochemical characterization of the nexin-dynein regulatory complex. *Mol Biol Cell* **26:** 294–304.

Olbrich H, Cremers C, Loges NT, Werner C, Nielsen KG, Marthin JK, Philipsen M, Wallmeier J, Pennekamp P, Menchen T, et al. 2015. Loss-of-function GAS8 mutations cause primary ciliary dyskinesia and disrupt the nexin-dynein regulatory complex. *Am J Hum Genet* **97:** 546–554.

Omoto CK, Gibbons IR, Kamiya R, Shingyoji C, Takahashi K, Witman GB. 1999. Rotation of the central pair microtubules in eukaryotic flagella. *Mol Biol Cell* **10:** 1–4.

Onoufriadis A, Paff T, Antony D, Shoemark A, Micha D, Kuyt B, Schmidts M, Petridi S, Dankert-Roelse JE, Haarman EG, et al. 2013. Splice-site mutations in the axonemal outer dynein arm docking complex gene CCDC114 cause primary ciliary dyskinesia. *Am J Hum Genet* **92:** 88–98.

Ooi CP, Rotureau B, Gribaldo S, Georgikou C, Julkowska D, Blisnick T, Perrot S, Subota I, Bastin P. 2015. The flagellar arginine kinase in *Trypanosoma brucei* is important for infection in tsetse flies. *PLoS ONE* **10:** e0133676.

O'Toole ET, Giddings TH Jr, Porter ME, Ostrowski LE. 2012. Computer-assisted image analysis of human cilia and *Chlamydomonas* flagella reveals both similarities and differences in axoneme structure. *Cytoskeleton (Hoboken)* **69:** 577–590.

Paps J, Medina-Chacon LA, Marshall W, Suga H, Ruiz-Trillo I. 2013. Molecular phylogeny of unikonts: New insights into the position of apusomonads and ancyromonads and the internal relationships of opisthokonts. *Protist* **164:** 2–12.

Parfrey LW, Lahr DJ, Knoll AH, Katz LA. 2011. Estimating the timing of early eukaryotic diversification with multigene molecular clocks. *Proc Natl Acad Sci* **108:** 13624–13629.

Patel-King RS, Gilberti RM, Hom EF, King SM. 2013. WD60/FAP163 is a dynein intermediate chain required for retrograde intraflagellar transport in cilia. *Mol Biol Cell* **24:** 2668–2677.

Pigino G, Maheshwari A, Bui KH, Shingyoji C, Kamimura S, Ishikawa T. 2012. Comparative structural analysis of eukaryotic flagella and cilia from *Chlamydomonas, Tetrahymena*, and sea urchins. *J Struct Biol* **178:** 199–206.

Pittis AA, Gabaldon T. 2016. Late acquisition of mitochondria by a host with chimaeric prokaryotic ancestry. *Nature* **531:** 101–104.

Rajagopalan V, Wilkes DE. 2016. Evolution of the dynein heavy chain family in ciliates. *J Eukaryot Microbiol* **63:** 138–141.

Ross L, Normark BB. 2015. Evolutionary problems in centrosome and centriole biology. *J Evol Biol* **28:** 995–1004.

Roy S. 2012. Cilia and Hedgehog: When and how was their marriage solemnized? *Differentiation* **83:** S43–S48.

Satir P, Mitchell DR, Jekely G. 2008. How did the cilium evolve? *Curr Top Dev Biol* **85:** 63–82.

Shih SM, Engel BD, Kocabas F, Bilyard T, Gennerich A, Marshall WF, Yildiz A. 2013. Intraflagellar transport drives flagellar surface motility. *eLife* **2:** e00744.

Soucy SM, Huang J, Gogarten JP. 2015. Horizontal gene transfer: Building the web of life. *Nat Rev Genet* **16:** 472–482.

Spang A, Saw JH, Jorgensen SL, Zaremba-Niedzwiedzka K, Martijn J, Lind AE, van ER, Schleper C, Guy L, Ettema TJ. 2015. Complex archaea that bridge the gap between prokaryotes and eukaryotes. *Nature* **521:** 173–179.

Sung CH, Leroux MR. 2013. The roles of evolutionarily conserved functional modules in cilia-related trafficking. *Nat Cell Biol* **15:** 1387–1397.

Takada S, Wilkerson CG, Wakabayashi K, Kamiya R, Witman GB. 2002. The outer dynein arm-docking complex: Composition and characterization of a subunit (Oda1) necessary for outer arm assembly. *Mol Biol Cell* **13:** 1015–1029.

Tombes RM, Brokaw CJ, Shapiro BM. 1987. Creatine kinase-dependent energy transport in sea urchin spermatozoa. Flagellar wave attenuation and theoretical analysis of high energy phosphate diffusion. *Biophys J* **52:** 75–86.

Turk E, Wills AA, Kwon T, Sedzinski J, Wallingford JB, Stearns T. 2015. ζ-Tubulin is a member of a conserved tubulin module and is a component of the centriolar basal foot in multiciliated cells. *Curr Biol* **25:** 2177–2183.

Valentine MS, Rajendran A, Yano J, Weeraratne SD, Beisson J, Cohen J, Koll F, Van HJ. 2012. *Paramecium* BBS genes are key to presence of channels in cilia. *Cilia* **1:** 1–16.

van Breugel M, Hirono M, Andreeva A, Yanagisawa HA, Yamaguchi S, Nakazawa Y, Morgner N, Petrovich M, Ebong IO, Robinson CV, et al. 2011. Structures of SAS-6 suggest its organization in centrioles. *Science* **331:** 1196–1199.

van Dam TJ, Townsend MJ, Turk M, Schlessinger A, Sali A, Field MC, Huynen MA. 2013. Evolution of modular intraflagellar transport from a coatomer-like progenitor. *Proc Natl Acad Sci* **110:** 6943–6948.

Wickstead B, Gull K. 2007. Dyneins across eukaryotes: A comparative genomic analysis. *Traffic* **8:** 1708–1721.

Wilkes DE, Watson HE, Mitchell DR, Asai DJ. 2008. Twenty-five dyneins in *Tetrahymena*: A re-examination of the multidynein hypothesis. *Cell Motil Cytoskeleton* **65:** 342–351.

Winey M, O'Toole E. 2014. Centriole structure. *Philos Trans R Soc Lond B Biol Sci* **369:** 20130457.

Wirschell M, Olbrich H, Werner C, Tritschler D, Bower R, Sale WS, Loges NT, Pennekamp P, Lindberg S, Stenram U, et al. 2013. The nexin-dynein regulatory complex subunit DRC1 is essential for motile cilia function in algae and humans. *Nat Genet* **45:** 262–268.

Yamamoto R, Yanagisawa HA, Yagi T, Kamiya R. 2006. A novel subunit of axonemal dynein conserved among lower and higher eukaryotes. *FEBS Lett* **580:** 6357–6360.

Yamamoto R, Yanagisawa HA, Yagi T, Kamiya R. 2008. Novel 44-kilodalton subunit of axonemal dynein conserved from *Chlamydomonas* to mammals. *Eukaryot Cell* **7:** 154–161.

Yamamoto R, Song K, Yanagisawa HA, Fox L, Yagi T, Wirschell M, Hirono M, Kamiya R, Nicastro D, Sale WS. 2013. The MIA complex is a conserved and novel dynein regulator essential for normal ciliary motility. *J Cell Biol* **201:** 263–278.

Yanagisawa HA, Mathis G, Oda T, Hirono M, Richey EA, Ishikawa H, Marshall WF, Kikkawa M, Qin H. 2014. FAP20 is an inner junction protein of doublet microtubules essential for both the planar asymmetrical waveform and stability of flagella in *Chlamydomonas*. *Mol Biol Cell* **25:** 1472–1483.

Yoshiba S, Shiratori H, Kuo IY, Kawasumi A, Shinohara K, Nonaka S, Asai Y, Sasaki G, Belo JA, Sasaki H, et al. 2012. Cilia at the node of mouse embryos sense fluid flow for left-right determination via Pkd2. *Science* **338:** 226–231.

Index